797,885 Books

are available to read at

www.ForgottenBooks.com

Forgotten Books' App
Available for mobile, tablet & eReader

ISBN 978-1-330-30834-9
PIBN 10021809

This book is a reproduction of an important historical work. Forgotten Books uses state-of-the-art technology to digitally reconstruct the work, preserving the original format whilst repairing imperfections present in the aged copy. In rare cases, an imperfection in the original, such as a blemish or missing page, may be replicated in our edition. We do, however, repair the vast majority of imperfections successfully; any imperfections that remain are intentionally left to preserve the state of such historical works.

Forgotten Books is a registered trademark of FB &c Ltd.
Copyright © 2015 FB &c Ltd.
FB &c Ltd, Dalton House, 60 Windsor Avenue, London, SW19 2RR.
Company number 08720141. Registered in England and Wales.

For support please visit www.forgottenbooks.com

1 MONTH OF FREE READING

at

www.ForgottenBooks.com

By purchasing this book you are eligible for one month membership to ForgottenBooks.com, giving you unlimited access to our entire collection of over 700,000 titles via our web site and mobile apps.

To claim your free month visit:

www.forgottenbooks.com/free21809

* Offer is valid for 45 days from date of purchase. Terms and conditions apply.

English
Français
Deutsche
Italiano
Español
Português

www.forgottenbooks.com

Mythology Photography **Fiction** Fishing Christianity **Art** Cooking Essays Buddhism Freemasonry Medicine **Biology** Music **Ancient Egypt** Evolution Carpentry Physics Dance Geology **Mathematics** Fitness Shakespeare **Folklore** Yoga Marketing **Confidence** Immortality Biographies Poetry **Psychology** Witchcraft Electronics Chemistry History **Law** Accounting **Philosophy** Anthropology Alchemy Drama Quantum Mechanics Atheism Sexual Health **Ancient History** **Entrepreneurship** Languages Sport Paleontology Needlework Islam **Metaphysics** Investment Archaeology Parenting Statistics Criminology **Motivational**

A THEORY OF
NATURAL PHILOSOPHY

PUT FORWARD AND EXPLAINED BY
ROGER JOSEPH BOSCOVICH, S.J.

LATIN—ENGLISH EDITION

FROM THE TEXT OF THE
FIRST VENETIAN EDITION
PUBLISHED UNDER THE PERSONAL
SUPERINTENDENCE OF THE AUTHOR
IN 1763

WITH
A SHORT LIFE OF BOSCOVICH

CHICAGO LONDON
OPEN COURT PUBLISHING COMPANY
1922

PRINTED IN GREAT BRITAIN
BY
BUTLER & TANNER, FROME, ENGLAND

Copyright

PREFACE

HE text presented in this volume is that of the Venetian edition of 1763. This edition was chosen in preference to the first edition of 1758, published at Vienna, because, as stated on the title-page, it was the first edition (revised and enlarged) issued under the personal superintendence of the author.

In the English translation, an endeavour has been made to adhere as closely as possible to a literal rendering of the Latin; except that the somewhat lengthy and complicated sentences have been broken up. This has made necessary slight changes of meaning in several of the connecting words. This will be noted especially with regard to the word "adeoque", which Boscovich uses with a variety of shades of meaning, from "indeed", "also" or "further", through "thus", to a decided "therefore", which would have been more correctly rendered by "ideoque". There is only one phrase in English that can also take these various shades of meaning, viz., "and so"; and this phrase, for the use of which there is some justification in the word "adeo" itself, has been usually employed.

The punctuation of the Latin is that of the author. It is often misleading to a modern reader and even irrational; but to have recast it would have been an onerous task and something characteristic of the author and his century would have been lost.

My translation has had the advantage of a revision by Mr. A. O. Prickard, M.A., Fellow of New College, Oxford, whose task has been very onerous, for he has had to watch not only for flaws in the translation, but also for misprints in the Latin. These were necessarily many; in the first place, there was only one original copy available, kindly loaned to me by the authorities of the Cambridge University Library; and, as this copy could not leave my charge, a type-script had to be prepared from which the compositor worked, thus doubling the chance of error. Secondly, there were a large number of misprints, and even omissions of important words, in the original itself; for this no discredit can be assigned to Boscovich; for, in the printer's preface, we read that four presses were working at the same time in order to take advantage of the author's temporary presence in Venice. Further, owing to almost insurmountable difficulties, there have been many delays in the production of the present edition, causing breaks of continuity in the work of the translator and reviser; which have not conduced to success. We trust, however, that no really serious faults remain.

The short life of Boscovich, which follows next after this preface, has been written by Dr. Branislav Petronievíc, Professor of Philosophy at the University of Belgrade. It is to be regretted that, owing to want of space requiring the omission of several addenda to the text of the *Theoria* itself, a large amount of interesting material collected by Professor Petronievíc has had to be left out.

The financial support necessary for the production of such a costly edition as the present has been met mainly by the Government of the Kingdom of Serbs, Croats and Slovenes; and the subsidiary expenses by some Jugo-Slavs interested in the publication.

After the "Life," there follows an "Introduction," in which I have discussed the ideas of Boscovich, as far as they may be gathered from the text of the *Theoria* alone; this also has been cut down, those parts which are clearly presented to the reader in Boscovich's own Synopsis having been omitted. It is a matter of profound regret to everyone that this discussion comes from my pen instead of, as was originally arranged, from that of the late Philip E. P. Jourdain, the well-known mathematical logician; whose untimely death threw into my far less capable hands the responsible duties of editorship.

I desire to thank the authorities of the Cambridge University Library, who time after time over a period of five years have forwarded to me the original text of this work of Boscovich. Great credit is also due to the staff of Messrs. Butler & Tanner, Frome, for the care and skill with which they have carried out their share of the work; and my special thanks for the unfailing painstaking courtesy accorded to my demands, which were frequently not in agreement with trade custom.

J. M. CHILD.

MANCHESTER UNIVERSITY,
December, 1921.

LIFE OF ROGER JOSEPH BOSCOVICH

By BRANISLAV PETRONIEVIĆ

HE Slav world, being still in its infancy, has, despite a considerable number of scientific men, been unable to contribute as largely to general science as the other great European nations. It has, nevertheless, demonstrated its capacity of producing scientific works of the highest value. Above all, as I have elsewhere indicated,[*] it possesses Copernicus, Lobachevski, Mendeljev, and Boscovich.

In the following article, I propose to describe briefly the life of the Jugo-Slav, Boscovich, whose principal work is here published for the sixth time; the first edition having appeared in 1758, and others in 1759, 1763, 1764, and 1765. The present text is from the edition of 1763, the first Venetian edition, revised and enlarged.

On his father's side, the family of Boscovich is of purely Serbian origin, his grandfather, Boško, having been an orthodox Serbian peasant of the village of Orakova in Herzegovina. His father, Nikola, was first a merchant in Novi Pazar (Old Serbia), but later settled in Dubrovnik (Ragusa, the famous republic in Southern Dalmatia), whither his father, Boško, soon followed him, and where Nikola became a Roman Catholic. Pavica, Boscovich's mother, belonged to the Italian family of Betere, which for a century had been established in Dubrovnik and had become Slavonicized—Bara Betere, Pavica's father, having been a poet of some reputation in Ragusa.

Roger Joseph Boscovich (Rudjer Josif Boškovic', in Serbo-Croatian) was born at Ragusa on September 18th, 1711, and was one of the younger members of a large family. He received his primary and secondary education at the Jesuit College of his native town; in 1725 he became a member of the Jesuit order and was sent to Rome, where from 1728 to 1733 he studied philosophy, physics and mathematics in the Collegium Romanum. From 1733 to 1738 he taught rhetoric and grammar in various Jesuit schools; he became Professor of mathematics in the Collegium Romanum, continuing at the same time his studies in theology, until in 1744 he became a priest and a member of his order.

In 1736, Boscovich began his literary activity with the first fragment, "De Maculis Solaribus," of a scientific poem, "De Solis ac Lunæ Defectibus"; and almost every succeeding year he published at least one treatise upon some scientific or philosophic problem. His reputation as a mathematician was already established when he was commissioned by Pope Benedict XIV to examine with two other mathematicians the causes of the weakness in the cupola of St. Peter's at Rome. Shortly after, the same Pope commissioned him to consider various other problems, such as the drainage of the Pontine marshes, the regularization of the Tiber, and so on. In 1756, he was sent by the republic of Lucca to Vienna as arbiter in a dispute between Lucca and Tuscany. During this stay in Vienna, Boscovich was commanded by the Empress Maria Theresa to examine the building of the Imperial Library at Vienna and the cupola of the cathedral at Milan. But this stay in Vienna, which lasted until 1758, had still more important consequences; for Boscovich found time there to finish his principal work, *Theoria Philosophiæ Naturalis*; the publication was entrusted to a Jesuit, Father Scherffer, Boscovich having to leave Vienna, and the first edition appeared in 1758, followed by a second edition in the following year. With both of these editions, Boscovich was to some extent dissatisfied (see the remarks made by the printer who carried out the third edition at Venice, given in this volume on page 3); so a third edition was issued at Venice, revised, enlarged and rearranged under the author's personal superintendence in 1763. The revision was so extensive that as the printer remarks, "it ought to be considered in some measure as a first and original edition"; and as such it has been taken as the basis of the translation now published. The fourth and fifth editions followed in 1764 and 1765.

One of the most important tasks which Boscovich was commissioned to undertake was that of measuring an arc of the meridian in the Papal States. Boscovich had designed to take part in a Portuguese expedition to Brazil on a similar errand; but he was per-

[*] *Slav Achievements in Advanced Science*, London, 1917.

suaded by Pope Benedict XIV, in 1750, to conduct, in collaboration with an English Jesuit, Christopher Maire, the measurements in Italy. The results of their work were published, in 1755, by Boscovich, in a treatise, *De Litteraria Expeditione per Pontificiam, &c.*; this was translated into French under the title of *Voyage astronomique et géographique dans l'État de l'Église*, in 1770.

By the numerous scientific treatises and dissertations which he had published up to 1759, and by his principal work, Boscovich had acquired so high a reputation in Italy, nay in Europe at large, that the membership of numerous academies and learned societies had already been conferred upon him. In 1760, Boscovich, who hitherto had been bound to Italy by his professorship at Rome, decided to leave that country. In this year we find him at Paris, where he had gone as the travelling companion of the Marquis Romagnosi. Although in the previous year the Jesuit order had been expelled from France, Boscovich had been received on the strength of his great scientific reputation. Despite this, he did not feel easy in Paris; and the same year we find him in London, on a mission to vindicate the character of his native place, the suspicions of the British Government, that Ragusa was being used by France to fit out ships of war, having been aroused; this mission he carried out successfully. In London he was warmly welcomed, and was made a member of the Royal Society. Here he published his work, *De Solis ac Lunæ defectibus*, dedicating it to the Royal Society. Later, he was commissioned by the Royal Society to proceed to California to observe the transit of Venus; but, as he was unwilling to go, the Society sent him to Constantinople for the same purpose. He did not, however, arrive in time to make the observation; and, when he did arrive, he fell ill and was forced to remain at Constantinople for seven months. He left that city in company with the English ambassador, Porter, and, after a journey through Thrace, Bulgaria, and Moldavia, he arrived finally at Warsaw, in Poland; here he remained for a time as the guest of the family of Poniatowski. In 1762, he returned from Warsaw to Rome by way of Silesia and Austria. The first part of this long journey has been described by Boscovich himself in his *Giornale di un viaggio da Constantinopoli in Polonia*—the original of which was not published until 1784, although a French translation had appeared in 1772, and a German translation in 1779.

Shortly after his return to Rome, Boscovich was appointed to a chair at the University of Pavia; but his stay there was not of long duration. Already, in 1764, the building of the observatory of Brera had been begun at Milan according to the plans of Boscovich; and in 1770, Boscovich was appointed its director. Unfortunately, only two years later he was deprived of office by the Austrian Government which, in a controversy between Boscovich and another astronomer of the observatory, the Jesuit Lagrange, took the part of his opponent. The position of Boscovich was still further complicated by the disbanding of his company; for, by the decree of Clement V, the Order of Jesus had been suppressed in 1773. In the same year Boscovich, now free for the second time, again visited Paris, where he was cordially received in official circles. The French Government appointed him director of "Optique Marine," with an annual salary of 8,000 francs; and Boscovich became a French subject. But, as an ex-Jesuit, he was not welcomed in all scientific circles. The celebrated d'Alembert was his declared enemy; on the other hand, the famous astronomer, Lalande, was his devoted friend and admirer. Particularly, in his controversy with Rochon on the priority of the discovery of the micrometer, and again in the dispute with Laplace about priority in the invention of a method for determining the orbits of comets, did the enmity felt in these scientific circles show itself. In Paris, in 1779, Boscovich published a new edition of his poem on eclipses, translated into French and annotated, under the title, *Les Eclipses*, dedicating the edition to the King, Louis XV.

During this second stay in Paris, Boscovich had prepared a whole series of new works, which he hoped would have been published at the Royal Press. But, as the American War of Independence was imminent, he was forced, in 1782, to take two years' leave of absence, and return to Italy. He went to the house of his publisher at Bassano; and here, in 1785, were published five volumes of his optical and astronomical works, *Opera pertinentia ad opticam et astronomiam*.

Boscovich had planned to return through Italy from Bassano to Paris; indeed, he left Bassano for Venice, Rome, Florence, and came to Milan. Here he was detained by illness and he was obliged to ask the French Government to extend his leave, a request that was willingly granted. His health, however, became worse; and to it was added a melancholia. He died on February 13th, 1787.

The great loss which Science sustained by his death has been fitly commemorated in the eulogium by his friend Lalande in the French Academy, of which he was a member; and also in that of Francesco Ricca at Milan, and so on. But it is his native town, his beloved Ragusa, which has most fitly celebrated the death of the greatest of her sons in the eulogium of the poet, Bernardo Zamagna.[a] This magnificent tribute from his native town was entirely deserved by Boscovich, both for his scientific works, and for his love and work for his country.

Boscovich had left his native country when a boy, and returned to it only once afterwards, when, in 1747, he passed the summer there, from June 20th to October 1st; but he often intended to return. In a letter, dated May 3rd, 1774, he seeks to secure a pension as a member of the Jesuit College of Ragusa; he writes: "I always hope at last to find my true peace in my own country and, if God permit me, to pass my old age there in quietness."

Although Boscovich has written nothing in his own language, he understood it perfectly; as is shown by the correspondence with his sister, by certain passages in his Italian letters, and also by his *Giornale* (p. 31; p. 59 of the French edition). In a dispute with d'Alembert, who had called him an Italian, he said: "we will notice here in the first place that our author is a Dalmatian, and from Ragusa, not Italian; and that is the reason why Marucelli, in a recent work on Italian authors, has made no mention of him."[b] That his feeling of Slav nationality was strong is proved by the tributes he pays to his native town and native land in his dedicatory epistle to Louis XV.

Boscovich was at once philosopher, astronomer, physicist, mathematician, historian, engineer, architect, and poet. In addition, he was a diplomatist and a man of the world; and yet a good Catholic and a devoted member of the Jesuit order. His friend, Lalande, has thus sketched his appearance and his character: "Father Boscovich was of great stature; he had a noble expression, and his disposition was obliging. He accommodated himself with ease to the foibles of the great, with whom he came into frequent contact. But his temper was a trifle hasty and irascible, even to his friends—at least his manner gave that impression—but this solitary defect was compensated by all those qualities which make up a great man. . . . He possessed so strong a constitution that it seemed likely that he would have lived much longer than he actually did; but his appetite was large, and his belief in the strength of his constitution hindered him from paying sufficient attention to the danger which always results from this." From other sources we learn that Boscovich had only one meal daily, déjeûner.

Of his ability as a poet, Lalande says: "He was himself a poet like his brother, who was also a Jesuit. . . . Boscovich wrote verse in Latin only, but he composed with extreme ease. He hardly ever found himself in company without dashing off some impromptu verses to well-known men or charming women. To the latter he paid no other attentions, for his austerity was always exemplary. . . . With such talents, it is not to be wondered at that he was everywhere appreciated and sought after. Ministers, princes and sovereigns all received him with the greatest distinction. M. de Lalande witnessed this in every part of Italy where Boscovich accompanied him in 1765."

Boscovich was acquainted with several languages—Latin, Italian, French, as well as his native Serbo-Croatian, which, despite his long absence from his country, he did not forget. Although he had studied in Italy and passed the greater part of his life there, he had never penetrated to the spirit of the language, as his Italian biographer, Ricca, notices. His command of French was even more defective; but in spite of this fact, French men of science urged him to write in French. English he did not understand, as he confessed in a letter to Priestley; although he had picked up some words of polite conversation during his stay in London.

His correspondence was extensive. The greater part of it has been published in the *Mémoirs de l'Académie Jougo-Slave* of Zagrab, 1887 to 1912.

---

[a] Oratio in funere R. J. Boscovichii . . . a Bernardo Zamagna.
[b] *Voyage Astronomique*, p. 750; also on pp. 707 seq.
[c] *Journal des Sçavans*, Février, 1792, pp. 113–118.

INTRODUCTION

LTHOUGH the title to this work to a very large extent correctly describes the contents, yet the argument leans less towards the explanation of a theory than it does towards the logical exposition of the results that must follow from the acceptance of certain fundamental assumptions, more or less generally admitted by natural philosophers of the time. The most important of these assumptions is the doctrine of Continuity, as enunciated by Leibniz. This doctrine may be shortly stated in the words: "Everything takes place by degrees"; or, in the phrase usually employed by Boscovich: "Nothing happens *per saltum*." The second assumption is the axiom of Impenetrability; that is to say, Boscovich admits as axiomatic that no two material points can occupy the same spatial, or local, point simultaneously. Clerk Maxwell has characterized this assumption as "an unwarrantable concession to the vulgar opinion." He considered that this axiom is a prejudice, or prejudgment, founded on experience of bodies of sensible size. This opinion of Maxwell cannot however be accepted without dissection into two main heads. The criticism of the axiom itself would appear to carry greater weight against Boscovich than against other philosophers; but the assertion that it is a prejudice is hardly warranted. For, Boscovich, in accepting the truth of the axiom, has no *experience* on which to found his acceptance. His material points have *absolutely no magnitude*; they are Euclidean points, "having no parts." There is, therefore, no *reason* for assuming, by a sort of induction (and Boscovich never makes an induction without expressing the reason why such induction can be made), that two material points cannot occupy the same local point simultaneously; that is to say, there cannot have been a prejudice in favour of the acceptance of this axiom, *derived from experience of bodies of sensible size;* for, since the material points are non-extended, they do not *occupy space*, and cannot therefore exclude another point from *occupying the same space*. Perhaps, we should say the reason is not the same as that which makes it impossible for bodies of sensible size. The acceptance of the axiom by Boscovich is purely theoretical; in fact, it constitutes practically the whole of the theory of Boscovich. On the other hand, for this very reason, there are no readily apparent grounds for the acceptance of the axiom; and no serious arguments can be adduced in its favour; Boscovich's own line of argument, founded on the idea that infinite improbability comes to the same thing as impossibility, is given in Art. 361. Later, I will suggest the probable source from which Boscovich derived his idea of impenetrability as applying to points of matter, as distinct from impenetrability for bodies of sensible size.

Boscovich's own idea of the merit of his work seems to have been chiefly that it met the requirements which, in the opinion of Newton, would constitute "a mighty advance in philosophy." These requirements were the "derivation, from the phenomena of Nature, of two or three general principles; and the explanation of the manner in which the properties and actions of all corporeal things follow from these principles, even if the causes of those principles had not at the time been discovered." Boscovich claims in his preface to the first edition (Vienna, 1758) that he has gone far beyond these requirements; in that he has reduced all the principles of Newton to a single principle—namely, that given by his Law of Forces.

The occasion that led to the writing of this work was a request, made by Father Scherffer, who eventually took charge of the first Vienna edition during the absence of Boscovich; he suggested to Boscovich the investigation of the centre of oscillation. Boscovich applied to this investigation the principles which, as he himself states, "he lit upon so far back as the year 1745." Of these principles he had already given some indication in the dissertations *De Viribus vivis* (published in 1745), *De Lege Virium in Natura existentium* (1755), and others. While engaged on the former dissertation, he investigated the production and destruction of velocity in the case of impulsive action, such as occurs in direct collision. In this, where it is to be noted that bodies of sensible size are under consideration, Boscovich was led to the study of the distortion and recovery of shape which occurs on impact; he came to the conclusion that, owing to this distortion and recovery of shape, there was produced by the impact a *continuous* retardation of the relative velocity during the *whole time* of impact, which was finite; in other words, the Law of Continuity, as enunciated by

Leibniz, was observed. It would appear that at this time (1745) Boscovich was concerned mainly, if not solely, with the *facts* of the change of velocity, and not with the *causes* for this change. The title of the dissertation, *De Viribus vivis*, shows however that a secondary consideration, of almost equal importance in the development of the Theory of Boscovich, also held the field. The natural philosophy of Leibniz postulated monads, without parts, extension or figure. In these features the monads of Leibniz were similar to the material points of Boscovich; but Leibniz ascribed to his monads [1] perception and appetition in addition to an equivalent of inertia. They are centres of force, and the force exerted is a *vis viva*. Boscovich opposes this idea of a "living," or "lively" force; and in this first dissertation we may trace the first ideas of the formulation of his own material points. Leibniz denies action at a distance; with Boscovich it is the fundamental characteristic of a material point.

The principles developed in the work on collisions of bodies were applied to the problem of the centre of oscillation. During the latter investigation Boscovich was led to a theorem on the mutual forces between the bodies forming a system of three; and from this theorem there followed the natural explanation of a whole sequence of phenomena, mostly connected with the idea of a statical moment; and his initial intention was to have published a dissertation on this theorem and deductions from it, as a specimen of the use and advantage of his principles. But all this time these principles had been developing in two directions, mathematically and philosophically, and by this time included the fundamental notions of the law of forces for material points. The essay on the centre of oscillation grew in length as it proceeded; until, finally, Boscovich added to it all that he had already published on the subject of his principles and other matters which, as he says, "obtruded themselves on his notice as he was writing." The whole of this material he rearranged into a more logical (but unfortunately for a study of development of ideas, non-chronological) order before publication.

As stated by Boscovich, in Art. 164, the whole of his Theory is contained in his statement that: "*Matter is composed of perfectly indivisible, non-extended, discrete points.*" To this assertion is conjoined the axiom that no two material points can be in the same point of space at the same time. As stated above, in opposition to Clerk Maxwell, this is no matter of prejudice. Boscovich, in Art. 361, gives his own reasons for taking this axiom as part of his theory. He lays it down that the number of material points is finite, whereas the number of local points is an infinity of three dimensions; hence it is infinitely improbable, i.e., impossible, that two material points, without the action of a directive mind, should ever encounter one another, and thus be in the same place at the same time. He even goes further; he asserts elsewhere that no material point ever returns to any point of space in which it has ever been before, or in which any other material point has ever been. Whether his arguments are sound or not, the matter does not rest on a prejudgment formed from experience of bodies of sensible size; Boscovich has convinced himself by such arguments of the truth of the principle of Impenetrability, and lays it down as axiomatic; and upon this, as one of his foundations, builds his complete theory. The consequence of this axiom is immediately evident; there can be no such thing as contact between any two material points; two points cannot be contiguous or, as Boscovich states, no two points of matter can be in mathematical contact. For, since material points have no dimensions, if, to form an imagery of Boscovich's argument, we take two little squares ABDC, CDFE to represent two points in mathematical contact along the side CD, then CD must also coincide with AB, and EF with CD; that is the points which we have supposed to be contiguous must also be coincident. This is contrary to the axiom of Impenetrability; and hence material points must be separated always by a finite interval, no matter how small. This finite interval however has no minimum; nor has it, on the other hand, on account of the infinity of space, any maximum, except under certain hypothetical circumstances which may possibly exist. Lastly, these points of matter float, so to speak, in an absolute void.

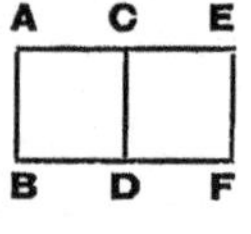

Every material point is exactly like every other material point; each is postulated to have an inherent propensity (*determinatio*) to remain in a state of rest or uniform motion in a straight line, whichever of these is supposed to be its initial state, so long as the point is not subject to some external influence. Thus it is endowed with an equivalent of inertia as formulated by Newton; but as we shall see, there does not enter the Newtonian idea of inertia as a characteristic of *mass*. The propensity is akin to the characteristic ascribed to the monad by Leibniz; with this difference, that it is not a symptom of activity, as with Leibniz, but one of inactivity.

[1] See Bertrand Russell, *Philosophy of Leibniz*; especially p. 91 for connection between Boscovich and Leibniz.

Further, according to Boscovich, there is a mutual *vis* between every *pair* of points, the magnitude of which depends only on the distance between them. At first sight, there would seem to be an incongruity in this supposition; for, since a point has no magnitude, it cannot have any mass, considered as "quantity of matter"; and therefore, if the slightest "force" (according to the ordinary acceptation of the term) existed between two points, there would be an infinite acceleration or retardation of each point relative to the other. If, on the other hand, we consider with Clerk Maxwell that each point of matter has a definite small mass, this mass must be finite, no matter how small, and not infinitesimal. For the mass of a point is the whole mass of a body, divided by the number of points of matter composing that body, which are all exactly similar; and this number Boscovich asserts is finite. It follows immediately that the density of a material point must be infinite, since the volume is an infinitesimal of the third order, if not of an infinite order, i.e., zero. Now, infinite density, if not to all of us, to Boscovich at least is unimaginable. Clerk Maxwell, in ascribing mass to a Boscovichian point of matter, seems to have been obsessed by a prejudice, that very prejudice which obsesses most scientists of the present day, namely, that there can be no force without mass. He understood that Boscovich ascribed to each pair of points a mutual attraction or repulsion; and, in consequence, prejudiced by Newton's Laws of Motion, he ascribed mass to a material point of Boscovich.

This apparent incongruity, however, disappears when it is remembered that the word *vis*, as used by the mathematicians of the period of Boscovich, had many different meanings; or rather that its meaning was given by the descriptive adjective that was associated with it. Thus we have *vis viva* (later associated with energy), *vis mortua* (the antithesis of vis viva, as understood by Leibniz), *vis acceleratrix* (acceleration), *vis motrix* (the real equivalent of force, since it varied with the mass directly), *vis descensiva* (moment of a weight hung at one end of a lever), and so on. Newton even, in enunciating his law of universal gravitation, apparently asserted nothing more than the fact of gravitation—a propensity for approach—according to the inverse square of the distance: and Boscovich imitates him in this. The mutual *vires*, ascribed by Boscovich to his pairs of points, are really accelerations, i.e. tendencies for mutual approach or recession of the two points, depending on the distance between the points at the time under consideration. Boscovich's own words, as given in Art. 9, are: "Censeo igitur bina quæcunque materiæ puncta determinari æque in aliis distantiis ad mutuum accessum, in aliis ad recessum mutuum, *quam ipsam determinationem apello vim*." The cause of this determination, or propensity, for approach or recession, which in the case of bodies of sensible size is more correctly called "force" (*vis motrix*), Boscovich does not seek to explain; he merely postulates the propensities. The measures of these propensities, i.e., the accelerations of the relative velocities, are the ordinates of what is usually called his curve of forces. This is corroborated by the statement of Boscovich that the areas under the arcs of his curve are proportional to squares of velocities; which is in accordance with the formula we should now use for the area under an "acceleration-space" graph (Area $= \int f.ds = \int \frac{dv}{dt}.ds = \int v.dv$). See Note (f) to Art. 118, where it is evident that the word *vires*, translated "forces," strictly means "accelerations;" see also Art. 64.

Thus it would appear that in the Theory of Boscovich we have something totally different from the monads of Leibniz, which are truly centres of force. Again, although there are some points of similarity with the ideas of Newton, more especially in the postulation of an acceleration of the relative velocity of every pair of points of matter due to and depending upon the relative distance between them, without any endeavour to explain this acceleration or gravitation; yet the Theory of Boscovich differs from that of Newton in being purely kinematical. His material point is defined to be without parts, i.e., it has *no volume*; as such it can have *no mass*, and can exert *no force*, as we understand such terms. The sole characteristic that has a finite measure is the relative acceleration produced by the simultaneous existence of two points of matter; and this acceleration depends solely upon the distance between them. The Newtonian idea of mass is replaced by something totally different; it is a mere number, without "dimension"; the "mass" of a body is simply the number of points that are combined to "form" the body.

Each of these points, if sufficiently close together, will exert on another point of matter, at a relatively much greater distance from every point of the body, the same acceleration very approximately. Hence, if we have two small bodies A and B, situated at a distance s from one another (the wording of this phrase postulates that the points of each body are very close together as compared with the distance between the bodies): and if the number of points in A and B are respectively a and b, and f is the mutual acceleration between any pair of material points at a distance s from one another; then, each point of A will give to each point of B an acceleration f. Hence, the body A will give to each point of B, and therefore to the whole of B, an acceleration equal to af. Similarly the body B will give to the body A an acceleration equal to *bf*. Similarly, if we placed a third body, C, at a distance *s* from A and B, the body A would give the body C an acceleration equal to *af*, and the body B would give the body C an acceleration equal to *bf*. That is, the accelerations given to a standard body C are proportional to the "number of points" in the bodies producing these accelerations; thus, *numerically*, the "mass" of Boscovich comes to the same thing as the "mass" of Newton. Further, the acceleration given by C to the bodies A and B is the same for either, namely, *cf*; from which it follows that all bodies have their velocities of fall towards the earth equally accelerated, apart from the resistance of the air; and so on. But the term "force," as the cause of acceleration is not applied by Boscovich to material points; nor is it used in the Newtonian sense at all. When Boscovich investigates the attraction of "bodies," he introduces the idea of a cause, but then only more or less as a convenient phrase. Although, as a philosopher, Boscovich denies that there is any possibility of a fortuitous circumstance (and here indeed we may admit a prejudice derived from experience; for he states that what we call fortuitous is merely something for which we, in our limited intelligence, can assign no cause), yet with him the existent thing is *motion* and not *force*. The latter word is merely a convenient phrase to describe the "product" of "mass" and "acceleration."

To sum up, it would seem that the curve of Boscovich is an acceleration-interval graph; and it is a mistake to refer to his cosmic system as a system of "force-centres." His material points have zero volume, zero mass, and exert zero force. In fact, if one material point alone existed outside the mind, and there were no material point forming part of the mind, then this single external point could in no way be perceived. In other words, a single point would give no sense-datum apart from another point; and thus single points might be considered as not perceptible in themselves, but as becoming so in relation to other material points. This seems to be the logical deduction from the strict sense of the definition given by Boscovich; what Boscovich himself thought is given in the supplements that follow the third part of the treatise. Nevertheless, the phraseology of "attraction" and "repulsion" is so much more convenient than that of "acceleration of the velocity of approach" and "acceleration of the velocity of recession," that it will be used in what follows: as it has been used throughout the translation of the treatise.

There is still another point to be considered before we take up the study of the Boscovich curve; namely, whether we are to consider Boscovich as, consciously or unconsciously, an atomist in the strict sense of the word. The practical test for this question would seem to be simply whether the divisibility of matter was considered to be limited or unlimited. Boscovich himself appears to be uncertain of his ground, hardly knowing which point of view is the logical outcome of his definition of a material point. For, in Art. 394, he denies infinite divisibility; but he admits infinite componibility. The denial of infinite divisibility is necessitated by his denial of "anything infinite in Nature, or in extension, or a self-determined infinitely small." The admission of infinite componibility is necessitated by his definition of the material point; since it has no parts, a fresh point can always be placed between any two points without being contiguous to either. Now, since he denies the existence of the infinite and the infinitely small, the attraction or repulsion between two points of matter (except at what he calls the limiting intervals) must be finite: hence, since the attractions of masses are all by observation finite, it follows that the number of points in a mass must be finite. To evade the difficulty thus raised, he appeals to the scale of integers, in which there is no infinite number: but, as he says, the scale of integers is a sequence of *numbers increasing indefinitely, and having no last term*. Thus, into any space, however small, there may be crowded an indefinitely great number of material points; this number can be still further increased to any extent; and yet the number of points finally obtained is always finite. It would, again, seem that the system of Boscovich was not a material system, but a system of relations; if it were not for the fact that he asserts, in Art. 7, that his view is that "the Universe does not consist of vacuum interspersed amongst matter, but that matter is interspersed in a vacuum and floats in it." The whole question is still further complicated by his remark, in Art. 393, that in the continual division of a body, "as soon as we reach intervals less than the distance between two material points, further sections will cut empty intervals and not matter"; and yet he has postulated that there is no minimum value to the interval between two material points. Leaving, however, this question of the philosophical standpoint of Boscovich to be decided by the reader, after a study of the supplements that follow the third part of the treatise, let us now consider the curve of Boscovich.

Boscovich, from experimental data, gives to his curve, when the interval is large, a branch asymptotic to the axis of intervals; it approximates to the "hyperbola" $x^2y = c$, in which x represents the interval between two points, and y the *vis* corresponding to that interval, which we have agreed to call an attraction, meaning thereby, not a force, but an acceleration of the velocity of approach. For small intervals he has as yet no knowledge of the quality or quantity of his ordinates. In Supplement IV, he gives some very ingenious arguments against forces that are attractive at very small distances and increase indefinitely, such as would be the case where the law of forces was represented by an inverse power of the interval, or even where the force varied inversely as the interval. For the inverse fourth or higher power, he shows that the attraction of a sphere upon a point on its surface would be less than the attraction of a part of itself on this point; for the inverse third power, he considers orbital motion, which in this case is an equiangular spiral motion, and deduces that after a finite time the particle must be nowhere at all. Euler, considering this case, asserted that on approaching the centre of force the particle must be annihilated; Boscovich, with more justice, argues that this law of force must be impossible. For the inverse square law, the limiting case of an elliptic orbit, when the transverse velocity at the end of the major axis is decreased indefinitely, is taken; this leads to rectilinear motion of the particle to the centre of force and a return from it; which does not agree with the otherwise proved oscillation *through* the centre of force to an equal distance on either side.

Now it is to be observed that this supplement is quoted from his dissertation *De Lege Virium in Natura existentium*, which was published in 1755; also that in 1743 he had published a dissertation of which the full title is: *De Motu Corporis attracti in centrum immobile viribus decrescentibus in ratione distantiarum reciproca duplicata in spatiis non resistentibus*. Hence it is not too much to suppose that somewhere between 1741 and 1755 he had tried to find a means of overcoming this discrepancy; and he was thus led to suppose that, in the case of rectilinear motion under an inverse square law, there was a departure from the law on near approach to the centre of force; that the attraction was replaced by a repulsion increasing indefinitely as the distance decreased; for this obviously would lead to an oscillation to the centre and back, and so come into agreement with the limiting case of the elliptic orbit. I therefore suggest that *it was this consideration that led Boscovich to the doctrine of Impenetrability*. However, in the treatise itself, Boscovich postulates the axiom of Impenetrability as applying in general, and thence argues that the force at infinitely small distances must be repulsive and increasing indefinitely. Hence the ordinate to the curve near the origin must be drawn in the opposite direction to that of the ordinates for sensible distances, and the area under this branch of the curve must be indefinitely great. That is to say, the branch must be asymptotic to the axis of ordinates; Boscovich however considers that this does not involve an infinite ordinate at the origin, because the interval between two material points is never zero; or, vice versa, since the repulsion increases indefinitely for very small intervals, the velocity of relative approach, no matter how great, of two material points is always destroyed before actual contact; which necessitates a finite interval between two material points, and the impossibility of encounter under any circumstances: the interval however, since a velocity of mutual approach may be supposed to be of any magnitude, can have no minimum. Two points are said to be in physical contact, in opposition to mathematical contact, when they are so close together that this great mutual repulsion is sufficiently increased to prevent nearer approach.

Since Boscovich has these two asymptotic branches, and he postulates Continuity, there must be a continuous curve, with a one-valued ordinate for any interval, to represent the "force" at all other distances; hence the curve must cut the axis at some point in between, or the ordinate must become infinite. He does not lose sight of this latter possibility, but apparently discards it for certain mechanical and physical reasons. Now, it is known that as the degree of a curve rises, the number of curves of that degree increases very rapidly; there is only one of the first degree, the conic sections of the second degree, while Newton had found over three-score curves with equations of the third degree, and nobody had tried to find all the curves of the fourth degree. Since his curve is not one of the known curves, Boscovich concludes that the degree of its equation is very high, even if it is not transcendent. But the higher the degree of a curve, the greater the number of possible intersections with a given straight line; that is to say, it is highly probable that there are a great many intersections of the curve with the axis; i.e., points giving zero action for material points situated at the corresponding distance from one another. Lastly, since the ordinate is one-valued, the equation of the curve, as stated in Supplement III, must be of the form $P-Qy = 0$, where P and Q are functions of x alone. Thus we have a curve winding about the axis for intervals that are very small and developing finally into the hyperbola of the third degree for sensible intervals. This final branch, however, cannot be exactly this hyperbola; for, Boscovich argues, if any finite arc of the curve ever coincided exactly with the hyperbola of the third degree, it would be a breach of continuity if it ever departed from it. Hence he concludes that the inverse square law is observed approximately only, even at large distances.

As stated above, the possibility of other asymptotes, parallel to the asymptote at the origin, is not lost sight of. The consequence of one occurring at a very small distance from the origin is discussed in full. Boscovich, however, takes great pains to show that all the phenomena discussed can be explained on the assumption of a number of points of intersection of his curve with the axis, combined with different characteristics of the arcs that lie between these points of intersection. There is, however, one suggestion that is very interesting, especially in relation to recent statements of Einstein and Weyl. Suppose that beyond the distances of the solar system, for which the inverse square law obtains approximately at least, the curve of forces, after touching the axis (as it may do, since it does not coincide exactly with the hyperbola of the third degree), goes off to infinity in the positive direction; or suppose that, after cutting the axis (as again it may do, for the reason given above), it once more begins to wind round the axis and finally has an asymptotic attractive branch. Then it is evident that the universe in which we live is a self-contained cosmic system; for no point within it can ever get beyond the distance of this further asymptote. If in addition, beyond this further asymptote, the curve had an asymptotic repulsive branch and went on as a sort of replica of the curve already obtained, then no point outside our universe could ever enter within it. Thus there is a possibility of infinite space being filled with a succession of cosmic systems, each of which would never interfere with any other; indeed, a mind existing in any one of these universes could never perceive the existence of any other universe except that in which it existed. Thus space might be in reality infinite, and yet never could be perceived except as finite.

The use Boscovich makes of his curve, the ingenuity of his explanations and their logic, the strength or weakness of his attacks on the theories of other philosophers, are left to the consideration of the reader of the text. It may, however, be useful to point out certain matters which seem more than usually interesting. Boscovich points out that no philosopher has attempted to prove the existence of a centre of gravity. It would appear especially that he is, somehow or other, aware of the mistake made by Leibniz in his early days (a mistake corrected by Huygens according to the statement of Leibniz), and of the use Leibniz later made of the principle of moments; Boscovich has apparently considered the work of Pascal and others, especially Guldinus; it looks almost as if (again, somehow or other) he had seen some description of "The Method" of Archimedes. For he proceeds to define the centre of gravity *geometrically*, and to prove that there is always a centre of gravity, or rather a geometrical centroid; whereas, even for a triangle, there is no centre of magnitude, with which Leibniz seems to have confused a centroid before his conversation with Huygens. This existence proof, and the deductions from it, are necessary foundations for the centrobaryc analysis of Leibniz. The argument is shortly as follows: Take a plane outside, say to the right of, all the points of all the bodies under consideration; find the sum of all the distances of all the points from this plane; divide this sum by the number of points; draw a plane to the left of and parallel to the chosen plane, at a distance from it equal to the quotient just found. Then, observing algebraic sign, this is a plane such that the sum of the distances of all the points from it is zero; i.e., the sum of the distances of all the points on one side of this plane is equal arithmetically to the sum of the distances of all the points on the other side. Find a similar plane of equal distances in another direction; this intersects the first plane in a straight line. A third similar plane cuts this straight line in a point; this point is the centroid; it has the unique property that all planes through it are planes of equal distances. If some of the points are conglomerated to form a particle, the sum of the distances for each of the points is equal to the distance of the particle multiplied by the number of points in the particle, i.e., by the mass of the particle. Hence follows the theorem for the statical moment for lines and planes or other surfaces, as well as for solids that have weight.

Another interesting point, in relation to recent work, is the subject-matter of Art. 230–236; where it is shown that, due solely to the mutual forces exerted on a third point by two points separated by a proper interval, there is a series of orbits, approximately confocal ellipses, in which the third point is in a state of steady motion; these orbits are alternately stable and stable. If the steady motion in a stable orbit is disturbed, by a sufficiently great difference of the velocity being induced by the action of a fourth point passing sufficiently near the third point, this third point will leave its orbit and immediately take up another stable orbit, after some initial oscillation about it. This elegant little theorem does not depend in any way on the exact form of the curve of forces, *so long as there are portions of the curve winding about the axis for very small intervals between the points.*

It is sufficient, for the next point, to draw the reader's attention to Art. 266–278, on collision, and to the articles which follow on the agreement between resolution and composition of forces as a working hypothesis. From what Boscovich says, it would appear that philosophers of his time were much perturbed over the idea that, when a force was resolved into two forces at a sufficiently obtuse angle, the force itself might be less than either of the resolutes. Boscovich points out that, in his Theory, there is no resolution, only composition; and therefore the difficulty does not arise. In this connection he adds that there are no signs in Nature of anything approaching the ***vires vivæ*** of Leibniz.

In Art. 294 we have Boscovich's contribution to the controversy over the correct measure of the "quantity of motion"; but, as there is no attempt made to follow out the change in either the velocity or the square of the velocity, it cannot be said to lead to anything conclusive. As a matter of fact, Boscovich uses the result to prove the non-existence of ***vires vivæ***.

In Art. 298–306 we have a mechanical exposition of reflection and refraction of light. This comes under the section on Mechanics, because with Boscovich light is matter moving with a very high velocity, and therefore reflection is a case of impact, in that it depends upon the destruction of the whole of the perpendicular velocity upon entering the "surface" of a denser medium, the surface being that part of space in front of the physical surface of the medium in which the particles of light are near enough to the denser medium to feel the influence of the last repulsive asymptotic branch of the curve of forces. If this perpendicular velocity is not all destroyed, the particle enters the medium, and is refracted; in which case, the existence of a sine law is demonstrated. It is to be noted that the "fits" of alternate attraction and repulsion, postulated by Newton, follow as a natural consequence of the winding portion of the curve of Boscovich.

In Art. 328–346 we have a discussion of the centre of oscillation, and the centre of percussion is investigated as well for masses in a plane perpendicular to the axis of rotation, and masses lying in a straight line, where each mass is connected with the different centres. Boscovich deduces from his theory the theorems, amongst others, that the centres of suspension and oscillation are interchangeable, and that the distance between them is equal to the distance of the centre of percussion from the axis of rotation; he also gives a rule for finding the simple equivalent pendulum. The work is completed in a letter to Fr. Scherffer, which is appended at the end of this volume.

In the third section, which deals with the application of the Theory to Physics, we naturally do not look for much that is of value. But, in Art. 505, Boscovich evidently has the correct notion that sound is a longitudinal vibration of the air or some other medium; and he is able to give an explanation of the propagation of the disturbance purely by means of the mutual forces between the particles of the medium. In Art. 507 he certainly states that the cause of heat is a "vigorous internal motion"; but this motion is that of the "particles of fire," if it is a motion; an alternative reason is however given, namely, that it may be a "fermentation of a sulphurous substance with particles of light." "Cold is a lack of this substance, or of a motion of it." No attention will be called to this part of the work, beyond an expression of admiration for the great ingenuity of a large part of it.

There is a metaphysical appendix on the seat of the mind, and its nature, and on the existence and attributes of God. This is followed by two short discussions of a philosophical nature on Space and Time. Boscovich does not look on either of these as being in themselves existent; his entities are modes of existence, temporal and local. These three sections are full of interest for the modern philosophical reader.

Supplement V is a theoretical proof, purely derived from the theory of mutual actions between points of matter, of the law of the lever; this is well worth study.

There are two points of historical interest beyond the study of the work of Boscovich that can be gathered from this volume. The first is that at this time it would appear that the nature of negative numbers and quantities was not yet fully understood. Boscovich, to make his curve more symmetrical, continues it to the left of the origin as a reflection in the axis of ordinates. It is obvious, however, that, if distances to the left of the origin stand for intervals measured in the opposite direction to the ordinary (remembering that of the two points under consideration one is supposed to be at the origin), then the force just the other side of the axis of ordinates must be repulsive; but the repulsion is in the opposite direction to the ordinary way of measuring it, and therefore should appear on the curve represented by an ordinate of attraction. Thus, the curve of Boscovich, if completed, should have point symmetry about the origin, and not line symmetry about the axis of ordinates. Boscovich, however, avoids this difficulty, intentionally or unintentionally, when showing how the equation to the curve may be obtained, by taking $z = x^2$ as his variable, and P and Q as functions of z, in the equation $P-Qy = 0$, referred to above. ***Note.***—In this connection (p. 410, Art. 25, l. 5), Boscovich has apparently made a slip over the negative sign: as the intention is clear, no attempt has been made to amend the Latin.

The second point is that Boscovich does not seem to have any idea of integrating between limits. He has to find the area, in Fig. 1 on p. 134, bounded by the axes, the curve and the ordinate ***ag***; this he does by the use of the calculus in Note (l) on p. 141. He assumes that the equation of the curve is $x^m y^n = 1$, and obtains the integral $\frac{n}{n-m} xy + A$, where A is the constant of integration. He states that, if n is greater than m, $A = 0$, *being the initial area at the origin.* He is then faced with the necessity of making the area infinite when $n = m$, and still more infinite when $n < m$. He says : "The area is infinite, when $n = m$, because this makes the divisor zero ; and thus the area becomes still more infinite if $n < m$." Put into symbols, the argument is : Since $n-m < 0$, $\frac{n}{n-m} > \frac{n}{0} > \infty$. The historically interesting point about this is that it represents the persistance of an error originally made by Wallis in his *Arithmetica Infinitorum* (it was Wallis who invented the sign ∞ to stand for "simple infinity," the value of $1/0$, and hence of $n/0$). Wallis had justification for his error, if indeed it was an error in his case ; for his exponents were characteristics of certain infinite series, and he could make his own laws about these so that they suited the geometrical problems to which they were applied ; it was not necessary that they should obey the laws of inequality that were true for ordinary numbers. Boscovich's mistake is, of course, that of assuming that the constant is zero in every case ; and in this he is probably deceived by using the formula $\frac{n}{n-m} xy + A$, instead of $\frac{n}{n-m} x^{n/(n-m)} + A$, for the area. From the latter it is easily seen that since the initial area is zero, we must have $A = \frac{n}{m-n} 0^{n/(n-m)}$. If n is equal to or greater than m, the constant A is indeed zero ; but if n is less than m, the constant is infinite. The persistence of this error for so long a time, from 1655 to 1758, during which we have the writings of Newton, Leibniz, the Bernoullis and others on the calculus, seems to lend corroboration to a doubt as to whether the integral sign was properly understood as a summation between limits, and that this sum could be expressed as the difference of two values of the same function of those limits. It appears to me that this point is one of very great importance in the history of the development of mathematical thought.

Some idea of how prolific Boscovich was as an author may be gathered from the catalogue of his writings appended at the end of this volume. This catalogue has been taken from the end of the original first Venetian edition, and brings the list up to the date of its publication, 1763. It was felt to be an impossible task to make this list complete up to the time of the death of Boscovich ; and an incomplete continuation did not seem desirable. Mention must however be made of one other work of Boscovich at least ; namely, a work in five quarto volumes, published in 1785, under the title of *Opera pertinentia ad Opticam et Astronomiam.*

Finally, in order to bring out the versatility of the genius of Boscovich, we may mention just a few of his discoveries in science, which seem to call for special attention. In astronomical science, he speaks of the use of a telescope filled with liquid for the purpose of measuring the aberration of light ; he invented a prismatic micrometer contemporaneously with Rochon and Maskelyne. He gave methods for determining the orbit of a comet from three observations, and for the equator of the sun from three observations of a "spot"; he carried out some investigations on the orbit of Uranus, and considered the rings of Saturn. In what was then the subsidiary science of optics, he invented a prism with a variable angle for measuring the refraction and dispersion of different kinds of glass ; and put forward a theory of achromatism for the objectives and oculars of the telescope. In mechanics and geodesy, he was apparently the first to solve the problem of the "body of greatest attraction"; he successfully attacked the question of the earth's density ; and perfected the apparatus and advanced the theory of the measurement of the meridian. In mathematical theory, he seems to have recognized, before Lobachevski and Bolyai, the impossibility of a proof of Euclid's "parallel postulate"; and considered the theory of the logarithms of negative numbers.

J. M. C.

N.B.—The page numbers on the left-hand pages of the index are the pages of the original Latin Edition of 1763 ; they correspond with the clarendon numbers inserted throughout the Latin text of this edition.

CORRIGENDA

Attention is called to the following important corrections, omissions, and alternative renderings; misprints volving a single letter or syllable only are given at the end of the volume.

27, l. 8, *for* in one plane *read* in the same direction
47, l. 62, *literally* on which . . . is exerted
49, l. 33, *for* just as . . . is *read* so that . . . may be
53, l. 9, *after* a line *add* but not parts of the line itself
61, Art. 47, *Alternative rendering:* These instances make good the same point as water making its way through the pores of a sponge did for impenetrability;
67, l. 5, *for* it is allowable for me *read* I am disposed; *unless in the original* libet *is taken to be a misprint for licet*
73, l. 26, *after* nothing *add* in the strict meaning of the term
85, l. 27, *after* conjunction *add* of the same point of space
91, l. 25, *Alternative rendering:* and these properties might distinguish the points even in the view of the followers of Leibniz
l. 5 from bottom, *Alternative rendering:* Not to speak of the actual form of the leaves present in the seed
115, l. 25, *after* the left *add* but that the two outer elements do not touch each other
l. 28, *for* two little spheres *read* one little sphere
117, l. 41, *for* precisely *read* abstractly
125, l. 29, *for* ignored *read* urged in reply
126, l. 6 from bottom, *it is possible that* acquirere *is intended for* acquiescere, *with a corresponding change in the translation*
129, Art. 162, marg. note, *for* on what they may be founded *read* in what it consists.
167, Art. 214, l. 2 of marg. note, *transpose* by *and* on
footnote, l. 1, *for* be at *read* bisect it at
199, l. 24, *for* so that *read* just as
233, l. 4 from bottom, *for* base to the angle *read* base to the sine of the angle
last line, *after* vary *insert* inversely
307, l. 5 from end, *for* motion, as (with fluids) takes place *read* motion from taking place
323, l. 39, *for* the agitation will *read* the fluidity will
345, l. 32, *for* described *read* destroyed
357, l. 44, *for* others *read* some, others of others
l. 5 from end, *for* fire *read* a fiery *and insert a comma before* substance

THEORIA

HILOSOPHIÆ NATURALIS

TYPOGRAPHUS
VENETUS
LECTORI

PUS, quod tibi offero, jam ab annis quinque Viennæ editum, quo plausu exceptum sit per Europam, noveris sane, si Diaria publica perlegeris, inter quæ si, ut omittam cætera, consulas ea, quæ in Bernensi pertinent ad initium anni 1761 ; videbis sane quo id loco haberi debeat. Systema continet Naturalis Philosophiæ omnino novum, quod jam ab ipso Auctore suo vulgo *Boscovichianum* appellant. Id quidem in pluribus Academiis jam passim publice traditur, nec tantum in annuis thesibus, vel dissertationibus impressis, ac propugnatis exponitur, sed & in pluribus elementaribus libris pro juventute instituenda editis adhibetur, exponitur, & a pluribus habetur pro archetypo. Verum qui omnem systematis compagem, arctissimum partium nexum mutuum, fœcunditatem summam, ac usum amplissimum ac omnem, quam late patet, Naturam ex unica simplici lege virium derivandam intimius velit conspicere, ac contemplari, hoc Opus consulat, necesse est.

Hæc omnia me permoverant jam ab initio, ut novam Operis editionem curarem : accedebat illud, quod Viennensia exemplaria non ita facile extra Germaniam itura videbam, & quidem nunc etiam in reliquis omnibus Europæ partibus, utut expetita, aut nuspiam venalia prostant, aut vix uspiam : systema vero in Italia natum, ac ab Auctore suo pluribus hic apud nos jam dissertationibus adumbratum, & casu quodam Viennæ, quo se ad breve tempus contulerat, digestum, ac editum, Italicis potissimum typis, censebam, per universam Europam disseminandum. Et quidem editionem ipsam e Viennensi exemplari jam tum inchoaveram ; cum illud mihi constitit, Viennensem editionem ipsi Auctori, post cujus discessum suscepta ibi fuerat, summopere displicere : innumera obrepsisse typorum menda : esse autem multa, inprimis ea, quæ Algebraicas formulas continent, admodum inordinata, & corrupta : ipsum eorum omnium correctionem meditari, cum nonnullis mutationibus, quibus Opus perpolitum redderetur magis, & vero etiam additamentis.

Illud ergo summopere desideravi, ut exemplar acquirerem ab ipso correctum, & auctum ac ipsum editioni præsentem haberem, & curantem omnia per sese. At id quidem per hosce annos obtinere non licuit, eo universam fere Europam peragrante ; donec demum ex tam longa peregrinatione redux huc nuper se contulit, & toto adstitit editionis tempore, ac præter correctores nostros omnem ipse etiam in corrigendo diligentiam adhibuit ; quanquam is ipse haud quidem sibi ita fidit, ut nihil omnino effugisse censeat, cum ea sit humanæ mentis conditio, ut in eadem re diu satis intente defigi non possit.

Hæc idcirco ut prima quædam, atque originaria editio haberi debet, quam qui cum Viennensi contulerit, videbit sane discrimen. E minoribus mutatiunculis multæ pertinent ad expolienda, & declaranda plura loca ; sunt tamen etiam nonnulla potissimum in paginarum fine exigua additamenta, vel mutatiunculæ exiguæ factæ post typographicam constructionem idcirco tantummodo, ut lacunulæ implerentur quæ aliquando idcirco supererant, quod plures phyliræ a diversis compositoribus simul adornabantur, & quatuor simul præla sudabant; quod quidem ipso præsente fieri facile potuit, sine ulla perturbatione sententiarum, & ordinis.

THE PRINTER AT VENICE
TO
THE READER

OU will be well aware, if you have read the public journals, with what applause the work which I now offer to you has been received throughout Europe since its publication at Vienna five years ago. Not to mention others, if you refer to the numbers of the Berne *Journal* for the early part of the year 1761, you will not fail to see how highly it has been esteemed. It contains an entirely new system of Natural Philosophy, which is already commonly known as the *Boscovichian theory*, from the name of its author, As a matter of fact, it is even now a subject of public instruction in several Universities in different parts; it is expounded not only in yearly theses or dissertations, both printed & debated; but also in several elementary books issued for the instruction of the young it is introduced, explained, & by many considered as their original. Any one, however, who wishes to obtain more detailed insight into the whole structure of the theory, the close relation that its several parts bear to one another, or its great fertility & wide scope for the purpose of deriving the whole of Nature, in her widest range, from a single simple law of forces; any one who wishes to make a deeper study of it must perforce study the work here offered.

All these considerations had from the first moved me to undertake a new edition of the work; in addition, there was the fact that I perceived that it would be a matter of some difficulty for copies of the Vienna edition to pass beyond the confines of Germany—indeed, at the present time, no matter how diligently they are inquired for, they are to be found on sale nowhere, or scarcely anywhere, in the rest of Europe. The system had its birth in Italy, & its outlines had already been sketched by the author in several dissertations published here in our own land; though, as luck would have it, the system itself was finally put into shape and published at Vienna, whither he had gone for a short time. I therefore thought it right that it should be disseminated throughout the whole of Europe, & that preferably as the product of an Italian press. I had in fact already commenced an edition founded on a copy of the Vienna edition, when it came to my knowledge that the author was greatly dissatisfied with the Vienna edition, taken in hand there after his departure; that innumerable printer's errors had crept in; that many passages, especially those that contain Algebraical formulæ, were ill-arranged and erroneous; lastly, that the author himself had in mind a complete revision, including certain alterations, to give a better finish to the work, together with certain additional matter.

That being the case, I was greatly desirous of obtaining a copy, revised & enlarged by himself; I also wanted to have him at hand whilst the edition was in progress, & that he should superintend the whole thing for himself. This, however, I was unable to procure during the last few years, in which he has been travelling through nearly the whole of Europe; until at last he came here, a little while ago, as he returned home from his lengthy wanderings, & stayed here to assist me during the whole time that the edition was in hand. He, in addition to our regular proof-readers, himself also used every care in correcting the proof; even then, however, he has not sufficient confidence in himself as to imagine that not the slightest thing has escaped him. For it is a characteristic of the human mind that it cannot concentrate long on the same subject with sufficient attention.

It follows that this ought to be considered in some measure as a first & original edition; any one who compares it with that issued at Vienna will soon see the difference between them. Many of the minor alterations are made for the purpose of rendering certain passages more elegant & clear; there are, however, especially at the foot of a page, slight additions also, or slight changes made after the type was set up, merely for the purpose of filling up gaps that were left here & there—these gaps being due to the fact that several sheets were being set at the same time by different compositors, and four presses were kept hard at work together. As he was at hand, this could easily be done without causing any disturbance of the sentences or the pagination.

Inter mutationes occurret ordo numerorum mutatus in paragraphis : nam numerus 82 de novo accessit totus : deinde is, qui fuerat 261 discerptus est in 5 ; demum in Appendice post num. 534 factæ sunt & mutatiunculæ nonnullæ, & additamenta plura in iis, quæ pertinent ad sedem animæ.

Supplementorum ordo mutatus est itidem ; quæ enim fuerant 3, & 4, jam sunt 1, & 2 : nam corum usus in ipso Opere ante alia occurrit. Illi autem, quod prius fuerat primum, nunc autem est tertium, accessit in fine Scholium tertium, quod pluribus numeris complectitur dissertatiunculam integram de argumento, quod ante aliquot annos in Parisiensi Academia controversiæ occasionem exhibuit in Encyclopedico etiam dictionario attactum, in qua dissertatiuncula demonstrat Auctor non esse, cur ad vim exprimendam potentia quæpiam distantiæ adhibeatur potius, quam functio.

Accesserunt per totum Opus notulae marginales, in quibus eorum, quæ pertractantur argumenta exponuntur brevissima, quorum ope unico obtutu videri possint omnia, & in memoriam facile revocari.

Postremo loco ad calcem Operis additus est fusior catalogus eorum omnium, quæ huc usque ab ipso Auctore sunt edita, quorum collectionem omnem expolitam, & correctam, ac eorum, quæ nondum absoluta sunt, continuationem meditatur, aggressurus illico post suum regressum in Urbem Romam, quo properat. Hic catalogus impressus fuit Venetisis ante hosce duos annos in reimpressione ejus poematis de Solis ac Lunæ defectibus. Porro eam omnium suorum Operum Collectionem, ubi ipse adornaverit, typis ego meis excudendam suscipiam, quam magnificentissime potero.

Hæc erant, quæ te monendum censui ; tu laboribus nostris fruere, & vive ſelix.

THE PRINTER AT VENICE TO THE READER

Among the more important alterations will be found a change in the order of numbering the paragraphs. Thus, Art. 82 is additional matter that is entirely new; that which was formerly Art. 261 is now broken up into five parts; &, in the Appendix, following Art. 534, both some slight changes and also several additions have been made in the passages that relate to the Seat of the Soul.

The order of the Supplements has been altered also: those that were formerly numbered III and IV are now I and II respectively. This was done because they are required for use in this work before the others. To that which was formerly numbered I, but is now III, there has been added a third scholium, consisting of several articles that between them give a short but complete dissertation on that point which, several years ago caused a controversy in the University of Paris, the same point being also discussed in the *Dictionnaire Encyclopédique*. In this dissertation the author shows that there is no reason why any one power of the distance should be employed to express the force, in preference to a function.

Short marginal summaries have been inserted throughout the work, in which the arguments dealt with are given in brief; by the help of these, the whole matter may be taken in at a glance and recalled to mind with ease.

Lastly, at the end of the work, a somewhat full catalogue of the whole of the author's publications up to the present time has been added. Of these publications the author intends to make a full collection, revised and corrected, together with a continuation of those that are not yet finished; this he proposes to do after his return to Rome, for which city he is preparing to set out. This catalogue was printed in Venice a couple of years ago in connection with a reprint of his essay in verse on the eclipses of the Sun and Moon. Later, when his revision of them is complete, I propose to undertake the printing of this complete collection of his works from my own type, with all the sumptuousness at my command.

Such were the matters that I thought ought to be brought to your notice. May you enjoy the fruit of our labours, & live in happiness.

EPISTOLA AUCTORIS DEDICATORIA

PRIMÆ EDITIONIS VIENNENSIS

AD CELSISSIMUM TUNC PRINCIPEM ARCHIEPISCOPUM VIENNENSEM, NUNC PRÆTEREA ET CARDINALEM EMINENTISSIMUM, ET EPISCOPUM VACCIENSEM CHRISTOPHORUM E COMITATIBUS DE MIGAZZI

ABIS veniam, Princeps Celsissime, si forte inter assiduas sacri regiminis curas importunus interpellator advenio, & libellum Tibi offero mole tenuem, nec arcana Religionis mysteria, quam in isto tanto constitutus fastigio administras, sed Naturalis Philosophiæ principia continentem. Novi ego quidem, quam totus in eo sis, ut, quam geris, personam sustineas, ac vigilantissimi sacrorum Antistitis partes agas. Videt utique Imperialis hæc Aula, videt universa Regalis Urbs, & ingenti admiratione defixa obstupescit, qua diligentia, quo labore tanti Sacerdotii munus obire pergas. Vetus nimirum illud celeberrimum *age, quod agis*, quod ab ipsa Tibi juventute, cum primum, ut Te Romæ dantem operam studiis cognoscerem, mihi fors obtigit, altissime jam insederat animo, id in omni reliquo amplissimorum munerum Tibi commissorum cursu hæsit firmissime, atque idipsum inprimis adjectum tam multis & dotibus, quas a Natura uberrime congestas habes, & virtutibus, quas tute diuturna Tibi exercitatione, atque assiduo labore comparasti, sanctissime observatum inter tam varias forenses, Aulicas, Sacerdotales occupationes, istos Tibi tam celeres dignitatum gradus quodammodo veluti coacervavit, & omnium una tam populorum, quam Principum admirationem excitavit ubique, conciliavit amorem ; unde illud est factum, ut ab aliis alia Te, sublimiora semper, atque honorificentiora munera quodammodo velut avulsum, atque abstractum rapuerint. Dum Romæ in celeberrimo illo, quod Auditorum Rotæ appellant, collegio toti Christiano orbi jus diceres, accesserat Hetrusca Imperialis Legatio apud Romanum Pontificem exercenda ; cum repente Mechliniensi Archiepiscopo in amplissima illa administranda Ecclesia Adjutor datus, & destinatus Successor, possessione præstantissimi muneris vixdum capta, ad Hispanicum Regem ab Augustissima Romanorum Imperatrice ad gravissima tractanda negotia Legatus es missus, in quibus cum summa utriusque Aulæ approbatione versatum per annos quinque ditissima Vacciensis Ecclesia adepta est ; atque ibi dum post tantos Aularum strepitus ea, qua Christianum Antistitem decet, & animi moderatione, & demissione quadam, atque in omne hominum genus charitate, & singulari cura, ac diligentia Religionem administras, & sacrorum exceres curam ; non ea tantum urbs, atque ditio, sed universum Hungariæ Regnum, quanquam exterum hominem, non ut civem suum tantummodo, sed ut Parentem amantissimum habuit, quem adhuc ereptum sibi dolet, & angitur ; dum scilicet minore, quam unius anni intervallo ab Ipsa Augustissima Imperatrice ad Regalem hanc Urbem, tot Imperatorum sedem, ac Austriacæ Dominationis caput, dignum tantis dotibus explicandis theatrum, eocatum videt, atque in hac Celsissima Archiepiscopali Sede, accedente Romani Pontificis Auctoritate collocatum ; in qua Tu quidem personam itidem, quam agis, diligentissime sustinens, totus es in gravissimis Sacerdotii Tui expediendis negotiis, in iis omnibus, quæ ad sacra pertinent, curandis vel per Te ipsum usque adeo, ut sæpe, raro admodum per

AUTHOR'S EPISTLE DEDICATING

THE FIRST VIENNA EDITION

TO

CHRISTOPHER, COUNT DE MIGAZZI, THEN HIS HIGHNESS THE PRINCE ARCHBISHOP OF VIENNA, AND NOW ALSO IN ADDITION HIS EMINENCE THE CARDINAL, BISHOP OF VACZ

OU will pardon me, Most Noble Prince, if perchance I come to disturb at an inopportune moment the unremitting cares of your Holy Office, & offer you a volume so inconsiderable in size; one too that contains none of the inner mysteries of Religion, such as you administer from the highly exalted position to which you are ordained; one that merely deals with the principles of Natural Philosophy. I know full well how entirely your time is taken up with sustaining the reputation that you bear, & in performing the duties of a highly conscientious Prelate. This Imperial Court sees, nay, the whole of this Royal City sees, with what care, what toil, you exert yourself to carry out the duties of so great a sacred office, & stands wrapt with an overwhelming admiration. Of a truth, that well-known old saying, "*What you do, DO*," which from your earliest youth, when chance first allowed me to make your acquaintance while you were studying in Rome, had already fixed itself deeply in your mind, has remained firmly implanted there during the whole of the remainder of a career in which duties of the highest importance have been committed to your care. Your strict observance of this maxim in particular, joined with those numerous talents so lavishly showered upon you by Nature, & those virtues which you have acquired for yourself by daily practice & unremitting toil, throughout your whole career, forensic, courtly, & sacerdotal, has so to speak heaped upon your shoulders those unusually rapid advances in dignity that have been your lot. It has aroused the admiration of all, both peoples & princes alike, in every land; & at the same time it has earned for you their deep affection. The consequence was that one office after another, each ever more exalted & honourable than the preceding, has in a sense seized upon you & borne you away a captive. Whilst you were in Rome, giving judicial decisions to the whole Christian world in that famous College, the Rota of Auditors, there was added the duty of acting on the Tuscan Imperial Legation at the Court of the Roman Pontiff. Suddenly you were appointed coadjutor to the Archbishop of Malines in the administration of that great church, & his future successor. Hardly had you entered upon the duties of that most distinguished appointment, than you were despatched by the August Empress of the Romans as Legate on a mission of the greatest importance. You occupied yourself on this mission for the space of five years, to the entire approbation of both Courts, & then the wealthy church of Vacz obtained your services. Whilst there, the great distractions of a life at Court being left behind, you administer the offices of religion & discharge the sacred rights with that moderation of spirit & humility that befits a Christian prelate, in charity towards the whole race of mankind, with a singularly attentive care. So that not only that city & the district in its see, but the whole realm of Hungary as well, has looked upon you, though of foreign race, as one of her own citizens; nay, rather as a well beloved father, whom she still mourns & sorrows for, now that you have been taken from her. For, after less than a year had passed, she sees you recalled by the August Empress herself to this Imperial City, the seat of a long line of Emperors, & the capital of the Dominions of Austria, a worthy stage for the display of your great talents; she sees you appointed, under the auspices of the authority of the Roman Pontiff, to this exalted Archiepiscopal see. Here too, sustaining with the utmost diligence the part you play so well, you throw yourself heart and soul into the business of discharging the weighty duties of your priesthood, or in attending to all those things that deal with the sacred rites with your own hands: so much so that we often see you officiating, & even administering the Sacraments, in our

hæc nostra tempora exemplo, & publico operatum, ac ipsa etiam Sacramenta administrantem videamus in templis, & Tua ipsius voce populos, e superiore loco docentum audiamus, atque ad omne virtutum genus inflammantem.

Novi ego quidem hæc omnia; novi hanc indolem, hanc animi constitutionem; nec sum tamen inde absterritus, ne, inter gravissimas istas Tuas Sacerdotales curas, Philosophicas hasce meditationes meas, Tibi sisterem, ac tantulæ libellum molis bomini ad tantum culmen evecto porrigerem, ac Tuo vellem Nomine insignitum. Quod enim ad primum pertinet caput, non Theologicas tantum, sed Philosophicas etiam perquisitiones Christiano Antistite ego quidem dignissimas esse censeo, & universam Naturæ contemplationem omnino arbitror cum Sacerdotii sanctitate penitus consentire. Mirum enim, quam belle ab ipsa consideratione Naturæ ad cælestium rerum contemplationem disponitur animus, & ad ipsum Divinum tantæ molis Conditorem assurgit, infinitam ejus Potentiam Sapientiam, Providentiam admiratus, quæ erumpunt undique, & utique se produnt.

Est autem & illud, quod ad supremi sacrorum Moderatoris curam pertinet providere, ne in prima ingenuæ juventutis institutione, quæ semper a naturalibus studiis exordium ducit, prava teneris mentibus irrepant, ac perniciosa principia, quæ sensim Religionem corrumpant, & vero etiam evertant penitus, ac eruant a fundamentis; quod quidem jam dudum tristi quodam Europæ fato passim evenire cernimus, gliscente in dies malo, ut fucatis quibusdam, profecto perniciosissimis, imbuti principiis juvenes, tum demum sibi sapere videantur, cum & omnem animo religionem, & Deum ipsum sapientissimum Mundi Fabricatorem, atque Moderatorem sibi mente excusserint. Quamobrem qui veluti ad tribunal tanti Sacerdotum Principis Universæ Physicæ Theoriam, & novam potissimum Theoriam sistat, rem is quidem præstet æquissimam, nec alienum quidpiam ab ejus munere Sacerdotali offerat, sed cum eodem apprime consentiens.

Nec vero exigua libelli moles deterrere me debuit, ne cum eo ad tantum Principem accederem. Est ille quidem satis tenuis libellus, at non & tenuem quoque rem continet. Argumentum pertractat sublime admodum, & nobile, in quo illustrando omnem ego quidem industriam collocavi, ubi si quid præstitero, si minus infiliciter me gessero, nemo sane me impudentiæ arguat, quasi vilem aliquam, & tanto indignam fastigio rem offeram. Habetur in eo novum quoddam Universæ Naturalis Philosophiæ genus a receptis huc usque, usitatisque plurimam discrepans, quanquam etiam ex iis, quæ maxime omnium per hæc tempora celebrantur, casu quodam præcipua quæque mirum sane in modum compacta, atque inter se veluti coagmentata conjunguntur ibidem, uti sunt simplicia atque inextensa Leibnitianorum clementa, cum Newtoni viribus inducentibus in aliis distantiis accessum mutuum, in aliis mutuum recessum, quas vulgo attractiones, & repulsiones appellant: casu, inquam: neque enim ego conciliandi studio hinc, & inde decerpsi quædam ad arbitrium selecta, quæ utcumque inter se componerem, atque compaginarem: sed omni præjudicio seposito, a principiis exorsus inconcussis, & vero etiam receptis communiter, legitima ratiocinatione usus, & continuo conclusionum nexu deveni ad legem virium in Natura existentium unicam, simplicem, continuam, quæ mihi & constitutionem elementorum materiæ, & Mechanicæ leges, & generales materiæ ipsius proprietates, & præcipua corporum discrimina, sua applicatione ita exhibuit, ut eadem in iis omnibus ubique se prodat uniformis agendi ratio, non ex arbitrariis hypothesibus, & fictitiis commentationibus, sed ex sola continua ratiocinatione deducta. Ejusmodi autem est omnis, ut eas ubique vel definiat, vel adumbret combinationes elementorum, quæ ad diversa præstanda phænomena sunt adhibendæ, ad quas combinationes Conditoris Supremi consilium, & immensa Mentis Divinæ vis ubique requiritur, quæ infinitos casus perspiciat, & ad rem aptissimos seligat, ac in Naturam inducat.

Id mihi quidem argumentum est operis, in quo Theoriam meam expono, comprobo, vindico: tum ad Mechanicam primum, deinde ad Physicam applico, & uberrimos usus expono, ubi brevi quidem libello, sed admodum diuturnas annorum jam tredecim meditationes complector meas, eo plerumque tantummodo rem deducens, ubi demum cum

churches (a somewhat unusual thing at the present time), and also hear you with your own voice exhorting the people from your episcopal throne, & inciting them to virtue of every kind.

I am well aware of all this; I know full well the extent of your genius, & your constitution of mind; & yet I am not afraid on that account of putting into your hands, amongst all those weighty duties of your priestly office, these philosophical meditations of mine; nor of offering a volume so inconsiderable in bulk to one who has attained to such heights of eminence; nor of desiring that it should bear the hall-mark of your name. With regard to the first of these heads, I think that not only theological but also philosophical investigations are quite suitable matters for consideration by a Christian prelate; & in my opinion, a contemplation of all the works of Nature is in complete accord with the sanctity of the priesthood. For it is marvellous how exceedingly prone the mind becomes to pass from a contemplation of Nature herself to the contemplation of celestial things, & to give honour to the Divine Founder of such a mighty structure, lost in astonishment at His infinite Power & Wisdom & Providence, which break forth & disclose themselves in all directions & in all things.

There is also this further point, that it is part of the duty of a religious superior to take care that, in the earliest training of ingenuous youth, which always takes its start from the study of the wonders of Nature, improper ideas do not insinuate themselves into tender minds; or such pernicious principles as may gradually corrupt the belief in things Divine, nay, even destroy it altogether, & uproot it from its very foundations. This is what we have seen for a long time taking place, by some unhappy decree of adverse fate, all over Europe; and, as the canker spreads at an ever increasing rate, young men, who have been made to imbibe principles that counterfeit the truth but are actually most pernicious doctrines, do not think that they have attained to wisdom until they have banished from their minds all thoughts of religion and of God, the All-wise Founder and Supreme Head of the Universe. Hence, one who so to speak sets before the judgment-seat of such a prince of the priesthood as yourself a theory of general Physical Science, & more especially one that is new, is doing nothing but what is absolutely correct. Nor would he be offering him anything inconsistent with his priestly office, but on the contrary one that is in complete harmony with it.

Nor, secondly, should the inconsiderable size of my little book deter me from approaching with it so great a prince. It is true that the volume of the book is not very great, but the matter that it contains is not unimportant as well. The theory it develops is a strikingly sublime and noble idea; & I have done my very best to explain it properly. If in this I have somewhat succeeded, if I have not failed altogether, let no one accuse me of presumption, as if I were offering some worthless thing, something unworthy of such distinguished honour. In it is contained a new kind of Universal Natural Philosophy, one that differs widely from any that are generally accepted & practised at the present time; although it so happens that the principal points of all the most distinguished theories of the present day, interlocking and as it were cemented together in a truly marvellous way, are combined in it; so too are the simple unextended elements of the followers of Leibniz, as well as the Newtonian forces producing mutual approach at some distances & mutual separation at others, usually called attractions and repulsions. I use the words "it so happens" because I have not, in eagerness to make the whole consistent, selected one thing here and another there, just as it suited me for the purpose of making them agree & form a connected whole. On the contrary, I put on one side all prejudice, & started from fundamental principles that are incontestable, & indeed are those commonly accepted; I used perfectly sound arguments, & by a continuous chain of deduction I arrived at a single, simple, continuous law for the forces that exist in Nature. The application of this law explained to me the constitution of the elements of matter, the laws of Mechanics, the general properties of matter itself, & the chief characteristics of bodies, in such a manner that the same uniform method of action in all things disclosed itself at all points; being deduced, not from arbitrary hypotheses, and fictitious explanations, but from a single continuous chain of reasoning. Moreover it is in all its parts of such a kind as defines, or suggests, in every case, the combinations of the elements that must be employed to produce different phenomena. For these combinations the wisdom of the Supreme Founder of the Universe, & the mighty power of a Divine Mind are absolutely necessary; naught but one that could survey the countless cases, select those most suitable for the purpose, and introduce them into the scheme of Nature.

This then is the argument of my work, in which I explain, prove & defend my theory; then I apply it, in the first instance to Mechanics, & afterwards to Physics, & set forth the many advantages to be derived from it. Here, although the book is but small, I yet include the well-nigh daily meditations of the last thirteen years, carrying on my conclu-

communibus Philosophorum consentio placitis, & ubi ea, quæ habemus jam pro compertis, ex meis etiam deductionibus sponte fluunt, quod usque adeo voluminis molem contraxit. Dederam ego quidem dispersa dissertatiunculis variis Theoriæ meæ quædam velut specimina, quæ inde & in Italia Professores publicos nonnullos adstipulatores est nacta, & jam ad exteras quoque gentes pervasit; sed ea nunc primum tota in unum compacta, & vero etiam plusquam duplo aucta, prodit in publicum, quem laborem postremo hoc mense, molestioribus negotiis, quæ me Viennam adduxerant, & curis omnibus exsolutus suscepi, dum in Italiam rediturus opportunam itineri tempus inter assiduas nives opperior, sed omnem in eodem adornando, & ad communem mediocrum etiam Philosophorum captum accommodando diligentiam adhibui.

Inde vero jam facile intelliges, cur ipsum laborem meum ad Te deferre, & Tuo nuncupare Nomini non dubitaverim. Ratio ex iis, quæ proposui, est duplex: primo quidem ipsum argumenti genus, quod Christianum Antistitem non modo non dedecet, sed etiam apprime decet: tum ipsius argumenti vis, atque dignitas, quæ nimirum confirmat, & erigit nimium fortasse impares, sed quantum fieri per me potuit, intentos conatus meos; nam quidquid eo in genere meditando assequi possum, totum ibidem adhibui, ut idcirco nihil arbitrer a mea tenuitate proferri posse te minus indignum, cui ut aliquem offerrem laborum meorum fructum quantumcunque, exposcebat sane, ac ingenti clamore quodam efflagitabat tanta erga me humanitas Tua, qua jam olim immerentem complexus Romæ, hic etiam fovere pergis, nec in tanto dedignatus fastigio, omni benevolentiæ significatione prosequeris. Accedit autem & illud, quod in hisce terris vix adhuc nota, vel etiam ignota penitus Theoria mea Patrocinio indiget, quod, si Tuo Nomine insignata prodeat in publicum, obtinebit sane validissimum, & secura vagabitur: Tu enim illam, parente velut hic orbatam suo, in dies nimirum discessuro, & quodammodo veluti posthumam post ipsum ejus discessum typis impressam, & in publicum prodeuntem tueberis, fovebisque.

Hæc sunt, quæ meum Tibi consilium probent, Princeps Celsissime: Tu, qua soles humanitate auctorem excipere, opus excipe, & si forte adhuc consilium ipsum Tibi visum fuerit improbandum; animum saltem æquus respice obsequentissimum Tibi, ac devinctissimum. Vale.

Dabam Viennæ in Collegio Academico Soc. JESU
Idibus Febr. MDCCLVIII.

sions for the most part only up to the point where I finally agreed with the opinions commonly held amongst philosophers, or where theories, now accepted as established, are the natural results of my deductions also; & this has in some measure helped to diminish the size of the volume. I had already published some instances, so to speak, of my general theory in several short dissertations issued at odd times; & on that account the theory has found some supporters amongst the university professors in Italy, & has already made its way into foreign countries. But now for the first time is it published as a whole in a single volume, the matter being indeed more than doubled in amount. This work I have carried out during the last month, being quit of the troublesome business that brought me to Vienna, and of all other cares; whilst I wait for seasonable time for my return journey through the everlasting snow to Italy. I have however used my utmost endeavours in preparing it, and adapting it to the ordinary intelligence of philosophers of only moderate attainments.

From this you will readily understand why I have not hesitated to bestow this book of mine upon you, & to dedicate it to you. My reason, as can be seen from what I have said, was twofold; in the first place, the nature of my theme is one that is not only not unsuitable, but is suitable in a high degree, for the consideration of a Christian priest; secondly, the power & dignity of the theme itself, which doubtless gives strength & vigour to my efforts—perchance rather feeble, but, as far as in me lay, earnest. Whatever in that respect I could gain by the exercise of thought, I have applied the whole of it to this matter; & consequently I think that nothing less unworthy of you can be produced by my poor ability; & that I should offer to you some such fruit of my labours was surely required of me, & as it were clamorously demanded by your great kindness to me; long ago in Rome you had enfolded my unworthy self in it, & here now you continue to be my patron, & do not disdain, from your exalted position, to honour me with every mark of your goodwill. There is still a further consideration, namely, that my Theory is as yet almost, if not quite, unknown in these parts, & therefore needs a patron's support; & this it will obtain most effectually, & will go on its way in security if it comes before the public franked with your name. For you will protect & cherish it, on its publication here, bereaved as it were of that parent whose departure in truth draws nearer every day; nay rather posthumous, since it will be seen in print only after he has gone.

Such are my grounds for hoping that you will approve my idea, most High Prince. I beg you to receive the work with the same kindness as you used to show to its author; &, if perchance the idea itself should fail to meet with your approval, at least regard favourably the intentions of your most humble & devoted servant. Farewell.

University College of the Society of Jesus,
VIENNA,
February 13th, 1758.

AD LECTOREM

EX EDITIONE VIENNENSI

ABES, amice Lector, Philosophiæ Naturalis Theoriam ex unica lege virium deductam, quam & ubi jam olim adumbraverim, vel etiam ex parte explicaverim, & qua occasione nunc uberius pertractandum, atque augendam etiam, susceperim, invenies in ipso primæ partis exordio. Libuit autem hoc opus dividere in partes tres, quarum prima continet explicationem Theoriæ ipsius, ac ejus analyticam deductionem, & vindicationem: secunda applicationem satis uberem ad Mechanicam; tertia applicationem ad Physicam.

Porro illud inprimis curandum duxi, ut omnia, quam liceret, dilucide exponerentur, nec sublimiore Geometria, aut Calculo indigerent. Et quidem in prima, ac tertia parte non tantum nullæ analyticæ, sed nec geometricæ demonstrationes occurrunt, paucissimis quibusdam, quibus indigeo, rejectis in adnotatiunculas, quas in fine paginarum quarundam invenies. Quædam autem admodum pauca, quæ majorem Algebræ, & Geometriæ cognitionem requirebant, vel erant complicatiora aliquando, & alibi a me jam edita, in fine operis apposui, quæ Supplementorum *appellavi nomine, ubi & ea addidi, quæ sentio de spatio, ac tempore, Theoriæ meæ consentanea, ac edita itidem jam alibi. In secunda parte, ubi ad Mechanicam applicatur Theoria, a geometricis, & aliquando etiam ab algebraicis demonstrationibus abstinere omnino non potui; sed eæ ejusmodi sunt, ut vix unquam requirant aliud, quam Euclideam Geometriam, & primas Trigonometriæ notiones maxime simplices, ac simplicem algorithmum.*

In prima quidem parte occurrunt Figuræ geometricæ complures, quæ prima fronte videbuntur etiam complicatæ rem ipsam intimius non perspectanti; verum eæ nihil aliud exhibent, nisi imaginem quandam rerum, quæ ipsis oculis per ejusmodi figuras sistuntur contemplandæ. Ejusmodi est ipsa illa curva, quæ legem virium exhibet. Invenio ego quidem inter omnia materiæ puncta vim quandam mutuam, quæ a distantiis pendet, & mutatis distantiis mutatur ita, ut in aliis attractiva sit, in aliis repulsiva, sed certa quadam, & continua lege. Leges ejusmodi variationis binarum quantitatum a se invicem pendentium, uti hic sunt distantia, & vis, exprimi possunt vel per analyticam formulam, vel per geometricam curvam; sed illa prior expressio & multo plures cognitiones requirit ad Algebram pertinentes, & imaginationem non ita adjuvat, ut hæc posterior, qua idcirco sum usus in ipsa prima operis parte, rejecta in Supplementa formula analytica, quæ & curvam, & legem virium ab illa expressam exhibeat.

Porro huc res omnis reducitur. Habetur in recta indefinita, quæ axis dicitur, punctum quoddam, a quo abscissa ipsius rectæ segmenta referunt distantias. Curva linea protenditur secundum rectam ipsam, circa quam etiam serpit, & eandem in pluribus secat punctis: rectæ a fine segmentorum erectæ perpendiculariter usque ad curvam, exprimunt vires, quæ majores sunt, vel minores, prout ejusmodi rectæ sunt itidem majores, vel minores; ac eædem ex attractivis migrant in repulsivis, vel vice versa, ubi illæ ipsæ perpendiculares rectæ directionem mutant, curva ab altera axis indefiniti plaga migrante ad alteram. Id quidem nullas requirit geometricas demonstrationes, sed meram cognitionem vocum quarundam, quæ vel ad prima pertinent Geometriæ elementa, & notissimæ sunt, vel ibi explicantur, ubi adhibentur. Notissima autem etiam est significatio vocis Asymptotus, *unde & crus asymptoticum curvæ appellatur; dicitur nimirum recta asymptotus cruris cujuspiam curvæ, cum ipsa recta in infinitum producta, ita ad curvilineum arcum productum itidem in infinitum semper accedit magis, ut distantia minuatur in infinitum, sed nusquam penitus evanescat, illis idcirco nunquam invicem convenientibus.*

Consideratio porro attenta curvæ propositæ in Fig. 1, *& rationis, qua per illam exprimitur*

THE PREFACE TO THE READER

THAT APPEARED IN THE VIENNA EDITION

EAR Reader, you have before you a Theory of Natural Philosophy deduced from a single law of Forces. You will find in the opening paragraphs of the first section a statement as to where the Theory has been already published in outline, & to a certain extent explained ; & also the occasion that led me to undertake a more detailed treatment & enlargement of it. For I have thought fit to divide the work into three parts ; the first of these contains the exposition of the Theory itself, its analytical deduction & its demonstration ; the second a fairly full application to Mechanics ; & the third an application to Physics.

The most important point, I decided, was for me to take the greatest care that everything, as far as was possible, should be clearly explained, & that there should be no need for higher geometry or for the calculus. Thus, in the first part, as well as in the third, there are no proofs by analysis ; nor are there any by geometry, with the exception of a very few that are absolutely necessary, & even these you will find relegated to brief notes set at the foot of a page. I have also added some very few proofs, that required a knowledge of higher algebra & geometry, or were of a rather more complicated nature, all of which have been already published elsewhere, at the end of the work ; I have collected these under the heading *Supplements* ; & in them I have included my views on Space & Time, which are in accord with my main Theory, & also have been already published elsewhere. In the second part, where the Theory is applied to Mechanics, I have not been able to do without geometrical proofs altogether ; & even in some cases I have had to give algebraical proofs. But these are of such a simple kind that they scarcely ever require anything more than Euclidean geometry, the first and most elementary ideas of trigonometry, and easy analytical calculations.

It is true that in the first part there are to be found a good many geometrical diagrams, which at first sight, before the text is considered more closely, will appear to be rather complicated. But these present nothing else but a kind of image of the subjects treated, which by means of these diagrams are set before the eyes for contemplation. The very curve that represents the law of forces is an instance of this. I find that between all points of matter there is a mutual force depending on the distance between them, & changing as this distance changes ; so that it is sometimes attractive, & sometimes repulsive, but always follows a definite continuous law. Laws of variation of this kind between two quantities depending upon one another, as distance & force do in this instance, may be represented either by an analytical formula or by a geometrical curve ; but the former method of representation requires far more knowledge of algebraical processes, & does not assist the imagination in the way that the latter does. Hence I have employed the latter method in the first part of the work, & relegated to the Supplements the analytical formula which represents the curve, & the law of forces which the curve exhibits.

The whole matter reduces to this. In a straight line of indefinite length, which is called the axis, a fixed point is taken ; & segments of the straight line cut off from this point represent the distances. A curve is drawn following the general direction of this straight line, & winding about it, so as to cut it in several places. Then perpendiculars that are drawn from the ends of the segments to meet the curve represent the forces ; these forces are greater or less, according as such perpendiculars are greater or less ; & they pass from attractive forces to repulsive, and vice versa, whenever these perpendiculars change their direction, as the curve passes from one side of the axis of indefinite length to the other side of it. Now this requires no geometrical proof, but only a knowledge of certain terms, which either belong to the first elementary principles of geometry, & are thoroughly well known, or are such as can be defined when they are used. The term *Asymptote* is well known, and from the same idea we speak of the branch of a curve as being asymptotic ; thus a straight line is said to be the asymptote to any branch of a curve when, if the straight line is indefinitely produced, it approaches nearer and nearer to the curvilinear arc which is also prolonged indefinitely in such manner that the distance between them becomes indefinitely diminished, but never altogether vanishes, so that the straight line & the curve never really meet.

A careful consideration of the curve given in Fig. 1, & of the way in which the relation

nexus inter vires, & distantias, est utique admodum necessaria ad intelligendam Theoriam ipsam, cujus ea est præcipua quædam veluti clavis, sine qua omnino incassum tentarentur cetera; sed & ejusmodi est, ut tironum, & sane etiam mediocrium, immo etiam longe infra mediocritatem collocatorum, captum non excedat, potissimum si viva accedat Professoris vox mediocriter etiam versati in Mechanica, cujus ope, pro certo habeo, rem ita patentem omnibus reddi posse, ut ii etiam, qui Geometriæ penitus ignari sunt, paucorum admodum explicatione vocabulorum accidente, eam ipsis oculis intueantur omnino perspicuam.

In tertia parte supponuntur utique nonnulla, quæ demonstrantur in secunda; sed ea ipsa sunt admodum pauca, & iis, qui geometricas demonstrationes fastidiunt, facile admodum exponi possunt res ipsæ ita, ut penitus etiam sine ullo Geometriæ adjumento percipiantur, quanquam sine iis ipsa demonstratio haberi non poterit; ut idcirco in eo differre debeat is, qui secundam partem attente legerit, & Geometriam calleat, ab eo, qui eam omittat, quod ille primus veritates in tertia parte adhibitis, ac ex secunda erutas, ad explicationem Physicæ, intuebitur per evidentiam ex ipsis demonstrationibus haustam, hic secundus easdem quodammodo per fidem Geometris adhibitam credet. Hujusmodi inprimis est illud, particulam compositam ex punctis etiam homogeneis, præditis lege virium proposita, posse per solam diversam ipsorum punctorum dispositionem aliam particulam per certum intervallum vel perpetuo attrahere, vel perpetuo repellere, vel nihil in eam agere, atque id ipsum viribus admodum diversis, & quæ respectu diversarum particularum diversæ sint, & diversæ respectu partium diversarum ejusdem particulæ, ac aliam particulam alicubi etiam urgeant in latus, unde plurium phænomenorum explicatio in Physica sponte fluit.

Verum qui omnem Theoriæ, & deductionum compagem aliquanto altius inspexerit, ac diligentius perpenderit, videbit, ut spero, me in hoc perquisitionis genere multo ulterius progressum esse, quam olim Newtonus ipse desideravit. Is enim in postremo Opticæ questione prolatis iis, quæ per vim attractivam, & vim repulsivam, mutata distantia ipsi attractivæ succedentem, explicari poterant, hæc addidit: "Atque hæc quidem omnia si ita sint, jam Natura universa valde erit simplex, & consimilis sui, perficiens nimirum magnos omnes corporum cælestium motus attractione gravitatis, quæ est mutua inter corpora illa omnia, & minores fere omnes particularum suarum motus alia aliqua vi attrahente, & repellente, quæ est inter particulas illas mutua." Aliquanto autem inferius de primigeniis particulis agens sic habet: "Porro videntur mihi hæ particulæ primigeniæ non modo in se vim inertiæ habere, motusque leges passivas illas, quæ ex vi ista necessario oriuntur; verum etiam motum perpetuo accipere a certis principiis actuosis, qualia nimirum sunt gravitas, & causa fermentationis, & cohærentia corporum. Atque hæc quidem principia considero non ut occultas qualitates, quæ ex specificis rerum formis oriri fingantur, sed ut universales Naturæ leges, quibus res ipsæ sunt formatæ. Nam principia quidem talia revera existere ostendunt phænomena Naturæ, licet ipsorum causæ quæ sint, nondum fuerit explicatum. Affirmare, singulas rerum species specificis præditas esse qualitatibus occultis, per quas eae vim certam in agendo habent, hoc utique est nihil dicere: at ex phænomenis Naturæ duo, vel tria derivare generalia motus principia, & deinde explicare, quemadmodum proprietates, & actiones rerum corporearum omnium ex istis principiis consequantur, id vero magnus esset factus in Philosophia progressus, etiamsi principiorum istorum causæ nondum essent cognitæ. Quare motus principia supradicta proponere non dubito, cum per Naturam universam latissime pateant."

Hæc ibi Newtonus, ubi is quidem magnos in Philosophia progressus facturum arbitratus est eum, qui ad duo, vel tria generalia motus principia ex Naturæ phænomenis derivata phænomenorum explicationem reduxerit, & sua principia protulit, ex quibus inter se diversis eorum aliqua tantummodo explicari posse censuit. Quid igitur, ubi & ea ipsa tria, & alia præcipua quæque, ut ipsa etiam impenetrabilitas, & impulsio reducantur ad principium unicum legitima ratiocinatione deductum? At id per meam unicam, & simplicem virium legem præstari, patebit sane consideranti operis totius Synopsim quandam, quam hic subjicio; sed multo magis opus ipsum diligentius pervolventi.

THE PRINTER AT VENICE
TO
THE READER

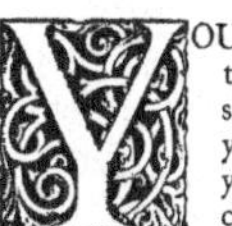

OU will be well aware, if you have read the public journals, with what applause the work which I now offer to you has been received throughout Europe since its publication at Vienna five years ago. Not to mention others, if you refer to the numbers of the Berne *Journal* for the early part of the year 1761, you will not fail to see how highly it has been esteemed. It contains an entirely new system of Natural Philosophy, which is already commonly known as the *Boscovichian theory*, from the name of its author, As a matter of fact, it is even now a subject of public instruction in several Universities in different parts; it is expounded not only in yearly theses or dissertations, both printed & debated; but also in several elementary books issued for the instruction of the young it is introduced, explained, & by many considered as their original. Any one, however, who wishes to obtain more detailed insight into the whole structure of the theory, the close relation that its several parts bear to one another, or its great fertility & wide scope for the purpose of deriving the whole of Nature, in her widest range, from a single simple law of forces; any one who wishes to make a deeper study of it must perforce study the work here offered.

All these considerations had from the first moved me to undertake a new edition of the work; in addition, there was the fact that I perceived that it would be a matter of some difficulty for copies of the Vienna edition to pass beyond the confines of Germany—indeed, at the present time, no matter how diligently they are inquired for, they are to be found on sale nowhere, or scarcely anywhere, in the rest of Europe. The system had its birth in Italy, & its outlines had already been sketched by the author in several dissertations published here in our own land; though, as luck would have it, the system itself was finally put into shape and published at Vienna, whither he had gone for a short time. I therefore thought it right that it should be disseminated throughout the whole of Europe, & that preferably as the product of an Italian press. I had in fact already commenced an edition founded on a copy of the Vienna edition, when it came to my knowledge that the author was greatly dissatisfied with the Vienna edition, taken in hand there after his departure; that innumerable printer's errors had crept in; that many passages, especially those that contain Algebraical formulæ, were ill-arranged and erroneous; lastly, that the author himself had in mind a complete revision, including certain alterations, to give a better finish to the work, together with certain additional matter.

That being the case, I was greatly desirous of obtaining a copy, revised & enlarged by himself; I also wanted to have him at hand whilst the edition was in progress, & that he should superintend the whole thing for himself. This, however, I was unable to procure during the last few years, in which he has been travelling through nearly the whole of Europe; until at last he came here, a little while ago, as he returned home from his lengthy wanderings, & stayed here to assist me during the whole time that the edition was in hand. He, in addition to our regular proof-readers, himself also used every care in correcting the proof; even then, however, he has not sufficient confidence in himself as to imagine that not the slightest thing has escaped him. For it is a characteristic of the human mind that it cannot concentrate long on the same subject with sufficient attention.

It follows that this ought to be considered in some measure as a first & original edition; any one who compares it with that issued at Vienna will soon see the difference between them. Many of the minor alterations are made for the purpose of rendering certain passages more elegant & clear; there are, however, especially at the foot of a page, slight additions also, or slight changes made after the type was set up, merely for the purpose of filling up gaps that were left here & there—these gaps being due to the fact that several sheets were being set at the same time by different compositors, and four presses were kept hard at work together. As he was at hand, this could easily be done without causing any disturbance of the sentences or the pagination.

nexus inter vires, & distantias, est utique admodum necessaria ad intelligendam Theoriam ipsam, cujus ea est præcipua quædam veluti clavis, sine qua omnino incassum tentarentur cetera; sed & ejusmodi est, ut tironum, & sane etiam mediocrium, immo etiam longe infra mediocritatem collocatorum, captum non excedat, potissimum si viva accedat Professoris vox mediocriter etiam versati in Mechanica, cujus ope, pro certo habeo, rem ita patentem omnibus reddi posse, ut ii etiam, qui Geometriæ penitus ignari sunt, paucorum admodum explicatione vocabulorum accidente, eam ipsis oculis intueantur omnino perspicuam.

In tertia parte supponuntur utique nonnulla, quæ demonstrantur in secunda; sed ea ipsa sunt admodum pauca, & iis, qui geometricas demonstrationes fastidiunt, facile admodum exponi possunt res ipsæ ita, ut penitus etiam sine ullo Geometriæ adjumento percipiantur, quanquam sine iis ipsa demonstratio haberi non poterit; ut idcirco in eo differre debeat is, qui secundam partem attente legerit, & Geometriam calleat, ab eo, qui eam omittat, quod ille primus veritates in tertia parte adhibitis, ac ex secunda erutas, ad explicationem Physicæ, intuebitur per evidentiam ex ipsis demonstrationibus haustam, hic secundus easdem quodammodo per fidem Geometris adhibitam credet. Hujusmodi inprimis est illud, particulam compositam ex punctis etiam homogeneis, præditis lege virium proposita, posse per solam diversam ipsorum punctorum dispositionem aliam particulam per certum intervallum vel perpetuo attrahere, vel perpetuo repellere, vel nihil in eam agere, atque id ipsum viribus admodum diversis, & quæ respectu diversarum particularum diversæ sint, & diversæ respectu partium diversarum ejusdem particulæ, ac aliam particulam alicubi etiam urgeant in latus, unde plurium phænomenorum explicatio in Physica sponte fluit.

Verum qui omnem Theoriæ, & deductionum compagem aliquanto altius inspexerit, ac diligentius perpenderit, videbit, ut spero, me in hoc perquisitionis genere multo ulterius progressum esse, quam olim Newtonus ipse desideravit. Is enim in postremo Opticæ questione prolatis iis, quæ per vim attractivam, & vim repulsivam, mutata distantia ipsi attractivæ succedentem, explicari poterant, hæc addidit: " Atque hæc quidem omnia si ita sint, jam Natura universa valde erit simplex, & consimilis sui, perficiens nimirum magnos omnes corporum cælestium motus attractione gravitatis, quæ est mutua inter corpora illa omnia, & minores fere omnes particularum suarum motus alia aliqua vi attrahente, & repellente, quæ est inter particulas illas mutua." Aliquanto autem inferius de primigeniis particulis agens sic habet: " Porro videntur mihi hæ particulæ primigeniæ non modo in se vim inertiæ habere, motusque leges passivas illas, quæ ex vi ista necessario oriuntur; verum etiam motum perpetuo accipere a certis principiis actuosis, qualia nimirum sunt gravitas, & causa fermentationis, & cohærentia corporum. Atque hæc quidem principia considero non ut occultas qualitates, quæ ex specificis rerum formis oriri fingantur, sed ut universales Naturæ leges, quibus res ipsæ sunt formatæ. Nam principia quidem talia revera existere ostendunt phænomena Naturæ, licet ipsorum causæ quæ sint, nondum fuerit explicatum. Affirmare, singulas rerum species specificis præditas esse qualitatibus occultis, per quas eae vim certam in agendo habent, hoc utique est nihil dicere: at ex phænomenis Naturæ duo, vel tria derivare generalia motus principia, & deinde explicare, quemadmodum proprietates, & actiones rerum corporearum omnium ex istis principiis consequantur, id vero magnus esset factus in Philosophia progressus, etiamsi principiorum istorum causæ nondum essent cognitæ. Quare motus principia supradicta proponere non dubito, cum per Naturam universam latissime pateant."

Hæc ibi Newtonus, ubi is quidem magnos in Philosophia progressus facturum arbitratus est eum, qui ad duo, vel tria generalia motus principia ex Naturæ phænomenis derivata phænomenorum explicationem reduxerit, & sua principia protulit, ex quibus inter se diversis eorum aliqua tantummodo explicari posse censuit. Quid igitur, ubi & ea ipsa tria, & alia præcipua quæque, ut ipsa etiam impenetrabilitas, & impulsio reducantur ad principium unicum legitima ratiocinatione deductum? At id per meam unicam, & simplicem virium legem præstari, patebit sane consideranti operis totius Synopsim quandam, quam hic subjicio; sed multo magis opus ipsum diligentius pervolventi.

THE PRINTER AT VENICE
TO
THE READER

OU will be well aware, if you have read the public journals, with what applause the work which I now offer to you has been received throughout Europe since its publication at Vienna five years ago. Not to mention others, if you refer to the numbers of the Berne *Journal* for the early part of the year 1761, you will not fail to see how highly it has been esteemed. It contains an entirely new system of Natural Philosophy, which is already commonly known as the *Boscovichian theory*, from the name of its author, As a matter of fact, it is even now a subject of public instruction in several Universities in different parts; it is expounded not only in yearly theses or dissertations, both printed & debated; but also in several elementary books issued for the instruction of the young it is introduced, explained, & by many considered as their original. Any one, however, who wishes to obtain more detailed insight into the whole structure of the theory, the close relation that its several parts bear to one another, or its great fertility & wide scope for the purpose of deriving the whole of Nature, in her widest range, from a single simple law of forces; any one who wishes to make a deeper study of it must perforce study the work here offered.

All these considerations had from the first moved me to undertake a new edition of the work; in addition, there was the fact that I perceived that it would be a matter of some difficulty for copies of the Vienna edition to pass beyond the confines of Germany—indeed, at the present time, no matter how diligently they are inquired for, they are to be found on sale nowhere, or scarcely anywhere, in the rest of Europe. The system had its birth in Italy, & its outlines had already been sketched by the author in several dissertations published here in our own land; though, as luck would have it, the system itself was finally put into shape and published at Vienna, whither he had gone for a short time. I therefore thought it right that it should be disseminated throughout the whole of Europe, & that preferably as the product of an Italian press. I had in fact already commenced an edition founded on a copy of the Vienna edition, when it came to my knowledge that the author was greatly dissatisfied with the Vienna edition, taken in hand there after his departure; that innumerable printer's errors had crept in; that many passages, especially those that contain Algebraical formulæ, were ill-arranged and erroneous; lastly, that the author himself had in mind a complete revision, including certain alterations, to give a better finish to the work, together with certain additional matter.

That being the case, I was greatly desirous of obtaining a copy, revised & enlarged by himself; I also wanted to have him at hand whilst the edition was in progress, & that he should superintend the whole thing for himself. This, however, I was unable to procure during the last few years, in which he has been travelling through nearly the whole of Europe; until at last he came here, a little while ago, as he returned home from his lengthy wanderings, & stayed here to assist me during the whole time that the edition was in hand. He, in addition to our regular proof-readers, himself also used every care in correeting the proof; even then, however, he has not sufficient confidence in himself as to imagine that not the slightest thing has escaped him. For it is a characteristic of the human mind that it cannot concentrate long on the same subject with sufficient attention.

It follows that this ought to be considered in some measure as a first & original edition; any one who compares it with that issued at Vienna will soon see the difference between them. Many of the minor alterations are made for the purpose of rendering certain passages more elegant & clear; there are, however, especially at the foot of a page, slight additions also, or slight changes made after the type was set up, merely for the purpose of filling up gaps that were left here & there—these gaps being due to the fact that several sheets were being set at the same time by different compositors, and four presses were kept hard at work together. As he was at hand, this could easily be done without causing any disturbance of the sentences or the pagination.

Inter mutationes occurret ordo numerorum mutatus in paragraphis : nam numerus 82 de novo accessit totus : deinde is, qui fuerat 261 discerptus est in 5 ; demum in Appendice post num. 534 factæ sunt & mutatiunculæ nonnullæ, & additamenta plura in iis, quæ pertinent ad sedem animæ.

Supplementorum ordo mutatus est itidem ; quæ enim fuerant 3, & 4, jam sunt 1, & 2 : nam eorum usus in ipso Opere ante alia occurrit. Illi autem, quod prius fuerat primum, nunc autem est tertium, accessit in fine Scholium tertium, quod pluribus numeris complectitur dissertatiunculam integram de argumento, quod ante aliquot annos in Parisiensi Academia controversiæ occasionem exhibuit in Encyclopedico etiam dictionario attactum, in qua dissertatiuncula demonstrat Auctor non esse, cur ad vim exprimendam potentia quæpiam distantiæ adhibeatur potius, quam functio.

Accesserunt per totum Opus notulae marginales, in quibus corum, quæ pertractantur argumenta exponuntur brevissima, quorum ope unico obtutu videri possint omnia, & in memoriam facile revocari.

Postremo loco ad calcem Operis additus est fusior catalogus eorum omnium, quæ huc usque ab ipso Auctore sunt edita, quorum collectionem omnem expolitam, & correctam, ac eorum, quæ nondum absoluta sunt, continuationem meditatur, aggressurus illico post suum regressum in Urbem Romam, quo properat. Hic catalogus impressus fuit Venetisis ante bosce duos annos in reimpressione ejus poematis de Solis ac Lunæ defectibus. Porro eam omnium suorum Operum Collectionem, ubi ipse adornaverit, typis ego meis excudendam suscipiam, quam magnificentissime potero.

Hæc erant, quæ te monendum censui ; tu laboribus nostris fruere, & vive ſelix.

THE PREFACE TO THE READER

THAT APPEARED IN THE VIENNA EDITION

DEAR Reader, you have before you a Theory of Natural Philosophy deduced from a single law of Forces. You will find in the opening paragraphs of the first section a statement as to where the Theory has been already published in outline, & to a certain extent explained; & also the occasion that led me to undertake a more detailed treatment & enlargement of it. For I have thought fit to divide the work into three parts; the first of these contains the exposition of the Theory itself, its analytical deduction & its demonstration; the second a fairly full application to Mechanics; & the third an application to Physics.

The most important point, I decided, was for me to take the greatest care that everything, as far as was possible, should be clearly explained, & that there should be no need for higher geometry or for the calculus. Thus, in the first part, as well as in the third, there are no proofs by analysis; nor are there any by geometry, with the exception of a very few that are absolutely necessary, & even these you will find relegated to brief notes set at the foot of a page. I have also added some very few proofs, that required a knowledge of higher algebra & geometry, or were of a rather more complicated nature, all of which have been already published elsewhere, at the end of the work; I have collected these under the heading *Supplements*; & in them I have included my views on Space & Time, which are in accord with my main Theory, & also have been already published elsewhere. In the second part, where the Theory is applied to Mechanics, I have not been able to do without geometrical proofs altogether; & even in some cases I have had to give algebraical proofs. But these are of such a simple kind that they scarcely ever require anything more than Euclidean geometry, the first and most elementary ideas of trigonometry, and easy analytical calculations.

It is true that in the first part there are to be found a good many geometrical diagrams, which at first sight, before the text is considered more closely, will appear to be rather complicated. But these present nothing else but a kind of image of the subjects treated, which by means of these diagrams are set before the eyes for contemplation. The very curve that represents the law of forces is an instance of this. I find that between all points of matter there is a mutual force depending on the distance between them, & changing as this distance changes; so that it is sometimes attractive, & sometimes repulsive, but always follows a definite continuous law. Laws of variation of this kind between two quantities depending upon one another, as distance & force do in this instance, may be represented either by an analytical formula or by a geometrical curve; but the former method of representation requires far more knowledge of algebraical processes, & does not assist the imagination in the way that the latter does. Hence I have employed the latter method in the first part of the work, & relegated to the Supplements the analytical formula which represents the curve, & the law of forces which the curve exhibits.

The whole matter reduces to this. In a straight line of indefinite length, which is called the axis, a fixed point is taken; & segments of the straight line cut off from this point represent the distances. A curve is drawn following the general direction of this straight line, & winding about it, so as to cut it in several places. Then perpendiculars that are drawn from the ends of the segments to meet the curve represent the forces; these forces are greater or less, according as such perpendiculars are greater or less; & they pass from attractive forces to repulsive, and vice versa, whenever these perpendiculars change their direction, as the curve passes from one side of the axis of indefinite length to the other side of it. Now this requires no geometrical proof, but only a knowledge of certain terms, which either belong to the first elementary principles of geometry, & are thoroughly well known, or are such as can be defined when they are used. The term *Asymptote* is well known, and from the same idea we speak of the branch of a curve as being asymptotic; thus a straight line is said to be the asymptote to any branch of a curve when, if the straight line is indefinitely produced, it approaches nearer and nearer to the curvilinear arc which is also prolonged indefinitely in such manner that the distance between them becomes indefinitely diminished, but never altogether vanishes, so that the straight line & the curve never really meet.

A careful consideration of the curve given in Fig. 1, & of the way in which the relation

nexus inter vires, & distantias, est utique admodum necessaria ad intelligendam Theoriam ipsam, cujus ea est præcipua quædam veluti clavis, sine qua omnino incassum tentarentur cetera; sed & ejusmodi est, ut tironum, & sane etiam mediocrium, immo etiam longe infra mediocritatem collocatorum, captum non excedat, potissimum si viva accedat Professoris vox mediocriter etiam versati in Mechanica, cujus ope, pro certo habeo, rem ita patentem omnibus reddi posse, ut ii etiam, qui Geometriæ penitus ignari sunt, paucorum admodum explicatione vocabulorum accidente, eam ipsis oculis intueantur omnino perspicuam.

In tertia parte supponuntur utique nonnulla, quæ demonstrantur in secunda; sed ea ipsa sunt admodum pauca, & iis, qui geometricas demonstrationes fastidiunt, facile admodum exponi possunt res ipsæ ita, ut penitus etiam sine ullo Geometriæ adjumento percipiantur, quanquam sine iis ipsa demonstratio haberi non poterit; ut idcirco in eo differre debeat is, qui secundam partem attente legerit, & Geometriam calleat, ab eo, qui eam omittat, quod ille primus veritates in tertia parte adhibitis, ac ex secunda erutas, ad explicationem Physicæ, intuebitur per evidentiam ex ipsis demonstrationibus haustam, hic secundus easdem quodammodo per fidem Geometris adhibitam credet. Hujusmodi inprimis est illud, particulam compositam ex punctis etiam homogeneis, præditis lege virium proposita, posse per solam diversam ipsorum punctorum dispositionem aliam particulam per certum intervallum vel perpetuo attrahere, vel perpetuo repellere, vel nihil in eam agere, atque id ipsum viribus admodum diversis, & quæ respectu diversarum particularum diversæ sint, & diversæ respectu partium diversarum ejusdem particulæ, ac aliam particulam alicubi etiam urgeant in latus, unde plurium phænomenorum explicatio in Physica sponte fluit.

Verum qui omnem Theoriæ, & deductionum compagem aliquanto altius inspexerit, ac diligentius perpenderit, videbit, ut spero, me in hoc perquisitionis genere multo ulterius progressum esse, quam olim Newtonus ipse desideravit. Is enim in postremo Opticæ questione prolatis iis, quæ per vim attractivam, & vim repulsivam, mutata distantia ipsi attractivæ succedentem, explicari poterant, hæc addidit: "Atque hæc quidem omnia si ita sint, jam Natura universa valde erit simplex, & consimilis sui, perficiens nimirum magnos omnes corporum cælestium motus attractione gravitatis, quæ est mutua inter corpora illa omnia, & minores fere omnes particularum suarum motus alia aliqua vi attrahente, & repellente, quæ est inter particulas illas mutua." Aliquanto autem inferius de primigeniis particulis agens sic habet: "Porro videntur mihi hæ particulæ primigeniæ non modo in se vim inertiæ habere, motusque leges passivas illas, quæ ex vi ista necessario oriuntur; verum etiam motum perpetuo accipere a certis principiis actuosis, qualia nimirum sunt gravitas, & causa fermentationis, & cohærentia corporum. Atque hæc quidem principia considero non ut occultas qualitates, quæ ex specificis rerum formis oriri fingantur, sed ut universales Naturæ leges, quibus res ipsæ sunt formatæ. Nam principia quidem talia revera existere ostendunt phænomena Naturæ, licet ipsorum causæ quæ sint, nondum fuerit explicatum. Affirmare, singulas rerum species specificis præditas esse qualitatibus occultis, per quas eae vim certam in agendo habent, hoc utique est nihil dicere: at ex phænomenis Naturæ duo, vel tria derivare generalia motus principia, & deinde explicare, quemadmodum proprietates, & actiones rerum corporearum omnium ex istis principiis consequantur, id vero magnus esset factus in Philosophia progressus, etiamsi principiorum istorum causæ nondum essent cognitæ. Quare motus principia supradicta proponere non dubito, cum per Naturam universam latissime pateant."

Hæc ibi Newtonus, ubi is quidem magnos in Philosophia progressus facturum arbitratus est eum, qui ad duo, vel tria generalia motus principia ex Naturæ phænomenis derivata phænomenorum explicationem reduxerit, & sua principia protulit, ex quibus inter se diversis eorum aliqua tantummodo explicari posse censuit. Quid igitur, ubi & ea ipsa tria, & alia præcipua quæque, ut ipsa etiam impenetrabilitas, & impulsio reducantur ad principium unicum legitima ratiocinatione deductum? At id per meam unicam, & simplicem virium legem præstari, patebit sane consideranti operis totius Synopsim quandam, quam hic subjicio; sed multo magis opus ipsum diligentius pervolventi.

between the forces & the distances is represented by it, is absolutely necessary for the understanding of the Theory itself, to which it is as it were the chief key, without which it would be quite useless to try to pass on to the rest. But it is of such a nature that it does not go beyond the capacity of beginners, not even of those of very moderate ability, or of classes even far below the level of mediocrity; especially if they have the additional assistance of a teacher's voice, even though he is only moderately familiar with Mechanics. By his help, I am sure, the subject can be made clear to every one, so that those of them that are quite ignorant of geometry, given the explanation of but a few terms, may get a perfectly good idea of the subject by ocular demonstration.

In the third part, some of the theorems that have been proved in the second part are certainly assumed, but there are very few such; &, for those who do not care for geometrical proofs, the facts in question can be quite easily stated in such a manner that they can be completely understood without any assistance from geometry, although no real demonstration is possible without them. There is thus bound to be a difference between the reader who has gone carefully through the second part, & who is well versed in geometry, & him who omits the second part; in that the former will regard the facts, that have been proved in the second part, & are now employed in the third part for the explanation of Physics, through the evidence derived from the demonstrations of these facts, whilst the second will credit these same facts through the mere faith that he has in geometricians. A specially good instance of this is the fact, that a particle composed of points quite homogeneous, subject to a law of forces as stated, may, merely by altering the arrangement of those points, either continually attract, or continually repel, or have no effect at all upon, another particle situated at a known distance from it; & this too, with forces that differ widely, both in respect of different particles & in respect of different parts of the same particle; & may even urge another particle in a direction at right angles to the line joining the two, a fact that readily gives a perfectly natural explanation of many physical phenomena.

Anyone who shall have studied somewhat closely the whole system of my Theory, & what I deduce from it, will see, I hope, that I have advanced in this kind of investigation much further than Newton himself even thought open to his desires. For he, in the last of his "Questions" in his *Opticks*, after stating the facts that could be explained by means of an attractive force, & a repulsive force that takes the place of the attractive force when the distance is altered, has added these words :—"Now if all these things are as stated, then the whole of Nature must be exceedingly simple in design, & similar in all its parts, accomplishing all the mighty motions of the heavenly bodies, as it does, by the attraction of gravity, which is a mutual force between any two bodies of the whole system; and Nature accomplishes nearly all the smaller motions of their particles by some other force of attraction or repulsion, which is mutual between any two of those particles." Farther on, when he is speaking about elementary particles, he says :—"Moreover, it appears to me that these elementary particles not only possess an essential property of inertia, & laws of motion, though only passive, which are the necessary consequences of this property; but they also constantly acquire motion from the influence of certain active principles such as, for instance, gravity, the cause of fermentation, & the cohesion of solids. I do not consider these principles to be certain mysterious qualities feigned as arising from characteristic forms of things, but as universal laws of Nature, by the influence of which these very things have been created. For the phenomena of Nature show that these principles do indeed exist, although their nature has not yet been elucidated. To assert that each & every species is endowed with a mysterious property characteristic to it, due to which it has a definite mode in action, is really equivalent to saying nothing at all. On the other hand, to derive from the phenomena of Nature two or three general principles, & then to explain how the properties & actions of all corporate things follow from those principles, this would indeed be a mighty advance in philosophy, even if the causes of those principles had not at the time been discovered. For these reasons I do not hesitate in bringing forward the principles of motion given above, since they are clearly to be perceived throughout the whole range of Nature."

These are the words of Newton, & therein he states his opinion that he indeed will have made great strides in philosophy who shall have reduced the explanation of phenomena to two or three general principles derived from the phenomena of Nature; & he brought forward his own principles, themselves differing from one another, by which he thought that some only of the phenomena could be explained. What then if not only the three he mentions, but also other important principles, such as impenetrability & impulsive force, be reduced to a single principle, deduced by a process of rigorous argument! It will be quite clear that this is exactly what is done by my single simple law of forces, to anyone who studies a kind of synopsis of the whole work, which I add below; but it will be far more clear to him who studies the whole work with some earnestness.

SYNOPSIS TOTIUS OPERIS

EX EDITIONE VIENNENSI

PARS I

*1 PRIMIS sex numeris exhibeo, quando, & qua occasione Theoriam meam invenerim, ac ubi hucusque de ea egerim in dissertationibus jam editis, quid ea commune habeat cum Leibnitiana, quid cum Newtoniana Theoria, in quo ab utraque discrepet, & vero etiam utrique præstet: addo, quid alibi promiserim pertinens ad æquilibrium, & oscillationis centrum, & quemadmodum iis nunc inventis, ac ex unico simplicissimo, ac elegantissimo theoremate profluentibus omnino sponte, cum dissertatiunculam brevem meditarer, jam eo consilio rem aggressus; repente mihi in opus integrum justæ molis evaserit tractatio.

7 Tum usque ad num. 11 expono Theoriam ipsam: materiam constantem punctis prorsus simplicibus, indivisibilibus, & inextensis, ac a se invicem distantibus, quæ puncta habeant singula vim inertiæ, & præterea vim activam mutuam pendentem a distantiis, ut nimirum, data distantia, detur & magnitudo, & directio vis ipsius, mutata autem distantia, mutetur vis ipsa, quæ, imminuta distantia in infinitum, sit repulsiva, & quidem excrescens in infinitum: aucta autem distantia, minuatur, evanescat, mutetur in attractivam crescentem primo, tum decrescentem, evanescentem, abeuntem iterum in repulsivam, idque per multas vices, donec demum in majoribus distantiis abeat in attractivam decrescentem ad sensum in ratione reciproca duplicata distantiarum; quem nexum virium cum distantiis, & vero etiam carum transitum a positivis ad negativas, sive a repulsivis ad attractivas, vel vice versa, oculis ipsis propono in vi, qua binæ elastri cuspides conantur ad es invicem accedere, vel a se invicem recedere, prout sunt plus justo distractæ, vel contractæ.

11 Inde ad num. 16 ostendo, quo pacto id non sit aggregatum quoddam virium temere coalescentium, sed per unicam curvam continuam exponatur ope abscissarum exprimentium distantias, & ordinatarum exprimentium vires, cujus curvæ ductum, & naturam expono, ac ostendo, in quo differat ab hyperbola illa gradus tertii, quæ Newtonianum gravitatem exprimit: ac demum ibidem & argumentum, & divisionem propono operis totius.

16 Hisce expositis gradum facio ad exponendam totam illam analysim, qua ego ad ejusmodi Theoriam deveni, & ex qua ipsam arbitror directa, & solidissima ratiocinatione deduci totam. Contendo nimirum usque ad numerum 19 illud, in collisione corporum debere vel haberi compenetrationem, vel violari legem continuitatis, velocitate mutata per saltum, si cum inæqualibus velocitatibus deveniant ad immediatum contactum, quæ continuitatis lex cum (ut evinco) debeat omnino observari, illud infero, antequam ad contactum deveniant corpora, debere mutari corum velocitates per vim quandam, quæ sit par extinguendæ velocitati, vel velocitatum differentiæ, cuivis utcunque magnæ.

19 A num. 19 ad 28 expendo effugium, quo ad eludendam argumenti mei vim utuntur ii, qui negant corpora dura, qua quidem responsione uti non possunt Newtoniani, & Corpusculares generaliter, qui elementares corporum particulas assumunt prorsus duras: qui autem omnes utcunque parvas corporum particulas molles admittunt, vel elasticas, difficultatem non effugiunt, sed transferunt ad primas superficies, vel puncta, in quibus committeretur omnino saltus, & lex continuitatis violaretur: ibidem quendam verborum lusum evolvo frustra adhibitum ad eludendam argumenti mei vim.

* Series numerorum, quibus tractari incipiunt, quæ sunt in textu.

SYNOPSIS OF THE WHOLE WORK

(FROM THE VIENNA EDITION)

PART I

N the first six articles, I state the time at which I evolved my Theory, what led me to it, & where I have discussed it hitherto in essays already pub lished: also what it has in common with the theories of Leibniz and Newton; in what it differs from either of these, & in what it is really superior to them both. In addition I state what I have published elsewhere about equilibrium & the centre of oscillation; & how, having found out that these matters followed quite easily from a single theorem of the most simple & elegant kind, I proposed to write a short essay thereon; but when I set to work to deduce the matter from this principle, the discussion, quite unexpectedly to me, developed into a whole work of considerable magnitude. 1*

From this until Art. 11, I explain the Theory itself: that matter is unchangeable, and consists of points that are perfectly simple, indivisible, of no extent, & separated from one another; that each of these points has a property of inertia, & in addition a mutual active force depending on the distance in such a way that, if the distance is given, both the magnitude & the direction of this force are given; but if the distance is altered, so also is the force altered; & if the distance is diminished indefinitely, the force is repulsive, & in fact also increases indefinitely; whilst if the distance is increased, the force will be diminished, vanish, be changed to an attractive force that first of all increases, then decreases, vanishes, is again turned into a repulsive force, & so on many times over; until at greater distances it finally becomes an attractive force that decreases approximately in the inverse ratio of the squares of the distances. This connection between the forces & the distances, & their passing from positive to negative, or from repulsive to attractive, & conversely, I illustrate by the force with which the two ends of a spring strive to approach towards, or recede from, one another, according as they are pulled apart, or drawn together, by more than the natural amount. 7

From here on to Art. 16 I show that it is not merely an aggregate of forces combined haphazard, but that it is represented by a single continuous curve, by means of abscissæ representing the distances & ordinates representing the forces. I expound the construction & nature of this curve; & I show how it differs from the hyperbola of the third degree which represents Newtonian gravitation. Finally, here too I set forth the scope of the whole work & the nature of the parts into which it is divided. 11

These statements having been made, I start to expound the whole of the analysis, by which I came upon a Theory of this kind, & from which I believe I have deduced the whole of it by a straightforward & perfectly rigorous chain of reasoning. I contend indeed, from here on until Art. 19, that, in the collision of solid bodies, either there must be compenetration, or the Law of Continuity must be violated by a sudden change of velocity, if the bodies come into immediate contact with unequal velocities. Now since the Law of Continuity must (as I prove that it must) be observed in every case, I infer that, before the bodies reach the point of actual contact, their velocities must be altered by some force which is capable of destroying the velocity, or the difference of the velocities, no matter how great that may be. 16

From Art. 19 to Art. 28 I consider the artifice, adopted for the purpose of evading the strength of my argument by those who deny the existence of hard bodies; as a matter of fact this cannot be used as an argument against me by the Newtonians, or the Corpuscularians in general, for they assume that the elementary particles of solids are perfectly hard. Moreover, those who admit that all the particles of solids, however small they may be, are soft or elastic, yet do not escape the difficulty, but transfer it to prime surfaces, or points; & here a sudden change would be made & the Law of Continuity violated. In the same connection I consider a certain verbal quibble, used in a vain attempt to foil the force of my reasoning. 19

* These numbers are the numbers of the articles, in which the matters given in the text are first discussed.

28 Sequentibus num. 28 & 29 binas alias responsiones rejicio aliorum, quarum altera, ut mei argumenti vis elidatur, affirmat quispiam, prima materiæ clementa compenetrari, alter dicuntur materiæ puncta adhuc moveri ad se invicem, ubi localiter omnino quiescunt, & contra primum effugium evinco impenetrabilitatem ex inductione; contra secundum expono æquivocationem quandam in significatione vocis *motus*, cui æquivocationi totum innititur.

30 Hinc num. 30, & 31 ostendo, in quo a Mac-Laurino dissentiam, qui considerata eadem, quam ego contemplatus sum, collisione corporum, conclusit, continuitatis legem violari, cum ego eandem illæsam esse debere ratus ad totam devenerim Theoriam meam.

32 Hic igitur, ut meæ deductionis vim exponam, in ipsam continuitatis legem inquiro, ac a num. 32 ad 38 expono, quid ipsa sit, quid mutatio continua per gradus omnes intermedios, quæ nimirum excludat omnem saltum ab una magnitudine ad aliam sine transitu per 39 intermedias, ac Geometriam etiam ad explicationem rei in subsidium advoco: tum eam probo primum ex inductione, ac in ipsum inductionis principium inquirens usque ad num. 44, exhibeo, unde habeatur ejusdem principii vis, ac ubi id adhiberi possit, rem ipsam illustrans exemplo impenetrabilitatis erutæ passim per inductionem, donec demum ejus vim 45 applicem ad legem continuitatis demonstrandam: ac sequentibus numeris casus evolvo quosdam binarum classium, in quibus continuitatis lex videtur lædi nec tamen læditur.

48 Post probationem principii continuitatis petitam ab inductione, aliam num. 48 ejus probationem aggredior metaphysicam quandam, ex necessitate utriusque limitis in quantitatibus realibus, vel seriebus quantitatum realium finitis, quæ nimirum nec suo principio, nec suo fine carere possunt. Ejus rationis vim ostendo in motu locali, & in Geometria 52 sequentibus duobus numeris: tum num. 52 expono difficultatem quandam, quæ petitur ex eo, quod in momento temporis, in quo transitur a *non esse* ad *esse*, videatur juxta ejusmodi Theoriam debere simul haberi ipsum *esse*, & *non esse*, quorum alterum ad finem præcedentis seriei statuum pertinet, alterum ad sequentis initium, ac solutionem ipsius fuse evolvo, Geometria etiam ad rem oculo ipsi sistendam vocata in auxilium.

63 Num. 63, post epilogum eorum omnium, quæ de lege continuitatis sunt dicta, id principium applico ad excludendum saltum immediatum ab una velocitate ad aliam, sine transitu per intermedias, quod & inductionem læderet pro continuitate amplissimam, & induceret pro ipso momento temporis, in quo fieret saltus, binas velocitates, ultimam nimirum seriei præcedentis, & primam novæ, cum tamen duas simul velocitates idem mobile habere omnino non possit. Id autem ut illustrem, & evincam, usque ad num. 72 considero velocitatem ipsam, ubi potentialem quandam, ut appello, velocitatem ab actuali secerno, & multa, quæ ad ipsarum naturam, ac mutationes pertinent, diligenter evolvo, nonnullis etiam, quæ inde contra meæ Theoriæ probationem objici possunt, dissolutis.

His expositis concludo jam illud ex ipsa continuitate, ubi corpus quodpiam velocius movetur post aliud lentius, ad contactum immediatum cum illa velocitatum inæqualitate deveniri non posse, in quo scilicet contactu primo mutaretur vel utriusque velocitas, vel alterius, per saltum, sed debere mutationem velocitatis incipere ante contactum ipsum. 73 Hinc num. 73 infero, debere haberi mutationis causam, quæ appelletur vis: tum num. 74 74 hanc vim debere esse mutuam, & agere in partes contrarias, quod per inductionem evinco, 75 & inde infero num. 75, appellari posse repulsivam ejusmodi vim mutuam, ac ejus legem exquirendam propono. In ejusmodi autem perquisitione usque ad num. 80 invenio illud, debere vim ipsam imminutis distantiis crescere in infinitum ita ut par sit extinguendæ velocitati utcunque magnæ; tum & illud, imminutis in infinitum etiam distantiis, debere in infinitum augeri, in maximis autem debere esse e contrario attractivam, uti est gravitas: inde vero colligo limitem inter attractionem, & repulsionem: tum sensim plures, ac etiam plurimos ejusmodi limites invenio, sive transitus ab attractione ad repulsionem, & vice versa, ac formam totius curvæ per ordinatas suas exprimentis virium legem determino.

28 In the next articles, 28 & 29, I refute a further pair of arguments advanced by others; in the first of these, in order to evade my reasoning, someone states that there is compenetration of the primary elements of matter; in the second, the points of matter are said to be moved with regard to one another, even when they are absolutely at rest as regards position. In reply to the first artifice, I prove the principle of impenetrability by induction; & in reply to the second, I expose an equivocation in the meaning of the term *motion*, an equivocation upon which the whole thing depends.

30 Then, in Art. 30, 31, I show in what respect I differ from Maclaurin, who, having considered the same point as myself, came to the conclusion that in the collision of bodies the Law of Continuity was violated; whereas I obtained the whole of my Theory from the assumption that this law must be unassailable.

32 At this point therefore, in order that the strength of my deductive reasoning might be shown, I investigate the Law of Continuity; and from Art. 32 to Art. 38, I set forth its nature, & what is meant by a continuous change through all intermediate stages, such as to exclude any sudden change from any one magnitude to another except by a passage through intermediate stages; & I call in geometry as well to help my explanation of the matter. 39 Then I investigate its truth first of all by induction; &, investigating the principle of induction itself, as far as Art. 44, I show whence the force of this principle is derived, & where it can be used. I give by way of illustration an example in which impenetrability is derived entirely by induction; & lastly I apply the force of the principle to demonstrate the Law of Continuity. 45 In the articles that follow I consider certain cases of two kinds, in which the Law of Continuity appears to be violated, but is not however really violated.

48 After this proof of the principle of continuity procured through induction, in Art. 48, I undertake another proof of a metaphysical kind, depending upon the necessity of a limit on either side for either real quantities or for a finite series of real quantities; & indeed it is impossible that these limits should be lacking, either at the beginning or the end. I demonstrate the force of this reasoning in the case of local motion, & also in geometry, in the next two articles. 52 Then in Art. 52 I explain a certain difficulty, which is derived from the fact that, at the instant at which there is a passage from *non-existence* to *existence*, it appears according to a theory of this kind that we must have at the same time both *existence* and *non-existence*. For one of these belongs to the end of the antecedent series of states, & the other to the beginning of the consequent series. I consider fairly fully the solution of this problem; and I call in geometry as well to assist in giving a visual representation of the matter.

63 In Art. 63, after summing up all that has been said about the Law of Continuity, I apply the principle to exclude the possibility of any sudden change from one velocity to another, except by passing through intermediate velocities; this would be contrary to the very full proof that I give for continuity, as it would lead to our having two velocities at the instant at which the change occurred. That is to say, there would be the final velocity of the antecedent series, & the initial velocity of the consequent series; in spite of the fact that it is quite impossible for a moving body to have two different velocities at the same time. Moreover, in order to illustrate & prove the point, from here on to Art. 72, I consider velocity itself; and I distinguish between a potential velocity, as I call it, & an actual velocity; I also investigate carefully many matters that relate to the nature of these velocities & to their changes. Further, I settle several difficulties that can be brought up in opposition to the proof of my Theory, in consequence.

This done, I then conclude from the principle of continuity that, when one body with a greater velocity follows after another body having a less velocity, it is impossible that there should ever be absolute contact with such an inequality of velocities; that is to say, a case of the velocity of each, or of one or the other, of them being changed suddenly at the instant of contact. I assert on the other hand that the change in the velocities must begin before contact. 73 Hence, in Art. 73, I infer that there must be a cause for this change: which is to be called "force." 74 Then, in Art. 74, I prove that this force is a mutual one, & that it acts in opposite directions; the p oof is by induction. 75 From this, in Art. 75, I infer that such a mutual force may be said to be repulsive; & I undertake the investigation of the law that governs it. Carrying on this investigation as far as Art. 80, I find that this force must increase indefinitely as the distance is diminished, in order that it may be capable of destroying any velocity, however great that velocity may be. Moreover, I find that, whilst the force must be indefinitely increased as the distance is indefinitely decreased, it must be on the contrary attractive at very great distances, as is the case for gravitation. Hence I infer that there must be a limit-point forming a boundary between attraction & repulsion; & then by degrees I find more, indeed very many more, of such limit-points, or points of transition from attraction to repulsion, & from repulsion to attraction; & I determine the form of the entire curve, that expresses by its ordinates the law of these forces.

81 Eo usque virium legem deduco, ac definio; tum num. 81 eruo ex ipsa lege constitutionem elementorum materiæ, quæ debent esse simplicia, ob repulsionem in minimis distantiis in immensum auctam; nam ea, si forte ipsa elementa partibus constarent, nexum omnem dissolveret. Usque ad num. 88 inquiro in illud, an hæc elementa, ut simplicia esse debent, ita etiam inextensa esse debeant, ac exposita illa, quam virtualem extensionem appellant, eandem excludo inductionis principio, & difficultatem evolvo tum eam, quæ peti possit ab exemplo ejus generis extensionis, quam in anima indivisibili, & simplice per aliquam corporis partem divisibilem, & extensam passim admittunt: vel omnipræsentiæ Dei: tum eam, quæ peti possit ab analogia cum quiete, in qua nimirum conjungi debeat unicum spatii punctum cum serie continua momentorum temporis, uti in extensione virtuali unicum momentum temporis cum serie continua punctorum spatii conjungeretur, ubi ostendo, nec quietem omnimodam in Natura haberi usquam, nec adesse semper omnimodam inter 88 tempus, & spatium analogiam. Hic autem ingentem colligo ejusmodi determinationis fructum, ostendens usque ad num. 91, quantum prosit simplicitas, indivisibilitas, inextensio elementorum materiæ, ob summotum transitum a vacuo continuo per saltum ad materiam continuam, ac ob sublatum limitem densitatis, quæ in ejusmodi Theoria ut minui in infinitum potest, ita potest in infinitum etiam augeri, dum in communi, ubi ad contactum deventum est, augeri ultra densitas nequaquam potest, potissimum vero ob sublatum omne continuum coexistens, quo sublato & gravissimæ difficultates plurimæ evanescunt, & infinitum actu existens habetur nullum, sed in possibilibus tantummodo remanet series finitorum in infinitum producta.

91 His definitis, inquiro usque ad num. 99 in illud, an ejusmodi elementa sint censenda homogenea, an heterogenea: ac primo quidem argumentum pro homogeneitate saltem in eo, quod pertinet ad totam virium legem, invenio in homogenietate tanta primi cruris repulsivi in minimis distantiis, ex quo pendet impenetrabilitas, & postremi attractivi, quo gravitas exhibetur, in quibus omnis materia est penitus homogenea. Ostendo autem, nihil contra ejusmodi homogenietatem evinci ex principio Leibnitiano indiscernibilium, nihil ex inductione, & ostendo, unde tantum proveniat discrimen in compositis massulis, ut in frondibus, & foliis; ac per inductionem, & analogiam demonstro, naturam nos ad homogeneitatem elementorum, non ad heterogeneitatem deducere.

100 Ea ad probationem Theoriæ pertinent; qua absoluta, antequam inde fructus colligantur multiplices, gradum hic facio ad evolvendas difficultates, quæ vel objectæ jam sunt, vel objici posse videntur mihi, primo quidem contra vires in genere, tum contra meam hanc expositam, comprobatamque virium legem, ac demum contra puncta illa indivisibilia, & inextensa, quæ ex ipsa ejusmodi virium lege deducuntur.

101 Primo quidem, ut iis etiam faciam satis, qui inani vocabulorum quorundam sono perturbantur, a num. 101 ad 104 ostendo, vires basce non esse quoddam occultarum qualitatum genus, sed patentem sane Mechanismum, cum & idea earum sit admodum distincta, & existentia, ac lex positive comprobata; ad Mechanicam vero pertineat omnis 104 tractatio de Motibus, qui a datis viribus etiam sine immediato impulsu oriuntur. A num. 104 ad 106 ostendo, nullum committi saltum in transitu a repulsionibus ad attractiones, 106 & vice versa, cum nimirum per omnes intermedias quantitates is transitus fiat. Inde vero ad objectiones gradum facio, quæ totam curvæ formam impetunt. Ostendo nimirum usque ad num. 116, non posse omnes repulsiones a minore attractione desumi; repulsiones ejusdem esse seriei cum attractionibus, a quibus differant tantummodo ut minus a majore, sive ut negativum a positivo; ex ipsa curvarum natura, quæ, quo altioris sunt gradus, eo in pluribus punctis rectam secare possunt, & eo in immensum plures sunt numero; haberi potius, ubi curva quæritur, quæ vires exprimat, indicium pro curva ejus naturæ, ut rectam in plurimis punctis secet, adeoque plurimos secum afferat virium transitus a repulsivis ad attractivas, quam pro curva, quæ nusquam axem secans attractiones solas, vel solas pro distantiis omnibus repulsiones exhibeat: sed vires repulsivas, & multiplicitatem transituum esse positive probatam, & deductam totam curvæ formam, quam itidem ostendo, non esse ex arcubus natura diversis temere coalescentem, sed omnino simplicem, atque eam ipsam

So far I have been occupied in deducing and settling the law of these forces. Next, in Art. 81, I derive from this law the constitution of the elements of matter. These must be quite simple, on account of the repulsion at very small distances being immensely great; for if by chance those elements were made up of parts, the repulsion would destroy all connections between them. Then, as far as Art. 88, I consider the point, as to whether these elements, as they must be simple, must therefore be also of no extent; &, having explained what is called "virtual extension," I reject it by the principle of induction. I then consider the difficulty which may be brought forward from an example of this kind of extension; such as is generally admitted in the case of the indivisible and one-fold soul pervading a divisible & extended portion of the body, or in the case of the omnipresence of God. Next I consider the difficulty that may be brought forward from an analogy with rest; for here in truth one point of space must be connected with a continuous series of instants of time, just as in virtual extension a single instant of time would be connected with a continuous series of points of space. I show that there can neither be perfect rest anywhere in Nature, nor can there be at all times a perfect analogy between time and space. In this connection, I also gather a large harvest from such a conclusion as this; showing, as far as Art. 91, the great advantage of simplicity, indivisibility, & non-extension in the elements of matter. For they do away with the idea of a passage from a continuous vacuum to continuous matter through a sudden change. Also they render unnecessary any limit to density: this, in a Theory like mine, can be just as well increased to an indefinite extent, as it can be indefinitely decreased: whilst in the ordinary theory, as soon as contact takes place, the density cannot in any way be further increased. But, most especially, they do away with the idea of everything continuous coexisting; & when this is done away with, the majority of the greatest difficulties vanish. Further, nothing infinite is found actually existing; the only thing possible that remains is a series of finite things produced indefinitely.

81

88

These things being settled, I investigate, as far as Art. 99, the point as to whether elements of this kind are to be considered as being homogeneous or heterogeneous. I find my first evidence in favour of homogeneity—at least as far as the complete law of forces is concerned—in the equally great homogeneity of the first repulsive branch of my curve of forces for very small distances, upon which depends impenetrability, & of the last attractive branch, by which gravity is represented. Moreover I show that there is nothing that can be proved in opposition to homogeneity such as this, that can be derived from either the Leibnizian principle of "indiscernibles," or by induction. I also show whence arise those differences, that are so great amongst small composite bodies, such as we see in boughs & leaves; & I prove, by induction & analogy, that the very nature of things leads us to homogeneity, & not to heterogeneity, for the elements of matter.

91

These matters are all connected with the proof of my Theory. Having accomplished this, before I start to gather the manifold fruits to be derived from it, I proceed to consider the objections to my theory, such as either have been already raised or seem to me capable of being raised; first against forces in general, secondly against the law of forces that I have enunciated & proved, & finally against those indivisible, non-extended points that are deduced from a law of forces of this kind.

100

First of all then, in order that I may satisfy even those who are confused over the empty sound of certain terms, I show, in Art. 101 to 104, that these forces are not some sort of mysterious qualities; but that they form a readily intelligible mechanism, since both the idea of them is perfectly distinct, as well as their existence, & in addition the law that governs them is demonstrated in a direct manner. To Mechanics belongs every discussion concerning motions that arise from given forces without any direct impulse. In Art. 104 to 106, I show that no sudden change takes place in passing from repulsions to attractions or *vice versa*; for this transition is made through every intermediate quantity. Then I pass on to consider the objections that are made against the whole form of my curve. I show indeed, from here on to Art. 116, that all repulsions cannot be taken to come from a decreased attraction; that repulsions belong to the self-same series as attractions, differing from them only as less does from more, or negative from positive. From the very nature of the curves (for which, the higher the degree, the more points there are in which they can intersect a right line, & vastly more such curves there are), I deduce that there is more reason for assuming a curve of the nature of mine (so that it may cut a right line in a large number of points, & thus give a large number of transitions of the forces from repulsions to attractions), than for assuming a curve that, since it does not cut the axis anywhere, will represent attractions alone, or repulsions alone, at all distances. Further, I point out that repulsive forces, and a multiplicity of transitions are directly demonstrated, & the whole form of the curve is a matter of deduction; & I also show that it is not formed of a number of arcs differing in nature connected together haphazard;

101

104

106

simplicitatem in Supplementis evidentissime demonstro, exhibens methodum, qua deveniri possit ad æquationem ejusmodi curvæ simplicem, & uniformem ; licet, ut hic ostendo, ipsa illa lex virium possit mente resolvi in plures, quæ per plures curvas exponantur, a quibus tamen omnibus illa reapse unica lex, per unicam illam continuam, & in se simplicem curvam componatur.

121 A num. 121 refello, quæ objici possunt a lege gravitatis decrescentis in ratione reciproca duplicata distantiarum, quæ nimirum in minimis distantiis attractionem requirit crescentem in infinitum. Ostendo autem, ipsam non esse uspiam accurate in ejusmodi ratione, nisi imaginarias resolutiones exhibeamus ; nec vero ex Astronomia deduci ejusmodi legem prorsus accurate servatam in ipsis Planetarum, & Cometarum distantiis, sed ad summum ita 124 proxime, ut differentia ab ea lege sit perquam exigua : ac a num. 124 expendo argumentum, quod pro ejusmodi lege desumi possit ex eo, quod cuipiam visa sit omnium optima, & idcirco electa ab Auctore Naturæ, ubi ipsum Optimismi principium ad trutinam revoco, ac excludo, & vero illud etiam evinco, non esse, cur omnium optima ejusmodi lex censeatur : in Supplementis vero ostendo, ad quæ potins absurda deducet ejusmodi lex, & vero etiam aliæ plures attractionis, quæ imminutis in infinitum distantiis excrescat in infinitum.

131 Num. 131 a viribus transeo ad elementa, & primum ostendo, cur punctorum inextensorum ideam non habeamus, quod nimirum eam haurire non possumus per sensus, quos solæ massæ, & quidem grandiores, afficiunt, atque idcirco eandem nos ipsi debemus per reflexionem efformare, quod quidem facile possumus. Ceterum illud ostendo, me non inducere primum in Physicam puncta indivisibilia, & inextensa, cum eo etiam Leibnitianæ monades recidant, sed sublata extensione continua difficultatem auferre illam omnem, quæ jam olim contra Zenonicos objecta, nunquam est satis soluta, qua fit, ut extensio continua ab inextensis effici omnino non possit.

140 Num. 140 ostendo, inductionis principium contra ipsa nullam habere vim, ipsorum autem existentiam vel inde probari, quod continuitas se se ipsam destruat, & ex ea assumpta probetur argumentis a me institutis hoc ipsum, prima elementa esse indivisibilia, & inextensa, 143 nec ullum haberi extensum continuum. A num. 143 ostendo, ubi continuitatem admittam, nimirum in solis motibus ; ac illud explico, quid mihi sit spatium, quid tempus, quorum naturam in Supplementis multo uberius expono. Porro continuitatem ipsam ostendo a natura in solis motibus obtineri accurate, in reliquis affectari quodammodo ; ubi & exempla quædam evolvo continuitatis primo aspectu violatæ, in quibusdam proprietatibus luminis, ac in aliis quibusdam casibus, in quibus quædam crescunt per additionem partium, non (ut ajunt) per intussumptionem.

153 A num. 153 ostendo, quantum hæc mea puncta a spiritibus differant ; ac illud etiam evolvo, unde fiat, ut in ipsa idea corporis videatur includi extensio continua, ubi in ipsam idearum nostrarum originem inquiro, & quæ inde præjudicia profluant, expono. Postremo 165 autem loco num. 165 innuo, qui fieri possit, ut puncta inextensa, & a se invicem distantia, in massam coalescant, quantum libet, cohærentem, & iis proprietatibus præditam, quas in corporibus experimur, quod tamen ad tertiam partem pertinet, ibi multo uberius pertractandum ; ac ibi quidem primam hanc partem absolvo.

PARS II

166 Num. 166 hujus partis argumentum propono ; sequenti vero 167, quæ potissimum in curva virium consideranda sint, enuncio. Eorum considerationem aggressus, primo quidem 168 usque ad num. 172 in ipsos arcus inquiro, quorum alii attractivi, alii repulsivi, alii asymptotici, ubi casuum occurrit mira multitudo, & in quibusdam consectaria notatu digna, ut & illud, cum ejus formæ curva plurium asymptotorum esse possit, Mundorum prorsus similium seriem posse oriri, quorum alter respectu alterius vices agat unius, & indissolubilis

but that it is absolutely one-fold. This one-fold character I demonstrate in the Supplements in a very evident manner, giving a method by which a simple and uniform equation may be obtained for a curve of this kind. Although, as I there point out, this law of forces may be mentally resolved into several, and these may be represented by several corresponding curves, yet that law, actually unique, may be compounded from all of these together by means of the unique, continuous & one-fold curve that I give.

121 In Art. 121, I start to give a refutation of those objections that may be raised from a consideration of the fact that the law of gravitation, decreasing in the inverse duplicate ratio of the distances, demands that there should be an attraction at very small distances, & that it should increase indefinitely. However, I show that the law is nowhere exactly in conformity with a ratio of this sort, unless we add explanations that are merely imaginative; nor, I assert, can a law of this kind be deduced from astronomy, that is followed with perfect accuracy even at the distances of the planets & the comets, but one merely that is at most so very nearly correct, that the difference from the law of inverse squares is very slight. 124 From Art. 124 onwards, I examine the value of the argument that can be drawn in favour of a law of this sort from the view that, as some have thought, it is the best of all, & that on that account it was selected by the Founder of Nature. In connection with this I examine the principle of Optimism, & I reject it; moreover I prove conclusively that there is no reason why this sort of law should be supposed to be the best of all. Further in the Supplements, I show to what absurdities a law of this sort is more likely to lead; & the same thing for other laws of an attraction that increases indefinitely as the distance is diminished indefinitely.

131 In Art. 131 I pass from forces to elements. I first of all show the reason why we may not appreciate the idea of non-extended points; it is because we are unable to perceive them by means of the senses, which are only affected by masses, & these too must be of considerable size. Consequently we have to build up the idea by a process of reasoning; & this we can do without any difficulty. In addition, I point out that I am not the first to introduce indivisible & non-extended points into physical science; for the "monads" of Leibniz practically come to the same thing. But I show that, by rejecting the idea of continuous extension, I remove the whole of the difficulty, which was raised against the disciples of Zeno in years gone by, & has never been answered satisfactorily; namely, the difficulty arising from the fact that by no possible means can continuous extension be made up from things of no extent.

140 In Art. 140 I show that the principle of induction yields no argument against these indivisibles; rather their existence is demonstrated by that principle, for continuity is self-contradictory. On this assumption it may be proved, by arguments originated by myself, that the primary elements are indivisible & non-extended, & that there does not exist anything possessing the property of continuous extension. 143 From Art. 143 onwards, I point out the only connection in which I shall admit continuity, & that is in motion. I state the idea that I have with regard to space, & also time: the nature of these I explain much more fully in the Supplements. Further, I show that continuity itself is really a property of motions only, & that in all other things it is more or less a false assumption. Here I also consider some examples in which continuity at first sight appears to be violated, such as in some of the properties of light, & in certain other cases where things increase by addition of parts, and not by intussumption, as it is termed.

153 From Art. 153 onwards, I show how greatly these points of mine differ from object-souls. I consider how it comes about that continuous extension seems to be included in the very idea of a body; & in this connection, I investigate the origin of our ideas & I explain the prejudgments that arise therefrom. 165 Finally, in Art. 165, I lightly sketch what might happen to enable points that are of no extent, & at a distance from one another, to coalesce into a coherent mass of any size, endowed with those properties that we experience in bodies. This, however, belongs to the third part; & there it will be much more fully developed. This finishes the first part.

PART II

166 In Art. 166 I state the theme of this second part; and in Art. 167 I declare what matters are to be considered more especially in connection with the curve of forces. 168 Coming to the consideration of these matters, I first of all, as far as Art. 172, investigate the arcs of the curve, some of which are attractive, some repulsive and some asymptotic. Here a marvellous number of different cases present themselves, & to some of them there are noteworthy corollaries; such as that, since a curve of this kind is capable of possessing a considerable number of asymptotes, there can arise a series of perfectly similar cosmi, each of which will act upon all the others as a single inviolate elementary system. From Art. 172

172 elementi. Ad. num. 179 areas contemplor arcubus clausas, quæ respondentes segmento axis cuicunque, esse possunt magnitudine utcunque magnæ, vel parvæ, sunt autem mensura
179 incrementi, vel decrementi quadrati velocitatum. Ad num. 189 inquiro in appulsus curvæ ad axem, sive is ibi secetur ab eadem (quo casu habentur transitus vel a repulsione ad attractionem, vel ab attractione ad repulsionem, quos dico limites, & quorum maximus est in tota mea Theoria usus), sive tangatur, & curva retro redeat, ubi etiam pro appulsibus considero recessus in infinitum per arcus asymptoticos, & qui transitus, sive limites, oriantur inde, vel in Natura admitti possint, evolvo.

189 Num. 189 a consideratione curvæ ad punctorum combinationem gradum facio, ac primo quidem usque ad num. 204 ago de systemate duorum punctorum, ea pertractans, quæ pertinent ad eorum vires mutuas, & motus, sive sibi relinquantur, sive projiciantur utcunque, ubi & conjunctione ipsorum exposita in distantiis limitum, & oscillationibus variis, sive nullam externam punctorum aliorum actionem sentiant, sive perturbentur ab eadem, illud innuo in antecessum, quanto id usui futurum sit in parte tertia ad exponenda cohæsionis varia genera, fermentationes, conflagrationes, emissiones vaporum, proprietates luminis, elasticitatem, mollitiem.

204 Succedit a Num. 204 ad 239 multo uberior consideratio trium punctorum, quorum vires generaliter facile definiuntur data ipsorum positione quacunque: verum utcunque data positione, & celeritate nondum a Geometris inventi sunt motus ita, ut generaliter pro casibus omnibus absolvi calculus possit. Vires igitur, & variationem ingentem, quam diversæ pariunt combinationes punctorum, utut tantummodo numero trium, persequor
209 usque ad num. 209. Hinc usque ad num. 214 quædam evolvo, quæ pertinent ad vires ortas in singulis ex actione composita reliquorum duorum, & quæ tertium punctum non ad accessum urgeant, vel recessum tantummodo respectu eorundem, sed & in latus, ubi & soliditatis imago prodit, & ingens sane discrimen in distantiis particularum perquam exiguis ac summa in maximis, in quibus gravitas agit, conformitas, quod quanto itidem ad Naturæ
214 explicationem futurum sit usui, significo. Usque ad num. 221 ipsis etiam oculis contemplandum propono ingens discrimen in legibus virium, quibus bina puncta agunt in tertium, sive id jaceat in recta, qua junguntur, sive in recta ipsi perpendiculari, & eorum intervallum secante bifariam, constructis ex data primigenia curva curvis vires compositas exhibentibus:
221 tum sequentibus binis numeris casum evolvo notatu dignissimum, in quo mutata sola positione binorum punctorum, punctum tertium per idem quoddam intervallum, situm in eadem distantia a medio eorum intervallo, vel perpetuo attrahitur, vel perpetuo repellitur, vel nec attrahitur, nec repellitur; cujusmodi discrimen cum in massis haberi debeat multo
222 majus, illud indico, num. 222, quantus inde itidem in Physicam usus proveniat.

223 Hic jam num. 223 a viribus binorum punctorum transeo ad considerandum totum ipsorum systema, & usque ad num. 228 contemplor tria puncta in directum sita, ex quorum mutuis viribus relationes quædam exurgunt, quæ multo generaliores redduntur inferius, ubi in tribus etiam punctis tantummodo adumbrantur, quæ pertinent ad virgas rigidas, flexiles, elasticas, ac ad vectem, & ad alia plura, quæ itidem inferius, ubi de massis, multo generaliora
228 fiunt. Demum usque ad num. 238 contemplor tria puncta posita non in directum, sive in æquilibrio sint, sive in perimetro ellipsium quarundam, vel curvarum aliarum; in quibus mira occurrit analogia limitum quorundam cum limitibus, quos habent bina puncta in axe curvæ primigeniæ ad se invicem, atque ibidem multo major varietas casuum indicatur pro massis, & specimen applicationis exhibetur ad soliditatem, & liquationem per celerem
238 intestinum motum punctis impressum. Sequentibus autem binis numeris generalia quædam expono de systemate punctorum quatuor cum applicatione ad virgas solidas, rigidas, flexiles, ac ordines particularum varios exhibeo per pyramides, quarum infimæ ex punctis quatuor, superiores ex quatuor pyramidibus singulæ coalescant.

240 A num. 240 ad massas gradu facto usque a num. 264 considero, quæ ad centrum gravitatis pertinent, ac demonstro generaliter, in quavis massa esse aliquod, & esse unicum · ostendo, quo pacto determinari generaliter possit, & quid in methodo, quæ communiter adhibetur, desit ad habendam demonstrationis vim, luculenter expono, & suppleo, ac

to Art. 179, I consider the areas included by the arcs; these, corresponding to different segments of the axis, may be of any magnitude whatever, either great or small; moreover they measure the increment or decrement in the squares of the velocities. Then, on as far as Art. 189, I investigate the approach of the curve to the axis; both when the former is cut by the latter, in which case there are transitions from repulsion to attraction and from attraction to repulsion, which I call 'limits,' & use very largely in every part of my Theory; & also when the former is touched by the latter, & the curve once again recedes from the axis. I consider, too, as a case of approach, recession to infinity along an asymptotic arc; and I investigate what transitions, or limits, may arise from such a case, & whether such are admissible in Nature. 172 179

In Art. 189, I pass on from the consideration of the curve to combinations of points. First, as far as Art. 204, I deal with a system of two points. I work out those things that concern their mutual forces, and motions, whether they are left to themselves or projected in any manner whatever. Here also, having explained the connection between these motions & the distances of the limits, & different cases of oscillations, whether they are affected by external action of other points, or are not so disturbed, I make an anticipatory note of the great use to which this will be put in the third part, for the purpose of explaining various kinds of cohesion, fermentations, conflagrations, emissions of vapours, the properties of light, elasticity and flexibility. 189

There follows, from Art. 204 to Art. 239, the much more fruitful consideration of a system of three points. The forces connected with them can in general be easily determined for any given positions of the points; but, when any position & velocity are given, the motions have not yet been obtained by geometricians in such a form that the general calculation can be performed for every possible case. So I proceed to consider the forces, & the huge variation that different combinations of the points beget, although they are only three in number, as far as Art. 209. From that, on to Art. 214, I consider certain things that have to do with the forces that arise from the action, on each of the points, of the other two together, & how these urge the third point not only to approach, or recede from, themselves, but also in a direction at right angles; in this connection there comes forth an analogy with solidity, & a truly immense difference between the several cases when the distances are very small, & the greatest conformity possible at very great distances such as those at which gravity acts; & I point out what great use will be made of this also in explaining the constitution of Nature. Then up to Art. 221, I give ocular demonstrations of the huge differences that there are in the laws of forces with which two points act upon a third, whether it lies in the right line joining them, or in the right line that is the perpendicular which bisects the interval between them; this I do by constructing, from the primary curve, curves representing the composite forces. Then in the two articles that follow, I consider the case, a really important one, in which, by merely changing the position of the two points, the third point, at any and the same definite interval situated at the same distance from the middle point of the interval between the two points, will be either continually attracted, or continually repelled, or neither attracted nor repelled; & since a difference of this kind should hold to a much greater degree in masses, I point out, in Art. 222, the great use that will be made of this also in Physics. 204 209 214 221 222

At this point then, in Art. 223, I pass from the forces derived from two points to the consideration of a whole system of them; and, as far as Art. 228, I study three points situated in a right line, from the mutual forces of which there arise certain relations, which I return to later in much greater generality; in this connection also are outlined, for three points only, matters that have to do with rods, either rigid, flexible or elastic, and with the lever, as well as many other things; these, too, are treated much more generally later on, when I consider masses. Then right on to Art. 238, I consider three points that do not lie in a right line, whether they are in equilibrium, or moving in the perimeters of certain ellipses or other curves. Here we come across a marvellous analogy between certain limits and the limits which two points lying on the axis of the primary curve have with respect to each other; & here also a much greater variety of cases for masses is shown, & an example is given of the application to solidity, & liquefaction, on account of a quick internal motion being impressed on the points of the body. Moreover, in the two articles that then follow, I state some general propositions with regard to a system of four points, together with their application to solid rods, both rigid and flexible; I also give an illustration of various classes of particles by means of pyramids, each of which is formed of four points in the most simple case, & of four of such pyramids in the more complicated cases. 223 228 238

From Art. 240 as far as Art. 264, I pass on to masses & consider matters pertaining to the centre of gravity; & I prove that in general there is one, & only one, in any given mass. I show how it can in general be determined, & I set forth in clear terms the point that is lacking in the usual method, when it comes to a question of rigorous proof; this deficiency 240

exemplum profero quoddam ejusdem generis, quod ad numerorum pertinet multiplicationem, & ad virium compositionem per parallelogramma, quam alia methodo generaliore exhibeo analoga illi ipsi, qua generaliter in centrum gravitatis inquiro : tum vero ejusdem ope demonstro admodum expedite, & accuratissime celebre illud Newtoni theorema de statu centri gravitatis per mutuas internas vires numquam turbato.

264, 265, 266, 276, 277, 278, 279 Ejus tractionis fructus colligo plures : conservationem ejusdem quantitatis motuum in Mundo in eandem plagam num. 264, æqualitatem actionis, & reactionis in massis num. 265, collisionem corporum, & communicationem motus in congressibus directis cum eorum legibus, inde num. 276 congressus obliquos, quorum Theoriam a resolutione motuum reduco ad compositionem num. 277, quod sequenti numero 278 transfero ad incursum etiam in planum immobile ; ac a num. 279 ad 289 ostendo nullam haberi in Natura veram virium, aut motuum resolutionem, sed imaginariam tantummodo, ubi omnia evolvo, & explico casuum genera, quæ prima fronte virium resolutionem requirere videntur.

289 A num. 289 ad 297 leges expono compositionis virium, & resolutionis, ubi & illud notissimum, quo pacto in compositione decrescat vis, in resolutione crescat, sed in illa priore conspirantium summa semper maneat, contrariis elisis ; in hac posteriore concipiantur tantummodo binæ vires contrariæ adjectæ, quæ consideratio nihil turbet phænomena ; unde fiat, ut nihil inde pro virium vivarum Theoria deduci possit, cum sine iis explicentur omnia, ubi plura itidem explico ex iis phænomenis, quæ pro ipsis viribus vivis afferri solent.

297 A num. 297 occasione inde arrepta aggredior quædam, quæ ad legem continuitatis pertinent, ubique in motibus sancte servatam, ac ostendo illud, idcirco in collisionibus corporum, ac in motu reflexo, leges vulgo definitas, non nisi proxime tantummodo observari, & usque ad num. 307 relationes varias persequor angulorum incidentiæ, & reflexionis, sive vires constanter in accessu attrahant, vel repellant constanter, sive jam attrahant, jam repellant : ubi & illud considero, quid accidat, si scabrities superficiei agentis exigua sit, quid, si ingens, ac clementa profero, quæ ad luminis reflexionem, & refractionem explicandam, definiendamque ex Mechanica requiritur, relationem itidem vis absolutæ ad relativam in obliquo gravium descensu, & nonnulla, quæ ad oscillationum accuratiorem Theoriam necessaria sunt, prorsus elementaria, diligenter expono.

307, 313 A num. 307 inquiro in trium massarum systema, ubi usque ad num. 313 theoremata evolvo plura, quæ pertinent ad directionem virium in singulis compositarum e binis reliquarum actionibus, ut illud, eas directiones vel esse inter se parallelas, vel, si utrinque indefinite producantur, per quoddam commune punctum transire omnes : tum usque ad 321 theoremata alia plura, quæ pertinent ad earumdem compositarum virium rationem ad se invicem, ut illud & simplex, & elegans, binarum massarum vires acceleratrices esse semper in ratione composita ex tribus reciprocis rationibus, distantiæ ipsarum a massa tertia, sinus anguli, quem singularum directio continet cum sua ejusmodi distantia, & massæ ipsius eam habentis compositam vim, ad distantiam, sinum, massam alteram ; vires autem motrices habere tantummodo priores rationes duas elisa tertia

321 Eorum theorematum fructum colligo deducens inde usque ad num. 328, quæ ad æquilibrium pertinent divergentium utcumque virium, & ipsius æquilibrii centrum, ac nisum centri in fulcrum, & quæ ad præponderantiam, Theoriam extendens ad casum etiam, quo massae non in se invicem agant mutuo immediate, sed per intermedias alias, quæ nexum concilient, & virgarum nectentium suppleant vices, ac ad massas etiam quotcunque, quarum singulas cum centro conversionis, & alia quavis assumpta massa connexas concipio, unde principium momenti deduco pro machinis omnibus : tum omnium vectium genera evolvo, ut & illud, facta suspensione per centrum gravitatis haberi æquilibrium, sed in ipso centro debere sentiri vim a fulcro, vel sustinente puncto, æqualem summæ ponderum totius systematis, unde demum pateat ejus ratio, quod passim sine demonstratione assumitur, nimirum systemate quiescente, & impedito omni partium motu per æquilibrium, totam massam concipi posse ut in centro gravitatis collectam.

I supply, & I bring forward a certain example of the same sort, that deals with the multiplication of numbers, & to the composition of forces by the parallelogram law; the latter I prove by another more general method, analogous to that which I use in the general investigation for the centre of gravity. Then by its help I prove very expeditiously & with extreme rigour that well-known theorem of Newton, in which he affirmed that the state of the centre of gravity is in no way altered by the internal mutual forces.

I gather several good results from this method of treatment. In Art. 264, the conservation of the same quantity of motion in the Universe in one plane; in Art. 265 the equality of action and reaction amongst masses; then the collision of solid bodies, and the communication of motions in direct impacts & the laws that govern them, & from that, in Art. 276, oblique impacts; in Art. 277 I reduce the theory of these from resolution of motions to compositions, & in the article that follows, Art. 278, I pass to impact on to a fixed plane; from Art. 279 to Art. 289 I show that there can be no real resolution of forces or of motions in Nature, but only a hypothetical one; & in this connection I consider & explain all sorts of cases, in which at first sight it would seem that there must be resolution. 264 265 266 276 277 278 279

From Art. 289 to Art. 297, I state the laws for the composition & resolution of forces; here also I give the explanation of that well-known fact, that force decreases in composition, increases in resolution, but always remains equal to the sum of the parts acting in the same direction as itself in the first, the rest being equal & opposite cancel one another; whilst in the second, all that is done is to suppose that two equal & opposite forces are added on, which supposition has no effect on the phenomena. Thus it comes about that nothing can be deduced from this in favour of the Theory of living forces, since everything can be explained without them; in the same connection, I explain also many of the phenomena, which are usually brought forward as evidence in favour of these 'living forces.' 289

In Art. 297, I seize the opportunity offered by the results just mentioned to attack certain matters that relate to the law of continuity, which in all cases of motion is strictly observed; & I show that, in the collision of solid bodies, & in reflected motion, the laws, as usually stated, are therefore only approximately followed. From this, as far as Art. 307, I make out the various relations between the angles of incidence & reflection, whether the forces, as the bodies approach one another, continually attract, or continually repel, or attract at one time & repel at another. I also consider what will happen if the roughness of the acting surface is very slight, & what if it is very great. I also state the first principles, derived from mechanics, that are required for the explanation & determination of the reflection & refraction of light; also the relation of the absolute to the relative force in the oblique descent of heavy bodies; & some theorems that are requisite for the more accurate theory of oscillations; these, though quite elementary, I explain with great care. 297

From Art. 307 onwards, I investigate the system of three bodies; in this connection, as far as Art. 313, I evolve several theorems dealing with the direction of the forces on each one of the three compounded from the combined actions of the other two; such as the theorem, that these directions are either all parallel to one another, or all pass through some one common point, when they are produced indefinitely on both sides. Then, as far as Art. 321, I make out several other theorems dealing with the ratios of these same resultant forces to one another; such as the following very simple & elegant theorem, that the accelerating forces of two of the masses will always be in a ratio compounded of three reciprocal ratios; namely, that of the distance of either one of them from the third mass, that of the sine of the angle which the direction of each force makes with the corresponding distance of this kind, & that of the mass itself on which the force is acting, to the corresponding distance, sine and mass for the other: also that the motive forces only have the first two ratios, that of the masses being omitted. 307 313

I then collect the results to be derived from these theorems, deriving from them, as far as Art. 328, theorems relating to the equilibrium of forces diverging in any manner, & the centre of equilibrium, & the pressure of the centre on a fulcrum. I extend the theorem relating to preponderance to the case also, in which the masses do not mutually act upon one another in a direct manner, but through others intermediate between them, which connect them together, & supply the place of rods joining them; and also to any number of masses, each of which I suppose to be connected with the centre of rotation & some other assumed mass, & from this I derive the principles of moments for all machines. Then I consider all the different kinds of levers; one of the theorems that I obtain is, that, if a lever is suspended from the centre of gravity, then there is equilibrium; but a force should be felt in this centre from the fulcrum or sustaining point, equal to the sum of the weights of the whole system; from which there follows most clearly the reason, which is everywhere assumed without proof, why the whole mass can be supposed to be collected at its centre of gravity, so long as the system is in a state of rest & all motions of its parts are prohibited by equilibrium. 321

328 A num. 328 ad 347 deduco ex iisdem theorematis, quæ pertinent ad centrum oscillationis quotcunque massarum, sive sint in eadem recta, sive in plano perpendiculari ad axem rotationis ubicunque, quæ Theoria per systema quatuor massarum, excolendum aliquanto diligentius, uberius promoveri deberet & extendi ad generalem habendum solidorum nexum, 344 qua re indicata, centrum itidem percussionis inde evolvo, & ejus analogiam cum centro oscillationis exhibeo.

347 Collecto ejusmodi fructu ex theorematis pertinentibus ad massas tres, innuo num. 347, quæ mihi communia sint cum ceteris omnibus, & cum Newtonianis potissimum, pertinentia ad summas virium, quas habet punctum, vel massa attracta, vel repulsa a punctis singulis 348 alterius massæ; tum a num. 348 ad finem hujus partis, sive ad num. 358, expono quædam, quæ pertinent ad fluidorum Theoriam, & primo quidem ad pressionem, ubi illud innuo demonstratum a Newtono, si compressio fluidi sit proportionalis vi comprimenti, vires repulsivas punctorum esse in ratione reciproca distantiarum, ac vice versa: ostendo autem illud, si eadem vis sit insensibilis, rem, præter alias curvas, exponi posse per Logisticam, & in fluidis gravitate nostra terrestri præditis pressiones haberi debere ut altitudines; deinde vero attingo illa etiam, quæ pertinent ad velocitatem fluidi erumpentis e vase, & expono, quid requiratur, ut ea sit æqualis velocitati, quæ acquiretur cadendo per altitudinem ipsam, quemadmodum videtur res obtingere in aquæ effluxu: quibus partim expositis, partim indicatis, hanc secundam partem concludo.

PARS III

358 Num. 358 propono argumentum hujus tertiæ partis, in qua omnes e Theoria mea 360 generales materiæ proprietates deduco, & particulares plerasque: tum usque ad num. 371 ago aliquanto fusius de impenetrabilitate, quam duplicis generis agnosco in meis punctorum inextensorum massis, ubi etiam de ea apparenti quadam compenetratione ago, ac de luminis transitu per substantias intimas sine vera compenetratione, & mira quædam phænomena 371 huc pertinentia explico admodum expedite. Inde ad num. 375 de extensione ago, quæ mihi quidem in materia, & corporibus non est continua, sed adhuc eadem præbet phænomenæ sensibus, ac in communi sententia; ubi etiam de Geometria ago, quæ vim suam in 375 mea Theoria retinet omnem: tum ad num. 383 figurabilitatem persequor, ac molem, massam, densitatem singillatim, in quibus omnibus sunt quædam Theoriæ meæ propria 383 scitu non indigna. De Mobilitate, & Motuum Continuitate, usque ad num. 388 notatu 388 digna continentur: tum usque ad num. 391 ago de æqualitate actionis, & reactionis, cujus consectaria vires ipsas, quibus Theoria mea innititur, mirum in modum confirmant. 391 Succedit usque ad num. 398 divisibilitas, quam ego ita admitto, ut quævis massa existens numerum punctorum realium habeat finitum tantummodo, sed qui in data quavis mole possit esse utcunque magnus; quamobrem divisibilitati in infinitum vulgo admissæ substituo componibilitatem in infinitum, ipsi, quod ad Naturæ phænomena explicanda 398 pertinet, prorsus æquivalentem. His evolutis addo num. 398 immutabilitatem primorum materiæ elementorum, quæ cum mihi sint simplicia prorsus, & inextensa, sunt utique immutabilia, & ad exhibendam perennem phænomenorum seriem aptissima.

399 A num. 399 ad 406 gravitatem deduco ex mea virium Theoria, tanquam ramum quendam e communi trunco, ubi & illud expono, qui fieri possit, ut fixæ in unicam massam 406 non coalescant, quod gravitas generalis requirere videretur. Inde ad num. 419 ago de cohæsione, qui est itidem veluti alter quidam ramus, quam ostendo, nec in quiete consistere, nec in motu conspirante, nec in pressione fluidi cujuspiam, nec in attractione maxima in contactu, sed in limitibus inter repulsionem, & attractionem; ubi & problema generale propono quoddam huc pertinens, & illud explico, cur massa fracta non iterum coalescat, cur fibræ ante fractionem distendantur, vel contrahantur, & innuo, quæ ad cohæsionem pertinentia mihi cum reliquis Philosophis communia sint.

419 A cohæsione gradum facio num. 419 ad particulas, quæ ex punctis cohærentibus efformantur, de quibus ago usque ad num. 426, & varia persequor earum discrimina:

From Art. 328 to Art. 347, I deduce from these same theorems, others that relate to the centre of oscillation of any number of masses, whether they are in the same right line, or anywhere in a plane perpendicular to the axis of rotation; this theory wants to be worked somewhat more carefully with a system of four bodies, to be gone into more fully, & to be extended so as to include the general case of a system of solid bodies; having stated this, I evolve from it the centre of percussion, & I show the analogy between it & the centre of oscillation. 328 344

I obtain all such results from theorems relating to three masses. After that, in Art. 347, I intimate the matters in which I agree with all others, & especially with the followers of Newton, concerning sums of forces, acting on a point, or an attracted or repelled mass, due to the separate points of another mass. Then, from Art. 348 to the end of this part, i.e., as far as Art. 359, I expound certain theorems that belong to the theory of fluids; & first of all, theorems with regard to pressure, in connection with which I mention that one which was proved by Newton, namely, that, if the compression of a fluid is proportional to the compressing force, then the repulsive forces between the points are in the reciprocal ratio of the distances, & conversely. Moreover, I show that, if the same force is insensible, then the matter can be represented by the logistic & other curves; also that in fluids subject to our terrestrial gravity pressures should be found proportional to the depths. After that, I touch upon those things that relate to the velocity of a fluid issuing from a vessel; & I show what is necessary in order that this should be equal to the velocity which would be acquired by falling through the depth itself, just as it is seen to happen in the case of an efflux of water. These things in some part being explained, & in some part merely indicated, I bring this second part to an end. 347 348

PART III

In Art. 358, I state the theme of this third part; in it I derive all the general & most of the special, properties of matter from my Theory. Then, as far as Art. 371, I deal somewhat more at length with the subject of impenetrability, which I remark is of a twofold kind in my masses of non-extended points; in this connection also, I deal with a certain apparent case of compenetrability, & the passage of light through the innermost parts of bodies without real compenetration; I also explain in a very summary manner several striking phenomena relating to the above. From here on to Art. 375, I deal with extension; this in my opinion is not continuous either in matter or in solid bodies, & yet it yields the same phenomena to the senses as does the usually accepted idea of it; here I also deal with geometry, which conserves all its power under my Theory. Then, as far as Art. 383, I discuss figurability, volume, mass & density, each in turn; in all of these subjects there are certain special points of my Theory that are not unworthy of investigation. Important theorems on mobility & continuity of motions are to be found from here on to Art. 388; then, as far as Art. 391, I deal with the equality of action & reaction, & my conclusions with regard to the subject corroborate in a wonderful way the hypothesis of those forces, upon which my Theory depends. Then follows divisibility, as far as Art. 398; this principle I admit only to the extent that any existing mass may be made up of a number of real points that are finite only, although in any given mass this finite number may be as great as you please. Hence for infinite divisibility, as commonly accepted, I substitute infinite multiplicity; which comes to exactly the same thing, as far as it is concerned with the explanation of the phenomena of Nature. Having considered these subjects I add, in Art. 398, that of the immutability of the primary elements of matter; according to my idea, these are quite simple in composition, of no extent, they are everywhere unchangeable, & hence are splendidly adapted for explaining a continually recurring set of phenomena. 358 360 371 375 383 388 391 398

From Art. 399 to Art. 406, I derive gravity from my Theory of forces, as if it were a particular branch on a common trunk; in this connection also I explain how it can happen that the fixed stars do not all coalesce into one mass, as would seem to be required under universal gravitation. Then, as far as Art. 419, I deal with cohesion, which is also as it were another branch; I show that this is not dependent upon quiescence, nor on motion that is the same for all parts, nor on the pressure of some fluid, nor on the idea that the attraction is greatest at actual contact, but on the limits between repulsion and attraction. I propose, & solve, a general problem relating to this, namely, why masses, once broken, do not again stick together, why the fibres are stretched or contracted before fracture takes place; & I intimate which of my ideas relative to cohesion are the same as those held by other philosophers. 399 406

In Art. 419, I pass on from cohesion to particles which are formed from a number of cohering points; & I consider these as far as Art. 426, & investigate the various distinctions 419

ostendo nimirum, quo pacto varias induere possint figuras quascunque, quarum tenacissime sint; possint autem data quavis figura discrepare plurimum in numero, & distributione punctorum, unde & oriantur admodum inter se diversæ vires unius particulæ in aliam, ac itidem diversæ in diversis partibus ejusdem particulæ respectu diversarum partium, vel etiam respectu ejusdem partis particulæ alterius, cum a solo numero, & distributione punctorum pendeat illud, ut data particula datam aliam in datis earum distantiis, & superficierum locis, vel attrahat, vel repellat, vel respectu ipsius sit prorsus iners: tum illud addo, particulas eo difficilius dissolubiles esse, quo minores sint; debere autem in gravitate esse penitus uniformes, quæcunque punctorum dispositio habeatur, & in aliis proprietatibus plerisque debere esse admodum (uti observamus) diversas, quæ diversitas multo major in majoribus massis esse debeat.

426 A num. 426 ad 446 de solidis, & fluidis, quod discrimen itidem pertinet ad varia cohæsionum genera; & discrimen inter solida, & fluida diligenter expono, horum naturam potissimum repetens ex motu faciliori particularum in gyrum circa alias, atque id ipsum ex viribus circumquaque æqualibus; illorum vero ex inæqualitate virium, & viribus quibusdam in latus, quibus certam positionem ad se invicem servare debeant. Varia autem distinguo fluidorum genera, & discrimen profero inter virgas rigidas, flexiles, elasticas, fragiles, ut & de viscositate, & humiditate ago, ac de organicis, & ad certas figuras determinatis corporibus, quorum efformatio nullam habet difficultatem, ubi una particula unam aliam possit in certis tantummodo superficiei partibus attrahere, & proinde cogere ad certam quandam positionem acquirendam respectu ipsius, & retinendam. Demonstro autem & illud, posse admodum facile ex certis particularum figuris, quarum ipsæ tenacissimæ sint, totum etiam Atomistarum, & Corpuscularium systema a mea Theoria repeti ita, ut id nihil sit aliud, nisi unicus itidem hujus veluti trunci fœcundissimi ramus e diversa cohæsionis ratione prorumpens. Demum ostendo, cur non quævis massa, utut constans ex homogeneis punctis, & circa se maxime in gyrum mobilibus, fluida sit; & fluidorum resistentiam quoque attingo, in ejus leges inquirens.

446 A num. 446 ad 450 ago de iis, quæ itidem ad diversa pertinent soliditatis genera, nimirum de elasticis, & mollibus, illa repetens a magna inter limites proximos distantia, qua fiat, ut puncta longe dimota a locis suis, idem ubique genus virium sentiant, & proinde se ad priorem restituant locum; hæc a limitum frequentia, atque ingenti vicinia, qua fiat, ut ex uno ad alium delata limitem puncta, ibi quiescant itidem respective, ut prius. Tum vero de ductilibus, & malleabilibus ago, ostendens, in quo a fragilibus discrepent: ostendo autem, hæc omnia discrimina a densitate nullo modo pendere, ut nimirum corpus, quod multo sit altero densius, possit tam multo majorem, quam multo minorem soliditatem, & cohæsionem habere, & quævis ex proprietatibus expositis æque possit cum quavis vel majore, vel minore densitate componi.

450 Num. 450 inquiro in vulgaria quatuor elementa; tum a num. 451 ad num. 467 persequor
452 chemicas operationes; num. 452 explicans dissolutionem, 453 præcipitationem, 454, & 455 commixtionem plurium substantiarum in unam: tum num. 456, & 457 liquationem binis methodis, 458 volatilizatiōnem, & effervescentiam, 461 emissionem effluviorum, quæ e massa constanti debeat esse ad sensum constans, 462 ebullitionem cum variis evaporationum generibus; 463 deflagrationem, & generationem aeris; 464 crystallizationem cum certis figuris; ac demum ostendo illud num. 465, quo pacto possit fermentatio desinere; & num. 466, quo pacto non omnia fermentescant cum omnibus.

467 A fermentatione num. 467 gradum facio ad ignem, qui mihi est fermentatio quædam substantiæ lucis cum sulphurea quadam substantia, ac plura inde consectaria deduco usque
471 ad num. 471; tum ab igne ad lumen ibidem transeo, cujus proprietates præcipuas, ex
472 quibus omnia lucis phænomena oriuntur, propono num. 472, ac singulas a Theoria mea deduco, & fuse explico usque ad num. 503, nimirum emissionem num. 473, celeritatem 474, propagationem rectilineam per media homogenea, & apparentem tantummodo compenetrationem a num. 475 ad 483, pellucidatem, & opacitatem num. 483, reflexionem ad angulos æquales inde ad 484, refractionem ad 487, tenuitatem num. 487, calorem, & ingentes intestinos motus allapsu tenuissimæ lucis genitos, num. 488, actionem majorem corporum oleosorum, & sulphurosorum in lumen num. 489: tum num. 490 ostendo, nullam resist-

between them. I show how it is possible for various shapes of all sorts to be assumed, which offer great resistance to rupture; & how in a given shape they may differ very greatly in the number & disposition of the points forming them. Also that from this fact there arise very different forces for the action of one particle upon another, & also for the action of different parts of this particle upon other different parts of it, or on the same part of another particle. For that depends solely on the number & distribution of the points, so that one given particle either attracts, or repels, or is perfectly inert with regard to another given particle, the distances between them and the positions of their surfaces being also given. Then I state in addition that the smaller the particles, the greater is the difficulty in dissociating them; moreover, that they ought to be quite uniform as regards gravitation, no matter what the disposition of the points may be; but in most other properties they should be quite different from one another (which we observe to be the case); & that this difference ought to be much greater in larger masses.

426 From Art. 426 to Art. 446, I consider solids & fluids, the difference between which is also a matter of different kinds of cohesion. I explain with great care the difference between solids & fluids; deriving the nature of the latter from the greater freedom of motion of the particles in the matter of rotation about one another, this being due to the forces being nearly equal; & that of the former from the inequality of the forces, and from certain lateral forces which help them to keep a definite position with regard to one another. I distinguish between various kinds of fluids also, & I cite the distinction between rigid, flexible, elastic & fragile rods, when I deal with viscosity & humidity; & also in dealing with organic bodies & those solids bounded by certain fixed figures, of which the formation presents no difficulty; in these one particle can only attract another particle in certain parts of the surface, & thus urge it to take up some definite position with regard to itself, & keep it there. I also show that the whole system of the Atomists, & also of the Corpuscularians, can be quite easily derived by my Theory, from the idea of particles of definite shape, offering a high resistance to deformation; so that it comes to nothing else than another single branch of this so to speak most fertile trunk, breaking forth from it on account of a different manner of cohesion. Lastly, I show the reason why it is that not every mass, in spite of its being constantly made up of homogeneous points, & even these in a high degree capable of rotary motion about one another, is a fluid. I also touch upon the resistance of fluids, & investigate the laws that govern it.

446 From Art. 446 to Art. 450, I deal with those things that relate to the different kinds of solidity, that is to say, with elastic bodies, & those that are soft. I attribute the nature of the former to the existence of a large interval between the consecutive limits, on account of which it comes about that points that are far removed from their natural positions still feel the effects of the same kind of forces, & therefore return to their natural positions; & that of the latter to the frequency & great closeness of the limits, on account of which it comes about that points that have been moved from one limit to another, remain there in relative rest as they were to start with. Then I deal with ductile and malleable solids, pointing out how they differ from fragile solids. Moreover I show that all these differences are in no way dependent on density; so that, for instance, a body that is much more dense than another body may have either a much greater or a much less solidity and cohesion than another; in fact, any of the properties set forth may just as well be combined with any density either greater or less.

450 In Art. 450 I consider what are commonly called the "four elements"; then from 452 Art. 451 to Art. 467, I treat of chemical operations; I explain solution in Art. 452, precipitation in Art. 453, the mixture of several substances to form a single mass in Art. 454, 455, liquefaction by two methods in Art. 456, 457, volatilization & effervescence in Art. 458, emission of effluvia (which from a constant mass ought to be approximately constant) in Art. 461, ebullition & various kinds of evaporation in Art. 462, deflagration & generation of gas in Art. 463, crystallization with definite forms of crystals in Art. 464; & lastly, I show, in Art. 465, how it is possible for fermentation to cease, & in Art. 466, how it is that any one thing does not ferment when mixed with any other thing.

467 From fermentation I pass on, in Art. 467, to fire, which I look upon as a fermentation of some substance in light with some sulphureal substance; & from this I deduce several 471 propositions, up to Art. 471. There I pass on from fire to light, the chief properties of 472 which, from which all the phenomena of light arise, I set forth in Art. 472; & I deduce & fully explain each of them in turn as far as Art. 503. Thus, emission in Art. 473, velocity in Art. 474, rectilinear propagation in homogeneous media, & a compenetration that is merely apparent, from Art. 475 on to Art. 483, pellucidity & opacity in Art. 483, reflection at equal angles to Art. 484, & refraction to Art. 487, tenuity in Art. 487, heat & the great internal motions arising from the smooth passage of the extremely tenuous light in Art. 488, the greater action of oleose & sulphurous bodies on light in Art. 489. Then I

entiam veram pati, ac num. 491 explico, unde sint phosphora, num. 492 cur lumen cum majo e obliquitate incidens reflectatur magis, num. 493 & 494 unde diversa refrangibilitas ortum ducat, ac num. 495, & 496 deduco duas diversas dispositiones ad æqualia redeuntes intervalla, unde num. 497 vices illas a Newtono detectas facilioris reflexionis, & facilioris transmissus eruo, & num. 498 illud, radios alios debere reflecti, alios transmitti in appulsu ad novum medium, & eo plures reflecti, quo obliquitas incidentiæ sit major, ac num. 499 & 500 expono, unde discrimen in intervallis vicium, ex quo uno omnis naturalium colorum pendet Newtoniana Theoria. Demum num. 501 miram attingo crystalli Islandicæ proprietatem, & ejusdem causam, ac num. 502 diffractionem expono, quæ est quædam inchoata refractio, sive reflexio.

503 504 507 511 514 Post lucem ex igne derivatam, quæ ad oculos pertinet, ago brevissime num. 503 de sapore, & odore, ac sequentibus tribus numeris de sono: tum aliis quator de tactu, ubi etiam de frigore, & calore: deinde vero usque ad num. 514 de electricitate, ubi totam Frankliniana Theoriam ex meis principiis explico, eandem ad bina tantummodo reducens principia, quæ ex mea generali virium Theoria eodem fere pacto deducuntur, quo præcipitationes, atque dissolutiones. Demum num. 514, ac 515 magnetismum persequor, tam directionem explicans, quam attractionem magneticam.

516 Hisce expositis, quæ ad particulares etiam proprietates pertinent, iterum a num. 516 ad finem usque generalem corporum complector naturam, & quid materia sit, quid forma, quæ censeri debeant essentialia, quæ accidentialia attributa, adeoque quid transformatio sit, quid alteratio, singillatim persequor, & partem hanc tertiam Theoriæ meæ absolvo.

De Appendice ad Metaphysicam pertinente innuam hic illud tantummodo, me ibi exponere de anima illud inprimis, quantum spiritus a materia differat, quem nexum anima habeat cum corpore, & quomodo in ipsum agat: tum de Deo, ipsius & existentiam me pluribus evincere, quæ nexum habeant cum ipsa Theoria mea, & Sapientiam inprimis, ac Providentiam, ex qua gradum ad revelationem faciendum innuo tantummodo. Sed hæc in antecessum veluti delibasse sit satis.

show, in Art. 490, that it suffers no real resistance, & in Art. 491 I explain the origin of bodies emitting light, in Art. 492 the reason why light that falls with greater obliquity is reflected more strongly, in Art. 493, 494 the origin of different degrees of refrangibility, & in Art. 495, 496 I deduce that there are two different dispositions recurring at equal intervals; hence, in Art. 497, I bring out those alternations, discovered by Newton, of easier reflection & easier transmission, & in Art. 498 I deduce that some rays should be reflected & others transmitted in the passage to a fresh medium, & that the greater the obliquity of incidence, the greater the number of reflected rays. In Art. 499, 500 I state the origin of the difference between the lengths of the intervals of the alternations; upon this alone depends the whole of the Newtonian theory of natural colours. Finally, in Art. 501, I touch upon the wonderful property of Iceland spar & its cause, & in Art. 502 I explain diffraction, which is a kind of imperfect refraction or reflection.

After light derived from fire, which has to do with vision, I very briefly deal with taste & smell in Art. 503, & of sound in the three articles that follow next. Then, in the next four articles, I consider touch, & in connection with it, cold & heat also. After that, as far as Art. 514, I deal with electricity; here I explain the whole of the Franklin theory by means of my principles; I reduce this theory to two principles only, & these are derived from my general Theory of forces in almost the same manner as I have already derived precipitations & solutions. Finally, in Art. 514, 515, I investigate magnetism, explaining both magnetic direction & attraction.

503
504
507
511
514

These things being expounded, all of which relate to special properties, I once more consider, in the articles from 516 to the end, the general nature of bodies, what matter is, its form, what things ought to be considered as essential, & what as accidental, attributes; and also the nature of transformation and alteration are investigated, each in turn; & thus I bring to a close the third part of my Theory.

516

I will mention here but this one thing with regard to the appendix on Metaphysics; namely, that I there expound more especially how greatly different is the soul from matter, the connection between the soul & the body, & the manner of its action upon it. Then with regard to God, I prove that He must exist by many arguments that have a close connection with this Theory of mine; I especially mention, though but slightly, His Wisdom and Providence, from which there is but a step to be made towards revelation. But I think that I have, so to speak, given my preliminary foretaste quite sufficiently.

[1] # PHILOSOPHIÆ NATURALIS THEORIA

PARS I

Theoriæ expositio, analytica deductio, & vindicatio.

Cujusmodi systema, Theoria exhibeat.

1. IRIUM mutuarum Theoria, in quam incidi jam ab Anno 1745, dum e notissimis principiis alia ex aliis consectaria eruerem, & ex qua ipsam simplicium materiæ elementorum constitutionem deduxi, systema exhibet medium inter Leibnitianum, & Newtonianum, quod nimirum & ex utroque habet plurimum, & ab utroque plurimum dissidet; at utroque in immensum simplicius, proprietatibus corporum generalibus sane omnibus, & [2] peculiaribus quibusque præcipuis per accuratissimas demonstrationes deducendis est profecto mirum in modum idoneum.

In quo conveniat cum systemate Newtoniano, & Leibnitiano.

2. Habet id quidem ex Leibnitii Theoria elementa prima simplicia, ac prorsus inextensa: habet ex Newtoniano systemate vires mutuas, quæ pro aliis punctorum distantiis a se invicem aliæ sint; & quidem ex ipso itidem Newtono non ejusmodi vires tantummodo, quæ ipsa puncta determinent ad accessum, quas vulgo attractiones nominant; sed etiam ejusmodi, quæ determinent ad recessum, & appellantur repulsiones: atque id ipsum ita, ut, ubi attractio desinat, ibi, mutata distantia, incipiat repulsio, & vice versa, quod nimirum Newtonus idem in postrema Opticæ Quæstione proposuit, ac exemplo transitus a positivis ad negativa, qui habetur in algebraicis formulis, illustravit. Illud autem utrique systemati commune est cum hoc meo, quod quævis particula materiæ cum aliis quibusvis, utcunque remotis, ita connectitur, ut ad mutationem utcunque exiguam in positione unius cujusvis, determinationes ad motum in omnibus reliquis immutentur, & nisi forte elidantur omnes oppositæ, qui casus est infinities improbabilis, motus in iis omnibus aliquis inde ortus habeatur.

In quo differat a Leibnitiano & ipsi præstet.

3. Distat autem a Leibnitiana Theoria longissime, tum quia nullam extensionem continuam admittit, quæ ex contiguis, & se contingentibus inextensis oriatur: in quo quidem difficultas jam olim contra Zenonem proposita, & nunquam sane aut soluta satis, aut solvenda, de compenetratione omnimoda inextensorum contiguorum, candem vim adhuc habet contra Leibnitianum systema: tum quia homogeneitatem admittit in clementis, omni massarum discrimine a sola dispositione, & diversa combinatione derivato, ad quam homogeneitatem in clementis, & discriminis rationem in massis, ipsa nos Naturæ analogia ducit, ac chemicæ resolutiones inprimis, in quibus cum ad adeo pauciora numero, & adeo minus inter se diversa principiorum genera, in compositorum corporum analysi deveniatur, id ipsum indicio est, quo ulterius promoveri possit analysis, eo ad majorem simplicitatem, & homogeneitatem devenire debere, adeoque in ultima demum resolutione ad homogeneitatem, & simplicitatem summam, contra quam quidem indiscernibilium principium, & principium rationis sufficientis usque adeo a Leibnitianis deprædicata, meo quidem judicio, nihil omnino possunt.

In quo differat a Newtoniano & ipsi præstet.

4. Distat itidem a Newtoniano systemate quamplurimum, tum in eo, quod ea, quæ Newtonus in ipsa postremo Quæstione Opticæ conatus est explicare per tria principia, gravitatis, cohæsionis, fermentationis, immo & reliqua quamplurima, quæ ab iis tribus principiis omnino non pendent, per unicam explicat legem virium, expressam unica, & ex pluribus inter se commixtis non composita algebraica formula, vel unica continua geometrica curva: tum in eo, quod in mi-[3]-nimis distantiis vires admittat non positivas, sive attractivas, uti Newtonus, sed negativas, sive repulsivas, quamvis itidem eo majores in

A THEORY OF NATURAL PHILOSOPHY

PART I

Exposition, Analytical Derivation & Proof of the Theory

1. HE following Theory of mutual forces, which I lit upon as far back as the year 1745, whilst I was studying various propositions arising from other very well-known principles, & from which I have derived the very constitution of the simple elements of matter, presents a system that is midway between that of Leibniz & that of Newton; it has very much in common with both, & differs very much from either; &, as it is immensely more simple than either, it is undoubtedly suitable in a marvellous degree for deriving all the general properties of bodies, & certain of the special properties also, by means of the most rigorous demonstrations.

The kind of system the Theory presents.

2. It indeed holds to those simple & perfectly non-extended primary elements upon which is founded the theory of Leibniz; & also to the mutual forces, which vary as the distances of the points from one another vary, the characteristic of the theory of Newton; in addition, it deals not only with the kind of forces, employed by Newton, which oblige the points to approach one another, & are commonly called attractions; but also it considers forces of a kind that engender recession, & are called repulsions. Further, the idea is introduced in such a manner that, where attraction ends, there, with a change of distance, repulsion begins; this idea, as a matter of fact, was suggested by Newton in the last of his 'Questions on Optics', & he illustrated it by the example of the passage from positive to negative, as used in algebraical formulæ. Moreover there is this common point between either of the theories of Newton & Leibniz & my own; namely, that any particle of matter is connected with every other particle, no matter how great is the distance between them, in such a way that, in accordance with a change in the position, no matter how slight, of any one of them, the factors that determine the motions of all the rest are altered; &, unless it happens that they all cancel one another (& this is infinitely improbable), some motion, due to the change of position in question, will take place in every one of them.

What there is in it common to the systems of Newton & Leibniz.

3. But my Theory differs in a marked degree from that of Leibniz. For one thing, because it does not admit the continuous extension that arises from the idea of consecutive, non-extended points touching one another; here, the difficulty raised in times gone by in opposition to Zeno, & never really or satisfactorily answered (nor can it be answered), with regard to compenetration of all kinds with non-extended consecutive points, still holds the same force against the system of Leibniz. For another thing, it admits homogeneity amongst the elements, all distinction between masses depending on relative position only, & different combinations of the elements; for this homogeneity amongst the elements, & the reason for the difference amongst masses, Nature herself provides us with the analogy. Chemical operations especially do so; for, since the result of the analysis of compound substances leads to classes of elementary substances that are so comparatively few in number, & still less different from one another in nature; it strongly suggests that, the further analysis can be pushed, the greater the simplicity, & homogeneity, that ought to be attained; thus, at length, we should have, as the result of a final decomposition, homogeneity & simplicity of the highest degree. Against this homogeneity & simplicity, the principle of indiscernibles, & the doctrine of sufficient reason, so long & strongly advocated by the followers of Leibniz, can, in my opinion at least, avail in not the slightest degree.

How it differs from, & surpasses, the theory of Leibniz.

4. My Theory also differs as widely as possible from that of Newton. For one thing, because it explains by means of a single law of forces all those things that Newton himself, in the last of his 'Questions on Optics', endeavoured to explain by the three principles of gravity, cohesion & fermentation; nay, & very many other things as well, which do not altogether follow from those three principles. Further, this law is expressed by a single algebraical formula, & not by one composed of several formulæ compounded together; or by a single continuous geometrical curve. For another thing, it admits forces that at very small distances are not positive or attractive, as Newton supposed, but negative or repul-

How it differs from, & surpasses, the theory of Newton.

infinitum, quo distantiæ in infinitum decrescant. Unde illud necessario consequitur, ut nec cohæsio a contactu immediato oriatur, quam ego quidem longe aliunde desumo; nec ullus immediatus, &, ut illum appellare soleo, mathematicus materiæ contactus habeatur, quod simplicitatem, & inextensionem inducit elementorum, quæ ipse variarum figurarum voluit, & partibus a se invicem distinctis composita, quamvis ita cohærentia, ut nulla Naturæ vi dissolvi possit compages, & adhæsio labefactari, quæ adhæsio ipsi, respectu virium nobis cognitarum, est absolute infinita.

Ubi de ipsa ctum ante; & quid promissum.

5. Quæ ad ejusmodi Theoriam pertinentia hucusque sunt edita, continentur dissertationibus meis, *De viribus vivis*, edita Anno 1745, *De Lumine* A. 1748, *De Lege Continuitatis* A. 1754, *De Lege virium in natura existentium* A. 1755, *De divisibilitate materiæ, & principiis corporum* A. 1757, ac in meis *Supplementis* Stayanæ Philosophiæ versibus traditæ, cujus primus Tomus prodiit A. 1755: eadem autem satis dilucide proposuit, & amplissimum ipsius per omnem Physicam demonstravit usum vir e nostra Societate doctissimus Carolus Benvenutus in sua *Physicæ Generalis Synopsi* edita Anno 1754. In ea Synopsi proposuit idem & meam deductionem æquilibrii binarum massarum, viribus parallelis animatarum, quæ ex ipsa mea Theoria per notissimam legem compositionis virium, & æqualitatis inter actionem, & reactionem, fere sponte consequitur, cujus quidem in supplementis illis § 4. ad lib. 3. mentionem feci, ubi & quæ in dissertatione *De centro Gravitatis* edideram, paucis proposui; & de centro oscillationis agens, protuli aliorum methodos præcipuas quasque, quæ ipsius determinationem a subsidiariis tantummodo principiis quibusdam repetunt. Ibidem autem de æquilibrii centro agens illud affirmavi: *In Natura nullæ sunt rigidæ virgæ, inflexiles, & omni gravitate, ac inertia carentes, adeoque nec revera ullæ leges pro iis conditæ; & si ad genuina, & simplicissima naturæ principia, res exigatur, invenietur, omnia pendere a compositione virium, quibus in se invicem agunt particulæ materiæ; a quibus nimirum viribus omnia Naturæ phænomena proficiscuntur.* Ibidem autem exhibitis aliorum methodis ad centrum oscillationis pertinentibus, promisi, me in quarto ejusdem Philosophiæ tomo ex genuinis principiis investigaturum, ut æquilibrii, sic itidem oscillationis centrum.

Qua occasione hoc de ipsa conscriptum opus.

6. Porro cum nuper occasio se mihi præbuisset inquirendi in ipsum oscillationis centrum ex meis principiis, urgente Scherffero nostro viro doctissimo, qui in eodem hoc Academico Societatis Collegio nostros Mathesim docet; casu incidi in theorema simplicisimum sane, & admodum elegans, quo trium massarum in se mutuo agentium comparantur vires, [4] quod quidem ipsa fortasse tanta sua simplicitate effugit hucusque Mechanicorum oculos; nisi forte ne effugerit quidem, sed alicubi jam ab alio quopiam inventum, & editum, me, quod admodum facile fieri potest, adhuc latuerit, ex quo theoremate & æquilibrium, ac omne vectium genus, & momentorum mensura pro machinis, & oscillationis centrum etiam pro casu, quo oscillatio fit in latus in plano ad axem oscillationis perpendiculari, & centrum percussionis sponte fluunt, & quod ad sublimiores alias perquisitiones viam aperit admodum patentem. Cogitaveram ego quidem initio brevi dissertatiuncula hoc theorema tantummodo edere cum consectariis, ac breve Theoriæ meæ specimen quoddam exponere; sed paullatim excrevit opusculum, ut demum & Theoriam omnem exposuerim ordine suo, & vindicarim, & ad Mechanicam prius, tum ad Physicam fere universam applicaverim, ubi & quæ maxima notatu digna erant, in memoratis dissertationibus ordine suo digessi omnia, & alia adjeci quamplurima, quæ vel olim animo conceperam, vel modo sese obtulerunt scribenti, & omnem hanc rerum farraginem animo pervolventi.

Prima elementa indivisibilia inextensa, nec contigua.

7. Prima elementa materiæ mihi sunt puncta prorsus indivisibilia, & inextensa, quæ in immenso vacuo ita dispersa sunt, ut bina quævis a se invicem distent per aliquod intervallum, quod quidem indefinite augeri potest, & minui, sed penitus evanescere non potest, sine conpenetratione ipsorum punctorum: eorum enim contiguitatem nullam admitto possibilem; sed illud arbitror omnino certum, si distantia duorum materiæ punctorum sit nulla, idem prorsus spatii vulgo concepti punctum indivisibile occupari ab utroque debere, &

sive; although these also become greater & greater indefinitely, as the distances decrease indefinitely. From this it follows of necessity that cohesion is not a consequence of immediate contact, as I indeed deduce from totally different considerations; nor is it possible to get any immediate or, as I usually term it, mathematical contact between the parts of matter. This idea naturally leads to simplicity & non-extension of the elements, such as Newton himself postulated for various figures; & to bodies composed of parts perfectly distinct from one another, although bound together so closely that the ties could not be broken or the adherence weakened by any force in Nature; this adherence, as far as the forces known to us are concerned, is in his opinion unlimited.

When & where I have already dealt with this theory; & a promise that I made.

5. What has already been published relating to this kind of Theory is contained in my dissertations, *De Viribus vivis*, issued in 1745, *De Lumine*, 1748, *De Lege Continuitatis*, 1754, *De Lege virium in natura existentium*, 1755, *De divisibilitate materiæ, & principiis corporum*, 1757, & in my *Supplements* to the philosophy of Benedictus Stay, issued in verse, of which the first volume was published in 1755. The same theory was set forth with considerable lucidity, & its extremely wide utility in the matter of the whole of Physics was demonstrated, by a learned member of our Society, Carolus Benvenutus, in his *Physicæ Generalis Synopsis* published in 1754. In this synopsis he also at the same time gave my deduction of the equilibrium of a pair of masses actuated by parallel forces, which follows quite naturally from my Theory by the well-known law for the composition of forces, & the equality between action & reaction; this I mentioned in those Supplements, section 4 of book 3, & there also I set forth briefly what I had published in my dissertation *De centro Gravitatis*. Further, dealing with the centre of oscillation, I stated the most noteworthy methods of others who sought to derive the determination of this centre from merely subsidiary principles. Here also, dealing with the centre of equilibrium, I asserted:—"*In Nature there are no rods that are rigid, inflexible, totally devoid of weight & inertia; & so, neither are there really any laws founded on them. If the matter is worked back to the genuine & simplest natural principles, it will be found that everything depends on the composition of the forces with which the particles of matter act upon one another; & from these very forces, as a matter of fact, all phenomena of Nature take their origin.*" Moreover, here too, having stated the methods of others for the determination of the centre of oscillation, I promised that, in the fourth volume of the Philosophy, I would investigate by means of genuine principles, such as I had used for the centre of equilibrium, the centre of oscillation as well.

The occasion that led to my writing this work on the matter.

6. Now, lately I had occasion to investigate this centre of oscillation, deriving it from my own principles, at the request of Father Scherffer, a man of much learning, who teaches mathematics in this College of the Society. Whilst doing this, I happened to hit upon a really most simple & truly elegant theorem, from which the forces with which three masses mutually act upon one another are easily to be found; this theorem, perchance owing to its extreme simplicity, has escaped the notice of mechanicians up till now (unless indeed perhaps it has not escaped notice, but has at some time previously been discovered & published by some other person, though, as may very easily have happened, it may not have come to my notice). From this theorem there come, as the natural consequences, the equilibrium & all the different kinds of levers, the measurement of moments for machines, the centre of oscillation for the case in which the oscillation takes place sideways in a plane perpendicular to the axis of oscillation, & also the centre of percussion; it opens up also a beautifully clear road to other and more sublime investigations. Initially, my idea was to publish in a short esssay merely this theorem & some deductions from it, & thus to give some sort of brief specimen of my Theory. But little by little the essay grew in length, until it ended in my setting forth in an orderly manner the whole of the theory, giving a demonstration of its truth, & showing its application to Mechanics in the first place, and then to almost the whole of Physics. To it I also added not only those matters that seemed to me to be more especially worth mention, which had all been already set forth in an orderly manner in the dissertations mentioned above, but also a large number of other things, some of which had entered my mind previously, whilst others in some sort obtruded themselves on my notice as I was writing & turning over in my mind all this conglomeration of material.

The primary elements are indivisible, non-extended & they are not contiguous.

7. The primary elements of matter are in my opinion perfectly indivisible & non-extended points; they are so scattered in an immense vacuum that every two of them are separated from one another by a definite interval; this interval can be indefinitely increased or diminished, but can never vanish altogether without compenetration of the points themselves; for I do not admit as possible any immediate contact between them. On the contrary I consider that it is a certainty that, if the distance between two points of matter should become absolutely nothing, then the very same indivisible point of space, according to the usual idea of it, must be occupied by both together, & we have true

haberi veram, ac omnimodam conpenetrationem. Quamobrem non vacuum ego quidem admitto disseminatum in materia, sed materiam in vacuo disseminatam, atque innatantem.

Eorum inertiæ vis cujusmodi.

8. In hisce punctis admitto determinationem perseverandi in eodem statu quietis, vel motus uniformis in directum (a) in quo semel sint posita, si seorsum singula in Natura existant; vel si alia alibi extant puncta, componendi per notam, & communem methodum compositionis virium, & motuum, parallelogrammorum ope, præcedentem motum cum mo-[5]-tu quem determinant vires mutuæ, quas inter bina quævis puncta agnosco a distantiis pendentes, & iis mutatis mutatas, juxta generalem quandam omnibus communem legem. In ea determinatione stat illa, quam dicimus, inertiæ vis, quæ, an a libera pendeat Supremi Conditoris lege, an ab ipsa punctorum natura, an ab aliquo iis adjecto, quodcunque, istud sit, ego quidem non quæro; nec vero, si velim quærere, inveniendi spem habeo; quod idem sane censeo de ea virium lege, ad quam gradum jam facio.

Eorundem vires mutuæ in aliis distantiis attractivæ, in aliis repulsivæ: virium ejusmodi exempla.

9. Censeo igitur bina quæcunque materiæ puncta determinari æque in aliis distantiis ad mutuum accessum, in aliis ad recessum mutuum, quam ipsam determinationem appello vim, in priore casu attractivam, in posteriore repulsivam, eo nomine non agendi modum, sed ipsam determinationem exprimens, undecunque proveniat, cujus vero magnitudo mutatis distantiis mutetur & ipsa secundum certam legem quandam, quæ per geometricam lineam curvam, vel algebraicam formulam exponi possit, & oculis ipsis, uti moris est apud Mechanicos repræsentari. Vis mutuæ a distantia pendentis, & ea variata itidem variatæ, atque ad omnes in immensum & magnas, & parvas distantias pertinentis, habemus exemplum in ipsa Newtoniana generali gravitate mutata in ratione reciproca duplicata distantiarum, quæ idcirco numquam e positiva in negativam migrare potest, adeoque ab attractiva ad repulsivam, sive a determinatione ad accessum ad determinationem ad recessum nusquam migrat. Verum in elastris inflexis habemus etiam imaginem ejusmodi vis mutuæ variatæ secundum distantias, & a determinatione ad recessum migrantis in determinationem ad accessum, & vice versa. Ibi enim si duæ cuspides, compresso elastro, ad se invicem accedant, acquirunt determinationem ad recessum, eo majorem, quo magis, compresso elastro, distantia decrescit; aucta distantia cuspidum, vis ad recessum minuitur, donec in quadam distantia evanescat, & fiat prorsus nulla; tum distantia adhuc aucta, incipit determinatio ad accessum, quæ perpetuo eo magis crescit, quo magis cuspides a se invicem recedunt: ac si e contrario cuspidum distantia minuatur perpetuo; determinatio ad accessum itidem minuetur, evanescet, & in determinationem ad recessum mutabitur. Ea determinatio oritur utique non ab immediata cuspidum actione in se invicem, sed a natura, & forma totius intermediæ laminæ plicatæ; sed hic physicam rei causam non moror, & solum persequor exemplum determinationis ad accessum, & recessum, quæ determinatio in aliis distantiis alium habeat nisum, & migret etiam ab altera in alteram.

Virium earundem lex.

10. Lex autem virium est ejusmodi, ut in minimis distantiis sint repulsivæ, atque eo majores in infinitum, quo distantiæ ipsæ minuuntur in infinitum, ita, ut pares sint extinguen-[6]-dæ cuivis velocitati utcunque magnæ, cum qua punctum alterum ad alterum possit accedere, antequam corum distantia evanescat; distantiis vero auctis minuuntur ita, ut in quadam distantia perquam exigua evadat vis nulla: tum adhuc, aucta distantia, mutentur in attractivas, primo quidem crescentes, tum decrescentes, evanescentes, abeuntes in repulsivas, eodem pacto crescentes, deinde decrescentes, evanescentes, migrantes iterum in attractivas, atque id per vices in distantiis plurimis, sed adhuc perquam exiguis, donec, ubi ad aliquanto majores distantias ventum sit, incipiant esse perpetuo attractivæ, & ad sensum reciproce

(a) *Id quidem respectu ejus spatii, in quo continemur nos, & omnia quæ nostris observari sensibus possunt, corpora; quod quiddam spatium si quiescat, nihil ego in ea re a reliquis differo; si forte moveatur motu quopiam, quem motum ex hujusmodi determinatione sequi debeant ipsa materiæ puncta; tum hæc mea erit quædam non absoluta, sed respectiva inertiæ vis, quam ego quidem exposui & in dissertatione* De Maris æstu *& in* Supplementis Stayanis *Lib. I.* § 13; *ubi etiam illud occurrit, quam ob causam ejusmodi respectivam inertiam excogitarim, & quibus rationibus evinci putem, absolutam omnino demonstrari non posse; sed ea huc non pertinent.*

compenetration in every way. Therefore indeed I do not admit the idea of vacuum interspersed amongst matter, but I consider that matter is interspersed in a vacuum & floats in it.

The nature of the force of inertia that they possess.

8. As an attribute of these points I admit an inherent propensity to remain in the same state of rest, or of uniform motion in a straight line, (a) in which they are initially set, if each exists by itself in Nature. But if there are also other points anywhere, there is an inherent propensity to compound (according to the usual well-known composition of forces & motions by the parallelogram law), the preceding motion with the motion which is determined by the mutual forces that I admit to act between any two of them, depending on the distances & changing, as the distances change, according to a certain law common to them all. This propensity is the origin of what we call the 'force of inertia'; whether this is dependent upon an arbitrary law of the Supreme Architect, or on the nature of points itself, or on some attribute of them, whatever it may be, I do not seek to know; even if I did wish to do so, I see no hope of finding the answer; and I truly think that this also applies to the law of forces, to which I now pass on.

The mutual forces between them are attractive at some distances & repulsive at others; examples of forces of this kind.

9. I therefore consider that any two points of matter are subject to a determination to approach one another at some distances, & in an equal degree recede from one another at other distances. This determination I call 'force'; in the first case 'attractive', in the second case 'repulsive'; this term does not denote the mode of action, but the propensity itself, whatever its origin, of which the magnitude changes as the distances change; this is in accordance with a certain definite law, which can be represented by a geometrical curve or by an algebraical formula, & visualized in the manner customary with Mechanicians. We have an example of a force dependent on distance, & varying with varying distance, & pertaining to all distances either great or small, throughout the vastness of space, in the Newtonian idea of general gravitation that changes according to the inverse squares of the distances: this, on account of the law governing it, can never pass from positive to negative; & thus on no occasion does it pass from being attractive to being repulsive, i.e., from a propensity to approach to a propensity to recession. Further, in bent springs we have an illustration of that kind of mutual force that varies according as the distance varies, & passes from a propensity to recession to a propensity to approach, and vice versa. For here, if the two ends of the spring approach one another on compressing the spring, they acquire a propensity for recession that is the greater, the more the distance diminishes between them as the spring is compressed. But, if the distance between the ends is increased, the force of recession is diminished, until at a certain distance it vanishes and becomes absolutely nothing. Then, if the distance is still further increased, there begins a propensity to approach, which increases more & more as the ends recede further & further away from one another. If now, on the contrary, the distance between the ends is continually diminished, the propensity to approach also diminishes, vanishes, & becomes changed into a propensity to recession. This propensity certainly does not arise from the immediate action of the ends upon one another, but from the nature & form of the whole of the folded plate of metal intervening. But I do not delay over the physical cause of the thing at this juncture; I only describe it as an example of a propensity to approach & recession, this propensity being characterized by one endeavour at some distances & another at other distances, & changing from one propensity to another.

The law of forces for the points.

10. Now the law of forces is of this kind; the forces are repulsive at very small distances, & become indefinitely greater & greater, as the distances are diminished indefinitely, in such a manner that they are capable of destroying any velocity, no matter how large it may be, with which one point may approach another, before ever the distance between them vanishes. When the distance between them is increased, they are diminished in such a way that at a certain distance, which is extremely small, the force becomes nothing. Then as the distance is still further increased, the forces are changed to attractive forces; these at first increase, then diminish, vanish, & become repulsive forces, which in the same way first increase, then diminish, vanish, & become once more attractive; & so on, in turn, for a very great number of distances, which are all still very minute: until, finally, when we get to comparatively great distances, they begin to be continually attractive & approxi-

(a) *This indeed holds true for that space in which we, and all bodies that can be observed by our senses, are contained. Now, if this space is at rest, I do not differ from other philosophers with regard to the matter in question; but if perchance space itself moves in some way or other, what motion ought these points of matter to comply with owing to this kind of propensity? In that case this force of inertia that I postulate is not absolute, but relative; as indeed I explained both in the dissertation* De Maris Aestu, *and also in the* Supplements to *Stay's Philosophy, book* 1, *section* 13. *Here also will be found the conclusions at which I arrived with regard to relative inertia of this sort, and the arguments by which I think it is proved that it is impossible to show that it is generally absolute. But these things do not concern us at present.*

proportionales quadratis distantiarum, atque id vel utcunque augeantur distantiæ etiam in infinitum, vel saltem donec ad distantias deveniatur omnibus Planetarum, & Cometarum distantiis longe majores.

Legis simplicitas exprimibilis per continuam curvam.

11. Hujusmodi lex primo aspectu videtur admodum complicata, & ex diversis legibus temere inter se coagmentatis coalescens; at simplicissima, & prorsus incomposita esse potest, expressa videlicet per unicam continuam curvam, vel simplicem Algebraicam formulam, uti innui superius. Hujusmodi curva linea est admodum apta ad sistendam oculis ipsis ejusmodi legem, nec requirit Geometram, ut id præstare possit: satis est, ut quis eam intueatur tantummodo, & in ipsa ut in imagine quadam solemus intueri depictas res qualescunque, virium illarum indolem contempletur. In ejusmodi curva eæ, quas Geometræ abscissas dicunt, & sunt segmenta axis, ad quem ipsa refertur curva, exprimunt distantias binorum punctorum a se invicem: illæ vero, quæ dicuntur ordinatæ, ac sunt perpendiculares lineæ ab axe ad curvam ductæ, referunt vires: quæ quidem, ubi ad alteram jacent axis partem, exhibent vires attractivas; ubi jacent ad alteram, repulsivas, & prout curva accedit ad axem, vel recedit, minuuntur ipsæ etiam, vel augentur: ubi curva axem secat, & ab altera ejus parte transit ad alteram, mutantibus directionem ordinatis, abeunt ex positivis in negativas, vel vice versa: ubi autem arcus curvæ aliquis ad rectam quampiam axi perpendicularem in infinitum productam semper magis accedit ita ultra quoscumque limites, ut nunquam in eam recidat, quem arcum asymptoticum appellant Geometræ, ibi vires ipsæ in infinitum excrescunt.

Forma curvæ ipsius.

12. Ejusmodi curvam exhibui, & exposui in dissertationibus *De viribus vivis* a Num. 51, *De Lumine* Num. 5, *De Lege virium in Naturam existentium* a Num. 68, & in sua *Synopsi Physicæ Generalis* P. Benvenutus candem protulit a Num. 108. En brevem quandem ejus ideam. In Fig. 1, Axis C′AC habet in puncto A asymptotum curvæ rectilineam AB indefinitam, circa quam habentur bini curvæ rami hinc, & inde æquales, prorsus inter se, & similes, quorum alter DEFGHIKLMNOPQRSTV habet inprimis arcum ED [7] asymptoticum, qui nimirum ad partes BD, si indefinite producatur ultra quoscunque limites, semper magis accedit ad rectam AB productam ultra quoscunque limites, quin unquam ad candem deveniat; hinc vero versus DE perpetuo recidit ab cadam recta, immo etiam perpetuo versus V ab eadem recedunt arcus reliqui omnes, quin uspiam recessus mutetur in accessum. Ad axem C′C perpetuo primum accedit, donec ad ipsum deveniat alicubi in E; tum eodem ibi seeto progreditur, & ab ipso perpetuo recedit usque ad quandam distantiam F, postquam recessum in accessum mutat, & iterum ipsum axem secat in G, ac flexibus continuis contorquetur circa ipsum, quem pariter secat in punctis quamplurimis, sed paucas admodum ejusmodi sectiones figura exhibet, uti I, L, N, P, R. Demum is arcus desinit in alterum crus T*ps*V, jacens ex parte opposita axis respectu primi cruris, quod alterum crus ipsum habet axem pro asymptoto, & ad ipsum accedit ad sensum ita, ut distantiæ ab ipso sint in ratione reciproca duplicata distantiarum a recta BA.

Abscissæ exprimentes distantias, ordinatæ exprimentes vires.

13. Si ex quovis axis puncto *a*, *b*, *d*, erigatur usque ad curvam recta ipsi perpendicularis *ag*, *br*, *dh*, segmentum axis A*a*, A*b*, A*d*, dicitur abscissa, & refert distantiam duorum materiæ punctorum quorumcunque a se invicem; perpendicularis *ag*, *br*, *dh*, dicitur ordinata, & exhibet vim repulsivam, vel attractivam, prout jacet respectu axis ad partes D, vel oppositas.

Mutationes ordinatarum, & virium iis expressarum.

14. Patet autem, in ea curvæ forma ordinatam *ag* augeri ultra quoscunque limites, si abscissa A*a*, minuatur pariter ultra quoscunque limites; quæ si augeatur, ut abeat in A*b*, ordinata minuetur, & abibit in *br*, perpetuo imminutam in accessu *b* ad E, ubi evanescet tum aneta abscissa in A*d*, mutabit ordinata directionem in *dh*, ac ex parte opposita augebitur prius usque ad F, tum decrescet per *il* usque ad G, ubi evanescet, & iterum mutabit directionem regressa in *mn* ad illam priorem, donec post evanescentiam, & directionis mutationem factam in omnibus sectionibus I, L, N, P, R, fiant ordinatæ *op*, *vs*, directionis constantis, & decrescentes ad sensum in ratione reciproca duplicata abscissarum A*o*, A*v*. Quamobrem illud est manifestum, per ejusmodi curvam exprimi eas ipsas vires, initio

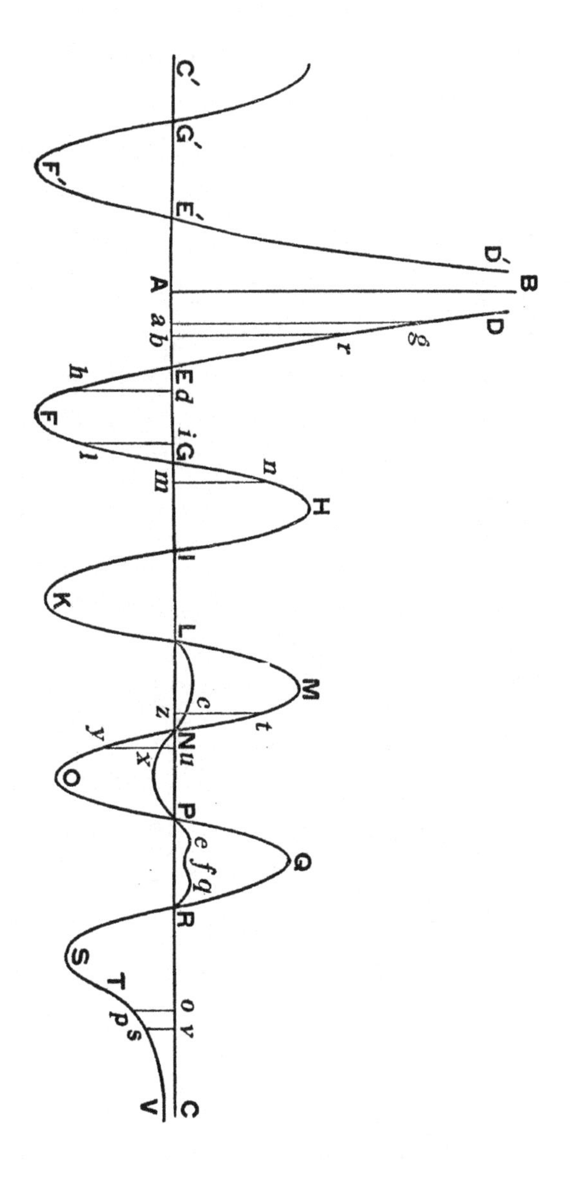
C′
G′
F′
E′
D′
B
A
a b
D
g
r
h
E d
F
i G
l
n
m
H
I
K
L
M
c
z
t
y
N u
x
O
P
e f q
Q
R
S
T
o v
p s
C
V

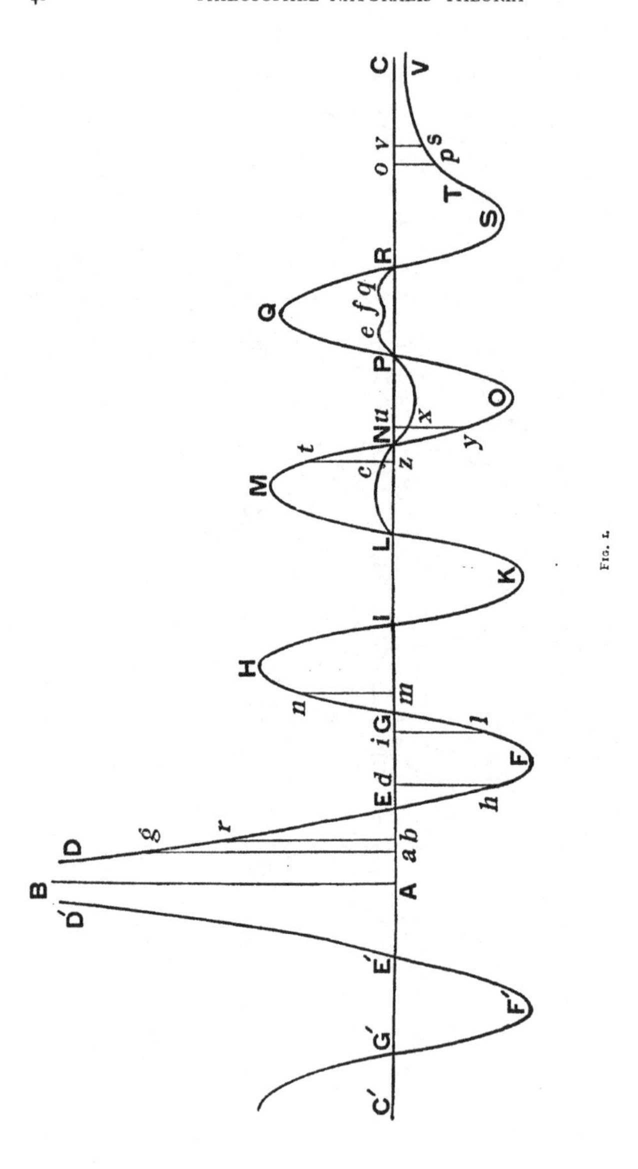

Fig. 1.

mately inversely proportional to the squares of the distances. This holds good as the distances are increased indefinitely to any extent, or at any rate until we get to distances that are far greater than all the distances of the planets & comets.

The simplicity of the law can be represented by means of a continuous curve.

11. A law of this kind will seem at first sight to be very complicated, & to be the result of combining together several different laws in a haphazard sort of way; but it can be of the simplest kind & not complicated in the slightest degree; it can be represented for instance by a single continuous curve, or by an algebraical formula, as I intimated above. A curve of this sort is perfectly adapted to the graphical representation of this sort of law, & it does not require a knowledge of geometry to set it forth. It is sufficient for anyone merely to glance at it, & in it, just as in a picture we are accustomed to view all manner of things depicted, so will he perceive the nature of these forces. In a curve of this kind, those lines, that geometricians call abscissæ, namely, segments of the axis to which the curve is referred, represent the distances of two points from one another; & those, which we called ordinates, namely, lines drawn perpendicular to the axis to meet the curve, represent forces. These, when they lie on one side of the axis represent attractive forces, and, when they lie on the other side, repulsive forces; & according as the curve approaches the axis or recedes from it, they too are diminished or increased. When the curve cuts the axis & passes from one side of it to the other, the direction of the ordinates being changed in consequence, the forces pass from positive to negative or vice versa. When any arc of the curve approaches ever more closely to some straight line perpendicular to the axis and indefinitely produced, in such a manner that, even if this goes on beyond all limits, yet the curve never quite reaches the line (such an arc is called asymptotic by geometricians), then the forces themselves will increase indefinitely.

The form of the curve.

12. I set forth and explained a curve of this sort in my dissertations *De Viribus vivis* (Art. 51), *De Lumine* (Art. 5), *De lege virium in Natura existentium* (Art. 68); and Father Benvenutus published the same thing in his *Synopsis Physicæ Generalis* (Art. 108). This will give you some idea of its nature in a few words.

In Fig. 1 the axis C′AC has at the point A a straight line AB perpendicular to itself, which is an asymptote to the curve; there are two branches of the curve, one on each side of AB, which are equal & similar to one another in every way. Of these, one, namely DEFGHIKLMNOPQRSTV, has first of all an asymptotic arc ED; this indeed, if it is produced ever so far in the direction ED, will approach nearer & nearer to the straight line AB when it also is produced indefinitely, but will never reach it; then, in the direction DE, it will continually recede from this straight line, & so indeed will all the rest of the arcs continually recede from this straight line towards V. The first arc continually approaches the axis C′C, until it meets it in some oint E; then it cuts it at this point & passes on, continually receding from the axis untiꝑit arrives at a certain distance given by the point F; after that the recession changes to an approach, & it cuts the axis once more in G; & so on, with successive changes of curvature, the curve winds about the axis, & at the same time cuts it in a number of points that is really large, although only a very few of the intersections of this kind, as I, L, N, P, R, are shown in the diagram. Finally the arc of the curve ends up with the other branch T*ps*V, lying on the opposite side of the axis with respect to the first branch; and this second branch has the axis itself as its asymptote, & approaches it approximately in such a manner that the distances from the axis are in the inverse ratio of the squares of the distances from the straight line AB.

The abscissæ represent distances, & the ordinates forces.

13. If from any point of the axis, such as *a*, *b*, or *d*, there is erected a straight line perpendicular to it to meet the curve, such as *ag*, *br*, or *dh* then the segment of the axis, A*a*, A*b*, or A*d*, is called the abscissa, & represents the distance of any two points of matter from one another; the perpendicular, *ag*, *br*, or *dh*, is called the ordinate, & this represents the force, which is repulsive or attractive, according as the ordinate lies with regard to the axis on the side towards D, or on the opposite side.

Change in the ordinates & the forces that they represent.

14. Now it is clear that, in a curve of this form, the ordinate *ag* will be increased beyond all bounds, if the abscissa A*a* is in the same way diminished beyond all bounds; & if the latter is increased and becomes A*b*, the ordinate will be diminished, & it will become *br*, which will continually diminish as *b* approaches to E, at which point it will vanish. Then the abscissa being increased until it becomes A*d*, the ordinate will change its direction as it becomes *dh*, & will be increased in the opposite direction at first, until the point F is reached, when it will be decreased through the value *il* until the point G is attained, at which point it vanishes; at the point G, the ordinate will once more change its direction as it returns to the position *mn* on the same side of the axis as at the start. Finally, after vanishing & changing direction at all points of intersection with the axis, such as I, L, N, P, R, the ordinates take the several positions indicated by *op*, *vs*: here the direction remains unchanged, & the ordinates decrease approximately in the inverse ratio of the squares of the abscissæ A*o*, A*v*. Hence it is perfectly evident that, by a curve of this kind, we can

repulsivas, & imminutis in infinitum distantiis auctas in infinitum, auctis imminutas, tum evanescentes, abeuntes, mutata directione, in attractivas, ac iterum evenescentes, mutatasque per vices: donec demum in satis magna distantia evadant attractivæ ad sensum in ratione reciproca duplicata distantiarum.

Discrimen hu us legis virium a gravitate Newtoniana: ejus usus in Physica: ordo pertractandorum.

15. Hæc virium lex a Newtoniana gravitate differt in ductu, & progressu curvæ eam exprimentis quæ nimirum, ut in fig. 2, apud Newtonum est hyperbola DV gradus tertii, jacens tota citra axem, quem nuspiam sceat, jacentibus omni-[8]-bus ordinatis *vs*, *op*, *bt*, *ag* ex parte attractiva, ut idcirco nulla habeatur mutatio e positivo ad negativum, ex attractione in repulsiouem, vel vice versa; cæterum utraque per ductum exponitur curvæ continuæ habentis duo crura infinita asymptotica in ramis singulis utrinque in infinitum productis. Ex hujusmodi autem virium lege, & ex solis principiis Mechanicis notissimis, nimirum quod ex pluribus viribus, vel motibus componatur vis, vel motus quidam ope parallelogrammorum, quorum latera exprimant vires, vel motus componentes, & quod vires ejusmodi in punctis singulis, tempusculis singulis æqualibus, inducant velocitates, vel motus proportionales sibi, omnes mihi profluunt generales, & præcipuæ quæque particulares proprietates corporum,uti etiam superius innui, nec ad singulares proprietates derivandas in genere affirmo, eas haberi per diversam combinationem, sed combinationes ipsas evolvo, & geometrice demonstro, quæ e quibus combinationibus phænomena, & corporum species oriri debeant. Verum antequam ea evolvo in parte secunda, & tertia, ostendam in hac prima, qua via, & quibus positivis rationibus ad eam virium legem devenerim, & qua ratione illam elementorum materiæ simplicitatem eruerim, tum quæ difficultatem aliquam videantur habere posse, dissolvam.

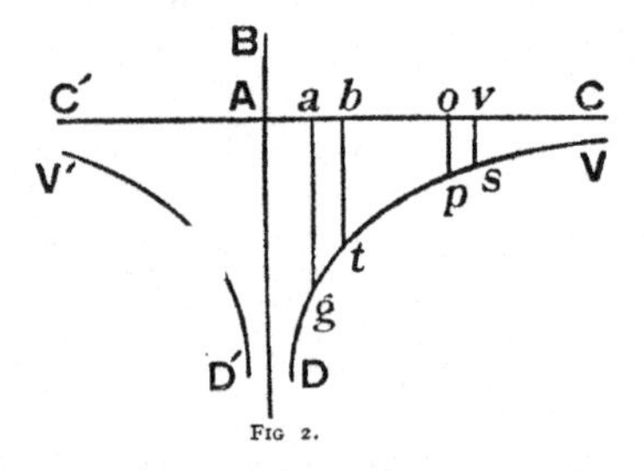

Fig 2.

Occasio inveniendæ Theoriæ ex consideratione impulsus.

16. Cum anno 1745 *De Viribus vivis* dissertationem conscriberem, & omnia, quæ a viribus vivis repetunt, qui Leibnitianam tuentur sententiam, & vero etiam plerique ex iis, qui per solam velocitatem vires vivas metiuntur, repeterem immediate a sola velocitate genita per potentiarum vires, quæ juxta communem omnium Mechanicorum sententiam velocitates vel generant, vel utcunque inducunt proportionales sibi, & tempusculis, quibus agunt, uti est gravitas, elasticitas, atque aliæ vires ejusmodi; cœpi aliquanto diligentius inquirere in eam productionem velocitatis, quæ per impulsum censetur fieri, ubi tota velocitas momento temporis produci creditur ab iis, qui idcirco percussionis vim infinities majorem esse censent viribus omnibus, quæ pressionem solam momentis singulis exercent. Statim illud mihi sese obtulit, alias pro percussionibus ejusmodi, quæ nimirum momento temporis finitam velocitatem inducant, actionum leges haberi debere.

Origo ejusdem ex oppositione impulsus immediati cum lege Continuitatis.

17. Verum re altius considerata, mihi illud incidit, si recta utamur ratiocinandi methodo, eum agendi modum submovendum esse a Natura, quæ nimirum eandem ubique virium legem, ac candem agendi rationem adhibeat: impulsum nimirum immediatum alterius corporis in alterum, & immediatam percussionem haberi non posse sine illa productione finitæ velocitatis facta momento temporis indivisibili, & hanc sine saltu quodam, & læsione illius, quam legem *Continuitatis* appellant, quam quidem legem in Natura existere, & quidem satis [9] valida ratione evinci posse existimabam. En autem ratiocinationem ipsam, qua tum quidem primo sum usus, ac deinde novis aliis, atque aliis meditationibus illustravi, ac confirmavi.

Læsio legis Continuitatis necessaria, si corpus velocius immediate incurrat in minus velox.

18. Concipiantur duo corpora æqualia, quæ moveantur in directum versus candem plagam, & id, quod præcedit, habeat gradus velocitatis 6, id vero, quod ipsum persequitur gradus 12. Si hoc posterius cum sua illa velocitate illæsa deveniat ad immediatum contactum cum illo priore; oportebit utique, ut ipso momento temporis, quo ad contactum devenerint, illud posterius minuat velocitatem suam, & illud primus suam augeat, utrumque per saltum, abeunte hoc a 12 ad 9, illo a 6 ad 9, sine ullo transitu per intermedios gradus 11, & 7; 10, & 8; $9\frac{1}{2}$, & $8\frac{1}{2}$, &c. Neque enim fieri potest, ut per aliquam utcunque exiguam continui

represent the forces in question, which are initially repulsive & increase indefinitely as the distances are diminished indefinitely, but which, as the distances increase, are first of all diminished, then vanish, then become changed in direction & so attractive, again vanish, & change their direction, & so on alternately; until at length, at a distance comparatively great they finally become attractive & are sensibly proportional to the inverse squares of the distance.

Difference between this law of forces & Newton's law of gravitation; its use in Physics; the order in which the subjects are to be taken.

15. This law of forces differs from the law of gravitation enunciated by Newton in the construction & development of the curve that represents it; thus, the curve given in Fig. 2, which is that according to Newton, is DV, a hyperbola of the third degree, lying altogether on one side of the axis, which it does not cut at any point; all the ordinates, such as *vs*, *op*, *bt*, *ag* lie on the side of the axis representing attractive forces, & therefore there is no change from positive to negative, i.e., from attraction to repulsion, or vice versa. On the other hand, each of the laws is represented by the construction of a continuous curve possessing two infinite asymptotic branches in each of its members, if produced to infinity on both sides. Now, from a law of forces of this kind, & with the help of well-known mechanical principles only, such as that a force or motion can be compounded from several forces or motions by the help of parallelograms whose sides represent the component forces or motions, or that the forces of this kind, acting on single points for single small equal intervals of time, produce in them velocities that are proportional to themselves; from these alone, I say, there have burst forth on me in a regular flood all the general & some of the most important particular properties of bodies, as I intimated above. Nor, indeed, for the purpose of deriving special properties, do I assert that they ought to be obtained owing to some special combination of points; on the contrary I consider the combinations themselves, & prove geometrically what phenomena, or what species of bodies, ought to arise from this or that combination. Of course, before I come to consider, both in the second part and in the third, all the matters mentioned above, I will show in this first part in what way, & by what direct reasoning, I have arrived at this law of forces, & by what argument I have made out the simplicity of the elements of matter; then I will give an explanation of every point that may seem to present any possible difficulty.

The occasion that led to the discovery of my Theory from the consideration of impulsive action.

16. In the year 1745, I was putting together my dissertation *De Viribus vivis*, & had derived everything that they who adhere to the idea of Leibniz, & the greater number of those who measure 'living forces' by means of velocity only, derive from these 'living forces'; as, I say I had derived everything directly & solely from the velocity generated by the forces of those influences, which, according to the generally accepted view taken by all Mechanicians, either generate, or in some way induce, velocities that are proportional to themselves & the intervals of time during which they act; take, for instance, gravity, elasticity, & other forces of the same kind. I then began to investigate somewhat more carefully that production of velocity which is thought to arise through impulsive action, in which the whole of the velocity is credited with being produced in an instant of time by those, who think, because of that, that the force of percussion is infinitely greater than all forces which merely exercise pressure for single instants. It immediately forced itself upon me that, for percussions of this kind, which really induce a finite velocity in an instant of time, laws for their actions must be obtained different from the rest.

The cause of the investigation was the opposition raised to the Law of Continuity by the idea of direct impulse.

17. However, when I considered the matter more thoroughly, it struck me that, if we employ a straightforward method of argument, such a mode of action must be withdrawn from Nature, which in every case adheres to one & the same law of forces, & the same mode of action. I came to the conclusion that really immediate impulsive action of one body on another, & immediate percussion, could not be obtained, without the production of a finite velocity taking place in an indivisible instant of time, & this would have to be accomplished without any sudden change or violation of what is called the *Law of Continuity*; this law indeed I considered as existing in Nature, & that this could be shown to be so by a sufficiently valid argument. The following is the line of argument that I employed initially; afterwards I made it clearer & confirmed it by further arguments & fresh reflection.

Violation of the Law of Continuity, if a body moving more swiftly comes into actual contact with another body moving more slowly.

18. Suppose there are two equal bodies, moving in the same straight line & in the same direction; & let the one that is in front have a degree of velocity represented by 6, & the one behind a degree represented by 12. If the latter, i.e., the body that was behind, should ever reach with its velocity undiminished, & come into absolute contact with, the former body which was in front, then in every case it would be necessary that, at the very instant of time at which this contact happened, the hindermost body should diminish its velocity, & the foremost body increase its velocity, in each case by a sudden change: one of them would pass from 12 to 9, the other from 6 to 9, without any passage through the intermediate degrees, 11 & 7, 10 & 8, 9½ & 8½, & so on. For it cannot possibly happen

temporis particulam ejusmodi mutatio fiat per intermedios gradus, durante contactu. Si enim aliquando alterum corpus jam habuit 7 gradus velocitatis, & alterum adhuc retinet 11; toto illo tempusculo, quod effluxit ab initio contactus, quando velocitates erant 12, & 6, ad id tempus, quo sunt 11, & 7, corpus secundum debuit moveri cum velocitate majore, quam primum, adeoque plus percurrere spatii, quam illud, & proinde anterior ejus superficies debuit transcurrere ultra illius posteriorem superficiem, & idcirco pars aliqua corporis sequentis cum aliqua antecedentis corporis parte compenetrari debuit, quod cum ob impenetrabilitatem, quam in materia agnoscunt passim omnes Physici, & quam ipsi tribuendam omnino esse, facile evincitur, fieri omnino non possit; oportuit sane, in ipso primo initio contactus, in ipso indivisibili momento temporis, quod inter tempus continuum præcedens contactum, & subsequens, est indivisibilis limes, ut punctum apud Geometras est limes indivisibilis inter duo continuæ lineæ segmenta, mutatio velocitatum facta fuerit per saltum sine transitu per intermedias, læsa penitus illa continuitatis lege, quæ itum ab una magnitudine ad aliam sine transitu per intermedias omnino vetat. Quod autem in corporibus æqualibus diximus de transitu immediato utriusque ad 9 gradus velocitatis, recurrit utique in iisdem, vel in utcunque inæqualibus de quovis alio transitu ad numeros quosvis. Nimirum ille posterioris corporis excessus graduum 6 momento temporis auferri debet, sive imminuta velocitate in ipso, sive aucta in priore, vel in altero imminuta utcunque, & aucta in altero, quod utique sine saltu, qui omissis infinitis intermediis velocitatibus habeatur, obtineri omnino non poterit.

Objectio petita a negatione durorum corporum.

19. Sunt, qui difficultatem omnem submoveri posse censeant, dicendo, id quidem ita se habere debere, si corpora dura habeantur, quæ nimirum nullam compressionem sentiant, nullam mutationem figuræ; & quoniam hæc a multis excluduntur penitus a Natura; dum se duo globi contingunt, introcessione, [10] & compressione partium fieri posse, ut in ipsis corporibus velocitas immutetur per omnes intermedios gradus transitu facto, & omnis argumenti vis eludatur.

Ea uti non posse, qui admittunt elementa solida, & dura.

20. At inprimis ea responsione uti non possunt, quicunque cum Newtono, & vero etiam cum plerisque veterum Philosophorum prima elementa materiæ omnino dura admittunt, & solida, cum adhæsione infinita, & impossibilitate absoluta mutationis figuræ. Nam in primis elementis illis solidis, & duris, quæ in anteriore adsunt sequentis corporis parte, & in præcedentis posteriore, quæ nimirum se mutuo immediate contingunt, redit omnis argumenti vis prorsus illæsa.

Extensionem continuam requirere primos poros, & parietes solidos, ac duros.

21. Deinde vero illud omnino intelligi sane non potest, quo pacto corpora omnia partes aliquas postremas circa superficiem non habeant penitus solidas, quæ idcirco comprimi omnino non possint. In materia quidem, si continua sit, divisibilitas in infinitum haberi potest, & vero etiam debet; at actualis divisio in infinitum difficultates secum trahit sane inextricabiles; qua tamen divisione in infinitum ii indigent, qui nullam in corporibus admittunt particulam utcunque exiguam compressionis omnis expertem penitus, atque incapacem. Ii enim debent admittere, particulam quamcunque actu interpositis poris distinctam, divisamque in plures pororum ipsorum velut parietes, poris tamen ipsis iterum distinctos. Illud sane intelligi non potest, qui fiat, ut, ubi e vacuo spatio transitur ad corpus, non aliquis continuus haberi debeat alicujus in se determinatæ crassitudinis paries usque ad primum porum, poris utique carens; vel quomodo, quod eodem recidit, nullus sit extimus, & superficiei externæ omnium proximus porus, qui nimirum si sit aliquis, parietem habeat utique poris expertem, & compressionis incapacem, in quo omnis argumenti superioris vis redit prorsus illæsa.

Læsio legis Continuitatis saltem in primis superficiebus, vel punctis.

22. At ea etiam, utcunque penitus inintelligibili, sententia admissa, redit omnis eadem argumenti vis in ipsa prima, & ultima corporum se immediate contingentium superficie, vel si nullæ continuæ superficies congruant, in lineis, vel punctis. Quidquid enim sit id, in quo contactus fiat, debet utique esse aliquid, quod nimirum impenetrabilitati occasionem præstet, & cogat motum in sequente corpore minui, in præcedente augeri; id, quidquid est, in quo exeritur impenetratibilitatis vis, quo fit immediatus contactus, id sane velocitatem mutare debet per saltum, sine transitu per intermedia, & in eo continuitatis lex abrumpi

that this kind of change is made by intermediate stages in some finite part, however small, of continuous time, whilst the bodies remain in contact. For if at any time the one body then had 7 degrees of velocity, the other would still retain 11 degrees; thus, during the whole time that has passed since the beginning of contact, when the velocities were respectively 12 & 6, until the time at which they are 11 & 7, the second body must be moved with a greater velocity than the first; hence it must traverse a greater distance in space than the other. It follows that the front surface of the second body must have passed beyond the back surface of the first body; & therefore some part of the body that follows behind must be penetrated by some part of the body that goes in front. Now, on account of impenetrability, which all Physicists in all quarters recognize in matter, & which can be easily proved to be rightly attributed to it, this cannot possibly happen. There really must be, in the commencement of contact, in that indivisible instant of time which is an indivisible limit between the continuous time that preceded the contact & that subsequent to it (just in the same way as a point in geometry is an indivisible limit between two segments of a continuous line), a change of velocity taking place suddenly, without any passage through intermediate stages; & this violates the Law of Continuity, which absolutely denies the possibility of a passage from one magnitude to another without passing through intermediate stages. Now what has been said in the case of equal bodies concerning the direct passing of both to 9 degrees of velocity, in every case holds good for such equal bodies, or for bodies that are unequal in any way, concerning any other passage to any numbers. In fact, the excess of velocity in the hindmost body, amounting to 6 degrees, has to be got rid of in an instant of time, whether by diminishing the velocity of this body, or by increasing the velocity of the other, or by diminishing somehow the velocity of the one & increasing that of the other; & this cannot possibly be done in any case, without the sudden change that is obtained by omitting the infinite number of intermediate velocities.

An objection derived from denying the existence of hard bodies.

19. There are some people, who think that the whole difficulty can be removed by saying that this is just as it should be, if hard bodies, such as indeed experience no compression or alteration of shape, are dealt with; whereas by many philosophers hard bodies are altogether excluded from Nature; & therefore, so long as two spheres touch one another, it is possible, by introcession & compression of their parts, for it to happen that in these bodies the velocity is changed, the passage being made through all intermediate stages; & thus the whole force of the argument will be evaded.

This reply cannot be made by those who admit solid & hard elements.

20. Now in the first place, this reply can not be used by anyone who, following Newton, & indeed many of the ancient philosophers as well, admit the primary elements of matter to be absolutely hard & solid, possessing infinite adhesion & a definite shape that it is perfectly impossible to alter. For the whole force of my argument then applies quite unimpaired to those solid and hard primary elements that are in the anterior part of the body that is behind, & in the hindmost part of the body that is in front; & certainly these parts touch one another immediately.

Continuous extension requires primary pores & walls bounding them, solid & hard.

21. Next it is truly impossible to understand in the slightest degree how all bodies do not have some of their last parts just near to the surface perfectly solid, & on that account altogether incapable of being compressed. If matter is continuous, it may & must be subject to infinite divisibility; but actual division carried on indefinitely brings in its train difficulties that are truly inextricable; however, this infinite division is required by those who do not admit that there are any particles, no matter how small, in bodies that are perfectly free from, & incapable of, compression. For they must admit the idea that every particle is marked off & divided up, by the action of interspersed pores, into many boundary walls, so to speak, for these pores; & these walls again are distinct from the pores themselves. It is quite impossible to understand why it comes about that, in passing from empty vacuum to solid matter, we are not then bound to encounter some continuous wall of some definite inherent thickness from the surface to the first pore, this wall being everywhere devoid of pores; nor why, which comes to the same thing in the end, there does not exist a pore that is the last & nearest to the external surface; this pore at least, if there were one, certainly has a wall that is free from pores & incapable of compression; & here then the whole force of the argument used above applies perfectly unimpaired.

Violation of the Law of Continuity takes place, at any rate, in prime surfaces or points.

22. Moreover, even if this idea is admitted, although it may be quite unintelligible, then the whole force of the same argument applies to the first or last surface of the bodies that are in immediate contact with one another; or, if there are no continuous surfaces congruent, then to the lines or points. For, whatever the manner may be in which contact takes place, there must be something in every case that certainly affords occasion for impenetrability, & causes the motion of the body that follows to be diminished, & that of the one in front to be increased. This, whatever it may be, from which the force of impenetrability is derived, at the instant at which immediate contact is obtained, must certainly change the velocity suddenly, & without any passage through intermediate stages; & by

debet, atque labefactari, si ad ipsum immediatum contactum illo velocitatum discrimine deveniatur. Id vero est sane aliquid in quacunque e sententiis omnibus continuam extensionem tribuentibus materiæ. Est nimirum realis affectio quædam corporis, videlicet ejus limes ultimus realis, superficies, realis superficiei limes linea, realis lineæ limes punctum, quæ affectiones utcunque in iis sententiis sint prorsus inseparabiles [11] ab ipso corpore, sunt tamen non utique intellectu confictæ, sed reales, quæ nimirum reales dimensiones aliquas habent, ut superficies binas, linea unam, ac realem motum, & translationem cum ipso corpore, cujus idcirco in iis sententiis debent, esse affectiones quædam, vel modi.

Objectio petita a voce *massa*, & *motus*, quæ superficiebus, & punctis non conveniant.

23. Est, qui dicat, nullum in iis committi saltum idcirco, quod censendum sit, nullum habere motum, superficiem, lineam, punctum, quæ massam habeant nullam. Motus, inquit, a Mechanicis habet pro mensura massam in velocitatem ductam; massa autem est superficies baseos ducta in crassitudinem, sive altitudinem, ex. gr. in prismatis. Quo minor est ejusmodi crassitudo, eo minor est massa, & motus, ac ipsa crassitudine evanescente, evanescat oportet & massa, & motus.

Responsionis initium: superficiem, lineam, punctum, posita extensione continua, esse aliquid.

24. Verum qui sic ratiocinatur, inprimis ludit in ipsis vocibus. Massam vulgo appellant quantitatem materiæ, & motum corporum metiuntur per massam ejusmodi, ac velocitatem. At quemadmodum in ipsa geometrica quantitate tria genera sunt quantitatum, corpus, vel solidum, quod trinam dimensionem habet, superficies quæ binas, linæ, quæ unicam, quibus accedit lineæ limes punctum, omni dimensione, & extensione carens; sic etiam in Physica habetur in communi corpus tribus extensionis speciebus præditum; superficies realis extimus corporis limes, prædita binis; linea, limes realis superficiei, habens unicam; & ejusdem lineæ indivisibilis limes punctum. Utrobique alterum alterius est limes, non pars, & quatuor diversa genera constituunt. Superficies est nihil corporeum, sed non & nihil superficiale, quin immo partes habet, & augeri potest, & minui; & eodem pacto linea in ratione quidem superficiei est nihil, sed aliquid in ratione lineæ; ac ipsum demum punctum est aliquid in suo genere, licet in ratione lineæ sit nihil.

Quo Pacto nomen *massæ* possit, *motus* debeat convenire superficiebus, lineis, punctis.

25. Hinc autem in iis ipsis massa quædam considerari potest duarum dimensionum, vel unius, vel etiam nullius continuæ dimensionis, sed numeri punctorum tantummodo, uti quantitas ejus generе designetur; quod si pro iis etiam usurpetur nomen massæ generaliter, motus quantitas definiri poterit per productum ex velocitate, & massa; si vero massæ nomen tribuendum sit soli corpori, tum motus quidem corporis mensura erit massa in velocitatem ducta; superficiei, lineæ, punctorum quotcunque motus pro mensura habebit quantitatem superficiei, vel lineæ, vel numerum punctorum in velocitatem ducta; sed motus utique iis omnibus speciebus tribuendus erit, eruntque quatuor motuum genera, ut quatuor sunt quantitatum, solidi, superficiei, lineæ, punctorum; ac ut altera harum erit nihil in alterius ratione, non in sua; ita alterius motus erit nihil in ratione alterius sed erit sane aliquid in ratione sui, non purum nihil.

Motum passim tribui punctis; fore, ut in eo lædatur Continuitatis lex.

[12] 26. Et quidem ipsi Mechanici vulgo motum tribuunt & superficiebus & lineis, & punctis, ac centri gravitatis motum ubique nominant Physici, quod centrum utique punctum est aliquod, non corpus trina præditum dimensione, quam iste ad motus rationem, & appellationem requirit, ludendo, ut ajebam, in verbis. Porro in ejusmodi motibus extimarum saltem superficierum, vel linearum, vel punctorum, saltus omnino committi debet, si ea ad contactum immediatum deveniant cum illo velocitatum discrimine, & continuitatis lex violari.

Fore, ut ea lædatur saltem in velocitate punctorum.

27. Verum hac omni disquisitione omissa de notione motus, & massæ, si factum ex velocitate, & massa, evanescente una e tribus dimensionibus, evanescit; remanet utique velocitas reliquarum dimensionum, quæ remanet, si eæ reapse remanent, uti quidem omnino remanent in superficie, & ejus velocitatis mutatio haberi deberet per saltum, ac in ea violari continuitatis lex jam toties memorata.

Objectio ex impenetrabilitate admissa in minimis particulis, & ejus confutatio.

28. Hæc quidem ita evidentia sunt, ut omnino dubitari non possit, quin continuitatis lex infringi debeat, & saltus in Naturam induci, ubi cum velocitatis discrimine ad se invicem accedant corpora, & ad immediatum contactum deveniant, si modo impenetrabilitas corporibus tribuenda sit, uti revera est. Eam quidem non in integris tantummodo corporibus, sed in minimis etiam quibusque corporum particulis, atque elementis agnoverunt Physici universi. Fuit sane, qui post meam editam Theoriam, ut ipsam vim mei argumenti

that the Law of Continuity must be broken & destroyed, if immediate contact is arrived at with such a difference of velocity. Moreover, there is in truth always something of this sort in every one of the ideas that attribute continuous extension to matter. There is some real condition of the body, namely, its last real boundary, or its surface, a real boundary of a surface, a line, & a real boundary of a line, a point; & these conditions, however inseparable they may be in these theories from the body itself, are nevertheless certainly not fictions of the brain, but real things, having indeed certain real dimensions (for instance, a surface has two dimensions, & a line one); they also have real motion & movement of translation along with the body itself; hence in these theories they must be certain conditions or modes of it.

Objection derived from the terms *mass* and *motion*, which do not accord with surfaces & points.

23. Someone may say that there is no sudden change made, because it must be considered that a surface, a line or a point, having no mass, cannot have any motion. He may say that motion has, according to Mechanicians, as its measure, the mass multiplied by the velocity; also mass is the surface of the base multiplied by the thickness or the altitude, as for instance in prisms. Hence the less the thickness, the less the mass & the motion; thus, if the thickness vanishes, then both the mass & therefore the motion must vanish as well.

Commencement of the answer to this; a surface, or a line, or a point, is something real, if continuous extension is supposed to exist.

24. Now the man who reasons in this manner is first of all merely playing with words. Mass is commonly called quantity of matter, & the motion of bodies is measured by mass of this kind & the velocity. But, just as in a geometrical quantity there are three kinds of quantities, namely, a body or a solid having three dimensions, a surface with two, & a line with one: to which is added the boundary of a line, a point, lacking dimensions altogether, & of no extension. So also in Physics, a body is considered to be endowed with three species of extension; a surface, the last real boundary of a body, to be endowed with two; a line, the real boundary of a surface, with one; & the indivisible boundary of the line, to be a point. In both subjects, the one is a boundary of the other, & not a part of it; & they form four different kinds. There is nothing solid about a surface; but that does not mean that there is also nothing superficial about it; nay, it certainly has parts & can be increased or diminished. In the same way a line is nothing indeed when compared with a surface, but a definite something when compared with a line; & lastly a point is a definite something in its own class, although nothing in comparison with a line.

The manner in which the term *mass* may, and the term *motus* is bound to, apply to surfaces, lines, & points.

25. Hence also in these matters, a mass can be considered to be of two dimensions, or of one, or even of no continuous dimension, but only numbers of points, just as quantity of this kind is indicated. Now, if for these also, the term mass is employed in a generalized sense, we shall be able to define the quantity of motion by the product of the velocity & the mass. But if the term mass is only to be used in connection with a solid body, then indeed the motion of a solid body will be measured by the mass multiplied by the velocity; but the motion of a surface, or a line, or any number of points will have as their measure the quantity of the surface, or line, or the number of the points, multiplied by the velocity. Motion at any rate will be ascribed in all these cases, & there will be four kinds of motion, as there are four kinds of quantity, namely, for a solid, a surface, a line, or for points; and, as each class of the latter will be as nothing compared with the class before it, but something in its own class, so the motion of the one will be as nothing compared with the motion of the other, but yet really something, & not entirely nothing, compared with those of its own class.

Motion is ascribed to points indiscriminately; the Law of Continuity is violated by doing so.

26. Indeed, Mechanicians themselves commonly ascribe motion to surfaces, lines & points, & Physicists universally speak of the motion of the centre of gravity; this centre is undoubtedly some point, & not a body endowed with three dimensions, which the objector demands for the idea & name of motion, by playing with words, as I said above. On the other hand, in this kind of motions of ultimate surfaces, or lines, or points, a sudden change must certainly be made, if they arrive at immediate contact with a difference of velocity as above, & the Law of Continuity must be violated.

It is at least a fact that this law is violated by the idea of the velocity of points.

27. But, omitting all debate about the notions of motion & mass, if the product of the velocity & the mass vanishes when one of the three dimensions vanish, there will still remain the velocity of the remaining dimensions; & this will persist so long as the dimensions persist, as they do persist undoubtedly in the case of a surface. Hence the change in its velocity must have been made suddenly, & thereby the Law of Continuity, which I have already mentioned so many times, is violated.

Objection derived from the admission of impenetrability in very small particles, & its refutation.

28. These things are so evident that it is absolutely impossible to doubt that the Law of Continuity is infringed, & that a sudden change is introduced into Nature, when bodies approach one another with a difference of velocity & come into immediate contact, if only we are to ascribe impenetrability to bodies, as we really should. And this property too, not in whole bodies only, but in any of the smallest particles of bodies, & in the elements as well, is recognized by Physicists universally. There was one, I must confess, who, after I

infringeret, affirmavit, minimas corporum particulas post contactum superficierum compenetrari non nihil, & post ipsam compenetrationem mutari velocitates per gradus. At id ipsum facile demonstrari potest contrarium illi inductioni, & analogiæ, quam unam habemus in Physica investigandis generalibus naturæ legibus idoneam, cujus inductionis vis quæ sit, & quibus in locis usum habeat, quorum locorum unus est hic ipse impenetrabilitatis ad minimas quasque particulas extendendæ, inferius exponam.

Objectio a voce *motus* assumpta pro mutatione; confutatio ex realitate motus localis.

29. Fuit itidem e Leibnitianorum familia, qui post evulgatam Theoriam meam censuerit, difficultatem ejusmodi amoveri posse dicendo, duas monades sibi etiam invicem occurrentes cum velocitatibus quibuscunque oppositis æqualibus, post ipsum contactum pergere moveri sine locali progressione. Eam progressionem, ajebat, revera omnino nihil esse, si a spatio percurso æstimetur, cum spatium sit nihil; motum utique perseverare, & extingui per gradus, quia per gradus extinguatur energia illa, qua in se mutuo agunt, sese premendo invicem. Is itidem ludit in voce *motus*, quam adhibet pro mutatione quacunque, & actione, vel actionis modo. Motus localis, & velocitas motus ipsius, sunt ea, quæ ego quidem adhibeo, & quæ ibi abrumpuntur per saltum. Ea, ut evidentissime constat, erant aliqua ante contactum, & post contactum mo-[13]-mento temporis in eo casu abrumpuntur; nec vero sunt nihil; licet spatium pure imaginarium sit nihil. Sunt realis affectio rei mobilis fundata in ipsis modis localiter existendi, qui modi etiam relationes inducunt distantiarum reales utique. Quod duo corpora magis a se ipsis invicem distent, vel minus; quod localiter celerius moveantur, vel lentius; est aliquid non imaginarie tantummodo, sed realiter diversum; in eo vero per immediatum contactum saltus utique induceretur in eo casu, quo ego superius sum usus.

Qui Continuitatis, legem summoverint.

30. Et sane summus nostri ævi Geometra, & Philosophus Mac-Laurinus, cum etiam ipse collisionem corporum contemplatus vidisset, nihil esse, quod continuitatis legem in collisione corporum facta per immediatum contactum conservare, ac tueri posset, ipsam continuitatis legem deferendam censuit, quam in eo casu omnino violari affirmavit in eo opere, quod de Newtoni Compertis inscripsit, lib. 1, cap. 4. Et sane sunt alii nonnulli, qui ipsam continuitatis legem nequaquam admiserint, quos inter Maupertuisius, vir celeberrimus, ac de Republica Litteraria optime meritus, absurdam etiam censuit, & quodammodo inexplicabilem. Eodem nimirum in nostris de corporum collisione contemplationibus devenimus Mac-Laurinus, & ego, ut viderimus in ipsa immediatum contactum, atque impulsionem cum continuitatis lege conciliari non posse. At quoniam de impulsione, & immediato corporum contactu ille ne dubitari quidem posse arbitrabatur, (nec vero scio, an alius quisquam omnem omnium corporum immediatum contactum subducere sit ausus antea, utcunque aliqui aeris velum, corporis nimirum alterius, in collisione intermedium retinuerint) continuitatis legem deseruit, atque infregit.

Theoriæ exortus, ea lege, uti fieri debet, retenta.

31. Ast ego cum ipsam continuitatis legem aliquanto diligentius considerarim, & fundamenta, quibus ea innititur, perpenderim, arbitratus sum, ipsam omnino e Natura submoveri non posse, qua proinde retenta contactum ipsum immediatum submovendum censui in collisionibus corporum, ac ea consectaria persecutus, quæ ex ipsa continuitate servata sponte profluebant, directa ratiocinatione delatus sum ad eam, quam superius exposui, virium mutuarum legem, quæ consectaria suo quæque ordine proferam, ubi ipsa, quæ ad continuitatis legem retinendam argumenta me movent, attigero.

Lex Continuitatis quid sit: discrimen inter statua, & incrementa.

32. Continuitatis lex, de qua hic agimus, in eo sita est, uti superius innui, ut quævis quantitas, dum ab una magnitudine ad aliam migrat, debeat transire per omnes intermedias ejusdem generis magnitudines. Solet etiam idem exprimi nominandi transitum per gradus intermedios, quos quidem gradus Maupertuisius ita accepit, quasi vero quædam exiguæ accessiones fierent momento temporis, in quo quidem is censuit violari jam necessario legem ipsam, quæ utcunque exiguo saltu utique violatur nihilo minus, quam maximo; cum nimi-[14]-rum magnum, & parvum sint tantummodo respectiva; & jure quidem id censuit; si nomine graduum incrementa magnitudinis cujuscunque momentanea intelligerentur.

had published my Theory, endeavoured to overcome the force of the argument I had used by asserting that the minute particles of the bodies after contact of the surfaces were subject to compenetration in some measure, & that after compenetration the velocities were changed gradually. But it can be easily proved that this is contrary to that induction & analogy, such as we have in Physics, one peculiarly adapted for the investigation of the general laws of Nature. What the power of this induction is, & where it can be used (one of the cases is this very matter of extending impenetrability to the minute particles of a body), I will set forth later.

Objection to the term *motus* being used for a change; refutation from the reality of local motion.

29. There was also one of the followers of Leibniz who, after I had published my Theory, expressed his opinion that this kind of difficulty could be removed by saying that two monads colliding with one another with any velocities that were equal & opposite would, after they came into contact, go on moving without any local progression. He added that that progression would indeed be absolutely nothing, if it were estimated by the space passed over, since the space was nothing; but the motion would go on & be destroyed by degrees, because the energy with which they act upon one another, by mutual pressure, would be gradually destroyed. He also is playing with the meaning of the term *motus*, which he uses both for any change, & for action & mode of action. Local motion, & the velocity of that motion are what I am dealing with, & these are here broken off suddenly. These, it is perfectly evident, were something definite before contact, & after contact in an instant of time in this case they are broken off. Not that they are nothing; although purely imaginary space is nothing. They are real conditions of the movable thing depending on its modes of extension as regards position; & these modes induce relations between the distances that are certainly real. To account for the fact that two bodies stand at a greater distance from one another, or at a less; or for the fact that they are moved in position more quickly, or more slowly; to account for this there must be something that is not altogether imaginary, but real & diverse. In this something there would be induced, in the question under consideration, a sudden change through immediate contact.

There are some who would deny the Law of Continuity.

30. Indeed the finest geometrician & philosopher of our times, Maclaurin, after he too had considered the collision of solid bodies & observed that there is nothing which could maintain & preserve the Law of Continuity in the collision of bodies accomplished by immediate contact, thought that the Law of Continuity ought to be abandoned. He asserted that, in general in the case of collision, the law was violated, publishing his idea in the work that he wrote on the discoveries of Newton, bk. 1, chap. 4. True, there are some others too, who would not admit the Law of Continuity at all; & amongst these, Maupertuis, a man of great reputation & the highest merit in the world of letters, thought it was senseless, & in a measure inexplicable. Thus, Maclaurin came to the same conclusion as myself with regard to our investigations on the collision of bodies; for we both saw that, in collision, immediate contact & impulsive action could not be reconciled with the Law of Continuity. But, whereas he came to the conclusion that there could be no doubt about the fact of impulsive action & immediate contact between the bodies, he impeached & abrogated the Law of Continuity. Nor indeed do I know of anyone else before me, who has had the courage to deny the existence of all immediate contact for any bodies whatever, although there are some who would retain a thin layer of air, (that is to say, of another body), in between the two in collision.

The origin of my Theory, retaining this Law, as should be done.

31. But I, after considering the Law of Continuity somewhat more carefully, & pondering over the fundamental ideas on which it depends, came to the conclusion that it certainly could not be withdrawn altogether out of Nature. Hence, since it had to be retained, I came to the conclusion that immediate contact in the collision of solid bodies must be got rid of; &, investigating the deductions that naturally sprang from the conservation of continuity, I was led by straightforward reasoning to the law that I have set forth above, namely, the law of mutual forces. These deductions, each set out in order, I will bring forward when I come to touch upon those arguments that persuade me to retain the Law of Continuity.

The nature of the Law of Continuity; distinction between states & increments.

32. The Law of Continuity, as we here deal with it, consists in the idea that, as I intimated above, any quantity, in passing from one magnitude to another, must pass through all intermediate magnitudes of the same class. The same notion is also commonly expressed by saying that the passage is made by intermediate stages or steps; these steps indeed Maupertuis accepted, but considered that they were very small additions made in an instant of time. In this he thought that the Law of Continuity was already of necessity violated, the law being indeed violated by any sudden change, no matter how small, in no less a degree than by a very great one. For, of a truth, large & small are only relative terms; & he rightly thought as he did, if by the name of steps we are to understand momentaneous

Verum id ita intelligendum est; ut singulis momentis singuli status respondeant; incrementa, vel decrementa non nisi continuis tempusculis.

Geometriæ usus ad eam exponendam: momenta punctis, tempora continua lineis expressa.

33. Id sane admodum facile concipitur ope Geometriæ. Sit recta quædam AB in fig. 3, ad quam referatur quædam alia linea CDE. Exprimat prior ex iis tempus, uti solet utique in ipsis horologiis circularis peripheria ab indicis cuspide denotata tempus definire. Quemadmodum in Geometria in lineis puncta sunt indivisibiles limites continuarum lineæ partium, non vero partes lineæ ipsius; ita in tempore distinguendæ erunt partes continui temporis respondentes ipsis lineæ partibus, continuæ itidem & ipsæ, a momentis, quæ sunt indivisibiles earum partium limites, & punctis respondent; nec inposterum alio sensu agens de tempore *momenti* nomen adhibebo, quam eo indivisibilis limitis; particulam vero temporis utcunque exiguam, & habitam etiam pro infinitesima, tempusculum appellabo.

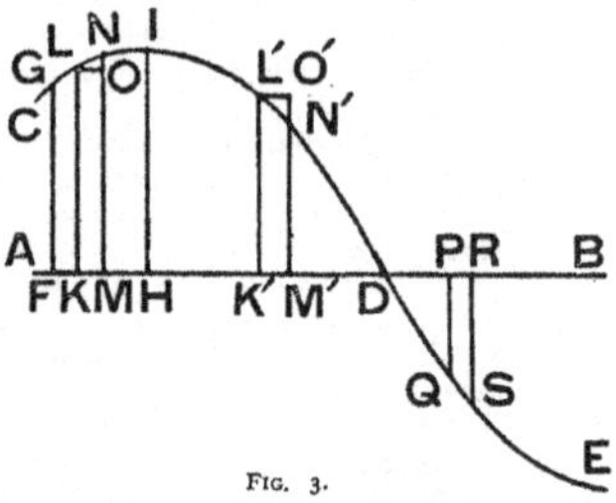

Fig. 3.

Fluxus ordinatæ transeuntis per magnitudines omnes intermedias.

34. Si jam a quovis puncto rectæ AB, ut F, H, erigatur ordinata perpendicularis FG, HI, usque ad lineam CD; ea poterit repræsentare quantitatem quampiam continuo variabilem. Cuicunque momento temporis F, H, respondebit sua ejus quantitatis magnitudo FG, HI; momentis autem intermediis aliis K, M, aliæ magnitudines, KL, MN, respondebunt; ac si a puncto G ad I continua, & finita abeat pars lineæ CDE, facile patet & accurate demonstrari potest, utcunque eadem contorqueatur, nullum fore punctum K intermedium, cui aliqua ordinata KL non respondeat; & e converso nullam fore ordinatam magnitudinis intermediæ inter FG, HI, quæ alicui puncto inter F, H intermedio non respondeat.

Idem in quantitate variabili expressa: æquivocatio in voce *gradus*.

35. Quantitas illa variabilis per hanc variabilem ordinatam expressa mutatur juxta continuitatis legem, quia a magnitudine FG, quam habet momento temporis F, ad magnitudinem HI, quæ respondet momento temporis H, transit per omnes intermedias magnitudines KL, MN, respondentes intermediis momentis K, M, & momento cuivis respondet determinata magnitudo. Quod si assumatur tempusculum quoddam continuum KM utcunque exiguum ita, ut inter puncta L, N arcus ipse LN non mutet recessum a recta AB in accessum; ducta LO ipsi parallela, habebitur quantitas NO, quæ in schemate exhibito est incrementum magnitudinis ejus quantitatis continuo variatæ. Quo minor est ibi temporis particula KM, eo minus est id incrementum NO, & illa evanescente, ubi congruant momenta K, M, hoc etiam evanescit. Potest quævis magnitudo KL, MN appellari status quidam variabilis illius quantitatis, & gradus nomine deberet potius in-[15]-telligi illud incrementum NO, quanquam aliquando etiam ille status, illa magnitudo KL nomine gradus intelligi solet, ubi illud dicitur, quod ab una magnitudine ad aliam per omnes intermedios gradus transeatur; quod quidem æquivocationibus omnibus occasionem exhibuit.

Status singulos momentis, incrementa vero utcunque parva tempusculis continuis respondere.

36. Sed omissis æquivocationibus ipsis, illud, quod ad rem facit, est accessio incrementorum facta non momento temporis, sed tempusculo continuo, quod est particula continui temporis. Utcunque exiguum sit incrementum ON, ipsi semper respondet tempusculum quoddam KM continuum. Nullum est in linea punctum M ita proximum puncto K, ut sit primum post ipsum; sed vel congruunt, vel intercipiunt lineolam continua bisectione per alia intermedia puncta perpetuo divisibilem in infinitum. Eodem pacto nullum est in tempore momentum ita proximum alteri præcedenti momento, ut sit primum post ipsum, sed vel idem momentum sunt, vel interjacet inter ipsa tempusculum continuum per alia intermedia momenta divisibile in infinitum; ac nullus itidem est quantitatis continuo variabilis status ita proximus præcedenti statui, ut sit primus post ipsum accessu aliquo momentaneo facto: sed differentia, quæ inter ejusmodi status est, debetur intermedio continuo tempusculo; ac data lege variationis, sive natura lineæ ipsam exprimentis, & quacunque utcunque exigua accessione, inveniri potest tempusculum continuum, quo ea accessio advenerit.

Transitus sine saltu, etiam a positivis ad negativa per nihilum, quod tamen non est vere nihilum, sed quidam realis status.

37. Atque sic quidem intelligitur, quo pacto fieri possit transitus per intermedias magnitudines omnes, per intermedios status, per gradus intermedios, quin ullus habeatur saltus utcunque exiguus momento temporis factus. Notari illud potest tantummodo, mutationem fieri alicubi per incrementa, ut ubi KL abit, in MN per NO; alicubi per decrementa, ut ubi K'L' abeat in N'M' per O'N'; quin immo si linea CDE, quæ legem

increments of any magnitude whatever. But the idea should be interpreted as follows single states correspond to single instants of time, but increments or decrements only to small intervals of continuous time.

Explanation by the use of geometry; instants represented by points, continuous intervals of time by lines.

33. The idea can be very easily assimilated by the help of geometry. Let AB be any straight line (Fig. 3), to which as axis let any other line CDE be referred. Let the first of them represent the time, in the same manner as it is customary to specify the time in the case of circular clocks by marking off the periphery with the end of a pointer. Now, just as in geometry, points are the indivisible boundaries of the continuous parts of a line, so, in time, distinction must be made between parts of continuous time, which correspond to these parts of a line, themselves also continuous, & instants of time, which are the indivisible boundaries of those parts of time, & correspond to points. In future I shall not use the term *instant* in any other sense, when dealing with time, than that of the indivisible boundary; & a small part of time, no matter how small, even though it is considered to be infinitesimal, I shall term a tempuscule, or small interval of time.

The flux of the ordinate as it passes through all intermediate values.

34. If now from any points F,H on the straight line AB there are erected at right angles to it ordinates FG, HI, to meet the line CD; any of these ordinates can be taken to represent a quantity that is continuously varying. To any instant of time F, or H, there will correspond its own magnitude of the quantity FG, or HI; & to other intermediate instants K, M, other magnitudes KL, MN will correspond. Now, if from the point G, there proceeds a continuous & finite part of the line CDE, it is very evident, & it can be rigorously proved, that, no matter how the curve twists & turns, there is no intermediate point K, to which some ordinate KL does not correspond; &, conversely, there is no ordinate of magnitude intermediate between FG & HI, to which there does not correspond a point intermediate between F & H.

The same holds good for the variable quantity so represented; equivocation in the use of the term *step*.

35. The variable quantity that is represented by this variable ordinate is altered in accordance with the Law of Continuity; for, from the magnitude FG, which it has at the instant of time F, to the magnitude HI, which corresponds to the instant H, it passes through all intermediate magnitudes KL, MN, which correspond to the intermediate instants K, M; & to every instant there corresponds a definite magnitude. But if we take a definite small interval of continuous time KM, no matter how small, so that between the points L & N the arc LN does not alter from recession from the line AB to approach, & draw LO parallel to AB, we shall obtain the quantity NO that in the figure as drawn is the increment of the magnitude of the continuously varying quantity. Now the smaller the interval of time KM, the smaller is this increment NO; & as that vanishes when the instants of time K, M coincide, the increment NO also vanishes. Any magnitude KL, MN can be called a state of the variable quantity, & by the name step we ought rather to understand the increment NO; although sometimes also the state, or the magnitude KL is accustomed to be called by the name step. For instance, when it is said that from one magnitude to another there is a passage through all intermediate stages or steps; but this indeed affords opportunity for equivocations of all sorts.

Single states correspond to instants, but increments however small to intervals of continuous time.

36. But, omitting all equivocation of this kind, the point is this: that addition of increments is accomplished, not in an instant of time, but in a small interval of continuous time, which is a part of continuous time. However small the increment ON may be, there always corresponds to it some continuous interval KM. There is no point M in the straight line AB so very close to the point K, that it is the next after it; but either the points coincide, or they intercept between them a short length of line that is divisible again & again indefinitely by repeated bisection at other points that are in between M & K. In the same way, there is no instant of time that is so near to another instant that has gone before it, that it is the next after it; but either they are the same instant, or there lies between them a continuous interval that can be divided indefinitely at other intermediate instants. Similarly, there is no state of a continuously varying quantity so very near to a preceding state that it is the next state to it, some momentary addition having been made; any difference that exists between two states of the same kind is due to a continuous interval of time that has passed in the meanwhile. Hence, being given the law of variation, or the nature of the line that represents it, & any increment, no matter how small, it is possible to find a small interval of continuous time in which the increment took place.

Passages without sudden change, from positive to negative through zero; zero however is not really nothing, but a certain real state.

37. In this manner we can understand how it is possible for a passage to take place through all intermediate magnitudes, through intermediate states, or through intermediate stages, without any sudden change being made, no matter how small, in an instant of time. It can merely be remarked that change in some places takes place by increments (as when KL becomes MN by the addition of NO), in other places by decrements (as when K'L'

variationis exhibit, alicubi secet rectam, temporis AB, potest ibidem evanescere magnitudo, ut ordinata M'N', puncto M' allapso ad D evanesceret, & deinde mutari in negativam PQ, RS, habentem videlicet directionem contrariam, quæ, quo magis ex oppositæ parte crescit, eo minor censetur in ratione priore, quemadmodum in ratione possessionis, vel divitiarum, pergit perpetuo se habere pejus, qui iis omnibus, quæ habebat, absumptis, æs alienum contrahit perpetuo majus. Et in Geometria quidem habetur a positivo ad negativa transitus, uti etiam in Algebraicis formulis, tam transeundo per nihilum, quam per infinitum, quos ego transitus persecutus sum partim in dissertatione adjecta meis *Sectionibus Conicis*, partim in *Algebra* § 14, & utrumque simul in dissertatione *De Lege Continuitatis*; sed in Physica, ubi nulla quantitas in infinitum excrescit, is casus locum non habet, & non, nisi transeundo per nihilum, transitus fit a positi-[16]-vis ad negativa, ac vice versa; quanquam, uti inferius innuam, id ipsum sit non nihilum revera in se ipso, sed realis quidem status, & habeatur pro nihilo in consideratione quadam tantummodo, in qua negativa etiam, qui sunt veri status, in se positivi, ut ut ad priorem seriem pertinentes negativo quodam modo, negativa appellentur.

Proponitur probanda existentia legis Continuitatis.

38. Exposita hoc pacto, & vindicata continuitatis lege, eam in Natura existere plerique Philosophi arbitrantur, contradicentibus nonnullis, uti supra innui. Ego, cum in eam primo inquirerem, censui, candem omitti omnino non posse; si eam, quam habemus unicam, Naturæ analogiam, & inductionis vim consulamus, ope cujus inductionis eam demonstrare conatus sum in pluribus e memoratis dissertationibus, ac eandem probationem adhibet Benvenutus in sua *Synopsi* Num. 119; in quibus etiam locis, prout diversis occasionibus conscripta sunt, repetuntur non nulla.

Ejus probatio ab inductione satis ampla.

39. Longum hic esset singula inde excerpere in ordinem redacta: satis crit exscribere dissertationis *De lege Continuitatis* numerum 138. Post inductionem petitam præcedente numero a Geometria, quæ nullum uspiam habet saltum, atque a motu locali, in quo nunquam ab uno loco ad alium devenitur, nisi ductu continuo aliquo, unde consequitur illud, distantiam a dato loco nunquam mutari in aliam, neque densitatem, quæ utique a distantiis pendet particularum in aliam, nisi transeundo per intermedias; fit gradus in eo numero ad motuum velocitates, & ductus, quæ magis hic ad rem faciunt, nimirum ubi de velocitate agimus non mutanda per saltum in corporum collisionibus. Sic autem habetur: " Quin immo in motibus ipsis continuitas servatur etiam in eo, quod motus omnes in lineis continuis fiunt nusquam abruptis. Plurimos ejusmodi motus videmus. Planetæ, & cometæ in lineis continuis cursum peragunt suum, & omnes retrogradationes fiunt paullatim, ac in stationibus semper exiguus quidem motus, sed tamen habetur semper, atque hine etiam dies paullatim per auroram venit, per vespertinum crepusculum abit, Solis diameter non per saltum, sed continuo motu supra horizontem ascendit, vel descendit. Gravia itidem oblique projecta in lineis itidem pariter continuis motus exercent suos, nimirum in parabolis, seclusa aeris resistentia, vel, ea considerata, in orbibus ad hyperbolas potius accedentibus, & quidem semper cum aliqua exigua obliquitate projiciuntur, cum infinities infinitam improbabilitatem habeat motus accurate verticalis inter infinities infinitas inclinationes, licet exiguas, & sub sensum non cadentes, fortuito obveniens, qui quidem motus in hypothesi Telluris motæ a parabolicis plurimum distant, & curvam continuam exhibent etiam pro casu projectionis accurate verticalis, quo, quiescente penitus Tellure, & nulla ventorum vi deflectente motum, haberetur [17] ascensus rectilineus, vel descensus. Immo omnes alii motus a gravitate pendentes, omnes ab elasticitate, a vi magnetica, continuitatem itidem servant; cum eam servent vires illæ ipsæ, quibus gignuntur. Nam gravitas, cum decrescat in ratione reciproca duplicata distantiarum, & distantiæ per saltum mutari non possint, mutatur per omnes intermedias magnitudines. Videmus pariter, vim magneticam a distantiis pendere lege continua; vim elasticam ab inflexione, uti in laminis, vel a distantia, ut in particulis aeris compressi. In iis, & omnibus ejusmodi viribus, & motibus, quos gignunt, continuitas habetur semper, tam in lineis quæ describuntur, quam in velocitatibus, quæ pariter per omnes intermedias magnitudines mutantur, ut videre est in pendulis, in ascensu corporum gravium,

becomes N′M′ by the subtraction of O′N′) ; moreover, if the line CDE, which represents the law of variation, cuts the straight AB, which is the axis of time, in any point, then the magnitude can vanish at that point (just as the ordinate M′N′ would vanish when the point M′ coincided with D), & be changed into a negative magnitude PQ, or RS, that is to say one having an opposite direction ; & this, the more it increases in the opposite sense, the less it is to be considered in the former sense (just as in the idea of property or riches, a man goes on continuously getting worse off, when, after everything he had has been taken away from him, he continues to get deeper & deeper into debt). In Geometry too we have this passage from positive to negative, & also in algebraical formulæ, the passage being made not only through nothing, but also through infinity ; such I have discussed, the one in a dissertation added to my *Conic Sections*, the other in my *Algebra* (§ 14), & both of them together in my essay *De Lege Continuitatis* ; but in Physics, where no quantity ever increases to an infinite extent, the second case has no place ; hence, unless the passage is made through the value nothing, there is no passage from positive to negative, or vice versa. Although, as I point out below, this nothing is not really nothing in itself, but a certain real state ; & it may be considered as nothing only in a certain sense. In the same sense, too, negatives, which are true states, are positive in themselves, although, as they belong to the first set in a certain negative way, they are called negative.

I propose to prove the existence of the Law of Continuity.

38. Thus explained & defended, the Law of Continuity is considered by most philosophers to exist in Nature, though there are some who deny it, as I mentioned above. I, when first I investigated the matter, considered that it was absolutely impossible that it should be left out of account, if we have regard to the unparalleled analogy that there is with Nature & to the power of induction ; & by the help of this induction I endeavoured to prove the law in several of the dissertations that I have mentioned, & Benvenutus also used the same form of proof in his *Synopsis* (Art. 119). In these too, as they were written on several different occasions, there are some repetitions.

Proof by induction sufficient for the purpose.

39. It would take too long to extract & arrange in order here each of the passages in these essays ; it will be sufficient if I give Art. 138 of the dissertation *De Lege Continuitatis.* After induction derived in the preceding article from geometry, in which there is no sudden change anywhere, & from local motion, in which passage from one position to another never takes place unless by some continuous progress (the consequence of which is that a distance from any given position can never be changed into another distance, nor the density, which depends altogether on the distances between the particles, into another density, except by passing through intermediate stages), the step is made in that article to the velocities of motions, & deductions, which have more to do with the matter now in hand, namely, where we are dealing with the idea that the velocity is not changed suddenly in the collision of solid bodies. These are the words : " Moreover in motions themselves continuity is preserved also in the fact that all motions take place in continuous lines that are not broken anywhere. We see a great number of motions of this kind. The planets & the comets pursue their courses, each in its own continuous line, & all retrogradations are gradual ; & in stationary positions the motion is always slight indeed, but yet there is always some ; hence also daylight comes gradually through the dawn, & goes through the evening twilight, as the diameter of the sun ascends above the horizon, not suddenly, but by a continuous motion, & in the same manner descends. Again heavy bodies projected obliquely follow their courses in lines also that are just as continuous ; namely, in parabolæ, if we neglect the resistance of the air, but if that is taken into account, then in orbits that are more nearly hyperbolæ. Now, they are always projected with some slight obliquity, since there is an infinitely infinite probability against accurate vertical motion, from out of the infinitely infinite number of inclinations (although slight & not capable of being observed), happening fortuitously. These motions are indeed very far from being parabolæ, if the hypothesis that the Earth is in motion is adopted. They give a continuous curve also for the case of accurate vertical projection, in which, if the Earth were at rest, & no wind-force deflected the motion, rectilinear ascent & descent would be obtained. All other motions that depend on gravity, all that depend upon elasticity, or magnetic force, also preserve continuity ; for the forces themselves, from which the motions arise, preserve it. For gravity, since it diminishes in the inverse ratio of the squares of the distances, & the distances cannot be changed suddenly, is itself changed through every intermediate stage. Similarly we see that magnetic force depends on the distances according to a continuous law ; that elastic force depends on the amount of bending as in plates, or according to distance as in particles of compressed air. In these, & all other forces of the sort, & in the motions that arise from them, we always get continuity, both as regards the lines which they describe & also in the velocities which are changed in similar manner through all intermediate magnitudes ; as is seen in pendulums, in the ascent of heavy

& in aliis mille ejusmodi, in quibus mutationes velocitatis fiunt gradatim, nec retro cursus reflectitur, nisi imminuta velocitate per omnes gradus. Ea diligentissime continuitatem servat omnia. Hinc nec ulli in naturalibus motibus habentur anguli, sed semper mutatio directionis fit paullatim, nec vero anguli exacti habentur in corporibus ipsis, in quibus utcunque videatur tenuis acies, vel cuspis, microscopii saltem ope videri solet curvatura, quam etiam habent alvei fluviorum semper, habent arborum folia, & frondes, ac rami, habent lapides quicunque, nisi forte alicubi cuspides continuæ occurrant, vel primi generis, quas Natura videtur affectare in spinis, vel secundi generis, quas videtur affectare in avium unguibus, & rostro, in quibus tamen manente in ipsa cuspide unica tangente continuitatem servari videbimus infra. Infinitum esset singula persequi, in quibus continuitas in Natura observatur. Satius est generaliter provocare ad exhibendum casum in Natura, in quo continuitas non servetur, qui omnino exhiberi non poterit."

Duplex inductionis genus, ubi & cur vim habeat inductio incompleta.

40. Inductio amplissima tum ex hisce motibus, ac velocitatibus, tum ex aliis pluribus exemplis, quæ habemus in Natura, in quibus ea ubique, quantum observando licet deprehendere, continuitatem vel observat accurate, vel affectat, debet omnino id efficere, ut ab ea ne in ipsa quidem corporum collisione recedamus. Sed de inductionis natura, & vi, ac ejusdem usu in Physica, libet itidem hic inserere partem numeri 134, & totum 135, dissertationis *De Lege Continuitatis*. Sic autem habent ibidem : " Inprimis ubi generales Naturæ leges investigantur, inductio vim habet maximam, & ad earum inventionem vix alia ulla superest via. Ejus ope extensionem, figurabilitem, mobilitatem, impenetrabilitatem corporibus omnibus tribuerunt semper Philosophi etiam veteres, quibus eodem argumento inertiam, & generalem gravitatem plerique e recentioribus addunt. Inductio, ut demonstrationis vim habeat, debet omnes singulares casus, quicunque haberi possunt percurrere. Ea in Natu-[18]-ræ legibus stabiliendis locum habere non potest. Habet locum laxior quædam inductio, quæ, ut adhiberi possit, debet esse ejusmodi, ut inprimis in omnibus iis casibus, qui ad trutinam ita revocari possunt, ut deprehendi debeat, an ea lex observetur, eadem in iis omnibus inveniatur, & ii non exiguo numero sint ; in reliquis vero, si quæ prima fronte contraria videantur, re accuratius perspecta, cum illa lege possint omnia conciliari ; licet, an eo potissimum pacto concilientur, immediate innotescere, nequaquam possit. Si eæ conditiones habeantur ; inductio ad legem stabiliendam censeri debet idonea. Sic quia videmus corpora tam multa, quæ habemus præ manibus, aliis corporibus resistere, ne in eorum locum adveniant, & loco cedere, si resistendo sint imparia, potius, quam eodem perstare simul ; impenetrabilitatem corporum admittimus ; nec obest, quod quædam corpora videamus intra alia, licet durissima, insinuari, ut oleum in marmora, lumen in crystalla, & gemmas. Videmus enim hoc phænomenum facile conciliari cum ipsa impenetrabilitate, dicendo, per vacuos corporum poros ea corpora permeare. (*Num.* 135). Præterea, quæcunque proprietates absolutæ, nimirum quæ relationem non habent ad nostros sensus, deteguntur generaliter in massis sensibilibus corporum, easdem ad quascunque utcunque exiguas particulas debemus transferre ; nisi positiva aliqua ratio obstet, & nisi sint ejusmodi, quæ pendeant a ratione totius, seu multitudinis, contradistincta a ratione partis. Primum evincitur ex eo, quod magna, & parva sunt respectiva, ac insensibilia dicuntur ea, quæ respectu nostræ molis, & nostrorum sensuum sunt exigua. Quare ubi agitur de proprietatibus absolutis non respectivis, quæcunque communia videmus in iis, quæ intra limites continentur nobis sensibiles, ea debemus censere communia etiam infra eos limites : nam ii limites respectu rerum, ut sunt in se, sunt accidentales, adeoque siqua fuisset analogiæ læsio, poterat illa multo facilius cadere intra limites nobis sensibiles, qui tanto laxiores sunt, quam infra eos, adeo nimirum propinquos nihilo. Quod nulla ceciderit, indicio est, nullam esse. Id indicium non est evidens, sed ad investigationis principia pertinet, quæ si juxta

bodies, & in a thousand other things of the same kind, where the changes of velocity occur gradually, & the path is not retraced before the velocity has been diminished through all degrees. All these things most strictly preserve continuity. Hence it follows that no sharp angles are met with in natural motions, but in every case a change of direction occurs gradually; neither do perfect angles occur in bodies themselves, for, however fine an edge or point in them may seem, one can usually detect curvature by the help of the microscope if nothing else. We have this gradual change of direction also in the beds of rivers, in the leaves, boughs & branches of trees, & stones of all kinds; unless, in some cases perchance, there may be continuous pointed ends, either of the first kind, which Nature is seen to affect in thorns, or of the second kind, which she is seen to do in the claws & the beak of birds; in these, however, we shall see below that continuity is still preserved, since we are left with a single tangent at the extreme end. It would take far too long to mention every single thing in which Nature preserves the Law of Continuity; it is more than sufficient to make a general statement challenging the production of a single case in Nature, in which continuity is not preserved; for it is absolutely impossible for any such case to be brought forward."

Induction of a two-fold kind: when & why incomplete induction has validity.

40. The effect of the very complete induction from such motions as these & velocities, as well as from a large number of other examples, such as we have in Nature, where Nature in every case, as far as can be gathered from direct observation, maintains continuity or tries to do so, should certainly be that of keeping us from neglecting it even in the case of collision of bodies. As regards the nature & validity of induction, & its use in Physics, I may here quote part of Art. 134 & the whole of Art. 135 from my dissertation *De Lege Continuitatis*. The passage runs thus: "Especially when we investigate the general laws of Nature, induction has very great power; & there is scarcely any other method beside it for the discovery of these laws. By its assistance, even the ancient philosophers attributed to all bodies extension, figurability, mobility, & impenetrability; & to these properties, by the use of the same method of reasoning, most of the later philosophers add inertia & universal gravitation. Now, induction should take account of every single case that can possibly happen, before it can have the force of demonstration; such induction as this has no place in establishing the laws of Nature. But use is made of an induction of a less rigorous type; in order that this kind of induction may be employed, it must be of such a nature that in all those cases particularly, which can be examined in a manner that is bound to lead to a definite conclusion as to whether or no the law in question is followed, in all of them the same result is arrived at; & that these cases are not merely a few. Moreover, in the other cases, if those which at first sight appeared to be contradictory, on further & more accurate investigation, can all of them be made to agree with the law; although, whether they can be made to agree in this way better than in any other whatever, it is impossible to know directly anyhow. If such conditions obtain, then it must be considered that the induction is adapted to establishing the law. Thus, as we see that so many of the bodies around us try to prevent other bodies from occupying the position which they themselves occupy, or give way to them if they are not capable of resisting them, rather than that both should occupy the same place at the same time, therefore we admit the impenetrability of bodies. Nor is there anything against the idea in the fact that we see certain bodies penetrating into the innermost parts of others, although the latter are very hard bodies; such as oil into marble, & light into crystals & gems. For we see that this phenomenon can very easily be reconciled with the idea of impenetrability, by supposing that the former bodies enter and pass through empty pores in the latter bodies (Art. 135). In addition, whatever absolute properties, for instance those that bear no relation to our senses, are generally found to exist in sensible masses of bodies, we are bound to attribute these same properties also to all small parts whatsoever, no matter how small they may be. That is to say, unless some positive reason prevents this; such as that they are of such a nature that they depend on argument having to do with a body as a whole, or with a group of particles, in contradistinction to an argument dealing with a part only. The proof comes in the first place from the fact that great & small are relative terms, & those things are called insensible which are very small with respect to our own size & with regard to our senses. Therefore, when we consider absolute, & not relative, properties, whatever we perceive to be common to those contained within the limits that are sensible to us, we should consider these things to be still common to those beyond those limits. For these limits, with regard to such matters as are self-contained, are accidental; & thus, if there should be any violation of the analogy, this would be far more likely to happen between the limits sensible to us, which are more open, than beyond them, where indeed they are so nearly nothing. Because then none did happen thus, it is a sign that there is none. This sign is not evident, but belongs to the principles of investigation, which generally proves successful if it is carried out in accordance with certain definite wisely

quasdam prudentes regulas fiat, successum habere solet. Cum id indicium fallere possit; fieri potest, ut committatur error, sed contra ipsum errorem habebitur præsumptio, ut etiam in jure appellant, donec positiva ratione evincatur oppositum. Hinc addendum fuit, *nisi ratio positiva obstet.* Sic contra hasce regulas peccaret, qui diceret, corpora quidem magna compenetrari, ac replicari, & inertia carere non posse, compenetrari tamen posse, vel replicari, vel sine inertia esse exiguas eorum partes. At si proprietas sit respectiva, respectu nostrorum sensuum, ex [19] eo, quod habeatur in majoribus massis, non debemus inferre, eam haberi in particulis minoribus, ut est hoc ipsum, esse sensibile, ut est, esse coloratas, quod ipsis majoribus massis competit, minoribus non competit; cum ejusmodi magnitudinis discrimen, accidentale respectu materiæ, non sit accidentale respectu ejus denominationis *sensibile, coloratum.* Sic etiam siqua proprietas ita pendet a ratione aggregati, vel totius, ut ab ea separari non possit; nec ea, ob rationem nimirum eandem, a toto, vel aggregato debet transferri ad partes. Est de ratione totius, ut partes habeat, nec totum sine partibus haberi potest. Est de ratione figurabilis, & extensi, ut habeat aliquid, quod ab alio distet, adeoque, ut habeat partes; hinc eæ proprietates, licet in quovis aggregato particularum materiæ, sive in quavis sensibili massa, inveniantur, non debent inductionis vi transferri ad particulas quascunque."

Et impenetrabilitatem, & continuitatem evinci per inductionem: ad ipsam quid requiratur.

41. Ex his patet, & impenetrabilitatem, & continuitatis legem per ejusmodi inductionis genus abunde probari, atque evinci, & illam quidem ad quascunque utcunque exiguas particulas corporum, hanc ad gradus utcunque exiguos momento temporis adjectos debere extendi. Requiritur autem ad hujusmodi inductionem primo, ut illa proprietas, ad quam probandam ea adhibetur, in plurimis casibus observetur, aliter enim probabilitas esset exigua; & ut nullus sit casus observatus, in quo evinci possit, eam violari. Non est necessarium illud, ut in iis casibus, in quibus primo aspectu timeri possit defectus proprietatis ipsius, positive demonstretur, eam non deficere; satis est, si pro iis casibus haberi possit ratio aliqua conciliandi observationem cum ipsa proprietate, & id multo magis, si in aliis casibus habeatur ejus conciliationis exemplum, & positive ostendi possit, eo ipso modo fieri aliquando conciliationem.

Ejus applicatio ad impenetrabilitatem.

42. Id ipsum fit, ubi per inductionem impenetrabilitas corporum accipitur pro generali lege Naturæ. Nam impenetrabilitatem ipsam magnorum corporum observamus in exemplis sane innumeris tot corporum, quæ pertractamus. Habentur quidem & casus, in quibus eam violari quis crederit, ut ubi oleum per ligna, & marmora penetrat, atque insinuatur, & ubi lux per vitra, & gemmas traducitur. At præsto est conciliatio phænomeni cum impenetrabilitate, petita ab eo, quod illa corpora, in quæ se ejusmodi substantiæ insinuant, poros habeant, quos eæ permeent. Et quidem hæc conciliatio exemplum habet manifestissimum in spongia, quæ per poros ingentes aqua immissa imbuitur. Poros marmorum illorum, & multo magis vitrorum, non videmus, ac multo minus videre possumus illud, non insinuari eas substantias nisi per poros. Hoc satis est reliquæ inductionis vi, ut dicere debeamus, eo potissimum pacto se rem habere, & ne ibi quidem violari generalem utique impenetrabilitatis legem.

Similis ad continuitatem: duo casuum genera, in quibus ea videatur lædi.

[20] 43. Eodem igitur pacto in lege ipsa continuitatis agendum est. Illa tam ampla inductio, quam habemus, debet nos movere ad illam generaliter admittendam etiam pro iis casibus, in quibus determinare immediate per observationes non possumus, an eadem habeatur, uti est collisio corporum; ac si sunt casus nonnulli, in quibus eadem prima fronte violari videatur; ineunda est ratio aliqua, qua ipsum phænomenum cum ea lege conciliari possit, uti revera potest. Nonnullos ejusmodi casus protuli in memoratis dissertationibus, quorum alii ad geometricam continuitatem pertinent, alii ad physicam. In illis prioribus non immorabor; neque enim geometrica continuitas necessaria est ad hanc physicam propugnandam, sed eam ut exemplum quoddam ad confirmationem quandam inductionis majoris adhibui. Posterior, ut sæpe & illa prior, ad duas classes reducitur; altera est eorum casuum, in quibus saltus videtur committi idcirco, quia nos per saltum omittimus intermedias quantitates: rem exemplo geometrico illustro, cui physicum adjicio.

chosen rules. Now, since the indication may possibly be fallacious, it may happen that an error may be made; but there is presumption against such an error, as they call it in law, until direct evidence to the contrary can be brought forward. Hence we should add: *unless some positive argument is against it.* Thus, it would be offending against these rules to say that large bodies indeed could not suffer compenetration, or enfolding, or be deficient in inertia, but yet very small parts of them could suffer penetration, or enfolding, or be without inertia. On the other hand, if a property is relative with respect to our senses, then, from a result obtained for the larger masses we cannot infer that the same is to be obtained in its smaller particles; for instance, that it is the same thing to be sensible, as it is to be coloured, which is true in the case of large masses, but not in the case of small particles; since a distinction of this kind, accidental with respect to matter, is not accidental with respect to the term *sensible* or *coloured.* So also if any property depends on an argument referring to an aggregate, or a whole, in such a way that it cannot be considered apart from the whole, or the aggregate; then, neither must it (that is to say, by that same argument), be transferred from the whole, or the aggregate, to parts of it. It is on account of its being a whole that it has parts; nor can there be a whole without parts. It is on account of its being figurable & extended that it has some thing that is apart from some other thing, & therefore that it has parts. Hence those properties, although they are found in any aggregate of particles of matter, or in any sensible mass, must not however be transferred by the power of induction to each & every particle."

Both impenetrability & continuity can be demonstrated by induction; what is required for this purpose.

41. From what has been said it is quite evident that both impenetrability & the Law of Continuity can be proved by a kind of induction of this type; & the former must be extended to all particles of bodies, no matter how small, & the latter to all additional steps, however small, made in an instant of time. Now, in the first place, to use this kind of induction, it is required that the property, for the proof of which it is to be used, must be observed in a very large number of cases; for otherwise the probability would be very small. Also it is required that no case should be observed, in which it can be proved that it is violated. It is not necessary that, in those cases in which at first sight it is feared that there may be a failure of the property, that it should be directly proved that there is no failure. It is sufficient if in those cases some reason can be obtained which will make the observation agree with the property; & all the more so, if in other cases an example of reconciliation can be obtained, & it can be positively proved that sometimes reconciliation can be obtained in that way.

Application of induction to impenetrability.

42. This is just what does happen, when the impenetrability of solid bodies is accepted as a law of Nature through inductive reasoning. For we observe this impenetrability of large bodies in innumerable examples of the many bodies that we consider. There are indeed also cases, in which one would think that it was violated, such as when oil penetrates wood and marble, & works its way through them, or when light passes through glasses & gems. But we have ready a means of making these phenomena agree with impenetrability, derived from the fact that those bodies, into which substances of this kind work their way, possess pores which they can permeate. There is a very evident example of this reconciliation in a sponge, which is saturated with water introduced into it by means of huge pores. We do not see the pores of the marble, still less those of glass; & far less can we see that these substances do not penetrate except by pores. It satisfies the general force of induction if we can say that the matter can be explained in this way better than in any other, & that in this case there is absolutely no contradiction of the general law of impenetrability.

Similar application to continuity; two classes of cases in which there seems to be violation.

43. In the same way, then, we must deal with the Law of Continuity. The full induction that we possess should lead us to admit in general this law even in those cases in which it is impossible for us to determine directly by observation whether the same law holds good, as for instance in the collision of bodies. Also, if there are some cases in which the law at first sight seems to be violated, some method must be followed, through which each phenomenon can be reconciled with the law, as is in every case possible. I brought forward several cases of this kind in the dissertations I have mentioned, some of which pertained to geometrical continuity, & others to physical continuity. I will not delay over the first of these: for geometrical continuity is not necessary for the defence of the physical variety; I used it as an example in confirmation of a wider induction. The latter, as well as very frequently the former, reduces to two classes; & the first of these classes is that class in which a sudden change seems to have been made on account of our having omitted the intermediate quantities with a jump. I give a geometrical illustration, and then add one in physics.

Exemplum geometricum primi generis, ubi nos intermedias magnitudines omittimus.

44. In axe curvæ cujusdam in fig. 4. sumantur segmenta AC, CE, EG æqualia, & erigantur ordinatæ AB, CD, EF, GH. Areæ BACD, DCEF, FEGH videntur continuæ cujusdam seriei termini ita, ut ab illa BACD ad DCEF, & inde ad FEGH immediate transeatur, & tamen secunda a prima, ut & tertia a secunda, differunt per quantitates finitas: si enim capiantur CI, EK æquales BA, DC, & arcus BD transferatur in IK; area DIKE erit incrementum secundæ supra primam, quod videtur immediate advenire totum absque eo, quod unquam habitum sit ejus dimidium, vel quævis alia pars incrementi ipsius; ut idcirco a prima ad secundam magnitudinem areæ itum sit sine transitu per intermedias. At ibi omittuntur a nobis termini intermedii, qui continuitatem servant; si enim *ac* æqualis AC motu continuo feratur ita, ut incipiendo ab AC desinat in CE; magnitudo areæ BACD per omnes intermedias *bacd* abit in magnitudinem DCEF sine ullo saltu, & sine ulla violatione continuitatis.

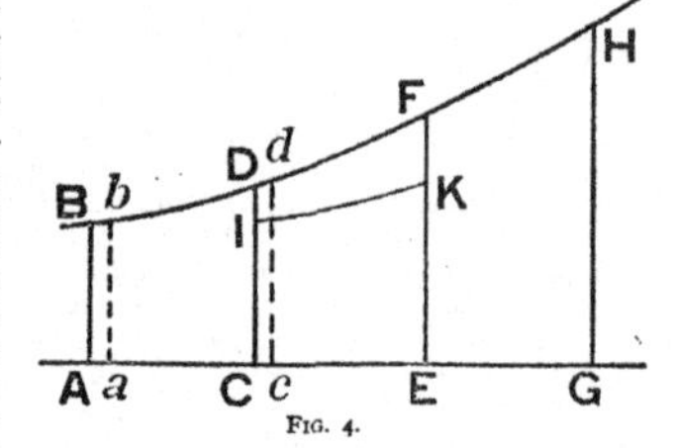

Fig. 4.

Quando id accidat exempla physica dierum, & oscillationum consequentium.

45. Id sane ubique accidit, ubi initium secundæ magnitudinis aliquo intervallo distat ab initio primæ; sive statim veniat post ejus finem, sive quavis alia lege ab ea disjungatur. Sic in physicis, si diem concipiamus intervallum temporis ab occasu ad occasum, vel etiam ab ortu ad occasum, dies præcedens a sequenti quibusdam anni temporibus differt per plura secunda, ubi videtur fieri saltus sine ullo intermedio die, qui minus differat. At seriem quidem continuam ii dies nequaquam constituunt. Concipiatur parallelus integer Telluris, in quo sunt continuo ductu disposita loca omnia, quæ candem latitudinem geographicam habent; ea singula loca suam habent durationem diei, & omnium ejusmodi dierum initia, ac fines continenter fluunt; donec ad eundem redeatur locum, cujus præ-[21]-cedens dies est in continua illa serie primus, & sequens postremus. Illorum omnium dierum magnitudines continenter fluunt sine ullo saltu: nos, intermediis omissis, saltum committimus non Natura. Atque huic similis responsio est ad omnes reliquos casus ejusmodi, in quibus initia, & fines continenter non fluunt, sed a nobis per saltum accipiuntur. Sic ubi pendulum oscillat in aere; sequens oscillatio per finitam magnitudinem distat a præcedente; sed & initium & finis ejus finito intervallo temporis distat a præcedentis initio, & fine, ac intermedii termini continua serie fluente a prima oscillatione ad secundam essent ii, qui haberentur, si primæ, & secundæ oscillationis arcu in æqualem partium numerum diviso, assumeretur via confecta, vel tempus in ea impensum, interjacens inter fines partium omnium proportionalium, ut inter trientem, vel quadrantem prioris arcus, & trientem, vel quadrantem posterioris, quod ad omnes ejus generis casus facile transferri potest, in quibus semper immediate etiam demonstrari potest illud, continuitatem nequaquam violari.

Exempla secundi generis, ubi mutatio sit celerrime, sed non momento temporis.

46. Secunda classis casuum est ea, in qua videtur aliquid momento temporis peragi, & tamen peragitur tempore successivo, sed perbrevi. Sunt, qui objiciant pro violatione continuitatis casum, quo quisquam manu lapidem tenens, ipsi statim det velocitatem quandam finitam: alius objicit aquæ e vase effluentis, foramine constituto aliquanto infra superficiem ipsius aquæ, velocitatem oriri momento temporis finitam. At in priore casu admodum evidens est, momento temporis velocitatem finitam nequaquam produci. Tempore opus est, utcunque brevissimo, ad excursum spirituum per nervos, & musculos, ad fibrarum tensionem, & alia ejusmodi: ac idcirco ut velocitatem aliquam sensibilem demus lapidi, manum retrahimus, & ipsum aliquandiu, perpetuo accelerantes, retinemus. Sic etiam, ubi tormentum bellicum exploditur, videtur momento temporis emitti globus, ac totam celeritatem acquirere; at id successive fieri, patet vel inde, quod debeat inflammari tota massa pulveris pyrii, & dilatari aer, ut elasticitate sua globum acceleret, quod quidem fit omnino per omnes gradus. Successionem multo etiam melius videmus in globo, qui ab elastro sibi relicto propellatur: quo elasticitas est major, eo citius, sed nunquam momento temporis velocitas in globum inducitur.

Applicatio ipsorum ad alia, nominatim ad effluxum aquæ e vase.

47. Hæc exempla illud præstant, quod aqua per poros spongiæ ingressa respectu impenetrabilitatis, ut ea responsione uti possimus in aliis casibus omnibus, in quibus accessio aliqua magnitudinis videtur fieri tota momento temporis; ut nimirum dicamus fieri tempore

Geometrical example of the first kind, where we omit intermediate magnitudes.

44. In the axis of any curve (Fig. 4) let there be taken the segments AC, CE, EG equal to one another ; & let the ordinates AB, CD, EF, GH be erected. The areas BACD, DCEF, FEGH seem to be terms of some continuous series such that we can pass directly from BACD to DCEF and then on to FEGH, & yet the second differs from the first, & also the third from the second, by a finite quantity. For if CI, EK are taken equal to BA, DC, & the arc BD is transferred to the position IK ; then the area DIKE will be the increment of the second area beyond the first ; & this seems to be directly arrived at as a whole without that which at any one time is considered to be the half of it, or indeed any other part of the increment itself : so that, in consequence, we go from the first to the second magnitude of area without passing through intermediate magnitudes. But in this case we omit intermediate terms which maintain the continuity ; for if *ac* is equal to AC, & this is carried by a continuous motion in such a way that, starting from the position AC it ends up at the position CE, then the magnitude of the area BACD will pass through all intermediate values such as *bacd* until it reaches the magnitude of the area DCEF without any sudden change, & hence without any breach of continuity.

When this will happen : physical examples in the case of consecutive days, or consecutive oscillations.

45. Indeed this always happens when the beginning of the second magnitude is distant by a definite interval from the beginning of the first ; whether it comes immediately after the end of the first or is disconnected from it by some other law. Thus in physics, if we look upon the day as the interval of time between sunset & sunset, or even between sunrise & sunset, the preceding day differs from that which follows it at certain times of the year by several seconds ; in which case we see that there is a sudden change made, without there being any intermediate day for which the change is less. But the fact is that these days do not constitute a continuous series. Let us consider a complete parallel of latitude on the Earth, along which in a continuous sequence are situated all those places that have the same geographical latitude. Each of these places has its own duration of the day, & the beginnings & ends of days of this kind change uninterruptedly ; until we get back again to the same place, where the preceding day is the first of that continuous series, & the day that follows is the last of the series. The magnitudes of all these days continuously alter without there being any sudden change : it was we who, by omitting the intermediates, made the sudden change, & not Nature. Similar to this is the answer to all the rest of the cases of the same kind, in which the beginnings & the ends do not change uninterruptedly, but are observed by us discontinuously. Similarly, when a pendulum oscillates in air, the oscillation that follows differs from the oscillation that has gone before by a finite magnitude. But both the beginning & the end of the second differs from the beginning & the end of the first by a finite interval of time ; & the intermediate terms in a continuously varying series from the first oscillation to the second would be those that would be obtained, if the arcs of the first & second oscillations were each divided into the same number of equal parts, & the path traversed (or the time spent in traversing the path) is taken between the ends of all these proportional paths ; such as that between the third or fourth part of the first arc & the third or fourth part of the second arc. This argument can be easily transferred so as to apply to all cases of this kind ; & in such cases it can always be directly proved that there is no breach of continuity.

Examples of the second class, in which the change is very rapid, but does not take place in an instant of time.

46. The second class of cases is that in which something seems to have been done in an instant of time, but still it is really done in a continuous, but very short, interval of time. There are some who bring forward, as an objection in favour of a breach of continuity, the case in which a man, holding a stone in his hand, gives to it a definite velocity all at once ; another raises an objection that favours a breach of continuity, in the case of water flowing from a vessel, where, if an opening is made below the level of the surface of the water, a finite velocity is produced in an instant of time. But in the first case it is perfectly clear that a finite velocity is in no wise produced in an instant of time. For there is need of time, although this is exceedingly short, for the passage of cerebral impulses through the nerves and muscles, for the tension of the fibres, and other things of that sort ; and therefore, in order to give a definite sensible velocity to the stone, we draw back the hand, and then retain the stone in it for some time as we continually increase its velocity forwards. So too when an engine of war is exploded, the ball seems to be driven forth and to acquire the whole of its speed in an instant of time. But that it is done continuously is clear, if only from the fact that the whole mass of the gunpowder has to be inflamed and the gas has to be expanded in order that it may accelerate the ball by its elasticity ; and this latter certainly takes place by degrees. The continuous nature of this is far better seen in the case of a ball propelled by releasing a spring ; here the stronger the elasticity, the greater the speed ; but in no case is the speed imparted to the ball in an instant of time.

Application of these to other cases ; particularly to the flow of water from a vessel.

47. These examples are superior to that of water entering through the pores of a sponge, which we employed in the matter of impenetrability ; so that we can make use of this reply in all other cases in which some addition to a magnitude seems to have taken place entirely in an instant of time. Thus, without doubt we may say that it takes place in an exceedingly

brevissimo, utique per omnes intermedias magnitudines, ac illæsa penitus lege continuitatis. Hinc & in aquæ effluentis exemplo res eodem redit, ut non unico momento, sed successivo aliquo tempore, & per [22] omnes intermedias magnitudines progignatur velocitas, quod quidem ita se habere optimi quique Physici affirmant. Et ibi quidem, qui momento temporis omnem illam velocitatem progigni, contra me affirmet, principium utique, ut ajunt, petat, necesse est. Neque enim aqua, nisi foramen aperiatur, operculo dimoto, effluet; remotio vero operculi, sive manu fiat, sive percussione aliqua, non potest fieri momento temporis, sed debet velocitatem suam acquirere per omnes gradus; nisi illud ipsum, quod quærimus, supponatur jam definitum, nimirum an in collisione corporum communicatio motus fiat momento temporis, an per omnes intermedios gradus, & magnitudines. Verum eo omisso, si etiam concipiamus momento temporis impedimentum auferri, non idcirco momento itidem temporis omnis illa velocitas produceretur; illa enim non a percussione aliqua, sed a pressione superincumbentis aquæ orta, oriri utique non potest, nisi per accessiones continuas tempusculo admodum parvo, sed non omnino nullo: nam pressio tempore indiget, ut velocitatem progignat, in communi omnium sententia.

Transitus ad metaphysicam probationem: limes in continuis unicus, ut in Geometria.

48. Illæsa igitur esse debet continuitatis lex, nec ad eam evertendam contra inductionem, tam uberem quidquam poterunt casus allati hucusque, vel iis similes. At ejusdem continuitatis aliam metaphysicam rationem adinveni, & proposui in dissertatione *De Lege Continuitatis*, petitam ab ipsa continuitatis natura, in qua quod Aristoteles ipse olim notaverat, communis esse debet limes, qui præcedentia cum consequentibus conjungit, qui idcirco etiam indivisibilis est in ea ratione, in qua est limes. Sic superficies duo solida dirimens & crassitudine caret, & est unica, in qua immediatus ab una parte fit transitus ad aliam; linea dirimens binas superficiei continuæ partes latitudine caret; punctum continuæ lineæ segmenta discriminans, dimensione omni: nec duo sunt puncta contigua, quorum alterum sit finis prioris segmenti, alterum initium sequentis, cum duo contigua indivisibilia, & inextensa haberi non possint sine compenetratione, & coalescentia quadam in unum.

Idem in tempore & in quavis serie continua: evidentius in quibusdam.

49. Eodem autem pacto idem debet accidere etiam in tempore, ut nimirum inter tempus continuum præcedens, & continuo subsequens unicum habeatur momentum, quod sit indivisibilis terminus utriusque; nec duo momenta, uti supra innuimus, contigua esse possint, sed inter quodvis momentum, & aliud momentum debeat intercedere semper continuum aliquod tempus divisibile in infinitum. Et eodem pacto in quavis quantitate, quæ continuo tempore duret, haberi debet series quædam magnitudinum ejusmodi, ut momento temporis cuivis respondeat sua, quæ præcedentem cum consequente conjungat, & ab illa per aliquam determinatam magnitudinem differat. Quin immo in illo quantitatum genere, in quo [23] binæ magnitudines simul haberi non possunt, id ipsum multo evidentius conficitur, nempe nullum haberi posse saltum immediatum ab una ad alteram. Nam illo momento temporis, quo deberet saltus fieri, & abrumpi series accessu aliquo momentaneo, deberent haberi duæ magnitudines, postrema seriei præcedentis, & prima seriei sequentis. Id ipsum vero adhuc multo evidentius habetur in illis rerum statibus, in quibus ex una parte quovis momento haberi debet aliquis status ita, ut nunquam sine aliquo ejus generis statu res esse possit; & ex alia duos simul ejusmodi status habere non potest.

Inde cur motus localis non fiat, nisi per lineam continuam.

50. Id quidem satis patebit in ipso locali motu, in quo habetur phænomenum omnibus sane notissimum, sed cujus ratio non ita facile aliunde redditur, inde autem patentissima est, Corpus a quovis loco ad alium quemvis devenire utique potest motu continuo per lineas quascunque utcunque contortas, & in immensum productas quaquaversum, quæ numero infinities infinitæ sunt: sed omnino debet per continuam aliquam abire, & nullibi interruptam. En inde rationem ejus rei admodum manifestam. Si alicubi linea motus abrumperetur; vel momentum temporis, quo esset in primo puncto posterioris lineæ, esset posterius eo momento, quo esset in puncto postremo anterioris, vel esset idem, vel anterius? In primo, & tertio casu inter ea momenta intercederet tempus aliquod continuum divisibile in infinitum per alia momenta intermedia, cum bina momenta temporis, in eo sensu accepta, in quo ego hic ea accipio, contigua esse non possint, uti superius exposui. Quamobrem in

short interval of time, and certainly passes through every intermediate magnitude, and that the Law of Continuity is not violated. Hence also in the case of water flowing from a vessel it reduces to the same example: so that the velocity is generated, not in a single instant, but in some continuous interval of time, and passes through all intermediate magnitudes; and indeed all the most noted physicists assert that this is what really happens. Also in this matter, should anyone assert in opposition to me that the whole of the speed is produced in an instant of time, then he must use a *petitio principii*, as they call it. For the water cannot flow out, unless the hole is opened, & the lid removed; & the removal of the lid, whether done by hand or by a blow, cannot be effected in an instant of time, but must acquire its own velocity by degrees; unless we suppose that the matter under investigation is already decided, that is to say, whether in collision of bodies communication of motion takes place in an instant of time or through all intermediate degrees and magnitudes. But even if that is left out of account, & if also we assume that the barrier is removed in an instant of time, none the more on that account would the whole of the velocity also be produced in an instant of time; for it is impossible that such velocity can arise, not from some blow, but from a pressure arising from the superincumbent water, except by continuous additions in a very short interval of time, which is however not absolutely nothing; for pressure requires time to produce velocity, according to the general opinion of everybody.

Passing to a metaphysical proof, we have a single limit in the case of continuous things, as in geometry.

48. The Law of Continuity ought then to be subject to no breach, nor will the cases hitherto brought forward, nor others like them, have any power at all to controvert this law in opposition to induction so copious. Moreover I discovered another argument, a metaphysical one, in favour of this continuity, & published it in my dissertation *De Lege Continuitatis*, having derived it from the very nature of continuity; as Aristotle himself long ago remarked, there must be a common boundary which joins the things that precede to those that follow; & this must therefore be indivisible for the very reason that it is a boundary. In the same way, a surface of separation of two solids is also without thickness & is single, & in it there is immediate passage from one side to the other; the line of separation of two parts of a continuous surface lacks any breadth; a point determining segments of a continuous line has no dimension at all; nor are there two contiguous points, one of which is the end of the first segment, & the other the beginning of the next; for two contiguous indivisibles, of no extent, cannot possibly be considered to exist, unless there is compenetration & a coalescence into one.

Similarly for time & any continuous series; more evident in some than in others.

49. In the same way, this should also happen with regard to time, namely, that between a preceding continuous time & the next following there should be a single instant, which is the indivisible boundary of either. There cannot be two instants, as we intimated above, contiguous to one another; but between one instant & another there must always intervene some interval of continuous time divisible indefinitely. In the same way, in any quantity which lasts for a continuous interval of time, there must be obtained a series of magnitudes of such a kind that to each instant of time there is its corresponding magnitude; & this magnitude connects the one that precedes with the one that follows it, & differs from the former by some definite magnitude. Nay even in that class of quantities, in which we cannot have two magnitudes at the same time, this very point can be deduced far more clearly, namely, that there cannot be any sudden change from one to another. For at that instant, when the sudden change should take place, & the series be broken by some momentary definite addition, two magnitudes would necessarily be obtained, namely, the last of the first series & the first of the next. Now this very point is still more clearly seen in those states of things, in which on the one hand there must be at any instant some state so that at no time can the thing be without some state of the kind, whilst on the other hand it can never have two states of the kind simultaneously.

Hence the reason why local motion only occurs in a continuous line.

50. The above will be sufficiently clear in the case of local motion, in regard to which the phenomenon is perfectly well known to all; the reason for it, however, is not so easily derived from any other source, whilst it follows most clearly from this idea. A body can get from any one position to any other position in any case by a continuous motion along any line whatever, no matter how contorted, or produced ever so far in any direction; these lines being infinitely infinite in number. But it is bound to travel by some continuous line, with no break in it at any point. Here then is the reason of this phenomenon quite clearly explained. If the motion in the line should be broken at any point, either the instant of time, at which it was at the first point of the second part of the line, would be after the instant, at which it was at the last point of the first part of the line, or it would be the same instant, or before it. In the first & third cases, there would intervene between the two instants some definite interval of continuous time divisible indefinitely at other intermediate instants; for two instants of time, considered in the sense in which I have

primo casu in omnibus iis infinitis intermediis momentis nullibi esset id corpus, in secundo casu idem esset eodem illo momento in binis locis, adeoque replicaretur; in terio haberetur replicatio non tantum respectu eorum binorum momentorum, sed omnium etiam intermediorum, in quibus nimirum omnibus id corpus esset in binis locis. Cum igitur corpus existens nec nullibi esse possit, nec simul in locis pluribus; illa viæ mutatio, & ille saltus haberi omnino non possunt.

Illustratio ejus argumenti ex Geometria: ratiocinatione metaphysica, pluribus exemplis.

51. Idem ope Geometriæ magis adhuc oculis ipsis subjicitur. Exponantur per rectam AB tempora, ac per ordinatas ad lineas CD, EF, abruptas alicubi, diversi status rei cujuspiam. Ductis ordinatis DG, EH, vel punctum H jaceret post G, ut in Fig. 5; vel cum ipso congrueret, ut in 6; vel ipsum præcederet, ut in 7. In primo casu nulla responderet ordinata omnibus punctis rectæ GH; in secundo binæ responderent GD, & HE eidem puncto G; in tertio vero binæ HI, & HE puncto H, binæ GD, GK puncto G, & binæ LM, LN

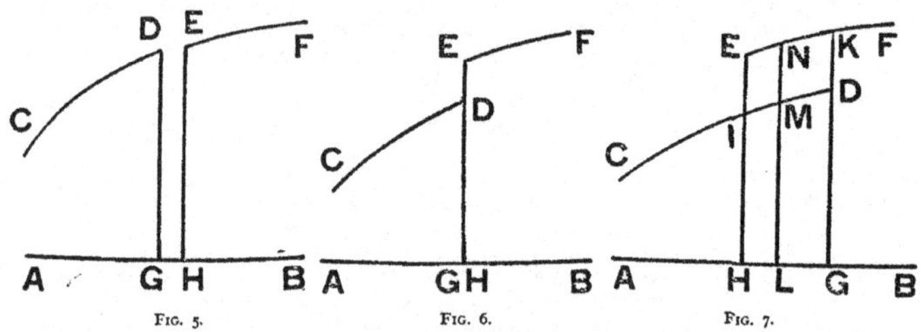

Fig. 5. Fig. 6. Fig. 7.

puncto cuivis intermedio L; nam ordinata est relatio quædam distantiæ, quam habet punctum curvæ cum puncto axis sibi respondente, adeoque ubi jacent in recta eadem perpendiculari axi bina curvarum puncta, habentur binæ ordinatæ respondentes eidem puncto axis. Quamobrem si nec o-[24]-mni statu carere res possit, nec haberi possint status simul bini; necessario consequitur, saltum illum committi non posse. Saltus ipse, si deberet accidere, uti vulgo fieri concipitur, accideret binis momentis G, & H, quæ sibi in fig. 6 immediate succederent sine ullo immediato hiatu, quod utique fieri non potest ex ipsa limitis ratione, qui in continuis debet esse idem, & antecedentibus, & consequentibus communis, uti diximus. Atque idem in quavis reali serie accidit; ut hic linea finita sine puncto primo, & postremo, quod sit ejus limes, & superficies sine linea esse non potest; unde fit, ut in casu figuræ 6 binæ ordinatæ necessario respondere debeant eidem puncto: ita in quavis finita reali serie statuum primus terminus, & postremus haberi necessario debent; adeoque si saltus fit, uti supra de loco diximus; debet eo momento, quo saltus confici dicitur, haberi simul status duplex; qui cum haberi non possit: saltus itidem ille haberi omnino non potest. Sic, ut aliis utamur exemplis, distantia unius corporis ab alio mutari per saltum non potest, nec densitas, quia duæ simul haberentur distantiæ, vel duæ densitates, quod utique sine replicatione haberi non potest; caloris itidem, & frigoris mutatio in thermometris, ponderis atmosphæræ mutatio in barometris, non fit per saltum, quia binæ simul altitudines mercurii in instrumento haberi deberent eodem momento temporis, quod fieri utique non potest; cum quovis momento determinato unica altitudo haberi debeat, ac unicus determinatus caloris gradus, vel frigoris; quæ quidem theoria innumeris casibus pariter aptari potest.

Objectio ab *esse*, & *non esse* conjungendis in creatione, & annihilatione, ac ejus solutio.

52. Contra hoc argumentum videtur primo aspectu adesse aliquid, quod ipsum prorsus evertat, & tamen ipsi illustrando idoneum est maxime. Videtur nimirum inde erui, impossibilem esse & creationem rei cujuspiam, & interitum. Si enim conjungendus est postremus terminus præcedentis seriei cum primo sequentis; in ipso transitu a *non esse* ad *esse*, vel vice versa, debebit utrumque conjungi, ac idem simul erit, & non erit, quod est absurdum. Responsio in promptu est. Seriei finitæ realis, & existentis, reales itidem, & existentes termini esse debent; non vero nihili, quod nullas proprietates habet, quas exigat, Hine si realium statuum seriei altera series realium itidem statuum succedat, quæ non sit communi termino conjuncta; bini eodem momento debebuntur status, qui nimirum sint bini limites earundem. At quoniam *non esse* est merum nihilum; ejusmodi series limitem nullum extremum requirit, sed per ipsum *esse* immediate, & directe excluditur. Quamobrem primo, & postremo momento temporis ejus continui, quo res est, crit utique, nec cum hoc *esse* suum *non esse* conjunget simul; at si densitas certa per horam duret, tum momento temporis in aliam mutetur duplam, duraturam itidem per alteram sequentem horam; momento temporis, [25] quod horas dirimit, binæ debebunt esse densitates simul, nimirum & simplex, & dupla, quæ sunt reales binarum realium serierum termini.

considered them, cannot be contiguous, as I explained above. Wherefore in the first case, at all those infinite intermediate instants the body would be nowhere at all; in the second case, it would be at the same instant in two different places & so there would be replication. In the third case, there would not only occur replication in respect of these two instants but for all those intermediate to them as well, in all of which the body would forsooth be in two places at the same time. Since then a body that exists can never be nowhere, nor in several places at one & the same time, there can certainly be no alteration of path & no sudden change.

Illustration of this argument from geometry; the line of reasoning being metaphysical, with several examples.

51. The same thing can be visualized better with the aid of Geometry. Let times be represented by the straight line AB, & diverse states of any thing by ordinates drawn to meet the lines CD, EF, which are discontinuous at some point. If the ordinates DG, EH are drawn, either the point H will fall after the point G, as in Fig. 5; or it will coincide with it, as in Fig. 6; or it will fall before it, as in Fig. 7. In the first case, no ordinate will correspond to any one of the points of the straight line GH; in the second case, GD and HE would correspond to the same point G; in the third case, two ordinates, HI, HE, would correspond to the same point H, two, GD, GK, to the same point G, and two, LM, LN, to any intermediate point L. Now the ordinate is some relation as regards distance, which a point on the curve bears to the point on the axis that corresponds with it; & thus, when two points of the curve lie in the same straight line perpendicular to the axis, we have two ordinates corresponding to the same point of the axis. Wherefore, if the thing in question can neither be without some state at each instant, nor is it possible that there should be two states at the same time, then it necessarily follows that the sudden change cannot be made. For this sudden change, if it is bound to happen, would take place at the two instants G & H, which immediately succeed the one the other without any direct gap between them; this is quite impossible, from the very nature of a limit, which should be the same for, & common to, both the antecedents & the consequents in a continuous set, as has been said. The same thing happens in any series of real things; as in this case there cannot be a finite line without a first & last point, each to be a boundary to it, neither can there be a surface without a line. Hence it comes about that in the case of Fig. 6 two ordinates must necessarily correspond to the same point. Thus, in any finite real series of states, there must of necessity be a first term & a last; & so if a sudden change is made, as we said above with regard to position, there must be at the instant, at which the sudden change is said to be accomplished, a twofold state at one & the same time. Now since this can never happen, it follows that this sudden change is also quite impossible. Similarly, to make use of other illustrations, the distance of one body from another can never be altered suddenly, no more can its density; for there would be at one & the same time two distances, or two densities, a thing which is quite impossible without replication. Again, the change of heat, or cold, in thermometers, the change in the weight of the air in barometers, does not happen suddenly; for then there would necessarily be at one & the same time two different heights for the mercury in the instrument; & this could not possibly be the case. For at any given instant there must be but one height, & but one definite degree of heat, & but one definite degree of cold; & this argument can be applied just as well to innumerable other cases.

A difficulty raised over the connecting together of *existence* & *non-existence* at the time of creation or annihilation; & its solution.

52. Against this argument it would seem at first sight that there is something ready to hand which overthrows it altogether; whilst as a matter of fact it is peculiarly fitted to exemplify it. It seems that from this argument it follows that both the creation of any thing, & its destruction, are impossible. For, if the last term of a series that precedes is to be connected with the first term of the series that follows, then in the passage from a state of existence to one of non-existence, or *vice versa*, it will be necessary that the two are connected together; & then at one & the same time the same thing will both exist & not exist, which is absurd. The answer to this is immediate. For the ends of a finite series that is real & existent must themselves be real & existent, not such as end up in absolute nothing, which has no properties. Hence, if to one series of real states there succeeds another series of real states also, which is not connected with it by a common term, then indeed there must be two states at the same instant, namely those which are their two limits. But since *non-existence* is mere nothing, a series of this kind requires no last limiting term, but is immediately & directly cut off by fact of *existence*. Wherefore, at the first & at the last instant of that continuous interval of time, during which the matter exists, it will certainly exist; & its *non-existence* will not be connected with its *existence* simultaneously. On the other hand if a given density persists for an hour, & then is changed in an instant of time into another twice as great, which will last for another hour; then in that instant of time which separates the two hours, there would have to be two densities at one & the same time, the simple & the double, & these are real terms of two real series.

Unde huc transferenda solutio ipsa.

53. Id ipsum in dissertatione *De lege virium in Natura existentium* satis, ni fallor, luculenter exposui, ac geometricis figuris illustravi, adjectis nonnullis, quæ eodem recidunt, & quæ in applicatione ad rem, de qua agimus, & in cujus gratiam hæc omnia ad legem continuitatis pertinentia allata sunt, proderunt infra; libet autem novem ejus dissertationis numeros huc transferre integros, incipiendo ab octavo, sed numeros ipsos, ut & schematum numeros mutabo hic, ut cum superioribus consentiant.

Solutio petita ex geometrico exemplo.

54. " Sit in fig. 8 circulus GMM'*m*, qui referatur ad datam reetam AB per ordinatas HM ipsi rectæ perpendiculares; uti itidem perpendiculares sint binæ tangentes EGF, E'G'F'. Concipiantur igitur recta quædam indefinita ipsi rectæ AB perpendicularis, motu quodam continuo delata ab A ad B. Ubi ea habuerit, positionem quamcumque CD, quæ præcedat tangentem EF, vel C'D', quæ consequatur tangentem E'F'; ordinata ad circulum nulla erit, sive erit impossibilis, & ut Geometræ loquuntur, imaginaria. Ubicunque autem ea sit inter binas tangentes EGF, E'G'F', in HI, H'I', occurret circulo in binis punctis M, *m*, vel M', *m'*, & habebitur valor ordinate HM, H*m*, vel H'M', H'*m'*. Ordinata quidem ipsa respondet soli intervallo EE': & si ipsa linea AB referat tempus; momentum E est limes inter tempus præcedens continuum AE, quo ordinata non est, & tempus continuum EE' subsequens, quo ordinata est; punctum E' est limes inter tempus præcedens EE', quo ordinata est, & subsequens E'B, quo non est. Vita igitur quædam ordinatæ est tempus EE'; ortus habetur in E, interitus in E'. Quid autem in ipso ortu, & interitu? Habetur-ne quoddam *esse* ordinatæ, an *non esse*? Habetur utique *esse*, nimirum EG, vel E'G', non autem *non esse*. Oritur tota finitæ magnitudinis ordinata EG, interit tota finitæ magnitudinis E'G', nec tamen ibi conjungit *esse*, & *non esse*, nec ullum absurdum secum trahit. Habetur momento E primus terminus seriei sequentis sine ultimo seriei præcedentis, & habetur momento E' ultimus terminus seriei præcedentis sine primo termino seriei sequentis."

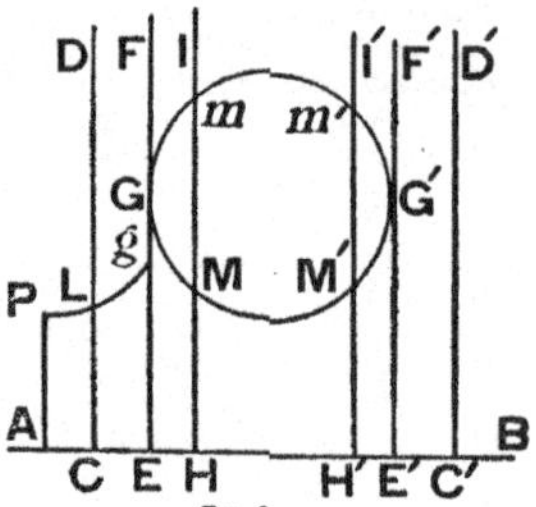

Fig. 8.

Solutio ex metaphysica consideratione.

55. " Quare autem id ipsum accidat, si metaphysica consideratione rem perpendimus, statim patebit. Nimirum veri nihili nullæ sunt veræ proprietates: entis realis veræ, & reales proprietates sunt. Quævis realis series initium reale debet, & finem, sive primum, & ultimum terminum. Id, quod non est, nullam habet veram proprietatem, nec proinde sui generis ultimum terminum, aut primum exigit. Series præcedens ordinatæ nullius, ultimum terminum non [26] habet, series consequens non habet primum: series realis contenta intervallo EE', & primum habere debet, & ultimum. Hujus reales termini terminum illum nihili per se se excludunt, cum ipsum *esse* per se excludat *non esse*."

Illustratio ulterior geometrica.

56. " Atque id quidem manifestum fit magis: si consideremus seriem aliquam præcedentem realem, quam exprimant ordinatæ ad lineam continuam PL*g*, quæ respondeat toti tempori AE ita, ut cuivis momento C ejus temporis respondeat ordinata CL. Tum vero si momento E debeat fieri saltus ab ordinata E*g* ad ordinatam EG: necessario ipsi momento E debent respondere binæ ordinatæ EG, E*g*. Nam in tota linea PL*g* non potest deesse solum ultimum punctum *g*; cum ipso sublato debeat adhuc illa linea terminum habere suum, qui terminus esset itidem punctum: id vero punctum idcirco fuisset ante contiguum puncto *g*, quod est absurdum, ut in eadem dissertatione *De Lege Continuitatis* demonstravimus. Nam inter quodvis punctum, & aliud punctum linea aliqua interjacere debet; quæ si non interjaceat; jam illa puncta in unicum coalescunt. Quare non potest deesse nisi lineola aliqua *g*L ita, ut terminus seriei præcedentis sit in aliquo momento C præcedente momentum E, & disjuncto ab eo per tempus quoddam continuum, in cujus temporis momentis omnibus ordinata sit nulla."

Applicatio ad creationem, & annihilationem.

57. " Patet igitur discrimen inter transitum a vero nihilo, nimirum a quantitate imaginaria, ad *esse*, & transitum ab una magnitudine ad aliam. In primo casu terminus nihili non habetur; habetur terminus uterque seriei veram habentis existentiam, & potest quantitas, cujus ea est series, oriri, vel occidere quantitate finita, ac per se excludere *non esse*. In secundo casu necessario haberi debet utriusque seriei terminus, alterius nimirum postremus, alterius primus. Quamobrem etiam in creatione, & in annihilatione potest quantitas oriri, vel interire magnitudine finita, & primum, ac ultimum *esse* erit quoddam *esse*, quod secum non conjunget una *non esse*. Contra vero ubi magnitudo realis ab una quantitate ad

53. I explained this very point clearly enough, if I mistake not, in my dissertation *De lege virium in Natura existentium*, & I illustrated it by geometrical figures; also I made some additions that reduced to the same thing. These will appear below, as an application to the matter in question; for the sake of which all these things relating to the Law of Continuity have been adduced. It is allowable for me to quote in this connection the whole of nine articles from that dissertation, beginning with Art. 8; but I will here change the numbering of the articles, & of the diagrams as well, so that they may agree with those already given.

The source from which the solution is to be borrowed.

54. "In Fig. 8, let GMM'*m* be a circle, referred to a given straight line AB as axis, by means of ordinates HM drawn perpendicular to that straight line; also let the two tangents EGF, E'G'F' be perpendiculars to the axis. Now suppose that an unlimited straight line perpendicular to the axis AB is carried with a continuous motion from A to B. When it reaches some such position as CD preceding the tangent EF, or as C'D' subsequent to the tangent E'F', there will be no ordinate to the circle, or it will be impossible &, as the geometricians call it, imaginary. Also, wherever it falls between the two tangents EGF, E'G'F', as at HI or H'I', it will meet the circle in two points, M, *m* or M', *m'*; & for the value of the ordinate there will be obtained HM & H*m*, or H'M' & H'*m'*. Such an ordinate will correspond to the interval EE' only; & if the line AB represents time, the instant E is the boundary between the preceding continuous time AE, in which the ordinate does not exist, & the subsequent continuous time EE', in which the ordinate does exist. The point E' is the boundary between the preceding time EE', in which the ordinate does exist, & the subsequent time E'B, in which it does not; the lifetime, as it were, of the ordinate, is EE'; its production is at E & its destruction at E'. But what happens at this production & destruction? Is it an *existence* of the ordinate, or a *non-existence*? Of a truth there is an *existence*, represented by EG & E'G', & not a *non-existence*. The whole ordinate EG of finite magnitude is produced, & the whole ordinate E'G' of finite magnitude is destroyed; & yet there is no connecting together of the states of *existence* & *non-existence*, nor does it bring in anything absurd in its train. At the instant E we get the first term of the subsequent series without the last term of the preceding series; & at the instant E' we have the last term of the preceding series without the first term of the subsequent series."

Solution derived from a geometrical example.

55. "The reason why this should happen is immediately evident, if we consider the matter metaphysically. Thus, to absolute nothing there belong no real properties; but the properties of a real absolute entity are also real. Any real series must have a real beginning & end, or a first term & a last. That which does not exist can have no true property; & on that account does not require a last term of its kind, or a first. The preceding series, in which there is no ordinate, does not have a last term; & the subsequent series has likewise no first term; whilst the real series contained within the interval EE' must have both a first term & a last term. The real terms of this series of themselves exclude the term of no value, since the fact of *existence* of itself excludes *non-existence*."

Solution from a metaphysical consideration.

56. "This indeed will be still more evident, if we consider some preceding series of real quantities, expressed by the ordinates to the curved line PL*g*; & let this curve correspond to the whole time AE in such a way that to every instant C of the time there corresponds an ordinate CL. Then, if at the instant E there is bound to be a sudden change from the ordinate E*g* to the ordinate EG, to that instant E there must of necessity correspond both the ordinates EG, E*g*. For it is impossible that in the whole line PL*g* the last point alone should be missing; because, if that point is taken away, yet the line is bound to have an end to it, & that end must also be a point; hence that point would be before & contiguous to the point *g*; & this is absurd, as we have shown in the same dissertation *De Lege Continuitatis*. For between any one point & any other point there must lie some line; & if such a line does not intervene, then those points must coalesce into one. Hence nothing can be absent, except it be a short length of line *g*L, so that the end of the series that precedes occurs at some instant, C, preceding the instant E, & separated from it by an interval of continuous time, at all instants of which there is no ordinate."

Further illustration by geometry.

57. "Evidently, then, there is a distinction between passing from absolute nothing, i.e., from an imaginary quantity, to a state of *existence*, & passing from one magnitude to another. In the first case the term which is naught is not reckoned in; the term at either end of a series which has real existence is given, & the quantity, of which it is the series, can be produced or destroyed, finite in amount; & of itself it will exclude *non-existence*. In the second case, there must of necessity be an end to either series, namely the last of the one series & the first of the other. Hence, in creation & annihilation, a quantity can be produced or destroyed, finite in magnitude; & the first & last state of *existence* will be a state of *existence* of some kind; & this will not associate with itself a state of *non-existence*. But, on the other hand, where a real magnitude is bound

Application to creation & annihilation.

aliam transire debet per saltum; momento temporis, quo saltus committitur, uterque terminus haberi deberet. Manet igitur illæsum argumentum nostrum metaphysicum pro exclusione saltus a creatione & annihilatione, sive ortu, & interitu."

Aliquando videri *nihilum* id, *quod* est aliq*u*id.

58. "At hic illud etiam notandum est; quoniam ad ortum, & interitum considerandum geometricas contemplationes assumpsimus, videri quidem prima fronte, aliquando etiam realis seriei terminum postremum esse nihilum; sed re altius considerata, non crit vere nihilum; sed status quidam itidem realis, & ejusdem generis cum præcedentibus, licet alio nomine insignitus."

Ordinatam nullam, ut & distantiam nullam existentium esse compenetrationem.

[27] 59. "Sit in Fig. 9. Linea AB, ut prius, ad quam linea quædam PL deveniat in G (pertinet punctum G ad lineam PL, E ad AB continuatas, & sibi occurrentes ibidem), & sive pergat ultra ipsam in GM′, sive retro resiliat per GM′. Recta CD habebit ordinatam CL, quæ evanescet, ubi puncto C abeunte in E, ipsa CD abibit in EF, tum in positione ulteriori rectæ perpendicularis HI, vel abibit in negativam HM, vel retro positiva regredietur in HM′. Ubi linea altera cum altera coit, & punctum E alterius cum alterius puncto G congreditur, ordinata CL videtur abire in nihilum ita, ut nihilum, quemadmodum & supra innuimus, sit limes quidam inter seriem ordinatarum positivarum CL, & negativarum HM; vel positivarum CL, & iterum positivarum HM′. Sed, si res altius consideretur ad metaphysicum conceptum reducta, in situ EF non habetur verum nihilum. In situ CD, HI habetur distantia quædam punctorum C, L; H, M: in situ EF habetur eorundem punctorum compenetratio. Distantia est relatio quædam binorum modorum, quibus bina puncta existunt; compenetratio itidem est relatio binorum modorum, quibus ea existunt, quæ compenetratio est aliquid reale ejusdem prorsus generis, cujus est distantia, constituta nimirum per binos reales existendi modos."

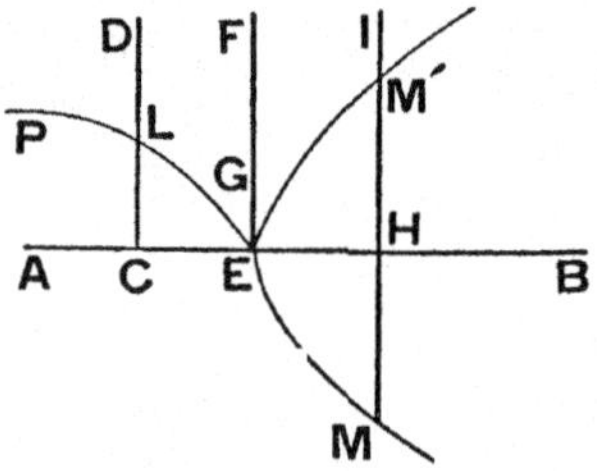

Fig. 9.

Ad idem perti*n*ere seriei realis ge*n*us eam dista*n*tiam nullam, & aliquam.

60. "Totum discrimen est in vocabulis, quæ nos imposuimus. Bini locales existendi modi infinitas numero relationes possunt constituere, alii alias. Hae omnes inter se & differunt, & tamen simul etiam plurimum conveniunt; nam reales sunt, & in quodam genere congruunt, quod nimirum sint relationes ortæ a binis localibus existendi modis. Diversa vero habent nomina ad arbitrarium instituta, cum aliæ ex ejusmodi relationibus, ut CL, dicantur distantiæ positivæ, relatio EG dicatur compenetratio, relationes HM dicantur distantiæ negativæ. Sed quoniam, ut a decem palmis distantiæ demptis 5, relinquuntur 5, ita demptis aliis 5, habetur nihil (non quidem verum nihil, sed nihil in ratione distantiæ a nobis ita appellatæ, cum remaneat compenetratio); ablatis autem aliis quinque, remanent quinque palmi distantiæ negativæ; ista omnia realia sunt, & ad idem genus pertinent; cum eodem prorsus modo inter se differant distantia palmorum 10 a distantia palmorum 5, hæc a distantia nulla, sed reali, quæ compenetrationem importat, & hæc a distantia negativa palmorum 5. Nam ex prima illa quantitate eodem modo devenitur ad basce posteriores per continuam ablationem palmorum 5. Eodem autem pacto infinitas ellipses, ab infinitis hyperbolis unica interjecta parabola discriminat, quæ quidem unica nomen peculiare sortita est, cum illas numero infinitas, & a se invicem admodum discrepantes unico vocabulo complectamur; licet altera magis oblonga ab altera minus oblonga plurimum itidem diversa sit."

Alia, quæ videntur nihil, & *sunt* aliquid: discrimen inter radicem imaginariam, & zero.

[28] 61. "Et quidem eodem pacto status quidam realis est quies, sive perseverantia in eodem modo locali existendi; status quidam realis est velocitas nulla puncti existentis. nimirum determinatio perseverandi in eodem loco; status quidam realis puncti existentis est vis nulla, nimirum determinatio retinendi præcedentem velocitatem, & ita porro; plurimum hæc discrepant a vero *non esse*. Casus ordinatæ respondentis lineæ EF in fig. 9, differt plurimum a casu ordinatæ circuli respondentis lineæ CD figuræ 8: in prima existunt puncta, sed compenetrata, in secunda alterum punctum impossible est. Ubi in solutione problematum devenitur ad quantitatem primi generis, problema determinationem peculiarem accipit; ubi devenitur ad quantitatem secundi generis, problema evadit impossibile; usque adeo in hoc secundo casu habetur verum nihilum, omni reali proprietate carens; in illo primo habetur aliquid realibus proprietatibus præditum, quod ipsis etiam solutionibus problematum, & constructionibus veras sufficit, & reales determinationes; cum realis, non imaginaria sit radix equationis cujuspiam, quæ sit = o, sive nihilo æqualis."

to pass suddenly from one quantity to another, then at the instant in which the sudden change is accomplished, both terms must be obtained. Hence, our argument on metaphysical grounds in favour of the exclusion of a sudden change from creation or annihilation, or production & destruction, remains quite unimpaired."

Sometimes what is really something appears to be nothing.

58. "In this connection the following point must be noted. As we have used geometrical ideas for the consideration of production & destruction, it seems also that sometimes the last term of a real series is nothing. But if we go deeper into the matter, we find that it is not in reality nothing, but some state that is also real and of the same kind as those that precede it, though designated by another name."

When the ordinate is nothing, just as when the distance between two existent things is nothing, there is compenetration.

59. "In Fig. 9, let AB be a line, as before, which some line PL reaches at G (where the point G belongs to the line PL, & E to the line AB, both being produced to meet one another at this point); & suppose that PL either goes on beyond the point as GM, or recoils along GM'. Then the straight line CD will contain the ordinate CL, which will vanish when, as the point C gets to E, CD attains the position EF; & after that, in the further position of the perpendicular straight line HI, will either pass on to the negative ordinate HM or return, once more positive, to HM'. Now when the one line meets the other, & the point E of the one coincides with the point G of the other, the ordinate CL seems to run off into nothing in such a manner that nothing, as we remarked above, is a certain boundary between the series of positive ordinates CL & the negative ordinates HM, or between the positive ordinates CL & the ordinates HM' which are also positive. But if the matter is more deeply considered & reduced to a metaphysical concept, there is not an absolute nothing in the position EF. In the position CD, or HI, we have given a certain distance between the points C,L, or H,M; in the position EF, there is compenetration of these points. Now distance is a relation between the modes of existence of two points; also compenetration is a relation between two modes of existence; & this compenetration is something real of the very same nature as distance, founded as it is on two real modes of existence."

This 'no' distance belongs to the same kind of series of real quantities as 'some' distance.

60. "The whole difference lies in the words that we have given to the things in question. Two local modes of existence can constitute an infinite number of relations, some of one sort & some of another. All of these differ from one another, & yet agree with one another in a high degree; for they are real & to a certain extent identical, since indeed they are all relations arising from a pair of local modes of existence. But they have different names assigned to them arbitrarily, so that some of the relations of this kind, as CL, are called positive distances, the relation EG is called compenetration, & relations like HM are called negative distances. But, just as when five palms of distance are taken away from ten palms, there are left five palms, so when five more are taken away, there is nothing left (& yet not really nothing, but nothing in comparison with what we usually call distance; for compenetration is left). Again, if we take away another five, there remain five palms of negative distance. All of these are real & belong to the same class; for they differ amongst themselves in exactly the same way, namely, the distance of ten palms from the distance of five palms, the latter from 'no' distance (which however is something real that denotes compenetration), & this again from a negative distance of five palms. For starting with the first quantity, the others that follow are obtained in the same manner, by a continual subtraction of five palms. In a similar manner a single intermediate parabola discriminates between an infinite number of ellipses & an infinite number of hyperbolas; & this single curve receives a special name, whilst under the one term we include an infinite number of them that to a certain extent are all different from one another, although one that is considerably elongated may be very different from another that is less elongated."

Other things that seem to be nothing, and yet are really something; distinction between an imaginary root & zero.

61. "In the same way, rest, i.e., a perseverance in the same mode of local existence, is some real state; so is 'no' velocity a real state of an existent point, namely, a propensity to remain in the same place; so also is 'no' force a real state of an existent point, namely, a propensity to retain the velocity that it has already; & so on. All these differ from a state of *non-existence* in the highest degree. The case of the ordinate corresponding to the line EF in Fig. 9 differs altogether from the case of the ordinate of the circle corresponding to the line CD in Fig. 8. In the first there exist two points, but there is compenetration of these points; in the other case, the second point cannot possibly exist. When, in the solution of problems, we arrive at a quantity of the first kind, the problem receives a special sort of solution; but when the result is a quantity of the second kind, the problem turns out to be incapable of solution. So much indeed that, in this second case, there is obtained a true nothing that lacks every real property; in the first case, we get something endowed with real properties, which also supplies true & real values to the solutions & constructions of the problems. For the root of any equation that = 0, or is equal to nothing, is something that is real, & is not an imaginary thing."

Conclusio pro solutione ejus objectionis.

62. " Firmum igitur manebit semper, & stabile, seriem realem quamcunque, quæ continuo tempore finito duret, debere habere & primum principium, & ultimum finem realem, sine ullo absurdo, & sine conjunctione sui *esse* cum *non esse*, si forte duret eo solo tempore : dum si præcedenti etiam exstitit tempore, habere debet & ultimum terminum seriei præcedentis, & primum sequentis, qui debent esse unicus indivisibilis communis limes, ut momentum est unicus indivisibilis limes inter tempus continuum præcedens, & subsequens. Sed hæc de ortu, & interitu jam satis."

Applicatio legis continuitatis ad collisionem corporum.

63. Ut igitur contrahamus jam vela, continuitatis lex & inductione, & metaphysico argumento abunde nititur, quæ idcirco etiam in velocitatis communicatione retineri omnino debet, ut nimirum ab una velocitate ad aliam numquam transeatur, nisi per intermedias velocitates omnes sine saltu. Et quidem in ipsis motibus, & velocitatibus inductionem habuimus num. 39, ac difficultates solvimus num. 46, & 47 pertinentes ad velocitates, quæ videri possent mutatæ per saltum. Quod autem pertinet ad metaphysicum argumentum, si toto tempore ante contactum subsequentis corporis superficies antecedens habuit 12 gradus velocitatis, & sequenti 9, saltu facto momentaneo ipso initio contactus ; in ipso momento ea tempora dirimente debuisset habere & 12, & 9 simul, quod est absurdum. Duas enim velocitates simul habere corpus non potest, quod ipsum aliquanto diligentius demonstrabo.

Duo velocitatum genera, potentialis, & actualis

64. Velocitatis nomen, uti passim usurpatur a Mechanicis, æquivocum est ; potest enim significare velocitatem actualem, quæ nimirum est relatio quædam in motu æquabili spatii percursi divisi per tempus, quo percurritur ; & potest significare [29] quandam, quam apto Scholiasticorum vocabulo potentialem appello, quæ nimirum est determinatio, ad actualem, sive determinatio, quam habet mobile, si nulla vis mutationem inducat, percurrendi motu æquabili determinatum quoddam spatium quovis determinato tempore, quæ quidem duo & in dissertatione *De Viribus Vivis*, & in Stayanis Supplementis distinxi, distinctione utique necessaria ad æquivocationes evitandas. Prima haberi non potest momento temporis, sed requirit tempus continuum, quo motus fiat, & quidem etiam motum æquabilem requirit ad accuratam sui mensuram ; secunda habetur etiam momento quovis determinata ; & hanc alteram intelligunt utique Mechanici, cum scalas geometricas efformant pro motibus quibuscunque difformibus, sive abscissa exprimente tempus, & ordinata velocitatem, utcunque etiam variatam, area exprimat spatium : sive abscissa exprimente itidem tempus, & ordinata vim, area exprimat velocitatem jam genitam, quod itidem in aliis ejusmodi scalis, & formulis algebraicis fit passim, hac potentiali velocitate usurpata, quæ sit tantummodo determinatio ad actualem, quam quidem ipsam intelligo, ubi in collisione corporum eam nego mutari posse per saltum ex hoc posteriore argumento.

Binas velocitates tum actuales, tum potentiales simul haberi non posse, ne detur, vel exigatur compenetratio.

65. Jam vero velocitates actuales non posse simul esse duas in eodem mobili, satis patet ; quia oporteret, id mobile, quod initio dati cujusdam temporis fuerit in dato spatii puncto, in omnibus sequentibus occupare duo puncta ejusdem spatii, ut nimirum spatium percursum sit duplex, alterum pro altera velocitate determinanda, adeoque requireretur actualis replicatio, quam non haberi uspiam, ex principio inductionis colligere sane possumus admodum facile. Cum nimirum nunquam videamus idem mobile simul ex eodem loco discedere in partes duas, & esse simul in duobis locis ita, ut constet nobis, utrobique esse illud idem. At nec potentiales velocitates duas simul esse posse, facile demonstratur. Nam velocitas potentialis est determinatio ad existendum post datum tempus continuum quodvis in dato quodam puncto spatii habente datam distantiam a puncto spatii, in quo mobile est eo temporis momento, quo dicitur habere illam potentialem velocitatem determinatam. Quamobrem habere simul illas duas potentiales velocitates est esse determinatum ad occupanda eodem momento temporis duo puncta spatii, quorum singula habeant suam diversam distantiam ab eo puncto spatii, in quo tum est mobile, quod est esse determinatum ad replicationem habendam momentis omnibus sequentis temporis. Dicitur utique idem mobile a diversis causis acquirere simul diversas velocitates, sed eæ componuntur in unicam ita, ut singulæ constituant statum mobilis, qui status respectu dispositionum, quas eo momento, in quo tum est, habet ipsum mobile, complectentium omnes circumstantias præteritas, & præsentes, est tantummodo conditionatus, non absolutus ; nimirum ut contineant determi-[30]-nationem, quam ex omnibus præteritis, & præsentibus circumstantiis haberet ad occupandum illud determinatum spatii punctum determinato illo momento

Conclusion in favour of a solution of this difficulty.

62. "Hence in all cases it must remain a firm & stable conclusion that any real series, which lasts for some finite continuous time, is bound to have a first beginning & a final end, without any absurdity coming in, & without any linking up of its *existence* with a state of *non-existence*, if perchance it lasts for that interval of time only. But if it existed at a previous time as well, it must have both a last term of the preceding series & a first term of the subsequent series; just as an instant is a single indivisible boundary between the continuous time that precedes & that which follows. But what I have said about production & destruction is already quite enough."

Application of the Law of Continuity to the collision of solid bodies.

63. But, to come back at last to our point, the Law of Continuity is solidly founded both on induction & on metaphysical reasoning; & on that account it should be retained in every case of communication of velocity. So that indeed there can never be any passing from one velocity to another except through all intermediate velocities, & then without any sudden change. We have employed induction for actual motions & velocities in Art. 39 & solved difficulties with regard to velocities in Art. 46, 47, in cases in which they might seem to be subject to sudden changes. As regards metaphysical argument, if in the whole time before contact the anterior surface of the body that follows had 12 degrees of velocity & in the subsequent time had 9, a sudden change being made at the instant of first contact; then at the instant that separates the two times, the body would be bound to have 12 degrees of velocity, & 9, at one & the same time. This is absurd; for a body cannot at the same time have two velocities, as I will now demonstrate somewhat more carefully.

Two kinds of velocity, potential & actual.

64. The term velocity, as it is used in general by Mechanicians is equivocal. For it may mean actual velocity, that is to say, a certain relation in uniform motion given by the space passed over divided by the time taken to traverse it. It may mean also something which, adopting a term used by the Scholastics, I call potential velocity. The latter is a propensity for actual velocity, or a propensity possessed by the movable body (should no force cause an alteration) for traversing with uniform motion some definite space in any definite time. I made the distinction between these two meanings, both in the dissertation *De Viribus Vivis* & in the Supplements to Stay's Philosophy; the distinction being very necessary to avoid equivocations. The former cannot be obtained in an instant of time, but requires continuous time for the motion to take place; it also requires uniform motion in order to measure it accurately. The latter can be determined at any given instant; & it is this kind that is everywhere intended by Mechanicians, when they make geometrical measured diagrams for any non-uniform velocities whatever. In which, if the abscissa represents time & the ordinate velocity, no matter how it is varied, then the area will express the distance passed over; or again, if the abscissa represents time & the ordinate force, then the area will represent the velocity already produced. This is always the case, for other scales of the same kind, whenever algebraical formulæ & this potential velocity are employed; the latter being taken to be but the propensity for actual velocity, such indeed as I understand it to be, when in collision of bodies I deny from the foregoing argument that there can be any sudden change.

It is impossible for a body to have two velocities, either actual or potential, unless it is given, or we are forced to admit, that there is compenetration.

65. Now it is quite clear that there cannot be two actual velocities at one & the same time in the same moving body. For, then it would be necessary that the moving body, which at the beginning of a certain time occupied a certain given point of space, should at all times afterwards occupy two points of that space; so that the space traversed would be twofold, the one space being determined by the one velocity & the other by the other. Thus an actual replication would be required; & this we can clearly prove in a perfectly simple way from the principle of induction. Because, for instance, we never see the same movable body departing from the same place in two directions, nor being in two places at the same time in such a way that it is clear to us that it is in both. Again, it can be easily proved that it is also impossible that there should be two potential velocities at the same time. For potential velocity is the propensity that the body has, at the end of any given continuous time, for existing at a certain given point of space that has a given distance from that point of space, which the moving body occupied at the instant of time in which it is said to have the prescribed potential velocity. Wherefore to have at one & the same time two potential velocities is the same thing as being prescribed to occupy at the same instant of time two points of space; each of which has its own distinct distance from that point of space that the body occupied at the start; & this is the same thing as prescribing that there should be replication at all subsequent instants of time. It is commonly said that a movable body acquires from different causes several velocities simultaneously; but these velocities are compounded into one in such a way that each produces a state of the moving body; & this state, with regard to the dispositions that it has at that instant (these include all circumstances both past & present), is only conditional, not absolute. That is to say, each involves the propensity which the body, on account of all past & present circumstances, would have for occupying that prescribed point of space at that particular

temporis ; nisi aliunde ejusmodi determinatio per conjunctionem alterius causæ, quæ tum agat, vel jam egerit, mutaretur, & loco ipsius alia, quæ composita dicitur, succederet. Sed status absolutus resultans ex omnibus eo momento præsentibus, & præteritis circumstantiis ipsius mobilis, est unica determinatio ad existendum pro quovis determinato momento temporis sequentis in quodam determinato puncto spatii, qui quidem status pro circumstantiis omnibus præteritis, & præsentibus est absolutus, licet sit itidem conditionatus pro futuris : si nimirum eædem, vel aliæ causæ agentes sequentibus momentis non mutent determinationem, & punctum illud loci, ad quod revera deveniri deinde debet dato illo momento temporis, & actu devenitur ; si ipsæ nihil aliud agant. Porro patet ejusmodi status ex omnibus præteritis, & præsentibus circumstantiis absolutos non posse eodem momento temporis esse duos sine determinatione ad replicationem, quam ille conditionatus status resultans e singulis componentibus velocitatibus non inducit ob id ipsum, quod conditionatus est. Jam vero si haberetur saltus a velocitate ex omnibus præteritis, & præsentibus circumstantiis exigente, ex. gr. post unum minutum, punctum spatii distans per palmos 6 ad exigentem punctum distans per palmos 9 ; deberet eo momento temporis, quo fieret saltus, haberi simul utraque determinatio absoluta respectu circumstantiarum omnium ejus momenti, & omnium præteritarum ; nam toto præcedenti tempore habita fuisset realis series statuum cum illa priore, & toto sequenti deberet haberi cum illa posteriore, adeoque eo momento, simul utraque, cum neutra series realis sine reali suo termino stare possit.

Quovis momento *punctum* existens debere habere statum *realem* ex *genere* velocitatis *potentialis*.

66. Præterea corporis, vel puncti existentis potest utique nulla esse velocitas actualis, saltem accurate talis ; si nimirum difformem habeat motum, quod ipsum etiam semper in Natura accidit, ut demonstrari posse arbitror, sed huc non pertinet ; at semper utique haberi debet aliqua velocitas potentialis, vel saltem aliquis status, qui licet alio vocabulo appellari soleat, & dici velocitas nulla, est tamen non nihilum quoddam, sed realis status, nimirum determinatio ad quietem, quanquam hanc ipsam, ut & quietem, ego quidem arbitrer in Natura reapse haberi nullam, argumentis, quæ in Stayanis Supplementis exposui in binis paragraphis de spatio, ac tempore, quos hic addam in fine inter nonnulla, quæ hic etiam supplementa appellabo, & occurrent primo, ac secundo loco. Sed id ipsum itidem nequaquam huc pertinet. Iis etiam penitus prætermissis, eruitur e reliquis, quæ diximus, admisso etiam ut existente, vel possibili in Natura motu uniformi, & quiete, utramque velocitatem habere conditiones necessarias ad [31] hoc, ut secundum argumentum pro continuitatis lege superius allatum vim habeat suam, nec ab una velocitate ad alteram abiri possit sine transitu per intermedias.

Non posse momento temporis transiri ab una velocitate ad aliam, demonstratur, & vindicatur.

67. Patet autem, hinc illud evinci, nec interire momento temporis posse, nec oriri velocitatem totam corporis, vel puncti non simul intereuntis, vel orientis, nec huc transferri posse, quod de creatione, & morte diximus ; cum nimirum ipsa velocitas nulla corporis, vel puncti existentis, sit non purum nihil, ut monui, sed realis quidam status, qui simul cum alio reali statu determinatæ illius intereuntis, vel orientis velocitatis deberet conjungi ; unde etiam fit, ut nullum effugium haberi possit contra superiora argumenta, dicendo, quando a 12 gradibus velocitatis transitur ad 9, durare utique priores 9, & interire reliquos tres, in quo nullum absurdum sit, cum nec in illorum duratione habeatur saltus, nec in saltu per interitum habeatur absurdi quidpiam, ejus exemplo, quod superius dictum fuit, ubi ostensum est, non conjungi *non esse* simul, & *esse*. Nam in primis 12 gradus velocitatis non sunt quid compositum e duodecim rebus inter se distinctis, atque disjunctis, quarum 9 manere possint, 3 interire, sed sunt unica determinatio ad existendum in punctis spatii distantibus certo intervallo, ut palmorum 12, elapsis datis quibusdam temporibus æqualibus quibusvis. Sic etiam in ordinatis GD, HE, quæ exprimunt velocitates in fig. 6, revera, in mea potissimuim Theoria, ordinata GD non est quædam pars ordinatæ HE communis ipsi usque ad D, sed sunt duæ ordinatæ, quarum prima constitit in relatione distantiæ, puncti curvæ D a puncto axis G, secunda in relatione puncti curvæ E a puncto axis H, quod est ibi idem, ac punctum G.

instant of time; were it not for the fact that that particular propensity is for other reasons altered by the conjunction of another cause, which acts at the time, or has already done so; & then another propensity, which is termed compound, will take the place of the former. But the absolute propensity, which arises from the combination of all the past & present circumstances of the moving body for that instant, is but a single propensity for existing at any prescribed instant of subsequent time in a certain prescribed point of space; & this state is absolute for all past & present circumstances, although it may be conditional for future circumstances. That is to say, if the same or other causes, acting during subsequent instants, do not change that propensity, & the point of space to which it ought to get thereafter at the given instant of time, & which it actually does reach if these causes have no other effect. Further, it is clear that we cannot have two such absolute states, arising from all past & present circumstances, at the same time without prescribing replication; & this the conditional state arising from each of the component velocities does not induce because of the very fact that it is conditional. If now there should be a jump from the velocity, arising out of all the past & present circumstances, which, after one minute for example, compels a point of space to move through 6 palms, to a velocity that compels the point to move through 9 palms; then, at the instant of time, in which the sudden change takes place, there would be each of two absolute propensities in respect of all the circumstances of that instant & all that had gone before, existing simultaneously. For in the whole of the preceding time there would have been a real series of states having the former velocity as a term, & in the whole of the subsequent time there must be one having the latter velocity as a term; hence at that particular instant each of them must occur at one & the same time, since neither real series can stand good without each having its own real end term.

At any instant an existing point must have a real state arising from a kind of potential velocity.

66. Again, it is at least possible that the actual velocity of a body, or of an existing point, may be nothing; that is to say, if the motion is non-uniform. Now, this always is the case in Nature; as I think can be proved, but it does not concern us at present. But, at any rate, it is bound to have some potential velocity, or at least some state, which, although usually referred to by another name, & the velocity stated to be nothing, yet is not definitely nothing, but is a real state, namely, a propensity for rest. I have come to the conclusion, however, that in Nature there is not really such a thing as this state, or absolute rest, from arguments that I gave in the Supplements to Stay's Philosophy in two paragraphs concerning space & time; & these I will add at the end of the work, amongst some matters, that I will call by the name of supplements in this work as well; they will be placed first & second amongst them. But that idea also does not concern us at present. Now, putting on one side these considerations altogether, it follows from the rest of what I have said that, if we admit both uniform motion & rest as existing in Nature, or even possible, then each velocity must have conditions that necessarily lead to the conclusion that according to the argument given above in support of the Law of Continuity it has its own corresponding force, & that no passage from one velocity to another can be made except through intermediate stages.

Rigorous proof that it is impossible to pass from one velocity to another in an instant of time.

67. Further, it is quite clear that from this it can be rigorously proved that the whole velocity of a body cannot perish or arise in an instant of time, nor for a point that does not perish or arise along with it; nor can our arguments with regard to production & destruction be made to refer to this. For, since that 'no' velocity of a body, or of an existing point, is not absolutely nothing, as I remarked, but is some real state; & this real state is bound to be connected with that other real state, namely, that of the prescribed velocity that is being created or destroyed. Hence it comes about that there can be no escape from the arguments I have given above, by saying that when the change from twelve degrees of velocity is made to nine degrees, the first nine at least endure, whilst the remaining three are destroyed; & then by asserting that there is nothing absurd in this, since neither in the duration of the former has there been any sudden change, nor is there anything absurd in the jump caused by the destruction of the latter, according to the instance of it given above, where it was shown that *non-existence* & *existence* must be disconnected. For in the first place those twelve degrees of velocity are not something compounded of twelve things distinct from, & unconnected with, one another, of which nine can endure & three can be destroyed; but are a single propensity for existing, after the lapse of any given number of equal times of any given length, in points of space at a certain interval, say twelve palms, away from the original position. So also, with regard to the ordinates GD, HE, which in Fig. 6. express velocities, it is the fact that (most especially in my Theory) the ordinate GD is not some part of the ordinate HE, common with it as far as the point D; but there are two ordinates, of which the first depends upon the relation of the distance of the point D of the curve from the point G on the axis, & the second upon the relation of the distance of point E on the curve from the point H on the axis, which is here the

Relationem distantiæ punctorum D, & G constituunt duo reales modi existendi ipsorum, relationem distantiæ punctorum D. & E duo reales modi existendi ipsorum, & relationem distantiæ punctorum H, & E duo reales modi existendi ipsorum. Hæc ultima relatio constat duobus modis realibus tantummodo pertinentibus ad puncta E, & H, vel G, & summa priorum constat modis realibus omnium trium, E, D, G. Sed nos indefinite concipimus possibilitatem omnium modorum realium intermediorum, ut infra dicemus, in qua præcisiva, & indefinita idea stat mihi idea spatii continui ; & intermedii modi possibiles inter G, & D sunt pars intermediorum inter E, & H. Præterea omissis etiam bisce omnibus ipse ille saltus a velocitate finita ad nullam, vel a nulla ad finitam, haberi non potest.

Cur adhibita collisio pergentium in eandem plagam pro Theoria deducenda.

68. Atque hinc ego quidem potuissem etiam adhibere duos globos æquales, qui sibi invicem occurrant cum velocitatibus æqualibus, quæ nimirum in ipso contactu deberent momento temporis interire ; sed ut basce ipsas considerationes evitarem de transitu a statu reali ad statum itidem realem, ubi a velocitate aliqua transitur ad velocitatem nullam ; adhibui potius [32] in omnibus dissertationibus meis globum, qui cum 12 velocitatis gradibus assequatur alterum præcedentem cum 6 ; ut nimirum abeundo ad velocitatem aliam quamcunque haberetur saltus ab una velocitate ad aliam, in quo evidentius esset absurdum.

Quo pacto mutata velocitate potentiali per saltum, non mutetur per saltum actualis.

69. Jam vero in hisce casibus utique haberi deberet saltus quidam, & violatio legis continuitatis, non quidem in velocitate actuali, sed in potentiali, si ad contactum deveniretur cum velocitatum discrimine aliquo determinato quocunque. In velocitate actuali, si eam metiamur spatio, quod conficitur, diviso per tempus, transitus utique fieret per omnes intermedias, quod sic facile ostenditur ope Geometriæ. In fig. 10 designent AB, BC bina tempora ante & post contactum, & momento quolibet H sit velocitas potentialis illa major HI, quæ æquetur velocitati primæ AD ; quovis autem momento Q posterioris temporis sit velocitas potentialis minor QR, quæ æquetur velocitati cuidam datæ CG. Assumpto quovis tempore HK determinatæ magnitudinis, area IHKL divisa per tempus HK, sive recta HI, exhibebit velocitatem actualem. Moveatur tempus HK versus B, & donec K adveniat ad B, semper eadem habebitur velocitatis mensura ; eo autem progresso in O ultra B, sed adhuc H existente in M citra B, spatium illi tempori respondens componetur ex binis MNEB, BFPO, quorum summa si dividatur per MO ; jam nec erit MN æqualis priori AD, nec BF, ipsa minor per datam quantitatem FE ; sed facile demonstrari potest (b), capta VE æquali IL, vel HK, sive MO, & ducta recta VF, quæ secet MN in X, quotum ex illo divisione prodeuntem fore MX, donec, abeunte toto illo tempore ultra B in QS, jam area QRTS divisa per tempus QS exhibeat velocitatem constantem QR.

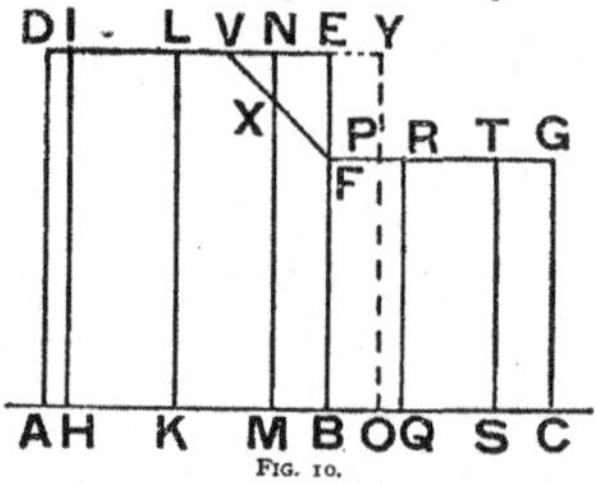

FIG. 10.

Irregularitas alia in expressione actualis velocitatis.

70. Patet igitur in ea consideratione a velocitate actuali præcedente HI ad sequentem QR transiri per omnes intermedias MX, quas continua recta VF definiet ; quanquam ibi etiam irregulare quid oritur inde, quod velocitas actualis XM diversa obvenire debeat pro diversa magnitudine temporis assumpti HK, quo nimirum assumpto majore, vel minore removetur magis, vel minus V ab E, & decrescit, vel crescit XM. Id tamen accidit in motibus omnibus, in quibus velocitas non manet eadem toto tempore, ut nimirum tum etiam, si velocitas aliqua actualis debeat agnosci, & determinari spatio diviso per tempus ; pro aliis, atque aliis temporibus assumptis pro mensura aliæ, atque aliæ velocitatis actualis mensuræ ob-[33]-veniant, secus ac accidit in motu semper æquabili, quam ipsam ob causam, velocitatis actualis in motu difformi nulla est revera mensura accurata, quod supra innui sed ejus idea præcisa, ac distincta æquabilitatem motus requirit, & idcirco Mechanici in difformibus motibus ad actualem velocitatem determinandam adhibere solent spatiolum infinitesimo tempusculo percursum, in quo ipso motum habent pro æquabili.

(b) *Si enim producatur OP usque ad NE in Υ, erit* $E\Upsilon = VN$, *ob* $VE = MO = N\Upsilon$. *Est autem* $VE : VN :: EF : NX$; *quare* $VN \times EF = VE \times NX$, *sive posito* $E\Upsilon$ *pro* VN, *&* MO *pro* VE, *erit* $E\Upsilon \times EF = MO \times NX$. *Totum MNΥO est* $MO \times MN$, *pars FEΥP est* $= E\Upsilon \times EF$. *Quare residuus gnomon NMOPFE est* $MO \times (MN - NX)$, *sive est* $MO \times MX$, *quo diviso per MO habetur MX.*

same as the point G. The relation of the distance between the points D & G is determined by the two real modes of existence peculiar to them, the relation of the distance between the points D & E by the two real modes of existence peculiar to them, & the relation of the distance between the points H & E by the two real modes of existence peculiar to them. The last of these relations depends upon the two real modes of existence that pertain to the points E & H (or G), & upon these alone; the sum of the first & second depends upon all three of the modes of the points E, D, & G. But we have some sort of ill-defined conception of the possibility of all intermediate real modes of existence, as I will remark later; & on this disconnected & ill-defined idea is founded my conception of continuous space; also the possible intermediate modes between G & D form part of those intermediate between E & H. Besides, omitting all considerations of this sort, that sudden change from a finite velocity to none at all, or from none to a finite, cannot happen.

Why the collision of bodies moving in the same direction is employed for the purpose of deducing my Theory.

68. Hence I might just as well have employed two equal balls, colliding with one another with equal velocities, which in truth at the moment of contact would have to be destroyed in an instant of time. But, in order to avoid the very considerations just stated with regard to the passage from a real state to another real state (when we pass from a definite velocity to none), I have preferred to employ in all my dissertations a ball having 12 degrees of velocity, which follows another ball going in front of it with 6 degrees; so that, by passing to some other velocity, there would be a sudden change from one velocity to another; & by this means the absurdity of the idea would be made more evident.

How, supposing that there were a sudden change in the potential velocity, there might not be a sudden change in the actual velocity.

69. Now, at least in such cases as these, there is bound to be some sudden change & a breach of the Law of Continuity, not indeed in the actual velocity, but in the potential velocity, if the collision occurs with any given difference of velocities whatever. In the actual velocity, measured by the space traversed divided by the time, the change will at any rate be through all intermediate stages; & this can easily be shown to be so by the aid of Geometry.

In Fig. 10 let AB, BC represent two intervals of time, respectively before & after contact; & at any instant let the potential velocity be the greater velocity HI, equal to the first velocity AD; & at any instant Q of the time subsequent to contact let the potential velocity be the less velocity QR, equal to some given velocity CG. If any prescribed interval of time HK be taken, the area IHKL divided by the time HK, i.e., the straight line HI, will represent the actual velocity. Let the time HK be moved towards B; then until K comes to B, the measure of the velocity will always be the same. If then, K goes on beyond B to O, whilst H still remains on the other side of B at M; then the space corresponding to that time will be composed of the two spaces MNEB, BFPO. Now, if the sum of these is divided by MO, the result will not be equal to either MN (which is equal to the first AD), or BF (which is less than MN by the given quantity FE). But it can easily be proved ( ) that, if VE is taken equal to IL, or HK, or MO, & the straight line VF is drawn to cut MN in X; then the quotient obtained by the division will be MX. This holds until, when the whole of the interval of time has passed beyond B into the position QS, the area QRTS divided by the time QS now represents a constant velocity equal to QR.

A further irregularity in the representation of actual velocity.

70. From the foregoing reasoning it is therefore clear that the change from the preceding actual velocity HI to the subsequent velocity QR is made through all intermediate velocities such as MX, which will be determined by the continuous straight line VF. There is, however, some irregularity arising from the fact that the actual velocity XM must turn out to be different for different magnitudes of the assumed interval of time HK. For, according as this is taken to be greater or less, so the point V is removed to a greater or less distance from E; & thereby XM will be decreased or increased correspondingly. This is the case, however, for all motions in which the velocity does not remain the same during the whole interval; as for instance in the case where, if any actual velocity has to be found & determined by the quotient of the space traversed divided by the time taken, far other & different measures of the actual velocities will arise to correspond with the different intervals of time assumed for their measurement; which is not the case for motions that are always uniform. For this reason there is no really accurate measure of the actual velocity in non-uniform motion, as I remarked above; but a precise & distinct idea of it requires uniformity of motion. Therefore Mechanicians in non-uniform motions, as a means to the determination of actual velocity, usually employ the small space traversed in an infinitesimal interval of time, & for this interval they consider that the motion is uniform.

(b) *For if OP be produced to meet NE in Υ, then $E\Upsilon = VN$; for $VE = MO = N\Upsilon$. Moreover $VE : VN = EF : NX$; and therefore $VN.EF = VE.NX$. Hence, replacing VN by $E\Upsilon$, and VE by MO, we have $E\Upsilon EF = MO.NX$. Now, the whole $MN\Upsilon O = MO.MN$, and the part $FE\Upsilon P = E\Upsilon.EF$. Hence the remainder (the gnomon NMOPFE) $= MO.(MN - NX) = MO.MX$: and this, on division by MO, will give MX.*

Concluditur ad contactum immediatum non posse deveniri cum differentia velocitatum.

71. At velocitas potentialis, quæ singulis momentis temporis respondet sua, mutaretur utique per saltum ipso momento B, quo deberet haberi & ultima velocitatum præcedentium BE, & prima sequentium BF, quod cum haberi nequeat, uti demonstratum est, fieri non potest per secundum ex argumentis, quæ adhibuimus pro lege continuitatis, ut cum illa velocitatum inæqualitate deveniatur ad immediatum contactum ; atque id ipsum excludit etiam inductio, quam pro lege continuitatis in ipsis quoque velocitatibus, atque motibus primo loco proposui.

Promovenda analysis eo excluso.

72. Atque hoc demum pacto illud constitit evidenter, non licere continuitatis legem deserere in collisione corporum, & illud admittere, ut ad contactum immediatum deveniatur cum illæsis binorum corporum velocitatibus integris. Videndum igitur, quid necessario consequi debeat, ubi id non admittatur, & hæc analysis ulterius promovenda.

Debere ante contactum haberi mutationem velocitatis, adeoque vim, quæ mutat.

73. Quoniam ad immediatum contactum devenire ea corpora non possunt cum præcedentibus velocitatibus ; oportet, ante contactum ipsum immediatum incipiant mutari velocitates ipsæ, & vel ea consequentis corporis minui, vel ea antecedentis augeri, vel utrumque simul. Quidquid accidat, habebitur ibi aliqua mutatio status, vel in altero corpore, vel in utroque, in ordine ad motum, vel quietem, adeoque habebitur aliqua mutationis causa, quæcunque illa sit. Causa vero mutans statum corporis in ordine ad motum, vel quietem, dicitur vis ; habebitur igitur vis aliqua, quæ effectum gignat, etiam ubi illa duo corpora nondum ad contactum devenerint.

Eam vim debere esse mutuam, & agere in partes oppositas.

74. Ad impediendam violationem continuitatis satis esset, si ejusmodi vis ageret in alterum tantummodo e binis corporibus, reducendo præcedentis velocitatem ad gradus 12, vel sequentis ad 6. Videndum igitur aliunde, an agere debeat in alterum tantummodo, an in utrumque simul, & quomodo. Id determinabitur per aliam Naturæ legem, quam nobis inductio satis ampla ostendit, qua nimirum evincitur, omnes vires nobis cognitas agere utrinque & æqualiter, & in partes oppositas, unde provenit principium, quod appellant actionis, & reactionis æqualium ; est autem fortasse quædam actio duplex semper æqualiter agens in partes oppositas. Ferrum, & magnes æque se mutuo trahunt ; elastrum binis globis æqualibus interjectum æque utrumque urget, & æqualibus velocitatibus propellit ; gravitatem ipsam generalem mutuam esse osten-[34]-dunt errores Jovis, ac Saturni potissimum, ubi ad se invicem accedunt, uti & curvatura orbitæ lunaris orta ex ejus gravitate in terram comparata cum æstu maris orto ex inæquali partium globi terraquei gravitate in Lunam. Ipsæ nostræ vires, quas nervorum ope exerimus, semper in partes oppositas agunt, nec satis valide aliquid propellimus, nisi pede humum, vel etiam, ut efficacius agamus, oppositum parietem simul repellamus. En igitur inductionem, quam utique ampliorem etiam habere possumus, ex qua illud pro eo quoque casu debemus inferre, eam ibi vim in utrumque corpus agere, quæ actio ad æqualitatem non reducet inæquales illas velocitates, nisi augeat præcedentis, minuat consequentis corporis velocitatem ; nimirum nisi in iis producat velocitates quasdam contrarias, quibus, si solæ essent, deberent a se invicem recedere : sed quia eæ componuntur cum præcedentibus ; hæc utique non recedunt, sed tantummodo minus ad se invicem accedunt, quam accederent.

Hinc dicendam esse repulsivam : quærendam ejus legem.

75. Invenimus igitur vim ibi debere esse mutuam, quæ ad partes oppositas agat, & quæ sua natura determinet per sese illa corpora ad recessum mutuum a se invicem. Hujusmodi igitur vis ex nominis definitione appellari potest vis repulsiva. Quærendum jam ulterius, qua lege progredi debeat, an imminutis in immensum distantiis ad datam quandam mensuram deveniat, an in infinitum excrescat ?

Ea vi debere totum velocitatum discrimen elidi ante contactum.

76. Ut in illo casu evitetur saltus ; satis est in allato exemplo ; si vis repulsiva, ad quam delati sumus, extinguat velocitatum differentiam illam 6 graduum, antequam ad contactum immediatum corpora devenirent : quamobrem possent utique devenire ad eum contactum eodem illo momento, quo ad æqualitatem velocitatum deveniunt. At si in alio quopiam casu corpus sequens impellatur cum velocitatis gradibus 20, corpore præcedente cum suis 6 ;

71. The potential velocity, each corresponding to its own separate instant of time, would certainly be changed suddenly at that instant of time B; & at this point we are bound to have both the last of the preceding velocities, BE, & the first of the subsequent velocities, BF. Now, since (as has been already proved) this is impossible, it follows from the second of the arguments that I used to prove the Law of Continuity, that it cannot come about that the bodies come into immediate contact with the inequality of velocities in question. This is also excluded by induction, such as I gave in the first place for the Law of Continuity, in the case also of these velocities & motions.

The *conclusion* is that immediate *contact* with a difference of velocities cannot be attained.

72. In this manner it is at length clearly established that it is not right to neglect the Law of Continuity in the collision of bodies, & admit the idea that they can come into immediate contact with the whole velocities of both bodies unaltered. Hence, we must now investigate the consequences that necessarily follow when this idea is not admitted; & the analysis must be carried further.

Immediate *contact* being *barred*, the analysis is to be carried further.

73. Since the bodies cannot come into immediate contact with the velocities they had at first, it is necessary that those velocities should commence to change before that immediate contact; & either that of the body that follows should be diminished, or that of the one going in front should be increased, or that both these changes should take place together. Whatever happens, there will be some change of state at the time, in one or other of the bodies, or in both, with regard to motion or rest; & so there must be some cause for this change, whatever it is. But a cause that changes the state of a body as regards motion or rest is called force. Hence there must be some force, which gives the effect, & that too whilst the two bodies have not as yet come into contact.

There must be then, before *contact*, a *change in* the velocity; & therefore some *force* that *causes* the *change*.

74. It would be enough, to avoid a breach of the Law of Continuity, if a force of this kind should act on one of the two bodies only, altering the velocity of the body in front to 12 degrees, or that of the one behind to 6 degrees. Hence we must find out, from other considerations, whether it should act on one of the two bodies only, or on both of them at the same time, & how. This point will be settled by another law of Nature, which sufficiently copious induction brings before us; that is, the law in which it is established that all forces that are known to us act on both bodies, equally, and in opposite directions. From this comes the principle that is called 'the principle of equal action & reaction'; perchance this may be a sort of twofold action that always produces its effect equally in opposite directions. Iron & a loadstone attract one another with the same strength; a spring introduced between two balls exerts an equal action on either ball, & generates equal velocities in them. That universal gravity itself is mutual is proved by the aberrations of Jupiter & of Saturn especially (not to mention anything else); that is to say, the way in which they err from their orbits & approach one another mutually. So also, when the curvature of the lunar orbit arising from its gravitation towards the Earth is compared with the flow of the tides caused by the unequal gravitation towards the Moon of different parts of the land & water that make up the Earth. Our own bodily forces, which produce their effect by the help of our muscles, always act in opposite dircetions; nor have we any power to set anything in motion, unless at the same time we press upon the earth with our feet or, in order to get a better purchase, upon something that will resist them, such as a wall opposite. Here then we have an induction, that can be made indeed more ample still; & from it we are bound in this case also to infer that the force acts on each of the two bodies. This action will not reduce to equality those two unequal velocities, unless it increases that of the body which is in front & diminishes that of the one which follows. That is to say, unless it produces in them velocities that are opposite in direction; & with these velocities, if they alone existed, the bodies would move away from one another. But, as they are compounded with those they had to start with, the bodies do not indeed recede from one another, but only approach one another less quickly than they otherwise would have done.

The *force must* be *mutual*, & act *in* opposite *directions*.

75. We have then found that the force must be a mutual force which acts in opposite directions; one which from its very nature imparts to those bodies a natural propensity for mutual recession from one another. Hence a force of this kind, from the very meaning of the term, may be called a repulsive force. We have now to go further & find the law that it follows, & whether, when the distances are indefinitely diminished, it attains any given measure, or whether it increases indefinitely.

Hence the *force* must be termed repulsive; the law *governing* it is now to be found.

76. In this case, in order that any sudden change may be avoided, it is sufficient, in the example under consideration, if the repulsive force, to which our arguments have led us, should destroy that difference of 6 degrees in the velocities before the bodies should have come into immediate contact. Hence they might possibly at least come into contact at the instant in which they attained equality between the velocities. But if in another case, say, the body that was behind were moving with 20 degrees of velocity, whilst the

The whole difference *between* the velocities *must* be destroyed by the force before contact.

tum vero ad contactum deveniretur cum differentia velocitatum majore, quam graduum 8. Nam illud itidem amplissima inductione evincitur, vires omnes nobis cognitas, quæ aliquo tempore agunt, ut velocitatem producant, agere in ratione temporis, quo agunt, & sui ipsius. Rem in gravibus oblique descendentibus experimenta confirmant; eadem & in elastris institui facile possunt, ut rem comprobent; ac id ipsum est fundamentum totius Mechanicæ, quæ inde motuum leges eruit, quas experimenta in pendulis, in projectis gravibus, in aliis pluribus comprobant, & Astronomia confirmat in cælestibus motibus. Quamobrem illa vis repulsiva, quæ in priore casu extinxit 6 tantummodo gradus discriminis, si agat breviore tempore in secundo casu, non poterit extinguere nisi pauciores, minore nimirum velocitate producta utrinque ad partes contrarias. At breviore utique tempore aget: nam cum majore velocitatum discrimine velocitas respectiva est major, ac proinde accessus celerior. [35] Extingueret igitur in secundo casu illa vis minus, quam 6 discriminis gradus, si in primo usque ad contactum extinxit tantummodo 6. Superessent igitur plures, quam 8; nam inter 20 & 6 erant 14, ubi ad ipsum deveniretur contactum, & ibi per saltum deberent velocitates mutari, ne compenetratio haberetur, ac proinde lex continuitatis violari. Cum igitur id accidere non possit; oportet, Natura incommodo caverit per ejusmodi vim, quæ in priore casu aliquanto ante contactum extinxerit velocitatis discrimen, ut nimirum imminutis in secundo casu adhuc magis distantiis, vis ulterior illud omne discrimen auferat, clisis omnibus illis 14 gradibus discriminis, qui habebantur.

Eam vim debere augeri in infinitum, imminutis, & quidem in infinitum, distantiis: habente virium curva aliquam asymptotum in origine abscissarum.

77. Quando autem huc jam delati sumus, facile est ulterius progredi, & illud considerare, quod in secundo casu accidit respectu primi, idem accidere aucta semper velocitate consequentis corporis in tertio aliquo respectu secundi, & ita porro. Debebit igitur ad omnem pro omni casu evitandum saltum Natura cavisse per ejusmodi vim, quæ imminutis distantiis crescat in infinitum, atque ita crescat, ut par sit extinguendæ cuicunque velocitati, utcunque magnæ. Devenimus igitur ad vires repulsivas imminutis distantiis crescentes in infinitum, nimirum ad arcum illum asymptoticum ED curvæ virium in fig. 1 propositum. Illud quidem ratiocinatione hactenus instituta immediate non deducitur, hujusmodi incrementa virium auctarum in infinitum respondere distantiis in infinitum imminutis. Posset pro hisce corporibus, quæ habemus præ manibus, quædam data distantia quæcunque esse ultimus limes virium in infinitum excrescentium, quo casu asymptotus AB non transiret per initium distantiæ binorum corporum, sed tanto intervallo post ipsum, quantus esset ille omnium distantiarum, quas remotiores particulæ possint acquirere a se invicem, limes minimus; sed aliquem demum esse debere extremum etiam asymptoticum arcum curvæ habentem pro asymptoto rectam transeuntem per ipsum initium distantiæ, sic evincitur; si nullus ejusmodi haberetur arcus; particulæ materiæ minores, & primo collocatæ in distantia minore, quam esset ille ultimus limes, sive illa distantia asymptoti ab initio distantiæ binorum punctorum materiæ, in mutuis incursibus velocitatem deberent posse mutare per saltum, quod cum fieri nequeat, debet utique aliquis esse ultimus asymptoticus arcus, qui asymptotum habeat transeuntem per distantiarum initium, & vires inducat imminutis in infinitum distantiis crescentes in infinitum ita, ut sint pares velocitati extinguendæ cuivis, utcunque magnæ. Ad summum in curva virium haberi possent plures asymptotici arcus, alii post alios, habentes ad exigua intervalla asymptotos inter se parallelas, qui casus itidem uberrimum aperit contemplationibus fœcundissimis campum, de quo aliquid inferius; sed aliquis arcus asympto-[36]-ticus postremus, cujusmodi est is, quem in figura 1 proposui, haberi omnino debet. Verum ea perquisitione hic omissa, pergendum est in consideratione legis virium, & curvæ eam exprimentis, quæ habentur auctis distantiis.

Vim in majoribus distantiis esse attractivam, curva secante axem in aliquo limite.

78. In primis gravitas omnium corporum in Terram, quam quotidie experimur, satis evincit, repulsionem illam, quam pro minimis distantiis invenimus, non extendi ad distantias quascunque, sed in magnis jam distantiis haberi determinationem ad accessum, quam vim attractivam nominavimus. Quin immo Keplerianæ leges in Astronomia tam feliciter a Newtono adhibitæ ad legem gravitatis generalis deducendam, & ad cometas etiam traductæ,

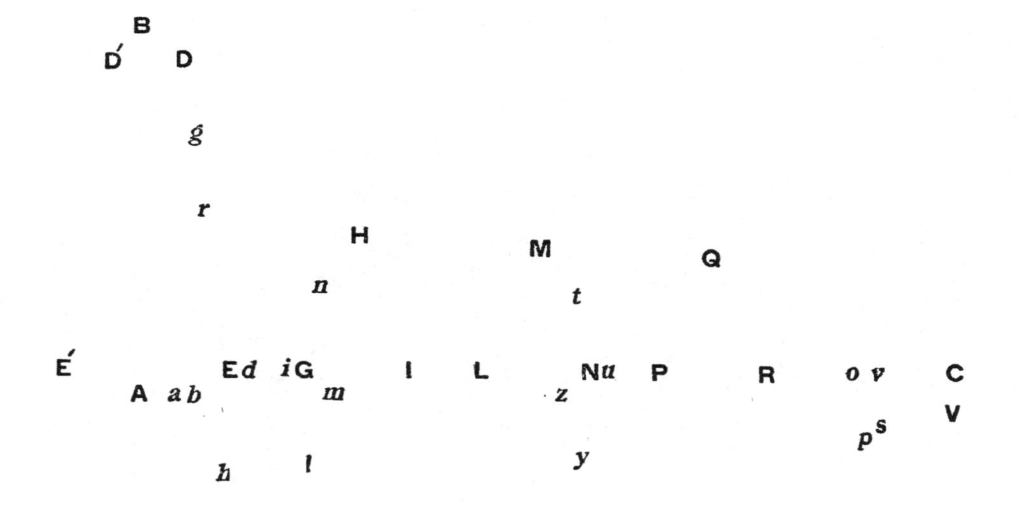

B
D′
D
g
r
H
M
Q
n
t
E′
A
a
b
E
d
i
G
m
I
L
z
N
u
P
R
o
v
C
V
p
S
h
l
y

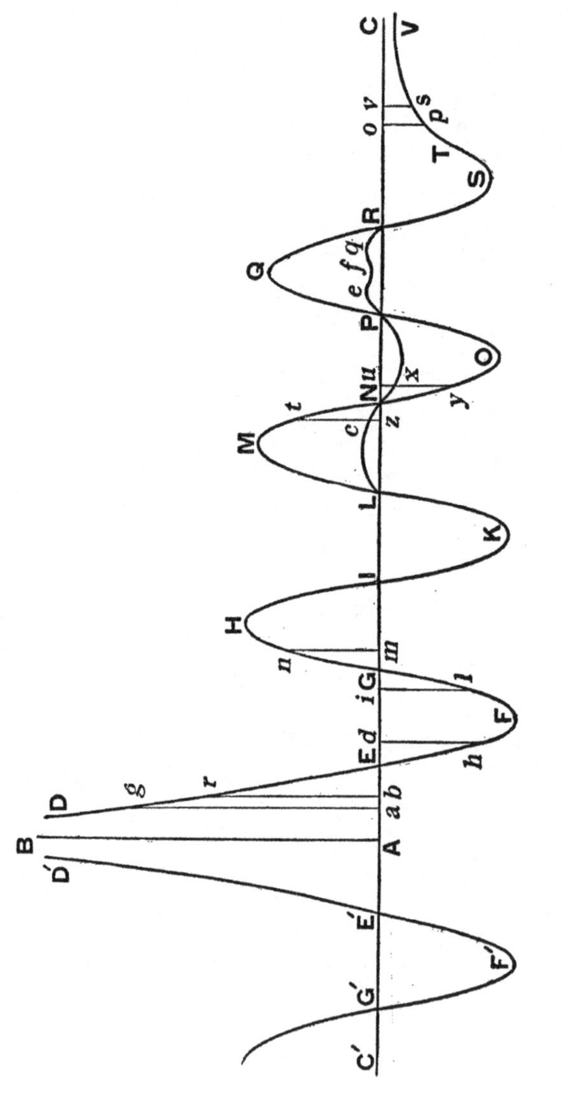

Fig. 1.

body in front still had its original 6 degrees; then they would come into contact with a difference of velocity greater than 8 degrees. For, it can also be proved by the fullest possible induction that all forces known to us, which act for any intervals of time so as to produce velocity, give effects that are proportional to the times for which they act, & also to the magnitudes of the forces themselves. This is confirmed by experiments with heavy bodies descending obliquely; the same things can be easily established in the case of springs so as to afford corroboration. Moreover it is the fundamental theorem of the whole of Mechanics, & from it are derived the laws of motion; these are confirmed by experiments with pendulums, projected weights, & many other things; they are corroborated also by astronomy in the matter of the motions of the heavenly bodies. Hence the repulsive force, which in the first case destroyed only 6 degrees difference of velocity, if it acts for a shorter time in the second case, will not be able to destroy aught but a less number of degrees, as the velocity produced in the two bodies in opposite directions is less. Now it certainly will act for a shorter time; for, owing to the greater difference of velocities, the relative velocity is greater & therefore the approach is faster. Hence, in the second case the force would destroy less than 6 degrees of the difference, if in the first case it had, just at contact, destroyed 6 degrees only. There would therefore be more than 8 degrees left over (for, between 20 & 6 there are 14) when contact happened, & then the velocities would have to be changed suddenly unless there was compenetration; & thereby the Law of Continuity would be violated. Since, then, this cannot be the case, Nature would be sure to guard against this trouble by a force of such a kind as that which, in the former case, extinguished the difference of velocity some time before contact; that is to say, so that, when the distances are still further diminished in the second case, a further force eliminates all that difference, all of the 14 degrees of difference that there were originally being destroyed.

The force must increase indefinitely, as the distances are diminished, also indefinitely; the curve of forces has an asymptote at the origin of abscissæ.

77. Now, after that we have been led so far, it is easy to go on further still & to consider that, what happens in the second case when compared with the first, will happen also in a third case, in which the velocity of the body that follows is once more increased, when compared with the second case; & so on, & so on. Hence, in order to guard against any sudden change at all in every case whatever, Nature will necessarily have taken measures for this purpose by means of a force of such a kind that, as the distances are diminished the force increases indefinitely, & in such a manner that it is capable of destroying any velocity, however great it may be. We have arrived therefore at repulsive forces that increase as the distances diminish, & increase indefinitely; that is to say, to the asymptotic arc, ED, of the curve of forces exhibited in Fig. 1. It is indeed true that by the reasoning given so far it is not immediately deduced that increments of the forces when increased to infinity correspond with the distances diminished to infinity. There may be for these bodies, such as we have in consideration, some fixed distance that acts as a boundary limit to forces that increase indefinitely; in this case the asymptote AB will not pass through the beginning of the distance between the two bodies, but at an interval after it as great as the least limit of all distances that particles, originally more remote, might acquire from one another. But, that there is some final asymptotic arc of the curve having for its asymptote the straight line passing through the very beginning of the distance, is proved as follows. If there were no arc of this kind, then the smaller particles of matter, originally collected at a distance less than this final limit would be, i.e., less than the distance of the asymptote from the beginning of the distance between the two points of matter, must be capable of having their velocities, on collision with one another, suddenly changed. Now, as this is impossible, then at any rate there must be some asymptotic arc, which has an asymptote passing through the very beginning of the distances; & this leads us to forces that, as the distances are indefinitely diminished, increase indefinitely in such a way that they are capable of destroying any velocity, no matter how large it may be. In general, in a curve of forces there may be several asymptotic arcs, one after the other, having at short intervals asymptotes parallel to one another; & this case also opens up a very rich field for fruitful investigations, about which I will say something later. But there must certainly be some one final asymptotic arc of the kind that I have given in Fig. 1. However, putting this investigation on one side, we must get on with the consideration of the law of forces, & the curve that represents them, which are obtained when the distances are increased.

The force at greater distances is attractive, the curve cutting the axis at some limit point.

78. First of all, the gravitation of all bodies towards the Earth, which is an everyday experience, proves sufficiently that the repulsion that we found for very small distances does not extend to all distances; but that at distances that are now great there is a propensity for approach, which we have called an attractive force. Moreover the Keplerian Laws in astronomy, so skilfully employed by Newton to deduce the law of universal gravitation, & applied even to the comets, show perfectly well that gravitation extends,

satis ostendunt, gravitatem vel in infinitum, vel saltem per totum planetarium, & cometarium systema extendi in ratione reciproca duplicata distantiarum. Quamobrem virium curva arcum habet aliquem jacentem ad partes axis oppositas, qui accedat, quantum sensu percipi possit, ad eam tertii gradus hyperbolam, cujus ordinatæ sunt in ratione reciproca duplicata distantiarum, qui nimirum est ille arcus STV figuræ 1. Ac illud etiam hinc patet, esse aliquem locum E, in quo curva ejusmodi axem secet, qui sit limes attractionum, & repulsionum, in quo ab una ad alteram ex iis viribus transitus fiat.

Plures esse debere, immo plurimos transitus, & limites.

79. Duos alios nobis indicat limites ejusmodi, sive alias duas intersectiones, ut G & I, phænomenum vaporum, qui oriuntur ex aqua, & acris, qui a fixis corporibus gignitur; cum in iis ante nulla particularum repulsio fuerit, quin immo fuerit attractio, ob cohærentiam, qua, una parte retracta, altera ipsam consequebatur, & in illa tanta expansione, & elasticitatis vi satis se manifesto prodat repulsio, ut idcirco a repulsione in minimis distantiis ad attractionem alicubi sit itum, tum inde iterum ad repulsionem, & iterum inde ad generalis gravitatis attractiones. Effervescentiæ, & fermentationes adeo diversæ, in quibus cum adeo diversis velocitatibus cunt, ac redeunt, & jam ad se invicem accedunt, jam recedunt a se invicem particulæ, indicant utique ejusmodi limites, atque transitus multo plures; sed illos prorsus evincunt substantiæ molles, ut cera, in quibus compressiones plurimæ acquiruntur cum distantiis admodum adversis, in quibus, tamen omnibus limites haberi debent; nam, anteriore parte ad se attracta, posteriores eam sequuntur, eadem propulsa, illæ recedunt, distantiis ad sensum non mutatis, quod ob illas repulsiones in minimis distantiis, quæ contiguitatem impediunt, fieri alio modo non potest, nisi si limites ibidem habeantur in iis omnibus distantiis inter attractiones, & repulsiones, quæ nimirum requiruntur ad hoc, ut pars altera alteram consequatur retractam, vel præcedat propulsam.

Hinc tota curvæ forma cum binis asymptotis, & pluribus flexibus, ac sectionibus.

80. Habentur igitur plurimi limites, & plurimi flexus curvæ hine, & inde ab axe præter duos arcus, quorum prior ED in infinitum protenditur, & asymptoticus est, alter STV, [37] si gravitas generalis in infinitum protenditur, est asymptoticus itidem, & ita accedit ad crus illud hyperbolæ gradus tertii, ut discrimen sensu percipi nequeat: nam cum ipso penitus congruere omnino non potest; non enim posset ab eodem deinde discedere, cum duarum curvarum, quarum diversa natura est, nulli arcus continui, utcunque exigui, possint penitus congruere, sed se tantummodo secare, contingere, osculari possint in punctis quotcunque, & ad se invicem accedere utcumque. Hine habetur jam tota forma curvæ virium, qualem initio proposui, directa ratiocinatione a Naturæ phænomenis, & genuinis principiis deducta. Remanet jam determinanda constitutio primorum elementorum materiæ ab iis viribus deducta, quo facto omnis illa Theoria, quam initio proposui, patebit, nec erit arbitraria quædam hypothesis, ac licebit progredi ad amovendas apparentes quasdam difficultates, & ad uberrimam applicationem ad omnem late Physicam qua exponendam, qua tantummodo, ne hoc opus plus æquo excrescat, indicandam.

Hinc elementorum primorum materiæ simplicitas carens partibus.

81. Quoniam, imminutis in infinitum distantiis, vis repulsiva augetur in infinitum; facile patet, nullam partem materiæ posse esse contiguam alteri parti: vis enim illa repulsiva protinus alteram ab altera removeret. Quamobrem necessario inde consequitur, prima materiæ elementa esse omnino simplicia, & a nullis contiguis partibus composita. Id quidem immediate, & necessario fluit ex illa constitutione virium, quæ in minimis distantiis sunt repulsivæ, & in infinitum excrescunt.

Solutio objectionis petitæ ex eo, quod vires repulsivas habere possent non puncta singula, se particulæ primigeniæ.

82. Objicit hic fortasse quispiam illud, fieri posse, ut particulæ primigeniæ materiæ sint compositæ quidem, sed nulla Naturæ vi divisibiles a se invicem, quarum altera tota respectu alterius totius habeat vires illas in minimis distantiis repulsivas, vel quarum pars quævis respectu reliquarum partium ejusdem particulæ non solum nullam habeat repulsivam vim, sed habeat maximam illam attractivam, quæ ad ejusmodi cohæsionem requiritur: eo pacto evitari debere quemvis immediatum impulsum, adeoque omnem saltum, & continuitatis læsionem. At in primis id esset contra homogeneitatem materiæ, de qua agemus infra: nam eadem materiæ pars in iisdem distantiis respectu quarundam paucissimarum partium, cum quibus particulam suam componit, haberet vim repulsivam, respectu autem

either to infinity or at least to the limits of the system including all the planets & comets, in the inverse ratio of the squares of the distances. Hence the curve will have an arc lying on the opposite side of the axis, which, as far as can be perceived by our senses, approximates to that hyperbola of the third degree, of which the ordinates are in the inverse ratio of the squares of the distances; & this indeed is the arc STV in Fig. 1. Now from this it is evident that there is some point E, in which a curve of this kind cuts the axis; and this is a limit-point for attractions and repulsions, at which the passage from one to the other of these forces is made.

There are bound to be many, nay, very many of these passages, with corresponding limit-points.

79. The phenomenon of vapour arising from water, & that of gas produced from fixed bodies lead us to admit two more of these limit-points, i.e., two other intersections, say, at G & I. Since in these there would be initially no repulsion, nay rather there would be an attraction due to cohesion, by which, when one part is retracted, another generally followed it: & since in the former, repulsion is clearly evidenced by the greatness of the expansion, & by the force of its elasticity; it therefore follows that there is, somewhere or other, a passage from repulsion at very small distances to attraction, then back again to repulsion, & from that back once more to the attractions of universal gravitation. Effervescences & fermentations of many different kinds, in which the particles go & return with as many different velocities, & now approach towards & now recede from one another, certainly indicate many more of these limit-points & transitions. But the existence of these limit-points is perfectly proved by the case of soft substances like wax; for in these substances a large number of compressions are acquired with very different distances, yet in all of these there must be limit-points. For, if the front part is drawn out, the part behind will follow; or if the former is pushed inwards, the latter will recede from it, the distances remaining approximately unchanged. This, on account of the repulsions existing at very small distances, which prevent contiguity, cannot take place in any way, unless there are limit-points there in all those distances between attractions & repulsions; namely, those that are requisite to account for the fact that one part will follow the other when the latter is drawn out, & will recede in front of the latter when that is pushed in.

Hence we get the whole form of the curve, with two asymptotes, many flexures & many intersections with the axis.

80. Therefore there are a large number of limit-points, & a large number of flexures on the curve, first on one side & then on the other side of the axis, in addition to two arcs, one of which, ED, is continued to infinity & is asymptotic, & the other, STV, is asymptotic also, provided that universal gravitation extends to infinity. It approximates to the form of the hyperbola of the third degree mentioned above so closely that the difference from it is imperceptible; but it cannot altogether coincide with it, because, in that case it would never depart from it. For, of two curves of different nature, there cannot be any continuous arcs, no matter how short, that absolutely coincide; they can only cut, or touch, or osculate one another in an indefinitely great number of points, & approximate to one another indefinitely closely. Thus we now have the whole form of the curve of forces, of the nature that I gave at the commencement, derived by a straightforward chain of reasoning from natural phenomena, & sound principles. It only remains for us now to determine the constitution of the primary elements of matter, derived from these forces; & in this manner the whole of the Theory that I enunciated at the start will become quite clear, & it will not appear to be a mere arbitrary hypothesis. We can proceed to remove certain apparent difficulties, & to apply it with great profit to the whole of Physics in general, explaining some things fully &, to prevent the work from growing to an unreasonable size, merely mentioning others.

The simplicity of the primary elements of matter; they are altogether without parts.

81. Now, because the repulsive force is indefinitely increased when the distances are indefinitely diminished, it is quite easy to see clearly that no part of matter can be contiguous to any other part; for the repulsive force would at once separate one from the other. Therefore it necessarily follows that the primary elements of matter are perfectly simple, & that they are not composed of any parts contiguous to one another. This is an immediate & necessary deduction from the constitution of the forces, which are repulsive at very small distances & increase indefinitely.

Solution of the objection derived from the assertion that single points cannot have repulsive forces, but that primary particles can have them.

82. Perhaps someone will here raise the objection that it may be that the primary particles of matter are composite, but that they cannot be disintegrated by any force in Nature; that one whole with regard to another whole may possibly have those forces that are repulsive at very small distances, whilst any one part with regard to any other part of the same particle may not only have no repulsive force, but indeed may have a very great attractive force such as is required for cohesion of this sort; that, in this way, we are bound to avoid all immediate impulse, & so any sudden change or breach of continuity. But, in the first place, this would be in opposition to the homogeneity of matter, which we will consider later; for the same part of matter, at the same distances with regard to those very few parts, along with which it makes up the particle, would have a repulsive

aliarum omnium attractivam in iisdem distantiis, quod analogiæ adversatur. Deinde si a Deo agente supra vires Naturæ sejungerentur illæ partes a se invicem, tum ipsius Naturæ vi in se invicem incurrerent; haberetur in carum collisione saltus naturalis, utut præsupponens aliquid factum vi agente supra Naturam. Demum duo tum cohæsionum genera deberent haberi in Natura admodum diversa, alterum per attractionem in minimis distantiis, alterum vero longe alio pacto in elementarium particularum massis, nimirum per limites cohæsionis; adeoque multo minus simplex, & minus uniformis evaderet Theoria.

An elementa sint extensa: argumenta pro virtuali eorum extensione.

[38] 83. Simplicitate & incompositione elementorum definita, dubitari potest, an ea sint etiam inextensa, an aliquam, utut simplicia, extensionem habeant ejus generis, quam virtualem extensionem appellant Scholastici. Fuerunt enim potissimum inter Peripateticos, qui admiserint clementa simplicia, & carentia partibus, atque ex ipsa natura sua prorsus indivisibilia, sed tamen extensa per spatium divisibile ita, ut alia aliis majus etiam occupent spatium, ac eo loco, quo unum stet, possint, eo remoto, stare simul duo, vel etiam plura; ac sunt etiamnum, qui ita sentiant. Sic etiam animam rationalem hominis utique prorsus indivisibilem censuerunt alii per totum corpus diffusam: alii minori quidem corporis parti, sed utique parti divisibili cuipiam, & extensæ, præsentem toti etiamnum arbitrantur. Deum autem ipsum præsentem ubique credimus per totum utique divisibile spatium, quod omnia corpora occupant, licet ipse simplicissimus sit, nec ullam prorsus compositionem admittat. Videtur autem sententia eadem inniti cuidam etiam analogiæ loci, ac temporis. Ut enim quies est conjunctio ejusdem puncti loci cum serie continua omnium momentorum ejus temporis, quo quies durat: sic etiam illa virtualis extensio est conjunctio unius momenti temporis cum serie continua omnium punctorum spatii, per quod simplex illud ens virtualiter extenditur; ut idcirco sicut illa quies haberi creditur in Natura, ita & hæc virtualis extensio debeat admitti, qua admissa poterunt utique illa primæ materiæ clementa esse simplicia, & tamen non penitus inextensa.

Excluditur virtualis extensio principio inductionis rite applicato.

84. At ego quidem arbitror, hanc itidem sententiam everti penitus eodem inductionis principio, ex quo alia tam multa hucusque, quibus usi sumus, deduximus. Videmus enim in his corporibus omnibus, quæ observare possumus, quidquid distinctum occupat locum, distinctum esse itidem ita, ut etiam satis magnis viribus adhibitis separari possint, quæ diversas occupant spatii partes, nec ullum casum deprehendimus, in quo magna hæc corpora partem aliquam habeant, quæ eodem tempore diversas spatii partes occupet, & eadem sit. Porro hæc proprietas ex natura sua ejus generis est, ut æque cadere possit in magnitudines, quas per sensum deprehendimus, ac in magnitudines, quæ infra sensuum nostrorum limites sunt; res nimirum pendet tantummodo a magnitudine spatii, per quod haberetur virtualis extensio, quæ magnitudo si esset satis ampla, sub sensus caderet. Cum igitur nunquam id comperiamus in magnitudinibus sub sensum cadentibus, immo in casibus innumeris deprehendamus oppositum: debet utique res transferri ex inductionis principio supra exposito ad minimas etiam quasque materiæ particulas, ut ne illæ quidem ejusmodi habeant virtualem extensionem.

Responsio ad exemplum animæ & Dei.

[39] 85. Exempla, quæ adduntur, petita ab anima rationali, & ab omnipræsentia Dei, nihil positive evincunt, cum ex alio entium genere petita sint; præterquam quod nec illud demonstrari posse censeo, animam rationalem non esse unico tantummodo, simplici, & inextenso corporis puncto ita præsentem, ut eundem locum obtineat, exerendo inde vires quasdam in reliqua corporis puncta rite disposita, in quibus viribus partim necessariis, & partim liberis, stet ipsum animæ commercium cum corpore. Dei autem præsentia cujusmodi sit, ignoramus omnino; quem sane extensum per spatium divisibile nequaquam dicimus, nec ab iis modis omnem excedentibus humanum captum, quibus ille existit, cogitat, vult, agit, ad humanos, ad materiales existendi, agendique modos, ulla esse potest analogia, & deductio.

Itidem ad analogiam cum quiete.

86. Quod autem pertinet ad analogiam cum quiete, sunt sane satis valida argumenta, quibus, ut supra innui, ego censeam, in Natura quietem nullam existere. Ipsam nec posse

force; but it would have an attractive force with regard to all others, at the very same distances; & this is in opposition to analogy. Secondly, if, due to the action of God surpassing the forces of Nature, those parts are separated from one another, then urged by the forces of Nature they would rush towards one another; & we should have, from their collision, a sudden change appertaining to Nature, although conveying a presumption that something was done by the action of a supernatural force. Lastly, with this idea, there would have to be two kinds of cohesion in Nature that were altogether different in constitution; one due to attraction at very small distances, & the other coming about in a far different way in the case of masses of elementary particles, that is to say, due to the limit-points of cohesion. Thus a theory would result that is far less simple & less uniform than mine.

Whether the elements are extended; certain arguments in favour of virtual extension.

83. Taking it for granted, then, that the elements are simple & non-composite, there can be no doubt as to whether they are also non-extended or whether, although simple, they have an extension of the kind that is termed virtual extension by the Scholastics. For there were some, especially among the Peripatetics, who admitted elements that were simple, lacking in all parts, & from their very nature perfectly indivisible; but, for all that, so extended through divisible space that some occupied more room than others; & such that in the position once occupied by one of them, if that one were removed, two or even more others might be placed at the same time; & even now there are some who are of the same opinion. So also some thought that the rational soul in man, which certainly is altogether indivisible, was diffused throughout the whole of the body; whilst others still consider that it is present throughout the whole of, indeed, a smaller part of the body, but yet a part that is at any rate divisible & extended. Further we believe that God Himself is present everywhere throughout the whole of the undoubtedly divisible space that all bodies occupy; & yet He is onefold in the highest degree & admits not of any composite nature whatever. Moreover, the same idea seems to depend on an analogy between space & time. For, just as rest is a conjunction with a continuous series of all the instants in the interval of time during which the rest endures; so also this virtual extension is a conjunction of one instant of time with a continuous series of all the points of space throughout which this one-fold entity extends virtually. Hence, just as rest is believed to exist in Nature, so also are we bound to admit virtual extension; & if this is admitted, then it will be possible for the primary elements of matter to be simple, & yet not absolutely non-extended.

Virtual extension is excluded by the principle of induction correctly applied.

84. But I have come to the conclusion that this idea is quite overthrown by that same principle of induction, by which we have hitherto deduced so many results which we have employed. For we see, in all those bodies that we can bring under observation, that whatever occupies a distinct position is itself also a distinct thing; so that those that occupy different parts of space can be separated by using a sufficiently large force; nor can we detect a case in which these larger bodies have any part that occupies different parts of space at one & the same time, & yet is the same part. Further, this property by its very nature is of the sort for which it is equally probable that it happens in magnitudes that we can detect by the senses & in magnitudes which are below the limits of our senses. In fact, the matter depends only upon the size of the space, throughout which the virtual extension is supposed to exist; & this size, if it were sufficiently ample, would become sensible to us. Since then we never find this virtual extension in magnitudes that fall within the range of our senses, nay rather, in innumerable cases we perceive the contrary; the matter certainly ought to be transferred by the principle of induction, as explained above, to any of the smallest particles of matter as well; so that not even they are admitted to have such virtual extension.

Reply to the parallel case of the Soul & God.

85. The illustrations that are added, derived from a consideration of the rational soul & the omnipresence of God, prove nothing positively; for they are derived from another class of entities, except that, I do not think that it can even be proved that the rational soul does not exist in merely a single, simple, & non-extended point of the body; so that it maintains the same position, & thence it puts forth some sort of force into the remaining points of the body duly disposed about it; & the intercommunication between the soul & the body consists of these forces, some of which are involuntary whilst others are voluntary. Further, we are absolutely ignorant of the nature of the presence of God; & in no wise do we say that He is really extended throughout divisible space; nor from those modes, surpassing all human intelligence, by which He exists, thinks, wills & acts, can any analogy or deduction be made which will apply to human or material modes of existence & action.

Again with regard to the analogy with rest.

86. Again, as regards the analogy with rest, we have arguments that are sufficiently strong to lead us to believe, as I remarked above, that there is no such thing in Nature as absolute rest. Indeed, I proved that such a thing could not be, by a direct argument

existere, argumento quodam positivo ex numero combinationum possibilium infinito contra alium finitum, demonstravi in Stayanis Supplementis, ubi de spatio, & tempore quæ juxta num. 66 occurrent infra Supplementorum § 1, & § 2; numquam vero eam existere in Natura, patet sane in ipsa Newtoniana sententia de gravitate generali, in qua in planetario systemate ex mutuis actionibus quiesçit tantummodo centrum commune gravitatis, punctum utique imaginarium, circa quod omnia planetarum, cometarumque corpora moventur, ut & ipse Sol; ac idem accidit fixis omnibus circa suorum systematum gravitatis centra; quin immo ex actione unius systematis in aliud utcunque distans, in ipsa gravitatis centra motus aliquis inducetur; & generalius, dum movetur quæcunque materiæ particula, uti luminis particula quæcunque; reliquæ omnes utcunque remotæ, quæ inde positionem ab illa mutant, mutant & gravitatem, ac proinde moventur motu aliquo exiguo, sed sane motu. In ipsa Telluris quiescentis sententia, quiescit quidem Tellus ad sensum, nec tota ab uno in alium transfertur locum; at ad quamcunque crispationem maris, rivuli decursum, muscæ volatum, æquilibrio dempto, trepidatio oritur, perquam exigua illa quidem, sed ejusmodi, ut veram quietem omnino impediat. Quamobrem analogia inde petita evertit potius virtualem ejusmodi simplicium elementorum extensionem positam in conjunctione ejusdem momenti temporis cum serie continua punctorum loci, quam comprobet.

In quo deficiat analogia loci, & temporis.

87. Sed nec ea ipsa analogia, si adesset, rem satis evinceret; cum analogiam inter tempus, & locum videamus in aliis etiam violari: nam in iis itidem paragraphis Supplementorum demonstravi, nullum materiæ punctum unquam redire ad punctum spatii quodcunque, in quo semel fuerit aliud materiæ punctum, ut idcirco duo puncta materiæ nunquam conjungant idem [40] punctum spatii ne cum binis quidem punctis temporis, dum quamplurima binaria punctorum materiæ conjungunt idem punctum temporis cum duobus punctis loci; nam utique coexistunt: ac præterea tempus quidem unicam dimensionem habet diuturnitatis, spatium vero habet triplicem, in longum, latum, atque profundum.

Inextensio utilis ad excludendum transitum momentaneum a densitate nulla ad summam.

88. Quamobrem illud jam tuto inferri potest, hæc primigenia materiæ elementa, non solum esse simplicia, ac indivisibilia, sed etiam inextensa. Et quidem hæc ipsa simplicitas, & inextensio elementorum præstabit commoda sane plurima, quibus eadem adhuc magis fulcitur, ac comprobatur. Si enim prima elementa materiæ sint quædam partes solidæ, ex partibus compositæ, vel etiam tantummodo extensæ virtualiter, dum a vacuo spatio motu continuo pergitur per unam ejusmodi particulam, fit saltus quidam momentaneus a densitate nulla, quæ habetur in vacuo, ad densitatem summam, quæ habetur, ubi ea particula spatium occupat totum. Is vero saltus non habetur, si elementa simplicia sint, & inextensa, ac a se invicem distantia. Tum enim omne continuum est vacuum tantummodo, & in motu continuo per punctum simplex fit transitus a vacuo continuo ad vacuum continuum. Punctum illud materiæ occupat unicum spatii punctum, quod punctum spatii est indivisibilis limes inter spatium præcedens, & consequens. Per ipsum non immoratur mobile continuo motu delatum, nec ad ipsum transit ab ullo ipsi immediate proximo spatii puncto, cum punctum puncto proximum, uti supra diximus, nullum sit; sed a vacuo continuo ad vacuum continuum transitur per ipsum spatii punctum a materiæ puncto occupatum.

Itidem ad hoc, ut densitatis augeri possit, ut potest minui in infinitum.

89. Accedit, quod in sententia solidorum, extensorumque elementorum habetur illud, densitatem corporis minui posse in infinitum, augeri autem non posse, nisi ad certum limitem in quo incrementi lex necessario abrumpi debeat. Primum constat ex eo, quod eadem particula continua dividi possit in particulas minores quotcunque, quæ idcirco per spatium utcunque magnum diffundi potest ita, ut nulla earum sit, quæ aliquam aliam non habeat utcunque libuerit parum a se distantem. Atque eo pacto aucta mole, per quam eadem illa massa diffusa sit, eaque aucta in ratione quacunque minuetur utique densitas in ratione itidem utcunque magna. Patet & alterum: ubi enim omnes particulæ ad contactum devenerint; densitas ultra augeri non poterit. Quoniam autem determinata quædam erit utique ratio spatii vacui ad plenum, nonnisi in ea ratione augeri poterit densitas, cujus augmentum, ubi ad contactum deventum fuerit, adrumpetur. At si elementa sint puncta penitus indivisibilia, & inextensa; uti augeri eorum distantia poterit in infinitum, ita utique poterit etiam minui pariter in ratione quacunque; cum

founded upon the infiniteness of a number of possible combinations as against the finiteness of another number, in the Supplements to Stay's Philosophy, in connection with space & time; these will be found later immediately after Art. 14 of the Supplements, §§ I and II. That it never does exist in Nature is really clear in the Newtonian theory of universal gravitation; according to this theory, in the planetary system the common centre of gravity alone is at rest under the action of the mutual forces; & this is an altogether imaginary point, about which all the bodies of the planets & comets move, as also does the sun itself. Moreover the same thing happens in the case of all the fixed stars with regard to the centres of gravity of their systems; & from the action of one system on another at any distance whatever from it, some motion will be imparted to these very centres of gravity. More generally, so long as any particle of matter, so long as any particle of light, is in motion, all other particles, no matter how distant, which on account of this motion have their distance from the first particle altered, must also have their gravitation altered, & consequently must move with some very slight motion, but yet a true motion. In the idea of a quiescent Earth, the Earth is at rest approximately, nor is it as a whole translated from place to place; but, due to any tremulous motion of the sea, the downward course of rivers, even to the fly's flight, equilibrium is destroyed & some agitation is produced, although in truth it is very slight; yet it is quite enough to prevent true rest altogether. Hence an analogy deduced from rest contradicts rather than corroborates virtual extension of the simple elements of Nature, on the hypothesis of a conjunction of the same instant of time with a continuous series of points of space.

Where the analogy of space and time fails.

87. But even if the foregoing analogy held good, it would not prove the matter satisfactorily; since we see that in other ways the analogy between space & time is impaired. For I proved, also in those paragraphs of the Supplements that I have mentioned, that no point of matter ever returned to any point of space, in which there had once been any other point of matter; so that two points of matter never connected the same point of space with two instants of time, let alone with more; whereas a huge number of pairs of points connect the same instant of time with two points of space, since they certainly coexist. Besides, time has but one dimension, duration; whilst space has three, length, breadth & depth.

Non-extension useful in excluding an instantaneous Passage from 'no' density to a very great one.

88. Therefore it can now be safely accepted that these primary elements of matter are not only simple & indivisible, but also that they are non-extended. Indeed this very simplicity & non-extension of the elements will prove useful in a really large number of cases for still further strengthening & corroborating the results already obtained. For if the primary elements were certain solid parts, themselves composed of parts or even virtually extended only, then, whilst we pass by a continuous motion from empty space through one particle of this kind, there would be a sudden change from a density that is nothing when the space is empty, to a density that is very great when the particle occupies the whole of the space. But there is not this sudden change if we assume that the elements are simple, non-extended & non-adjacent. For then the whole of space is merely a continuous vacuum, &, in the continuous motion by a simple point, the passage is made from continuous vacuum to continuous vacuum. The one point of matter occupies but one point of space; & this point of space is the indivisible boundary between the space that precedes & the space that follows. There is nothing to prevent the moving point from being carried through it by a continuous motion, nor from passing to it from any point of space that is in immediate proximity to it: for, as I remarked above, there is no point that is the next point to a given point. But from continuous vacuum to continuous vacuum the passage is made through that point of space which is occupied by the point of matter.

Also for the idea that density can be increased, as it can be decreased, indefinitely.

89. There is also the point, that arises in the theory of solid extended elements, namely that the density of a body can be diminished indefinitely, but cannot be increased except up to a certain fixed limit, at which the law of increase must be discontinuous. The first comes from the fact that this same continuous particle can be divided into any number of smaller particles; these can be diffused through space of any size in such a way that there is not one of them that does not have some other one at some little (as little as you will) distance from itself. In this way the volume through which the same mass is diffused is increased; & when that is increased in any ratio whatever, then indeed the density will be diminished in the same ratio, no matter how great the ratio may be. The second thing is also evident; for when the particles have come into contact, the density cannot be increased any further. Moreover, since there will undoubtedly be a certain determinate ratio for the amount of space that is empty compared with the amount of space that is full, the density can only be increased in that ratio; & the regular increase of density will be arrested when contact is attained. But if the elements are points that are perfectly indivisible & non-extended, then, just as their distances can be increased indefinitely,

in [41] ratione quacunque lineola quæcunque secari sane possit: adeoque uti nullus est limes raritatis anetæ, ita etiam nullus erit auctæ densitatis.

Et ad excludendum continuum extensum, & in infinitum in existentibus.

90. Sed & illud commodum accidet, quod ita omne continuum coexistens eliminabitur e Natura, in quo explicando usque adeo desudarunt, & fere incassum, Philosophi, nec idcirco divisio ulla realis entis in infinitum produci poterit, nec hærebitur, ubi quæratur, an numerus partium actu distinctarum, & separabilium, sit finitus, an infinitus; nec alia ejusmodi sane innumera, quæ in continui compositione usque adeo negotium facessunt Philosophis, jam habebuntur. Si enim prima materiæ elementa sint puncta penitus inextensa, & indivisibilia, a se invicem aliquo intervallo disjuncta; jam erit finitus punctorum numerus in quavis massa: nam distantiæ omnes finitæ erunt; infinitesimas enim quantitates in se determinatas nullas esse, satis ego quidem, ut arbitror, luculenter demonstravi & in dissertatione *De Natura, & Usu infinitorum, ac infinite parvorum*, & in dissertatione *De Lege Continuitatis*, & alibi. Intervallum quodcunque finitum erit, & divisibile utique in infinitum per interpositionem aliorum, atque aliorum punctorum, quæ tamen singula, ubi fuerint posita, finita itidem erunt, & aliis pluribus, finitis tamen itidem, ubi extiterint, locum reliquent, ut infinitum sit tantummodo in possibilibus, non autem in existentibus, in quibus possibilibus ipsis omnem possibilium seriem idcirco ego appellare soleo constantem terminis finitis in infinitum, quod quæcunque, quæ existant, finita esse debeant, sed nullus sit existentium finitus numerus ita ingens, ut alii, & alii majores, sed itidem finiti, haberi non possint, atque id sine ullo limite, qui nequeat præteriri. Hoc autem pacto, sublato ex existentibus omni actuali infinito, innumeræ sane difficultates auferentur.

Inextensionem admitti oportere; quærendum de homogeneitate.

91. Cum igitur & positivo argumento, a lege virium positive demonstrata desumpto, simplicitas, & inextensio primorum materiæ elementorum deducatur, & tam multis aliis vel indiciis fulciatur, vel emolumentis inde derivatis confirmetur; ipsa itidem admitti jam debet, ac supererit quærendum illud tantummodo, utrum hæc elementa homogenea censeri debeant, & inter se prorsus similia, ut ea initio assumpsimus, an vero heterogenea, ac dissimilia.

Homogeneitatem suaderi ab homogeneitate primi, & ultimi asymptotici cruris pro punctis omnibus.

92. Pro homogeneitate primorum materiæ elementorum illud est quoddam veluti principium, quod in simplicitate, & inextensione conveniant, ac etiam vires quasdam habeant utique omnia. Deinde curvam ipsam virium eandem esse omnino in omnibus illud indicat, vel etiam evincit, quod primum crus repulsivum impenetrabilitatem secum trahens, & postremum attractivum gravitatem definiens, omnino communia in omnibus sint: nam corpora omnia æque impenetrabilia sunt, & vero etiam æque gravia pro quantitate materiæ suæ, uti satis [42] evincit æqualis velocitas auri, & plumæ cadentis in Boyliano recipiente Si reliquus curvæ arcus intermedius esset difformis in diversis materiæ punctis; infinities probabilius esset, difformitatem extendi etiam ad crus primum, & ultimum, cum infinities plures sint curvæ, quæ, cum in reliquis differant partibus, differant plurimum etiam in hisce extremis, quam quæ in hisce extremis tantum modo tam arcte consentiant. Et hoc quidem argumento illud etiam colligitur, curvam virium in quavis directione ab eodem primo materiæ elemento, nimirum ab eodem materiæ puncto eandem esse, cum & primum impenetrabilitatis, & postremum gravitatis crus pro omnibus directionibus sit ad sensum idem. Cum primum in dissertatione *De Viribus Vivis* hanc Theoriam protuli, suspicabar diversitatem legis virium respondentis diversis directionibus; sed hoc argumento adi majorem simplicitatem, & uniformitatem deinde adductus sum. Diversitas autem legum virium pro diversis particulis, & pro diversis respectu ejusdem particulæ directionibus, habetur utique ex diverso numero, & positione punctorum eam componentium, qua de re inferius aliquid.

Nihil contra deduci ex principio indiscernibilium, & rationis sufficientis.

93. Nec vero huic homogeneitati opponitur inductionis principium, quo ipsam Leibnitiani oppugnare solent, nec principium rationis sufficientis, atque indiscernibilium, quod superius innui numero 3. Infinitam Divini Conditoris mentem, ego quidem omnino. arbitror, quod & tam multi Philosophi censuerunt, ejusmodi perspicacitatem habere, atque intuitionem quandam, ut ipsam etiam, quam individuationem appellant, omnino similium individuorum cognoscat, atque illa inter se omnino discernat. Rationis autem sufficientis

so also can they just as well be diminished in any ratio whatever. For it is certainly possible that a short line can be divided into parts in any ratio whatever; & thus, just as there is no limit to increase of rarity, so also there is none to increase of density.

Also for excluding the idea of a continuum in existing things, that is extended & infinite.

90. The theory of non-extension is also convenient for eliminating from Nature all idea of a coexistent continuum—to explain which philosophers have up till now laboured so very hard & generally in vain. Assuming non-extension, no division of a real entity can be carried on indefinitely; we shall not be brought to a standstill when we seek to find out whether the number of parts that are actually distinct & separable is finite or infinite; nor with it will there come in any of those other truly innumerable difficulties that, with the idea of continuous composition, have given so much trouble to philosophers. For if the primary elements of matter are perfectly non-extended & indivisible points separated from one another by some definite interval, then the number of points in any given mass must be finite; because all the distances are finite. I proved clearly enough, I think, in the dissertation *De Natura, & Usu infinitorum ac infinite parvorum*, & in the dissertation *De Lege Continuitatis*, & in other places, that there are no infinitesimal quantities determinate in themselves. Any interval whatever will be finite, & at least divisible indefinitely by the interpolation of other points, & still others; each such set however, when they have been interpolated, will be also finite in number, & leave room for still more; & these too, when they existed, will also be finite in number. So that there is only an infinity of possible points, but not of existing points; & with regard to these possible points, I usually term the whole series of possibles a series that ends at finite limits at infinity. This for the reason that any of them that exist must be finite in number; but there is no finite number of things that exist so great that other numbers, greater & greater still, but yet all finite, cannot be obtained; & that too without any limit, which cannot be surpassed. Further, in this way, by doing away with all idea of an actual infinity in existing things, truly countless difficulties are got rid of.

Non-extension must be admitted; we have now to investigate homogeneity.

91. Since therefore, by a direct argument derived from a law of forces that has been directly proved, we have both deduced the simplicity & non-extension of the primary elements of matter, & also we have strengthened the theory by evidence pointing towards it, or corroborated it by referring to the advantages to be derived from it; this theory ought now to be accepted as true. There only remains the investigation as to whether these elements ought to be considered to be homogeneous & perfectly similar to one another, as we assumed at the start, or whether they are really heterogeneous & dissimilar.

Homogeneity for all points to be advocated from a consideration of the homogeneity of the first & last asymptotic branches of the curve of forces.

92. In favour of the homogeneity of the primary elements of matter we have so to speak some foundation derived from the fact that all of them agree in simplicity & non-extension, & also that they are all endowed with forces of some sort. Now, that this curve of forces is exactly the same for all of them is indicated or even proved by the fact that the first repulsive branch necessitating impenetrability, & the last attractive branch determining gravitation, are exactly the same in all respects. For all bodies are equally impenetrable; & also all are equally heavy in proportion to the amount of matter contained in them, as is sufficiently proved by the equal velocity of the piece of gold & the feather when falling in Boyle's experiment. If the remaining intermediate arc of the curve were non-uniform for different points of matter, it would be infinitely more probable that the non-uniformity would extend also to the first & last branches also; for there are infinitely more curves which, when they differ in the remaining parts, also differ to the greatest extent in the extremes, than there are curves, which agree so closely only in these extremes. Also from this argument we can deduce that the curve of forces is indeed exactly the same from the same point of matter, in any direction whatever from the same primary element of matter; for both the first branch of impenetrability & the last branch of gravitation are the same, so far as we can perceive, for all directions. When I first published this Theory in my dissertation *De Viribus Vivis*, I was inclined to believe that there was a diversity in the law of forces corresponding to diversity of direction; but I was led by the argument given above to the greater simplicity & the greater uniformity derived therefrom. Further, diversity of the laws of forces for diverse particles, & for different directions with the same particle, is certainly to be obtained from the diverse number & position of the points composing it; about which I shall have something to say later.

Nothing to be brought against this from the doctrines of indiscernibles & 'sufficient reason.'

93. Nor indeed is there anything opposed to this idea of homogeneity to be derived from the principle of induction, by means of which the followers of Leibniz usually raise an objection to it; nor from the principle of sufficient reason, & of indiscernibles, that I mentioned above in Art. 3. I am indeed quite convinced, & a great many other philosophers too have thought, that the Infinite Will of the Divine Founder has a perspicacity & an intuition of such a nature that it takes cognizance of that which is called individuation amongst individuals that are perfectly similar, & absolutely

principium falsum omnino esse censeo, ac ejusmodi, ut omnem veræ libertatis ideam omnino tollat; nisi pro ratione, ubi agitur de voluntatis determinatione, ipsum liberum arbitrium, ipsa libera determinatio assumatur, quod nisi fiat in voluntate divina, quæcunque existunt, necessario existunt, & quæcunque non existunt, ne possibilia quidem erunt, vera aliqua possibilitate, uti facile admodum demonstratur; quod tamen si semel admittatur, mirum sane, quam prona demum ad fatalem necessitatem patebit via. Quamobrem potest divina voluntas determinari ex toto solo arbitrio suo ad creandum hoc individuum potius, quam illud ex omnibus omnino similibus, & ad ponendum quodlibet ex iis potius eo loco, quo ponit, quam loco alterius. Sed de rationis sufficientis principio hæc ipsa fusius pertractavi tum in aliis locis pluribus, tum in Stayanis Supplementis, ubi etiam illud ostendi, id principium nullum habere usum posse in iis ipsis casibus, in quibus adhibetur, & prædicari solet tantopere, atque id idcirco, quod nobis non innotescant rationes omnes, quas tamen oporteret utique omnes nosse ad hoc, ut eo principio uti possemus, affirmando, nullam esse rationem sufficientem pro hoc potius, quam pro illo [43] alio: sane in exemplo illo ipso, quod adhiberi solet, Archimedis hoc principio æquilibrium determinantis, ibidem ostendi, ex ignoratione causarum, sive rationum, quæ postea detectæ sunt, ipsum in suæ investigationis progressu errasse plurimum, deducendo per abusum ejus principii sphæricam figuram marium, ac Telluris.

Posse etiam puncta convenire in iis, differre in aliis.

94. Accedit & illud, quod illa puncta materiæ, licet essent prorsus similia in simplicitate, & extensione, ac mensura virium, pendentium a distantia, possent alias habere proprietates metaphysicas diversas inter se, nobis ignotas, quæ ipsa etiam apud ipsos Leibnitianos discriminarent.

Non valere hic principium inductionis a massis: eas deferre ex diversis combinationibus.

95. Quod autem attinet ad inductionem, quam Leibnitiani desumunt a dissimilitudine, quam observamus in rebus omnibus, cùm nimirum nusquam ex. gr. in amplissima silva reperire sit duo folia prorsus similia; ea sane me nihil movet; cum nimirum illud discrimen sit proprietas relativa ad rationem aggregati, & nostros sensus, quos singula materiæ clementa non afficiunt vi sufficiente ad excitandam in animo ideam, nisi multa sint simul, & in molem majorem excrescant. Porro scimus utique combinationes ejusdem numeri terminorum in immensum excrescere, si ille ipse numerus sit aliquanto major. Solis 24 litterulis Alphabeti diversimodo combinatis formantur voces omnes, quibus huc usque usa sunt omnia idiomata, quæ extiterunt, & quibus omnia illa, quæ possunt existere, uti possunt. Quid si numerus earum existeret tanto major, quanto major est numerus punctorum materiæ in quavis massa sensibili? Quod ibi diversus est litterarum diversarum ordo, id in punctis etiam prorsus homogeneis sunt positiones, & distantia, quibus variatis, variatur utique forma, & vis, qua sensus afficitur in aggregatis. Quanto major est numerus combinationum diversarum possibilium in massis sensibilibus, quam earum massarum, quas possumus observare, & inter se conferre (qui quidem ob distantias, & directiones in infinitum variabiles præscindendo ab æquilibrio virium, est infinitus, cum ipso æquilibrio est immensus); tanto major est improbabilitas duarum massarum omnino similium, quam omnium aliquantisper saltem inter se dissimilium.

Physica ratio discriminis in pluribus massis ut in foliis.

96. Et quidem accedit illud etiam, quod alicujus dissimilitudinis in aggregatis physicam quoque rationem cernimus in iis etiam casibus, in quibus maxime inter se similia esse deberent. Cum enim mutuæ vires ad distantias quascunque pertineant; status uniuscujusque puncti pendebit saltem aliquantisper a statu omnium aliorum punctorum, quæ sunt in Mundo. Porro utcunque puncta quædam sint parum a se invicem remota, uti sunt duo folia in eadem silva, & multo magis in eodem ramo; adhuc tamen non candem prorsus relationem distantiæ, & virium habent ad reliqua omnia materiæ puncta, quæ [44] sunt in Mundo, cum non eundem prorsus locum obtineant; & inde jam in aggregato discrimen aliquod oriri debet, quod perfectam similitudinem omnino impediat. Sed illud eam inducit magis, quod quæ maxime conferunt ad ejusmodi dispositionem, necessario respectu diversarum frondium diversa non nihil esse debeant. Omissa ipsa earum forma in semine, solares radii, humoris ad nutritionem necessarii quantitas, distantia, a qua debet is progredi, ut ad locum suum deveniat, aura ipsa, & agitatio inde orta, non sunt omnino similia, sed diversitatem aliquam habent, ex qua diversitas in massas inde efformatas redundat.

distinguishes them one from the other. Moreover, I consider that the principle of sufficient reason is altogether false, & one that is calculated to take away all idea of true freewill. Unless free choice or free determination is assumed as the basis of argument, in discussing the determination of will, unless this is the case with the Divine Will, then, whatever things exist, exist because they must do so, & whatever things do not exist will not even be possible, i.e., with any real possibility, as is very easily proved. Nevertheless, once this idea is accepted, it is truly wonderful how it tends to point the way finally to fatalistic necessity. Hence the Divine Will is able, of its own pleasure alone, to be determined to the creation of one individual rather than another out of a whole set of exactly similar things, & to the setting of any one of these in the place in which it puts it rather than in the place of another. But I have discussed these very matters more at length, besides several other places, in the Supplements to Stay's Philosophy; where I have shown that the principle cannot be employed in those instances in which it is used & generally so strongly asserted. The reason being that all possible reasons are not known to us; & yet they should certainly be known, to enable us to employ the principle by stating that there is no sufficient reason in favour of this rather than that other. In truth, in that very example of the principle generally given, namely, that of Archimedes' determination of equilibrium by means of it, I showed also that Archimedes himself had made a very big mistake in following out his investigation because of his lack of knowledge of causes or reasons that were discovered in later days, when he deduced a spherical figure for the seas & the Earth by an abuse of this principle.

It is possible for points also to agree in these properties but to disagree in others.

94. There is also this, that these points of matter, although they might be perfectly similar as regards simplicity & extension, & in having the measure of their forces dependent on their distances, might still have other metaphysical properties different from one another, & unknown to us; & these distinctions also are made by the followers of Leibniz.

The principle does not hold good here of induction from masses; they differ on account of different combinations of their parts.

95. As regards the induction which the followers of Leibniz make from the lack of similitude that we see in all things, (for instance such as that there never can be found in the largest wood two leaves exactly alike), their argument does not impress me in the slightest degree. For that distinction is a property that is concerned with reasoning for an aggregate, & also with our senses; & these senses single elements of matter cannot influence with sufficient force to excite an idea in the mind, except when there are many of them together at a time, & they develop into a mass of considerable size. Further it is well known that combinations of the same number of terms increase enormously, if that number itself increase a little. From the 24 letters of the alphabet alone, grouped together in different ways, are formed all the words that have hitherto been used in all expressions that have existed, or can possibly come into existence. What then if their number were increased to equal the number of points of matter in any sensible mass? Corresponding to the different order of the several letters in the one, we have in perfectly homogeneous points also different positions & distances; & if these are altered at least the form & the force, which affect our senses in the groups, are altered as well. How much greater is the number of different combinations that are possible in sensible masses than the number of those masses that we can observe & compare with one another (& this number, on account of the infinitely variable distances & directions of the forces, when equilibrium is precluded, is infinite, since including equilibrium it is very great); just so much greater is the improbability of two masses being exactly similar than of their being all at least slightly different from one another.

Physical reason for the difference in several masses, as in leaves.

96. There is also this point in addition; we discern a physical reason as well for some dissimilarity in groups for those cases too, in which they ought to be especially similar to one another. For since mutual forces pertain to all possible distances, the state of any one point will depend upon, at least in some slight degree, the state of all other points that are in the universe. Further, however short the distance between certain points may be, as of two leaves in the same wood, much more so on the same branch, still for all that they do not have quite the same relation as regards distance & forces as all the rest of the points of matter that are in the universe, because they do not occupy quite the same place. Hence in a group some distinction is bound to arise which will entirely prevent perfect similarity. Moreover this tendency is all the stronger, because those things which especially conduce to this sort of disposition must necessarily be somewhat different with regard to different leaves. For the form itself being absent in the seed, the rays of the sun, the quantity of moisture necessary for nutrition, the distance from which it has to proceed to arrive at the place it occupies, the air itself & the continual motion derived from this, these are not exactly similar, but have some diversity; & from this diversity there proceeds a diversity in the masses thus formed.

Similitudine qualicunque in aliquibus magis probari homogeneitatem, quam dissimilitudine heterogeneitatem.

97. Patet igitur, varietatem illam a numero pendere combinationum possibilium in numero punctorum necessario ad sensationem, & circumstantiarum, quæ ad formationem massæ sunt necessariæ, adeoque ejusmodi inductionem extendi ad elementa non posse. Quin immo illa tanta similitudo, quæ cum exigua dissimilitudine commixta invenitur in tam multis corporibus, indicat potius similitudinem ingentem in elementis. Nam ob tantum possibilium combinationum numerum, massæ elementorum etiam penitus homogeneorum debent a se invicem differre plurimum, adeoque si clementa heterogenea sint, in immensum majorem debent habere dissimilitudinem, quam ipsa prima elementa, ex quibus idcirco nullæ massæ, ne tantillum quidem, similes provenire deberent. Cum elementa multo minus dissimilia esse debeant, quam aggregata elementorum, multo magis ad elementorum homogeneitatem valere debet illa quæcunque similitudo, quam in corporibus observamus, potissimum in tam multis, quæ ad candem pertinent speciem, quam ad homogeneitatem eorundem tam exiguum illud discrimen, quod in aliis tam multis observatur. Rem autem penitus conficit illa tanta similitudo, qua superius usi sumus, in primo crure exhibente impenetrabilitatem, & in postremo exhibente gravitatem generalem, quæ crura cum ob hasce proprietates corporibus omnibus adeo generales, adeo inter se in omnibus similia sint, etiam reliqui arcus curvæ exprimentis vires omnimodam similitudinem indicant pro corporibus itidem omnibus.

Homogeneitatem ab analysi Naturæ insinuari: exemplum a libris, litteris, punctulis.

98. Superest, quod ad hanc rem pertinet, illud unum iterum hic monendum, quod ipsum etiam initio hujus Operis innui, ipsam Naturam, & ipsum analyseos ordinem nos ducere ad simplicitatem & homogeneitatem elementorum, cum nimirum, quo analysis promovetur magis, eo ad pauciora, & inter se minus discrepantia principia deveniatur, uti patet in resolutionibus Chemicis. Quam quidem rem ipsum litterarum, & vocum exemplum multo melius animo sistet. Fieri utique possent nigricantes litteræ, non duetu atramenti continuo, sed punctulis rotundis nigricantibus, & ita parum a se invicem remotis, ut intervalla non nisi ope microscopii discerni possent, & quidem ipsæ litterarum formæ pro typis fieri pos-[45]-sent ex ejusmodi rotundis sibi proximis cuspidibus constantes. Concipiatur ingens quædam bibliotheca, cujus omnes libri constent litteris impressis, ac sit incredibilis in ea multitudo librorum conscriptorum linguis variis, in quibus omnibus forma characterum sit eadem. Si quis scripturæ ejusmodi, & linguarum ignarus circa ejusmodi libros, quos omnes a se invicem discrepantes intueretur, observationem institueret cum diligenti contemplatione; primo quidem inveniret vocum farraginem quandam, quæ voces in quibusdam libris occurrerent sæpe, cum eædem in aliis nusquam apparent, & inde lexica posset quædam componere totidem numero, quot idiomata sunt, in quibus singulis omnes ejusdem idiomatis voces reperirentur, quæ quidem numero admodum pauca essent, discrimine illo ingenti tot, tam variorum librorum redacto ad illud usque adeo minus discrimen, quod contineretur lexicis illis, & haberetur in vocibus ipsa lexica constituentibus. At inquisitione promota, facile adverteret, omnes illas tam varias voces constare ex 24 tantummodo diversis litteris, discrimen aliquod inter se habentibus in duetu linearum, quibus formantur, quarum combinatio diversa pareret omnes illas voces tam varias, ut earum combinatio libros efformaret usque adeo magis a se invicem discrepantes. Et ille quidem si aliud quodcunque sine microscopio examen institueret, nullum aliud inveniret magis adhuc simile elementorum genus, ex quibus diversa ratione combinatis orirentur ipsæ litteræ; at microscopio arrepto, intueretur utique illam ipsam litterarum compositionem e punctis illis rotundis prorsus homogeneis, quorum sola diversa positio, ac distributio litteras exhiberet.

Applicatio exempli ad Naturæ analysim.

99. Hæc mihi quædam imago videtur esse eorum, quæ cernimus in Natura. Tam multi, tam varii illi libri corpora sunt, & quæ ad diversa pertinent regna, sunt tanquam diversis conscripta linguis. Horum omnium Chemica analysis principia quædam invenit minus inter se difformia, quam sint libri, nimirum voces. Hæ tamen ipsæ inter se habent discrimen aliquod, ut tam multas oleorum, terrarum, salium species eruit Chemica analysis e diversis corporibus. Ulterior analysis harum, veluti vocum, litteras minus adhuc inter se difformes inveniret, & ultima juxta Theoriam meam deveniret ad homogenea punctula, quæ ut illi circuli nigri litteras, ita ipsa diversas diversorum corporum particulas per solam dispositionem diversam efformarent: usque adeo analogia ex ipsa Naturæ consideratione

97. It is clear then that this variety depends on the number of possible combinations to be found for the number of points that are necessary to make the mass sensible, & of the circumstances that arenecessary for the formation of the mass; & so it is not possible that the induction should be extended to the elements. Nay rather, the great similarity that is found accompanied by some very slight dissimilarity in so many bodies points more strongly to the greatest possible similarity of the elements. For on account of the great number of the possible combinations, even masses of elements that are perfectly homogeneous must be greatly different from one another; & thus if the elements are heterogeneous, the masses must have an immensely greater dissimilarity than the primary elements themselves; & therefore no masses formed from these ought to come out similar, not even in the very slightest degree. Since the elements are bound to be much less dissimilar than aggregates formed from these elements, homogeneity of the elements must be indicated by that certain similarity that we observe in bodies, especially in so many of those that belong to the same species, far more strongly than heterogeneity of the elements is indicated by the slight differences that are observed in so many others. The whole discussion is made perfectly complete by that great similarity, which we made use of above, that exists in the first branch representing impenetrability, & in the last branch representing universal gravitation; for since these branches, on account of properties that are so general to all bodies, are so similar to one another in all cases, they indicate complete similarity of the remaining arc of the curve expressing the forces for all bodies as well.

Homogeneity is to be demonstrated from any sort of similitude in some cases more than heterogeneity from dissimilarity.

98. Naught that concerns this subject remains but for me to once more mention in this connection that one thing, which I have already remarked at the beginning of this work, namely, that Nature itself & the method of analysis lead us towards simplicity & homogeneity of the elements; since in truth the farther the analysis is pushed, the fewer the fundamental substances we arrive at & the less they differ from one another; as is to be seen in chemical experiments. This will be presented to the mind far more clearly by an illustration derived from letters & words. Suppose we have made black letters, not by drawing a continuous line with ink, but by means of little black dots which are at such small distances from one another that the intervals cannot be perceived except with the aid of a microscope—& indeed such forms of letters may be made as types from round points of this sort set close to one another. Now imagine that we have a huge library, all the books in it consisting of printed letters, & let there be an incredible multitude of books printed in various languages, in all which the form of the characters is the same. If anyone, who was ignorant of such compositions or languages, started on a careful study of books of this kind, all of which he would perceive differed from one another; then first of all he would find a medley of words, some of which occurred frequently in certain books whilst they never appeared at all in others. Hence he could compose lexicons, as many in number as there are languages; in each of these all words of the same language would be found, & these would indeed be very few in number; for the immense multiplicity of words in this numerous collection of books of so many kinds is now reduced to what is still a multiplicity, but smaller, than is contained in the lexicons & the words forming these lexicons. Now if he continued his investigation, he would easily perceive that the whole of these words of so many different kinds were formed from 24 letters only; that these differed in some sort from one another in the manner in which the lines forming them were drawn; that the different combinations of these would produce the whole of that great variety of words, & that combinations of these words would form books differing from one another still more widely. Now if he made yet another examination without the aid of a microscope, he would not find any other kind of elements that were more similar to one another than these letters, from a combination of which in different ways the letters themselves could be produced. But if he took a microscope, then indeed would he see the mode of formation of the letters from the perfectly homogeneous round points, by the different position & distribution of which the letters were depicted.

Homogeneity is suggested by an analysis of Nature; example taken from books, letters and dots.

99. This seems to me to be a sort of picture of what we perceive in Nature. Those books, so many in number & so different in character are bodies, & those which belong to the different kingdoms are written as it were in different tongues. Of all of these, chemical analysis finds out certain fundamental constituents that are less unlike one another than the books; these are the words. Yet these constituent substances have some sort of difference amongst themselves, & thus chemical analysis produces a large number of species of oils, earths & salts from different bodies. Further analysis of these, like that of the words, would disclose the letters that are still less unlike one another; & finally, according to my Theory, the little homogeneous points would be obtained. These, just as the little black circles formed the letters, would form the diverse particles of diverse bodies through diverse arrangement alone. So far then the analogy derived from such a

Application of the illustration to the analysis of Nature.

derivata non ad difformitatem, sed ad conformitatem elementorum nos ducit.

Transitus a pro-baſſone Theoriæ ad objectiones.

100. Atque hoc demum pacto ex principiis certis & vulgo receptis, per legitimam, consectariorum seriem devenimus ad omnem illam, quam initio proposui, Theoriam, nimirum ad legem virium mutuarum, & ad constitutionem primorum materiæ elementorum ex illa ipsa virium lege derivatorum. [46] Videndum jam superest, quam uberes inde fructus per universam late Physicam colligantur, explicatis per eam unam præcipuis corporum proprietatibus, & Naturæ phænomenis. Sed antequam id aggredior, præcipuas quasdam e difficultatibus, quæ contra Theoriam ipsam vel objectæ jam sunt, vel in oculos etiam sponte incurrunt, dissolvam, uti promisi.

Legem virium non inducere actionem in distans, nec esse occultam qualitatem.

101. Contra vires mutuas illud solent objicere, illas esse occultas quasdam qualitates, vel etiam actionem in distans inducere. His satisfit notione virium exhibita numero 8, & 9. Illud unum præterea hic addo, admodum manifestas eas esse, quarum idea admodum facile efformatur, quarum existentia positivo argumento evincitur, quarum effectus multiplices continuo oculis observantur. Sunt autem ejusmodi hæ vires. Determinationis ad accessum, vel recessum idea efformatur admodum facile. Constat omnibus, quid sit accedere, quid recedere; constat, quid sit esse indifferens, quid determinatum; adeoque & determinationis ad accessum, vel recessum habetur idea admodum sane distincta. Argumenta itidem positiva, quæ ipsius ejusmodi determinationis existentiam probant, superius prolata sunt. Demum etiam motus varii, qui ab ejusmodi viribus oriuntur, ut ubi corpus quoddam incurrit in aliud corpus, ubi partem solidi arreptam pars alia sequitur, ubi vaporum, vel elastrorum particulæ se invicem repellunt, ubi gravia descendunt, hi motus, inquam, quotidie incurrant in oculos. Patet itidem saltem in genere forma curvæ ejusmodi vires exprimentis. Hæc omnia non occultam, sed patentem reddunt ejusmodi virium legem.

Quid adhuc lateat: admittendam omnino: quo pacto evitetur hic actio in distans.

102. Sunt quidem adhuc quædam, quæ ad eam pertinent, prorsus incognita, uti est numerus, & distantia intersectionum curvæ cum axe, forma arcuum intermediorum, atque alia ejusmodi, quæ quidem longe superant humanum captum, & quæ ille solus habuit omnia simul præ oculis, qui Mundum condidit; sed id omnino nil officit. Nec sane id ipsum in causa esse debet, ut non admittatur illud, cujus existentiam novimus, & cujus proprietates plures, & effectus deprehendimus; licet alia multa nobis incognita eodem pertinentia supersint. Sic aurum incognitam, occultamque substantiam nemo appellarit, & multo minus ejusdem existentiam negabit idcirco, quod admodum probabile sit, plures alias latere ipsius proprietates, olim forte detegendas, uti aliæ tam multæ subinde detectæ sunt, & quia non patet oculis, qui sit particularum ipsum componentium textus, quid, & qua ratione Natura ad ejus compositionem adhibeat. Quod autem pertinet ad actionem in distans, id abunde ibidem prævenimus, cum inde pateat fieri posse, ut punctum quodvis in se ipsum agat, & ad actionis directionem, ac energiam determinetur ab altero puncto, vel ut Deus juxta liberam sibi legem a se in Natura condenda stabilitam motum progignat in utroque pun-[47]-cto. Illud sane mihi est evidens, nihilo magis occultam esse, vel explicatu, & captu difficilem productionem motus per hasce vires pendentes a certis distantiis, quam sit productio motus vulgo concepta per immediatum impulsum, ubi ad motum determinat impenetrabilitas, quæ itidem vel a corporum natura, vel a libera conditoris lege repeti debet.

Sine impulsione melius explicatam esse hucusque Naturam, & melius explicandam imposterum.

103. Et quidem hoc potius pacto, quam per impulsionem, in motuum causas, & leges inquirendum esse, illud etiam satis indicat, quod ubi huc usque, impulsione omissa, vires adhibitæ sunt a distantiis pendentes, ibi sane tantummodo accurate definita sunt omnia, atque determinata, & ad calculum redacta cum phænomenis congruunt ultra, quam sperare liceret, accuratissime. Ego quidem ejusmodi in explicando, ac determinando felicitatem nusquam alibi video in universa Physica, nisi tantummodo in Astronomia mechanica, quæ abjectis vorticibus, atque omni impulsione submota, per gravitatem generalem absolvit omnia, ac in Theoria luminis, & colorum, in quibus per vires in aliqua distantia agentes, & reflexionem, & refractionem, & diffractionem Newtonus exposuit, ac priorum duarum, potissimum leges omnes per calculum, & Geometriam determinavit, & ubi illa etiam, quæ ad diversas vices facilioris transmissus, & facilioris reflexionis, quas Physici passim relinquunt

consideration of Nature leads us not to non-uniformity but to uniformity of the elements.

Passing from the proof of the Theory to the consideration of objections against it.

100. Thus at length, from known principles that are commonly accepted, by a legitimate series of deductions, we have arrived at the whole of the Theory that I enunciated at the start; that is to say, at a law of mutual forces & the constitution of the primary elements of matter derived from that law of forces. Now it remains to be seen what a bountiful harvest is to be gathered throughout the wide field of general physics; for from this one theory we obtain explanations of all the chief properties of bodies, & of the phenomena of Nature. But before I go on to that, I will give solutions of a few of the principal difficulties that have been raised against the Theory itself, as well as some that naturally meet the eye, according to the promise I made.

The law of forces does not necessitate action at a distance, nor is it some mysterious quality.

101. The objection is frequently brought forward against mutual forces that they are some sort of mysterious qualities or that they necessitate action at a distance. This is answered by the idea of forces outlined in Art. 8, & 9. In addition, I will make just one remark, namely, that it is quite evident that these forces exist, that an idea of them can be easily formed, that their existence is demonstrated by direct reasoning, & that the manifold results that arise from them are a matter of continual ocular observation. Moreover these forces are of the following nature. The idea of a propensity to approach or of a propensity to recede is easily formed. For everybody knows what approach means, and what recession is; everybody knows what it means to be indifferent, & what having a propensity means; & thus the idea of a propensity to approach, or to recede, is perfectly distinctly obtained. Direct arguments, that prove the existence of this kind of propensity, have been given above. Lastly also, the various motions that arise from forces of this kind, such as when one body collides with another body, when one part of a solid is seized & another part follows it, when the particles of gases, & of springs, repel one another, when heavy bodies descend, these motions, I say, are of everyday occurrence before our eyes. It is evident also, at least in a general way, that the form of the curve represents forces of this kind. In all of these there is nothing mysterious; on the contrary they all tend to make the law of forces of this kind perfectly plain.

What is so far unknown; the theory to be admitted in all detail; the way in which the idea of action at a distance is eliminated.

102. There are indeed certain things that relate to the law of forces of which we are altogether ignorant, such as the number & distances of the intersections of the curve with the axis, the shape of the intervening arcs, & other things of that sort; these indeed far surpass human understanding, & He alone, Who founded the universe, had the whole before His eyes. But truly there is no reason on that account, why a thing, whose existence we fully recognize, & many of the properties & results of which are readily understood, should not be accepted; although certainly there do remain many other things pertaining to it that are unknown to us. For instance, nobody would call gold an unknown & mysterious substance, & still less would deny its existence, simply because it is quite probable that many of its properties are unknown to us, to be discovered perhaps in the future, as so many others have been already discovered from time to time, or because it is not visually apparent what is the texture of the particles composing it, or why & in what way Nature adopts that particular composition. Again, as regards action at a distance, we amply guard against this by the same means; for, if this is admitted, then it would be possible for any point to act upon itself, & to be determined as to its direction of action & energy apart from another point, or that God should produce in either point a motion according to some arbitrary law fixed by Him when founding the universe. To my mind indeed it is clear that motions produced by these forces depending on the distances are not a whit more mysterious, involved or difficult of understanding than the production of motion by immediate impulse as it is usually accepted; in which impenetrability determines the motion, & the latter has to be derived just the same either from the nature of solid bodies, or from an arbitrary law of the founder of the universe.

As far as we have gone, Nature has been more clearly explained without the idea of impulse; and what follows will be so too.

103. Now, that the investigation of the causes & laws of motion are better made by my method, than through the idea of impulse, is sufficiently indicated by the fact that, where hitherto we have omitted impulse & employed forces depending on the distances, only in this way has everything been accurately defined & determined, & when reduced to calculation everything agrees with the phenomena with far more accuracy than we could possibly have expected. Indeed I do not see anywhere such felicity in explaining & determining the matters of general physics, except only in celestial mechanics; in which indeed, rejecting the idea of vortices, & doing away with that of impulse entirely, Newton gave a solution of everything by means of universal gravitation; & in the theory of light & colours, where by means of forces acting at some distance he explained reflection, refraction & diffraction; &, especially in the two first mentioned, he determined all the laws by calculus & Geometry. Here also those things depending on alternate fits of easier transmission & easier reflection, which physicists everywhere leave almost

fere intactas, ac alia multa admodum feliciter determinantur, explicanturque, quod & ego præstiti in dissertatione *De Lumine*, & præstabo hic in tertia parte ; cum in ceteris Physicæ partibus plerumque explicationes habeantur subsidariis quibusdam principiis innixæ & vagæ admodum. Unde jam illud conjectare licet, si ab impulsione immediata penitus recedatur, & sibi constans ubique adhibeatur in Natura agendi ratio a distantiis pendens, multo sane facilius, & certius explicatum iri cetera ; quod quidem mihi omnino successit ut patebit inferius, ubi Theoriam ipsam applicavero ad Naturam.

Non fieri saltum in transitu a vi attractiva ad repulsivam.

104. Solent & illud objicere, in hac potissimo Theoria virium committi saltum illum, ad quem evitandum ea inprimis admittitur ; fieri enim transitum ab attractionibus ad repulsiones per saltum, ubi nimirum a minima ultima repulsione ad minimam primam attractionem transitur. At isti continuitatis naturam, quam supra exposuimus, nequaquam intelligunt. Saltus, cui evitando Theoria inducitur, in eo consistit, quod ab una magnitudine ad aliam eatur sine transitu per intermedias. Id quidem non accidit in casu exposito. Assumatur quæcunque vis repulsiva utcunque parva ; tum quæcunque vis attractiva. Inter eas intercedunt omnes vires repulsivæ minores usque ad *zero*, in quo habetur determinatio ad conservandum præcedentum statum quietis, vel motus uniformis in directum : tum omnes vires attractivæ a *ze*-[48]-*ro* usque ad eam determinatam vim, & omnino nullus erit ex hisce omnibus intermediis statibus, quem aliquando non sint habitura puncta, quæ a repulsione abeunt ad attractionem. Id ipsum facile erit contemplari in fig. 1, in qua a vi repulsiva *br* ad attractionem *dh* itur utique continuo motu puncti *b* ad *d* transeundo per omnes intermedias, & per ipsum *zero* in E sine ullo saltu ; cum ordinata in eo motu habitura sit omnes magnitudines minores priore *br* usque ad *zero* in E ; tum omnes oppositas majores usque ad posteriorem *dh*. Qui in ea veluti imagine mentis oculos defigat, is omnem apparentem difficultatem videbit plane sibi penitus evanescere.

Nullum esse postremum attractionis, & primum repulsionis gradum, qui si essent, adhuc transire per omnes intermedios.

105. Quod autem additur de postremo repulsionis gradu, & primo attractionis nihil sane probaret, quando etiam essent aliqui ii gradus postremi, & primi ; nam ab altero eorum transiretur ad alterum per intermedium illud *zero*, & ex eo ipso, quod illi essent postremus, ac primus, nihil omitteretur intermedium, quæ tamen sola intermedii omissio continuitatis legem evertit, & saltum inducit. Sed nec habetur ullus gradus postremus, aut primus, sicut nulla ibi est ordinata postrema, aut prima, nulla lineola omnium minima. Data quacunque lineola utcunque exigua, aliæ illa breviores habentur minores, ac minores ad infinitum sine ulla ultima, in quo ipso stat, uti supra etiam monuimus, continuitatis natura. Quamobrem qui primum, aut ultimum sibi confingit in lineola, in vi, in celeritatis gradu, in tempusculo, is naturam continuitatis ignorat, quam supra hic innui, & quam ego idcirco initio meæ dissertationis *De Lege Continuitatis* abunde exposui.

Objectio ab apparenti compositione curvæ, & duobus virium generibus.

106. Videri potest cuipiam saltem illud, ejusmodi legem virium, & curvam, quam in fig. 1 protuli, esse nimium complicatam, compositam, & irregularem, quæ nimirum coalescat ex ingenti numero arcuum jam attractivorum, jam repulsivorum, qui inter se nullo pacto cohæreant ; rem eo redire, ubi erat olim, cum apud Peripateticos pro singulis proprietatibus corporum singulæ qualitates distinctæ, & pro diversis speciebus diversæ formæ substantiales confingebantur ad arbitrium. Sunt autem, qui & illud addant, repulsionem, & attractionem esse virium genera inter se diversa ; satius esse, alteram tantummodo adhibere, & repulsionem explicare tantummodo per attractionem minorem.

Responsio : vim repulsivam positive demonstrari præter attractionem.

107. Inprimis quod ad hoc postremum pertinet, satis patet, per positivam meæ Theoriæ probationem immediate evinci repulsionem ita, ut a minore attractione repeti omnino non possit ; nam duæ materiæ particulæ si etiam solæ in Mundo essent, & ad se invicem cum aliqua velocitatum inæqualitate accederent, deberent utique ante contactum ad æqualitatem devenire vi, quæ a nulla attractione pendere posset.

untouched, & many other matters were most felicitously determined & explained by him; & also that which I enunciated in the dissertation *De Lumine*, & will repeat in the third part of this work. For in other parts of physics most of the explanations are independent of, & disconnected from, one another, being based on several subsidiary principles. Hence we may now conclude that if, relinquishing all idea of immediate impulses, we employ a reason for the action of Nature that is everywhere the same & depends on the distances, the remainder will be explained with far greater ease & certainty; & indeed it is altogether successful in my hands, as will be evident later, when I come to apply the Theory to Nature.

There is no sudden change in the transition from an attractive to a repulsive force.

104. It is very frequently objected that, in this Theory more especially, a sudden change is made in the forces, whilst the theory is to be accepted for the very purpose of avoiding such a thing. For it is said that the transition from attractions to repulsions is made suddenly, namely, when we pass from the last extremely minute repulsive force to the first extremely minute attractive force. But those who raise these objections in no wise understand the nature of continuity, as it has been explained above. The sudden change, to avoid which the Theory has been brought forward, consists in the fact that a passage is made from one magnitude to another without going through the intermediate stages. Now this kind of thing does not take place in the case under consideration. Take any repulsive force, however small, & then any attractive force. Between these two there lie all the repulsive forces that are less than the former right down to zero, in which there is the propensity for preserving the original state of rest or of uniform motion in a straight line; & also all the attractive forces from zero up to the prescribed attractive force, & there will be absolutely no one of all these intermediate states, which will not be possessed at some time or other by the points as they pass from repulsion to attraction. This can be readily understood from a study of Fig. 1, where indeed the passage is made from the repulsive force *br* to the attractive force *dh* by the continuous motion of a point from *b* to *d*; the passage is made through every intermediate stage, & through zero at E, without any sudden change. For in this motion there will be obtained as ordinates all magnitudes, less than the first one *br*, down to zero at E, & after that all magnitudes of opposite sign greater than zero as far as the last ordinate *dh*. Anyone, who will fix his intellectual vision on this as on a sort of pictorial illustration cannot fail to perceive for himself that all the apparent difficulty vanishes completely.

There is no last stage of attraction, and no first for repulsion; and even if there were, the passage would be made through all intermediate stages.

105. Further, as regards what is said in addition about the last stage of repulsion & the first stage of attraction, it would really not matter, even if there were these so called last & first stages; for, from one of them to the other the passage would be made through the one intermediate stage, namely zero; since it passes zero, & because they are the first & last, therefore no intermediate stage is omitted. Nevertheless the omission of this intermediate alone would upset the law of continuity, & introduce a sudden change. But, as a matter of fact, there cannot possibly be a last stage or a first; just as there cannot be a last ordinate or a first in the curve, that is to say, a short line that is the least of them all. Given any short line, no matter how short, there will be others shorter than it, less & less in infinite succession without any limit whatever; & in this, as we remarked also above, there lies the nature of continuity. Hence anyone who brings forward the idea of a first or a last in the case of a line, or a force, or a degree of velocity, or an interval of time, must be ignorant of continuity; this I have mentioned before in this work, & also for this very reason I explained it very fully at the beginning of my dissertation *De Lege Continuitatis*.

Objection raised against the apparent composite character of the curve, and the two kinds of forces.

106. It may seem to some that at least a law of forces of this nature, & the curve expressing it, which I gave in Fig. 1, is very complicated, composite & irregular, being indeed made up of an immense number of arcs that are alternately attractive & repulsive, & that these are joined together according to no definite plan; & that it reduces to the same thing as obtained amongst the ancients, since with the Peripatetics separate distinct qualities were invented for the several properties of bodies, & different substantial forms for different species. Moreover there are some who add that repulsion & attraction are kinds of forces that differ from one another; & that it would be quite enough to use only the latter, & to explain repulsion merely as a smaller attraction.

In reply; it is possible to prove directly the existence of a repulsive force apart from attraction.

107. First of all, as regards the last objection, it is clear enough from what has been directly proved in my Theory that the existence of repulsion has been rigorously demonstrated in such a way that it cannot possibly be derived from the idea of a smaller attraction. For two particles of matter, if they were also the only particles in the universe, & approached one another with some difference of velocity, would be bound to attain to an equality of velocity on account of a force which could not possibly be derived from an attraction of any kind.

Hinc nihil obstare, si diversi sint generis; sed esse ejusdem, uti sunt positiva, & negativa

108. Deinde vero quod pertinet ad duas diversas species attractionis, & repulsionis; id quidem licet ita se haberet, ni-[49]-hil sane obesset, cum positivo argumento evincatur & repulsio, & attractio, uti vidimus; at id ipsum est omnino falsum. Utraque vis ad eandem pertinet speciem, cum altera respectu alterius negativa sit, & negativa a positivis specie non differant. Alteram negativam esse respectu alterius, patet inde, quod tantummodo differant in directione, quæ in altera est prorsus opposita directioni alterius; in altera enim habetur determinatio ad accessum, in altera ad recessum, & uti recessus, & accessus sunt positivum, ac negativum; ita sunt pariter & determinationes ad ipsos. Quod autem negativum, & positivum ad eandem pertineant speciem, id sane patet vel ex eo principio: *magis, & minus non differunt specie.* Nam a positivo per continuam subtractionem, nimirum diminutionem, habentur prius minora positiva, tum *zero*, ac demum negativa, continuando subtractionem candem.

Probatio hujus a progressu, & regressu, in fluvio.

109. Id facile patet exemplis solitis. Eat aliquis contra fluvii directionem versus locum aliquem superiori alveo proximum, & singulis minutis perficiat remis, vel vento 100 hexapedas, dum a cursu fluvii retroagitur per hexapedas 40; is habet progressum hexapedarum 60 singulis minutis. Crescat autem continuo impetus fluvii ita, ut retroagatur per 50, tum per 60, 70, 80, 90, 100, 110, 120, &c. Is progredietur per 50, 40, 30, 20, 10, nihil; tum regredietur per 10, 20, quæ erunt negativa priorum; nam erat prius 100—50, 100—60, 100—70, 100—80, 100—90, tum 100—100 = 0, 100—110, = —10, 100—120 = —20, et ita porro. Continua imminutione, sive subtractione itum est a positivis in negativa, a progressu ad regressum, in quibus idcirco eadem species mansit, non duæ diversæ.

Probatio ex Algebra, & Geometria: applicatio ad omnes quantitates variabiles.

110. Idem autem & algebraicis formulis, & geometricis lineis satis manifeste ostenditur. Sit formula $10-x$, & pro x ponantur valores 6, 7, 8, 9, 10, 11, 12, &c.; valor formulæ exhibebit 4, 3, 2, 1, 0, —1, —2, &c., quod eodem redit, ubi erat superius in progressu, & regressu, qui exprimerentur simul per formulam $10-x$. Eadem illa formula per continuam mutationem valoris x migrat e valore positivo in negativum, qui æque ad eandem formulam pertinent. Eodem pacto in Geometria in fig. 11, si duæ lineæ MN, OP referantur invicem per ordinatas AB, CD, &c. parallelas inter se, secent autem se in E; continuo motu ipsius ordinatæ a positivo abitur in negativum, mutata directione AB, CD, quæ hic habentur pro positivis, in FG, HI, post evanescentiam in E. Ad eandem lineam continuam OEP æque pertinet omnis ea ordinatarum series, nec est altera linea, alter locus geometricus OE, ubi ordinatæ sunt positivæ, ac EP, ubi sunt negativæ. Jam vero variabilis quantitatis cujusvis natura, & lex plerumque per formulam aliquam analyticam, semper per ordinatas ad lineam aliquam exprimi potest; si [50] enim singulis ejus statibus ducatur perpendicularis respondens; vertices omnium ejusmodi perpendicularium erunt utique ad lineam quandam continuam. Si ea linea nusquam ad alteram abeat axis partem, si ea formula nullum valorem negativum habeat; illa etiam quantitas semper positiva manebit. Sed si mutet latus linea, vel formula valoris signum; ipsa illa quantitatis debebit itidem ejusmodi mutationem habere. Ut autem a formulæ, vel lineæ exprimentis natura, & positione respectu axis mutatio pendet; ita mutatio eadem a natura quantitatis illius pendebit; & ut nec duæ formulæ, nec duæ lineæ speciei diversæ sunt, quæ positiva exhibent, & negativa; ita nec in ea quantitate duæ erunt naturæ, duæ species, quarum altera exhibeat positiva, altera negativa, ut altera progressus, altera regressus; altera accessus, altera recessus; & hic altera attractiones, altera repulsiones exhibeat; sed eadem erit, unica, & ad eandem pertinens quantitatis speciem tota.

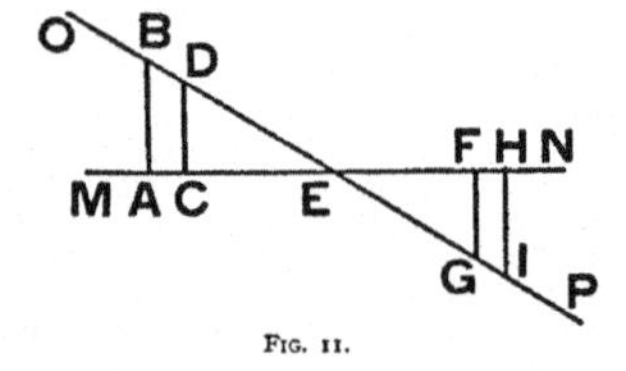

FIG. 11.

An habeatur transitus e positivis in negativa; investigatio ex sola curvarum natura.

111. Quin immo hic locum habet argumentum quoddam, quo usus sum in dissertatione *De Lege Continuitatis*, quo nimirum Theoria virium attractivarum, & repulsivarum pro diversis distantiis, multo magis rationi consentanea evincitur, quam Theoria virium tantummodo attractivarum, vel tantummodo repulsivarum. Fingamus illud, nos ignorare penitus, quodnam virium genus in Natura existat, an tantummodo attractivarum, vel repulsivarum tantummodo, an utrumque simul: hac sane ratiocinatione ad eam perquisitionem uti liceret. Erit utique aliqua linea continua, quæ per suas ordinatas ad axem exprimentem distantias, vires ipsas determinabit, & prout ipsa axem secuerit, vel non

Hence it does not matter if they are of different kinds; but as a matter of fact they are of the same kind, just as a positive and a negative are so.

108. Next, as regards attraction & repulsion being of different species, even if it were a fact that they were so, it would not matter in the slightest degree, since by rigorous argument the existence of both attraction & repulsion is proved, as we have seen; but really the supposition is untrue. Both kinds of force belong to the same species; for one is negative with regard to the other, & a negative does not differ in species from positives. That the one is negative with regard to the other is evident from the fact that they only differ in direction, the direction of one being exactly the opposite of the direction of the other; for in the one there is a propensity to approach, in the other a propensity to recede; & just as approach & recession are positive & negative, so also are the propensities for these equally so. Further, that such a negative & a positive belong to the same species, is quite evident from the principle *the greater & the less are not different in kind.* For from a positive by continual subtraction, or diminution, we first obtain less positives, then zero, & finally negatives, the same subtraction being continued throughout.

Demonstration by means of progressive and retrograde motion on a river.

109. The matter is easily made clear by the usual illustrations. Suppose a man to go against the current of a river to some place on the bank up-stream; & suppose that he succeeds in doing, either by rowing or sailing, 100 fathoms a minute, whilst he is carried back by the current of the river through 40 fathoms; then he will get forward a distance of 60 fathoms a minute. Now suppose that the strength of the current continually increases in such a way that he is carried back first 50, then 60, 70, 80, 90, 100, 110, 120, &c. fathoms per minute. His forward motion will be successively 50, 40, 30, 20, 10 fathoms per minute, then nothing, & then he will be carried backward through 10, 20, &c. fathoms a minute; & these latter motions are the negatives of the former. For first of all we had $100 - 50$, $100 - 60$, $100 - 70$, $100 - 80$, $100 - 90$, then $100 - 100$ (which $= 0$), then $100 - 110$ (which $= -10$), $100 - 120$ (which $= -20$), and so on. By a continual diminution or subtraction we have passed from positives to negatives, from a progressive to a retrograde motion; & therefore in these there was a continuance of the same species, and there were not two different species.

Proof from algebra and geometry; application to all variable quantities.

110. Further, the same thing is shown plainly enough by algebraical formulæ, & by lines in geometry. Consider the formula $10 - x$, & for x substitute the values, 6, 7, 8, 9, 10, 11, 12, &c.; then the value of the formula will give in succession 4, 3, 2, 1, 0, -1, -2, &c.; & this comes to the same thing as we had above in the case of the progressive & retrograde motion, which may be expressed by the formula $10 - x$, all together. This same formula passes, by a continuous change in the value of x, from a positive value to a negative, which equally belong to the same formula. In the same manner in geometry, in Fig. 11, if two lines MN, OP are referred to one another by ordinates AB, CD, & also cut one another in E; then by a continuous motion of the ordinate itself it passes from positive to negative, the direction of AB, CD, which are here taken to be positive, being changed to that of FG, HI, after evanescence at E. To the same continuous line OEP belongs equally the whole of this series of ordinates; & OE, where the ordinates are positive, is not a different line, or geometrical locus from EP, where the ordinates are negative. Now the nature of any variable quantity, & very frequently also the law, can be expressed by an algebraical formula, & can always be expressed by some line; for if a perpendicular be drawn to correspond to each separate state of the quantity, the vertices of all perpendiculars so drawn will undoubtedly form some continuous line. If the line never passes over to the other side of the axis, if the formula has no negative value, then also the quantity will always remain positive. But if the line changes side, or the formula the sign of its value, then the quantity itself must also have a change of the same kind. Further, as the change depends on the nature of the formula & the line expressing it, & its position with respect to the axis; so also the same change will depend on the nature of the quantity; & just as there are not two formulæ, or two lines of different species to represent the positives & the negatives, so also there will not be in the quantity two natures, or two species, of which the one yields positives & the other negatives, as the one a progressive & the other a retrograde motion, the one approach & the other recession, & in the matter under consideration the one will give attractions & the other repulsions. But it will be one & the same nature & wholly belonging to the same spec es of quantity.

Whether there can be a transition from positive to negative; investigation by means of the nature of the curve only.

111. Lastly, this is the proper place for me to bring forward an argument that I used in the dissertation *De Lege Continuitatis*; by it indeed it is proved that a theory of attractive & repulsive forces for different distances is far more reasonable than one of attractive forces only, or of repulsive forces only. Let us imagine that we are quite ignorant of the kind of forces that exist in Nature, whether they are only attractive or only repulsive, or both; it would be allowable to use the following reasoning to help us to investigate the matter. Without doubt there will be some continuous line which, by means of ordinates drawn from it to an axis representing distances, will determine the forces; & according

secuerit; vires erunt alibi attractivæ, alibi repulsivæ; vel ubique attractivæ tantum, aut repulsivæ tantum. Videndum igitur, an sit rationi consentaneum magis, lineam ejus naturæ, & positionis censere, ut axem alicubi secet, an ut non secet.

Transitum deduci ex eo, quod plures sint curvæ, quas recte secent, quam eæ, quas non secent.

112. Inter rectas axem rectilineum unica parallela ducta per quodvis datum punctum non secat, omnes aliæ numero infinitæ secant alicubi. Curvarum nulla est, quam infinitæ numero rectæ secare non possint; & licet aliquæ curvæ ejus naturæ sint, ut eas aliquæ rectæ non secent; tamen & eas ipsas aliæ infinitæ numero rectæ secant, & infinitæ numero curvæ, quod Geometriæ sublimioris peritis est notissimum, sunt ejus naturæ, ut nulla prorsus sit recta linea, a qua possint non secari. Hujusmodi ex. gr. est parabola illa, cujus ordinatæ sunt in ratione triplicata abscissarum. Quare infinitæ numero curvæ sunt, & infinitæ numero rectæ, quæ sectionem necessario habeant, pro quavis recta, quæ non habeat, & nulla est curva, quæ sectionem cum axe habere non possit. Ergo inter casus possibiles multo plures sunt ii, qui sectionem admittunt, quam qui ea careant; adeoque seclusis rationibus aliis omnibus, & sola casuum probabilitate, & rei [51] natura abstracte considerata, multo magis rationi consentaneum est, censere lineam illam, quæ vires exprimat, esse unam ex iis, quæ axem secant, quam ex iis, quæ non secant, adeoque & ejusmodi esse virium legem, ut attractiones, & repulsiones exhibeat simul pro diversis distantiis, quam ut alteras tantummodo referat; usque adeo rei natura considerata non solam attractionem, vel solam repulsionem, sed utramque nobis objicit simul.

Ulterior perquisitio: curvarum genera: quo altiores, eo in pluribus punctis secabiles a recta.

113. Sed eodem argumento licet ulterius quoque progredi, & primum etiam difficultatis caput amovere, quod a sectionum, & idcirco etiam arcuum jam attractivorum, jam repulsivorum multiplicitate desumitur. Curvas lineas Geometræ in quasdam classes dividunt ope analyseos, quæ earum naturam exprimit per illas, quas Analystæ appellant, æquationes, & quæ ad varios gradus ascendunt. Aequationes primi gradus exprimunt rectas; æquationes secundi gradus curvas primi generis; æquationes tertii gradus curvas secundi generis, atque ita porro; & sunt curvæ, quæ omnes gradus transcendunt finitæ Algebræ, & quæ idcirco dicuntur transcendentes. Porro illud demonstrant Geometræ in Analysi ad Geometriam applicata, lineas, quæ exprimuntur per æquationem primi gradus, posse secari a recta in unico puncto; quæ æquationem habent gradus secundi, tertii, & ita porro, secari posse a recta in punctis duobus, tribus, & ita porro: unde fit, ut curva noni, vel nonagesimi noni generis secari possit a recta in punctis decem, vel centum.

Quo altiores, eo itidem in immensum plures in eodem genere.

114. Jam vero curvæ primi generis sunt tantummodo tres conicæ sectiones, ellipis, parabola, hyperbola, adnumerato ellipsibus etiam circulo, quæ quidem veteribus quoque Geometris innotuerunt. Curvas secundi generis enumeravit Newtonus omnium primus, & sunt circiter octoginta; curvarum generis tertii nemo adhuc numerum exhibuit accuratum, & mirum sane, quantus sit is ipse illarum numerus. Sed quo altius assurgit curvæ genus, eo plures in eo genere sunt curvæ, progressione ita in immensum crescente, ut ubi aliquanto altius ascenderit genus ipsum, numerus curvarum omnem superet humanæ imaginationis vim. Idem nimirum ibi accidit, quod in combinationibus terminorum, de quibus supra mentionem fecimus, ubi diximus a 24 litterulis omnes exhiberi voces linguarum omnium, & quæ fuerunt, aut sunt, & quæ esse possunt.

Deductio inde plurimarum intersectionum, axis, & curvæ exprimentis vires.

115. Inde jam pronum est argumentationem hujusmodi instituere. Numerus linearum, quæ axem secare possint in punctis quamplurimis, est in immensum major earum numero, quæ non possint, nisi in paucis, vel unico: igitur ubi agitur de linea exprimente legem virium, ei, qui nihil aliunde seiat, in immensum probabilius crit, ejusmodi lineam esse ex prio-[52]-rum genere unam, quam ex genere posteriorum, adeoque ipsam virium naturam plurimos requirere transitus ab attractionibus ad repulsiones, & vice versa, quam paucos, vel nullum.

Curvam virium propositam posse esse simplicem: in quo sita sit curvarum simplicitas.

116. Sed omissa ista conjecturali argumentatione quadam, formam curvæ exprimentis vires positivo argumento a phænomenis Naturæ deducto nos supra determinavimus cum plurimis intersectionibus, quæ transitus ejusmodi quamplurimos exhibeant. Nec ejusmodi curva debet esse e pluribus arcubus temere compaginata, & compacta: diximus enim,

as it will cut the axis, or will not, the forces will be either partly attractive & partly repulsive, or everywhere only attractive or only repulsive. Accordingly it is to be seen if it is more reasonable to suppose that a line of this nature & position cuts the axis anywhere, or does not.

Intersection is to be inferred from the fact that there are more lines that cut a straight line than lines that do not.

112. Amongst straight lines there is only one, drawn parallel to the rectilinear axis, through any given point that does not cut the axis; all the rest (infinite in number) will cut it somewhere. There is no curve that an infinite number of straight lines cannot cut; & although there are some curves of such a nature that some straight lines do not cut them, yet there are an infinite number of other straight lines that do cut these curves; & there are an infinite number of curves, as is well-known to those versed in higher geometry, of such a nature that there is absolutely not a single straight line by which they cannot be cut. An example of this kind of curve is that parabola, in which the ordinates are in the triplicate ratio of the abscissæ. Hence there are an infinite number of curves & an infinite number of straight lines which necessarily have intersection, corresponding to any straight line that has not; & there is no curve that cannot have intersection with an axis. Therefore amongst the cases that are possible, there are far more curves that admit intersection than those that are free from it; hence, putting all other reasons on one side, & considering only the probability of the cases & the nature of the matter on its own merits, it is far more reasonable to suppose that the line representing the forces is one of those, which cut the axis, than one of those that do not cut it. Thus the law of forces is such that it yields both attractions & repulsions (for different distances), rather than such that it deals with either alone. Thus far the nature of the matter has been considered, with the result that it presents to us, not attraction alone, nor repulsion alone, but both of these together.

Further investigation; classes of curves; the higher their order, the more the points in which a straight line can cut them.

113. But we can also proceed still further adopting the same line of argument, & first of all remove the chief point of the difficulty, that is derived from the multiplicity of the intersections, & consequently also of the arcs alternately attractive & repulsive. Geometricians divide curves into certain classes by the help of analysis, which expresses their nature by what the analysts call equations; these equations rise to various degrees. Equations of the first degree represent straight lines, equations of the second degree represent curves of the first class, equations of the third degree curves of the second class, & so on. There are also curves which transcend all degrees of finite algebra, & on that account these are called transcendental curves. Further, geometricians prove, in analysis applied to geometry, that lines that are expressed by equations of the first degree can be cut by a straight line in one point only; those that have equations of the second, third, & higher degrees can be cut by a straight line in two, three, & more points respectively. Hence it comes about that a curve of the ninth, or the ninety-ninth class can be cut by a straight line in ten, or in a hundred, points.

As the class gets higher, the number of curves of that class becomes immensely greater.

114. Now there are only three curves of the first class, namely the conic sections, the parabola, the ellipse & the hyperbola; the circle is included under the name of ellipse; & these three curves were known to the ancient geometricians also. Newton was the first of all persons to enumerate the curves of the second class, & there are about eighty of them. Nobody hitherto has stated an exact number for the curves of the third class; & it is really wonderful how great is the number of these curves. Moreover, the higher the class of the curve becomes, the more curves there are in that class, according to a progression that increases in such immensity that, when the class has risen but a little higher, the number of curves will altogether surpass the fullest power of the human imagination. Indeed the same thing happens in this case as in combinations of terms; we mentioned the latter above, when we said that by means of 24 little letters there can be expressed all the words of all languages that ever have been, or are, or can be in the future.

Hence we deduce that there are very many intersections of the axis and the curve representing forces.

115. From what has been said above we are led to set up the following line of argument. The number of lines that can cut the axis in very many points is immensely greater than the number of those that can cut it in a few points only, or in a single point. Hence, when the line representing the law of forces is in question, it will appear to one, who otherwise knows nothing about its nature, that it is immensely more probable that the curve is of the first kind than that it is of the second kind; & therefore that the nature of the forces must be such as requires a very large number of transitions from attractions to repulsions & back again, rather than a small number or none at all.

It may be that the curve of forces is simple; the characteristic of simplicity in curves.

116. But, omitting this somewhat conjectural line of reasoning, we have already determined, by what has been said above, the form of the curve representing forces by a rigorous argument derived from the phenomena of Nature, & that there are very many intersections which represent just as many of these transitions. Further, a curve of this

notum esse Geometris, infinita esse curvarum genera, quæ ex ipsa natura sua debeant axem in plurimis secare punctis, adeoque & circa ipsum sinuari; sed præter hanc generalem responsionem desumptam a generali curvarum natura, in dissertatione *De Lege Virium in Natura existentium* ego quidem directe demonstravi, curvam illius ipsius formæ, cujusmodi ea est, quam in fig. 1 exhibui, simplicem esse posse, non ex arcubus diversarum curvarum compositam. Simplicem autem ejusmodi curvam affirmavi esse posse: eam enim simplicem appello, quæ tota est uniformis naturæ, quæ in Analysi exponi possit per æquationem non resolubilem in plures, e quarum multiplicatione eadem componatur cujuscunque demum ea curva sit generis, quotcunque habeat flexus, & contorsiones. Nobis quidem altiorum generum curvæ videntur minus simplices; quia nimirum nostræ humanæ menti, uti pluribus ostendi in dissertatione *De Maris Aestu*, & in Stayanis Supplementis, recta linea videtur omnium simplicissima, cujus congruentiam in superpositione intuemur mentis oculis evidentissime, & ex qua una omnem nos homines nostram derivamus Geometriam; ac idcirco, quæ lineæ a recta recedunt magis, & discrepant, illas habemus pro compositis, & magis ab ea simplicitate, quam nobis confinximus, recedentibus. At vero lineæ continuæ, & uniformis naturæ omnes in se ipsis sunt æque simplices; & aliud mentium genus, quod cujuspiam ex ipsis proprietatem aliquam æque evidenter intueretur, ac nos intuemur congruentiam rectarum, illas maxime simplices esse crederet curvas lineas, ex illa carum proprietate longe alterius Geometriæ sibi elementa conficeret, & ad illam ceteras referret lineas, ut nos ad rectam referimus; quæ quidem mentes si aliquam ex. gr. parabolæ proprietatem intime perspicerent, atque intuerentur, non illud quærerent, quod nostri Geometræ quærunt, ut parabolam rectificarent, sed, si ita loqui fas est, ut reetam *parabolarent.*

Problema continens naturam curvæ analytice exprimendam.

117. Et quidem analyseos ipsius profundiorem cognitionem requirit ipsa investigatio æquationis, qua possit exprimi curva ejus formae, quæ meam exhibet virium legem. Quamobrem hic tantummodo exponam conditiones, quas ipsa curva habere debet, & quibus æquatio ibi inventa satis facere [53] debeat. (c) Continetur autem id ipsum num. 75, illius dissertationis, ubi habetur hujusmodi Problema: *Invenire naturam curvæ, cujus abscissis exprimentibus distantias, ordinatæ exprimant vires, mutatis distantiis utcunque mutatas, & in datis quotcunque limitibus transeuntes e repulsivis in attractivas, ac ex attractivis in repulsivas, in minimis autem distantiis repulsivas, & ita crescentes, ut sint pares extinguendæ cuicunque velocitati utcunque magnæ.* Proposito problemate illud addo: *quoniam posuimus* mutatis distantiis utcunque mutatas, *complectitur propositio etiam rationem quæ ad rationem reciprocam duplicatam distantiarum accedat, quantum libuerit, in quibusdam satis magnis distantiis.*

Conditiones ejus problematis.

118. His propositis numero illo 75, sequenti numero propono sequentes sex conditiones, quæ requirantur, & sufficiant ad habendam curvam, quæ quæritur. *Primo: ut sit regularis, ac simplex, & non composita ex aggregato arcuum diversarum curvarum. Secundo: ut secet axem* C'AC *figuræ* 1. *tantum in punctis quibusdam datis ad binas distantias* AE', AE; AG', AG; *& ita porro æquales* (d) *hinc, & inde. Tertio: ut singulis abscissis respondeant singulæ ordinatæ.* (e) *Quarto: ut sumptis abscissis æqualibus hinc, & inde ab* A, *respondeant ordinatæ*

(c) *Qui velit ipsam rei determinationem videre, poterit hic in fine, ubi Supplementorum,* § 3. *exhibebitur solutio problematis, quæ in memorata dissertatione continetur a num.* 77. *ad* 110. *Sed & numerorum ordo, & figurarum mutabitur, ut cum reliquis hujusce operis cohæreat.*

Addetur præterea eidem §. *postremum scholium pertinens ad quæstionem agitatam ante hos aliquot annos Parisiis; an vis mutua inter materiæ particulas debeat omnino exprimi per solam aliquam distantiæ potentiam, an possit per aliquam ejus functionem; & constabit, posse utique per functionem, ut hic ego præsto, quæ uti superiore numero de curvis est dictum, est in se æque simplex etiam, ubi nobis potentias ad ejus expressionem adhibentibus videatur admodum composita.*

(d) *Id, ut & quarta conditio, requiritur, ut curva utrinque sit sui similis, quod ipsam magis uniformem reddit; quanquam de illo crure, quod est citra asymptotum* AB, *nihil est, quod soliciti simus; cum ob vim repulsivam imminutis distantiis ita in infinitum excrescentem, non possit abscissa distantiam exprimens unquam evadere* zero, *& abire in negativam.*

(e) *Nam singulis distantiis singulæ vires respondent.*

kind is not bound to be built up by connecting together a number of independent arcs. For, as I said, it is well known to Geometricians that there are an infinite number of classes of curves that, from their very nature, must cut the axis in a very large number of points, & therefore also wind themselves about it. Moreover, in addition to this general answer to the objector, derived from the general nature of curves, in my dissertation *De Lege Virium in Natura existentium*, I indeed proved in a straightforward manner that a curve, of the form that I have given in Fig. 1, might be simple & not built up of arcs of several different curves. Further, I asserted that a simple curve of this kind was perfectly feasible; for I call a curve simple, when the whole of it is of one uniform nature. In analysis, this can be expressed by an equation that is not capable of being resolved into several other equations, such that the former is formed from the latter by multiplication; & that too, no matter of what class the curve may be, or how many flexures or windings it may have. It is true that the curves of higher classes seem to us to be less simple; this is so because, as I have shown in several places in the dissertation *De Maris Aestu*, & the supplements to Stay's Philosophy, a straight line seems to our human mind to be the simplest of all lines; for we get a real clear mental perception of the congruence on superposition in the case of a straight line, & from this we human beings form the whole of our geometry. On this account, the more that lines depart from straightness & the more they differ, the more we consider them to be composite & to depart from that simplicity that we have set up as our standard. But really all lines that are continuous & of uniform nature are just as simple as one another. Another kind of mind, which might form an equally clear mental perception of some property of any one of these curves, as we do the congruence of straight lines, might believe these curves to be the simplest of all & from that property of these curves build up the elements of a far different geometry, referring all other curves to that one, just as we compare them with a straight line. Indeed, these minds, if they noticed & formed an extremely clear perception of some property of, say, the parabola, would not seek, as our geometricians do, to rectify the parabola; they would endeavour, if one may use the words, to *parabolify a straight line.*

Problem dealing with the analytical expression of the nature of the curve.

117. The investigation of the equation, by which a curve of the form that will represent my law of forces can be expressed, requires a deeper knowledge of analysis itself. Wherefore I will here do no more than set out the necessary requirements that the curve must fulfil & those that the equation thereby discovered must satisfy.(c) It is the subject of Art. 75 of the dissertation *De Lege Virium*, where the following problem is proposed. *Required to find the nature of the curve, whose abscissæ represent distances & whose ordinates represent forces that are changed as the distances are changed in any manner, & pass from attractive forces to repulsive, & from repulsive to attractive, at any given number of limit-points; further, the forces are repulsive at extremely small distances and increase in such a manner that they are capable of destroying any velocity, however great it may be.* To the problem as there proposed I now add the following:—*As we have used the words* are changed as the distances are changed in any manner, *the proposition includes also the ratio that approaches as nearly as you please to the reciprocal ratio of the squares of the distances, whenever the distances are sufficiently great.*

The conditions of the problem.

118. In addition to what is proposed in this Art. 75, I set forth in the article that follows it the following six conditions; these are the necessary and sufficient conditions for determining the curve that is required.

(i) *The curve is règular & simple, & not compounded of a number of arcs of different curves.*

(ii) *It shall cut the axis C'AC of Fig.* 1, *only in certain given points, whose distances, AE', AE, AG', AG, and so on, are equal* (d) *in pairs on each side of A* [see p. 80].

(iii) *To each abscissa there shall correspond one ordinate & one only.* (e)

(iv) *To equal abscissæ, taken one on each side of A, there shall correspond equal ordinates.*

(c) *Anyone who desires to see the solution of the problem will be able to do so at the end of this work; it will be found in* § 3 *of the Supplements; it is the solution of the problem, as it was given in the dissertation mentioned above, from Art.* 77 *to* 110. *But here both the numbering of the articles & of the diagrams have been changed, so as to agree with the rest of the work. In addition, at the end of this section, there will be found a final note dealing with a question that was discussed some years ago in Paris. Namely, whether the mutual force between particles of matter is bound to be expressible by some one power of the distance only, or by some function of the distance. It will be evident that at any rate it may be expressible by a function as I here assert; & that function, as has been stated in the article above, is perfectly simple in itself also; whereas, if we adhere to an expression by means of powers, the curve will seem to be altogether complex.*

(d) *This, & the fourth condition too, is required to make the curve symmetrical, thus giving it greater uniformity; although we are not concerned with the branch on the other side of the asymptote AB at all. For, on account of the repulsive force at very small distances increasing indefinitely in such a manner as postulated, it is impossible that the abscissa that represents the distance should ever become zero & then become negative.*

(e) *For to each distance one force, & and only one, corresponds.*

æquales. Quinto: ut habeant rectam AB *pro asymptoto, area asymptotica* BAED *existente* (f) *infinita. Sexto: ut arcus binis quibuscunque intersectionibus terminati possint variari, ut libuerit, & ad quascunque distantias recedere ab axe* C′AC, *ac accedere ad quoscunque quarumcunque curvarum arcus, quantum libuerit, eos secando, vel tangendo, vel osculando ubicunque, & quomodocunque libuerit.*

Curvæ virium resolutio in attractionem gravitatis Newtonianam, & aliam quandam vim.

[54] 119. Verum quod ad multiplicitatem virium pertinet, quas diversis jam Physici nominibus appellant, illud hic etiam notari potest, si quis singulas seorsim considerare velit, licere illud etiam, hanc curvam in se unicam per resolutionem virium cogitatione nostra, atque fictione quadam, dividere in plures. Si ex. gr. quis velit considerare in materia gravitatem generalem accurate reciprocam distantiarum quadratis; poterit sane is describere ex parte attractiva hyperbolam illam, quæ habeat accurate ordinatas in ratione reciproca duplicata distantiarum, quæ quidem erit quædam velut continuatio cruris VTS, tum singulis ordinatis *ag*, *db* curvæ virium expressæ in fig. 1. adjungere ordinatas hujus novæ hyperbolæ ad partes AB incipiendo a punctis curvæ *g*, *b*, & eo pacto orietur nova quædam curva, quæ versus partes *p*V coincidet ad sensum cum axe *o*C, in reliquis locis ab eo distabit, & contorquebitur etiam circa ipsum, si vertices F, K, O distiterint ab axe magis, quam distet ibidem hyperbola illa. Tum poterit dici, puncta omnia materiæ habere gravitatem decrescentem accurate in ratione reciproca duplicata distantiarum, & simul habere vim aliam expressam ab illa nova curva: nam idem erit, concipere simul hasce binas leges virium, ac illam præcedentem unicam, & iidem effectus orientur.

Hujus posterioris vis resolutio in alias plures.

120. Eodem pacto hæc nova curva potest dividi in alias duas, vel plures, concipiendo aliam quamcunque vim, ut ut accurate servantem quasdam determinatas leges, sed simul mutando curvam jam genitam, translatis ejus punctis per intervalla æqualia ordinatis respondentibus novæ legi assumptæ. Hoc pacto habebuntur plures etiam vires diversæ, quod aliquando, ut in resolutione virium accidere diximus, inserviet ad faciliorem determinationem effectuum, & ea erit itidem vera virium resolutio quædam; sed id omne erit nostræ mentis partus quidam; nam reipsa unica lex virium habebitur, quam in fig. 1. exposui, & quæ ex omnibus ejusmodi legibus componetur.

Non obesse theoriam gravitatis; cujus lex in minimis distantiis locum non habet.

121. Quoniam autem hic mentio injecta est gravitatis decrescentis accurate in ratione reciproca duplicata distantiarum; cavendum, ne cui difficultatem aliquam pariat illud, quod apud Physicos, & potissimum apud Astronomiæ mechanicæ cultores, habetur pro comperto, gravitatem decrescere in ratione reciproca duplicata distantiarum accurate, cum in hac mea Theoria lex virium discedat plurimum ab ipsa ratione reciproca duplicata distantiarum. Inprimis in minoribus distantiis vis integra, quam in se mutuo exercent particulæ, omnino plurimum discrepat a gravitate, quæ sit in ratione reciproca duplicata distantiarum. Nam & vapores, qui tantam exercent vim ad se expandendos, repulsionem habent utique in illis minimis distantiis a se invicem, non attractionem; & ipsa attractio, quæ in cohæsione se prodit, est illa quidem in immensum major, quam quæ ex generali gravitate consequitur; cum ex ipsis Newtoni compertis attractio gravitati respondens [55] in globos homogeneos diversarum diametrorum sit in eadem ratione, in qua sunt globorum diametri, adeoque vis ejusmodi in exiguam particulam est eo minor gravitate corporum in Terram, quo minor est diameter particulæ diametro totius Terræ, adeoque penitus insensibilis. Et idcirco Newtonus aliam admisit vim pro cohæsione, quæ decrescat in ratione majore, quam sit reciproca duplicata distantiarum; & multi ex Newtonianis admiserunt vim respondentem huic formulæ $\frac{a}{x^3} + \frac{b}{x^2}$, cujus prior pars respectu posterioris sit in immensum minor, ubi x sit in immensum major unitate assumpta; sit vero major, ubi x sit in immensum minor, ut idcirco in satis magnis distantiis evanescente ad sensum prima parte, vis remaneat quam proxime in ratione reciproca duplicata distantiarum x, in minimis vero distantiis sit quam proxime in ratione reciproca triplicata: usque adeo ne apud Newtonianos quidem servatur omnino accurate ratio duplicata distantiarum.

Ex planetarum aphelii erui eam legem quamproxime, non accurate.

122. Demonstravit quidem Newtonus, in ellipsibus planetariis, eam, quam Astronomi lineam apsidum nominant, & est axis ellipseos, habituram ingentem motum, si ratio virium a reciproca duplicata distantiarum aliquanto magis aberret, cumque ad sensum quiescant

(f) *Id requiritur, quia in Mechanica demonstratur, aream curvæ, cujus abscissæ exprimant distantias, & ordinatæ vires, exprimere incrementum, vel decrementum quadrati velocitatis: quare ut illæ vires sint pares extinguendæ velocitati cuivis utcunque magnæ, debet illa area esse omni finita major.*

(v) *The straight line AB shall be an asymptote, and the asymptotic area BAED shall be infinite.*(f)

(vi) *The arcs lying between any two intersections may vary to any extent, may recede to any distances whatever from the axis C'AC, and approximate to any arcs of any curves to any degree of closeness, cutting them, or touching them, or osculating them, at any points and in any manner.*

Resolution of the curve of forces into the Newtonian attraction of gravitation and some other force.

119. Now, as regards the multiplicity of forces which at the present time physicists call by different names, it can also here be observed that, if anyone wants to consider one of these separately, the curve though it is of itself quite one-fold can yet be divided into several parts by a sort of mental & fictitious resolution of the forces. Thus, for instance, if anyone wishes to consider universal gravitation of matter exactly reciprocal to the squares of the distances; he can indeed describe on the attractive side the hyperbola which has its ordinates accurately in the inverse ratio of the squares of the distances, & this will be as it were a continuation of the branch VTS. Then he can add on to every ordinate, such as *ag*, *dh*, the ordinates of this new hyperbola, in the direction of AB, starting in each case from points on the curve, as *g*,*h* ; & in this way there will be obtained a fresh curve, which for the part *p*V will approximately coincide with the axis *o*C, & for the remainder will recede from it & wind itself about it, if the vertices F,K,O are more distant from the axis than the corresponding point on the hyperbola. Then it can be stated that all points of matter have gravitation accurately decreasing in the inverse square of the distance, together with another force represented by this new curve. For it comes to the same thing to think of these two laws of forces acting together as of the single law already given; & the results that arise will be the same also.

The resolution of this latter force into several other forces.

120. In the same manner this new curve can be divided into two others, or several others, by considering some other force, in some way or other accurately obeying certain fixed laws, & at the same time altering the curve just obtained by translating the points of it through intervals equal to the ordinates corresponding to the new law that has been taken. In this manner several different forces will be obtained ; & this will be sometimes useful, as we mentioned that it would be in resolution of forces, for determining their effects more readily ; & will be a sort of true resolution of forces. But all this will be as it were only a conception of our mind ; for, in reality, there is a single law of forces, & that is the one which I gave in Fig. 1, & it will be the compounded resultant of all such forces as the above.

The theory of gravitation is not in opposition ; this law does not hold good at very small distances.

121. Moreover, since I here make mention of gravitation decreasing accurately in the inverse ratio of the squares of the distances, it is to be remarked that no one should make any difficulty over the fact that, amongst physicists & more especially those who deal with celestial mechanics, it is considered as an established fact that gravitation decreases accurately in the inverse ratio of the squares of the distances, whilst in my Theory the law of forces is very different from this ratio. Especially, in the case of extremely small distances, the whole force, which the particles exert upon one another, will differ very much in every case from the force of gravity, if that is supposed to be inversely proportional to the squares of these distances. For, in the case of gases, which exercise such a mighty force of self-expansion, there is certainly repulsion at those very small distances from one another, & not attraction ; again, the attraction that arises in cohesion is immensely greater than it ought to be according to the law of universal gravitation. Now, from the results obtained by Newton, the attraction corresponding to gravitation in homogeneous spheres of different diameters varies as the diameters of the spheres ; & therefore this kind of force for the case of a tiny particle is as small in proportion to the gravitation of bodies to the Earth as the diameter of the particle is small in proportion to the diameter of the whole Earth ; & is thus insensible altogether. Hence Newton admitted another force in the case of cohesion, decreasing in a greater ratio than the inverse square of the distances ; also many of the followers of Newton have admitted a force corresponding to the formula, $a'x^3 + b'x^2$; in this the first term is immensely less than the second, when x is immensely greater than some distance assumed as unit distance ; & immensely greater, when x is immensely less. By this means, at sufficiently great distances the first part practically vanishes & the force remains very approximately in the inverse ratio of the squares of the distances x ; whilst, at very small distances, it is very nearly in the inverse ratio of the cubes of the distances. Thus indeed, not even amongst the followers of Newton has the inverse ratio of the squares of the distances been altogether rigidly adhered to.

The law follows very nearly, but not accurately, from the aphelia of the planets.

122. Now Newton proved, in the case of planetary elliptic orbits, that that which Astronomers call the apsidal line, i.e., the axis of the ellipse, would have a very great motion, if the ratio of the forces varied to any great extent from the inverse ratio of the squares of the distances ; & since as far as could be observed the lines of apses were stationary

(f) *This is required because in Mechanics it is shown that the area of a curve, whose abscissæ represent distances & ordinates forces, represents the increase or decrease of the square of the velocity. Hence in order that the forces should be capable of destroying any velocity however great, this area must be greater than any finite area.*

in earum orbitis apsidum lineæ, intulit, eam rationem observari omnino in gravitate. At id nequaquam evincit, accurate servari illam legem, sed solum proxime, neque inde ullum efficax argumentum contra meam Theoriam deduci potest. Nam inprimis nec omnino quiescunt illæ apsidum lineæ, sive, quod idem est, aphelia planetarum, sed motu exiguo quidem, at non insensibili prorsus, moventur etiam respectu fixarum, adeoque motu non tantummodo apparente, sed vero. Tribuitur is motus perturbationi virium ortæ ex mutua planetarum actione in se invicem ; at illud utique huc usque nondum demonstratum est, illum motum accurate respondere actionibus reliquorum planetarum agentium in ratione reciproca duplicata distantiarum ; neque enim adhuc sine contemptibus pluribus, & approximationibus a perfectione, & exactitudine admodum remotis solutum est problema, quod appellant, trium corporum, quo quæratur motus trium corporum in se mutuo agentium in ratione reciproca duplicata distantiarum, & utcunque projectorum, ac illæ ipsæ adhuc admodum imperfectæ solutiones, quæ prolatæ huc usque sunt, inserviunt tantummodo particularibus quibusdam casibus, ut ubi unum corpus sit maximum, & remotissimum, quemadmodum Sol, reliqua duo admodum minora & inter se proxima, ut est Luna, ac Terra, vel remota admodum a majore, & inter se, ut est Jupiter, & Saturnus. Hinc nemo hucusque accuratum instituit, aut etiam instituere potuit calculum pro actione perturbativa omnium planetarum, quibus si accedat actio perturbativa cometarum, qui, nec scitur, quam multi sint, nec quam longe abeant ; multo adhuc magis evidenter patebit, nullum inde confici posse argumentum [56] pro ipsa penitus accurata ratione reciproca duplicata distantiarum.

Idem ex reliqua Astronomia : posse autem hanc legem accedere ad illam quantum libuerit.

123. Clairautius quidem in schediasmate ante aliquot annos impresso, crediderat, ex ipsis motibus lineæ apsidum Lunæ colligi sensibilem recessum a ratione reciproca duplicata distantiæ, & Eulerus in dissertatione *De Aberrationibus Jovis, & Saturni*, quæ premium retulit ab Academia Parisiensi an. 1748, censuit, in ipso Jove, & Saturno haberi recessum admodum sensibilem ab illa ratione ; sed id quidem ex calculi defectu non satis producti sibi accidisse Clairautius ipse agnovit, ac edidit ; & Eulero aliquid simile fortasse accidit : nec ullum habetur positivum argumentum pro ingenti recessu gravitatis generalis a ratione duplicata distantiarum in distantia Lunæ, & multo magis in distantia planetarum. Vero nec ullum habetur argumentum positivum pro ratione ita penitus accurata, ut discrimen sensum omnem prorsus effugiat. At & si id haberetur ; nihil tamen pati posset inde Theoria mea ; cum arcus ille meæ curvæ postremus VT possit accedere, quantum libuerit, ad arcum illius hyperbolæ, quæ exhibet legem gravitatis reciprocam quadratorum distantiæ, ipsam tangendo, vel osculando in punctis quotcunque, & quibuscunque ; adeoque ita possit accedere, ut discrimen in iis majoribus distantiis sensum omnem effugiat, & effectus nullum habeat sensibile discrimen ab effectu, qui responderet ipsi legi gravitatis ; si ea accurate servaret proportionem cum quadratis distantiarum reciproce sumptis.

Difficultas a Maupertuisiana perfectione maxima Newtonianæ legis.

124. Nec vero quidquam ipsi meæ virium Theoriæ obsunt meditationes Maupertuisii, ingeniosæ illæ quidem, sed meo judicio nequaquam satis conformes Naturæ legibus circa legem virium decrescentium in ratione reciproca duplicata distantiarum, cujus ille perfectiones quasdam persequitur, ut illam, quod in hac una integri globi habeant candem virium legem, quam singulæ particulæ. Demonstravit enim Newtonus, globos, quorum singuli paribus a centro distantiis homogenei sint, & quorum particulæ minimæ se attrahant in ratione reciproca duplicata distantiarum, se itidem attrahere in eadem ratione distantiarum reciproca duplicata. Ob basce perfectiones hujus Theoriæ virium ipse censuit hanc legem reciprocam duplicatam distantiarum ab Auctore Naturæ selectam fuisse, quam in Natura esse vellet.

Prima responsio : nec cognosci fines omnes, & perfectiones, ac seligi etiam minus Perfecta in gratiam perfectionum.

125. At mihi quidem inprimis nec unquam placuit, nec placebit sane unquam in investigatione Naturæ causarum finalium usus, quas tantummodo ad meditationem quandam, contemplationemque, usui esse posse abitror, ubi leges Naturæ aliunde innotuerint. Nam nec perfectiones omnes innotescere nobis possunt, qui intimas rerum naturas nequaquam inspicimus, sed externas tantummodo proprietates quasdam agnoscimus, & fines omnes, quos Naturæ Auctor sibi potuit [57] proponere, ac proposuit, dum Mundum conderet,

in the orbits of each, he deduced that the ratio of the inverse square of the distances was exactly followed in the case of gravitation. But he only really proved that that law was very approximately followed, & not that it was accurately so; nor from this can any valid argument against my Theory be brought forward. For, in the first place these lines of apses, or what comes to the same thing, the aphelia of the planets are not quite stationary; but they have some motion, slight indeed but not quite insensible, with respect to the fixed stars, & therefore move not only apparently but really. This motion is attributed to the perturbation of forces which arises from the mutual action of the planets upon one another. But the fact remains that it has never up till now been proved that this motion exactly corresponds with the actions of the rest of the planets, where this is in accordance with the inverse ratio of the squares of the distances. For as yet the problem of three bodies, as they call it, has not been solved except by much omission of small quantities & by adopting approximations that are very far from truth and accuracy; in this problem is investigated the motion of three bodies acting mutually upon one another in the inverse ratio of the squares of the distances, & projected in any manner. Moreover, even these still only imperfect solutions, such as up till now have been published, hold good only in certain particular cases; such as the case in which one of the bodies is very large & at a very great distance, the Sun for instance, whilst the other two are quite small in comparison & very near one another, as are the Earth and the Moon, or at a large distance from the greater & from one another as well, as Jupiter & Saturn. Hence nobody has hitherto made, nor indeed could anybody make, an accurate calculation of the disturbing influence of all the other planets combined. If to this is added the disturbing influence of the comets, of which we neither know the number, nor how far off they are; it will be still more evident that from this no argument can be built up in favour of a perfectly exact observance of the inverse ratio of the squares of the distances.

The same thing is to be deduced from the rest of astronomy; moreover this law of mine can approximate to the other as nearly as is desired.

123. Clairaut indeed, in a pamphlet printed several years ago, asserted his belief that he had obtained from the motions of the line of apses for the Moon a sensible discrepancy from the inverse square of the distance. Also Euler, in his dissertation *De Aberrationibus Jovis, & Saturni*, which carried off the prize given by the Paris Academy, considered that in the case of Jupiter & Saturn there was quite a sensible discrepancy from that ratio. But Clairaut found out, & proclaimed the fact, that his result was indeed due to a defect in his calculation which had not been carried far enough; & perhaps something similar happened in Euler's case. Moreover, there is no positive argument in favour of a large discrepancy from the inverse ratio of the squares of the distances for universal gravitation in the case of the distance of the Moon, & still more in the case of the distances of the planets. Neither is there any rigorous argument in favour of the ratio being so accurately observed that the difference altogether eludes all observation. But even if this were the case, my Theory would not suffer in the least because of it. For the last arc VT of my curve can be made to approximate as nearly as is desired to the arc of the hyperbola that represents the law of gravitation according to the inverse squares of the distances, touching the latter, or osculating it in any number of points in any positions whatever; & thus the approximation can be made so close that at these relatively great distances the difference will be altogether unnoticeable, & the effect will not be sensibly different from the effect that would correspond to the law of gravitation, even if that exactly conformed to the inverse ratio of the squares of the distances.

Objection arising from the greatest perfection, according to Maupertuis, of the Newtonian law.

124. Further, there is nothing really to be objected to my Theory on account of the meditations of Maupertuis; these are certainly most ingenious, but in my opinion in no way sufficiently in agreement with the laws of Nature. Those meditations of his, I mean, with regard to the law of forces decreasing in the inverse ratio of the squares of the distances; for which law he strives to adduce certain perfections as this, that in this one law alone complete spheres have the same law of forces as the separate particles of which they are formed. For Newton proved that spheres, each of which have equal densities at equal distances from the centre, & of which the smallest particles attract one another in the inverse ratio of the squares of the distances, themselves also attract one another in the same ratio of the inverse squares of the distances. On account of such perfections as these in this Theory of forces, Maupertuis thought that this law of the inverse squares of the distances had been selected by the Author of Nature as the one He willed should exist in Nature.

First reply to this; all the aims and perfections are not known; and even a less perfect might be selected for the sake of greater perfection.

125. Now, in the first place I was never satisfied, nor really shall I ever be satisfied, with the use of final causes in the investigation of Nature; these I think can only be employed for a kind of study & contemplation, in such cases as those in which the laws of Nature have already been ascertained from other methods. For we cannot possibly be acquainted with all perfections; for in no wise do we observe the inmost nature of things, but all we know are certain external properties. Nor is it at all possible for us to see & know all the intentions which the Author of Nature could and did set before Himself when He founded

videre, & nosse omnino non possumus. Quin immo cum juxta ipsos Leibnitianos inprimis, aliosque omnes defensores acerrimos principii rationis sufficientis, & Mundi perfectissimi, qui inde consequitur, multa quidem in ipso Mundo sint mala, sed Mundus ipse idcirco sit optimus, quod ratio boni ad malum in hoc, qui electus est, omnium est maxima; fieri utique poterit, ut in ea ipsius Mundi parte, quam hic, & nunc contemplamur, id, quod electum fuit, debuerit esse non illud bonum, in cujus gratiam tolerantur alia mala, sed illud malum, quod in aliorum bonorum gratiam toleratur. Quamobrem si ratio reciproca duplicata distantiarum esset omnium perfectissima pro viribus mutuis particularum, non inde utique sequeretur, eam pro Natura fuisse electam, & constitutam.

Eandem legem nec perfectam esse, nec in corporibus, non utique accurate sphæricis habere locum.

126. At nec revera perfectissima est, quin immo meo quidem judicio est omnino imperfecta, & tam ipsa, quam aliæ plurimæ leges, quæ requirunt attractionem imminutis distantiis crescentem in ratione reciproca duplicata distantiarum, ad absurda deducunt plurima, vel saltem ad inextricabiles difficultates, quod ego quidem tum alibi etiam, tum inprimis demonstravi in dissertatione *De Lege Virium in Natura existentium* a num. 59. (g) Accedit autem illud, quod illa, quæ videtur ipsi esse perfectio maxima, quod nimirum eandem sequantur legem globi integri, quam particulæ minimæ, nulli fere usui est in Natura; si res accurate ad exactitudinem absolutam exigatur; cum nulli in Natura sint accurate perfecti globi paribus a centro distantiis homogenei, nam præter non exiguam inæqualitatem interioris textus, & irregularitatem, quam ego quidem in Tellure nostra demonstravi in Opere, quod de *Litteraria Expeditione per Pontificiam ditionem* inscripsi, in reliquis autem planetis, & cometis suspicari possumus ex ipsa saltem analogia, præter scabritiem superficiei, quæ utique est aliqua, satis patet, ipsa rotatione circa proprium axem induci in omnibus compressionem aliquam, quæ ut ut exigua, exactam globositatem impedit, adeoque illam assumptam perfectionem maximam corrumpit. Accedit autem & illud, quod Newtoniana determinatio rationis reciprocæ duplicatæ distantiarum locum habet tantummodo in globis materia continua constantibus sine ullis vacuolis, qui globi in Natura non existunt, & multo minus a me admitti possunt, qui non vacuum tantummodo disseminatum in materia, ut Philosophi jam sane passim, sed materiam in immenso vacuo innatantem, & punctula a se invicem remota, ex quibus, qui apparentes globi fiant, illam habere proprietatem non possunt rationis reciprocæ duplicatæ distantiarum, adeoque nec illins perfectionis creditæ maxime perfectam, absolutamque applicationem.

Objectio ex præjudicio pro impulsione, & ex testimonio sensuum: responsio ad hanc posteriorem.

[58] 127. Demum & illud nonnullis difficultatem parit summam in hac Theoria Virium, quod censeant, phænomena omnia per impulsionem explicari debere, & immediatum contactum, quem ipsum credant evidenti sensuum testimonio evinci; hinc hujusmodi nostras vires *immechanicas* appellant, & eas, ut & Newtonianorum generalem gravitatem, vel idcirco rejiciunt, quod mechanicæ non sint, & mechanismum, quem Newtoniana labefactare coeperat, penitus evertant. Addunt autem etiam per jocum ex serio argumento petito a sensibus, baculo utendum esse ad persuadendum neganti contactum. Quod ad sensuum testimonium pertinet, exponam uberius infra, ubi de extensione agam, quæ eo in genere habeamus præjudicia, & unde: cum nimirum ipsis sensibus tribuamus id, quod nostræ ratiocinationis, atque illationis vitio est tribuendum. Satis erit hic monere illud, ubi corpus ad nostra organa satis accedat, vim repulsivam, saltem illam ultimam, debere in organorum ipsorum fibris excitare motus illos ipsos, qui excitantur in communi sententia ab impenetrabilitate, & contactu, adeoque eundem tremorem ad cerebrum propagari, & eandem excitari debere in anima perceptionem, quæ in communi sententia excitaretur; quam ob rem ab iis sensationibus, quæ in hac ipsa Theoria Virium haberentur, nullum utique argumentum desumi potest contra ipsam, quod ullam vim habeant utcunque tenuem.

Felicius explicari omnia sine impulsione: eam nusquam positive probari.

128. Quod pertinet ad explicationem phænomenorum per impulsionem immediatam, monui sane superius, quanto felicius, ea prorsus omissa, Newtonus explicarit Astronomiam, & Opticam; & patebit inferius, quanto felicius phænomena quæque præcipua sine ulla immediata impulsione explicentur. Cum iis exemplis, tum aliis, commendatur abunde ea ratio explicandi phænomena, quæ adhibet vires agentes in aliqua distantia. Ostendant

(g) *Quæ huc pertinent, & continentur novem numeris ejus Dissertationis incipiendo a* 59, *habentur in fine Supplem.* § 4.

the Universe. Nay indeed, since in the doctrine of the followers of Leibniz more especially, and of all the rest of the keenest defenders of the principle of sufficient reason, and a most perfect Universe which is a direct consequence of that idea, there may be many evils in the Universe, and yet the Universe may be the best possible, just because the ratio of good to evil, in this that has been chosen, is the greatest possible. It might certainly happen that in this part of the Universe, which here & now we are considering, that which was chosen would necessarily be not that goodness in virtue of which other things that are evil are tolerated, but that evil which is tolerated because of the other things that are good. Hence, even if the inverse ratio of the squares of the distances were the most perfect of all for the mutual forces between particles, it certainly would not follow from that fact that it was chosen and established for Nature.

This law is neither perfect, nor does it hold good for bodies that are not exactly spherical.

126. But this law as a matter of fact is not the most perfect of all; nay rather, in my opinion, it is altogether imperfect. Both it, & several other laws, that require attraction at very small distances increasing in the inverse ratio of the squares of the distances lead to very many absurdities; or at least, to insuperable difficulties, as I showed in the dissertation *De Lege Virium in Natura existentium* in particular, as well as in other places.(g) In addition there is the point that the thing, which to him seems to be the greatest perfection, namely, the fact that complete spheres obey the same law as the smallest particles composing them, is of no use at all in Nature; for there are in Nature no exactly perfect spheres having equal densities at equal distances from the centre. Besides the not insignificant inequality & irregularity of internal composition, of which I proved the existence in the Earth, in a work which I wrote under the title of *De Litteraria Expeditione per Pontificiam ditionem*, we can assume also in the remaining planets & the comets (at least by analogy), in addition to roughness of surface (of which it is sufficiently evident that at any rate there is some), that there is some compression induced in all of them by the rotation about their axes. This compression, although it is indeed but slight, prevents true sphericity, & therefore nullifies that idea of the greatest perfection. There is too the further point that the Newtonian determination of the inverse ratio of the squares of the distances holds good only in spheres made up of continuous matter that is free from small empty spaces; & such spheres do not exist in Nature. Much less can I admit such spheres; for I do not so much as admit a vacuum disseminated throughout matter, as philosophers of all lands do at the present time, but I consider that matter as it were swims in an immense vacuum, & consists of little points separated from one another. These apparent spheres, being composed of these points, cannot have the property of the inverse ratio of the squares of the distances; & thus also they cannot bear the true & absolute application of that perfection that is credited so highly.

Objection founded on a prejudice for impulse, and on the testimony of the senses; reply to this latter.

127. Finally, some persons raise the greatest objections to this Theory of mine, because they consider that all the phenomena must be explained by impulse and immediate contact; this they believe to be proved by the clear testimony of the senses. So they call forces like those I propose *non-mechanical*, and reject them, just as they also reject the universal gravitation of Newton, for the alleged reason that they are not mechanical, and overthrow altogether the idea of mechanism which the Newtonian theory had already begun to undermine. Moreover, they also add, by way of a joke in the midst of a serious argument derived from the senses, that a stick would be useful for persuading anyone who denies contact. Now as far as the evidence of the senses is concerned, I will set forth below, when I discuss extension, the prejudices that we may form in such cases, and the origin of these prejudices. Thus, for instance, we may attribute to the senses what really ought to be attributed to the imperfection of our reasoning and inference. It will be enough just for the present to mention that, when a body approaches close enough to our organs, my repulsive force (at any rate it is that finally), is bound to excite in the nerves of those organs the motions which, according to the usual idea, are excited by impenetrability and contact; & that thus the same vibrations are sent to the brain, and these are bound to excite the same perception in the mind as would be excited in accordance with the usual idea. Hence, from these sensations, which are also obtained in my Theory of Forces, no argument can be adduced against the theory, which will have even the slightest validity.

Everything is more happily explained without the idea of impulse; and the latter is nowhere rigorously proved to exist.

128. As regards the explanation of phenomena by means of immediate contact I, indeed, mentioned above how much more happily Newton had explained Astronomy and Optics by omitting it altogether; and it will be evident, in what follows, how much more happily every one of the important phenomena is explained without any idea of immediate contact. Both by these instances, and by many others, this method of explaining phenomena, by employing forces acting at a distance, is strongly recommended. Let objectors bring

(g) *That which refers to this point, & which is contained in nine articles of the dissertation commencing with Art. 59, is to be found at the end of this work as Supplement IV.*

isti vel unicum exemplum, in quo positive probare possint, per immediatam impulsionem communicari motum in Natura. Id sane ii præstabunt nunquam; cum oculorum testimonium ad excludendas distantias illas minimas, ad quas primum crus repulsivum pertinet, & contorsiones curvæ circa axem, quæ oculos necessario fugiunt, adhibere non possint; cum e contrario ego positivo argumento superius excluserim immediatum contactum omnem, & positive probaverim, ipsum, quem ii ubique volunt, haberi nusquam.

Vires hujus Theoriæ pertinere ad verum, nec occultum mechanismum.

129. De nominibus quidem non esset, cur solicitudinem haberem ullam; sed ut & in iisdem aliquid præjudicio cuidam, quod ex communi loquendi usu provenit, illud notandum duco, Mechanicam non utique ad solam impulsionem immediatam fuisse restrictam unquam ab iis, qui de ipsa tractarunt, sed ad liberos inprimis adhibitam contemplandos motus, qui independenter ab omni impulsione habeantur. Quæ Archimedes de æquilibrio tradidit, quæ Galilæus de li-[59]-bero gravium descensu, ac de projectis, quæ de centralibus in circulo viribus, & oscillationis centro Hugenius, quæ Newtonus generaliter de motibus in trajectoriis quibuscunque, utique ad Mechanicam pertinent, & Wolfiana & Euleriana, & aliorum Scriptorum Mechanica passim utique ejusmodi vires, & motus inde ortos contemplatur, qui fiant impulsione vel exclusa penitus, vel saltem mente seclusa. Ubicunque vires agant, quæ motum materiæ gignant, vel immutent, & leges expandantur, secundum quas velocitas oriatur, mutetur motus, ac motus ipse determinetur; id omne inprimis ad Mechanicam pertinet in admodum propria significatione acceptam. Quamobrem ii maxime ea ipsa propria vocum significatione abutuntur, qui impulsionem unicam ad Mechanismum pertinere arbitrantur, ad quem hæc virium genera pertinent multo magis, quæ idcirco appellari jure possunt vires *Mechanicæ*, & quidquid per illas fit, jure affirmari potest fieri per *Mechanismum*, nec vero incognitum, & occultum, sed uti supra demonstravimus, admodum patentem, a manifestum.

Discrimen inter contactum mathematicum, & physicum: hunc dici proprie contactum.

130. Eodem etiam pacto in omnino propria significatione usurpare licebit vocem *contactus;* licet intervallum semper remaneat aliquod; quanquam ego ad æquivocationes evitandas soleo distinguere inter contactum *Mathematicum*, in quo distantia sit prorsus nulla, & contactum *Physicum*, in quo distantia sensus effugit omnes, & vis repulsiva satis magna ulteriorem accessum per nostras vires inducendum impedit. Voces ab hominibus institutæ sunt ad significandas res corporeas, & corporum proprietates, prout nostris sensibus subsunt, iis, quæ continentur infra ipsos, nihil omnino curatis. Sic planum, sic læve proprie dicitur id, in quo nihil, quod sensu percipi possit, sinuetur, nihil promineat; quanquam in communi etiam sententia nihil sit in Natura mathematice planum, vel læve. Eodem pacto & nomen *contactus* ab hominibus institutum est, ad exprimendum *physicum* illum *contactum* tantummodo, sine ulla cura *contactus mathematici*, de quo nostri sensus sententiam ferre non possunt. Atque hoc quidem pacto si adhibeantur voces in propria significatione illa, quæ ipsarum institutioni respondeat; ne a vocibus quidem ipsis huic Theoriæ virium invidiam creare poterunt ii, quibus ipsa non placet.

Transitus ab objectionibus contra Theoriam virium ad objectiones contra puncta.

131. Atque hæc de iis, quæ contra ipsam virium legem a me propositam vel objecta sunt hactenus, vel objici possent, sint satis, ne res in infinitum excrescat. Nunc ad illa transibimus, quæ contra constitutionem elementorum materiæ inde deductam se menti offerunt, in quibus itidem, quæ maxime notatu digna sunt, persequar.

Objectio ab idea puncti inextensi, qua caremus: responsio: unde idea extensionis sit orta.

132. Inprimis quod pertinet ad hanc constitutionem elementorum materiæ, sunt sane multi, qui nullo pacto in animum sibi possint inducere, ut admittant puncta prorsus indi-[60]-visibilia, & inextensa, quod nullam se dicant habere posse corum ideam. At id hominum genus præjudiciis quibusdam tribuit multo plus æquo. Ideas omnes, saltem eas, quæ ad materiam pertinent, per sensus hausimus. Porro sensus nostri nunquam potuerunt percipere singula elementa, quæ nimirum vires exerunt nimis tenues ad movendas fibras, & propagandum motum ad cerebrum: massis indiguerunt, sive elementorum aggregatis, quæ ipsas impellerent collata vi. Hæc omnia aggregata constabant partibus, quarum partium extremæ sumptæ hinc, & inde, debebant a se invicem distare per aliquod intervallum, nec ita exiguum. Hine factum est, ut nullam unquam per sensus acquirere potuerimus ideam pertinentem ad materiam, quæ simul & extensionem, & partes, ac divisibilitatem non involverit. Atque idcirco quotiescunque punctum nobis animo sistimus, nisi reflexione utamur, habemus ideam globuli cujusdam perquam exigui, sed tamen globuli rotundi, habentis binas superficies oppositas distinctis.

forward but a single instance in which they can positively prove that motion in Nature is communicated by immediate impulse. Of a truth they will never produce one; for they cannot use the testimony of the eyes to exclude those very small distances to which the first repulsive branch of my curve refers & the windings about the axis; for these necessarily evade ocular observation. Whilst I, on the other hand, by the rigorous argument given above, have excluded all idea of immediate contact; & I have positively proved that the thing, which they wish to exist everywhere, as a matter of fact exists nowhere.

The forces in this Theory refer to a real and not to an occult mechanism.

129. There is no reason why I should trouble myself about nomenclature; but, as in that too there is something that, from the customary manner of speaking, gives rise to a kind of prejudice, I think it should be observed that Mechanics was certainly never restricted to immediate impulse alone by those who have dealt with it; but that in the first place it was employed for the consideration of free motions, such as exist quite independently of any impulse. The work of Archimedes on equilibrium, that of Galileo on the free descent of heavy bodies & on projectiles, that of Huygens on central forces in a circular orbit & on the centre of oscillation, what Newton proved in general for motion on all sorts of trajectories; all these certainly belong to the science of Mechanics. The Mechanics of Wolf, Euler & other writers in different lands certainly treats of such forces as these & the motions that arise from them, & these matters have been accomplished with the idea of impulse excluded altogether, or at least put out of mind. Whenever forces act, & there is an investigation of the laws in accordance with which velocity is produced, motion is changed, or the motion itself is determined; the whole of this belongs especially to Mechanics in a truly proper signification of the term. Hence, they greatly abuse the proper signification of terms, who think that impulse alone belongs to the science of Mechanics; to which these kinds of forces belong to a far greater extent. Therefore these forces may justly be called *Mechanical*; & whatever comes about through their action can be justly asserted to have come about through a *mechanism*; & one too that is not unknown or mysterious, but, as we proved above, perfectly plain & evident.

Distinction between mathematical and physical contact; the latter to be more properly called contact.

130. Also in the same way we may employ the term contact in an altogether special sense; the interval may always remain something definite. Although, in order to avoid ambiguity, I usually distinguish between *mathematical* contact, in which the distance is absolutely nothing, & *physical* contact, in which the distance is too small to affect our senses, and the repulsive force is great enough to prevent closer approach being induced by the forces we are considering. Words are formed by men to signify corporeal things & the properties of such, as far as they come within the scope of the senses; & those that fall beneath this scope are absolutely not heeded at all. Thus, we properly call a thing plane or smooth, which has no bend or projection in it that can be perceived by the senses; although, in the general opinion, there is nothing in Nature that is mathematically plane or smooth. In the same way also, the term contact was invented by men to express *physical contact* only, without any thought of *mathematical contact*, of which our senses can form no idea. In this way, indeed, if words are used in their correct sense, namely, that which corresponds to their original formation, those who do not care for my Theory of forces cannot from those words derive any objection against it.

Passing on from objections against my Theory of forces to objections against points.

131. I have now said sufficient about those objections that either up till now have been raised, or might be raised, against the law of forces that I have proposed; otherwise the matter would grow beyond all bounds. Now we will pass on to objections against the constitution of the elements of matter derived from it, which present themselves to the mind; & in these also I will investigate those that more especially seem worthy of remark.

Objection to the idea of non-extended points, which we postulate; reply; the origin of the idea of extension.

132. First of all, as regards the constitution of the elements of matter, there are indeed many persons who cannot in any way bring themselves into that frame of mind to admit the existence of points that are perfectly indivisible and non-extended; for they say that they cannot form any idea of such points. But that type of men pays more heed than is right to certain prejudices. We derive all our ideas, at any rate those that relate to matter, from the evidences of our senses. Further, our senses never could perceive single elements, which indeed give forth forces that are too slight to affect the nerves & thus propagate motion to the brain. The senses would need masses, or aggregates of the elements, which would affect them as a result of their combined force. Now all these aggregates are made up of parts; & of these parts the two extremes on the one side and on the other must be separated from one another by a certain interval, & that not an insignificant one. Hence it comes about that we could never obtain through the senses any idea relating to matter, which did not involve at the same time extension, parts & divisibility. So, as often as we thought of a point, unless we used our reflective w , we should get the idea of a sort of ball, exceedingly small indeed, but still a round ball, having two distinct and opposite faces.

Ideam puncti debere acquiri per reflexionem: quomodo ejus idea negativa acquiratur.

133. Quamobrem ad concipiendum punctum indivisibile, & inextensum; non debemus consulere ideas, quas immediate per sensus hausimus; sed eam nobis debemus efformare per reflexionem. Reflexione adhibita non ita difficulter efformabimus nobis ideam ejusmodi. Nam inprimis ubi & extensionem, & partium compositionem conceperimus; si utranque negemus; jam inextensi, & indivisibilis ideam quandam nobis comparabimus per negationem illam ipsam eorum, quorum habemus ideam; uti foraminis ideam habemus utique negando existentiam illius materiæ, quæ deest in loco foraminis.

Quomodo ejus idea positiva acquiri possit per limites, & limitum intersectionem.

134. Verum & positivam quandam indivisibilis, & inextensi puncti ideam poterimus comparare nobis ope Geometriæ, & ope illius ipsius ideæ extensi continui, quam per sensus hausimus, & quam inferius ostendemus, fallacem esse, ac fontem ipsum fallaciæ ejusmodi aperiemus, quæ tamen ipsa ad indivisibilium, & inextensorum ideam nos ducet admodum claram. Concipiamus planum quoddam prorsus continuum, ut mensam, longum ex. gr. pedes duos; atque id ipsum planum concipiamus secari transversum secundum longitudinem ita, ut tamen iterum post sectionem conjungantur partes, & se contingant. Sectio illa erit utique limes inter partem dexteram & sinistram, longus quidem pedes duos, quanta erat plani longitudo, at latitudinis omnino expers: nam ab altera parte immediate motu continuo transitur ad alteram, quæ, si illa sectio crassitudinem haberet aliquam, non esset priori contigua. Illa sectio est limes secundum crassitudinem inextensus, & indivisibilis, cui si occurrat altera sectio transversa eodem pacto indivisibilis, & inextensa; oportebit utique, intersectio utriusque in superficie plani concepti nullam omnino habeat extensionem in partem quamcumque. Id erit punctum peni-[61]-tus indivisibile, & inextensum, quod quidem punctum, translato plano, movebitur, & motu suo lineam describet, longam quidem, sed latitudinis expertem.

Natura inextensi, quod non potest esse inextenso contiguum in lineis.

135. Quo autem melius ipsius indivisibilis natura concipi possit; quærat a nobis quispiam, ut aliam faciamus ejus planæ massæ sectionem, quæ priori ita sit proxima, ut nihil prorsus inter utramque intersit. Respondebimus sane, id fieri non posse: vel enim inter novam sectionem, & veteram intercedet aliquid ejus materiæ, ex qua planum continuum constare concipimus, vel nova sectio congruet penitus cum præcedente. En quomodo ideam acquiremus etiam ejus naturæ indivisibilis illius, & inextensi, ut aliud indivisibile, & inextensum ipsi proximum sine medio intervallo non admittat, sed vel cum eo congruat, vel aliquod intervallum relinquat inter se, & ipsum. Atque hinc patebit etiam illud, non posse promoveri planum ipsum ita, ut illa sectio promoveatur tantummodo per spatium latitudinis sibi æqualis. Utcunque exiguus fuerit motus, jam ille novus sectionis locus distabit a præcedente per aliquod intervallum, cum sectio sectioni contigua esse non possit.

Eadem in punctis: idea puncti geometrici translata ad physicum, & materiale.

136. Hæc si ad concursum sectionum transferamus, habebimus utique non solum ideam puncti indivisibilis, & inextensi, sed ejusmodi naturæ puncti ipsius, ut aliud punctum sibi contiguum habere non possit, sed vel congruant, vel aliquo a se invicem intervallo distent. Et hoc pacto sibi & Geometræ ideam sui puncti indivisibilis, & inextensi, facile efformare possunt, quam quidem etiam efformant sibi ita, ut prima Euclidis definitio jam inde incipiat: *punctum est, cujus nulla pars est.* Post hujusmodi ideam acquisitam illud unum intererit inter geometricum punctum, & punctum physicum materiæ, quod hoc secundum habebit proprietates reales vis inertiæ, & virium illarum activarum, quæ cogent duo puncta ad se invicem accedere, vel a se invicem recedere, unde fiet, ut ubi satis accesserint ad organa nostrorum sensuum, possint in iis excitare motus, qui propagati ad cerebrum, perceptiones ibi eliciant in anima, quo pacto sensibilia erunt, adeoque materialia, & realia, non pure imaginaria.

Punctorum existentiam aliunde demonstrari: per ideam acquisitam ea tantum concipi.

137. En igitur per reflexionem acquisitam ideam punctorum realium, materialium, indivisibilium, inextensorum, quam inter ideas ab infantia acquisitas per sensus incassum quærimus. Idea ejusmodi non evincit eorum existentiam. Ipsam quam nobis exhibent positiva argumenta superius facta, quod nimirum, ne admittatur in collisione corporum saltus, quem & inductio, & impossibilitas binarum velocitatum diversarum habendarum omnino ipso momento, quo saltus fieret, excludunt, oportet admittere in materia vires, quæ repulsivæ sint in minimis distantiis, & iis in infinitum imminutis augeantur in infinitum;

133. Hence for the purpose of forming an idea of a point that is indivisible & non-extended, we cannot consider the ideas that we derive directly from the senses; but we must form our own idea of it by reflection. If we reflect upon it, we shall form an idea of this sort for ourselves without much difficulty. For, in the first place, when we have conceived the idea of extension and composition by parts, if we deny the existence of both, then we shall get a sort of idea of non-extension & indivisibility by that very negation of the existence of those things of which we already have formed an idea. For instance, we have the idea of a hole by denying the existence of matter, namely, that which is absent from the position in which the hole lies.

The idea of a point must be obtained by reflection; how a negative idea of it may be acquired.

134. But we can also get an idea of a point that is indivisible & non-extended, by the aid of geometry, and by the help of that idea of an extended continuum that we derive from the senses; this we will show below to be a fallacy, & also we will open up the very source of this kind of fallacy, which nevertheless will lead us to a perfectly clear idea of indivisible & non-extended points. Imagine some thing that is perfectly plane and continuous, like a table-top, two feet in length; & suppose that this plane is cut across along its length; & let the parts after section be once more joined together, so that they touch one another. The section will be the boundary between the left part and the right part; it will be two feet in length (that being the length of the plane before section), & altogether devoid of breadth. For we can pass straightaway by a continuous motion from one part to the other part, which would not be contiguous to the first part if the section had any thickness. The section is a boundary which, as regards breadth, is non-extended & indivisible; if another transverse section which in the same way is also indivisible & non-extended fell across the first, then it must come about that the intersection of the two in the surface of the assumed plane has no extension at all in any direction. It will be a point that is altogether indivisible and non-extended; & this point, if the plane be moved, will also move and by its motion will describe a line, which has length indeed but is devoid of breadth.

How a positive idea can be acquired by means of boundaries, and intersections of boundaries.

135. The nature of an indivisible itself can be better conceived in the following way. Suppose someone should ask us to make another section of the plane mass, which shall lie so near to the former section that there is absolutely no distance between them. We should indeed reply that it could not be done. For either between the new section & the old there would intervene some part of the matter of which the continuous plane was composed; or the new section would completely coincide with the first. Now see how we acquire an idea also of the nature of that indivisible and non-extended thing, which is such that it does not allow another indivisible and non-extended thing to lie next to it without some intervening interval; but either coincides with it or leaves some definite interval between itself & the other. Hence also it will be clear that it is not possible so to move the plane, that the section will be moved only through a space equal to its own breadth. However slight the motion is supposed to be, the new position of the section would be at a distance from the former position by some definite interval; for a section cannot be contiguous to another section.

The nature of a non-extended thing, which cannot lie next to an extended thing as far as lines are concerned.

136. If now we transfer these arguments to the intersection of sections, we shall truly have not only the idea of an indivisible & non-extended point, but also an idea of the nature of a point of this sort; which is such that it cannot have another point contiguous to it, but the two either coincide or else they are separated from one another by some interval. In this way also geometricians can easily form an idea of their own kind of indivisible & non-extended points; & indeed they do so form their idea of them, for the first definition of Euclid begins:—*A point is that which has no parts.* After an idea of this sort has been acquired, there is but one difference between a geometrical point & a physical point of matter; this lies in the fact that the latter possesses the real properties of a force of inertia and of the active forces that urge the two points to approach towards, or recede from, one another; whereby it comes about that when they have approached sufficiently near to the organs of our senses, they can excite motions in them which, when propagated to the brain, induce sensations in the mind, and in this way become sensible, & thus material and real, & not imaginary.

The same thing for points; the idea of a geometrical point transferred to a physical and material point.

137. See then how by reflection the idea of real, material, indivisible, non-extended points can be acquired; whilst we seek for it in vain amongst those ideas that we have acquired since infancy by means of the senses. But an idea of this sort about things does not prove that these things exist. That is just what the rigorous arguments given above point out to us; that is to say, because, in order that in the collision of solids a sudden change should not be admitted (which change both induction & the impossibility of there being two different velocities at the same instant in which the change should take place), it had to be admitted that in matter there were forces which are repulsive at very small distances, & that these increased indefinitely as the distances were diminished.

The existence of points must be otherwise demonstrated; they can merely be thought of through acquiring an idea of them.

unde fit, ut duæ particulæ materiæ sibi [62] invicem contiguæ esse non possint: nam illico vi illa repulsiva resilient a se invicem, ac particula iis constans statim disrumpetur, adeoque prima materiæ elementa non constant contiguis partibus, sed indivisibilia sunt prorsus, atque simplicia, & vero etiam ob inductionem separabilitatis, ac distinctionis eorum, quæ occupant spatii divisibilis partes diversas, etiam penitus inextensa. Illa idea acquisita per reflexionem illud præstat tantummodo, ut distincte concipiamus id, quod ejusmodi rationes ostendunt existere in Natura, & quod sine reflexione, & ope illins supellectilis tantummodo, quam per sensus nobis comparavimus ab ipsa infantia, concipere omnino non liceret.

Puncta simplicia, & inextensa ab aliis quoque admissa: sed iis præstare hanc eorum theoriam.

138. Ceterum simplicium, & inextensorum notionem non ego primus in Physicam induco. Eorum ideam habuerunt veteres post Zenonem, & Leibnitiani monades suas & simplices utique volunt, & inextensas; ego cum ipsorum punctorum contiguitatem auferam, & distantias velim inter duo quælibet materiæ puncta, maximum evito scopulum, in quem utrique incurrunt, dum ex ejusmodi indivisibilibus, & inextensis continuum extensum componunt. Atque ibi quidem in eo videntur mihi peccare utrique, quod cum simplicitate, & inextensione, quam iis elementis tribuunt, commiscent ideam illam imperfectam, quam sibi compararunt per sensus, globuli cujusdam rotundi, qui binas habeat superficies a se distinctas, utcumque interrogati, an id ipsum faciant, omnino sint negaturi. Neque enim aliter possent ejusmodi simplicibus inextensis implere spatium, nisi concipiendo unum elementum in medio duorum ab altero contactum ad dexteram, ab altero ad lævam, quin ea extrema se contingant; in quo, præter contiguitatem indivisibilium, & inextensorum impossibilem, uti supra demonstravimus, quam tamen coguntur admittere, si rem altius perpenderint; videbunt sane, se ibi illam ipsam globuli inter duos globulos interjacentis ideam admiscere.

Impugnatur conciliatio extensionis formatæ ab inextensis petita ab impenetrabilitate.

139. Nec ad indivisibilitatem, & inextensionem elementorum conjungendas cum continua extensione massarum ab iis compositarum prosunt ea, quæ nonnulli ex Leibnitianorum familia proferunt, de quibus egi in una adnotatiuncula adjecta num. 13. dissertationis *De Materiæ Divisibilitate, & Principiis Corporum*, ex qua, quæ eo pertinent, huc libet transferre. Sic autem habet: *Qui dicunt, monades non compenetrari, quia natura sua impenetrabiles sunt, ii difficultatem nequaquam amovent; nam si & natura sua impenetrabiles sunt, & continuum debent componere, adeoque contigua esse; compenetrabuntur simul, & non compenetrabuntur, quod ad absurdum deducit, & ejusmodi entium impossibilitatem evincit. Ex omnimodæ inextensionis, & contiguitatis notione evincitur, compenetrari debere argumento contra Zenonistas instituto per tot sæcula, & cui nunquam satis responsum est. Ex natura, quæ in* [63] *iis supponitur, ipsa compenetratio excluditur, adeoque habetur contradictio, & absurdum.*

Inductionem a sensibilibus compositis, & extensis haud valere contra puncta simplicia, & inextensa.

140. Sunt alii, quibus videri poterit, contra hæc ipsa puncta indivisibilia, & inextensa adhiberi posse inductionis principium, a quo continuitatis legem, & alias proprietates derivavimus supra, quæ nos ad hæc indivisibilia, & inextensa puncta deduxerunt. Videmus enim in materia omni, quæ se uspiam nostris objiciat sensibus, extensionem, divisibilitatem, partes; quamobrem hanc ipsam proprietatem debemus transferre ad elementa etiam per inductionis principium. Ita ii: at hanc difficultatem jam superius præoccupavimus, ubi egimus de inductionis principio. Pendet ea proprietas a ratione sensibilis, & aggregati, cum nimirum sub sensus nostros ne composita quidem, quorum moles nimis exigua sit, cadere possint. Hinc divisibilitatis, & extensionis proprietas ejusmodi est; ut ejus defectus, si habeatur alicubi is casus, ex ipsa earum natura, & sensuum nostrorum constitutione non possit cadere sub sensus ipsos, atque idcirco ad ejusmodi proprietates argumentum desumptum ab inductione nequaquam pertingit, ut nec ad sensibilitatem extenditur.

Per ipsam etiam exclusionem inextensi vi inductionis habitam ipsum extensum excludi.

141. Sed etiam si extenderetur, esset adhuc nostræ Theoriæ causa multo melior in eo, quod circa extensionem, & compositionem partium negativa sit. Nam eo ipso, quod continuitate admissa, continuitas elementorum legitima ratiocinatione excludatur, excludi omnino debet absolute; ubi quidem illud accidit, quod a Metaphysicis, & Geometris nonnullis animadversum est jam diu, licere aliquando demonstrare propositionem ex

From this it comes about that two particles of matter cannot be contiguous; for thereupon they would recoil from one another owing to that repulsive force, & a particle composed of them would at once be broken up. Thus, the primary elements of matter cannot be composed of contiguous parts, but must be perfectly indivisible & simple; and also on account of the induction from separability & the distinction between those that occupy different divisible parts of space, they must be perfectly non-extended as well. The idea acquired by reflection only yields the one result, namely, that through it we may form a clear conception of that which reasoning of this kind proves to be existent in Nature; of which, without reflection, using only the equipment that we have got together for ourselves by means of the senses from our infancy, we could not have formed any conception.

Simple and non-extended points are admitted by others as well; but my Theory about them is the best.

138. Besides, I was not the first to introduce the notion of simple non-extended points into physics. The ancients from the time of Zeno had an idea of them, & the followers of Leibniz indeed suppose that their monads are simple & non-extended. I, since I do not admit the contiguity of the points themselves, but suppose that any two points of matter are separated from one another, avoid a mighty rock, upon which both these others come to grief, whilst they build up an extended continuum from indivisible & non-extended things of this sort. Both seem to me to have erred in doing so, because they have mixed up with the simplicity & non-extension that they attribute to the elements that imperfect idea of a sort of round globule having two surfaces distinct from one another, an idea they have acquired through the senses; although, if they were asked if they had made this supposition, they would deny that they had done so. For in no other way can they fill up space with indivisible and non-extended things of this sort, unless by imagining that one element between two others is touched by one of them on the right & by the other on the left. If such is their idea, in addition to contiguity of indivisible & non-extended things (which is impossible, as I proved above, but which they are forced to admit if they consider the matter more carefully); in addition to this, I say, they will surely see that they have introduced into their reasoning that very idea of the two little spheres lying between two others.

The deduction from impenetrability of a conciliation of extension with its formation from non-extended things.

139. Those arguments that some of the Leibnitian circle put forward are of no use for the purpose of connecting indivisibility & non-extension of the elements with continuous extension of the masses formed from them. I discussed the arguments in question in a short note appended to Art. 13 of the dissertation *De Materiæ Divisibilitate and Principiis Corporum*; & I may here quote from that dissertation those things that concern us now. These are the words :—*Those, who say that monads cannot be compenetrated, because they are by nature impenetrable, by no means remove the difficulty. For, if they are both by nature impenetrable, & also at the same time have to make up a continuum, i.e., have to be contiguous, then at one & the same time they are compenetrated & they are not compenetrated; & this leads to an absurdity & proves the impossibility of entities of this sort. For, from the idea of non-extension of any sort, & of contiguity, it is proved by an argument instituted against the Zenonists many centuries ago that there is bound to be compenetration; & this argument has never been satisfactorily answered. From the nature that is ascribed to them, this compenetration is excluded. Thus there is a contradiction & an absurdity.*

Induction derived from things that are sensible, compound, and extended are of no avail for the purpose of opposing simple and non-extended things.

140. There are others, who will think that it is possible to employ, for the purpose of opposing the idea of these indivisible & non-extended points, the principle of induction, by which we derived the Law of Continuity & other properties, which have led us to these indivisible & non-extended points. For we perceive (so they say) in all matter, that falls under our notice in any way, extension, divisibility & parts. Hence we must transfer this property to the elements also by the principle of induction. Such is their argument. But we have already discussed this difficulty, when we dealt with the principle of induction. The property in question depends on a reasoning concerned with a sensible body, & one that is an aggregate; for, in fact, not even a composite body can come within the scope of our senses, if its mass is over-small. Hence the property of divisibility & extension is such that the absence of this property (if this case ever comes about), from the very nature of divisibility & extension, & from the constitution of our senses, cannot fall within the scope of those senses. Therefore an argument derived from induction will not apply to properties of this kind in any way, inasmuch as the extension does not reach the point necessary for sensibility.

Extension itself is excluded by the exclusion of non-extension, obtained by the force of induction.

141. But even if this point is reached, there would only be all the more reason for our Theory from the fact that it denies extension and composition by parts. For, from the very fact that, if continuity be admitted, continuity of the elements is excluded by legitimate argument, it follows that continuity ought to be absolutely excluded in all cases. For in that case we get an instance of the argument that has been observed by metaphysicists and some geometers for a very long time, namely, that a proposition may sometimes be

assumpta veritate contradictoriæ propositionis ; cum enim ambæ simul veræ esse non possint, si ab altera inferatur altera, hanc posteriorem veram esse necesse est. Sic nimirum, quoniam a continuitate generaliter assumpta defectus continuitatis consequitur in materiæ elementis, & in extensione, defectum hunc haberi vel inde eruitur : nec oberit quidquam principium inductionis physicæ, quod utique non est demonstrativum, nec vim habet, nisi ubi aliunde non demonstretur, casum illum, quem inde colligere possumus, improbabilem esse tantummodo, adhuc tamen haberi, uti aliquando sunt & falsa veris probabiliora.

Cujusmodi continuum in hac Theoria admittatur: quid sit spatium, & tempus.

142. Atque hic quidem, ubi de continuitate seipsam excludente mentio injecta est, notandum & illud, continuitatis legem a me admitti, & probari pro quantitatibus, quæ magnitudinem mutent, quas nimirum ab una magnitudine ad aliam censeo abire non posse, nisi transeant per intermedias, quod elementorum materiæ, quæ magnitudinem nec mutant, nec ullam habent variabilem, continuitatem non inducit, sed argumento superius facto penitus summovet. Quin etiam ego quidem continuum nullum agnosco coexistens, uti & supra monui ; nam nec spatium reale mihi est ullum continuum, sed [64] imaginarium tantummodo, de quo, uti & de tempore, quæ in hac mea Theoria sentiam, satis luculenter exposui in Supplementis ad librum I. Stayanæ Philosophiæ (h). Censeo nimirum quodvis materiæ punctum, habere binos reales existendi modos, alterum localem, alterum temporarium, qui num appellari debeant res, an tantummodo modi rei, ejusmodi litem, quam arbitror esse tantum de nomine, nihil omnino curo. Illos modos debere admitti, ibi ego quidem positive demonstro : eos natura sua immobiles esse, censeo ita, ut idcirco ejusmodi existendi modi per se inducant relationes prioris, & posterioris in tempore, ulterioris, vel citerioris in loco, ac distantiæ cujusdam determinatæ, & in spatio determinatæ positionis etiam, qui modi, vel eorum alter, necessario mutari debeant, si distantia, vel etiam in spatio sola mutetur positio. Pro quovis autem modo pertinente ad quodvis punctum, penes omnes infinitos modos possibiles pertinentes ad quodvis aliud, mihi est unus, qui cum eo inducat in tempore relationem coexistentiæ ita, ut existentiam habere uterque non possit, quin simul habeant, & coexistant ; in spatio vero, si existunt simul, inducant relationem compenetrationis, reliquis omnibus inducentibus relationem distantiæ temporariæ, vel localis, ut & positionis cujusdam localis determinatæ. Quoniam autem puncta materiæ existentia habent semper aliquam a se invicem distantiam, & numero finita sunt ; finitus est semper etiam localium modorum coexistentium numerus, nec ullum reale continuum efformat. Spatium vero imaginarium est mihi possibilitas omnium modorum localium confuse cognita, quos simul per cognitionem præcisivam concipimus, licet simul omnes existere non possint, ubi cum nulli sint modi ita sibi proximi, vel remoti, ut alii viciniores, vel remotiores haberi non possint, nulla distantia inter possibiles habetur, sive minima omnium, sive maxima. Dum animum abstrahimus ab actuali existentia, & in possibilium serie finitis in infinitum constante terminis mente secludimus tam minimæ, quam maximæ distantiæ limitem, ideam nobis efformamus continuitatis, & infinitatis in spatio, in quo idem spatii punctum appello possibilitatem omnium modorum localium, sive, quod idem est, realium localium punctorum pertinentium ad omnia materiæ puncta, quæ si existerent, compenetrationis relationem inducerent, ut eodem pacto idem nomino momentum temporis temporarios modos omnes, qui relationem inducunt coexistentiæ. Sed de utroque plura in illis dissertatiunculis, in quibus & analogiam persequor spatii, ac temporis multiplicem.

Ubi habeat continuitatem Natura ubi affectet tantummodo.

[65] 143. Continuitatem igitur agnosco in motu tantummodo, quod est successivum quid, non coexistens, & in eo itidem solo, vel ex eo solo in corporeis saltem entibus legem continuitatis admitto. Atque hinc patebit clarius illud etiam, quod superius innui, Naturam ubique continuitatis legem vel accurate observare, vel affectare saltem. Servat in motibus, & distantiis, affectat in aliis casibus multis, quibus continuitas, uti etiam supra definivimus, nequaquam convenit, & in aliis quibusdam, in quibus haberi omnino non potest continuitas, quæ primo aspectu sese nobis objicit res non aliquanto intimius inspectantibus, ac perpendentibus : ex. gr. quando Sol oritur supra horizontem, si concipiamus Solis discum

(h) *Binæ dissertatiunculæ, quæ huc pertinent, inde excerptæ habentur hic Supplementorum* § I, *&* 2, *quarum mentio facta est etiam superius num.* 66, *&* 86.

proved by assuming the truth of the contradictory proposition. For since both propositions cannot be true at the same time, if from one of them the other can be inferred, then the latter of necessity must be the true one. Thus, for instance, because it follows, from the assumption of continuity in general, that there is an absence of continuity in the elements of matter, & also in the case of extension, we come to the conclusion that there is this absence. Nor will any principle of physical induction be prejudicial to the argument, where the induction is not one that can be proved in every case; neither will it have any validity, except in the case where it cannot be proved in other ways that the conclusion that we can come to from the argument is highly improbable but yet is to be held as true; for indeed sometimes things that are false are more plausible than the true facts.

The sort of continuum that is admitted in this Theory; the nature of space and time.

142. Now, in this connection, whilst incidental mention has been made of the exclusion of continuity, it should be observed that the Law of Continuity is admitted by me, & proved for those quantities that change their magnitude, but which indeed I consider cannot pass from one magnitude to another without going through intermediate stages; but that this does not lead to continuity in the case of the elements of matter, which neither change their magnitude nor have anything variable about them; on the contrary it proves quite the opposite, as the argument given above shows. Moreover, I recognize no co-existing continuum, as I have already mentioned; for, in my opinion, space is not any real continuum, but only an imaginary one; & what I think about this, and about time as well, as far as this Theory is concerned, has been expounded clearly enough in the supplements to the first book of Stay's Philosophy.(h) For instance, I consider that any point of matter has two modes of existence, the one local and the other temporal; I do not take the trouble to argue the point as to whether these ought to be called things, or merely modes pertaining to a thing, as I consider that this is merely a question of terminology. That it is necessary that these modes be admitted, I prove rigorously in the supplements mentioned above. I consider also that they are by their very nature incapable of being displaced; so that, of themselves, such modes of existence lead to the relations of before & after as regards time, far & near as regards space, & also of a given distance & a given position in space. These modes, or one of them, must of necessity be changed, if the distance, or even if only the position in space is altered. Moreover, for any one mode belonging to any point, taken in conjunction with all the infinite number of possible modes pertaining to any other point, there is in my opinion one which, taken in conjunction with the first mode, leads as far as time is concerned to a relation of coexistence; so that both cannot have existence unless they have it simultaneously, i.e., they coexist. But, as far as space is concerned, if they exist simultaneously, the conjunction leads to a relation of compenetration. All the others lead to a relation of temporal or of local distance, as also of a given local position. Now since existent points of matter always have some distance between them, & are finite in number, the number of local modes of existence is also always finite; & from this finite number we cannot form any sort of real continuum. But I have an ill-defined idea of an imaginary space as a possibility of all local modes, which are precisely conceived as existing simultaneously, although they cannot all exist simultaneously. In this space, since there are not modes so near to one another that there cannot be others nearer, or so far separated that there cannot be others more so, there cannot therefore be a distance that is either the greatest or the least of all, amongst those that are possible. So long as we keep the mind free from the idea of actual existence &, in a series of possibles consisting of an indefinite number of finite terms, we mentally exclude the limit both of least & greatest distance, we form for ourselves a conception of continuity & infinity in space. In this, I define the same point of space to be the possibility of all local modes, or what comes to the same thing, of real local points pertaining to all points of matter, which, if they existed, would lead to a relation of compenetration; just as I define the same instant of time as all temporal modes, which lead to a relation of coexistence. But there is a fuller treatment of both these subjects in the notes referred to; & in them I investigate further the manifold analogy between space & time.

Where there is continuity in Nature; where Nature does no more than attempt to maintain it.

143. Hence I acknowledge continuity in motion only, which is something successive and not co-existent; & also in it alone, or because of it alone, in corporeal entities at any rate, lies my reason for admitting the Law of Continuity. From this it will be all the more clear that, as I remarked above, Nature accurately observes the Law of Continuity, or at least tries to do so. Nature observes it in motions & in distance, & tries to in many other cases, with which continuity, as we have defined it above, is in no wise in agreement; also in certain other cases, in which continuity cannot be completely obtained. This continuity does not present itself to us at first sight, unless we consider the subjects somewhat more deeply & study them closely. For instance, when the sun rises above the horizon,

(h) *The two notes, which refer to this matter, have been quoted in this work as supplements I & II: these have been already referred to in Arts. 66 & 86 above.*

ut continuum, & horizontem ut planum quoddam ; ascensus Solis fit per omnes magnitudines ita, ut a primo ad postremum punctum & segmenta solaris disci, & chordæ segmentorum crescant transeundo per omnes intermedias magnitudines. At Sol quidem in mea Theoria non est aliquid continuum, sed est aggregatum punctorum a se invicem distantium, quorum alia supra illud imaginarium planum ascendunt post alia, intervallo aliquo temporis interposito semper. Hinc accurata illa continuitas huic casui non convenit, & habetur tantummodo in distantiis punctorum singulorum componentium eam massam ab illo imaginario plano. Natura tamen etiam hic continuitatem quandam affectat, cum nimirum illa punctula ita sibi sint invicem proxima, & ita ubique dispersa, ac disposita, ut apparens quædam ibi etiam continuitas habeatur, ac in ipsa distributione, a qua densitas pendet, ingentes repentini saltus non fiant.

Exempla continuitatis apparentis tantum : unde ea ortum ducat.

144. Innumera ejus rei exempla liceret proferre, in quibus eodem pacto res pergit. Sic in fluviorum alveis, in frondium flexibus, in ipsis salium, & crystallorum, ac aliorum corporum angulis, in ipsis cuspidibus unguium, quæ acutissimæ in quibusdam animalibus apparent nudo oculo ; si microscopio adhibito inspiciantur ; nusquam cuspis abrupta prorsus, nusquam omnino cuspidatus apparet angulus, sed ubique flexus quidam, qui curvaturam habeat aliquam, & ad continuitatem videatur accedere. In omnibus tamen iis casibus vera continuitas in mea Theoria habetur nusquam ; cum omnia ejusmodi corpora constent indivisibilibus, & a se distantibus punctis, quæ continuam superficiem non efformant, & in quibus, si quævis tria puncta per rectas lineas conjuncta intelligantur ; triangulum habebitur utique cum angulis cuspidatis. Sed a motuum, & virium continuitate accurata etiam ejusmodi proximam continuitatem massarum oriri censeo, & a casuum possibilium multitudine inter se collata, quod ipsum innuisse sit satis.

Motuum omnium continuitas in lineis continuis nusquam interruptis, aut mutatis.

145. Atque hinc fiet manifestum, quid respondendum ad casus quosdam, qui eo pertinent, & in quibus violari quis crederet [66] continuitatis legem. Quando plano aliquo speculo lux excipitur, pars refringitur, pars reflectitur : in reflexione, & refractione, uti eam olim creditum est fieri, & etiamnum a nonnullis creditur, per impulsionem nimirum, & incursum immediatum, fieret violatio quædam continui motus mutata linea recta in aliam ; sed jam hoc Newtonus advertit, & ejusmodi saltum abstulit, explicando ea phænomena per vires in aliqua distantia agentes, quibus fit, ut quævis particula luminis motum incurvet paullatim in accessu ad superficiem reflectentem, vel refringentem ; unde accessuum, & recessuum lex, velocitas, directionum flexus, omnia juxta continuitatis legem mutantur. Quin in mea Theoria non in aliqua vicinia tantum incipit flexus ille, sed quodvis materiæ punctum a Mundi initio unicam quandam continuam descripsit orbitam, pendentem a continua illa virium lege, quam exprimit figura 1, quæ ad distantias quascunque protenditur ; quam quidem lineæ continuitatem nec liberæ turbant animarum vires, quas itidem non nisi juxta continuitatis legem exerceri a nobis arbitror ; unde fit, ut quemadmodum omnem accuratam quietem, ita omnem accurate rectilineum motum, omnem accurate circularem, ellipticum, parabolicum excludam ; quod tamen aliis quoque sententiis omnibus commune esse debet ; cum admodum facile sit demonstrare, ubique esse perturbationem quandam, & mutationum causas, quæ non permittant ejusmodi linearum nobis ita simplicium accuratas orbitas in motibus.

Apparens saltus in diffusione reflexi, ac refracti luminis.

146. Et quidem ut in iis omnibus, & aliis ejusmodi Natura semper in mea Theoria accuratissimam continuitatem observat, ita & hic in reflexionibus, ac refractionibus luminis. At est aliud ea in re, in quo continuitatis violatio quædam haberi videatur, quam, qui rem altius perpendat, credet primo quidem, servari itidem accurate a Natura, tum ulterius progressus, inveniet affectari tantummodo, non servari. Id autem est ipsa luminis diffusio, atque densitas. Videtur prima fronte discindi radius in duos, qui hiatu quodam intermedio a se invicem divellantur velut per saltum, alia parte reflexa, ali refracta, sine ullo intermedio flexu cujuspiam. Alius itidem videtur admitti ibidem saltus quidam : si enim radius integer excipiatur prismate ita, ut una pars reflectatur, alia transmittatur, & prodeat etiam e secunda superficie, tum ipsum prisma sensim convertatur ; ubi ad certum devenitur in conversione angulum, lux, quæ datam habet refrangibilitatem, jam non egreditur, sed reflectitur in totum ; ubi itidem videtur fieri transitus a prioribus angulis cum superficie semper minoribus, sed jacentibus ultra ipsam, ad angulum reflexionis æqualem angulo

if we think of the Sun's disk as being continuous, & the horizon as a certain plane; then the rising of the Sun is made through all magnitudes in such a way that, from the first to the last point, both the segments of the solar disk & the chords of the segments increase by passing through all intermediate magnitudes. But, in my Theory, the Sun is not something continuous, but is an aggregate of points separate from one another, which rise, one after the other, above that imaginary plane, with some interval of time between them in all cases. Hence accurate continuity does not fit this case, & it is only observed in the case of the distances from the imaginary plane of the single points that compose the mass of the Sun. Yet Nature, even here, tries to maintain a sort of continuity; for instance, the little points are so very near to one another, & so evenly spread & placed that, even in this case, we have a certain apparent continuity, and even in this distribution, on which the density depends, there do not occur any very great sudden changes.

Examples of continuity that is merely apparent; its origin.

144. Innumerable examples of this apparent continuity could be brought forward, in which the matter comes about in the same manner. Thus, in the channels of rivers, the bends in foliage, the angles in salts, crystals and other bodies, in the tips of the claws that appear to the naked eye to be very sharp in the case of certain animals; if a microscope were used to examine them, in no case would the point appear to be quite abrupt, or the angle altogether sharp, but in every case somewhat rounded, & so possessing a definite curvature & apparently approximating to continuity. Nevertheless in all these cases there is nowhere true continuity according to my Theory; for all bodies of this kind are composed of points that are indivisible & separated from one another; & these cannot form a continuous surface; & with them, if any three points are supposed to be joined by straight lines, then a triangle will result that in every case has three sharp angles. But I consider that from the accurate continuity of motions & forces a very close approximation of this kind arises also in the case of masses; &, if the great number of possible cases are compared with one another, it is sufficient for me to have just pointed it out.

The continuity of motions in continuous lines is nowhere interrupted or altered.

145. Hence it becomes evident how we are to refute certain cases, relating to this matter, in which it might be considered that the Law of Continuity was violated. When light falls upon a plane mirror, part is refracted & part is reflected. In reflection & refraction, according to the idea held in olden times, & even now credited by some people, namely, that it took place by means of impulse & immediate collision, there would be a breach of continuous motion through one straight line being suddenly changed for another. But already Newton has discussed this point, & has removed any sudden change of this sort, by explaining the phenomena by means of forces acting at a distance; with these it comes about that any particle of light will have its path bent little by little as it approaches a reflecting or refracting surface. Hence, the law of approach and recession, the velocity, the alteration of direction, all change in accordance with the Law of Continuity. Nay indeed, in my Theory, this alteration of direction does not only begin in the immediate neighbourhood, but any point of matter from the time that the world began has described a single continuous orbit, depending on the continuous law of forces, represented in Fig. 1, a law that extends to all distances whatever. I also consider that this continuity of path is undisturbed by any voluntary mental forces, which also cannot be exerted by us except in accordance with the Law of Continuity. Hence it comes about that, just as I exclude all idea of absolute rest, so I exclude all accurately rectilinear, circular, elliptic, or parabolic motions. This too ought to be the general opinion of all others; for it is quite easy to show that there is everywhere some perturbation, & reasons for alteration, which do not allow us to have accurate paths along such simple lines for our motions.

Apparent discontinuity in diffusion of reflected and refracted light.

146. Just as in all the cases I have mentioned, & in others like them, Nature always in my Theory observes the most accurate continuity, so also is this done here in the case of the reflection and refraction of light. But there is another thing in this connection, in which there seems to be a breach of continuity; & anyone who considers the matter fairly deeply, will think at first that Nature has observed accurate continuity, but on further consideration will find that Nature has only endeavoured to do so, & has not actually observed it; that is to say, in the diffusion of light, & its density. At first sight the ray seems to be divided into two parts, which leave a gap between them & diverge from one another as it were suddenly, the one part being reflected & the other part refracted without any intermediate bending of the path. It also seems that another sudden change must be admitted; for suppose that a beam of light falls upon a prism, & part of it is reflected & the rest is transmitted & issues from the second surface, and that then the prism is gradually rotated; when a certain angle of rotation is reached, light, having a given refrangibility, is no longer transmitted, but is totally reflected. Here also it seems that there is a sudden transition from the first case in which the angles made with the surface by the issuing rays are always less than the angle of incidence, & lie on the far side of the surface, to the latter case in which the angles of reflection are equal to

incidentiæ, & jacentem citra, sine ulla reflexione in angulis intermediis minoribus ab ipsa superficie ad ejusmodi finitum angulum.

Apparens conciliatio cum lege continuitatis per radios irregulariter dispersos.

147. Huic cuidam velut læsioni continuitatis videtur responderi posse per illam lucem quæ reflectitur, vel refrin-[67]-gitur irregulariter in quibusvis angulis. Jam olim enim observatum est illud, ubi lucis radius reflectitur, non reflecti totum ita, ut angulus reflexionis æquetur angulo incidentiæ, sed partem dispergi quaquaversus; quam ob causam si Solis radius in partem quandam speculi incurrat, quicunque est in conclavi, videt, qui sit ille locus, in quem incurrit radius, quod utique non fieret, nisi e solaribus illis directis radiis etiam ad oculum ipsius radii devenirent, egressi in omnibus iis directionibus, quæ ad omnes oculi positiones tendunt; licet ibi quidem satis intensum lumen non appareat, nisi in directione faciente angulum reflexionis æqualem incidentiæ, in qua resilit maxima luminis pars. Et quidem hisce radiis redeuntibus in angulis bisce inæqualibus egregie utitur Newtonus in fine Opticæ ad explicandos colores laminarum crassarum: & eadem irregularis dispersio in omnes plagas ad sensum habetur in tenui parte, sed tamen in aliqua, radii refracti. Hine inter vividum illum reflexum radium, & refractum, habetur intermedia omnis ejusmodi radiorum series in omnibus iis intermediis angulis prodeuntium, & sic etiam ubi transitur a refractione ad reflexionem in totum, videtur per bosce intermedios angulos res posse fieri citissimo transitu per ipsos, atque idcirco illæsa perseverare continuitas.

Cur ea apparens tantum: vera conciliatio per continuitatem viæ cujusvis punctilucis.

148. Verum si adhuc altius perpendatur res; patebit in illa intermedia serie non haberi accuratam continuitatem, sed apparentem quandam, quam Natura affectat, non accurate servat illæsam. Nam lumen in mea Theoria non est corpus quoddam continuum, quod diffundatur continuo per illud omne spatium, sed est aggregatum punctorum a se invicem disjunctorum, atque distantium, quorum quodlibet suam percurrit viam disjunctam a proximi via per aliquod intervallum. Continuitas servatur accuratissime in singulorum punctorum viis, non in diffusione substantiæ non continuæ, & quo pacto ea in omnibus iis motibus servetur, & mutetur, mutata inclinatione incidentiæ, via a singulis punctis descripta sine saltu, satis luculenter exposui in secunda parte meæ dissertationis *De Lumine* a num. 98. Sed hæc ad applicationem jam pertinent Theoriæ ad Physicam.

Quo pacto servetur continuitas in quibusdam casibus, in quibus videtur lædi.

149. Haud multum absimiles sunt alii quidam casus, in quibus singula continuitatem observant, non aggregatum utique non continuum, sed partibus disjunctis constans. Hujusmodi est ex. gr. altitudo cujusdam domus, quæ ædificatur de novo, cui cum series nova adjungitur lapidum determinatæ cujusdam altitudinis, per illam additionem repente videtur crescere altitudo domus, sine transitu per altitudines intermedias: & si dicatur id non esse Naturæ opus, sed artis; potest difficultas transferri facile ad Naturæ opera, ut ubi diversa inducuntur glaciei strata, vel in aliis incrustationibus, ac in iis omnibus casibus, in quibus incrementum fit per externam applicationem partium, ubi accessiones finitæ videntur acquiri simul totæ sine [68] transitu per intermedias magnitudines. In iis casibus continuitas servatur in motu singularum partium, quæ accedunt. Illæ per lineam quandam continuam, & continua velocitatis mutatione accedunt ad locum sibi deditum: quin immo etiam posteaquam eo advenerunt, pergunt adhuc moveri, & nunquam habent quietem nec absolutam, nec respectivam respectu aliarum partium, licet jam in respectiva positione sensibilem mutationem non subeant: parent nimirum adhuc viribus omnibus, quæ respondent omnibus materiæ punctis utcunque distantibus, & actio proximarum partium, quæ novam adhæsionem parit, est continuatio actionis, quam multo minorem exercebant, cum essent procul. Hoc autem, quod pertineant ad illam domum, vel massam, est aliquid non in se determinatum, quod momento quodam determinato fiat, in quo saltus habeatur, sed ab æstimatione quadam pendet nostrorum sensuum satis crassa; ut licet perpetuo accedant illæ partes, & pergant perpetuo mutare positionem respectu ipsius massæ; tum incipiant censeri ut pertinentes ad illam domum, vel massam: cum desinit respectiva mutatio esse sensibilis, quæ sensibilitatis cessatio fit ipsa etiam quodammodo per gradus omnes, & continuo aliquo tempore, non vero per saltum.

Generalis responsio ad casus similes inde eruta.

150. Hinc distinctius ibi licebit difficultatem omnem amovere dicendo, non servari mutationem continuam in magnitudinibus earum rerum, quæ continuæ non sunt, & magnitudinem non habent continuam, sed sunt aggregata rerum disjunctarum; vel in iis rebus, quæ a nobis ita censentur aliquod totum constituere, ut magnitudinem aggregati non

the angles of incidence & lie on the near side of the surface, without any reflection for rays at intermediate angles with the surface less than a certain definite angle.

Apparent reconciliation with the Law of Continuity effected by means of irregular dispersion.

147. It seems that an explanation of this apparent breach of continuity can be given by means of light that is reflected or refracted irregularly at all sorts of angles. For long ago it was observed that, when a ray of light is reflected, it is not reflected entirely in such a manner that the angle of reflection is equal to the angle of incidence, but that a part of it is dispersed in all directions. For this reason, if a ray of light from the Sun falls upon some part of a mirror, anybody who is in the room sees where the ray strikes the mirror; & this certainly would not be the case, unless some of the solar rays reached his eye directly issuing from the mirror in all those directions that reach to all positions that the eye might be in. Nevertheless, in this case the light does not appear to be of much intensity, unless the eye is in the position facing the angle of reflection equal to the angle of incidence, along which the greatest part of the light rebounds. Newton indeed employed in a brilliant way these rays that issue at irregular angles at the end of his Optics to explain the colours of solid laminæ. The same irregular dispersion in all directions takes place as far as can be observed in a small part, but yet in some part, of the refracted ray. Hence, in between the intense reflected & refracted rays, we have a whole series of intermediate rays of this sort issuing at all intermediate angles. Thus, when the transition is effected from refraction to total reflection, it seems that it can be done through these intermediate angles by an extremely rapid transition through them, & therefore continuity remains unimpaired.

Why this is only an apparent reconciliation; the true reconciliation is through the continuity of path for any point of light.

148. But if we inquire into the matter yet more carefully, it will be evident that in that intermediate series there is no accurate continuity, but only an apparent continuity; & this Nature tries to maintain, but does not accurately observe it unimpaired. For, in my Theory, light is not some continuous body, which is continuously diffused through all the space it occupies; but it is an aggregate of points unconnected with & separated from one another; & of these points, any one pursues its own path, & this path is separated from the path of the next point to it by a definite interval. Continuity is observed perfectly accurately for the paths of the several points, not in the diffusion of a substance that is not continuous; & the manner in which continuity is preserved in all these motions, & the path described by the several points is altered without sudden change, when the angle of incidence is altered, I have set forth fairly clearly in the second part of my dissertation *De Lumine*, Art. 98. But in this work these matters belong to the application of the Theory to physics.

The manner in which continuity is maintained in certain cases in which it is apparently impaired.

149. There are certain cases, not greatly unlike those already given, in which each part preserves continuity, but not so the whole, which is not continuous but composed of separate parts. For an instance of this kind, take the height of a new house which is being built; as a fresh layer of stones of a given height is added to it, the height of the house on account of that addition seems to increase suddenly without passing through intermediate heights. If it is said that that is not a work of Nature, but of art; then the same difficulty can easily be transferred to works of Nature, as when different strata of ice are formed, or in other incrustations, and in all cases in which an increment is caused by the external application of parts, where finite additions seem to be acquired all at once without any passage through intermediate magnitudes. In these cases the continuity is preserved in the motions of the separate parts that are added. These reach the place allotted to them along some continuous line & with a continuous change of velocity. Further, after they have reached it, they still continue to move, & never have absolute rest; no, nor even relative rest with respect to the other parts, although they do not now suffer a sensible change in their relative positions. Thus, they still submit to the action of all the forces that correspond to all points of matter at any distances whatever; and the action of the parts nearest to them, which produces a new adhesion, is the continuation of the action that they exert to a far smaller extent when they are some distance away. Moreover, in the fact that they belong to that house or mass, there is something that is not determinate in itself, because it happens at a determinate instant in which the sudden change takes place; but it depends on a somewhat rough assessment by our senses. So that, although these parts are continually being added, & continually go on changing their position with respect to the mass, they both begin to be thought of as belonging to that house or mass, & the relative change ceases to be sensible; also this cessation of sensibility itself also takes place to some extent through all stages, and in some continuous interval of time, & not by a sudden jump.

General refutation for similar cases derived from this.

150. From this consideration we may here in a clearer manner remove all difficulty by saying that a continuous change is not maintained in the magnitudes of those things, which are not themselves continuous, & do not possess continuous magnitude, but are aggregates of separate things. That is to say, in those things that are thus considered as forming a certain whole, in such a way that the magnitude of the aggregate is not determined

determinent distantiæ inter eadem extrema, sed a nobis extrema ipsa assumantur jam alia, jam alia, quæ censeantur incipere ad aggregatum pertinere, ubi ad quasdam distantias devenerint, quas ut ut in se juxta legem continuitatis mutatas, nos a reliquis divellimus per saltum, ut dicamus pertinere eas partes ad id aggregatum. Id accidit, ubi in objectis casibus accessiones partium novæ fiunt, atque ibi nos in usu vocabuli saltum facimus; ars, & Natura saltum utique habet nullum.

Alii casus in quibus læditur, alii, in quibus habetur solum proxima, non accurata continuitas.

151. Non idem contingit etiam, ubi plantæ, vel animantia crescunt, succo se insinuante per tubulos fibrarum, & procurrente, ubi & magnitudo computata per distantias punctorum maxime distantium transit per omnes intermedias; cum nimirum ipse procursus fiat per omnes intermedias distantias. At quoniam & ibi mutantur termini illi, qui distantias determinant, & nomen suscipiunt altitudinis ipsius plantæ; vera & accurata continuitas ne ibi quidem observatur, nisi tantummodo in motibus, & velocitatibus, ac distantiis singularum partium : quanquam ibi minus recedatur a continuitate accurata, quam in superioribus. In his autem, & in illis habetur ubique illa alia continuitas quædam apparens, & affectata tantummodo a Natura, quam intuemur etiam in progressu substantiarum, ut incipiendo ab inanima-[69]-tis corporibus progressu facto per vegetabilia, tum per quædam fere semianimalia torpentia, ac demum animalia perfectiora magis, & perfectiora usque ad simios bomini tam similes. Quoniam & harum specierum, ac existentium individuorum in quavis specie numerus est finitus, vera continuitas haberi non potest, sed ordinatis omnibus in seriem quandam, inter binas quasque intermedias species hiatus debet esse aliquis necessario, qui continuitatem abrumpat. In omnibus iis casibus habentur discretæ quædam quantitates, non continuæ; ut & in Arithmetica series ex. gr. naturalium numerorum non est continua, sed discreta; & ut ibi series ad continuam reducitur tantummodo, si generaliter omnes intermediæ fractiones concipiantur; sic & in superiore exemplo quædam velut continua series habebitur tantummodo; si concipiantur omnes intermediæ species possibiles.

Conclusio pertinens ad ea, quæ veram, & ea, quæ affectatam habent continuitatem.

152. Hoc pacto excurrendo per plurimos justmodi casus, in quibus accipiuntur aggregata rerum a se invicem certis intervallis distantium, & unum aliquid continuum non constituentium, nusquam accurata occurret continuitatis lex, sed per quandam dispersionem quodammodo affectata, & vera continuitas habebitur tantummodo in motibus, & in iis, quæ a motibus pendent, uti sunt distantiæ, & vires determinatæ a distantiis, & velocitates a viribus ortæ; quam ipsam ob causam ubi supra num. 39 inductionem pro lege continuitatis assumpsimus, exempla accepimus a motu potissimum, & ab iis, quæ cum ipsis motibus connectuntur, ac ab iis pendent.

Difficultates petitæ a discrimine debito inter materiam, & spiritum.

153. Sed jam ad aliam difficultatem gradum faciam, quæ non nullis negotium ingens facessit, & obvia est etiam, contra hanc indivisibilium, & inextensorum punctorum Theoriam; quod nimirum ea nullum habitura sint discrimen a spiritibus. Ajunt enim, si spiritus ejusmodi vires habeant, præstituros eadem phænomena, tolli nimirum corpus, & omnem corporeæ substantiæ notionem sublata extensione continua, quæ sit præcipua materiæ proprietas ita pertinens ad naturam ipsius; ut vel nihil aliud materia sit, nisi substantia prædita extensione continua; vel saltem idea corporis, & materiæ haberi non possit; nisi in ea includatur idea extensionis continuæ. Multa hic quidem congeruntur simul, quæ nexum aliquem inter se habent, quæ hic seorsum evolvam singula.

Differre hæc puncta a spiritibus per impenetrabilitatem, sensibilitatem, incapacitatem cogitationis.

154. Inprimis falsum omnino est, nullum esse horum punctorum discrimen a spiritibus. Discrimen potissimum materiæ a spiritu situm est in hisce duobus, quod materia est sensibilis, & incapax cogitationis, ac voluntatis, spiritus nostros sensus non afficit, & cogitare potest, ac velle. Sensibilitas autem non ab extensione continua oritur, sed ab impenetrabilitate, qua fit, ut nostrorum organorum fibræ tendantur a corporibus, quæ ipsis sistuntur, & motus ad cerebrum pro-[70]-pagetur. Nam si extensa quidem essent corpora, sed impenetrabilitate carerent; manu contrectata fibras non sisterent, nec motum ullum in iis progignerent, ac eadem radios non reflecterent, sed liberum intra se aditum luci præberent. Porro hoc discrimen utrumque manere potest integrum, & manet inter mea indivisibilia hæc puncta, & spiritus. Ipsa impenetrabilitatem habent, & sensus nostros afficiunt, ob illud primum crus asymptoticum exhibens vim illam repulsivam primam; spiritus autem, quos impenetrabilitate carere credimus, ejusmodi viribus itidem carent, & sensus nostros idcirco nequaquam afficiunt, nec oculis inspectantur, nec manibus palpari possunt. Deinde in meis hisce punctis ego nihil admitto aliud, nisi illam virium legem cum inertiæ vi conjunctam, adeoque illa volo prorsus incapacia cogitationis, & voluntatis.

by the distances between the same extremes all the time, but the extremes we take are different, one after another; & these are considered to begin to belong to the aggregate when they attain to certain distances from it; &, although in themselves changed in accordance with the Law of Continuity, we separate them from the rest in a discontinuous manner, by saying that these parts belong to the aggregate. This comes about, whenever in the cases under consideration fresh additions of parts take place; & then we make a discontinuity in the use of a term; art, as well as Nature, has no discontinuity.

Cases in which there is a breach of continuity; others in which the continuity is only very nearly, but not accurately, observed.

151. It is not the same thing however in the case of the growth of plants or animals, which is due to a life-principle insinuating itself into, & passing along the fine tubes of the fibres; here the magnitude, calculated by means of the distance between the points furthest from one another, passes through all intermediate distances; for the flow of the life-principle takes place indeed through all intermediate distances. But, since here also the extremes are changed, which determine the distances, & denominate the altitude of the plant; not even in this case is really accurate continuity observed, except only in the motions & velocities and distances of the separate parts; however there is here less departure from accurate continuity, than there was in the examples given above. In both there is indeed that kind of apparent continuity, which Nature does no more than try to maintain; such as we also see in the series of substantial things, which starting from inanimate bodies, continues through vegetables, then through certain sluggish semianimals, & lastly, through animals more & more perfect, up to apes that are so like to man. Also, since the number of these species, & the number of existent individuals of any species, is finite, it is impossible to have true continuity; but if they are all ordered in a series, between any two intermediate species there must necessarily be a gap; & this will break the continuity. In all these cases we have certain discrete, & not continuous, quantities; just as, for instance, the arithmetical series of the natural numbers is not continuous, but discrete. Also, just as the series is reduced to continuity only by mentally introducing in general all the intermediate fractions; so also, in the example given above a sort of continuous series is obtained, if & only if all intermediate possible species are so included.

Conclusion as regards those things that possess true continuity, and those that have a counterfeit continuity.

152. In the same way, if we examine a large number of cases of the same kind, in which aggregates of things are taken, separated from one another by certain definite intervals, & not composing a single continuous whole, an accurate continuity law will never be met with, but only a sort of counterfeit depending on dispersion. True continuity will only be obtained in motions, & in those things that depend on motions, such as distances & forces determined by distances, & velocities derived from such forces. It was for this very reason that, when we adopted induction for the proof of the Law of Continuity in Art. 39 above, we took our examples mostly from motion, & from those things which are connected with motions & depend upon them.

Objections derived from the distinction that has to be made between matter & spirit.

153. Now I will pass on to another objection, which some people have made a great to-do about, and which has also been raised in opposition to this Theory of indivisible & non-extended points; namely, that there will be no difference between my points & spirits. For, they say that, if spirits were endowed with such forces, they would show the same phenomena as bodies, & that bodies & all idea of corporeal substance would be done away with by denying continuous extension; for this is one of the chief properties of matter, so pertaining to Nature itself; so that either matter is nothing else but substance endowed with continuous extension, or the idea of a body and of matter cannot be obtained without the inclusion of the idea of continuous extension. Here indeed there are many matters all jumbled together, which have no connection with one another; these I will now separate & discuss individually.

These points differ from spirits on account of their impenetrability, their being sensible, & their incapacity for thought.

154. First of all it is altogether false that there is no difference between my points & spirits. The most important difference between matter & spirit lies in the two facts, that matter is sensible & incapable of thought, whilst spirit does not affect the senses, but can think or will. Moreover, sensibility does not arise from continuous extension, but from impenetrability, through which it comes about that the fibres of our organs are subjected to stress by bodies that are set against them & motions are thereby propagated to the brain. For if indeed bodies were extended, but lacked impenetrability, they would not resist the fibres of the hand when touched, nor produce in them any motion; nor would they reflect light, but allow it an uninterrupted passage through themselves. Further, it is possible that each of these distinctions should hold good independently; & they do so between these indivisible points of mine & spirits. My points have impenetrability & affect our senses, because of that first asymptotic branch representing that first repulsive force; but spirits, which we suppose to lack impenetrability, lack also forces of this kind, and therefore can in no wise affect our senses, nor be examined by the eyes, nor be felt by the hands. Then, in these points of mine, I admit nothing else but the law of forces conjoined with the force of inertia; & hence I intend them to be incapable

Quamobrem discrimen essentiæ illud utrumque, quod inter corpus, & spiritum agnoscunt omnes, id & ego agnosco, nec vero id ab extensione, & compositione continua desumitur, sed ab iis, quæ cum simplicitate, & inextensione æque conjungi possunt, & cohærere cum ipsis.

Si possibilis substantia prædita hisce viribus, & capax cogitationis; eam nec fore materiam, nec spiritum.

155. At si substantiæ capaces cogitationis & voluntatis haberent ejusmodi virium legem, an non eosdem præstarent effectus respectu nostrorum sensuum, quos ejusmodi puncta? Respondebo sane, me hic non quærere, utrum impenetrabilitas, & sensibilitas, quæ ab iis viribus pendent, conjungi possint cum facultate cogitandi, & volendi, quæ quidem quæstio eodem redit, ac in communi sententia de impenetrabilitate extensorum, ac compositorum relata ad vim cogitandi, & volendi. Illud ajo, notionem, quam habemus partim ex observationibus tam sensuum respectu corporum, quam intimæ conscientiæ respectu spiritus, una cum reflexione, partim, & vero etiam circa spiritus potissimum, ex principiis immediate revelatis, vel connexis cum principiis revelatis, continere pro materia impenetrabilitatem, & sensibilitatem, una cum incapacitate cogitationis, & pro spiritu incapacitatem afficiendi per impenetrabilitatem nostros sensus, & potentiam cogitandi, ac volendi, quorum priores illas ego etiam in meis punctis admitto, posteriores hasce in spiritibus; unde fit, ut mea ipsa puncta materialia sint, & eorum massæ constituant corpora a spiritibus longissime discrepantia. Si possibile sit illud substantiæ genus, quod & hujusmodi vires activas habeat cum inertia conjunctas, & simul cogitare possit, ac velle; id quidem nec corpus erit, nec spiritus, sed tertium quid, a corpore discrepans per capacitatem cogitationis, & voluntatis, discrepans autem a spiritu per inertiam, & vires basce nostras, quæ impenetrabilitatem inducunt. Sed, ut ajebam, ea quæstio huc non pertinet, & aliunde resolvi debet; ut aliunde utique debet resolvi quæstio, qua quæratur, an substantia extensa, & impenetrabilis [71] basce proprietates conjungere possit cum facultate cogitandi, volendique.

Nihil amitti, amisso argumento eorum, qui a compositione partium deducunt incapacitatem cogitationis.

156. Nec vero illud reponi potest, argumentum potissimum ad evincendum, materiam cogitare non posse, deduci ab extensione, & partium compositione, quibus sublatis, omne id fundamentum prorsus corruere, & ad materialismum sterni viam. Nam ego sane non video, quid argumenti peti possit ab extensione, & partium compositione pro incapacitate cogitandi, & volendi. Sensibilitas, præcipua corporum, & materiæ proprietas, quæ ipsam adeo a spiritibus discriminat, non ab extensione continua, & compositione partium pendet, uti vidimus, sed ab impenetrabilitate, quæ ipsa proprietas ab extensione continua, & compositione non pendet. Sunt qui adhibent hoc argumentum ad excludendam capacitatem cogitandi a materia, desumptum a compositione partium: si materia cogitaret; singulæ ejus partes deberent singulas cogitationis partes habere, adeoque nulla pars objectum perciperet; cum nulla haberet eam perceptionis partem, quam habet altera. Id argumentum in mea Theoria amittitur; at id ipsum, meo quidem judicio, vim nullam habet. Nam posset aliquis respondere, cogitationem totam indivisibilem existere in tota massa materiæ, quæ certa partium dispositione sit prædita, uti anima rationalis per tam multos Philosophos, ut ut indivisibilis, in omni corpore, vel saltem in parte corporis aliqua divisibili existit, & ad ejusmodi præsentiam præstandam certa indiget dispositione partium ipsius corporis, qua semel læsa per vulnus, ipsa non potest ultra ibi esse; atque ut viventis corporei, sive animalis rationalis natura, & determinatio habetur per materiam divisibilem, & certo modo constructam, una cum anima indivisibili; ita ibi per indivisibilem cogitationem inhærentem divisibili materiæ natura, & determinatio cogitantis haberetur. Unde aperte constat eo argumento amisso, nihil omnino amitti, quod jure dolendum sit.

Etiam si quidpiam amittatur; theoriam positive probari, & in ea manere summum discrimen inter materiam, & spiritum.

157. Sed quidquid de eo argumento censeri debeat, nihil refert, nec ad infirmandam Theoriam positivis, & validis argumentis comprobatam, ac e solidissimis principiis directa ratiocinatione deductam, quidquam potest unum, vel alterum argumentum amissum, quod ad probandam aliquam veritatem aliunde notam, & a revelatis principiis aut directe, aut indirecte confirmatam, ab aliquibus adhibeatur, quando etiam vim habeat aliquam, quam, uti ostendi, superius allatum argumentum omnino non habet. Satis est, si illa Theoria cum ejusmodi veritate conjungi possit, uti hæc nostra cum immaterialitate spirituum conjungitur optime, cum retineat pro materia inertiam, impenetrabilitatem, sensibilitatem, incapacitatem cogitandi, & pro spiritibus retineat incapacitatem afficiendi sensus nostros per impenetrabilitatem, & facultatem cogitandi, ac volendi. [72] Ego quidem in ipsius

of thought or will. Wherefore I also acknowledge each of those essential differences between matter and spirit, which are acknowledged by everyone; but by me it is not deduced from extension and continuous composition, but, just as correctly, from things that can be conjoined with simplicity & non-extension, & can combine with them.

If it were possible that there was a substance that was both endowed with these forces & was capable of thought, it would be neither matter nor spirit.

155. Now if there were substances capable of thought & will that also had a law of forces of this kind, is it possible that they would produce the same effects with respect to our senses, as points of this sort? Truly, I will answer that I do not seek to know in this connection, whether impenetrability & sensibility, which depend on these forces, can be conjoined with the faculty of thinking & willing; indeed this question comes to the same thing as the general idea of the relations of impenetrability of extended & composite things to the power of thinking & willing. I will say but this, that we form our ideas, partly from observations, of the senses in the case of bodies, & of the inner consciousness in the case of spirits, together with reflections upon them, partly, & indeed more especially in the case of spirits, from directly revealed principles, or matters closely connected with revealed principles; & these ideas involve for matter impenetrability, sensibility, combined with incapacity for thought, & for spirit an incapacity for affecting our senses by means of impenetrability, together with the capacity for thinking and willing. I admit the former of these in the case of my points, & the latter for spirits; so that these points of mine are material points, & masses of them compose bodies that are far different from spirits. Now if it were possible that there should be some kind of substance, which has both active forces of this kind together with a force of inertia & also at the same time is able to think and will; then indeed it will neither be body nor spirit, but some third thing, differing from a body in its capacity for thought & will, & also from spirit by possessing inertia and these forces of mine, which lead to compenetration. But as I was saying, that question does not concern me now, & the answer must be found by other means. So by other means also must the answer be found to the question, in which we seek to know whether a substance that is extended & impenetrable can conjoin these two properties with the faculty of thinking and willing.

Nothing is lost even if we dismiss the argument of those who deduce incapacity for thought from composition by parts.

156. Now it cannot be ignored that an argument of great importance in proving that matter is incapable of thought is deduced from extension & composition by parts; & if these are denied, the whole foundation breaks down, & the way is laid open to materialism. But really I do not see what in the way of argument can be derived from extension & composition by parts, to support incapacity for thinking and willing. Sensibility, the chief property of bodies & of matter, which is so much different from spirits, does not depend on continuous extension & composition by parts, as we have seen, but on impenetrability; & this latter property does not depend on continuous extension & composition. There are some, who use the following argument, derived from composition by parts, to exclude from matter the capacity for thought:—If matter were to think, then each of its parts would have a separate part of the thought, & thus no part would have perception of the object of thought; for no part can have that part of the perception that another part has. This argument is neglected in my Theory; but the argument itself, at least so I think, is unsound. For one can reply that the complete thought exists as an indivisible thing in the whole mass of matter, which is endowed with a certain arrangement of parts, in the same way as the rational soul in the opinion of so many philosophers exists, although it is indivisible, in the whole of the body, or at any rate in a certain divisible part of the body; & to maintain a presence of this kind there is need for a definite arrangement of the parts of the body, which if at any time impaired by a wound would no longer exist there. Thus, just as from the nature of a living body, or of a rational animal, determination arises from matter that is divisible & constructed on a definite plan, in conjunction with an indivisible mind; so also in this case by means of indivisible thought inherent in the nature of divisible matter, there is a propensity for thought. From this it is very plain that, if this argument is dismissed, there will be nothing neglected that we have any reason to regret.

Even if something is thus neglected, the Theory can be proved in a direct manner, & there will still remain in it the greatest difference between matter & spirit.

157. But whatever opinion we are to form about this argument, it makes no difference, nor can it weaken a Theory that has been corroborated by direct & valid arguments, & deduced from the soundest principles by a straightforward chain of reasoning, if we leave out one or other of the arguments, which have been used by some for the purpose of testing some truth that is otherwise known & confirmed by revealed principles either directly or indirectly; even when the argument has some validity, which, as I have shown, that adduced above has not in any way. It is sufficient if that theory can be conjoined with such a truth; just as this Theory of mine can be conjoined in an excellent manner with the immateriality of spirits. For it retains for matter inertia, impenetrability, sensibility, & incapacity for thinking, & for spirits it retains the incapacity for affecting our senses by impenetrability, & the faculty of thinking or willing. Indeed I assume the

materiæ, & corporeæ substantiæ definitione ipsa assumo incapacitatem cogitandi, & volendi, & dico corpus massam compositam e punctis habentibus vim inertiæ conjunctam cum viribus activis expressis in fig. 1, & cum incapacitate cogitandi, ac volendi, qua definitione admissa, evidens est, materiam cogitare non posse ; quæ erit metaphysica quædam conclusio, ea definitione admissa, certissima : tum ubi solæ rationes physicæ adhibeantur, dicam, hæc corpora, quæ meos afficiunt sensus, esse materiam, quod & sensus afficiant per illas utique vires, & non cogitent. Id autem deducam inde, quod nullum cogitationis indicium præstent ; quæ erit conclusio tantum physica, circa existentiam illius materiæ ita definitæ, æque physice certa, ac est conclusio, quæ dicat lapides non habere levitatem, quod nunquam eam prodiderint ascendendo sponte, sed semper e contrario sibi relicti descenderint.

Sensus omnino falli in illa tanta continuitate in extensionis, quam nobis ingerunt.

158. Quod autem pertinet ad ipsam corporum, & materiæ ideam, quæ videtur extensionem continuam, & contactum partium involvere, in eo videntur mihi quidem Cartesiani inprimis, qui tantopere contra præjudicia pugnare sunt visi, præjudiciis ipsis ante omnes alios indulsisse. Ideam corporum habemus per sensus ; sensus autem de continuitate accurata judicare omnino non possunt, cum minima intervalla sub sensus non cadant. Et quidem omnino certo deprehendimus illam continuitatem, quam in plerisque corporibus nobis objiciunt sensus nostri, nequaquam haberi. In metallis, in marmoribus, in vitris, & crystallis continuitas nostris sensibus apparet ejusmodi, ut nulla percipiamus in iis vacua spatiola, nullos poros, in quo tamen hallucinari sensus nostros manifesto patet, tum ex diversa gravitate specifica, quæ a diversa multitudine vacuitatum oritur utique, tum ex eo, quod per illa insinuentur substantiæ plures, ut per priora oleum diffundatur, per posteriora liberrime lux transeat, quod quidem indicat, in posterioribus hisce potissimum ingentem pororum numerum, qui nostris sensibus delitescunt

Fons præjudiciorum : haberi pro nullis in se, quæ sunt nulla in nostris sensibus : eorum exempla.

159. Quamobrem jam ejusmodi nostrorum sensuum testimonium, vel potius noster eorum ratiociniorum usus, in hoc ipso genere suspecta esse debent, in quo constat nos decipi. Suspicari igitur licet, exactam continuitatem sine ullis spatiolis, ut in majoribus corporibus ubique deest, licet sensus nostri illam videantur denotare, ita & in minimis quibusvis particulis nusquam haberi, sed esse illusionem quandam sensuum tantummodo, & quoddam figmentum mentis, reflexione vel non utentis, vel abutentis. Est enim solemne illud hominibus, atque usitatum, quod quidem est maximorum præjudiciorum fons, & origo præcipua, ut quidquid in nostris sensibus est nihil, habeamus pro nihilo absoluto. Sic utique per tot sæcula a multis est creditum, & nunc etiam a vulgo creditur, [73] quietem Telluris, & diurnum Solis, ac fixarum motum sensuum testimonio evinci, cum apud Philosophos jam constet, ejusmodi quæstionem longe aliunde resolvendam esse, quam per sensus, in quibus debent eædem prorsus impressiones fieri, sive stemus & nos, & Terra, ac moveantur astra, sive moveamur communi motu & nos, & Terra, ac astra consistant. Motum cognoscimus per mutationem positionis, quam objecti imago habet in oculo, & quietem per ejusdem positionis permanentiam. Tam mutatio, quam permanentia fieri possunt duplici modo : mutatio, primo si nobis immotis objectum moveatur ; & permanentia, si id ipsum stet : secundo, illa, si objecto stante moveamur nos ; hæc, si moveamur simul motu communi. Motum nostrum non sentimus, nisi ubi nos ipsi motum inducimus, ut ubi caput circumagimus, vel ubi curru delati succutimur. Idcirco habemus tum quidem motum ipsum pro nullo, nisi aliunde admoneamur de eodem motu per causas, quæ nobis sint cognitæ, ut ubi *provehimur portu*, quo casu vector, qui jam diu assuevit ideæ littoris stantis, & navis promotæ per remos, vel vela, corrigit apparentiam illius, *terræque urbesque recedunt*, & sibi, non illis, motum adjudicat.

Eorum correctio ubi deprehendatur, rem alio etiam modo cum sensuum apparentia conciliari posse.

160. Hine Philosophus, ne fallatur, non debet primis hisce ideis acquirere, quas e sensationibus haurimus, & ex illis deducere consectaria sine diligenti perquisitione, ac in ea quæ ab infantia deduxit, debet diligenter inquirere. Si inveniat, easdem illas sensuum perceptiones duplici modo æque fieri posse ; peccabit utique contra Logicæ etiam naturalis leges, si alterum modum præ altero pergat eligere, unice, quia alterum antea non viderat, & pro nullo habuerat, & idcirco alteri tantum assueverat. Id vero accidit in casu nostro :

incapacity for thinking & willing in the very definition of matter itself & corporeal substance; & I say that a body is a mass composed of points endowed with a force of inertia together with such active forces as are represented in Fig. 1, & an incapacity for thinking & willing. If this definition is taken, it is clear that matter cannot think; & this will be a sort of metaphysical conclusion, which will follow with absolute certainty from the acceptation of the definition. Again, where physical arguments are alone employed, I say that such bodies as affect our senses are matter, because they affect the senses by means of the forces under consideration, & do not think. I also deduce the same conclusion from the fact that they afford no evidence of thought. This will be a conclusion that is solely physical with regard to the existence of matter so defined; & it will be just as physically true as the conclusion that says that stones do not possess levity, deduced from the fact that they never display such a thing by an act of spontaneous ascent, but on the contrary always descend if left to themselves.

The senses are altogether at fault in the greatness of the continuity of extension that they force us to believe.

158. With regard to the idea of bodies & matter, which seems to involve continuous extension, it seems to me indeed that in this matter the Cartesians in particular, who have appeared to impugn prejudgments with so much vigour, have given themselves up to these prejudgments more than anyone else. We obtain the idea of bodies through the senses; and the senses cannot in any way judge on a matter of accurate continuity; for very small intervals do not fall within the scope of the senses. Indeed we quite take it for granted that the continuity, which our senses meet with in a large number of bodies, does not really exist. In metals, marble, glass & crystals there appears to our senses to be continuity, of such sort that we do not perceive in them any little empty spaces, or pores; but in this respect the senses have manifestly been deceived. This is clear, both from their different specific gravities, which certainly arises from the differences in the numbers of the empty spaces; & also from the fact that several substances will insinuate themselves through their substance. For instance, oil will diffuse itself through the former, & light will pass quite freely through the latter; & this indeed indicates, especially in the case of the latter, an immense number of pores; & these are concealed from our senses.

The origin of prejudgments; things considered as nothing, which are nothing so far as the senses are concerned; examples of these.

159. Hence such evidence of our senses, or rather our employment of such arguments, must now lie open to suspicion in that class, in which it is known that we have been deceived. We may then suspect that accurate continuity without the presence of any little empty spaces—such as is certainly absent from bodies of considerable size, although our senses seem to remark its presence—is also nowhere existent in any of their smallest particles; but that it is merely an illusion of the senses, & a sort of figment of the brain through its not using, or through misusing, reflection. For it is a customary thing for men (& a thing that is frequently done) to consider as absolutely nothing something that is nothing as far as the senses are concerned; & this indeed is the source & principal origin of the greatest prejudices. Thus for many centuries it was credited by many, & still is believed by the unenlightened, that the Earth is at rest, & that the daily motions of the Sun & the fixed stars is proved by the evidence of the senses; whilst among philosophers it is now universally accepted that such a question has to be answered in a far different manner from that by means of the senses. Exactly the same impressions are bound to be obtained, whether we & the Earth stand still & the stars are moved, or we & the Earth are moved with a common motion & the stars are at rest. We recognize motion by the change of position, which the image of an object has in the eye; and rest by the permanence of that position. Now both the change & the permanence can come about in two ways. Firstly, if we remain at rest, there is a change of position if the object is moved, & permanence if it too is at rest; secondly, if we move, there is a change if the object is at rest, & permanence if we & it move with a motion common to both. We do not feel ourselves moving, unless we ourselves induce the motion, as when we turn the head, or when we are jolted as we are borne in a vehicle. Hence we consider that the motion is nothing, unless we are made to notice in other ways that there is motion by causes that are known to us. Thus, when "*we leave the harbour*," a passenger who has for some time been accustomed to the idea of a shore remaining still, & of a ship being propelled by oars or sails, corrects the apparent motion of the shore; &, as "*the land & buildings recede*," he attributes the motion to himself and not to them.

Correction of those things, where it is known that the matter cannot be brought into agreement with what is apparent to the senses in some other way.

160. Hence, the philosopher, to avoid being led astray, must not seek to obtain from these primary ideas that we derive from the senses, or deduce from them, consequential theorems, without careful investigation; & he must carefully study those things that he has deduced from infancy. If he find that these very perceptions by the senses can come about in two ways, one of which is as probable as the other; then he will certainly commit an offence against the laws of natural logic, if he should proceed to choose one method in preference to the other, solely for the reason that previously he had not seen the one & took no account of it, & thus had become accustomed to the other. Now

sensationes habebuntur eædem, sive materia constet punctis prorsus inextensis, & distantibus inter se per intervalla minima, quæ sensum fugiant, ac vires ad illa intervalla pertinentes organorum nostrorum fibras sine ulla sensibili interruptione afficiant, sive continua sit, & per immediatum contactum agat. Patebit autem in tertia hujusce operis parte, quo pacto proprietates omnes sensibiles corporum generales, immo etiam ipsorum præcipua discrimina, cum punctis hisce indivisibilibus conveniant, & quidem multo sane melius, quam in communi sententia de continua extensione materiæ. Quamobrem errabit contra rectæ ratiocinationis usum, qui ex præjudicio ab hujusce conciliationis, & alterius hujusce sensationum nostrarum causæ ignoratione inducto, continuam extensionem ut proprietatem necessariam corporum omnino credat, & multo magis, qui censeat, materialis substantiæ ideam in ea ipsa continua extensione debere consistere.

Ordo idearum, quas hausimus circa corpus: primas habitas esse per tactum.

161. Verum quo magis evidenter constet horum præjudiciorum origo, afferam hic dissertationis *De Materiæ Divisibilita*-[74]-*te*, *& Principiis Corporum*, numeros tres incipiendo a 14, ubi sic: " utcunque demus, quod ego omnino non censeo, aliquas esse innatas ideas, & non per sensus acquisitas; illud procul dubio arbitror omnino certum, ideam corporis, materiæ, rei corporeæ, rei materialis, nos hausisse ex sensibus. Porro ideæ primæ omnium, quas circa corpora acquisivimus per sensus, fuerunt omnino eæ, quas in nobis tactus excitavit, & easdem omnium frequentissimas hausimus. Multa profecto in ipso materno utero se tactui perpetuo offerebant, antequam ullam fortasse saporum, aut odorum, aut sonorum, aut colorum ideam habere possemus per alios sensus, quarum ipsarum, ubi eas primum habere cœpimus, multo minor sub initium frequentia fuit. Ideæ autem, quas per tactum habuimus, ortæ sunt ex phænomenis hujusmodi. Experiebamur palpando, vel temere impingendo resistentiam vel a nostris, vel a maternis membris ortam, quæ cum nullam interruptionem per aliquod sensibile intervallum sensui objiceret, obtulit nobis ideam impenetrabilitatis, & extensionis continuæ: cumque deinde cessaret in eadem directione, alicubi resistentia, & secundum aliam directionem exerceretur; terminos ejusdem quantitatis concepimus, & figuræ ideam hausimus."

Quæ fuerint tum consideranda: infantia ad eas reflexiones, inepta: in quo ea sita sit.

162. " Porro oriebantur hæc phænomena a corporibus e materia jam efformatis, non a singulis materiæ particulis, e quibus ipsa corpora componebantur. Considerandum diligenter erat, num extensio ejusmodi esset ipsius corporis, non spatii cujusdam, per quod particulæ corpus efformantes diffunderentur: num ea particulæ ipsæ iisdem proprietatibus essent præditæ: num resistentia exerceretur in ipso contactu, an in minimis distantiis sub sensus non cadentibus vis aliqua impedimento esset, quæ id ageret, & resistentia ante ipsum etiam contactum sentiretur: num ejusmodi proprietates essent intrinsecæ ipsi materiæ, ex qua corpora componuntur, & necessariæ: an casu tantum aliquo haberentur, & ab extrinseco aliquo determinante. Hæc, & alia sane multa considerare diligentius oportuisset: sed erat id quidem tempus maxime caliginosum, & obscurum, ac reflexionibus minus obviis minime aptum. Præter organorum debilitatem, occupabat animum rerum novitas, phænomenorum paucitas, & nullus, aut certe satis tenuis usus in phænomenis ipsis inter se comparandis, & ad certas classes revocandis, ex quibus in eorum leges, & causas liceret inquirere & systema quoddam efformare, quo de rebus extra nos positis possemus ferre judicium. Nam in hac ipsa phænomenorum inopia, in hac efformandi systematis difficultate, in hoc exiguo reflexionum usu, magis etiam, quam in organorum imbecillitate, arbitror, sitam esse infantiam."

Præjudicia inde orta extensionis continuæ ut essentialis, odorum, &c., ut accidentalium.

[75] 163. " In hac tanta rerum caligine ea prima sese obtulerunt animo, quæ minus alta indagine, minus intentis reflexionibus indigebant, eaque ipsa ideis toties repetitis altius impressa sunt, & tenacius adhæserunt, & quendam veluti campum nacta prorsus vacuum, & adhuc immunem, suo quodammodo jure quandam veluti possessionem inierunt. Intervalla, quæ sub sensum nequaquam cadebant, pro nullis habita: ea, quorum ideæ semper simul conjunctæ excitabantur, habita sunt pro iisdem, vel arctissimo, & necessario nexu inter se conjunctis. Hinc illud effectum est, ut ideam extensionis continuæ, ideam

that is just what happens in the case under consideration. The same sensations will be experienced, whether matter consists of points that are perfectly non-extended & distant from one another by very small intervals that escape the senses, & forces pertaining to those intervals affect the nerves of our organs without any sensible interruption; or whether it is continuous and acts by immediate contact. Moreover it will be clearly shown, in the third part of this work, how all the general sensible properties of bodies, nay even the principal distinctions between them as well, will fit in with these indivisible points; & that too, in a much better way than is the case with the common idea of continuous extension of matter. Wherefore he will commit an offence against the use of true reasoning, who, from a prejudgment derived from this agreement & from ignorance of this alternative cause for our sensations, persists in believing that continuous extension is an absolutely necessary property of bodies; and much more so, one who thinks that the very idea of material substance must depend upon this very same continuous extension.

Order of the ideas which we obtain about bodies; the first ideas come through the sense of touch.

161. Now in order that the source of these prejudices may be the more clearly known, I will here quote, from the dissertation *De Materiæ Divisibilitate & Principii Corporum*, three articles, commencing with Art. 14, where we have:—" Even if we allow (a thing quite opposed to my way of thinking) that some ideas are innate & are not acquired through the senses, there is no doubt in my mind that it is quite certain that we derive the idea of a body, of matter, of a corporeal thing, or a material thing, through the senses. Further, the very first ideas, of all those which we have acquired about bodies through the senses, would be in every circumstance those which have excited our sense of touch, & these also are the ideas that we have derived on more occasions than any other ideas. Many things continually present themselves to the sense of touch actually in the very womb of our mothers, before ever perchance we could have any idea of taste, smell, sound, or colour, through the other senses; & of these latter, when first we commenced to have them, there were to start with far fewer occasions for experiencing them. Moreover the ideas which we have obtained through the sense of touch have arisen from phenomena of the following kind. We experienced a resistance on feeling, or on accidental contact with, an object; & this resistance arose from our own limbs, or from those of our mothers. Now, since this resistance offered no opposition through any interval that was perceptible to the senses, it gave us the idea of impenetrability & continuous extension; & then when it ceased in the original direction at any place & was exerted in some other direction, we conceived the boundaries of this quantity, & derived the idea of figure."

Such things demand reflection at the time; ineptitude of infancy for such reflection; on what they may be founded.

162. " Furthermore, these phenomena will have arisen from bodies already formed from matter, not from the single particles of matter of which the bodies themselves were composed. It would have to be considered carefully whether such extension was a property of the body itself, & not of some space through which the particles forming the body were diffused; whether the particles themselves were endowed with the same properties; whether the resistance was exerted only on actual contact, or whether, at very small distances such as did not fall within the scope of the senses, some force would act as a hindrance & produce the same effect, and resistance would be felt even before actual contact; whether properties of this kind would be intrinsic in the matter of which the bodies are composed, & necessary to its existence; or only possessed in certain cases, being due to some external influence. These, & very many other things, should have been investigated most carefully; but the period was indeed veiled in mist & obscurity to a great degree, & very little fitted for aught but the most easy thought. In addition to the weakness of the organs, the mind was occupied with the novelty of things & the rareness of the phenomena; & there was no, or certainly very little, use made of comparisons of these phenomena with one another, to reduce them to definite classes, from which it would be permissible to investigate their laws & causes & thus form some sort of system, through which we could bring the judgment to bear on matters situated outside our own selves. Now, in this very paucity of phenomena, in this difficulty in the matter of forming a system, in this slight use of the powers of reflection, to a greater extent even than in the lack of development of the organs, I consider that infancy consists."

Thence prejudgments are derived that continuity of extension is an essential, but that continuity of odours &c., is accidental.

163. " In this dense haze of things, the first that impressed themselves on the mind were those which required a less deep study & less intent investigation; & these, since the ideas were the more often renewed, made the greater impression & became fixed the more firmly in the mind, & as it were took possession of, so to speak, a land that they found quite empty & hitherto immune, by a sort of right of discovery. Intervals, which in no wise came within the scope of the senses, were considered to be nothing; those things, the ideas of which were always excited simultaneously & conjointly, were considered as identical, or bound up with one another by an extremely close & necessary bond. Hence the result is that we have formed the idea of continuous extension, the idea of

impenetrabilitatis prohibentis ulteriorem motum in ipso tantum contactu corporibus affinxerimus, & ad omnia, quæ ad corpus pertinent, ac ad materiam, ex qua ipsum constat, temere transtulerimus : quæ ipsa cum primum insedissent animo, cum frequentissimis, immo perpetuis phænomenis, & experimentis confirmarentur ; ita tenaciter sibi invicem adhæserunt, ita firmiter ideæ corporum immixta sunt, & cum ea copulata ; ut ea ipsa pro primis corporibus, & omnium corporearum rerum, nimirum etiam materiæ corpora componentis, ejusque partium proprietatibus maxime intrinsecis, & ad naturam, atque essentiam earundem pertinentibus, & tum habuerimus, & nunc etiam habeamus, nisi nos præjudiciis ejusmodi liberemus. Extensionem nimirum continuam, impenetrabilitatem ex contactu, compositionem ex partibus, & figuram, non solum naturæ corporum, sed etiam corporeæ materiæ, & singulis ejusdem partibus, tribuimus tanquam proprietates essentiales : cætera, quæ serins, & post aliquem reflectendi usum deprehendimus, colorem, saporem, odorem sonum, tanquam accidentales quasdam, & adventitias proprietates consideravimus."

Binæ propositiones dissertationis totam Theoriam continentis.

164. Ita ego ibi, ubi Theoriam virium deinde refero, quam supra hic exposui, ac ad præcipuas corporum proprietates applico, quas ex illa deduco, quod hic præstabo in parte tertia. Ibi autem ea adduxeram ad probandam primam e sequentibus propositionibus, quibus probatis & evincitur Theoria mea, & vindicatur : sunt autem hujusmodi : 1. *Nullo prorsus argumento evincitur materiam habere extensionem continuam, & non potius constare e punctis prorsus indivisibilibus a se per aliquod intervallum distantibus ; nec ulla ratio seclusis præjudiciis suadet extensionem ipsam continuam potius, quam compositionem e punctis prorsus indivisibilibus, inextensis, & nullum continuum extensum constituentibus.* 2. *Sunt argumenta, & satis valida illa quidem, quæ hanc compositionem e punctis indivisibilibus evincant extensioni ipsi continuæ præferri oportere.*

Quo pacto congeries punctorum coalescant in massas tenaces : transitus ad partem secundam.

165. At quodnam extensionis genus erit istud, quod e punctis inextensis, & spatio imaginario, sive puro nihilo [76] constat ? Quo pacto Geometria locum habere poterit, ubi nihil habetur reale continuo extensum ? An non punctorum ejusmodi in vacuo innatantium congeries erit, ut quædam nebula unico oris flatu dissolubilis prorsus sine ulla consistenti figura, solidate, resistentia ? Hæc quidem pertinent ad illud extensionis ,& cohæsionis genus, de quo agam in tertia parte, in qua Theoriam applicabo ad Physicam, ubi istis ipsis difficultatibus faciam satis. Interea hic illud tantummodo innuo in antecessum, me cohæsionem desumere a limitibus illis, in quibus curva virium ita sceat axem, ut a repulsione in minoribus distantiis transitus fiat ad attractionem in majoribus. Si enim duo puncta sint in distantia alicujus limitis ejus generis, & vires, quæ immutatis distantiis oriuntur, sint satis magnæ, curva secante axem ad angulum fere rectum, & longissime abeunte ab ipso ; ejusmodi distantiam ea puncta tuebuntur vi maxima ita, ut etiam insensibiliter compressa resistant ulteriori compressioni, ac distracta resistant ulteriori distractioni ; quo pacto si multa etiam puncta cohæreant inter se, tuebuntur utique positionem suam, & massam constituent formæ tenacissimam, ac eadem prorsus phænomena exhibentem, quæ exhiberent solidæ massulæ in communi sententia. Sed de hac re uberius, uti monui, in parte tertia : nunc autem ad secundam faciendus est gradus.

impenetrability preventing further motion only on the absolute contact of bodies; & then we have heedlessly transferred these ideas to all things that pertain to a solid body, and to the matter from which it is formed. Further, these ideas, from the time when they first entered the mind, would be confirmed by very frequent, not to say continual, phenomena & experiences. So firmly are they mutually bound up with one another, so closely are they intermingled with the idea of solid bodies & coupled with it, that we at the time considered these two things as being just the same as primary bodies, & as peculiarly intrinsic properties of all corporeal things, nay further, of the very matter from which bodies are composed, & of its parts; indeed we shall still thus consider them, unless we free ourselves from prejudgments of this nature. To sum up, we have attributed continuous extension, impenetrability due to actual contact, composition by parts, & shape, as if they were essential properties, not only to the nature of bodies, but also to corporeal matter & every separate part of it; whilst others, which we comprehend more deeply & as a consequence of some considerable use of thought, such as colour, taste, smell & sound, we have considered as accidental or adventitious properties."

A pair of propositions of the dissertation containing the whole of my Theory.

164. Such are the words I used; & then I stated the Theory of forces which I have expounded in the previous articles of this work, and I applied the theory to the principal properties of bodies, deducing them from it; & this I will set forth in the third part of the present work. In the dissertation I had brought forward the arguments quoted in order to demonstrate the truth of the first of the following theorems. If these theorems are established, then my Theory is proved & verified; they are as follows:— 1. *There is absolutely no argument that can be brought forward to prove that matter has continuous extension, & that it is not rather made up of perfectly indivisible points separated from one another by a definite interval; nor is there any reason apart from prejudice in favour of continuous extension in preference to composition from points that are perfectly indivisible, non-extended, & forming no extended continuum of any sort.* 2. *There are arguments, & fairly strong ones too, which will prove that this composition from indivisible points is preferable to continuous extension.*

The manner in which groups of points coalesce into tenacious masses: & then we pass on to the second part.

165. Now what kind of extension can that be which is formed out of non-extended points & imaginary space, i.e., out of pure nothing? How can Geometry be upheld if no thing is considered to be actually continuously extended? Will not groups of points, floating in an empty space of this sort be like a cloud, dissolving at a single breath, & absolutely without a consistent figure, or solidity, or resistance? These matters pertain to that kind of extension & cohesion, which I will discuss in the third part, where I apply my Theory to physics & deal fully with these very difficulties. Meanwhile I will here merely remark in anticipation that I derive cohesion from those limit-points, in which the curve of forces cuts the axis, in such a way that a transition is made from repulsion at smaller distances to attraction at greater distances. For if two points are at the distance that corresponds to that of any of the limit-points of this kind, & the forces that arise when the distances are changed are great enough (the curve cutting the axis almost at right angles & passing to a considerable distance from it), then the points will maintain this distance apart with a very great force; so that when they are insensibly compressed they will resist further compression, & when pulled apart they resist further separation. In this way also, if a large number of points cohere together, they will in every case maintain their several positions, & thus form a mass that is most tenacious as regards its form; & this mass will exhibit exactly the same phenomena as little solid masses, as commonly understood, exhibit. But I will discuss this more fully, as I have remarked, in the third part; for now we must pass on to the second part.

[77] PARS II

Theoriæ Applicato ad Mechanicam

Ante applicationem ad Mechanicam consideratio curvæ.

166. Considerabo in hac secunda parte potissimum generales quasdam leges æquilibrii & motus tam punctorum, quam massarum, quæ ad Mechanicam utique pertinent, & ad plurima ex iis, quæ in elementis Mechanicæ passim traduntur, ex unico principio, & adhibito constanti ubique agendi modo, demonstranda viam sternunt pronissimam. Sed prius præmittam nonnulla quæ pertinent ad ipsam virium curvam, a qua utique motuum, phænomena pendent omnia.

Quid in ea considerandum.

167. In ea curva consideranda sunt potissimum tria, arcus curvæ, area comprehensa inter axem, & arcum, quam generat ordinata continuo fluxu, ac puncta illa, in quibus curva secat axem.

Diversa arcuum genera: arcus asymptotici etiam numero infiniti.

168. Quod ad arcus pertinet, alii dici possunt repulsivi, & alii attractivi, prout nimirum jacent ad partes cruris asymptotici ED, vel ad contrarias, ac terminant ordinatas exhibentes vires repulsivas, vel attractivas. Primus arcus ED debet omnino esse asymptoticus ex parte repulsiva, & in infinitum productus: ultimus TV, si gravitas cum lege virium reciproca duplicata distantiarum protenditur in infinitum, debet itidem esse asymptoticus ex parte attractiva, & itidem natura sua in infinitum productus. Reliquos figura 1 exprimit omnes finitos. Verum curva Geometrica etiam ejus naturæ, quam exposuimus, posset habere alia itidem asymptotica crura, quot libuerit, ut si ordinata *mn* in H abeat in infinitum. Sunt nimirum curvæ continuæ, & uniformis naturæ, quæ asymptotos habent plurimas, & habere possunt etiam numero infinitas. (*i*)

Arcus intermedii.

[78] 169. Arcus intermedii, qui se contorquent circa axem, possunt etiam alicubi, ubi ad ipsum devenerint, retro redire, tangendo ipsum, atque id ex utralibet parte, & possent itidem ante ipsum contactum inflecti, & redire retro, mutando accessum in recessum, ut in fig. 1. videre est in arcu P*efq*R.

Arcus prostremus gravitatis fortasse non asymptoticus.

170. Si gravitas generalis legem vis proportionalis inverse quadrato distantiæ, quam non accurate servat, sed quamproxime, uti diximus in priore parte, retinet ad sensum non mutatam solum per totum planetarium, & cometarium systema, fieri utique poterit, ut curva virium non habeat illud postremum crus asymptoticum TV, habens pro asymptoto ipsam reetam AC, sed iterum secet axem, & se contorqueat circa ipsum.(*k*) Tum vero inter

---

(i) *Sit ex. gr. in fig. 12. cycloïs continua* CDEFGH *&c., quam generet punctum peripheriæ circuli continuo revoluti supra rectam* AB, *quæ natura sua protenditur utrinque in infinitum, adeoque in infinitis punctis* C, E, G, I, *&c. occurrit basi* AB. *Si ubicunque ducatur quævis ordinata* PQ, *productaturque in* R *ita, ut sit* PR *tertia post* PQ, *& datam quampiam rectam; punctum* R *erit ad curvam continuam constantem totidem ramis* MNO, VXY, *&c., quot erunt arcus Cycloidales* CDE, EFG, *&c., quorum ramorum singuli habebunt bina crura asymptotica, cum ordinata* PQ *in accessu ad omnia puncta,* C, E, G, &c. *decrescat ultra quoscunque limites, adeoque ordinata* PR *crescat ultra limites quoscunque. Erunt hic quidem omnes asymptoti* CK, EL, GS *&c. parallelæ inter se, & perpendiculares basi* AB, *quod in aliis curvis non est necessarium, cum etiam divergentes utcunque possint esse. Erunt autem & totidem numero, quot puncta illa* C, E, G *&c., nimirum infinitæ. Eodem autem pacto curvarum quarumlibet singuli occursus cum axe in curvis per eas hac eadem lege genitis bina crura asymptotica generant, cruribus ipsis jacentibus, vel, ut hic, ad eandem axis partem, ubi curva genetrix ab eo regreditur retro post appulsum, vel etiam ad partes oppositas, ubi curva genetrix ipsum secet, ac transiliat: cumque possit eadem curva altiorum generum secari in punctis plurimis a recta, vel contingi; poterunt utique haberi & rami asymptotici in curva eadem continua, quo libuerit dato numero.*

(*k*) *Nam ex ipsa Geometrica continuitate, quam persecutus sum in dissertatione* De Lege Continuitatis, *& in dissertatione* De Transformatione Locorum Geometricorum *adjecta Sectionibus Conicis, exhibui necessitatem generalem secundi illius cruris asymptotici redeuntis ex infinito. Quotiescunque enim curva aliqua saltem algebraica habet asymptoticum crus aliquod, debet necessario habere & alterum ipsi respondens, & habens pro asymptoto eandem rectam: sed id habere*

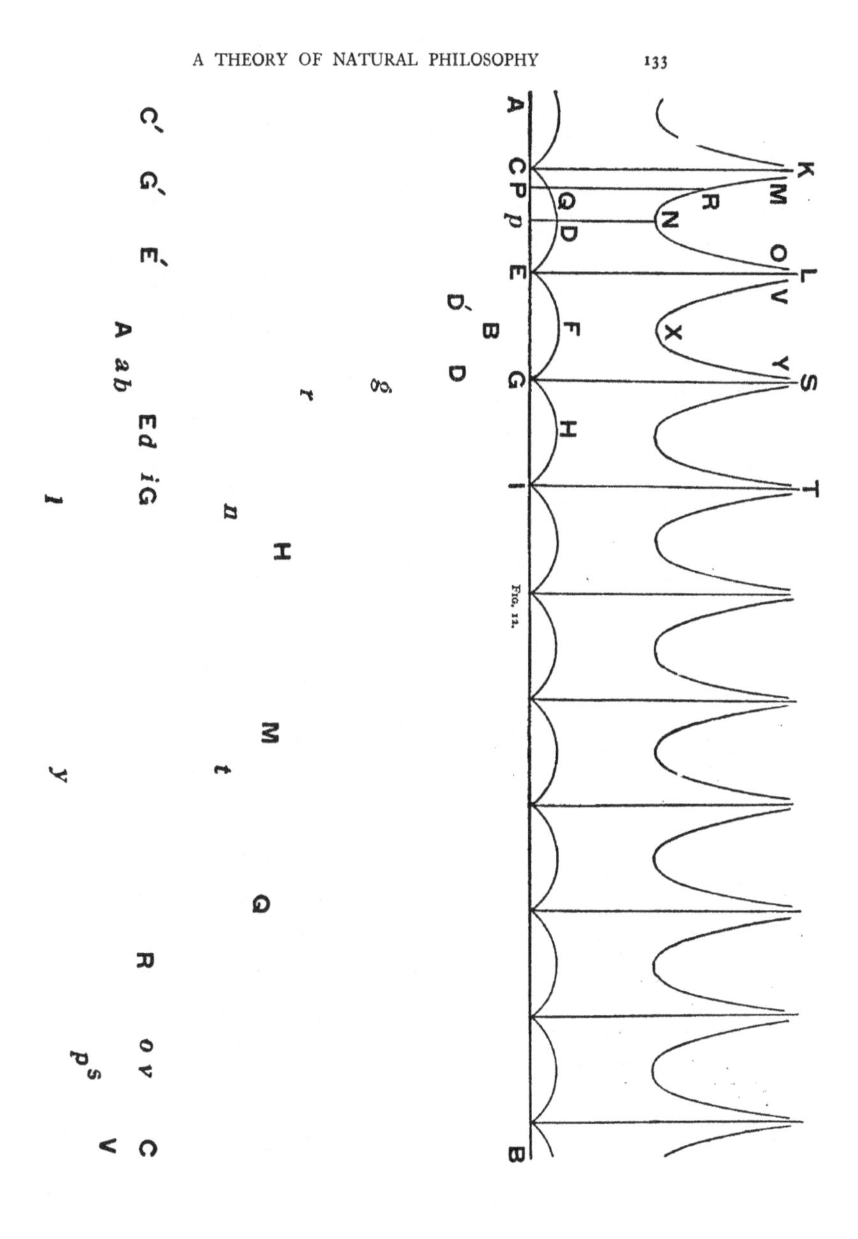

Fig. 12.

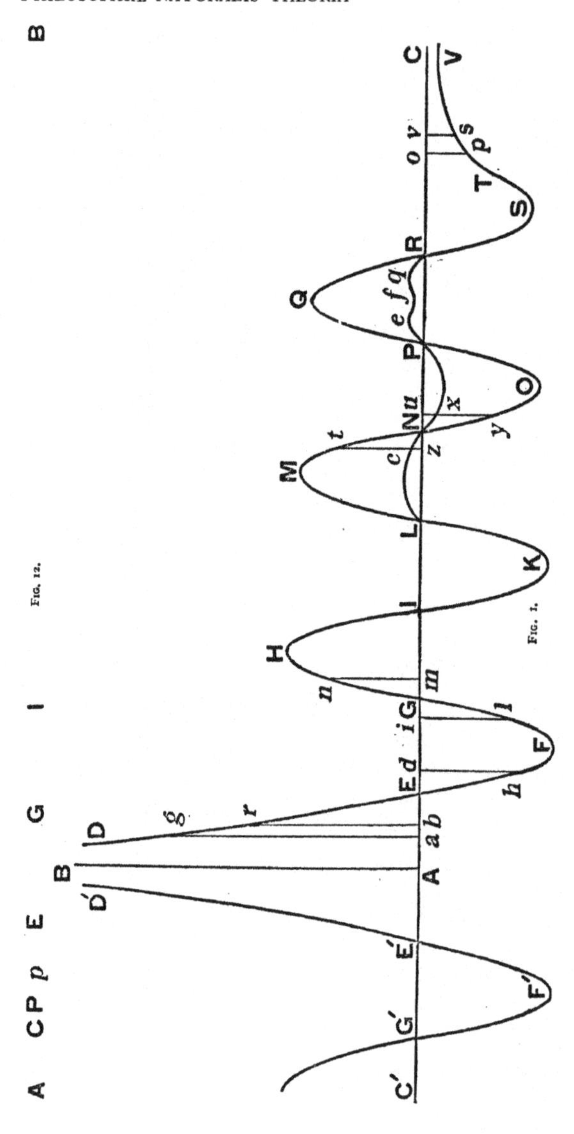

Fig. 12.

Fig. 1.

PART II

Application of the Theory to Mechanics

Consideration of the curve before proceeding with the application to Mechanics.

166. I will consider in this second part more especially certain general laws of equilibrium, & motions both of points & masses; these certainly belong to the science of Mechanics, & they smooth the path that is most favourable for proving very many of those theorems, that are everywhere expounded in the elements of Mechanics, from a single principle, & in every case by the constant employment of a single method of dealing with them. But, before I do that, I will call attention to a few points that pertain to the curve of forces itself, upon which indeed all the phenomena of motions depend.

The Points we have to consider with regard to it.

167. With regard to the curve, there are three points that are especially to be considered; namely, the arcs of the curve, the area included between the axis & the curve swept out by the ordinate by its continuous motion, & those points in which the curve cuts the axis.

The different kinds of arcs; asymptotic arcs may even be infinite in number.

168. As regards the arcs, some may be called repulsive, & others attractive, according indeed as they lie on the same side of the axis as the asymptotic branch ED or on the opposite side, & terminate ordinates that represent repulsive or attractive forces. The first arc ED must certainly be asymptotic on the repulsive side of the axis, & continued indefinitely. The last arc TV, if gravity extends to indefinite distances according to a law of forces in the inverse ratio of the squares of the distances, must also be asymptotic on the attractive side of the axis, & by its nature also continued indefinitely. All the remaining arcs are represented in Fig. 1 as finite. But a geometrical curve, of the kind that we have expounded, may also have other asymptotic branches, as many in number as one can wish; for instance, suppose the ordinate *mn* at H to go away to infinity. There are indeed curves, that are continuous & uniform, which have very many asymptotes, & such curves may even have an infinite number of asymptotes.(*i*)

Intermediate arcs.

169. The intermediate arcs, which wind about the axis, can also, at any point where they reach it, return backwards & touch it; and they can do this on either side of it; they may also be reflected and recede before actual contact, the approach being altered into a recession, as is to be seen in Fig. 1 with regard to the arc P*efq*R.

The ultimate arc representing gravity possibly not asymptotic.

170. If universal gravity obeys the law of a force inversely proportional to the square of the distance (which, as I remarked in the first part, it only obeys as nearly as possible, but not exactly), sensibly unchanged only throughout the planetary & cometary system, it will certainly be the case that the curve of forces will not have the last arm PV asymptotic with the straight line AC as the asymptote, but will again cut the axis & wind about it. (*k*) Then

---

(i) *Let, for example, in Fig.* 12, CDEFGH &c. *be a continuous cycloid, generated by a point on the circumference of a circle rolling continuously along the straight line* AB; *this by its nature extends on either side to infinity, & thus meets the base* AB *in an infinite number of points such as* C, E, G, I, &c. *If at every point there is drawn an ordinate such as* PQ, *and this is produced to* R, *so that* PR *is a third proportional to* PQ *& some given straight line; then the point* R *will trace out a continuous curve consisting of as many branches,* MNO, VXY, &c., *as there are cycloidal arcs,* CDE, EFG, &c.; *each of these branches will have a pair of asymptotic arms, since the ordinate* PQ *on approaching any one of the points* C,E,G, &c., *will decrease beyond all limits, & thus the ordinate* PR *will increase beyond all limits. In this curve then there will be* CK, EL, GS, &c., *all asymptotes parallel to one another & perpendicular to the base* AB; *this is not necessarily the case in other curves, since they may be also inclined to one another in any manner. Further they will be as many in number as there are points such as* C, E, G, &c., *that is to say, infinite. Again, in a similar way, the several intersections of any curves you please with the axis give rise to a pair of asymptotic arms in curves derived from them according to the same law; and these arms lie, either on the same side of the axis, as in this case, where the original curve leaves the axis once more after approaching it, or indeed on opposite sides of the axis, where the original curve cuts & crosses it. Also, since it is possible for the same curve of higher orders to be cut in a large number of points, or to be touched, there will possibly be also asymptotic arms in this same continuous curve equal to any given number you please.*

(k) *For, from the principle of geometrical continuity itself, which I discussed in my dissertation* De Lege Continuitatis *and in the dissertation* De Transformatione Locorum Geometricorum *appended to my* Sectionum Conicarum Elementa, *I showed the necessity for the second asymptotic arm returning from infinity. For as often as an algebraical curve has at least one asymptotic arm, it must also have another that corresponds to it & has the same straight line*

alios casus innumeros, qui haberi possent, unum censeo speciminis gratia hic non omittendum; incredibile enim est, quam ferax casuum, quorum singuli sunt notatu dignissimi, unica etiam hujusmodi curva esse possit.

Series curvarum similium, cum serie Mundorum magnitudine proportionalium.

171. Si in fig. 14 in axe C'C sint segmenta AA', A'A'' numero quocunque, quorum posteriora sint in immensum majora respectu præcedentium, & per singula transeant, asympto-[79]-ti AB, A'B', A''B'' perpendiculares axi; possent inter binas quasque asymptotos esse curvæ ejus formæ, quam in fig. 1 habuimus, & quæ exhibetur hic in DEFI &c., D'E'F'I', &c., in quibus primum crus ED esset asymptoticum repulsivum, postremum SV attractivum, in singulis vero intervallum EN, quo arcus curvæ contorquetur, sit perquam exiguum respectu intervalli circa S, ubi arcus diutissime perstet proximus hyperbolæ habenti ordinatas in ratione reciproca duplicata distantiarum, tum vero vel immediate abiret in arcum asymptoticum attractivum, vel iterum contorqueretur utcunque usque ad ejusmodi asymptoticum attractivum arcum, habente utroque asymptotico arcu arcam infinitam; in eo casu collocato quocunque punctorum numero inter binas quascunque asymptotos, vel inter binaria quotlibet, & rite ordinato, posset exurgere quivis, ut ita dicam, Mundorum numerus, quorum singuli essent inter se simillimi, vel dissimillimi, prout arcus EF&cN, E'F'&cN' essent inter se similes, vel dissimiles, atque id ita, ut quivis ex iis nullum haberet commercium cum quovis alio; cum nimirum nullum punctum posset egredi ex spatio incluso iis binis arcubus, hinc repulsivo, & inde attractivo; & ut omnes Mundi minorum dimensionum simul sumpti vices agerent unius puncti respectu proxime majoris, qui constaret ex ejusmodi massulis respectu sui tanquam punctualibus, dimensione nimirum omni singulorum, respectu ipsius, & respectu distantiarum, ad quas in illo devenire possint, fere nulla; unde & illud consequi posset, ut quivis ex ejusmodi tanquam Mundis nihil ad sensum perturbaretur a motibus, & viribus Mundi illius majoris, sed dato quovis utcunque magno tempore totus Mundus inferior vires sentiret a quovis puncto materiæ extra ipsum posito accedentes, quantum libuerit, ad æquales, & parallelas quæ idcirco nihil turbarent respectivum ipsius statum internum.

Omissis sublimioribus, progressus ad areas.

172. Sed ea jam pertinent ad applicationem ad Physicam, quæ quidem hic innui tantummodo, ut pateret, quam multa notatu dignissima considerari ibi possent, & quanta sit hujusce campi fœcunditas, in quo combinationes possibiles, & possibiles formæ sunt sane infinities infinitæ, quarum, quæ ab humana mente perspici utcunque possunt, ita sunt paucæ respectu totius, ut haberi possint pro mero nihilo, quas tamen omnes unico intuitu præsentes vidit, qui Mundum condidit, DEUS. Nos in iis, quæ consequentur, simpliciora tantummodo quædam plerumque consectabimur, quæ nos ducant ad phænomena iis conformia, quæ in Natura nobis pervia intuemur, & interea progrediemur ad areas arcubus respondentes.

Guicunque axis segmento posse aream respondere utcunque magnam vel parvam partis secundæ demonstratio.

173. Aream curvæ propositæ cuicunque, utcunque exiguo, axis segmento respondentem posse esse utcunque magnam, & arcam respondentem cuicunque, utcunque magno, [80] posse esse utcunque parvam, facile patet. Sit in fig. 15, MQ segmentum axis utcunque parvum, vel magnum; ac detur area utcunque magna, vel parva. Ea applicata ad MQ exhibebit quandam altitudinem MN ita, ut, dueta NR parallela MQ, sit MNRQ æqualis areæ datæ, adeoque assumpta QS dupla QR, area trianguli MSQ crit itidem æqualis areæ datæ. Jam vero pro secundo casu satis patet, posse curvam transire infra reetam NR, uti transit XZ, cujus area idcirco esset minor, quam area MNRQ; nam esset ejus pars.

potest vel ex eadem parte, vel ex opposita; & crus ipsum jacere potest vel ad easdem plagas partis utriuslibet cum priore crure, vel ad oppositas, adeoque cruris redeuntis ex infinito positiones quatuor esse possunt. Si in fig. 13 *crus* ED *abeat in infinitum, existente asymptoto* ACA', *potest regredi ex parte* A *vel ut* HI, *quod crus jacet ad eandem plagam, vel ut* KL, *quod jacet ad oppositam; & ex parte* A', *vel ut* MN, *ex eadem plaga, vel ut* OP, *ex opposita. In posteriore ex iis duabus dissertationibus profero exempla omnium ejusmodi regressuum; ac secundi, & quarti casus exempla exhibet etiam superior genesis, si curva generans contingat axem, vel secet, ulterius progressa respectu ipsius. Inde autem fit, ut crura asymptotica rectilineam habentia asymptotum esse non possint, nisi numero pari, ut & radices imaginariæ in æquationibus algebraicis.*

Verum hic in curva virium, in qua arcus semper debet progredi, ut singulis distantiis, sive abscissis, singulæ vires, sive ordinatæ respondeant, casus primus, & tertius haberi non possunt. Nam ordinata RQ *cruris* DE *occurreret alicubi in* S, S' *cruribus etiam* HI, MN; *adeoque relinquentur soli quartus, & secundus, quorum usus erit infra.*

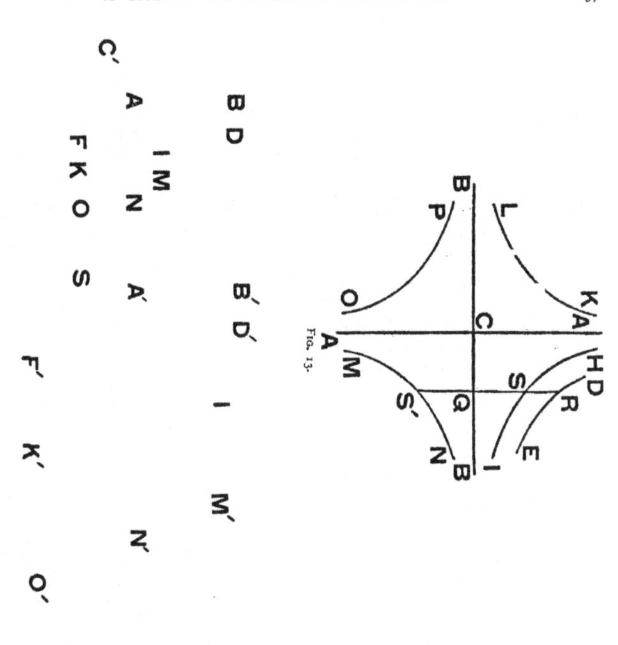

Fig. 13.

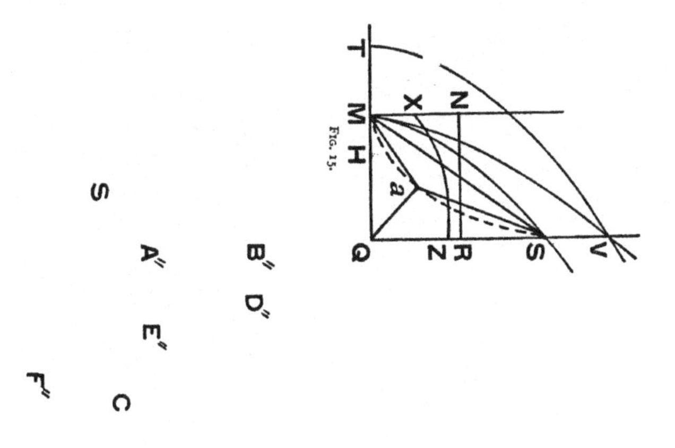

Fig. 15.

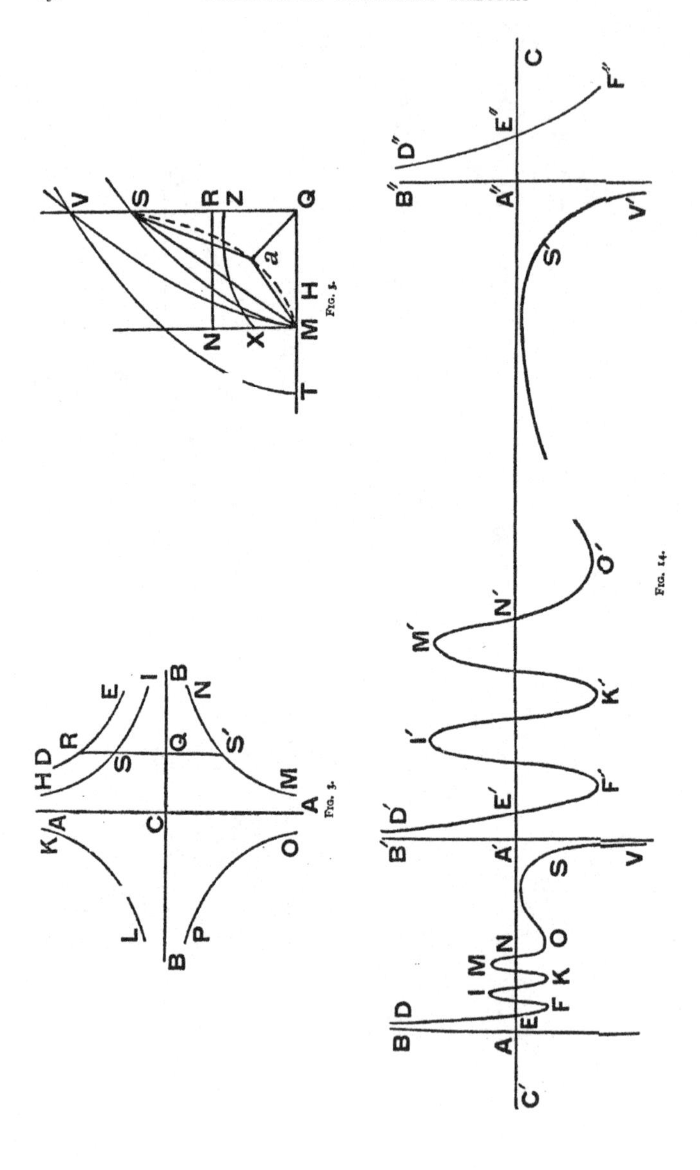

Fig. 5.

Fig. 3.

Fig. 14.

there is one, out of an innumerable number of other cases that may possibly happen, which I think for the sake of an example should not be omitted here; for it is incredible how prolific in cases, each of which is well worth mentioning, a single curve of this kind can be.

A series of similar curves, with a series of universes proportional in magnitude.

171. If, in Fig. 14, there are any number of segments AA′, A′A″, of which each that follows is immensely great with regard to the one that precedes it; & if through each point there passes an asymptote, such as AB, A′B′, A″B″, perpendicular to the axis; then between any two of these asymptotes there may be curves of the form given in Fig. 1. These are represented in Fig. 14 by DEFI &c., D′E′F′I′ &c.; & in these the first arm E would be asymptotic & repulsive, & the last SV attractive. In each the interval EN, where the arc of the curve is winding, is exceedingly small compared with the interval near S, where the arc for a very long time continues closely approximating to the form of the hyperbola having its ordinates in the inverse ratio of the squares of the distances; & then, either goes off straightway into an asymptotic & attractive arm, or once more winds about the axis until it becomes an asymptotic attractive arc of this kind, the area corresponding to either asymptotic arc being infinite. In such a case, if a number of points are assembled between any pair of asymptotes, or between any number of pairs you please, & correctly arranged, there can, so to speak, arise from them any number of universes, each of them being similar to the other, or dissimilar, according as the arcs EF . . . . N, E′F′ . . . . N′ are similar to one another, or dissimilar; & this too in such a way that no one of them has any communication with any other, since indeed no point can possibly move out of the space included between these two arcs, one repulsive & the other attractive; & such that all the universes of smaller dimensions taken together would act merely as a single point compared with the next greater universe, which would consist of little point-masses, so to speak, of the same kind compared with itself, that is to say, every dimension of each of them, compared with that universe & with respect to the distances to which each can attain within it, would be practically nothing. From this it would also follow that any one of these universes would not be appreciably influenced in any way by the motions & forces of that greater universe; but in any given time, however great, the whole inferior universe would experience forces, from any point of matter placed without itself, that approach as near as possible to equal & parallel forces; these therefore would have no influence on its relative internal state.

Leaving out more abstruse matters, we pass on to areas.

172. Now these matters really belong to the application of the Theory to physics; & indeed I only mentioned them here to show how many things there may be well worth considering in that section, & how great is the fertility of this field of investigation, in which possible combinations & possible forms are truly infinitely infinite; of these, those that can be in any way comprehended by the human intelligence are so few compared with the whole, that they can be considered as a mere nothing. Yet all of them were seen in clear view at one gaze by GOD, the Founder of the World. We, in what follows, will for the most part investigate only certain of the more simple matters which will lead us to phenomena in conformity with those things that we contemplate in Nature as far as our intelligence will carry us; meanwhile we will proceed to the areas corresponding to the arcs.

To any segment of the axis there may correspond any area, however great or however small; proof of the second part of this assertion.

173. It is easily shown that the area corresponding to any segment of the axis, however small, can be anything, no matter how great; & the area corresponding to any segment, however great, can be anything, no matter how small. In Fig. 15, let MQ be a segment of the axis, no matter how small, or great; & let an area be given, no matter how great, or small. If this area is applied to MQ a certain altitude MN will be given, such that, if NR is drawn parallel to MQ, then MNRQ will be equal to the given area; & thus, if QS is taken equal to twice QR, the area of the triangle MSQ will also be equal to the given area. Now, for the second case it is sufficiently evident that a curve can be drawn below the straight line NR, in the way XZ is shown, the area under which is less than the area MNRQ;

---

as its asymptote; & this can take place with either the same part of the line or with the other part; also the arm itself can lie either on the same side of either of the two parts, or on the opposite side. Thus there may be four positions of the arm that returns from infinity. If, in Fig. 13, the arm ED goes off to infinity, the asymptote being ACA, *it may return from the direction of* A, *either like* HI, *where the arm lies on the same side of the asymptote or as* KL *which lies on the opposite side of it; or from the direction of* A′, *either as* MN, *on the same side, or as,* DP, *on the opposite side. In the second of these two dissertations, I have given examples of all regressions of this sort; & the method of generation given above will yield examples of the second & fourth cases, if the generating curve touches the axis, or cuts it & passes over beyond it. Further, it thus comes about that asymptotic arms having a rectilinear asymptote cannot exist except in pairs, just like imaginary roots in algebraical equations.*

But here in the curve of forces, in which the arc must always proceed in such a manner that to each distance or abscissa there corresponds a single force or ordinate, the first & third cases cannot occur. For the ordinate RQ *of the arm* DE *would meet somewhere, in* S, S′, *the branches* HI, MN *as well. Hence only the fourth & second cases are left; & these we will make use of later.*

Quin immo licet ordinata QV sit utcunque magna; facile patet, posse arcum M*a*V ita accedere ad rectas MQ, QV; ut area inclusa iis rectis, & ipsa curva, minuatur infra quoscunque determinatos limites. Potest enim jacere totus arcus intra duo triangula Q*a*M, Q*a*V, quorum altitudines cum minui possint, quantum libuerit, stantibus basibus MQ, QV, potest utique area ultra quoscunque limites imminui. Posset autem ea area esse minor quacunque data; etiamsi QV esset asymptotus, qua de re paullo inferius.

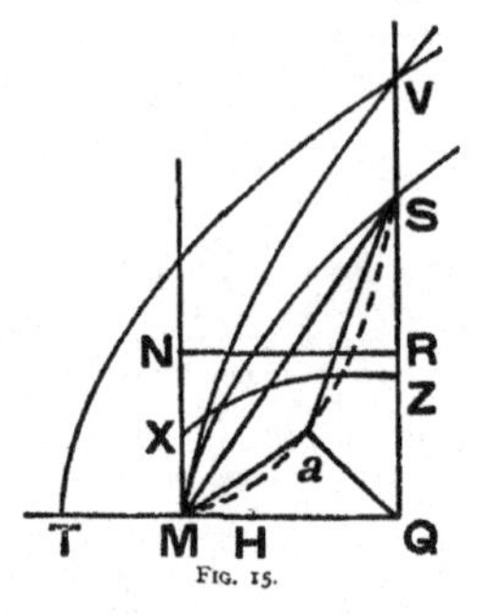

Fig. 15.

Demonstratio primæ.

174. Pro primo autem casu vel curva secet axem extra MQ, ut in T, vel in altero extremo, ut in M; fieri poterit, ut ejus arcus TV, vel MV transeat per aliquod punctum V jacens ultra S, vel etiam per ipsum S ita, ut curvatura illum ferat, quemadmodum figura exhibet, extra triangulum MSQ, quo casu patet, arcam curvæ respondentem intervallo MQ fore majorem, quam sit area trianguli MSQ, adeoque quam sit area data; crit enim ejus trianguli area pars areæ pertinentis ad curvam. Quod si curva etiam secaret alicubi axem, ut in H inter M, & Q, tum vero fieri posset, ut area respondens alteri e segmentis MH, QH esset major, quam area data. simul, & area alia assumpta, qua area assumpta esset minor area respondens segmento, alteri adeoque excessus prioris supra posteriorem remaneret major, quam area data.

Aream asymptoticam posse esse infinitam, vel finitam magnitudinis cujuscunque.

175. Area asymptotica clausa inter asymptotum, & ordinatam quamvis, ut in fig. 1 BA*ag*, potest esse vel infinita, vel finita magnitudinis cujusvis ingentis, vel exiguæ. Id quidem etiam geometrice demonstrari potest, sed multo facilius demonstratur calculo integrali admodum elementari; & in Geometriæ sublimioris clementis habentur theoremata, ex quibus id admodum facile deducitur (*l*). Generaliter nimi-[81]-rum area ejusmodi est infinita; si ordinata crescit in ratione reciproca abscissarum simplici, aut majore: & est finita; si crescit in ratione multiplicata minus, quam per unitatem.

Areas exprimere incrementa, vel decrementa quadrati velocitatis.

176. Hoc, quod de areis dictum est, necessarium fuit ad applicationem ad Mechanicam, ut nimirum habeatur scala quædam velocitatum, quæ in accessu puncti cujusvis ad aliud punctum, vel recessu generantur, vel eliduntur; prout ejus motus conspiret directione vis, vel sit ipsi contrarius. Nam, quod innuimus & supra in adnot. (*f*) ad num. 118., ubi vires exprimuntur per ordinatas, & spatia per abscissas, area, quam texit ordinata, exprimit incrementum, vel decrementum quadrati velocitatis, quod itidem ope Geometriæ demonstratur facile, & demonstravi tam in dissertatione *De Viribus Vivis*, quam in Stayanis Supplementis; sed multo facilius res conficitur ope calculi integralis. (*m*)

(l) *Sit* Aa *in Fig.* 1 $=x$, $ag=y$; *ac sit* $x^m y^n = 1$; *erit* $y = x^{-m/n}$, $y\,dx$ *elementum areæ* $= x^{-m/n}dx$, *cujus integrale* $\frac{n}{n-m}x^{\frac{n-m}{n}} + A$, *addita constanti* A, *sive ob* $x^{-m/n}=y$, *habebitur* $\frac{n}{n-m}xy + A$. *Quoniam incipit area in* A, *in origine abscissarum; si* $n-m$ *fuerit numerus positivus, adeoque* n *major, quam* m; *area erit finita, ac valor* $A=0$; *area vero erit ad rectangulum* $Aa \times ag$, *ut in ad* $n-m$, *quod rectangulum, cum* ag *possit esse magna, & parva, ut libuerit, potest esse magnitudinis cujusvis. Is valor fit infinitus, si facto* $m=n$, *divisor evadat* $=0$; *adeoque multo magis fit infinitus valor areæ, si* m *sit major, quam* n. *Unde constat, aream fore infinitam, quotiescunque ordinatæ crescent in ratione reciproca simplici, & majore; secus fore finitam.*

(m) *Sit* u *vis,* c *celeritas,* t *tempus,* s *spatium: erit* $u\,dt = dc$, *cum celeritatis incrementum sit proportionale vi, & tempusculo; ac erit* $c\,dt = ds$, *cum spatiolum confectum respondeat velocitati, & tempusculo. Hinc eruitur* $dt = \frac{dc}{u}$, *& pariter* $dt = \frac{ds}{c}$, *adeoque* $\frac{dc}{u} = \frac{ds}{c}$, *&* $c\,dc = u\,ds$. *Porro* $2c\,dc$ *est incrementum quadrati velocitatis* cc, *&* $u\,ds$ *in hypothesi, quod ordinata sit* u, *& spatium* s *sit abscissa, est areola respondens spatiolo* ds *confecto. Igitur incrementum quadrati velocitatis conspirante vi, adeoque decrementum vi contraria, respondet areæ respondenti spatiolo percurso quovis infinitesimo tempusculo; & proinde tempore etiam quovis finito incrementum, vel decrementum quadrati velocitatis respondet areæ pertinenti ad partem axis referentem spatium percursum.*

Hinc autem illud sponte consequitur: si per aliquod spatium vires in singulis punctis eædem permaneant, mobile autem adveniat cum velocitate quavis ad ejus initium; differentiam quadrati velocitatis finalis a quadrato velocitatis initialis fore semper eandem, quæ idcirco erit tota velocitas finalis in casu, in quo mobile initio illius spatii haberet velocitatem nullam. Quare, quod nobis erit inferius usui, quadratum velocitatis finalis, conspirante vi cum directione motus, æquabitur binis quadratis binarum velocitatum, ejus, quam habuit initio, & ejus, quam acquisivisset in fine, si initio ingressum fuisset sine ulla velocitate.

for it is part of it. Again, although the ordinate QV may be of any size, however great, it is easily shown that an arc M*a*V can approach so closely to the straight lines MQ, QV that the area included between these lines & the curve shall be diminished beyond any limits whatever. For it is possible for the curve to lie within the two triangles Q*a*M, Q*a*V; & since the altitudes of these can be diminished as much as you please, whilst the bases MQ, QV remain the same, therefore the area can indeed be diminished beyond all limits whatever. Moreover it is possible for this area to be less than any given area, even although QV should be an asymptote; we will consider this a little further on.

Proof of the first part.

174. Again, for the first case, either the curve will cut the axis beyond MQ, as at T, or at either end, as at M. Then it is possible for it to happen that an arc of it, TV or MV, will pass through some point V lying beyond S, or even through S itself, in such a way that its curvature will carry it, as shown in the diagram, outside the triangle MSQ; in this case it is clear that the area of the curve corresponding to the interval MQ will be greater than the area of the triangle MSQ, & therefore greater than the given area, for the area of this triangle is part of the area belonging to the curve. But if the curve should even cut the axis anywhere, as at H, between M & Q, then it would be possible for it to come about that the area corresponding to one of the two segments MH, QH would be greater than the given area together with some other assumed area; & that the area corresponding to the other segment should be less than this assumed area; and thus the excess of the former over the latter would remain greater than the given area.

An asymptotic area may be either infinite or equal to any finite area whatever.

175. An asymptotic area, bounded by an asymptote & any ordinate, like BA*ag* in Fig. 1, can be either infinite, or finite of any magnitude either very great or very small. This can indeed be also proved geometrically, but it can be demonstrated much more easily by an application of the integral calculus that is quite elementary; & in the elements of higher geometry theorems are obtained from which it is derived quite easily. (*l*) In general, it is true, an area of this kind is infinite; namely when the ordinate increases in the simple inverse ratio of the abscissæ, or in a greater ratio; and it is finite, if it increases in this ratio multiplied by something less than unity.

The areas represent the increments or decrements of the square of the velocit

176. What has been said with regard to areas was a necessary preliminary to the application of the Theory to Mechanics; that is to say, in order that we might obtain a diagrammatic representation of the velocities, which, on the approach of any point to another point, or on recession from it, are produced or destroyed, according as its motion is in the same direction as the direction of the force, or in the opposite direction. For, as we also remarked above, in note (*f*) to Art. 118, when the forces are represented by ordinates & the distances by abscissae, the area that the ordinate sweeps out represents the increment or decrement of the square of the velocity. This can also be easily proved by the help of geometry; & I gave the proof both in the dissertation *De Viribus Vivis* & in the Supplements to Stay's Philosophy; but the matter is much more easily made out by the aid of the integral calculus.(*m*)

(l) *In Fig.* 1 *let* Aa $= x$, $ag = y$; *& let* $x^m y^n = 1$. *Then will* $y = x^{-m/n}$, *& the element of area* $y\,dx = x^{-m/n}\,dx$: *the integral of this is* $\frac{n}{n-m} x^{(n-m)/n} + A$, *where a constant* A *is added; or, since* $x^{-m/n} = y$, *we shall have* $\frac{n}{n-m} xy + A$. *Now, since the area is initially* A, *at the origin of the abscissæ, if* $n-m$ *happened to be a positive number, & thus* n *greater than* m, *then the area will be finite, & the value of* A *will be* $= 0$. *Also the area will be to the rectangle* Aa.ag *as* n *is to* $n-m$; *& this rectangle, since* ag *can be either great or small, as you please, may be of any magnitude whatever. The value is infinite, if by making* m *equal to* n *the divisor becomes equal to zero; & thus the value of the area becomes all the more infinite, if* m *is greater than* n. *Hence it follows that the area will be infinite, whenever the ordinates increase in a simple inverse ratio, or in a greater ratio; otherwise it will be finite.*

(m) *Let* u *be the force,* c *the velocity,* t *the time, &* s *the distance. Then will* $u\,dt = dc$, *since the increment of the velocity is proportional to the force, & to the small interval of time. Also* $c\,dt = ds$, *since the distance traversed corresponds with the velocity & the small interval of time. Hence it follows that* $dt = dc/u$, *& similarly* $dt = ds/c$, *& therefore* $dc/u = ds/c$, *&* $c\,dc = u\,ds$. *Further,* $2c\,dc$ *is the increment of the square of the velocity* c^2, *&* $u\,ds$, *on the hypothesis that the ordinate represents* u, *& the abscissa the distance* s, *is the small area corresponding to the small distance traversed. Hence the increment of the square of the velocity, when in the direction of the force, & the decrement when opposite in direction to the force, is represented by the area corresponding to* ds, *the small distance traversed in any infinitely short time. Hence also, in any finite interval of time, the increment or decrement of the square of the velocity will be represented by the area corresponding to that part of the axis which represents the distance traversed.*

Hence also it follows immediately that, if through any distance the force on each of the points remains as before, but the moving body arrives at the beginning of it with any velocity, then the difference between the square of the final velocity & the square of the initial velocity will always be the same; & this therefore will be the total final velocity, in the case where the moving body had no velocity at the beginning of the distance. Hence, the square of the final velocity, when the motion is in the same direction as the force, will be equal to the sum of the squares of the velocity which it had at the beginning & of the velocity it would have acquired at the end, if it had at the beginning started without any velocity; a theorem that we shall make use of later.

Atque id ipsum, licet segmenta axis sint dimidia spatiorum percursorum a singulis punctis.

177. Duo tamen hic tantummodo notanda sunt; primo quidem illud: si duo puncta ad se invicem accedant, vel a se invicem recedant in ea recta, quæ ipsa conjungit, segmenta illius [82] axis, qui exprimit distantias, non expriment spatium confectum; nam moveri debebit punctum utrumque: adhuc tamen illa segmenta erunt proportionalia ipsi spatio confecto, eorum nimirum dimidio; quod quidem satis est ad hoc, ut illæ areæ adhuc sint proportionales incrementis, vel decrementis quadrati velocitatum, adeoque ipsa exprimant.

Si areæ sint partim attractivæ, partim repulsivæ, assumendam esse differentiam earundem.

178. Secundo loco notandum illud, ubi areæ respondentes dato cuipiam spatio sint partim attractivæ, partim repulsivæ, earum differentiam, quæ oritur subtrahendo summam omnium repulsivarum a summa attractivarum, vel vice versa, exhibituram incrementum illud, vel decrementum quadrati velocitatis; prout directio motus respectivi conspiret cum vi, vel oppositam habeat directionem. Quamobrem si interea, dum per aliquod majus intervallum a se invicem recesserunt puncta, habuerint vires directionis utriusque; ut innotescat, an celeritas creverit, an decreverit & quantum; erit investigandum, an areæ omnes attractivæ simul, omnes repulsivas simul superent, an deficiant, & quantum; inde enim, & a velocitate, quæ habebatur initio, erui poterit quod quæritur.

Appulsus ad axem curvæ secantis, vel tangentis: sectionum seu limitum duo genera.

179. Hæc quidem de arcubus, & areis; nunc aliquanto diligentius considerabimus illa axis puncta, ad quæ curva appellit. Ea puncta vel sunt ejusmodi, ut in iis curva axem secet, cujusmodi in fig. 1 sunt E,G,I, &c., vel ejusmodi, ut in iis ipsa curva axem contingat tantummodo. Primi generis puncta sunt ea, in quibus fit transitus a repulsionibus ad attractiones, vel vice versa, & hæc ego appello limites, quod nimirum sint inter eas oppositarum directionum vires. Sunt autem hi limites duplicis generis: in aliis, aucta distantia, transitur a repulsione ad attractionem: in aliis contra ab attractione ad repulsionem. Prioris generis sunt E,I,N,R; posterioris G,L,P: & quoniam, posteaquam ex parte repulsiva in una sectione curva transiit ad partem attractivam; in proxime sequenti sectione debet necessario ex parte attractiva transire ad repulsivam, ac vice versa; patet, limites fore alternatim prioris illius, & hujus posterioris generis.

In quo conveniant inter se, in quo differant: limites cohæsionis, & non cohæsionis.

180. Porro limites prioris generis, a limitibus posterioris ingens habent inter se discrimen. Habent illi quidem hoc commune, ut duo puncta collocata in distantia unius limitis cujuscunque nullam habeant mutuam vim, adeoque si respective quiescebant, pergant itidem respective quiescere. At si ab illa respectiva quiete dimoveantur; tum vero in limite primi generis ulteriori dimotioni resistent, & conabuntur priorem distantiam recuperare, ac sibi relicta ad illam ibunt; in limite vero secundi generis, utcunque parum dimota, sponte magis fugient, ac a priore distantia statim recedent adhuc magis. Nam si distantia minuatur; habebunt in limite prioris generis vim repulsivam, quæ obstabit uteriori accessui, & urgebit puncta ad mutuum recessum, quem sibi relicta acquirent, [83] adeoque tendent ad illam priorem distantiam: at in limite secundi generis habebunt attractionem, qua adhuc magis ad se accedent, adeoque ab illa priore distantia, quæ erat major, adhuc magis sponte fugient. Pariter si distantia augeatur, in primo limitum genere a vi attractiva, quæ habetur statim in distantia majore; habebitur resistentia ad ulteriorem recessum, & conatus ad minuendam distantiam, ad quam recuperandam sibi relicta tendent per accessum; at in limitibus secundi generis orietur repulsio, qua sponte se magis adhuc fugient, adeoque a minore illa priore distantia sponte magis recedent. Hine illos prioris generis limites, qui mutuæ positionis tenaces sunt, ego quidem appellavi *limites cohæsionis*, & secundi generis limites appellavi *limites non cohæsionis*.

Duo genera contactuum.

181. Illa puncta, in quibus curva axem tangit, sunt quidem terminus quidam virium, quæ ex utraque parte, dum ad ea acceditur, decrescunt ultra quoscunque limites, ac demum ibidem evanescunt; sed in iis non transitur ab una virium directione ad aliam. Si contactus fiat ab areu repulsivo; repulsiones evanescunt, sed post contactum remanent itidem repulsiones; ac si ab arcu attractivo, attractionibus evanescentibus attractiones iterum immediate succedunt. Duo puncta collocata in ejusmodi distantia respective quiescunt;

The same result holds good even when the segments of the axis are the halves of the distances traversed by single points.

177. However, there are here two things that want noting only. The first of them is this, that if two points approach one another or recede from one another in the straight line joining them, the segments of the axis, which expresses distances, do not represent the distances traversed; for both points will have to move. Nevertheless the segments will still be proportional to the distance traversed, namely, the half of it; & this indeed is sufficient for the areas to be still proportional to the increments or decrements of the squares of the velocities, & thus to represent them.

If the areas are partly attractive & partly repulsive, their difference must be taken.

178. In the second place it is to be noted that, where the areas corresponding to any given interval are partly attractive & partly repulsive, their difference, obtained by subtracting the sum of all those that are repulsive from the sum of those that are attractive, or vice versa, will represent the increment, or the decrement, of the square of the velocity, according as the direction of relative motion is in the same direction as the force, or in the opposite direction. Hence, if, during the time that the points have receded from one another by some considerable interval, they had forces in each direction; then in order to ascertain whether the velocity had been increased or decreased, & by how much, it will have to be considered whether all the attractive areas taken together are greater or less than all the repulsive areas taken together, & by how much. For from this, & from the velocity which initially existed, it will be possible to deduce what is required.

Approach of the curve to the axis when it cuts or touches it; two kinds of intersections or limit-points.

179. So much for the arcs & the areas; now we must consider in a rather more careful manner those points of the axis to which the curve approaches. These points are either such that the curve cuts the axis in them, for instance, the points E, G, I, &c. in Fig. 1; or such that the curve only touches the axis at the points. Points of the first kind are those in which there is a transition from repulsions to attractions, or vice-versa; & these I call limit-points or boundaries, since indeed they are boundaries between the forces acting in opposite directions. Moreover these limit-points are twofold in kind; in some, when the distance is increased, there is a transition from repulsion to attraction; in others, on the contrary, there is a transition from attraction to repulsion. The points E, I, N, R are of the first kind, and G, L, P are of the second kind. Now, since at one intersection, the curve passes from the repulsive part to the attractive part, at the next following intersection it is bound to pass from the attractive to the repulsive part, & vice versa. It is clear then that the limit-points will be alternately of the first & second kinds.

In what they agree & in what they differ: the limit-points of cohesion & of non-cohesion.

180. Further, there is a distinction between limit-points of the first & those of the second kind. The former kind have this property in common; namely that, if two points are situated at a distance from one another equal to the distance of any one of these limit-points from the origin, they will have no mutual force; & thus, if they are relatively at rest with regard to one another, they will continue to be relatively at rest. Also, if they are moved apart from this position of relative rest, then, for a limit-point of the first kind, they will resist further separation & will strive to recover the original distance, & will attain to it if left to themselves; but, in a limit-point of the second kind, however small the separation, they will of themselves seek to get away from one another & will immediately depart from the original distance still more. For, if the distance is diminished, they will have, in a limit-point of the first kind, a repulsive force, which will impede further approach & impel the points to mutual recession, & this they will acquire if left to themselves; thus they will endeavour to maintain the original distance apart. But in a limit-point of the second kind they will have an attraction, on account of which they will approach one another still more; & thus they will seek to depart still further from the original distance, which was a greater one. Similarly, if the distance is increased, in limit-points of the first kind, due to the attractive force which is immediately obtained at this greater distance, there will be a resistance to further recession, & an endeavour to diminish the distance; & they will seek to recover the original distance if left to themselves by approaching one another. But, in limit-points of the second class, a repulsion is produced, owing to which they try to get away from one another still further; & thus of themselves they will depart still more from the original distance, which was less. On this account indeed I have called those limit-points of the first kind, which are tenacious of mutual position, *limit-points of cohesion*, & I have termed limit-points of the second kind *limit-points of non-cohesion*.

Two kinds of contact.

181. Those points in which the curve touches the axis are indeed end-terms of series of forces, which decrease on both sides, as approach to these points takes place, beyond all limits, & at length vanish there; but with such points there is no transition from one direction of the forces to the other. If contact takes place with a repulsive arc, the repulsion vanishes, but after contact remains still a repulsion. If it takes place with an attractive arc, attraction follows on immediately after a vanishing attraction. Two points situated such a distance remain in a state of relative rest; but in the first case they will

sed in primo casu resistunt soli compressioni, non etiam distractioni, & in secundo resistunt huic soli, non illi.

Limites cohæsionis validi, vel languidi pro forma curvæ prope sectionem.

182. Limites cohæsionis possunt esse validissimi, & languidissimi. Si curva ibi quasi ad perpendiculum secat axem, & ab eo longissime recedit; sunt validissimi: si autem ipsum secet in angulo perquam exiguo, & parum ab ipso recedat; erunt languidissimi. Primum genus limitum cohæsionis exhibet in fig. 1 arcus tNy, secundum cNx. In illo assumptis in axe Nz, Nu utcunque exiguis, possunt vires zt, uy, & areæ Nzt, Nuy esse utcumque magnæ, adeoque, mutatis utcunque parum distantiis, possunt haberi vires ab ordinatis expressæ utcunque magnæ, quæ vi comprimenti, vel distrahenti, quantum libuerit, valide resistant, vel areæ utcunque magnæ, quæ velocitates quantumlibet magnas respectivas elidant, adeoque sensibilis mutatio positionis mutuæ impediri potest contra utcunque magnam vel vim prementem, vel celeritatem ab aliorum punctorum actionibus impressam. In hoc secundo genere limitum cohæsionis, assumptis etiam majoribus segmentis Nz, Nu, possunt & vires zc,ux, & areæ Nzc, Nux, esse quantum libuerit exiguæ, & idcirco exigua itidem, quantum libuerit, resistentia, quæ mutationem vetet.

Posse limites esse quotcunque numero, utcunque proximos, vel remotos invicem, & respectu originis abscissarum, positos ordine quocunque.

183. Possunt autem hi limites esse quocunque, utcunque magno numero; cum demonstratum sit, posse curvam in quotcunque, & quibuscunque punctis axem secare. Possunt idcirco etiam esse utcunque inter se proximi, vel remoti, ut [84] alicubi intervallum inter duos proximos limites sit etiam in quacunque ratione majus, quam sit distantia præcedentis ab origine abscissarum A; alibi in intervallo vel exiguo, vel ingenti sint quamplurimi inter se ita proximi, ut a se invicem distent minus, quam pro quovis assumpto, aut dato intervallo. Id evidenter fluit ex eo ipso, quod possint sectiones curvæ cum axe haberi quotcunque, & ubicunque. Sed ex eo, quod arcus curvæ ubicunque possint habere positiones quascunque, cum ad datas curvas accedere possint, quantum libuerit, sequitur, quod limites ipsi cohæsionis possint alii aliis esse utcunque validiores, vel languidiores, atque id quocunque ordine, vel sine ordine ullo; ut nimirum etiam sint in minoribus distantiis alicubi limites validissimi, tum in majoribus languidiores, deinde itidem in majoribus multo validiores, & ita porro; cum nimirum nullus sit nexus necessarius inter distantiam limitis ab origine abscissarum, & ejus validitatem pendentem ab inclinatione, & recessu arcus secantis respectu axis, quod probe notandum est, futurum nimirum usui ad ostendendum, tenacitatem, sive cohæsionem, a densitate non pendere.

Quæ positio rectæ tangentis curvam in limite rarissima, quæ frequentissima. Arcus exigui hinc & inde æquales, & similes.

184. In utroque limitum genere fieri potest, ut curva in ipso occursu cum axe pro tangente habeat axem ipsum, ut habeat ordinatam, ut aliam rectam aliquam inclinatam. In primo casu maxime ad axem accedit, & initio saltem languidissimus est limes; in secundo maxime recedit, & initio saltem est validissimus; sed hi casus debent esse rarissimi, si uspiam sunt: nam cum ibi debeat & axem secare curva, & progredi, adeoque secari in puncto eodem ab ordinata producta, debebit habere flexum contrarium, sive mutare directionem flexus, quod utique fit, ubi curva & rectam tangit simul, & secat. Rarissimos tamen debere esse ibi hos flexus, vel potius nullos, constat ex eo, quod flexus contrarii puncta in quovis finito arcu datæ curvæ cujusvis numero finito esse debent, ut in Theoria curvarum demonstrari potest, & alia puncta sunt infinita numero, adeoque illa cadere in intersectiones est infinities improbabilius. Possunt tamen sæpe cadere prope limites: nam in singulis contorsionibus curvæ saltem singuli flexus contrarii esse debent. Porro quamcunque directionem habuerit tangens, si accipiatur exiguus arcus hinc, & inde a limite, vel maxime accedet ad rectam, vel habebit curvaturam ad sensum æqualem, & ad sensum æquali lege progredientem utrinque, adeoque vires in æquali distantia exigua a limite erunt ad sensum hinc, & inde æquales; sed distantiis auctis poterunt & diu æqualitatem retinere, & cito etiam ab ea recedere.

Transitus per infinitum cruribus asymptoticis.

185. Hi quidem sunt limites per intersectionem curvæ cum axe, viribus evanescentibus in ipso limite. At possunt [85] esse alii limites, ac transitus ab una directione virium ad aliam non per evanescentiam, sed per vires auctas in infinitum, nimirum per asymptoticos

resist compression only, & not separation; and in the second case the latter only, but not the former.

182. Limit-points may be either very strong or very weak. If the curve cuts the axis at the point almost at right angles, & goes off to a considerable distance from it, they are very strong. But if it cuts the axis at a very small angle & recedes from it but little, then they will be very weak. The arc *t*N*y* in Fig. 1 represents the first kind of limit-points of cohesion, and the arc *c*N*x* the second kind. At the point N, if N*z*, N*u* are taken along the axis, no matter how small, the forces *zt*, *uy*, & the areas N*zt*, N*uy* may be of any size whatever; & thus, if the distances are changed ever so little, it is possible that there will be forces represented by ordinates ever so great; & these will strongly resist the compressing or separating force, be it as great as you please; also that we shall have areas, ever so large, that will destroy the relative velocities, no matter how great they may be. Thus, a sensible change of relative position will be hindered in opposition to any impressed force, however great, or against a velocity generated by the actions upon them of other points. In the second kind of limit-points of cohesion, if also segments N*z*, N*u* are taken of considerable size even, then it is possible for both the forces *zc*, *ux*, & the areas N*zc*, N*ux* to be as small as you please; & therefore also the resistance that opposes the change will be as small as you please.

The limit-points of cohesion are strong or weak according to the form of the curve near the point of intersection.

183. Moreover, there can be any number of these limit-points, no matter how great; for it has been proved that the curve can cut the axis in any number of points, & anywhere. Therefore it is possible for them to be either close to or remote from one another, without any restriction whatever, so that the interval between any two consecutive limit-points at any place shall even bear to the distance of the first of the two from A, the origin of abscissae, a ratio that is greater than unity. In other words, in any interval, either very small or very large, there may be an exceedingly large number of them so close to one another, that they are less distant from one another than they are from any chosen or given interval. This evidently follows from the fact that the intersections of the curve with the axis can happen any number of times & anywhere. Again, from the fact that arcs of the curve can anywhere, owing to their being capable of approximating as closely as you please to given curves, have any positions whatever, it follows that these limit-points of cohesion can be some of them stronger than others, or weaker, in any manner; & that too, in any order, or without order. So that, for instance, we may have at small distances anywhere very strong limit-points, then at greater distances weaker ones, & then again at still greater distances much stronger ones, & so on. That is to say, since there is no necessary connection between the distance of a limit-point from the origin of abscissæ and its strength, which depends on the inclination of the intersecting arc & the distance it recedes from the axis. It is well that this should be made a note of; for indeed it will be used later to prove that tenacity or cohesion does not depend on density.

The limit-points are indefinite as regards number, their proximity to or remoteness from one another, & the order of their occurrence with respect to the origin of abscissæ.

184. In each of these kinds of limit-points it may happen that the curve, where it meets the axis, may have the axis itself as its tangent, or the ordinate, or any other straight line inclined to the axis. In the first case it approximates very closely to the axis, & close to the point at any rate it is a very weak limit-point; in the second case, it departs from the axis very sharply, & close to the point at any rate it is a very strong limit-point. But these two cases must be of very rare occurrence, if indeed they ever occur. For, since at the point the curve is bound to cut the axis & go on, & thus be cut in the same point by the ordinate produced, it is bound to have contrary-flexure; that is to say, a change in the direction of its curvature, such as always takes place at a point where the curve both touches a straight line & cuts it at the same time. Yet, that these flexures must occur very rarely at such points, or rather never occur at all, is evident from the fact that in any finite arc of any given curve the number of points of contrary-flexure must be finite, as can be proved in the theory of curves; & other points are infinite in number; hence that the former should happen at the points of intersection with the axis is infinitely improbable. On the other hand they may often fall close to the limit-points; for in each winding of the curve about the axis there must be at least one point of contrary-flexure. Further, whatever the direction of the tangent, if a very small arc of the curve is taken on each side of the limit-point, this arc will either approximate very closely to the straight line, or will have its curvature the same very nearly, & will proceed very nearly according to the same law on each side; & thus the forces, at equal small distances on each side of the limit-point will be very nearly equal to one another; but when the distances are increased, they can either maintain this equality, for some considerable time, or indeed very soon depart from it.

What position of the straight line touching the curve at a limit-point is most unusual, & what most frequent; small arcs on each side of the limit-point are equal & similar.

185. The limit-points so far discussed are those obtained through the intersection of the curve with the axis, where the forces vanish at the limit-point. But there may be other limit-points; the transition from one direction of the forces to another

Passage through infinity for asymptotic branches.

curvæ arcus. Diximus supra num. 168. adnot. (i), quando crus asymptoticum abit in infinitum, debere ex infinito regredi crus aliud habens pro asymptoto eandem reetam, & posse regredi cum quatuor diversis positionibus pendentibus a binis partibus ipsius rectæ, & binis plagis pro singulis rectæ partibus; sed cum nostra curva debeat semper progredi, diximus, relinqui pro ea binas ex ejusmodi quatuor positionibus pro quovis crure abeunte in infinitum, in quibus nimirum regressus fiat ex plaga opposita. Quoniam vero, progrediente curva, abire potest in infinitum tam crus repulsivum, quam crus attractivum; jam iterum fiunt casus quatuor possibiles, quos exprimunt figuræ 16, 17, 18, & 19, in quibus omnibus est axis ACB, asymptotus DCD′, crus recedens in infinitum EKF, regrediens ex infinito GMH.

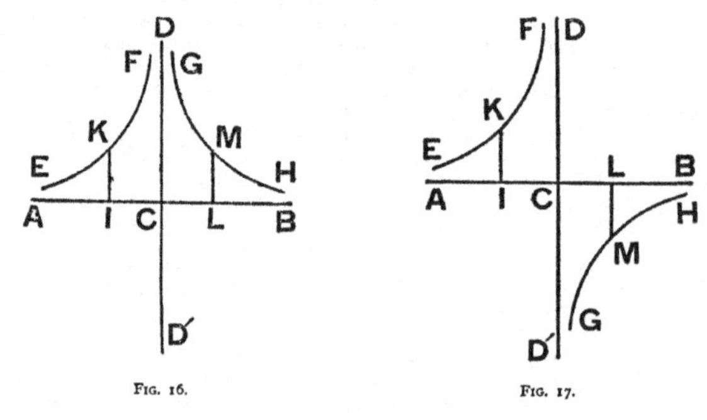

Fig. 16. Fig. 17.

Quatuor eorum genera; bini respondentes contactibus, bini limitibus, alter cohæsionis, alter non cohæsionis.

186. In fig. 16. cruri repulsivo EKF succedit itidem repulsivum GMH; in fig. 17 repulsivo attractivum; in 18 attractivo attractivum; in 19 attractivo repulsivum. Primus & tertius casus respondent contactibus. Ut enim in illis evanescebat vis; sed directionem non mutabat; ita & hic abit quidem in infinitum, sed directionem non mutat. Repulsioni IK in fig. 16 succedit repulsio LM; & attractioni in fig. 18 attractio. Quare ii casus non habent limites quosdam. Secundus, & quartus habent utique limites; nam in fig. 17 repulsioni IK succedit attractio LM; & in fig. 19 attractioni repulsio; atque idcirco secundus continet limitem *cohæsionis*, quartus limitem *non cohæsionis*.

Nullum in Natura admittendum præter postremum, nec vero eum ipsum utcunque.

187. Ex istis casibus a nostra curva censeo removendos esse omnes præter solum quartum; & in hoc ipso removenda omnia crura, in quibus ordinata crescit in ratione minus, quam simplici reciproca distantiarum a limite. Ratio excludendi est, ne haberi aliquando vis infinita possit, quam & per se se absurdam censeo, & idcirco præterea, quod infinita vis natura sua velocitatem infinitam requirit a se generandam finito tempore. Nam in primo, & secundo casu punctum collocatum in ea distantia ab alio puncto, quam habet I, ab origine abscissarum, abiret ad C per omnes gradus virium auctarum in infinitum, & in C deberet habere vim infinitam; in tertio vero idem accideret puncto collocato in distantia, quam habet L. At in quarto casu accessum ad C prohibet ex parte I attractio IK, & ex parte L repulsio LM. Sed quoniam, si eæ crescant in ratione reciproca minus, quam simplici distantiarum CI,CL; area FKICD, vel GMLCD erit finita, adeoque punctum impulsum versus C velocitate majore, quam quæ respondeat illi areæ, debet transire per omnes virium magnitudines usque ad vim absolute infinitam in C, quæ ibi [86] præterea & attractiva esse deberet, & repulsiva, limes videlicet omnium & attractivarum, & repulsivarum; idcirco ne hic quidem casus admitti debet, nisi cum hac conditione, ut ordinata crescat in ratione reciproca simplici distantiarum a C, vel etiam majore, ut nimirum area infinita evadat, & accessum a puncto C prohibeat.

may occur, not with evanescence of the forces, but through the forces increasing indefinitely, that is to say through asymptotic arcs of the curve. We said above, in Note (*i*) to Art. 168, when an asymptotic arm goes off to infinity, there must be another asymptotic arm returning from infinity having the same straight line for an asymptote; & it may return in four different positions, which depend on the two parts of the straight line & the two sides of each part of the straight line. But, since our curve must always go forward, we said that for it there remained only two out of these four positions, for any arm going off to infinity; that is to say, those in which the return is made on the opposite side of the straight line. However, since, whilst the curve goes forward, either a repulsive or an attractive arm can go off to infinity, here again we must have four possible cases, represented in Figs. 16, 17, 18, 19, in all of which ACB is the axis, DCD′ the asymptote, EKF the arm going off to infinity, & GMH the arm returning from infinity.

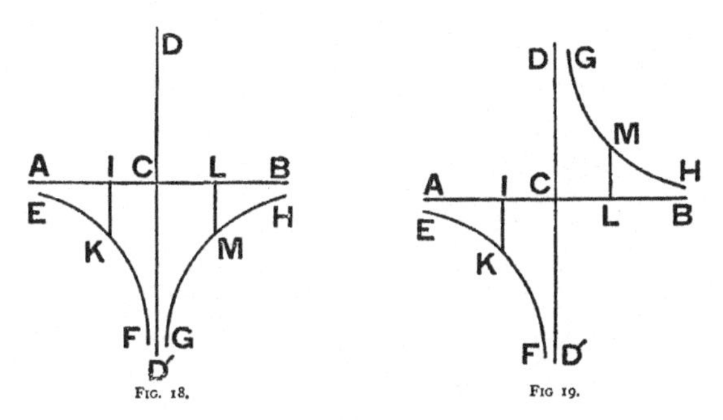

FIG. 18.

FIG 19.

Four kinds of them; two corresponding to contact, & two to limit-points, of which the one is a limit-point of cohesion & the other of non-cohesion.

186. In Fig. 16, to a repulsive arm EKF there succeeds an arm that is also repulsive; in Fig. 17, to a repulsive succeeds an attractive; in Fig. 18, to an attractive succeeds an attractive; and in Fig. 19, to an attractive succeeds a repulsive. The first & third cases correspond to contacts. For, just as in contact, the force vanished, but did not change its direction, so here also the force indeed becomes infinite but does not change its direction. In Fig. 16, to the repulsion IK there succeeds the repulsion LM, & in Fig. 18 to an attraction an attraction; & thus these two cases cannot have any limit-points. But the second & fourth cases certainly have limit-points; for, in Fig. 17, to the repulsion IK there succeeds the attraction LM, & in Fig. 19 to an attraction a repulsion; & thus the second case contains a limit-point of cohesion, & the fourth a limit-point of non-cohesion.

None of these except the last admissible in Nature; & not even that in general.

187. Out of these cases I think that all except the last must be barred from our curve; & even with that all arms must be rejected for which the ordinates increase in a ratio less than the simple reciprocal of the distances from the limit-point. My reasons for excluding these are to avoid the possibility of there being at any time an infinite force (which of itself I consider to be impossible), & because, in addition to that, an infinite force, by its very nature necessitates the creation by it of an infinite velocity in a finite time. For, in the first & second cases, a point, situated at the distance from another point equal to that which I has from the origin of abscissæ, would go off to C through all stages of forces increased indefinitely, & at C would be bound to have an infinite force. In the third case, too, the same thing would happen to a point situated at a distance equal to that of L. Now, in the fourth case, the approach to C is restrained, from the side of I by the attraction IK, & from the side of L by the repulsion LM. However, since, if these forces increase in a ratio that is less than the simple reciprocal ratio of the distances CI, CL, then the area FKICD or the area GMLCD will be finite; & thus the point, being impelled towards C with a velocity that is greater than that corresponding to the area, must pass through all magnitudes of the forces up to a force that is absolutely infinite at C; and this force must besides be both attractive & repulsive, the limit so to speak of all attractive & repulsive forces. Hence not even this case is admissible, unless with the condition that the ordinate increases in the simple reciprocal ratio of the distances from C, or in a greater; that is to say, the area must turn out to be infinite and so restrain the approach towards the point C.

Transitus per eum limitem impossibilis: in quibus distantiis constet, eum non haberi.

188. Quando habeatur hic quartus casus in nostra curva cum ea conditione; tum quidem nullum punctum collocatum ex altera parte puncti C poterit ad alteram transilire, quacunque velocitate ad accessum impellatur versus alterum punctum, vel ad recessum ab ipso, impediente transitum area repulsiva infinita, vel infinita attractiva. Inde vero facile colligitur, eum casum non haberi saltem in ea distantia, quæ a diametris minimarum particularum conspicuarum per microscopia ad maxima protenditur fixarum intervalla nobis conspicuarum per telescopia: lux enim liberrime permeat intervallum id omne. Quamobrem si ejusmodi limites asymptotici sunt uspiam, debent esse extra nostræ sensibilitatis sphæram, vel ultra omnes telescopicas fixas, vel citra microscopicas moleculas.

Transitus ad puncta materiæ, & massas.

189. Expositas hisce, quæ ad curva virium pertinebant, aggrediar simpliciora quædam, quæ maxime notatu digna sunt, ac pertinent ad combinationem punctorum primo quidem duorum, tum trium, ac deinde plurium in massa etiam coalescentium, ubi & vires mutuas, & motus quosdam, & vires, quas in alia exercent puncta, considerabimus.

Quies in limitibus; motus puncti positi extra ipsos.

190. Duo puncta posita in distantia æquali distantiæ limitis cujuscunque ab origine abscissarum, ut in fig. 1. AE, AG, AI, &c, (immo etiam si curva alicubi axem tangat, æquali distantiæ contactus ab eodem), ac ibi posita sine ulla velocitate, quiescent, ut patet, quia nullam habebunt ibi vim mutuam: posita vero extra ejusmodi limites, incipient statim ad se invicem accedere, vel a se invicem recedere per intervalla æqualia, prout fuerint sub arcu attractivo, vel repulsivo. Quoniam autem vis manebit semper usque ad proximum limitem directionis ejusdem; pergent progredi in ea recta, quæ ipsa urgebat prius, usque ad distantiam limitis proximi, motu semper accelerato, juxta legem expositam num. 176, ut nimirum quadrata velocitatum integrarum, quæ acquisitæ jam sunt usque ad quodvis momentum (nam velocitas initio ponitur nulla) respondeant areis clausis inter ordinatam respondentem puncto axis terminanti abscissam, quæ exprimebat distantiam initio motus, & ordinatam respondentem puncto axis terminanti abscissam, quæ exprimit distantiam pro eo sequenti momento. Atque id quidem, licet interea occurrat contactus aliquis; quamvis enim in eo vis sit nulla, tamen superata distantia per velocitatem jam acquisitam, statim habentur iterum [87] vires ejusdem directionis, quæ habebatur prius, adeoque perget acceleratio prioris motus.

Motus post proximum limitem superatum, & oscillatio.

191. Proximus limes crit ejus generis, cujus generis diximus limites cohæsionis, in quo nimirum si distantia per repulsionem augebatur, succedet attractio; si vero minuebatur per attractionem, succedet e contrario repulsio, adeoque in utroque casu limes erit ejusmodi, ut in distantiis minoribus repulsionem, in majoribus attractionem secum ferat. In eo limite in utroque casu recessus mutui, vel accessus ex præcedentibus viribus, incipiet, velocitas motus minui vi contraria priori, sed motus in eadem directione perget; donec sub sequenti arcu obtineatur area curvæ æqualis illi, quam habebat prior arcus ab initio motus usque ad limitem ipsum. Si ejusmodi æqualitas obtineatur alicubi sub areu sequente; ibi, extincta omni præcedenti velocitate, utrumque punctum retro reflectet cursum; & si prius accedebant, incipient a se invicem recedere; si recedebant, incipient accedere, atque id recuperando per eosdem gradus velocitates, quas amiserant, usque ad limitem, quem fuerant prætergressa; tum amittendo, quas acquisiverant usque ad distantiam, quam habuerant initio; viribus nimirum iisdem occurrentibus in ingressu, & areolis curvæ iisdem per singula tempuscula exhibentibus quadratorum velocitatis incrementa, vel decrementa eadem, quæ fuerant antea decrementa, vel incrementa. Ibi autem iterum retro cursum reflectent, & oscillabunt circa illum cohæsionis limitem, quem fuerant prætergressa, quod facient hinc, & inde perpetuo, nisi aliorum externorum punctorum viribus perturbentur, habentia velocitatem maximam in plagam utramlibet in distantia ipsius illius limitis cohæsionis.

Casus oscillationis majoris trans plures limites.

192. Quod si ubi primum transgressa sunt proximum limitem cohæsionis, offendant arcum ita minus validum præcedente, qui arcus nimirum ita minorem concludat arcam, quam præcedens, ut tota ejus area sit æqualis, vel etiam minor, quam illa præcedentis arcus area, quæ habetur ab ordinata respondente distantiæ habitæ initio motus, usque ad

Passage through a limit-point of this kind is impossible; distances at which it is certain that there are no such limit-points.

188. When, if ever, this fourth case occurs in our curve, then indeed no point situated on either side of the point C will be able to pass through it to the other side, no matter what the velocity with which it is impelled to approach towards, or recede from, the other point; for the infinite repulsive area, or the infinite attractive area, will prevent such passage. Now, it can easily be derived from this, that this case cannot happen at any rate in the distance lying between the diameters of the smallest particles visible under the microscope & the greatest distances of the stars visible to us through the telescope; for light passes with the greatest freedom through the whole of this interval. Therefore, if there are ever any such asymptotic limit-points, they must be beyond the scope of our senses, either superior to all telescopic stars, or inferior to microscopic molecules.

We now pass on to points of matter, & masses.

189. Having thus set forth these matters relating to the curve of forces, I will now discuss some of the simpler things that are more especially worth mentioning with regard to combination of points; & first of all I will consider a combination of two points, then of three, & then of many, coalescing into masses; & with them we will discuss their mutual forces, & certain motions, and forces, which they exercise on other points.

Rest at limit-points; motion of a point situated without them.

190. Two points situated at a distance apart equal to the distance of any limit-point from the origin of abscissæ, like AE, AG, AI, &c. in Fig. 1 (or indeed also where the curve touches the axis anywhere, equal to the distance of the point of contact from the origin), & placed in that position without any velocity, will be relatively at rest; this is evident from the fact that they have then no mutual force; but if they are placed at any other distance, they will immediately commence to move towards one another or away from one another through equal intervals, according as they lie below an attractive or a repulsive arc. Moreover, as the force always remains the same in direction as far as the next following limit-point, they continue to move in the same straight line which contained them initially as far as the distance apart equal to the distance of the next limit-point from the origin, with a motion that is continually accelerated according to the law given in Art. 176; that is to say, in such a manner that the squares of the whole velocities which have been already acquired up to any instant (for the velocity at the commencement is supposed to be nothing) will correspond to the areas included between the ordinate corresponding to the point of the axis terminating the abscissa which the distance traversed since motion began and the ordinate corresponding to the point on the axis terminating the abscissa which expresses the distance for the next instant after it. This is still the case, even if a contact should occur in the meantime. For, although at a point where contact occurs the force is nothing, yet, this distance being passed by the velocity already acquired, immediately afterwards there will be forces having the same direction as before; and thus the acceleration of the former motion will proceed.

Motion after the next limit-point is passed; oscillation.

191. The next limit-point will be one of the kind we have called limit-points of cohesion, namely, one in which, if the distance is increased by repulsion, then attraction follows; but if the distance is diminished by attraction, then on the contrary repulsion will follow; & thus, in either case, the limit-point will be of such a kind, that it gives a repulsion at smaller distances & an attraction at larger. In this limit-point, in either case, the separation or approach, due to the forces that have preceded, will be changed, & the velocity of motion will begin to be diminished by a force opposite to the original force, but the motion will continue in the same direction; until an area of the curve under the arc that follows the limit-point becomes equal to the area under the former arc from the commencement of the motion as far as the limit-point. If equality of this kind is obtained somewhere under the subsequent arc, then, the whole of the preceding velocity being destroyed, both the points will return along their paths; & if at the start they approached one another, they will now begin to recede from one another, or if they originally receded from one another, they will now commence to approach; and as they do this, they will regain by the same stages the velocities which they lost, as far as the limit-point which they passed through; then they will lose those which they had acquired, until they reach the distance apart which they had at the commencement. That is to say, the same forces occur on the return path, & the same little areas of the curve for the several short intervals of time represent increments or decrements of the squares of the velocities which are the same as were formerly decrements or increments. Then again they will once more retrace their paths, & they will oscillate about the limit-point of cohesion which they had passed through; & this they will do, first on this side & then on that, over & over again, unless they are disturbed by forces due to other points outside them; & their greatest velocity in either direction will occur at a distance apart equal to that of the limit-point of cohesion from the origin.

The case of a larger oscillation through several limit-points.

192. But if, when they first passed through the nearest limit-point of cohesion, they happened to come to an arc representing forces so much weaker than those of the preceding arc that the whole area of it was equal to, or even less than, the area of the preceding arc, reckoning from the ordinate corresponding to the distance apart at the commencement

limitem ipsum; tum vero devenient ad distantiam alterius limitis proximi priori, qui idcirco crit limes non cohæsionis. Atque ibi quidem in casu æqualitatis illarum arearum consistent, velocitatibus prioribus clisis, & nulla vi gignente novas. At in casu, quo tota illa area sequentis arcus fuerit minor, quam illa pars areæ præcedentis, appellent ad distantiam ejus limitis motu quidem retardato, sed cum aliqua velocitate residua, quam distantiam idcirco prætergressa, & nacta vires directionis mutatæ jam conspirantes cum directione sui motus, non, ut ante, oppositas, accelerabunt motum usque ad distantiam limitis proxime sequentis, quam prætergressa procedent, sed motu retardato, ut in priore; & si area sequentis arcus non sit par extinguendæ ante suum finem toti [88] velocitati, quæ fuerat residua in appulsu ad distantiam limitis præcedentis non cohæsionis, & quæ acquisita est in arcu sequenti usque ad limitem cohæsionis proximum; tum puncta appellent ad distantiam limitis non cohæsionis sequentis, ac vel ibi sistent, vel progredientur itidem, eritque semper reciprocatio quædam motus perpetuo accelerati, tum retardati; donec deveniatur ad arcum ita validum, nimirum qui concludat ejusmodi arcam, ut tota velocitas acquisita extinguatur: quod si accidat alicubi, & non accidat in distantia alicujus limitis; cursum reflectent retro ipsa puncta, & oscillabunt perpetuo.

Velocitatis mutationes alternæ: ubi ea habeat maximum, & minimum ubi extingui possit

193. Porro in hujusmodi motu patet illud, dum itur a distantia limitis cohæsionis ad distantiam limitis non cohæsionis, velocitatem semper debere augeri; tum post transitum per ipsam debere minui, usque ad appulsum ad distantiam limitis non cohæsionis, adeoque habebitur semper in ipsa velocitate aliquod *maximum* in appulsu ad distantiam limitis cohæsionis, & *minimum* in appulsu ad distantiam limitis non cohæsionis. Quamobrem poterit quidem sisti motus in distantia limitis hujus secundi generis; si sola existant illa duo puncta, nec ullum externum punctum turbet illorum motum: sed non poterit sisti in distantia limitis illius primi generis; cum ad ejusmodi distantias deveniatur semper motu accelerato. Præterea patet & illud, si ex quocunque loco impellantur velocitatibus æqualibus vel alterum versus alterum, vel ad partes oppositas, debere haberi reciprocationes easdem auctis semper æque velocitatibus utriusque, dum itur versus distantiam limitis primi generis, & imminutis, dum itur versus distantiam limitis secundi generis.

Circa quos limites oscillatio major esse debeat, & unde pendeat ejus magnitudo.

194. Patet & illud, si a distantia limitis primi generis dimoveantur vi aliqua, vel non ita ingenti velocitate impressa, oscillationem fore perquam exiguam, saltem si quidam validus fuerit limes; nam velocitas incipiet statim minui, & ei vi statim vis contraria invenietur, ac puncta parum dimota a loco suo, tum sibi relicta statim retro cursum reflectent. At si dimoveantur a distantia limitis secundi generis vi utcunque exigua; oscillatio crit multo major, quia necessario debebunt progredi ultra distantiam sequentis limitis primi generis, post quem motus primo retardari incipiet. Quin immo si arcus proximus hinc, & inde ab ejusmodi limite secundi generis concluserit arcam ingentem, ac majorem pluribus sequentibus contrariæ directionis, vel majorem excessu corundem supra areas interjacentes directionis suæ; tum vero oscillatio poterit esse ingens: nam fieri poterit, ut transcurrantur hinc, & inde limites plurimi, antequam deveniatur ad arcum ita validum, ut velocitatem omnem elidat, & motum retro reflectat. Ingens itidem oscillatio esse poterit, si cum ingenti vi dimoveantur puncta a distantia limitum generis utriuslibet; ac res tota pendet a velocitate initiali, & ab areis, quæ post oc-[89]-currunt, & quadratum velocitatis vel augent, vel minuunt quantitate sibi proportionali.

Accessum debere sisti saltem a primo arcu repulsivo, recessum posse haberi in infinitum: casus notabilis exiguæ differentiæ velocitatis ingentis.

195. Utcunque magna sit velocitas, qua dimoveantur a distantia limitum illa duo puncta, utcunque validos inveniant arcus conspirantes cum velocitatis directione, si ad se invicem accedunt, debebunt utique alicubi motum retro reflectere, vel saltem sistere, quia saltem advenient ad distantias illas minimas, quæ respondent arcui asymptotico, cujus area est capax extinguendæ cujuscunque velocitatis utcunque magnæ. At si recedant a se invicem, fieri potest, ut deveniant ad arcum aliquem repulsivum validissimum, cujus area sit major, quam omnis excessus sequentium arearum attractivarum supra repul-

of the motion up to the limit-point; then indeed they will arrive at a distance apart equal to that of the limit-point next following the first one, which will therefore be a limit-point of non-cohesion. Here they will stop, in the case of equality between the areas in question; for the preceding velocities have been destroyed & no fresh ones will be generated. But in the case when the whole of the area under the second arc is less than the said part of the first area, they will reach a distance apart equal to that of the limit-point with a motion that is certainly diminished; but some velocity will be left, & this distance will therefore be passed, & the points, coming under the influence of forces changed in direction so that they now act in the same sense as their own motion, will accelerate their motion as far as the next following limit-point; & having passed through this they will go on, but with retarded motion as in the first case. Then, if the area of the subsequent arc is not capable before it ends of destroying the whole of the velocity which remained on attaining the distance of the preceding limit-point of non-cohesion, & that which was acquired in the arc that followed it up to the next limit-point of cohesion, then the points will move to a distance apart equal to that of the next following limit-point from the origin, & will either stop there or proceed; & there will always be a repetition of the motion, continually accelerated & retarded. Until at length it comes to an arc so strong, that is to say, one under which the area is such, that the whole velocity acquired is destroyed; & when this happens anywhere, & does not happen at a distance equal to that of any limit-point, then the points will retrace their paths & oscillate continuously.

Alternate changes of velocity; where it has a maximum value, & a minimum value; where it may be destroyed.

193. Further in this kind of motion it is clear that along the path from the distance of a limit-point of cohesion to a limit-point of non-cohesion the velocity is bound to be always increasing; then after passing through the latter it must decrease up to its arrival at the distance of a limit-point of non-cohesion. Thus, there will always be in the velocity a *maximum* on arrival at a distance equal to that of a limit-point of cohesion, & a *minimum* on arrival at a distance of a limit-point of non-cohesion. Hence indeed the motion may possibly cease at a limit-point of this second kind, if the two points exist by themselves, & no other point influences their motion from without. But it cannot cease at a distance of a limit-point of the first kind; for it will always arrive at distances of this kind with an accelerated motion. Moreover it is also clear that, if they are urged from any given position with equal velocities, either towards one another or in opposite directions, the same alternations must be had as before, the velocities being increased equally for each point whilst they are moving up to a distance of a limit-point of the first kind, & diminished whilst they are moving up to a distance of a limit-point of the second kind.

The limit-points about which the oscillation must be larger; & the thing on which its magnitude depends.

194. It is evident also that, if the points are moved from a distance apart equal to that of a limit-point of the first kind by some force (especially when the velocity thus impressed is not extremely great), then the oscillation will be exceedingly small, at least so long as the limit-point is a fairly strong one. For the velocity will commence to be diminished immediately, & to the force another force will be obtained at once, acting in opposition to it; & the points, being moved but little from their original position, will immediately afterwards retrace their paths if left to themselves. But if they are moved from a distance apart equal to that of a limit-point of the second kind by any force, no matter how small, then the oscillation will be much greater; for, of necessity, they are bound to go on beyond the distance equal to that of the next following limit-point of the first kind; & not until this has been done, will the motion begin to be retarded. Nay, if the next arc on each side of such a limit-point of the second kind should include a very large area, and one that is greater than several of those subsequent to them, which are opposite in direction, or greater than the excess of these over the intervening areas that are in the same direction, then indeed the oscillation may be exceedingly large. For it may be that very many limit-points on either side are traversed before an arc is arrived at, which is sufficiently strong to destroy the whole of the velocity & reverse the direction of motion. A very large oscillation will also be possible, if the points are moved from a distance apart equal to that of a limit-point of either kind by an exceedingly large force. The whole thing depends on the initial velocity & the areas which occur subsequently, & either increase or decrease the square of the velocity by a quantity that is proportional to the areas themselves.

Approach is bound to cease at any rate owing to the first repulsive arc, but separation can go on indefinitely; a noteworthy case of the very small difference for a very great velocity.

195. However great the velocity may be, with which the two points are moved from a distance equal to that of any limit-point, no matter how strong are the arcs they come upon, which are in the same direction as that of the velocity; yet, if they approach one another, they are bound somewhere to have their motion reversed, or at least to come to rest; for, at all events, they must finally attain to those very small distances that correspond to an asymptotic arm, the area of which is capable of destroying any velocity whatever, no matter how great. But, if they recede from one another, it may happen that they come to some very strong repulsive arc, the area of which is greater than the whole of the excess of the subsequent attractive arcs above those that are repulsive, as far as the very weak

sivas, usque ad languidissimum illum arcum postremi cruris gravitatem exhibentis. Tum vero motus acquisitus ab illo arcu nunquam poterit a sequentibus sisti, & puncta illa recedent a se invicem in immensum : quin immo si ille arcus repulsivus cum sequentibus repulsivis ingentem habeat areæ excessum supra arcus sequentes attractivos ; cum ingenti velocitate pergent puncta in immensum recedere a se invicem ; & licet ad initium ejus tam validi arcus repulsivi deveniant puncta cum velocitatibus non parum diversis ; tamen velocitates recessuum post novum ingens illud augmentum crunt parum admodum discrepantes a se invicem : nam si ingentis radicis quadrato addatur quadratum radicis multo minoris, quamvis non exiguæ ; radix extracta ex summa parum admodum differet a radice priore.

Demonstratio admodum simplex.

196. Id quidem ex Euclidea etiam Geometria manifestum fit. Sit in fig. 20 AB linea longior, cui addatur ad perpendiculum BC, multo minor, quam fit ipsa ; tum centro A, intervallo AC, fiat semicirculus occurrens AB hinc, & inde in E, D. Quadrato AB addendo quadratum BC habetur quadratum AC, sive AD ; & tamen hæc excedit præcedentem radicem AB per solam BD, quæ semper est minor, quam BC, & est ad ipsam, ut est ipsa ad totam BE. Exprimat AB velocitatem, quam in punctis quiescentibus gigneret arcus ille repulsivus per suam aream, una cum differentia omnium sequentium arcuum repulsivorum supra omnes sequentes attractivos : exprimat autem BC velocitatem, cum qua advenitur ad distantiam respondentem initio ejus arcus : exprimet AC velocitatem, quæ habebitur, ubi jam distantia evasit major, & vis insensibilis, ac ejus excessus supra priorem AB crit BD, exiguus sane etiam respectu BC, si BC fuerit exigua respectu AB, adeoque multo magis respectu AB ; & ob candem rationem perquam exigua area sequentis cruris attractivi ingentem illam jam acquisitam velocitatem nihil ad sensum mutabit, quæ permanebit ad sensum eadem post recessum in immensum.

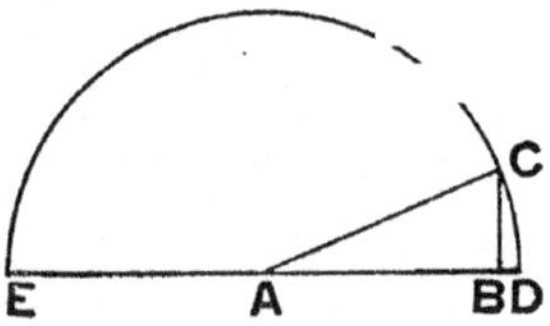

FIG. 20.

Quid accidat binis punctis, cum sunt sola, quid possit accidere actionibus aliorum externis.

197. Hæc accident binis punctis sibi relictis, vel impulsis [90] in recta, qua junguntur, cum oppositis velocitatibus æqualibus, quo casu etiam facile demonstratur, punctum, quod illorum distantiam bifariam secat, debere quiescere ; nunquam in hisce casibus poterit motus extingui in adventu ad distantiam limitis cohæsionis, & multo minus poterunt ea bina puncta consistere extra distantiam limitis cujuspiam, ubi adhuc habeatur vis aliqua vel attractiva, vel repulsiva. Verum si alia externa puncta agant in illa, poterit res multo aliter se habere. Ubi ex. gr. a se recedunt, & velocitates recessus augeri deberent in accessu ad distantiam limitis cohæsionis ; potest externa compressio illam velocitatem minuere, & extinguere in ipso appulsu ad ejusmodi distantiam. Potest externa compressio cogere illa puncta manere immota etiam in ea distantia, in qua se validissime repellunt, uti duæ cuspides elastri manu compressæ detinentur in ea distantia, a qua sibi relictæ statim recederent : & simile quid accidere potest vi attractivæ per vires externas distrahentes.

Si limites sint a se invicem remoti, mutata multum distantia rediri retro : secus, si sint proximi.

198. Tum vero diligenter notandum discrimen inter casus varios, quos inducit varia arcuum curvæ natura. Si puncta sint in distantia alicujus limitis cohæsionis, circa quem sint arcus amplissimi, ita, ut proximi limites plurimum inde distent, & multo magis etiam, quam sit tota distantia proximi citerioris limitis ab origine abscissarum ; tum poterunt externa vi comprimente, vel distrahente redigi ad distantiam multis vicibus minorem, vel majorem priore ita, ut semper adhuc conentur se restituere ad priorem positionem recedendo, vel accedendo, quod nimirum semper adhuc sub arcu repulsivo permaneat, vel attractivo. At si ibi frequentissimi limites, curva sæpissime secante axem ; tum quidem post compressionem, vel distractionem ab externa vi factam, poterunt sisti in multo minore, vel majore distantia, & adhuc esse in distantia alterius limitis cohæsionis sine ullo conatu ad recuperandum priorem locum.

Superiorum usus in Physica.

199. Hæc omnia aliquanto fusius considerare libuit, quia in applicatione ad Physicam magno usui crunt infra hæc ipsa, & multo magis bisce similia, quæ massis respondent habentibus utique multo uberiores casus, quam bina tantummodo habeant puncta. Illa ingens agitatio cum oscillationibus variis, & motibus jam acceleratis, jam retardatis, jam retro reflexis, fermentationes, & conflagrationes exhibebit : ille egressus ex ingenti arcu

arc of the last branch which represents gravity. Then indeed the velocity acquired through that arc can never be stopped by the subsequent arcs, & the points will recede from one another to an immense distance. Nay further, if that repulsive arc taken together with the subsequent repulsive arcs has a very great excess of area over the subsequent attractive arcs, then the points will continue to recede to an immense distance from one another with a very great velocity ; &, although points arrive at this repulsive arc, which is so strong, with considerably different velocities, yet the velocities after this fresh & exceedingly great increase will be very little different from one another. For, if to the square of a very great number there is added the square of a number that is much less, although not in itself very small, the square root of the sum differs very little from the first number.

The demonstration is perfectly simple.

196. This indeed is very evident from Euclidean geometry even. In Fig. 20, let AB be a fairly long line, to which is added, perpendicular to it, BC, which is much less than AB. Then, with centre A, & radius AC, describe a semicircle meeting AB on either side in E & D. On adding the square on BC to the square on AB, we get the square on AC or AD ; & yet this exceeds the former root AB by BD only, which is always less than BC, bearing the same ratio to it as BC bears to the whole length BE. Suppose that AB represents the velocity which the repulsive arc, owing to the area under it, would generate in points initially at rest, together with the difference for all the subsequent repulsive arcs over all the subsequent attractive arcs ; also let BC represent the velocity with which the distance corresponding to the beginning of this arc is reached ; then AC will represent the velocity which is obtained when the distance has already become of considerable amount, & the force insensible. Now the excess of this above the former velocity AB will be represented by BD ; & this is really very small compared with BC, if BC were very small compared with AB ; & therefore much more so with regard to AB. For the same reason, the very small area under the subsequent attractive branch will not sensibly change the very great velocity acquired so far ; this will remain sensibly the same after recession to a huge distance.

What may happen to two points when they are by themselves : what may happen to them when under the actions of other points external to them.

197. These things will take place in the case of two points left to themselves, or impelled along the straight line joining them with velocities that are equal & opposite ; in such a case it can be easily proved that the middle point of the distance between them is bound to remain at rest. The motion in the cases we have discussed can never be destroyed altogether on arrival at a distance equal to that of a limit-point of cohesion, & much less will the two points be able to stop at a distance apart that is not equal to that of some limit-point, as far as which there is some force acting, either attractive or repulsive. But if other external points act upon them, we may have altogether different results. For instance, in a case where they recede from one another, & the velocities would therefore be bound to be increased as they approached a distance equal to that of a limit-point of cohesion, an external compression may diminish that velocity, & completely destroy it as it approaches the distance of that limit-point. An external compression may even force the points to remain motionless at a distance for which they repel one another very strongly ; just as the two ends of a spring compressed by the hands are kept at a distance from which if left to themselves they will immediately depart. A similar thing may come about in the case of an attractive force when there are external tensile forces.

If the limit-points are far apart, there is a tendency to return if the distance suffers any considerable change; but this is not the case when the limit-points are very close together.

198. Now, a careful note must be made of the distinctions between the various cases, which arise from the various natures of the arcs of the curve. If our points are at a distance of any limit-point of cohesion, on each side of which the arcs are very wide, so that the nearest limit-points are very far distant from it, & also much more so than the nearest limit-point to the left is distant from the origin of abscissæ ; they may, under the action of an external force causing either compression or tension, be reduced after many alternations to a distance, either less, or greater, than the original distance, in such a way that they will always strive however to revert to their old position by receding from or approaching towards one another ; for indeed they will still always remain under a repulsive, or an attractive arc. But if, near the limit-point in question, the limit-points on either side occur at very frequent intervals ; then indeed, after compression, or separation, caused by an external force, they may stop at a much less, or a much greater, distance apart, & still be at a distance equal to that of another limit-point of cohesion, without there being any endeavour to revert to their original position.

The use of the above facts in physics.

199. All these considerations I have thought it a good thing to investigate somewhat at length ; for they will be of great service later in the application of the Theory to physics, both these considerations, & others like them to an even greater degree ; namely those that correspond to masses, for which indeed there are far more cases than for a system of only two points. The great agitation, with its various oscillations & motions that are sometimes accelerated, sometimes retarded, & sometimes reversed, will represent fermentations

repulsivo cum velocitatibus ingentibus, quæ ubi jam ad ingentes deventum est distantias, parum admodum a se invicem differant, nec ad sensum mutentur quidquam per immensa intervalla, luminis emissionem, & propagationem uniformem, ac ferme eandem celeritatem in quovis ejusdem speciei radio fixarum, Solis, flammæ, cum exiguo discrimine inter diversos coloratos radios; illa vis permanens post compressionem ingentem, vel diffractionem elasticitati explicandæ in-[91]-serviet; quies ob frequentiam limitum, sine conatu ad priorem recuperandam figuram, mollium corporum ideam suggeret; quæ quidem hic innuo in antecessum, ut magis hæreant animo, prospicienti jam hinc insignes eorum usus.

Motus binorum punctorum oblique projectorum.

200. Quod si illa duo puncta projiciantur oblique motibus contrariis, & æqualibus per directiones, quæ cum recta jungente ipsa illa duo puncta angulos æquales efficiant; tum vero punctum, in quo recta illa conjungens secatur bifariam, manebit immotum; ipsa autem duo puncta circa id punctum gyrabunt in curvis lineis æqualibus, & contrariis, quæ data lege virium per distantias ab ipso puncto illo immoto (uti daretur, data nostra curva virium figuræ 1, cujus nimirum abscissæ exprimunt distantias punctorum a se invicem, adeoque eorum dimidiæ distantias a puncto illo medio immoto) invenitur solutione problematis a Newtono jam olim soluti, quod vocant *inversum problema virium centralium*, cujus problematis generalem solutionem & ego exhibui syntheticam eodem cum Newtoniana recidentem, sed non nihil expolitam, in Stayanis Supplementis ad lib. § 19.

Casus, in quo duo puncta debeant describere spirales circa medium immotum.

201. Hic illud notabo tantummodo, inter infinita curvarum genera, quæ describi possunt, cum nulla sit curva, quæ assumpto quovis puncto pro centro virium describi non possit cum quadam virium lege, quæ definitur per Problema directum virium centralium, esse innumeras, quæ in se redeant, vel in spiras contorqueantur. Hinc fieri potest, ut duo puncta delata sibi obviam e remotissimis regionibus, sed non accurate in ipsa recta, quæ illa jungit (qui quidem casus accurati occursus in ea recta est infinities improbabilior casu deflexionis cujuspiam, cum sit unicus possibilis contra infinitos), non recedant retro, sed circa punctum spatii medium immotum gyrent perpetuo sibi deinceps semper proxima, intervallo etiam sub sensus non cadente; qui quidem casus itidem diligenter notandi sunt, cum sint futuri usui, ubi de cohæsione, & mollibus corporibus agendum erit.

Theorema de statu puncti medii, & generaliter in massis centri gravitatis perseverante.

202. Si utcunque alio modo projiciantur bina puncta velocitatibus quibuscunque; potest facile ostendi illud: punctum, quod est medium in recta jungente ipsa, debere quiescere, vel progredi uniformiter in directum, & circa ipsum vel quietum, vel uniformiter progrediens, debere haberi vel illas oscillationes, vel illarum curvarum descriptiones. Verum id generalius pertinet ad massas quotcunque, & quascunque, quarum commune gravitatis centrum vel quiescit, vel progreditur uniformiter in directum a viribus mutuis nihil turbatum. Id theorema Newtonus proposuit, sed non satis demonstravit. Demonstrationem accuratissimam, ac generalem simul, & non per casuum inductionem tantummodo, inveni, ac in dissertatione *De Centro Gravitatis* proposui, quam ipsam demonstrationem hic etiam inferius exhibebo.

Accessum alterius e binis ad planum quodvis alterius æquari recessui ex vi mutua.

[92] 203. Interea hic illud postremo loco adnotabo, quod pertinet ad duorum punctorum motum ibi usui futurum: si duo puncta moveantur viribus mutuis tantummodo, & ultra ipsa assumatur planum quodcunque; accessus alterius ad illud planum secundum directionem quamcunque, æquabitur recessui alterius. Id sponte consequitur ex eo, quod eorum absoluti motus sint æquales, & contrarii; cum inde fiat, ut ad directionem aliam quamcunque redacti æquales itidem maneant, & contrarii, ut erant ante. Sed de æquilibrio, & motibus duorum punctorum jam satis.

Transitus ad systema trium punctorum: bina generalia problemata.

204. Deveniendo ad systema trium punctorum, uti etiam pro punctis quotcunque, res, si generaliter pertractari deberet, reduceretur ad hæc duo problemata, quorum alterum pertinet ad vires, & alterum ad motus: 1. *Data positione, & distantia mutua eorum punctorum, invenire magnitudinem, & directionem vis, qua urgetur quodvis ex ipsis, compositæ a viribus, quibus urgetur a reliquis, quarum singularum virium lex communis datur per curvam figuræ primæ.* 2. *Data illa lege virium figuræ primæ invenire motus eorum punctorum, quorum singula cum datis velocitatibus projiciantur ex datis locis cum datis directionibus.* Primum facile solvi potest, & potest etiam ope curvæ figuræ 1 determinari lex virium

& conflagrations. The starting forth from a very large repulsive arc with very great velocities, which, as soon as very great distances have been reached, are very little different from one another; nor are they sensibly changed in the slightest degree for very great intervals; this will represent the emission & uniform propagation of light, & the approximately equal velocities in any ray of the same kind from the stars, the sun, and a flame, with a very slight difference between rays of different colours. The force persisting after compression, or separation, will serve to explain elasticity. The lack of motion due to the frequent occurrence of limit-points, without any endeavour towards recovering the original configuration, will suggest the idea of soft bodies. I mention these matters here in anticipation, in order that they may the more readily be assimilated by a mind that already sees from what has been said that there is an important use for them.

The motion of two points projected obliquely.

200. But if the two points are projected obliquely with velocities that are equal and opposite to one another, in directions making equal angles with the straight line joining the two points; then, the point in which the straight line joining them is bisected will remain motionless; the two points will gyrate about this middle point in equal curved paths in opposite directions. Moreover, if the law of forces is given in terms of the distances from that motionless point (as it will be given when our curve of forces in Fig. 1 is given, where the abscissæ represent the distances of the points from one another, & therefore the halves of these abscissæ represent the distances from the motionless middle point), then we arrive at a solution of the problem already solved by Newton some time ago, which is called the *inverse problem of central forces*. Of this problem I also gave a general synthetic solution that was practically the same thing as that of Newton, not altogether devoid of neatness, in the Supplements to Stay's Philosophy, Book 3, Art. 19.

The case in which the two points are bound to describe spirals about the motionless middle point.

201. At present I will only remark that, amongst the infinite number of different curves that can be described, there are an innumerable number which will either re-enter their paths, or wind in spirals; for there is no curve that, having taken any point whatever for the centre of forces, cannot be described with some law of forces, which is determined by the direct problem of central forces. Hence it may happen that two points approaching one another from a long way off, but not exactly in the straight line joining them—and the case of accurate approach along the straight line joining them is infinitely more improbable than the case in which there is some deviation, since the former is only one possible case against an infinite number of others—then the points will not reverse their motion and recede, but will gyrate about a motionless middle point of space for evermore, always remaining very near to one another, the distance between them not being appreciable by the senses. These cases must be specially noted; for they will be of use when we come to consider cohesion & soft bodies.

Theorem on the steady state of the central point &, more generally, of the centre of gravity in the case of masses.

202. If two points are projected in any manner whatever with any velocities whatever, it can readily be proved that the middle point of the line joining them must remain at rest or move uniformly in a straight line; and that about this point, whether it is at rest or is moving uniformly, the oscillations or descriptions of the curved paths, referred to above, must take place. But this, more generally, is a property relating to masses, of any number or kind, for which the common centre of gravity is either at rest or moves uniformly in a straight line, in no wise disturbed by the mutual forces. This theorem was enunciated by Newton, but he did not give a satisfactory proof of it. I have discovered a most rigorous demonstration, & one that is at the same time general, & I gave it in the dissertation *De Centro Gravitatis*; this demonstration I will also give here in the articles that follow.

The approach of one of the two points towards any plane is equal to the recession of the other from it, on account of the mutual force.

203. Lastly, I will here mention in passing something that refers to the motion of two points, which will be of use later, in connection with that subject. If two points move subject to their mutual forces only, & any plane is taken beyond them both, then the approach of one of them to that plane, measured in any direction, will be equal to the recession of the other. This follows immediately from the fact that their absolute motions are equal & opposite; for, on that account, it comes about that the resolved parts in any other direction also remain equal & opposite, as they were to start with. However, I have said enough for the present about the equilibrium & motions of two points.

Extension to a system of three points; two general problems.

204. When we come to consider systems of three points, as also systems of any number of points, the whole matter in general will reduce to these two problems, of which the one refers to forces and the other to motions. 1. *Being given the position and the mutual distance of the points, it is required to find the magnitude and direction of the force, to which any one of them is subject; this force being the resultant of the forces due to the remaining points, and each of these latter being found by a general law which is given by the curve of Fig. 1.* 2. *Being given the law of forces represented by Fig. 1, it is required to find the motions of the points, when each of them is projected with known velocities from given initial positions in given directions.* The first of these problems is easily solved; and also, by the aid of

generaliter pro omnibus distantiis assumptis in quavis recta positionis datæ, a que id tam geometrice determinando per puncta curvas, quæ ejusmodi legem exhibeant, ac determinent sive magnitudinem vis absolutæ, sive magnitudines binarum virium, in quas ea concipiatur resoluta, & quarum altera sit perpendicularis datæ illi rectæ, altera secundum illam agat; quam exhibendo tres formulas analyticas, quæ id præstent. Secundum omnino generaliter acceptum, & ita, ut ipsas curvas describendas liceat definire in quovis casu vel constructione, vel caculo, superat (licet puncta sint tantummodo tria) vires methodorum adhuc cognitarum: & si pro tribus punctis substituantur tres massæ punctorum, est illud ipsum celeberrimum problema quod appellant trium corporum, usque adeo quæsitum per hæc nostra tempora, & non nisi pro peculiaribus quibusdam casibus, & cum ingentibus limitationibus, nec adhuc satis promoto ad accurationem calculo, solutum a paucissimis nostri ævi Geometris primi ordinis, uti diximus num. 122.

Theorema de motu puncti habentis actionem cum aliis binis.

205. Pro hoc secundo casu illud est notissimum, si tria puncta sint in fig. 21 A, C, B, & distantia AB duorum divisa semper bifariam in D, ac ducta CD, & assumpto ejus triente DE, utcunque moveantur eadem puncta motibus compositis a projectionibus quibuscunque, & mutatis viribus; punctum E debere vel quiescere semper, vel progredi in directum motu uniformi. Pendet id a generali theoremate de centro gravitatis, cujus & superius injecta est mentio, & de quo age-[93]-mus infra pro massis quibuscunque. Hinc si sibi relinquantur, accedet C ad E, & rectæ AB punctum medium D ibit ipsi obviam versus ipsum cum velocitate dimidia ejus, quam ipsum habebit, vel contra recedent, vel hinc, aut inde movebuntur in latus, per lineas tamen similes, atque ita, ut C, & D semper respectu puncti E immoti ex adverso sint, in quo motu tam directio rectæ AB, quam directio rectæ CD, & ejus inclinatio ad AB, plerumque mutabitur.

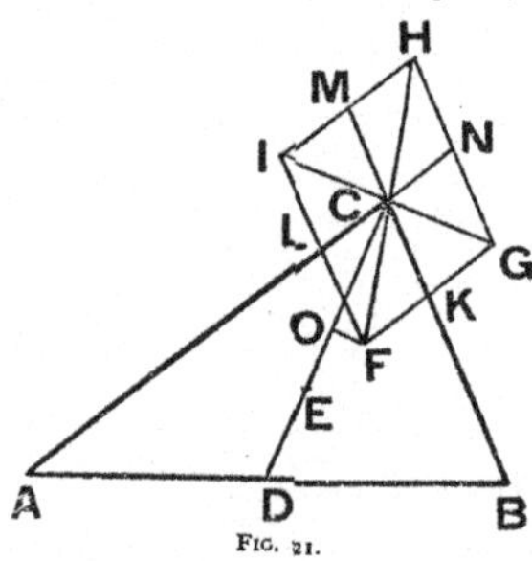

FIG. 21.

Determinatio vis ejusdem compositæ e binis viribus.

206. Quod pertinet ad inveniendam vim pro quacunque positione puncti C respectu punctorum A, & B, ea facile sic invenietur. In fig. 1 assumendæ essent abscissæ in axe æquales rectis AC, BC figuræ 21, & erigendæ ordinatæ ipsis respondentes, quæ vel ambæ essent ex parte attractiva, vel ambæ ex parte repulsiva; vel prima attractiva, & secunda repulsiva; vel prima repulsiva & secunda attractiva. In primo casu sumendæ essent CL, CK ipsis æquales (figura 21 exhibet minores, ne nimis excrescat) versus A, & B; in secundo CN, CM ad partes oppositas A,B: in tertio CL versus A, & CM ad partes oppositas B; in quarto CN ad partes oppositas A, & CK versus B. Tam completo parallelogrammo LCKF, vel MCNH, vel LCMI, vel KCNG, diameter CF, vel CH, vel CI, vel CG exprimeret directionem, & magnitudinem vis compositæ, qua urgetur C a reliquis binis punctis.

Methodus construendi curvam, quæ generaliter exprimat vim ejusmodi.

207. Hinc si assumantur ad arbitrium duo loca quæcunque punctorum A, & B, ad quæ referendum sit tertium C; ducta quavis recta DEC indefinita, ex quovis ejus puncto posset erigi recta ipsi perpendicularis, & æqualis illi diametro, ut CF in primo casu, ac haberetur curva exprimens vim absolutam puncti in eo siti, & solicitati a viribus, quas habet cum ipsis A, & B. Sed satis esset binas curvas construere, alteram, quæ exprimeret vim redactam ad directionem DC per perpendiculum FO, ut CO; alteram, quæ exprimeret vim perpendicularem OF: nam eo pacto haberentur etiam directiones vis absolutæ ab iis compositæ per ejusmodi binas ordinatas. Oporteret autem ipsam ordinatam curvæ utriuslibet assumere ex altera plaga ipsius CD, vel ex altera opposita; prout CO jaceret versus D, vel ad plagam oppositam pro prima curva; & prout OF jaceret ad alteram partem rectæ DC, vel ad oppositam, pro secunda.

Expressio magis generalis per superficiem.

208. Hoc pacto datis locis A, B pro singulis rectis egressis e puncto medio D duæ haberentur diversæ curvæ, quæ diversas admodum exhiberent virium leges; ac si quæreretur locus geometricus continuus, qui exprimeret simul omnes ejusmodi leges pertinentes ad omnes ejusmodi curvas, sive indefinite exhiberet omnes vires pertinentes ad omnia

the curve given in Fig. 1, the law of forces can be determined in general for any assumed distances along any straight line given in position. Moreover, this can be effected either by constructing geometrically curves through sets of points, which represent a law of this sort & give either the magnitude of the absolute force, or the magnitudes of the pair of forces into which it may be considered to be resolved, the one acting perpendicularly to the given straight line & the other in its direction ; or else by writing down three analytical formulæ, which will represent its value. The second, if treated perfectly generally, & in such a manner that the curves to be described can be assigned in any case whatever, either by construction or by calculation, is (even when there are only three points in question) beyond the power of all methods known hitherto. Further, if instead of three points we have three masses of points, then we have the well-known problem that is called "the problem of three bodies." The solution of this problem is still sought after in our own times ; & has only been solved in certain special cases, with great limitations by a very few of the geometricians of our age belonging to the highest rank, & even then with insufficient accuracy of calculation ; as was pointed out in Art. 122.

Theorem with regard to the motion of a point under the action of two other points.

205. As for this second case, it is very well known that, if in Fig. 21, A,C,B, are three points, & the distance between two of them, A & B, is always bisected at D, & CD is joined, & DE is taken equal to one third of DC, then, however these points move under the influence of the forces compounded from the forces of any projection whatever & the mutual forces, the point E must always remain at rest or proceed in a straight line with uniform motion. This depends on a general theorem with regard to the centre of gravity, about which passing mention has already been made, & with which we shall deal in what follows for the case of any masses whatever. From this it follows that, if they are left to themselves, the point C will approach the point E, & D, the middle point of the straight line AB, will move in the opposite direction towards E with half the velocity of C ; or, on the contrary, both C & D will recede from E ; or they will move, one in one direction & the other in the opposite direction ; nevertheless they will follow similar paths, in such a manner that C & D will always be on opposite sides of the stable point E ; & in this motion, the direction of the straight line AB, that of the straight line DE, & the inclination of the latter to AB will usually be altered.

Determination of this force that is compounded from two forces.

206. As regards the determination of the force for any position of the point C with regard to the points A & B, that is easily effected in the following manner. Take, in Fig. 1, abscissæ measured along the axis equal to the straight lines AC & BC of Fig. 21 ; draw the ordinates corresponding to them, which may be either both on the attractive side of the axis, or both on the repulsive side ; or the first on the attractive & the second on the repulsive ; or the first on the repulsive & the second on the attractive side. In the first case, take CL, CK, equal to these ordinates (in Fig. 21 they are reduced so as to prevent the figure from being too large) ; let them be taken in the direction of A & B ; similarly, in the second case, take CN & CM in the opposite directions to those of A & B ; and, in the third case, take CL in the direction of A, & CM in the direction opposite to that of B ; whilst, in the fourth case, take CN in the direction opposite to that of A, & CK in the direction of B. Then, completing the parallelogram LCKF, or MCNH, or LCMI, or KCNG, the diagonal CF, or CH, or CI, or CG, will represent the direction & the magnitude of the resultant force, which is exerted upon the point C by the remaining two points.

The method of constructing a curve which will in general express a force of this sort.

207. Hence, if any two positions are taken at random as those of the points A & B, & to these the third point C is referred ; & if any straight line DEC is drawn of indefinite length ; then from any point of it a straight line can be erected perpendicular to it, & equal to the diagonal of the parallelogram, for instance CF in the first case. From these perpendiculars a curve will be obtained, which will represent the absolute force on a point situated in the straight line DEC, & under the action of the forces exerted upon it by the points A & B. However, it would be more satisfactory if two curves were constructed ; one of which would represent the force resolved along the direction DC by means of a perpendicular FO, such as CO ; & the other to represent the perpendicular force OF. For, in this way, we should also obtain the directions of the absolute forces compounded from these resolved parts, by means of the two ordinates of this kind. Moreover, we ought to take these ordinates of either of the curves on the one side or the other of the straight line CD, according as CO would be towards D, or away from it, in the first curve, & according as OF would be away from the straight line CD, on the one side or on the other, in the second curve.

A more general expression by means of a surface.

208. In this way, given the positions of A & B, for each straight line drawn through the point D, we should obtain distinct curves ; & these would represent altogether different laws of forces. If then a continuous geometrical locus is required, which would simultaneously represent all the laws of this kind relating to every curve of this sort, or express in general all the forces pertaining to all points such as C, wherever they might

puncta C, ubicunque collocata; oporteret erigere in omnibus punctis C rectas normales plano ACB, alteram æqualem CO, [94] alteram OF, & vertices ejusmodi normalium determinarent binas superficies quasdam continuas, quarum altera exhiberet vires in directione CD attractivas ad D, vel repulsivas respectu ipsius, prout, cadente O citra, vel ultra C, normalis illa fuisset erecta supra, vel infra planum; & altera pariter vires perpendiculares. Ejusmodi locus geometricus, si algebraice tractari deberet, esset ex iis, quos Geometræ tractant tribus indeterminatis per unicam æquationem inter se connexis; ac data æquatione ad illam primam curvam figuræ 1, posset utique inveniri tam æquatio ad utramlibet curvam respondentem singulis rectis DC, constans binis tantum indeterminatis, quam æquatio determinans utramlibet superficiem simul indefinite per tres indeterminatas. (n)

Methodus determinandi vim compositam ex viribus respicientibus puncta quotcunque.

[95] 209. Si pro duobus punctis tantummodo agentibus in tertium daretur numerus quicunque punctorum positorum in datis locis, ac agentium in idem punctum, posset utique constructione simili inveniri vis, qua singula agunt in ipsum collocatum in quovis assumpto loci puncto, ac vis ex ejusmodi viribus composita definiretur tam directione, quam magnitudine, per notam virium compositionem. Posset etiam analysis adhiberi ad exprimendas curvas per æquationes duarum indeterminatarum pro rectis quibuscunque, & (o) si omnia puncta jaceant in eodem plano, superficies per æquationem trium. [96] Mirum autem, quanta inde diversarum legum combinatio oriretur. Sed & ubi duo tantummodo puncta agant in tertium, incredibile dictu est, quanta diversitas legum, & curvarum inde erumpat. Manente etiam distantia AB, leges pertinentes ad diversas inclinationes rectæ DC ad AB, admodum diversæ obveniunt inter se: mutata vero punctorum A, B distantia

---

(n) *Stantibus in fig. 22 punctis* ADBCKFLO, *ut in fig. 21, ducantur perpendicula* BP, AQ *in* CD, *quæ dabuntur data inclinatione* DC, & *punctis* B, A, *ac pariter dabuntur* & DP, DQ. *Dicatur præterea* DC $= x$, & *dabuntur analytice* CQ, CP. *Quare ob angulos rectos* P, Q, *dabuntur etiam analytice* CB, CA. *Denominentur* CK$=u$, CL $=z$, CF $=y$. *Quoniam datur* AB, & *dantur analytice* AC, CB; *dabitur analytice ex applicatione Algebræ ad Trigonometriam sinus anguli* ACB *per* x, & *datas quantitates, qui est idem, ac sinus anguli* CKF *complementi ad duos rectos. Datur autem idem ex datis analytice valoribus* CK $= u$, KF $=$ CL $= z$, CF $= y$; *quare habetur ibi una æquatio per* x, y, z, u, & *constantes. Si præterea valor* CB *ponatur pro valore abscissae in æquatione curvæ figuræ* 1; *acquiritur altera æquatio per valores* CK, CB, *sive per* x,u, & *constantes. Eodem pacto invenietur ope æquationis curvæ figuræ* 1 *tertia æquatio per* AC, & CL, *adeoque per* x, z, & *constantes. Quare jam habebuntur æquationes tres per* x,u,z,y, & *constantes, quæ, eliminatis* u, & z, *reducentur ad unicam per* x,y, & *constantes, ac ea primam illam curvam definiet.*

Quod si quæratur æquatio ad secundam curvam, cujus ordinata est CO, *vel tertiam, cujus ordinata* OF, *inveniri itidem poterit. Nam datur analytice sinus anguli* DCB $= \frac{BP}{CB}$, & *in triangulo* FCK *datur analytice sinus* FCK $= \frac{FK}{CF} \times$ *sin* CKF. *Quare datur analytice etiam sinus differentiæ* OCF, *adeoque* & *ejus cosinus,* & *inde, ac ex* CF *datur analytice* OF, *vel* CO. *Si igitur altera ex illis dicatur* p, *acquiritur nova æquatio, cujus ope una cum superioribus eliminari poterit præterea una alia indeterminata; adeoque eliminata* CF $= y$, *habebitur unica æquatio per* x,p, & *constantes, quæ exhibebit utramlibet e reliquis curvis determinantibus legem virium* CO, *vel* OF.

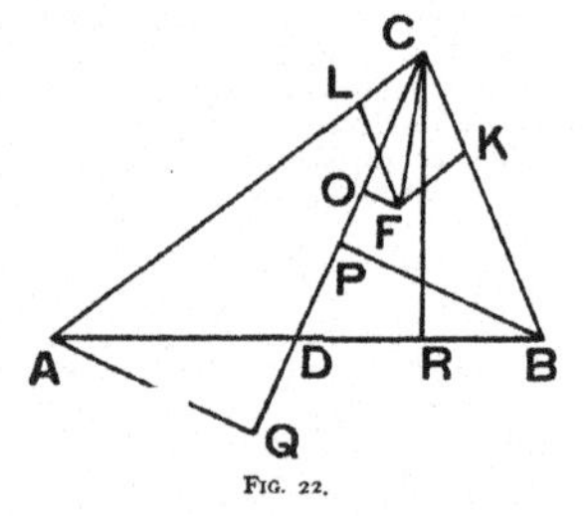

Fig. 22.

Pro æquatione cum binis indeterminatis, quæ exhibebit locum ad superficiem, ducatur CR *perpendicularis ad* AB, & *dicatur* DR $= x$, RC $= q$, *denominatis, ut prius,* CK $= u$, CL $= z$, CF $= v$; & *quoniam dantur* AD, DB; *dabuntur analytice per* x, & *constantes* AR, RB, *adeoque per* x, q, & *constantes* AC, CB, & *factis omnibus reliquis, ut prius, habebuntur quatuor æquationes per* x,q,u,z,y,p, & *constantes, quæ eliminatis valoribus* u,z,y, *reducentur ad unicam datam per constantes,* & *tres indeterminatas* x,p,q, *sive* DR, RC, & CO, *vel* OF, *quæ exhibebit quæsitum locum ad superficiem.*

Calculus quidem esset immensus, sed patet methodus, qua deveniri possit ad æquationem quæsitam. Mirum autem, quanta curvarum, & *superficierum, adeoque* & *legum virium varietas obvenerit, mutata tantummodo distantia* AB *binorum punctorum agentium in tertium, qua mutata, mutatur tota lex,* & *æquatio.*

(o) *Hæc conditio punctorum jacentium in eodem plano necessaria fuit pro loco ad superficiem,* & *pro æquatione, quæ legem virium exhibeat per æquationem indeterminatarum tantummodo trium: at si puncta sint plura,* & *in eodem plano non jaceant, quod punctis tantummodo tribus accidere omnino non potest; tum vero locus ad superficiem,* & *æquatio trium indeterminatarum non sufficit, sed ad eam generaliter exprimendam legem Geometria omnis est incapax,* & *analysis indiget æquatione indeterminatarum quatuor. Primum patet ex eo, quod si manentibus punctis* A, B, *exeat punctum* C *ex dato quodam plano, pro quo constructus sit locus ad superficiem; liceret convertere circa rectam* AB *planum illud cum superficie curva legem virium determinante, donec ad punctum* C *deveniret planum ipsum: tum enim erecto perpendiculo usque ad superficiem illam curvam, definiretur per ipsum vis agens secundum rectam* CD, *vel ipsi perpendicularis, prout locus ille ad curvam superficiem constructus fuerit pro altera ex iis.*

be situated; we should have to erect at every point C normals to the plane ACB, one of them equal to CO & the other to OF. The ends of these normals would determine two continuous surfaces; & of these, the one would represent the forces in the direction CD, attractive or repulsive with respect to the point D, according as the normal was erected above or below this plane, whether C fell on the near side or on the far side of D; & similarly the other would represent the perpendicular forces. A geometrical locus of this kind, if it has to be treated algebraically, is such as geometricians deal with by means of three unknowns connected together by a single equation; &, if the equation to the primary curve of Fig. 1 is given, it would in all cases be possible to find, not only the equations to the two curves corresponding to each & every straight line DC, involving only two unknowns, but also the equations for both the surfaces corresponding to the general determination, by means of three unknowns.(n)

The method of determining the force compounded from the forces due to any number of points. The great number & variety of laws.

209. If instead of only two points acting upon a third we are given any number of points situated in given positions, & acting on the same point, it would be possible, by a similar construction in each case, to find the force, with which each acts on the point situated in any chosen position; & the force compounded from forces of this kind would be determined, both in position & magnitude, by the well-known method for composition of forces. Also analysis could be employed to represent the curves by equations involving two unknowns for any straight lines; & (o) provided that all the points were in the same plane, the surface could be represented by an equation involving three unknowns. But it is marvellous what a huge number of different laws arise. But, indeed, it is incredible, even when there are only two points acting on a third, how great a number of different laws & curves are produced in this way. Even if the distance AB remains the same, the laws with respect to different inclinations of the straight line CD to the straight line AB, come out quite different to one another. But when the distance of the points A & B from

---

(n) *In Fig. 22, let the points* A,D,B,C,K,F,L,O *be in the same positions as in Fig. 21, & let* BP, AQ *be drawn perpendicular to* CD; *then these will be known, if the inclination of* CD *& the positions of* A *&* B *are known: & so also will* DP *&* DQ *be known. Further, suppose* DC $= x$, *then* CQ *&* CP *will be given analytically. Hence on account of the right angles at P & Q,* CB *&* CA *will also be given analytically. Suppose* CK $= u$, CL $= z$, CF $= y$. *Since* AB *is known, &* AC, CB *are given analytically, by an application of algebra to trigonometry, the sine of the angle* ACB *is also known analytically in terms of x & known quantities; & this is the same thing as the sine of the supplementary angle* CKF. *Moreover the same thing will be given in terms of the known analytical values of* CK $= u$, KF $=$ CL $= z$, CF $= y$. *Hence there is obtained in this case an equation involving x,y,z,u, & constants. If, in addition, the value* CB *is substituted for the value of the abscissa in the equation of the curve in Fig. 1, another equation will be obtained in terms of the values of* CK, CB, *i.e. in terms of x, u, & constants. In a similar way by the help of the equation of the curve of Fig. 1, there can be found a third equation in terms of* AC *&* CL, *i.e., in terms of x,z, & constants. Now, snce there will be thus obtained three equations in terms of x,y,z,u, & constants, these, on eliminating u,z, will reduce to a single equation involving x,y, & constants; & this will be the equation defining the first curve.*

Again, if the equation to the second curve is required, of which the ordinate is CO, *or of a third curve for which the ordinate is* CF, *it will be possible to find either of these as well. For the sine of the angle* DCB *is analytically given, being equal to* BP/CB; *& from the triangle* FCK, *the sine of the angle* FCK *is given, being equal to sin* CKF.(FK/CF). *Therefore the sine of the difference* OCF *is also given analytically, & therefore also its cosine; & from this & the value of* CF, *the value of* OF *or* CO *will be given analytically. If then one or the other of them is denoted by p, a new equation will be obtained: by the help of this & one of the equations given above, another of the unknowns can be eliminated. If then, we eliminate* CF $= y$, *a single equation will be obtained in terms of x,p, & constants, which will be that of one or other of the remaining curves determining the law of forces for* CO *or* OF.

For an equation in three unknowns, which will represent the surface, draw CR *perpendicular to* AB, *& let* DR$=x$ *&* RC $= q$; *&, as before, let* CK $= u$, CL $= z$, CF $= y$. *Then, since* AD, DB *are given,* AR *&* RB *are also given analytically in terms of x & constants: & therefore* AC *&* CB *are given in terms of x,q, & constants: & if all the rest of the work is done as before, four equations will be obtained in terms of x,q,u,z,y,p, & constants. These, on eliminating the values u,z,y, will reduce to a single equation in terms of constants & the three unknowns x,p,q, or* DR, RC, *&* CO *or* OF; *this equation will represent the surface required.*

The calculation would indeed be enormous; but the method, by which the required equation might be obtained is perfectly clear. But it is wonderful what a great number of curves & surfaces, & therefore of laws of force, would be met with, if merely the distance between A *&* B, *the two points which act upon the third, is changed; for if this alone is changed, the whole law is altered & so too is the equation.*

(o) *This condition, that the points should all lie in the same plane, is necessary for the determination of the surface, & for the equation, whick will express the law of the forces by an equation involving only three unknowns. If the points are numerous, & they do not all lie in the same plane (which is quite impossible in the case of only three points), then indeed a surface locus, & an equation in three unknowns, will not be sufficient; indeed, to express the law generally, the whole of geometry is powerless, & analysis requires an equation in four unknowns. The first point is clear from the fact that if, whilst the points* A *&* B *remain where tkey were, the point* C *moves out of the given plane, witk regard to whick the construction for the surface locus was made, it would be right to rotate about the straight line* AB *that plane together with its curved surface, which determines the law of forces, until the plane passes through the point* C. *For then, if a perpendicular is drawn to meet the curved surface, this would define the force acting along the straight line* CD, *or perpendicular to it, according as the locus to the curved surface had been constructed for the one or for the other of them.*

a se invicem, leges etiam pertinentes ad eandem inclinationem DC differunt inter se plurimum; & infinitum esset singula persequi; quanquam carum variationum cognitio, si obtineri utcunque posset, mirum in modum vires imaginationis extenderet, & objiceret discrimina quamplurima scitu dignissima, & maximo futura usui, atque incredibilem Theoriæ fœcunditatem ostenderet.

Vis in latus in exiguis distantiis, ac ejus usus pro solidis: in magnis nulla: in iis summa virium simplicium.

210. Ego hic simpliciora quædam, ac faciliora, & usum habitura in sequentibus, ac in applicatione ad Physicam inprimis attingam tantummodo; sed interea quod ad generalem pertinet determinationem expositam, duo adnotanda proponam. Primo quidem in ipsa trium punctorum combinatione occurrit jam hic nobis præter vim determinantem ad accessum, & recessum, vis urgens in latus, ut in fig. 21, præter vim CF, vel CH, vis CI, vel CG. Id erit infra magno usui ad explicanda solidorum phænomena, in quibus, inclinato fundo virgæ solidæ, tota virga, & ejus vertex moventur in latus, ut certam ad basim positionem acquirant. Deinde vero illud: hæc omnia curvarum, & legum discrimina tam quæ [97] pertinent ad diversas directiones rectarum DC, data distantia punctorum A, B, quam quæ pertinent ad diversas distantias ipsorum punctorum A, B, data etiam directione DC, ac basce vires in latus haberi debere in exiguis illis distantiis, in quibus curva figuræ 1 circa axem contorquetur, ubi nimirum mutata parum admodum distantia, vires singulorem punctorum mutantur plurimum, & e repulsivis etiam abeunt in attractivas, ac vice versa, & ubi respectu alterius puncti haberi possit attractio, respectu alterius repulsio, quod utique requiritur, ut vis dirigatur extra angulum ACB, & extra ipsi ad verticem oppositum. At in majoribus distantiis, in quibus jam habetur illud postremum crus figuræ 1 exprimens arcum attractivum ad sensum in ratione reciproca duplicata distantiarum, vis in punctum C a punctis A, B inter se proximis, utcunque ejusmodi distantia mutetur, & quæcunque fuerit inclinatio CD ad AB, erit semper ad sensum eadem, directa ad sensum ad punctum D, ad sensum proportionalis reciproce quadrato distantiæ DC ab ipso puncto D, & ad sensum dupla ejus, quam in curva figuræ 1 requireret distantia DC.

At secundum sit manifestum ex eo, quod si puncta agentia sint etiam omnia in eodem plano, & punctum, cujus vis composita quæritur, in quavis recta posita extra ipsum planum, relationes omnes distantiarum a reliquis punctis, ac directionum, a quibus pendent vires singulorum, & compositio ipsarum virium, longe aliæ essent, ac in quavis recta in eodem plano posita, uti facile videre est. Hinc pro quovis puncto loci ubicunque assumpto sua responderet vis composita, & quarta aliqua plaga, seu dimensio, præter longum, latum, & profundum, requireretur ad ducendas ex omnibus punctis spatii rectas iis viribus proportionales, quarum rectarum vertices locum continuum aliquem exhiberent determinantem virium legem.

Sed quod Geometria non assequitur, assequeretur quarta alia dimensio mente concepta, ut si conciperetur spatium totum plenum materia continua, quod in mea sententia cogitatione tantummodo effingi potest, & ea esset in omnibus spatii punctis densitatis diversæ, vel diversi pretii; tum illa diversa densitas, vel illud pretium, vel quidpiam ejusmodi, exhibere posset legem virium ipsi respondentium, quæ nimirum ipsi essent proportionales. Sed ibi iterum ad determinandam directionem vis compositæ non esset satis resolutio in duas vires, alteram secundum rectam transeuntem per datum punctum; alteram ipsi perpendicularem; ed requirerentur tres, nimirum vel omnes secundum tres datas directiones, vel tendentes per rectas, quæ per data tria puncta transeant, vel quavis alia certa lege definitas: adeoque tria loca ejusmodi ad spatium, quarta aliqua dimensione, vel qualitate affectum requirerentur, quæ tribus ejusmodi plusquam Geometricis legibus vis compositæ legem definirent, tum quod pertinet ad ejus magnitudinem, tum quod ad directionem.

Verum quod non assequitur Geometria, assequeretur Analysis ope æquationis quatuor indeterminatarum: si enim conciperetur planum, quod libuerit, ut ACB, *& in eo quævis recta* AB, *ac in ipsa recta quodvis punctum* D; *tum quovis hujus segmento* DR *appellato* x, *quavis recta* RC *ipsi perpendiculari y, quavis tertia perpendiculari ad totum planum z, per hasce tres indeterminatas involveretur positio puncti spatii cujuscumque, in quo collocatum esset punctum materiæ, cujus vis quæritur.*

Punctorum agentium utcunque collocatorum ubicunque vel intra id planum, vel extra, possent definiri positiones per ejusmodi tres rectas, datas utique pro singulis, si eorum positiones dentur. Per eas, & per illas x,y,z, posset utique haberi distantia cujuscumque ex iis punctis agentibus, & positione datis, a puncto indefinite accepto; adeoque ope æquationis figuræ 1 *posset haberi analytice per æquationes quasdam, ut supra, vis ad singula agentia puncta pertinens, & per easdem rectas ejus etiam directio resoluta in tres parallelas illis x,y,z. Hinc haberetur analytice omnium summa pro singulis ejusmodi directionibus per aliam æquationem derivatam ab ejus summæ denominatione, ea nimirum facta = u, ac expunctis omnibus subsidiariis valoribus, methodo non absimili ei, quam adhibuimus superius pro loco ad superficiem, deveniretur ad unam æquationem constitutam illis quatuor indeterminatis x,y,z,u, & constantibus; ac tres ejusmodi æquationes pro tribus directionibus vim omnem compositam definirent. Sed hæc innuisse sit satis, quæ nimirum & altiora sunt, & ob ingentem complicationem casuum, ac nostræ humanæ mentis imbecillitatem nulli nobis inferius futura sunt usui.*

one another is also changed, the laws corresponding to the same inclination of DC are altogether different to one another; & it would be an interminable task to consider them all, case by case. However, a comprehensive insight into their variations, if it could be obtained, would enlarge the powers of imagination to a marvellous extent; it would bring to the notice a very large number of characteristics that would be well worth knowing & most useful for further work; & it would give a demonstration of the marvellous fertility of my Theory.

The lateral force at very small distances, & its use in the consideration of solids; the absence of this force at great distances, the sum of the simple forces in the latter case.

210. First of all, therefore, I will here only deal slightly with certain of the more simple cases, such as will be of use in what follows, & later when considering the application to Physics. But meanwhile, I will enunciate two theorems, applying to the general determination set forth above, which should be noted. Firstly, in the case of the combination of three points, we have here already met with, in addition to a force inducing approach & recession, i.e., in Fig. 21, in addition to a force CF or CH, a force CI or CG, urging the point C to one side. This will be of great service to us in explaining certain phenomena of solids; for instance, the fact that, if the bottom of a solid rod is inclined, the whole rod, including its top, is moved to one side & takes up a definite position with respect to the base. Secondly, there is the fact that we are bound to have all these differences of curves & laws, not only those corresponding to different directions of the straight lines DC when the distance between the points A & B is given, but also those corresponding to different distances of the points A & B when the direction of DC is given; & that we are bound to have these lateral forces for very small distances, for which the curve in Fig. 1 twists about the axis; for then indeed, if the change in distance is very slight, the change in the forces corresponding to the several points is very great, & even passes from repulsion to attraction & vice versa; & also there may be attraction for one point & repulsion for another; & this must be the case if the direction of the force has to be without the angle ACB, or the angle vertically opposite to it. But, at distances that are fairly large, for which we have already seen that there is a final branch of the curve of Fig. 1 that represents attraction approximately in the ratio of the inverse square of the distance, the force on the point C, due to two points A & B very near to one another, will be approximately the same, no matter how this distance may be altered, or what the inclination of CD to AB may be; its direction is approximately towards D; & its magnitude will be approximately in inverse proportion to the square of DC, its distance from the point D; that is to say, it will be approximately double of that to which in Fig. 1 the distance DC would correspond.

The second point is evident from the fact that, if all the points acting are all in the same plane, & the point for which the resultant force is required, lies in any straight line situated without that plane, even then all the relations between the distances from the remaining points as well as between their directions, will be altogether different from those for any straight line situated in the same plane, as can be easily seen. Hence, for any point of space chosen at random there would be a corresponding force; & a fourth region, or dimension, in addition to length, breadth, & depth, would be required, in order to draw through each point of space straight lines proportional to these forces, the ends of which straight lines would give a continuous locus determining the law for the forces.

But what can not be attained by the use of geometry, could be attained, by imagining another, a fourth, dimension (just as if the whole of space were imagined to be full of continuous matter, which in my opinion can only be a mental fiction); & this would be of different density, or different value, at all points of space. Then the different density, or value, or something of that kind, might represent the law of forces corresponding to it, these indeed being proportional to it. But here again, in order to find the direction of the resultant force, resolution into two forces, the one along the straight line passing through the given point, & the other perpendicular to it, would not be sufficient. Three resolved parts would be required, either all in three given directions, or along straight lines passing through three given points, or defined by some other fixed law. Thus, three regions of this kind in space possessed of some fourth dimension or quality would be required; & these would define, by three ultra-geometrical laws of this sort, the law of the resultant force both as regards magnitude & direction.

But what cannot be obtained with the help of geometry could be obtained by the aid of analysis by employing an equation with four unknowns. For, if we take any arbitrary plane, as ACB, & in it any straight line AB, & in this straight line any point D; then, calling any segment of it x, any straight line perpendicular to it y, & any third straight line perpendicular to the whole plane z, there would be contained in these three unknowns the position of any point in space, at which is situated a point of matter, for which the force is required.

The positions of the acting points, however & wherever they may be situated, either within the plane or without it, could be defined by three straight lines of this sort; & these would in all cases be known for each point, if the positions of the points are given. By means of these, & the former straight lines denoted by x,y,z, there could be obtained in all cases the distance of each of the acting points, that are given in position, from any point assumed indefinitely. Thus by the help of the equation to the curve of Fig. 1, there could be obtained analytically, by means of certain equations similar to those above, the force corresponding to each of the acting points; also from the same straight lines, its direction as well, by resolving along three parallels to x, y, & z. Hence there could be obtained analytically the sum of all of them for each of these directions, by means of another equation derived from the symbol used for the sum (for instance, let this be called u); &, eliminating all the subsidiary values, by a method not unlike that which was used above for the surface locus, we should arrive at a single equation in terms of the four unknowns, x, y, z, u, & constants. Three equations of this sort, one for each of the three directions, would determine the resultant force completely. But let it suffice merely to have mentioned these things; for indeed they are too abstruse, &, on account of the enormous complexity of cases, & the disability of the human intelligence, will not be of any use to us later.

Demonstratio postremi theorematis.

211. Id quidem facile demonstratur. Si enim AB respectu DC sit perquam exigua, angulus ACB crit perquam exiguus, & a recta CD ad sensum bifariam sectus : distantiæ AC, CB crunt ad se invicem ad sensum in ratione æqualitatis, adeoque & vires CL, CK ambæ attractivæ debebunt ad sensum æquales esse inter se, & proinde LCKF ad sensum rhombus, diametro CF ad sensum secante angulum LCK bifariam, quæ rhombi proprietas est, & ipsa CF congruente cum CO, ac (ob angulum FCK insensibilem, & CKF ad sensum æqualem duobus rectis) æquali ad sensum binis CK, KF, sive CK, CL, simul sumptis ; quæ singulæ cum sint quam proxime in ratione reciproca duplicata distantiarum CB, BA ; erunt & eadem, & earum summa ad sensum in ratione reciproca duplicata distantiæ CD.

Discrimen ingens virium, quas massula exercet in massulam proximam, conformitas summa in remotarum viribus, quæ sunt directe ut massæ, & reciproce, ut quadrata distantiarum.

212. Porro id quidem commune est etiam massulis constantibus quocunque punctorum numero. Mutata illarum combinatione, vis composita a viribus singulorum agens in punctum distans a massula ipsa per intervallum perquam exiguum, nimirum ejusmodi, in quo curva figuræ 1 circa axem contorquetur, debet mutare plurimum tam intensitatem suam, quam directionem, & fieri utique potest, quod infra etiam in aliquo simpliciore casu trium punctorum videbimus, ut in alia combinatione punctorum massulæ pro eadem distantia a medio repulsiones prævaleant, in alia attractiones, in alia oriatur vis in latus ad perpendiculum, ac in eadem constitutione massulæ pro diversis directionibus admodum diversæ sint vires pro eadem etiam distantia a medio. At in magnis illis distantiis, in quibus singulorum punctorum vires jam attractivæ sunt omnes, & directiones, ob molem massulæ tam exiguam respectu ingentis distantiæ, ad sensum conspirant, vis com-[98] -posita ex omnibus dirigetur necessario ad punctum aliquod intra massulam situm, adeoque ad sensum ejus directio erit eadem, ac directio rectæ tendentis ad mediam massulam, & æquabitur vis ipsa ad sensum summæ virium omnium punctorum constituentium ipsam massulam, adeoque erit attractiva semper, & ad sensum proportionalis in diversis etiam massulis numero punctorum directe, & quadrato distantiæ a medio massulæ ipsius reciproce ; sive generaliter erit in ratione composita ex directa simplici massarum, & reciproca duplicata distantiarum. Multo autem majus erit discrimen in exiguis illis distantiis, si non unicum punctum a massula illa solicitetur, sed massula alia, cujus vis componatur e singulis viribus singulorum suorum punctorum, quod tamen in massula etiam respectu massulæ admodum remotæ evanescet, singulis ejus punctis vires habentibus ad sensum æquales, & agentes in eadem ad sensum directione ; unde fiet, ut vis motrix ejus massulæ solicitatæ, orta ab actionibus illius alterius remotæ massulæ, sit ad sensum proportionalis numero punctorum, quæ habet ipsa, numero corum, quæ habet altera, & quadrato distantiæ, quæcunque sit diversa dispositio punctorum in utralibet, quicunque numerus.

Unde necessaria omnium corporum uniformitas in gravitate, difformitas in aliis innumeris proprietatibus.

213. Mirum sane, quantum in applicatione ad Physicam hæc animadversio habitura sit usum ; nam inde constabit, cur omnia corporum genera gravitatem acceleratricem habeant proportionalem massæ, in quam tendunt, & quadrato distantiæ, adeoque in superficie Terræ aurum, & pluma cum æquali celeritate descendant seclusa resistentia, vim autem totam, quam etiam pondus appellamus, proportionalem præterea massæ suæ, adeoque in ordine ad gravitatem nullum sit discrimen, quæcunque differentia habeatur inter corpora, quæ gravitant, & in quæ gravitant, sed ad solam demum massam, & distantiam res omnis deveniat ; at in iis proprietatibus, quæ pendent a minimis distantiis, in quibus nimirum fiunt reflexionis lucis, & refractiones cum separatione colorum pro visu, vellicationes fibrarum palati pro gustu, incursus odoriferarum particularum pro odoratu, tremor communicatus particulis acris proximis, & propagatus usque ad tympanum auriculare pro auditu, asperitas, ac aliæ sensibiles ejusmodi qualitates pro tactu, tot cohæsionum tam diversa genera, secretiones, nutritionesque, fermentationes, conflagrationes, displosiones, dissolutiones. præcipitationes, ac alii effectus Chemici omnes, & mille alia ejusmodi, quæ diversa corpora a se invicem discernunt, in iis, inquam, tantum sit discrimen, & vires tam variæ, ac tam

Proof of the latter theorem.

211. The latter theorem can be easily demonstrated. For, if AB is very small compared with DC, the angle ACB will be very small, & will be very nearly bisected by the straight line CD. The distances AC, CB will be approximately equal to one another; & thus the forces CL, CK, which are both attractive, must be approximately equal to one another. Hence, LCKF is approximately a rhombus, & the diagonal CF very nearly bisects the angle LCK, that being a property of a rhombus; CF will fall along CO, &, because the angle FCK is exceedingly small & CKF very nearly two right angles, CF will be very nearly equal to CK & KF, or CK & CL, taken together. Now each of these are as nearly as possible in the inverse ratio of the square of the distances CB, CA; & these will be the same, & their sum therefore approximately inversely proportional to the square of the distance DC.

There is a huge difference in the forces which a small mass exerts on a small mass that is very near to it; but the greatest possible uniformity in the forces due to remote masses; these vary directly as the masses, & inversely as the square of the distances.

212. Further this theorem is also true in general for little masses consisting of points, whatever their number may be. The force compounded from the several forces acting on a point, whose distance from the mass is very small, i.e., such a distance as that for which, in Fig. 1, the curve is twisted about the axis, must be altered very greatly if the combination of the points is altered; & this is so, both as regards its intensity, & as regards its direction. It may even happen, as will be seen later in the more simple case of three points, that in one combination of the points forming the little mass, & for one & the same distance from the mean point, repulsions will preponderate, in another case attractions, & in another case there will arise a perpendicular lateral force. Also for the same constitution of the mass, for the same distance from the mean point, there may be altogether different forces for different directions. But, for considerable distances, where the forces due to the several points are now attractive, & their directions practically coincide owing to the dimensions of the little mass being so small compared with the greatness of the distance, the force compounded from all of them will necessarily be directed towards some point within the mass itself; & thus its direction will be approximately the same as the straight line drawn through the mean centre of the mass; & the force itself will be equal approximately to the sum of all the forces due to the points composing the little mass. Hence, it will always be an attractive force; & in different masses, it will be approximately proportional to the number of points directly, & to the square of the distance from the mean centre of the mass inversely. That is, in general, it will be in the ratio compounded of the simple direct ratio of the masses & the inverse duplicate ratio of the distances. Further, the differences will be far greater, in the case of very small distances, if not a single point alone, but another mass, is under the action of the little mass under consideration; for in this case, the force is compounded from the several forces on each of the points that constitute it; & yet these differences will also disappear in the case of a mass acted on by a mass considerably remote from it, since each of the points composing it is under the influence of forces that are approximately equal & act in practically the same direction. Hence it comes about that the motive force of the mass acted upon, which is produced by the action of the other mass remote from it, is approximately proportional to the number of points in itself, to the number of points in the other mass, & to the square of the distance between them, whatever the difference in the disposition of the points, or their number, may be for either mass.

Hence we have necessarily, for all bodies, uniformity in the case of gravity, & non-uniformity in the case of numerous other properties.

213. It is indeed wonderful what great use can be made of this consideration in the application of my Theory to Physics; for, from it it will be clear why all classes of bodies have an accelerating gravity, proportional to the mass on which they act, & to the square of the distance [inversely]; & hence that, on the surface of the Earth, a piece of gold & a feather will descend with equal velocity, when the resistance of the air is eliminated. It will be clear also that the whole force, which we call the weight, is in addition proportional to the mass itself; & thus, without exception, there is no difference as regards gravity, no matter what difference there may be between the bodies which gravitate, or towards which they gravitate; the whole matter reducing finally to a consideration of mass & distance alone. However, for those properties that depend on very small distances, for instance, where we have reflection of light, & refraction with separation of colours, with regard to sight, the titillation of the nerves of the palate, with regard to taste, the inrush of odoriferous particles where smell is concerned, the quivering motion communicated to the nearest particles of the air & propagated onwards till it reaches the drum of the ear for sound, roughness & other such qualities as may be felt in the case of touch, the large number of kinds of cohesion that are so different from one another, secretion, nutrition, fermentation, conflagration, explosion, solution, precipitation, & all the rest of the effects met with in Chemistry, & a thousand other things of the same sort, which distinguish different bodies from one another; for these, I say, the differences become as great, the forces and the motions become as different, as the differences in the phenomena,

varii motus, qui tam varia phænomena, & omnes specificas tot corporum differentias inducunt, consensu Theoriæ hujus cum omni Natura sane admirabili. Sed hæc, quæ huc usque dicta sunt ad massas pertinent, & ad amplicationem ad Physicam: interea peculiaria quædam persequar ex innumeris iis, quæ per-[99]-tinent ad diversas leges binorum punctorum agentium in tertium.

Vis in duo puncta puncti positi in recta jungente ipsa, vel in recta secante hanc bifariam, & ad angulos rectos directa secundum eandem rectam.

214. Si libeat considerare illas leges, quæ oriuntur in recta perpendiculari ad AB dueta per D, vel in ipsa AB hine, & inde producta, inprimis facile est videre illud, directionem vis compositæ utrobique fore eandem cum ipsa recta sine ulla vi in latus, & sine ulla declinatione a recta, quæ tendit ad ipsum D, vel ab ipso. Pro recta AB res constat per sese; nam vires illæ, quæ ad bina ea puncta pertinent, vel habebunt directionem candem, vel oppositas, jacente ipso tertio puncto in directum cum utroque e prioribus: unde fit, ut vis composita æquetur summæ, vel differentiæ virium singularum componentium, quæ in eadem recta remaneat. Pro recta perpendiculari facile admodum demonstratur. Si enim in fig. 23 recta DC fuerit perpendicularis ad AB sectam bifariam in D, crunt AC, BC æquales inter se. Quare vires, quibus C agitatur ab A, & B, æquales crunt, & proinde vel ambæ attractivæ, ut CL, CK, vel ambæ repulsivæ, ut CN, CM. Quare vis composita CF, vel CH, erit diameter rhombi, adeoque secabit bifariam angulum LCK, vel NCM; quos angulos cum bifariam secet etiam recta DC, ob æqualitatem triangulorum DCA, DCB, patet, ipsas CF, CH debere cum eadem congruere. Quamobrem in hisce casibus evanescit vis illa perpendicularis FO, quæ in præcedentibus binis figuris habebatur, ac in iis per unicam æquationem res omnis absolvitur (p), quarum ea, quæ ad posteriorem casum pertinet, admodum facile invenitur.

Constructio curvæ exhibentis legem casus posterioris.

215. Legem pro recta perpendiculari rectæ jungenti duo puncta, & æque distanti ab utroque exhibet fig. 24, quæ vitandæ confusionis causa exhibetur, ubi sub numero 24 habetur littera B, sed quod ad ejus constructionem pertinet, habetur separatim, ubi sub num. 24 habetur littera A; ex quibus binis figuris fit unica; si puncta XYEAE′ censeantur utrobique eadem. In ea X, Y sunt duo materiæ puncta, & ipsam XY recta CC′ secat bifariam in A. Curva, quæ vires compositas ibi exhibet per ordinatas, constructa est ex fig. 1, quod fieri potest, inveniendo vires singulas singulorum punctorum, tum vim compositam ex iis more consueto juxta [100] generalem constructionem numeri 205; sed etiam sic facilius idem præstatur; centro Y intervallo cujusvis abscissæ *Ad* figuræ 1 inveniatur in figura 24 sub littera A in recta CC′ punctum *d*, sumaturque *de* versus Y æqualis ordinatæ *dh* figuræ 1, ductoque *ea* perpendiculo in CA, erigatur eidem CA itidem perpendicularis *dh* dupla *da* versus plagam electam ad arbitrium pro attractionibus, vel versus oppositam, prout illa ordinata in fig. 1 attractionem, vel repulsionem expresserit, & erit punctum *h* ad curvam exprimentem legem virium, qua punctum ubicunque collocatum in recta C′C solicitatur a binis X, Y.

Constructionis demonstratio.

216. Demonstratio facilis est: si enim ducatur *d*X, & in ea sumatur *dc* æqualis *de*, ac compleatur rhombus *debc*; patet fore ejus verticem *b* in recta *d*A secante angulum X*d*Y bifariam, cujus diameter *db* exprimet vim compositam a binis *de*, *dc*, quæ bifariam secaretur a diametro altera *ec*, & ad angulos rectos, adeoque in ipso illo puncto *a*; & *dh*, dupla *da*, æquabitur *db* exprimenti vim, quæ respectu A crit attractiva, vel repulsiva, prout illa *dh* figuræ 1 fuerit itidem attractiva, vel repulsiva.

Plures ejus curvæ proprietates.

217. Porro ex ipsa constructione patet, si centro Y, intervallis AE, AG, AI figuræ 1 inveniantur in recta CAC′ hujus figuræ positæ sub littera B puncta E, G, I, &c, ea fore limites respectu novæ curvæ; & eodem pacto reperiri posse limites E′, G′, I′, &c. ex parte opposita A; in iis enim punctis evanescente *de* figuræ ejusdem positæ sub A, evadit nulla *da*, & *db*. Notandum tamen, ibi in figura posita sub B mutari plagam attractivam in

---

(p) *Ducta enim* LK *in Fig.* 23. *ipsam* FC *secabit alicubi in I bifariam, & ad angulos rectos ex rhombi natura. Dicatur* $CD = x$, $CF = y$, $DB = a$, *& erit* $CB = \sqrt{aa+xx}$, *&* $CD = x.CB = \sqrt{aa + xx} :: CI = \frac{1}{2}y.CK = \frac{y}{2x}\sqrt{aa + xx}$, *quo valore posito in æquatione curvæ figuræ* 1 *pro valore ordinatæ, &* $\sqrt{aa + xx}$ *pro valore abscissæ, habebitur immediate æquatio nova per* x, y, *& constantes, quæ ejusmodi curvam determinabit.*

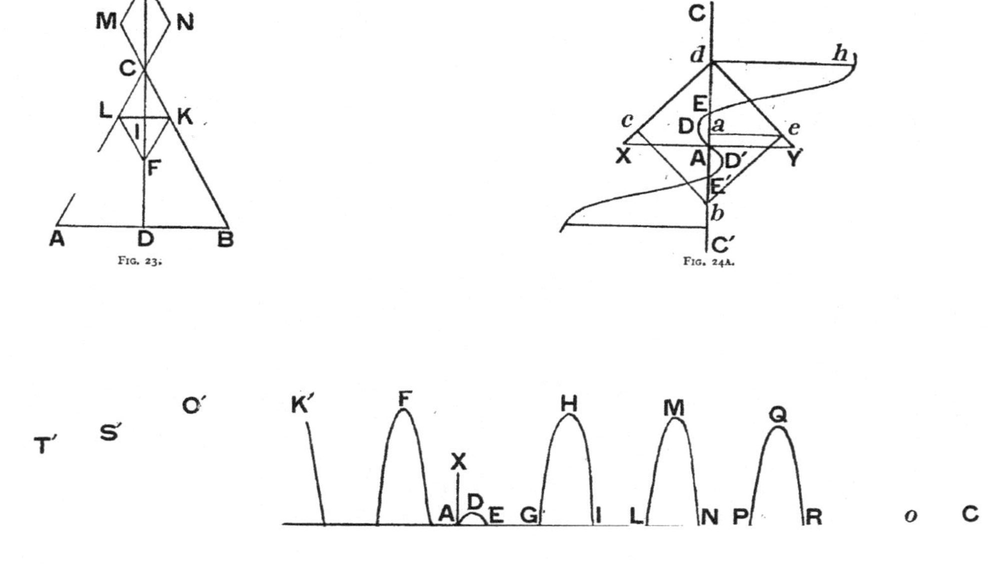

FIG. 23.

FIG. 24A.

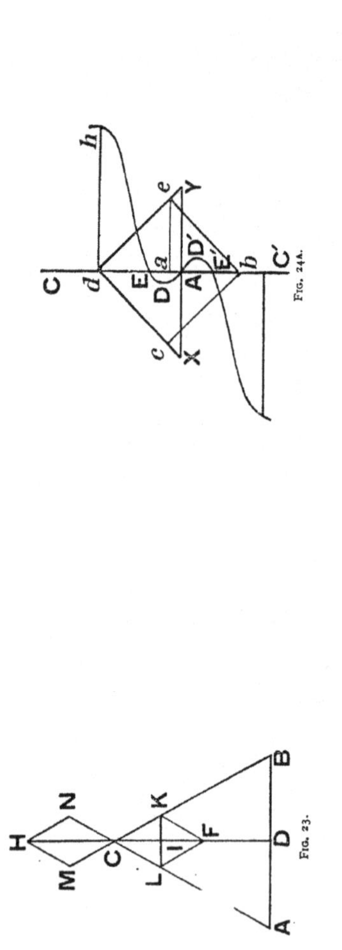

Fig. 23.

Fig. 24A.

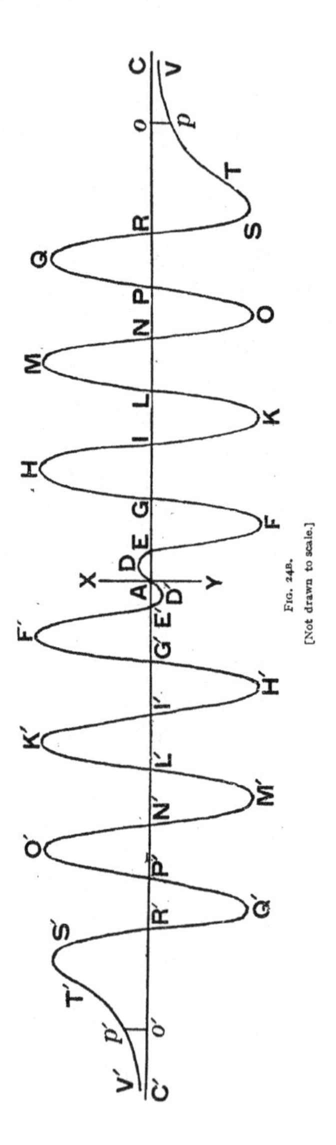

Fig. 24B.
[Not drawn to scale.]

& all the specific differences between the large number of bodies which they yield; the agreement between the Theory & the whole of Nature is truly remarkable. But what has so far been said refers to masses, & to the application of the Theory to Physics. Before we come to this, however, I will discuss certain particular cases, out of an innumerable number of those which refer to the different laws concerning the action of two points on a third.

The force exerted by two points on a point situated on the straight line joining them, or in the straight line which bisects it at right angles.

214. If we wish to consider the laws that arise in the case of a straight line drawn through D perpendicular to AB, or in the case of AB itself produced on either side, first of all it is easily seen that the direction of the resultant force in either case will coincide with the line itself without any lateral force or any declination from the straight line which is drawn towards or away from D. In the case of AB itself the matter is self-evident; for the forces which pertain to the two points either have the same direction as one another, or are opposite in direction, since the third point lies in the same straight line as each of the two former points. Whence it comes about that the resultant force is equal to the sum, or the difference, of the two component forces; & it will be in the same straight line as they. In the case of the line at right angles, the matter can be quite easily demonstrated. For, if in Fig. 23 the straight line DC were perpendicular to AB, passing through its middle point, then will AC, BC be equal to one another. Hence, the forces, by which C is influenced by A & B, will also be equal; secondly, they will either be both attractive, as CL, CK, or they will be both repulsive, as CN, CM. Hence the resultant force, CF, or CH, will be the diagonal of a rhombus, & thus it will bisect the angle LCK, or NCM. Now since these angles are also bisected by the straight line DC, on account of the equality of the triangles DCA, DCB, it is evident that CF, CH must coincide with DC. Therefore, in these cases the perpendicular force FO, which was obtained in the two previous figures, will vanish. Also in these cases, the whole matter can be represented by a single equation (*p*); & the one, which refers to the latter case, can be found quite easily.

Construction for the curve giving the law in the second case.

215. The law in the case of the straight line perpendicular to the straight line joining the two points, & equally distant from each, is graphically given in Fig. 24; to avoid confusion the curve itself is given in Fig. 24B, whilst the construction for it is given separately in Fig. 24A. These two figures are but one & the same, if the points X,Y,E,A,E′ are supposed to be the same in both. Then, in the figure, X,Y are two points of matter, & the straight line CC′ bisects XY at A. The curve, which here gives the resultant forces by means of the ordinates drawn to it, is constructed from that of Fig. 1: & this can be done, by finding the forces for the points, each for each, then the force compounded from them in the usual manner according to the general construction given in Art. 205. But the same thing can be more easily obtained thus;—With centre Y, & radius equal to any abscissa A*d* in Fig. 1, construct a point *d* in the straight line CC′, of Fig. 24A, & mark off *de* towards Y equal to the ordinate *db* in Fig. 1; draw *ea* perpendicular to CA, & erect a perpendicular, *dh*, to the same line CA also, so that $dh = 2ae$; this perpendicular should be drawn towards the side of CA which is chosen at will to represent attractions, or towards the opposite side, according as the ordinate in Fig. 1 represents an attraction or a repulsion; then the point *h* will be a point on the curve expressing the law of forces, with which a point situated anywhere on the line CC′ will be influenced by the two points X & Y.

Proof of the foregoing construction.

216. The demonstration is easy. For, if *d*X is drawn, & in it *dc* is taken equal to *de*, & the rhombus *debc* is completed, then it is clear that the point *b* will fall on the straight line *d*A bisecting the angle X*d*Y; & the diagonal of this rhombus represents the resultant of the two forces *de*, *dc*. Now, this diagonal is bisected at right angles by the other diagonal *ec*, & thus, at the point *a* in it. Also *dh*, being double of *da*, will be equal to *db*, which expresses the resultant force; this will be attractive with respect to A, or repulsive, according as the ordinate *dh* in Fig. 1 is also attractive or repulsive.

Further Properties of a curve of this sort.

217. Further, from the construction, it is evident that, if with centre Y & radii respectively equal to AE, AG, AI in Fig. 1, there are found in the straight line CAC′ of Fig. 24B the points E, G, I, &c, then these will be limit-points for the new curve; & that in the same way limit-points E′, G′, I′, &c. may be found on the opposite side of A. For, since at these points, in Fig. 24A, *de* vanishes, it follows that *da* & *db* become nothing also. Yet it must be noted that, in this case, in Fig. 24B, there is a change from the attractive

---

(p) *For, if in Fig. 23, LK is drawn, it will cut FC somewhere, in I say; & it will be at right angles to it on account of the nature of a rhombus. Suppose* $CD = x$, $CF = y$, $DB = a$; *then* $CB = \sqrt{(a^2 + x^2)}$, *& we have*

$$CD\ (or\ x) : CB\ (or\ \sqrt{(a^2 + x^2)}) = CI\ (or\ \tfrac{1}{2}y) : CK,\ \therefore\ CK = y.\sqrt{(a^2 + x^2)}/2x;$$

& if this value is substituted in the equation of the curve in Fig. 1 instead of the ordinate, & $\sqrt{(a^2 + x^2)}$ *for the abscissa, we shall get straightaway a new equation in* x, y, *& constants; & this will determine a curve of the kind under consideration.*

repulsivam, & vice versa; nam in toto tractu CA vis attractiva ad A habet directionem CC′, & in tractu AC′ vis itidem attractiva ad A habet directionem oppositam C′C. Deinde facile patebit, vim in A fore nullam, ubi nimirum oppositæ vires se destruent, adeoque ibi debere curvam axem secare; ac licet distantiæ AX, AY fuerint perquam exiguæ, ut idcirco repulsiones singulorum punctorum evadant maximæ; tamen prope A vires erunt perquam exiguæ ob inclinationes duarum virium ad XY ingentes, & contrarias; & si ipsæ AY, AX fuerint non majores, quam sit AE figuræ 1; postremus arcus EDA erit repulsivus; secus si fuerint majores, quam AE, & non majores, quam AG, atque ita porro; cum vires in exigua distantia ab A debeant esse ejus directionis, quam in fig. 1 requirunt abscissæ paullo majores, quam sit hæc YA. Postrema crura T*p*V,T′*p*′V′, patet, fore attractiva; & si in figura 1 fuerint asymptotica, fore asymptotica etiam hic; sed in A nullum erit asymptoticum crus.

Constructio curvæ exhibentis legem casus prioris.

218. At curva quæ exhibet in fig. 25 legem virium pro recta CC′ transeunte per duo puncta X, Y, est admodum diversa a priore. Ea facile construitur: satis est pro quovis ejus puncto *d* assumere in fig. 1 duas abscissæ æquales, alteram Y*d* hujus figuræ, alteram X*d* ejusdem, & sumere hic *db* æqualem [101] summæ, vel differentiæ binarum ordinatarum pertinentium ad eas abscissas, prout fuerint ejusdem directionis, vel contrariæ, & eam ducere ex parte attractiva, vel repulsiva, prout ambæ ordinatæ figuræ 1, vel earum major, attractiva fuerit, vel repulsiva. Habebitur autem asymptotus *b*Y*c*, & ultra ipsam crus asymptoticum DE, citra ipsam autem crus itidem asymptoticum *dg* attractivum respectu A, cui attractivum, sed directionis mutatæ respectu CC′, ut in fig. superiore diximus, ad partes oppositas A debet esse aliud *g*′*d*′, habens asymptotum *c*′*b*′ transeuntem per X; ac utrumque crus debet continuari usque ad A, ubi curva secabit axem. Hoc postremum patet ex eo, quod vires oppositæ in A debeant elidi; illud autem prius ex eo, quod si *a* sit prope Y, & ad ipsum in infinitum accedat, repulsio ab Y crescat in infinitum, vi, quæ provenit ab X, manente finita; adeoque tam summa, quam differentia debet esse vis repulsiva respectu Y, & proinde attractiva respectu A, quæ imminutis in infinitum distantiis ab Y augebitur in infinitum. Quare ordinata *ag* in accessu ad *b*Y*c* crescet in infinitum; unde consequitur, arcum *gd* fore asymptoticum respectu Y*c*; & eadem erit ratio pro *a*′*g*′, & areu *g*′*d*′ respectu *b*′X*c*′.

Ejus curvæ proprietates: discrimina pro mutata distantia punctorum: collatio cum curva casus alterius.

219. Poterit autem etiam arcus curvæ interceptus asymptotis *b*Y*c*, *b*′X*c*′ sive cruribus *dg*, *d*′*g*′ secare alicubi axem, ut exhibet figura 26; quin immo & in locis pluribus, si nimirum AY sit satis major, quam AE figuræ 1, ut ab Y habeatur alicubi citra A attractio, & ab X repulsio, vel ab X repulsio major, quam repulsio ab Y. Ceterum sola inspectione postremarum duarum figurarum patebit, quantum discrimen inducat in legem virium, vel curvam, sola distantia punctorum X, Y. Utraque enim figura derivata est a figura 1, & in fig. 25 assumpta est XY æqualis AE figuræ 1, in fig. 26 æqualis AI, ejusdem quæ variatio usque adeo mutavit figuræ genitæ ductum; & assumptis aliis, atque aliis distantiis pnnctorum X, Y, aliæ, atque aliæ curvæ novæ provenirent, quæ inter se collatæ, & cum illis, quæ habentur in recta CAC′ perpendiculari ad XAY, uti est in fig. 24; ac multo magis cum iis, quæ pertinentes ad alias rectas mente concipi possunt, satis confirmant id, quod supra innui de tanta multitudine, & varietate legum provenientium a sola etiam duorum punctorum agentium in tertium dispositione diversa; ut & illud itidem patet ex sola etiam harum trium curvarum delineatione, quanta sit ubique conformitas in arcu illo attractivo T*p*V, ubique conjuncta cum tanto discrimine in areu se circa axem contorquente.

Tria genera hujus casus notatu dignissima.

220. Verum ex tanto discriminum numero unum seligam maxime notatu dignum, & maximo nobis usui futurum inferius. Sit in fig. 27C—′AC axis idem, ac in fig. 1, & quinque arcus consequenter accepti alicubi GHI, IKL, LMN, NOP, PQR sint æquales prorsus inter se, ac similes. Ponantur autem bina puncta B′, B hine, & inde ab A in fig. 28 [102] ad intervallum æquale dimidiæ amplitudini unius e quinque iis arcubus, uti uni GI, vel IL; in fig. 29 ad intervallum æquale integræ ipsi amplitudini; in fig. 30 ad intervallum æquale duplæ; sint autem puncta L, N in omnibus hisce figuris eadem, & quæratur, quæ futura sit vis in quovis puncto *g* in intervallo LN in hisce tribus positionibus punctorum B′, B.

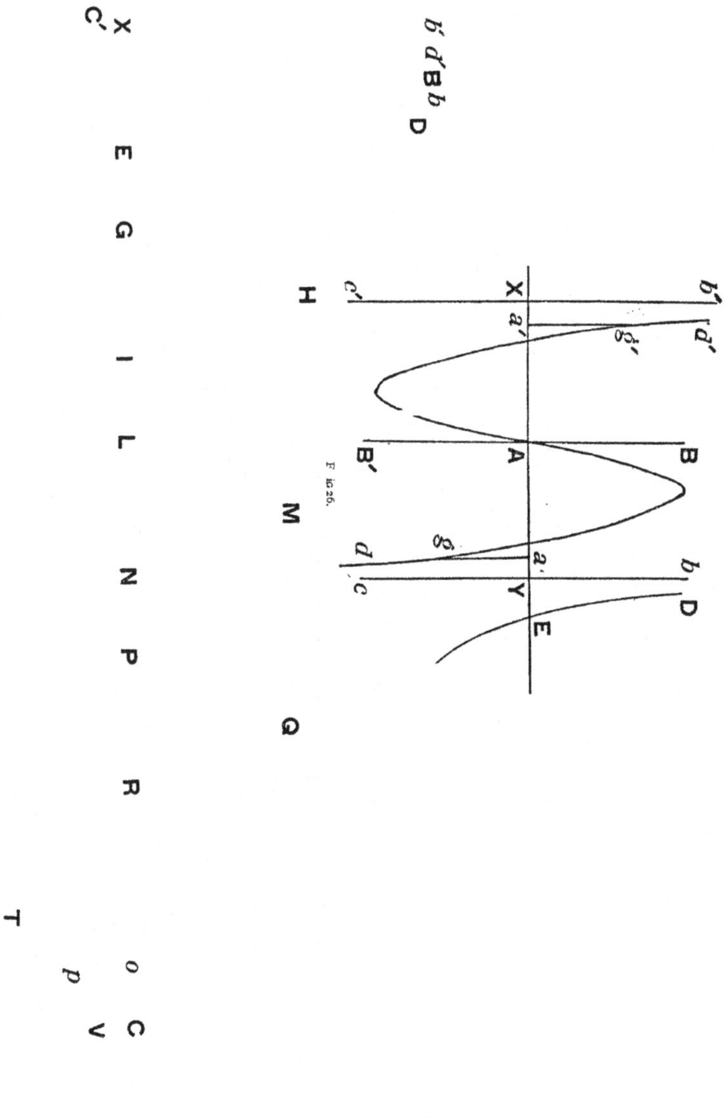

Fig. 26.

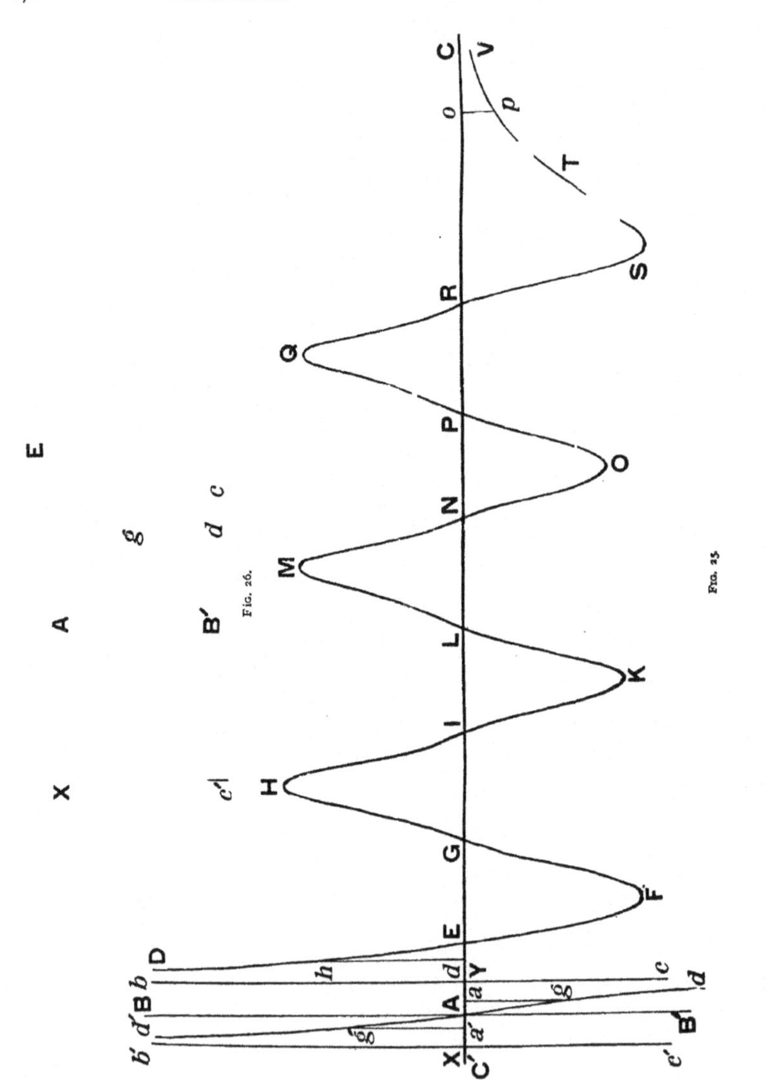

Fig. 26.

Fig. 25

side to the repulsive side, & vice versa. For along the whole portion CA, the force of attraction towards A has the direction CC′, whilst for the portion AC′, the force of attraction also towards A has the direction C′C. Secondly, it will be clearly seen that the force at A will be nothing; for there indeed the forces, being equal & opposite, cancel one another, & so the curve cuts the axis there; & although the distances AX, AY would be very small, & thus the repulsions due to each of the two points would be immensely great, nevertheless, close to A, the resultants would be very small, on account of the inclinations of the two forces to XY being extremely great & oppositely inclined. Also if AY, AX were not greater than AE in Fig. 1, the last arc would be repulsive; & attractive, if they were greater than AE, but not greater than AG, & so on; for the forces at very small distances from A must have their directions the same as that required in Fig. 1 for abscissæ that are slightly greater than YA. The final branches T*p*V, T′*p*′V′ will plainly be attractive; &, if in Fig. 1 they were asymptotic, they would also be asymptotic in this case; but there will not be an asymptotic branch at A.

Construction for the curve expressing the law for the first case.

218. But the curve, in Fig. 25, which expresses the law of forces for the straight line CC′, when it passes through the points X,Y, is quite different from the one just considered. It is easily constructed; it is sufficient, for any point *d* upon it, to take, in Fig. 1, two abscissæ, one equal to Y*d*, & the other equal to X*d*; & then, for Fig. 25, to take *db* equal to the sum or the difference of the two ordinates corresponding to these abscissæ, according as they are in the same direction or in opposite directions; &, according as each ordinate, or the greater of the two, in Fig. 1, is attractive or repulsive, to draw *db* on the attractive or repulsive side of CC′. Moreover there will be obtained an asymptote *b*Y*c*; on the far side of this there will be an asymptotic branch DE, & on the near side of it there will also be an asymptotic branch *dg*, which will be attractive with respect to A; & with respect to this part, there must be another branch *g*′*d*′, which is attractive but, since the direction with regard to CC′ is altered, as we mentioned in the case of the preceding figure, falling on the opposite side of CC′; this has an asymptote *c*′*b*′ passing through X. Also each branch must be continuous up to the point A, where it cuts the axis. This last fact is evident from the consideration that the equal & opposite forces at A must cancel one another; & the former is clear from the fact that, if *a* is very near to Y, & approaches indefinitely near to it, the repulsion due to Y increases indefinitely, whilst the force due to X remains finite. Thus, both the sum & the difference must be repulsive with respect to Y, & therefore attractive with respect to A; & this, as the distance from Y is diminished indefinitely, will increase indefinitely. Hence the ordinate *ag*, when approaching *b*Y*c*, increases indefinitely: & it thus follows that the arc *gd* will be asymptotic with respect to Y*c*; & the reasoning will be the same for *a*′*g*′, & the arc *g*′*d*′, with respect to *b*′X*c*′.

The properties of this curve; differences corresponding to changed distances between the points; comparison with the curve obtained in the other case.

219. Again, it is even possible that the arc intercepted between the asymptotes *b*Y*c*, *b*′X*c*′, i.e., between the branches *dg*, *d*′*g*′, to cut the axis somewhere, as is shown in Fig. 26; nay rather, it may cut it in more places than one, for instance, if AY is sufficiently greater than AE in Fig. 1; so that, at some place on the near side of A, there is obtained an attraction from the point Y & a repulsion from the point X, or a repulsion from X greater than the repulsion from Y. Besides, by a mere inspection of the last two figures, it will be evident how great a difference in the law of forces, & the curve, may be derived from the mere distance apart of the points X & Y. For both figures are derived from Fig. 1, &, in Fig. 25, XY is taken equal to AE in Fig. 1, whilst, in Fig. 26, it is taken equal to AI of Fig. 1; & this variation alone has changed the derived figure to such a degree as is shown. If other distances, one after another, are taken for the points X & Y, fresh curves, one after the other, will be produced. If these are compared with one another, & with those that are obtained for a straight line CAC′ perpendicular to XAY, like the one in Fig 24, nay, far more, if they are compared with those, referring to other straight lines, that can be imagined, will sufficiently confirm what has been said above with regard to the immense number & variety of the laws arising from a mere difference of disposition of the two points that act on the third. Also, from the drawing of merely these three curves, it is plainly seen what great uniformity there is in all cases for the attractive arc T*p*V, combined always with a great dissimilarity for the arc that is twisted about the axis.

Three classes of this case that are well worth remark.

220. But I will select, from this great number of different cases, one which is worth notice in a high degree, which also will be of the greatest service to us later. In Fig. 27, let CAC′ be the same axis as in Fig. 1, & let the five arcs, GHI, IKL, LMN, NOP, PQR taken consecutively anywhere along it, be exactly equal & like one another. Moreover, in Fig. 28, let the two points B & B′, one on each side of A, be taken at a distance equal to half the width of one of these five arcs, i.e., half of the one GL, or LI; in Fig. 29, at a distance equal to the whole of this width; &, in Fig. 30, at a distance equal to double the width; also let the points L,N be the same in all these figures. It is required to find the force at any point *g* in the interval LN, for these three positions of the points B & B′.

Determinatio vis compositæ in iisdem.

221. Si in Fig. 27 capiantur hinc, & inde ab ipso *g* intervalla æqualia intervallis AB', AB reliquarum trium figurarum ita, ut *ge*, *gi* respondeant figuræ 28; *gc*, *gm* figuræ 29; *ga*, *go* figuræ 30; patet, intervallum *ei* fore æquale amplitudini LN, adeoque L*e*, N*i* æquales fore dempto communi L*i*, sed puncta *e*, *i* debere cadere sub arcus proximos directionum contrariarum; ob arcuum vero æqualitatem fore æqualem vim *ef* vi contrariæ *il*, adeoque in fig. 28 vim ab utraque compositam, respondentem puncto *g*, fore nullam. At quoniam *gc*, *gm* integræ amplitudini æquantur; cadent puncta *c*, *m* sub arcus IKL, NOP, conformes etiam directione inter se, sed directionis contrariæ respectu arcus LMN, eruntque æquales *m*N, *c*I ipsi *g*L, adeoque attractiones *mn*, *cd*, & repulsioni *gh* æquales, & inter se; ac idcirco in figura 29 habebitur vis attractiva *gh* composita ex iis binis dupla repulsivæ figuræ 27. Demum cum *ga*, *go* sint æquales duplæ amplitudini, cadent puncta *a*, *o* sub arcus GHI, PQR conformis directionis inter se, & cum areu LMN, eruntque pariter binæ repulsiones *ab*, *op* æquales repulsioni *gh*, & inter se. Quare vis ex iis compositae pro fig. 30 erit repulsio *gh* dupla repulsionis *gh* figuræ 27, & æqualis attractioni figuræ 29.

In alia dispositione vim in tractu continuo nullam, in alia attractionem, in alia repulsionem, manente distantia; usus in Physica summus.

222. Inde igitur jam patet, loci geometrici exprimentis vim compositam, qua bina puncta B', B agunt in tertium, partem, quæ respondet intervallo eidem LN, fore in prima e tribus corum positionibus propositis ipsum axem LN, in secunda arcum attractivum LMN, in tertia repulsivum, utroque recedente ab axe ubique duplo plus, quam in fig. 27; ac pro quovis situ puncti *g* in toto intervallo LN in primo e tribus casibus fore prorsus nullam, in secundo fore attractionem, in tertio repulsionem æqualem ei, quam bina puncta B', B exercerent in tertium punctum situm in *g*, si collocarentur simul in A, licet in omnibus hisce casibus distantia puncti ejusdem *g* a medio systematis eorundem duorum punctorum, sive a centro particulæ constantis iis duobus punctis sit omnino eadem. Possunt autem in omnibus hisce casibus puncta B', B esse simul in arctissimis limitibus cohæsionis inter se, adeoque particulam quandam constantis positionis constituere. Aequalitas ejusmodi accurata inter arcus, & amplitudines, ac limitum distantias in figura 1 non dabitur uspiam; cum nullus arcus curvæ derivatæ utique continuæ, deductæ nimirum certa lege a curva continua, possit congruere accurate cum recta; at poterunt ea omnia ad æqualitatem accedere, quantum [103] libuerit; poterunt hæc ipsa discrimina haberi ad sensum per tractus continuos aliis modis multo adhuc pluribus, immo etiam pluribus in immensum, ubi non duo tantummodo puncta, sed immensus corum numerus constituat massulas, quæ in se agant, & ut in hoc simplicissimo exemplo deprompto e solo trium punctorum systemate, multo magis in systematis magis compositas, & plures idcirco variationes admittentibus, in eadem centrorum distantia, pro sola varia positione punctorum componentium massulas ipsas vel a se mutuo repelli, vel se mutuo attrahere, vel nihil ad sensum agere in se invicem. Quod si ita res habet, nihil jam mirum accidet, quod quædam substantiæ inter se commixtæ ingentem acquirant intestinarum partium motum per effervescentiam, & fermentationem, quæ deinde cesset, particulis post novam commixtionem respective quiescentibus; quod ex eodem cibo alia per secretionem repellantur, alia in succum nutrititium convertantur, ex quo ad eandem præterfluente distantiam alia aliis partibus solidis adhæreant, & per alias valvulas transmittantur, aliis libere progredientibus. Sed adhuc multa supersunt notatu dignissima, quæ pertinent ad ipsum etiam adeo simplex trium punctorum systema.

Alius casus vis nullius trium punctorum positorum in directum ex distantiis limitum: tres alii in quorum binis vis nulla ex elisione contrariarum.

223. Jaceant in figura 31 tria puncta A,D,B, in directum: ea poterunt respective quiescere, si omnibus mutuis viribus careant, quod fieret, si tres distantiæ AD, DB, AB omnes essent distantiæ limitum; sed potest haberi etiam quies respectiva per elisionem contrariarum virium. Porro virium mutuarum casus diversi tres esse poterunt: vel enim punctum medium D ab utroque extremorum A, B attrahitur, vel ab utroque repellitur, vel ab altero attrahitur, ab altero repellitur. In hoc postremo casu, patet, non haberi quietem respectivam; cum debeat punctum medium moveri versus extremum attrahens recedendo simul ab altero extremo repellente. In reliquis binis casibus poterit utique

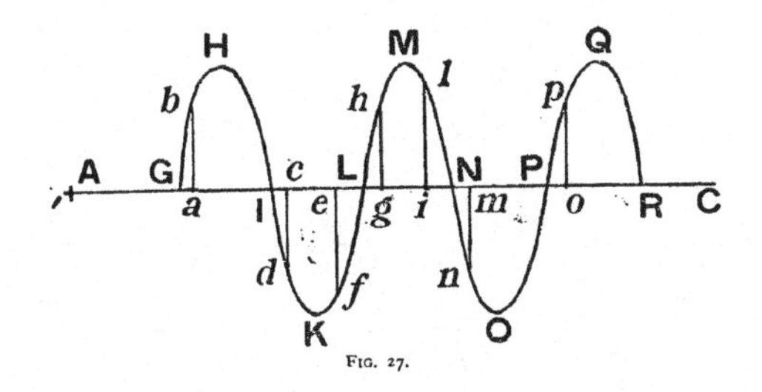

Fig. 27.

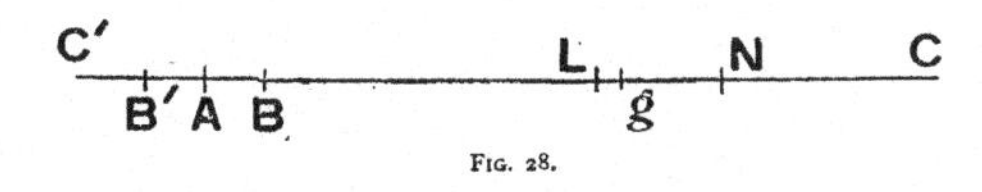

Fig. 28.

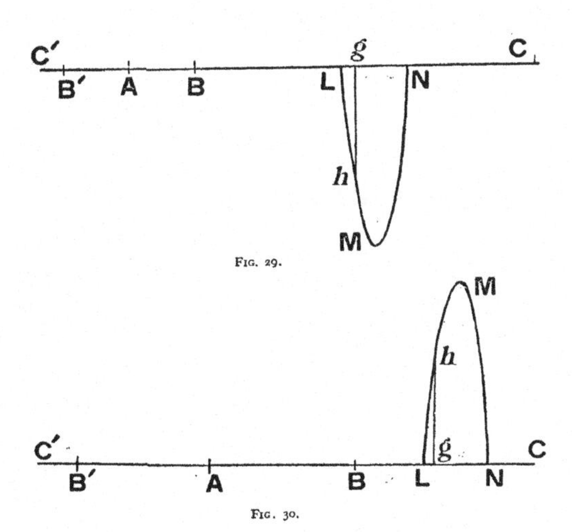

Fig. 29.

Fig. 30.

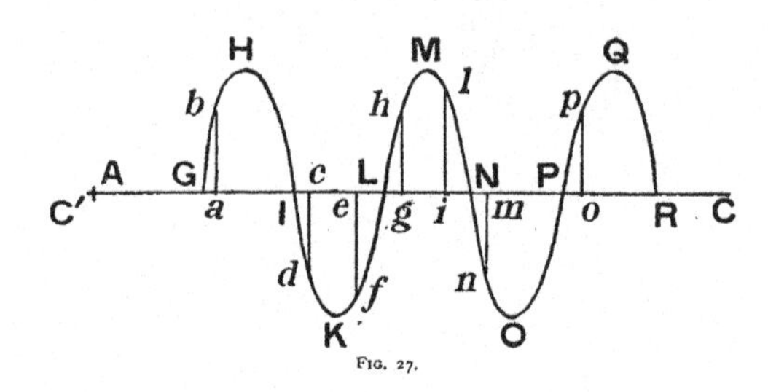

Fig. 27.

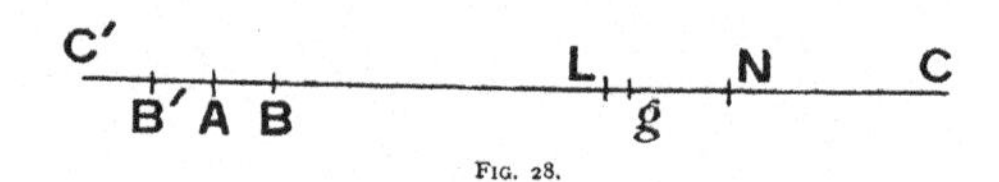

Fig. 28.

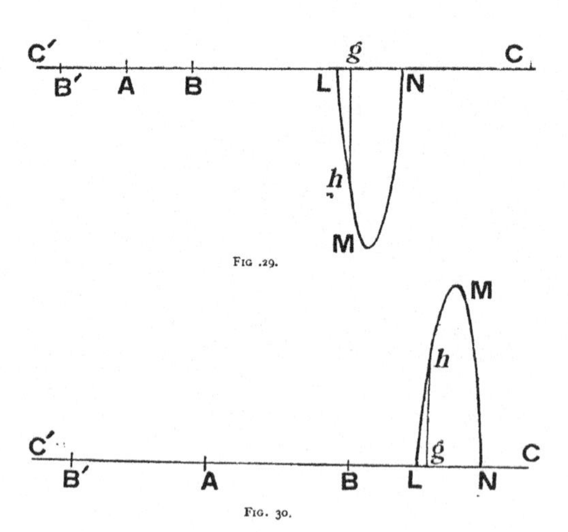

Fig. 29.

Fig. 30.

Determination of the resultant force in these three cases.

221. If, in Fig. 27, we take, on either side of this point *g*, intervals that are equal to the intervals AB′, AB of the other three figures; so that *ge*, *gi* correspond to Fig. 28; *gc*, *gm* to Fig. 29; & *ga*, *go* to Fig. 30; then it is plain that the interval *ei* will be equal to the width LN, & thus, taking away the common part L*i*, we have L*e* & N*i* equal to one another, but the points *e* & *i* must fall under successive arcs of opposite directions. Now, on account of the equality of the arcs, the force *ef* will be equal to the opposite force *il*; thus, in Fig. 28, the force compounded from the two, corresponding to the point *g*, will be nothing. Again, in Fig. 29, since *gc*, *gm* are each equal to the whole width of an arc, the points *c* & *m* fall under arcs IKL, NOP, which lie in the same direction as one another, but in the opposite direction to the arc LMN. Hence, *m*N, *c*I will be equal to *g*L; & thus the attractions *mn*, *cd* will be equal to the repulsion *gh*, & to one another. Therefore, in Fig. 29, we shall have an attractive force, compounded of these two, which is double of the repulsive force in Fig. 27. Lastly, in Fig. 30, since *ga*, *go* are equal to double the width of an arc, the points *a* & *o* will fall beneath arcs GHI, PQR, lying in the same direction as one another, & as that of the arc LMN as well. As before, the two repulsions, *ab*, *op* will be equal to the repulsion *gh*, & to one another. Hence, in Fig. 30, the force compounded from the two of them will be a repulsion *gh* which is double of the repulsion *gh* in Fig. 27, & equal to the attraction in Fig. 29.

In one arrangement, there is, in a continuous region, no force at all, in another there is an attraction & in a third a repulsion the distance remaining constant; this result is of the highest utility in Physics.

222. Therefore, from the preceding article, it is now evident that the part of the geometrical locus representing the resultant force, with which two points B′, B act upon a third, corresponding to the same interval LN, will be the axis LN itself in the first of the three stated positions of the points; in the second position it will be an attractive arc LMN, & in the third a repulsive arc; each of these will recede from the axis at all points along it to twice the corresponding distance shown in Fig. 27. So, for any position of the point *g* in the whole interval LN, the force will be nothing at all in the first of the three cases, an attraction in the second, & a repulsion in the third. This latter will be equal to that which the two points B′, B would exert on the third point, if they were both situated at the same time at the point A. And yet, in all these three cases, the distance of the point *g* under consideration remains absolutely the same, measured from the centre of the system of the same two points, or from the mean centre of a particle formed from them. Moreover, in all three cases, the points B′,B may be in the positions defining the strongest limits of cohesion with regard to one another, & so constitute a particle fixed in position. Now we never can have such accurate equality as this between the arcs, the widths, & the distances of the limit-points; for no arc of the derived curve, which is everywhere continuous because it is obtained by a given law from a continuous curve, can possibly coincide accurately with a straight line; but there could be an approximation to equality for all of them, to any degree desired. The same distinctions could be obtained, approximately for continuous regions in very many more different ways, nay the number of ways is immeasurable; in which the number of points constituting the little masses is not two only, but a very large number, acting upon one another; &, as in this very simple case derived from a consideration of a single system of three points, so, much more in systems that are more complex & on that account admitting of more variations, corresponding to a single variation of the points composing the masses, whilst the distance between the masses themselves remains the same, there may be either mutual repulsion, mutual attraction, or no mutual action to any appreciable extent. But, that being the case, there is nothing wonderful in the fact that certain substances, when mixed together, acquire a huge motion of their inmost parts, as in effervescence & fermentation; this motion ceasing & the particles attaining relative rest after rearrangement. There is nothing wonderful in the fact that from the same food some things are repelled by secretion, whilst others are converted into nutritious juices; & that from these juices, though flowing past at exactly the same distances, some things adhere to some solid parts & some to others; that some are transmitted through certain little passages, some through others, whilst some pass along uninterruptedly. However, there yet remain many things with regard to this ever so simple system of three points; & these are well worth our attention.

Another instance of no force in the case of three points situated in a straight line at the distances corresponding to limit-points. Three others, in two of which the absence of a resultant force arises from an elimination of equal & opposite forces.

223. In Fig. 31, let A,D,B be three points in a straight line. These will be at rest with regard to one another if they lack all mutual forces; & this would be the case, if the three distances AD, DB, AB were all distances corresponding to limit-points. In addition, relative rest could be obtained owing to elimination of equal & opposite forces. Further, there will be three different cases with regard to the mutual forces. For, either the middle point D is attracted by each of the outside points A & B, or it is repelled by each of them, or it is attracted by one of them & repelled by the other. In the last case, it is evident that relative rest could not obtain; for the middle point must then be moved towards the outside point that is attracting it, & recede from the other outside point which is repelling it at the same time. But in the other two cases, it is at least possible that there may be

res haberi : nam vires attractivæ, vel repulsivæ, quas habet medium punctum, possunt esse æquales ; tum autem extrema puncta debebunt itidem attrahi a medio in primo casu, repelli in secundo ; quæ si se invicem e contrario æque repellant in casu primo, attrahant in secundo ; poterunt mutuæ vires elidi omnes.

In eorum altero nisus ad recuperandam positionem, in altero ad magis ab ea recedendum, si incipiant inde removeri.

224. Adhuc autem ingens est discrimen inter hosce binos casus. Si nimirum puncta illa a directione rectæ lineæ quidquam removeantur, ut nimirum medium punctum D distet jam non nihil a recta AB, delatam in C, in secundo casu adhuc magis sponte recedet inde, & in primo accedet iterum ; vel si vi aliqua externa urgeatur, conabitur recuperare positionem priorem, & ipsi urgenti vi resistet. Nam binæ repulsiones CM, CN adhuc habebuntur in secundo casu in ipso primo recessu a D (licet eæ mutatis jam satis distantiis BD, AD inBC, AC, evadere possint attractiones) & vim com-[104]-ponent directam per CH contrariam directioni tendenti ad rectam AB. At in primo casu habebuntur attractiones CL, CK, quæ component vim CF directam versus AB, quo casu attractio AP cum repulsione AR, et attractio BV, cum repulsione BS component vires AQ, BT, quibus puncta A, B ibunt obviam puncto C redeunti ad rectam transituram per illud punctum E, quod est in triente rectæ DC, & de quo supra mentionem fecimus num. 205.

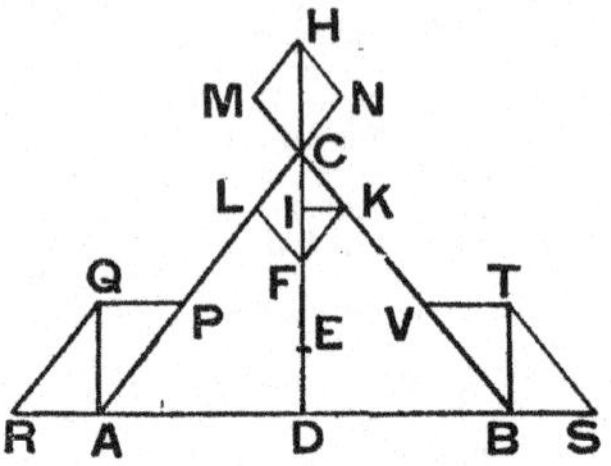

Fig. 31.

Theoria generalior indicata : trium punctorum jacentium in directum : vis maxima ad conservandam distantiam.

225. Hæc Theoria generaliter etiam non rectilineæ tantum, sed & cuivis positioni trium massarum applicari potest, ac applicabitur infra, ubi etiam generale simplicissimum, ac fœcundissimum theorema eruetur pro comparatione virium inter se : sed hic interea evolvemus nonnulla, quæ pertinent ad simpliciorem hunc casum trium punctorum. Inprimis fieri utique potest, ut ejusmodi tria puncta positionem ad sensum rectilineam retineant cum prioribus distantiis, utcunque magna fuerit vis, quæ illa dimovere tentet, vel utcunque magna velocitas impressa fuerit ad ea e suo respectivo statu deturbanda. Nam vires ejusmodi esse possunt, ut tam in eadem directione ipsius rectæ, quam in directione ad eam perpendiculari, adeoque in quavis obliqua etiam, quæ in eas duas resolvi cogitatione potest, validissimus exurgat conatus ad redeundum ad priorem locum, ubi inde discesserint puncta. Contra vim impressam in directione ejusdem rectæ satis est, si pro puncto medio attractio plurimum crescat, aucta distantia ab utrolibet extremo, & plurimum decrescat eadem imminuta ; ac pro utrovis puncto extremo satis est, si repulsio decrescat plurimum aucta distantia ab extremo, & attractio plurimum crescat, aucta distantia a medio, quod secundum utique fiet, cum, ut dictum est, debeat attractio medii in ipsum crescere, aucta distantia. Si hæc ita se habuerint, ac vice versa ; differentia virium vi extrinsecæ resistet, sive ea tenet contrahere, sive distrahere puncta, & si aliquod ex iis velocitatem in ea directione acquisiverit utcunque magnam, poterit differentia virium esse tanta, ut extinguat ejusmodi respectivam velocitatem tempusculo, quantum libuerit, parvo, & post percursum spatiolum, quantum libuerit, exiguum.

Quid ubi vis externa urgeat in latus : idea virgæ rigidæ, & virgæ flexilis.

226. Quod si vis urgeat perpendiculariter, ut ex.gr. punctum medium D moveatur per rectam DC perpendicularem ad AB ; tum vires CK, CL possunt utique esse ita validæ, ut vis composita CF sit post recessum, quantum libuerit, exiguum satis magna ad ejusmodi vim elidendam, vel ad extinguendam velocitatem impressam. In casu vis, quæ constanter urgeat, & punctum D versus C, & puncta A, B ad partes oppositas, habebitur inflexio ; ac in casu vis, quæ agat in eadem directione rectæ jungentis puncta, habebitur contractio, seu distractio ; sed vires resistentes ipsis poterunt esse ita validæ, ut & inflexio, & contractio, vel distractio, sint prorsus insensibiles ; [105] ac si actione externa velocitas imprimatur punctis ejusmodi, quæ flexionem, vel contractionem, aut distractionem inducat, tum ipsa puncta permittantur sibi libera ; habebitur oscillatio quædam, angulo jam in alteram plagam obverso, jam in alteram oppositam, ac longitudine ejus veluti virgæ constantis iis tribus punctis jam aucta, jam imminuta, fieri poterit ; ut oscillatio ipsa sensum omnem effugiat, quod quidem exhibebit nobis ideam virgæ, quam vocamus rigidam, & solidam, contractionis nimirum, & dilatationis incapacem, quas proprietates nulla virga in Natura

[The reader should draw a more general figure for Art. 224 & 227, taking AD, DB unequal and CD not at right angles to AB.]

relative rest; for the attractive, or repulsive, forces which are acting on the middle point may be equal. But then, in these cases, the outside points must be respectively attracted, or repelled by the middle point; & if they are equally & oppositely repelled by one another in the first case, & attracted by one another in the second case, then it will be possible for all the mutual forces to cancel one another.

In one of these cases there is an endeavour towards a recovery of position, & in the other towards a further recession from it, if they are initially moved out of that position.

224. Further, there is also a very great difference between these two cases. For instance, if the points are moved a small distance out of the direct straight line, so that the middle point D, say, is now slightly off the straight line AB, being transferred to C, then, if left to itself, it will recede still further from it in the first case, & will approach it once more in the second case. Or, if it is acted on by some external force, it will endeavour to recover its position & will resist the force acting on it. For two repulsions, CM, CN, will at first be obtained in the second case, at the first instant of motion from the position D; although indeed these may become attractions when the distances BD, AD are sufficiently altered into the distances BC, AC. These will give a resultant force acting along CH in a direction away from the straight line AB. But in the first case we shall have two attractions CL, CK; & these will give a force directed towards AB. In this case, the attraction AP combined with the repulsion AR, & the attraction BV combined with the repulsion BY, will give resultant forces, AQ, BT, under the action of which the points A,B will move in the opposite direction to that of the point C, as it returns to the straight line passing through that point E, which is a third of the way along the straight line DC, of which mention was made above in Art. 205.

Enunciation of a more general theory for three points lying in a straight line; possibility of a very great force tending to conservation of distance.

225. This Theory can also be applied more generally, to include not only a position of the three points in a straight line but also any position whatever. This application will be made in what follows, where also a general theorem, of a most simple & fertile nature will be deduced for comparison of forces with one another. But for the present we will consider certain points that have to do with this more simple case of three points. First of all, it may come about that three points of this kind may maintain a position practically in a straight line, no matter how great the force tending to drive them from it may be, or no matter how great a velocity may be impressed upon them for the purpose of disturbing them from their relative positions. For there may be forces of such a kind that both in the direction of the straight line, & perpendicular to it, & hence in any oblique direction which may be mentally resolved into the former, there may be produced an extremely strong endeavour towards a return to the initial position as soon as the points had departed from it. To counterbalance the force impressed in the direction of the same straight line itself, it is sufficient if the attraction for the middle point should increase by a large amount when the distance from either of the outside points is increased, & should be decreased by a large amount if this distance is decreased. For either of the outside points it is sufficient if the repulsion should greatly decrease, as the distance is increased, from the outside point, and the attraction should greatly increase, as the distance is increased, from the middle point; & this second requirement will be met in every case, since, as has been said, and attraction on it of the middle point will necessarily increase when the distance is increased. If matters should turn out to be as stated, or vice versa, then the difference of the forces will resist the external force, whether it tries to bring the points together or to drive them apart; & if any one of them should have acquired a velocity in the direction of the straight line, no matter how great, there will be a possibility that the difference of the forces may be so great that it will destroy any relative velocity of this kind, in any interval of time, no matter how short the time assigned may be; & this, after passing over any very small assigned space, no matter how small.

What happens if the external force does not act along the straight line; idea of a rigid, & of a non-rigid rod.

226. But if the force acts perpendicularly, so that, for instance, the point D moves along the line DC perpendicular to AB, then the forces CK, CL, can in any case be so strong that the resultant force CF may become, after a recession of any desired degree of smallness, large enough to eliminate any force of this kind, or to destroy any impressed velocity. In the case of a force continually urging the point D towards C, & the points A & B in the opposite direction, there will be some bending; & in the case of a force acting in the same direction as the straight line joining the two points, there will be some contraction or distraction. But the forces resisting them may be so strong that the bending, the contraction, or the distraction will be altogether inappreciable. If by external action a velocity is impressed on points of this kind, & if this induces bending, contraction or distraction, & if the points are then left to themselves, there will be produced an oscillation, in which the angle will jut out first on one side & then on the other side; & the length of, so to speak, the rod consisting of the three points will be at one time increased & at another decreased; & it may possibly be the case that the oscillation will be totally unappreciable; & this indeed will give us the idea of a rod, such as we call rigid & solid, incapable of being contracted or dilated; these properties are possessed by no rod in Nature perfectly

habet accurata tales, sed tantummodo ad sensum. Quod si vires sint aliquanto debiliores, tum vero & inflexio ex vi externa mediocri, & oscillatio, ac tremor erunt majores, & jam hine ex simplicissimo trium punctorum systemate habebitur species quædam satis idonea ad sistendum animo discrimen, quod in Natura observatur quotidie oculis, inter virgas rigidas, ac eas, quæ sunt flexiles, & ex elasticitate trementes.

Systemate inflexo per vires parallelas vis puncti medii contraria extremis, & æqualis eorum summæ.

227. Ibidem si binæ vires, ut AQ, BT fuerint perpendiculares ad AB, vel etiam utcunque parallelæ inter se, tertia quoque crit parallela illis, & æqualis carum summæ, sed directionis contrariæ. Ducta enim CD parallela iis, tum ad illam KI parallela BA, crit ob CK, VB æquales, triangulum CIK æquale simili BTV, sive TBS, adeoque CI æqualis BT, IK æqualis BS, sive AR, vel QP. Quare si sumpta IF æquali AQ ducatur KF ; crit triangulum FIK æquale AQP, ac proinde FK æqualis, & parallela AP, sive LC, & CLFK parallelogrammum, ac CF, diameter ipsius, exprimet vim puncti C utique parallelam viribus AQ, BT, & æqualem carum summæ, sed directionis contrariæ. Quoniam vero est SB ad BT, ut BD ad DC ; ac AQ ad AR, ut DC ad DA ; crit ex æqualitate perturbata AQ ad BT, ut BD ad DA, nimirum vires in A, & B in ratione reciproca distantiarum AD, DB a recta CD ducta per C secundum directionem virium.

Postremum theorema generale, ubi etiam tria puncta non jaceant in directum.

228. Ea, quæ hoc postremo numero demonstravimus, æque pertinent ad actiones mutuas trium punctorum habentium positionem mutuam quamcunque, etiam si a rectilinea recedat quantumlibet ; nam demonstratio generalis est : sed ad massas utcunque inæquales, & in se agentes viribus etiam divergentibus, multo generalius traduci possunt, ac traducentur inferius, & ad æquilibrii leges, & vectem, & centra oscillationis ac percussionis nos deducent. Sed interea pergemus alia nonnulla persequi pertinentia itidem ad puncta tria, quæ in directum non jaceant.

Æquilibrium trium punctorum non in directum jacentium impossibile sine vi externa, nisi sint in distantiis limitum : cum iis qui nisus ad retinendam formam systematis.

229. Si tria puncta non jaceant in directum, tum vero sine externis viribus non poterunt esse in æquilibrio ; nisi omnes tres distantiæ, quæ latera trianguli constituunt, sint distantiæ limitum figuræ 1. Cum enim vires illæ mutuæ non habeant [106] directiones oppositas ; sive unica vis ab altero e reliquis binis punctis agat in tertium punctum, sive ambæ ; haberi debebit in illo tertio puncto motus, vel in recta, quæ jungit ipsum cum puncto agente, vel in diagonali parallelogrammi, cujus latera binas illas exprimant vires. Quamobrem si assumantur in figura 1 tres distantiæ limitum ejusmodi, ut nulla ex iis sit major reliquis binis simul sumptis, & ex ipsis constituatur triangulum, ac in singulis angulorum cuspidibus singula materiæ puncta collocentur ; habebitur systema trium punctorum quiescens, cujus punctis singulis si imprimantur velocitates æquales, & parallelæ ; habebitur systema progrediens quidem, sed respective quiescens ; adeoque istud etiam systema habebit ibi suum quemdam limitem, sed horum quoque limitum duo genera erunt : ii, qui orientur ab omnibus tribus limitibus cohæsionis, crunt ejusmodi, ut mutata positione, conentur ipsam recuperare, cum debeant conari recuperare distantias : ii vero, in quibus etiam una e tribus distantiis fuerit distantia limitis non cohæsionis, crunt ejusmodi, ut mutata positione : ab ipsa etiam sponte magis discedat systema punctorum eorundem. Sed consideremus jam casus quosdam peculiares, & elegantes, & utiles, qui huc pertinent.

Elegans theoria puncti siti in perimetro ellipsis binis aliis occupantibus foco : vis nulla in verticibus axium.

230. Sint in fig. 32 tria puncta A,E,B ita collocata, ut tres distantiæ AB, AE, BE sint distantiæ limitum cohæsionis, & postremæ duæ sint æquales. Focis A, B concipiatur ellipsis transiens per E, cujus axis transversus sit FO, conjugatus EH, centrum D : sit in fig. 1 AN æqualis semiaxi transverso hujus DO, sive BE, vel AE, ac sit DB hic minor, quam in fig. 1 amplitudo proximorum arcuum LN, NP, & sint in eadem fig. 1 arcus ipsi NM, NO similes, & æquales ita, ut ordinatæ *uy*, *zt*, æque distantes ab N, sint inter se æquales. Inprimis si punctum materiæ sit hic in E ; nullum ibi habebit vim, cum AE, BE sint æquales distantiæ AN limitis N figuræ 1 ; ac eadem est ratio pro puncto collocato in H. Quod si fuerit in O, itidem crit in æquilibrio. Si enim assumantur in fig. 1 Az, Au æquales hisce BO, AO ; crunt Nz, Nu illius æquales DB, DA hujus, adeoque & inter se. Quare & vires illius *zt*, *uy* erunt æquales inter se, quæ cum pariter oppositæ directionis sint, se mutuo elident ; ac eadem ratio est pro collocatione in F. Attrahetur hic utique A, & repelletur B ab O ; sed si limes, qui respondet distantiæ AB, sit satis validus ; ipsa puncta nihil ad sensum discedent a focis

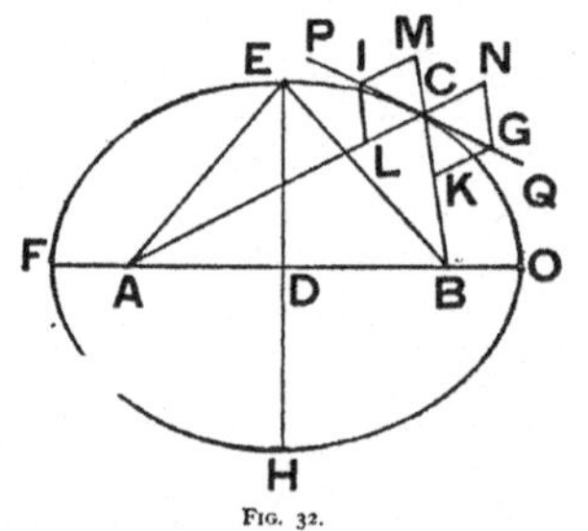

Fig. 32.

accurately, but only approximately. But if the forces are somewhat more feeble, then indeed, under the action of a moderate external force, the bending, the oscillation & the vibration will all be greater; & from this extremely simple system of three points we now obtain several kinds of cases that are adapted to giving us a mental conception of the differences, that meet our eyes every day in Nature, between rigid rods & those that are flexible & elastically tremulous.

In a system distorted by parallel forces the force on the middle point is in the opposite direction to that of the outside forces, and equal to their sum.

227. At the same time, if the two forces, represented by AQ, BT, were perpendicular to AB, or parallel to one another in any manner, then the third force would also be parallel to them, equal to their sum, but in the opposition direction. For, if CD is drawn parallel to the forces, & KI parallel to BA to meet CD in I, then, since CK & VB are equal to one another, the triangle CIK will be equal to the similar triangle BTV, or to the triangle TBS; & therefore CI will be equal to BT, IK to BS or AR or QP. Hence if IF is taken equal to AQ & KF is drawn, then the triangle FIK will be equal to AQP, & thus FK will be equal & parallel to AP or LC, CLFK will be a parallelogram, & its diagonal CF will represent the force for the point C, in every case parallel to the forces AQ, BT, & equal to their sum, but opposite in direction. But, because SB : BT : : BD : DC, & AQ : AR : : DC : DA; hence, *ex æquali* we have AQ : BT : : BD : DA, that is to say, the forces on A & B are in the inverse ratio of the distances AD & DB, drawn from the straight line CD in the direction of the forces.

The last theorem in general, even when the three points do not lie in a straight line.

228. What has been proved in the last article applies equally to the mutual actions of three points having any relative positions whatever, even if it departs from a rectilinear position to any extent you may please. For the demonstration is general; &, further, the results can be deduced much more generally for masses that are in every manner unequal, & that act upon one another even with diverging forces; & they will be thus deduced later; & these will lead us to the laws of equilibrium, the lever, & the centres of oscillation & percussion. But meanwhile we will go straight on with our consideration of some matters relating in the same manner to three points, which do not lie in a straight line.

Equilibrium of three points that do not lie in a straight line is impossible without the presence of an external force, unless the points are at distances corresponding to limit-points; the endeavour, in this case, to conserve the form of the system.

229. If the three points do not lie in a straight line, then indeed without the presence of an external force they cannot be in equilibrium; unless all three distances, which form the sides of the triangle, are those corresponding to the limit-points in Fig. 1. For, since the mutual forces do not have opposite directions, either a single force from one of the remaining two points acts on the third, or two such forces. Hence there must be for that third point some motion, either in the direction of the straight line joining it to the acting point, or along the diagonal of the parallelogram whose sides represent those two forces. Therefore, if in Fig. 1 we take three limit-distances of such a kind, that no one of them is greater than the other two taken together, & if from them a triangle is formed & at each vertical angle a material point is situated, then we shall have a system of three points at rest. If to each point of the system there is given a velocity, and these are all equal & parallel to one another, we shall have a system which moves indeed, but which is relatively at rest. Thus also that system will have a certain limit of its own; moreover, of such limits there are also two kinds. Namely, those that arise from all three limit-points being those of cohesion which will be such that, if the relative position is altered, they will strive to recover it; for they are bound to try to restore the distances. Secondly, those in which one of the three distances corresponds to a limit-point of non-cohesion, which will be such that, if the relative position is altered, the system will of its own accord depart still more from it. However, let us now consider certain special cases, that are both elegant & useful, for which this is the appropriate place.

An elegant theory for a point situated in the perimeter of an ellipse, each of the other two being placed in a focus; no force at the ends of the axes.

230. In Fig. 32, let the three points A,E,B be so placed that the three distances AB, AE, BE correspond to limit-points of cohesion, & let the two last be equal to one another. Suppose that an ellipse, whose foci are A & B, passes through E; let the transverse axis of this be FO, & the conjugate axis EH, & the centre D. In Fig. 1, let AN be equal to the transverse semiaxis DO of Fig. 32, that is equal to BE or AE; also in the latter figure let DB be less than the width of the successive arcs LN, NP of Fig. 1; also, in Fig. 1, let the arcs NM, NO be similar & equal, so that the ordinates uy, zt, which are equidistant from N, are equal to one another. Then, first of all, if in Fig. 32, the point of matter is situated at E, there will be no force upon it; for AE, BE are equal to the distance AN of the limit-point N in Fig. 1; & the argument is the same for a point situated at H. Further, if it is at O, it will in like manner be in equilibrium. For, if in Fig. 1 we take Az, Au equal to AO, BO of Fig. 32, then Nz, Nu of the former figure will be equal to DB, DA of the latter; & thus equal also to one another. Hence also the forces in that figure, zt & uy, will be equal to one another; & since they are likewise opposite in direction, they will cancel one another; & the argument is the same for a point situated at F. Here in every case A is attracted & B is repelled from O; but if the limit-point, which corresponds to the distance AB is strong enough, the points will not depart to any appreciable extent

ellipseos, in quibus fuerant collocata, vel si debeant discedere ob limitem minus validum, considerari poterunt per externam vim ibidem immota, ut contemplari liceat solam relationem tertii puncti ad illa duo.

In reliquis puncti perimetri vis directa per ipsam perimetrum versus vertices axis conjugati.

231. Manet igitur immotum, ac sine vi, punctum collocatum tam in verticibus axis conjugati ejus ellipseos, quam in verticibus axis transversi; & si ponatur in quovis puncto C [107] perimetri ejus ellipseos, tum ob AC, CB simul æquales in ellipsi axi transverso, sive duplo semiaxi DO; crit AC tanto longior, quam ipsa DO, quanto BC brevior; adeoque si jam in fig. 1 sint A*u*, A*z* æquales hisce AC, BC; habebuntur ibi utique *uy*, *zt* itidem æquales inter se. Quare hic attractio CL æquabitur repulsioni CM, & LIMC erit rhombus, in quo inclinatio IC secabit bifariam angulum LCM; ac proinde si ea utrinque producatur in P, & Q; angulus ACP, qui est idem, ac LCI, crit æqualis angulo BCQ, qui est ad verticem oppositus angulo ICM. Quæ cum in ellipsi sit notissima proprietas tangentis relatæ ad focos; crit ipsa PQ tangens. Quamobrem dirigetur vis puncti C in latus secundum tangentem, sive secundum directionem arcus elliptici, atque id, ubicunque fuerit punctum in perimetro ipsa, versus verticem propiorem axis conjugati, & sibi relictum ibit per ipsam perimetrum versus eum verticem, nisi quatenus ob vim centrifugam motum non nihil adhuc magis incurvabit.

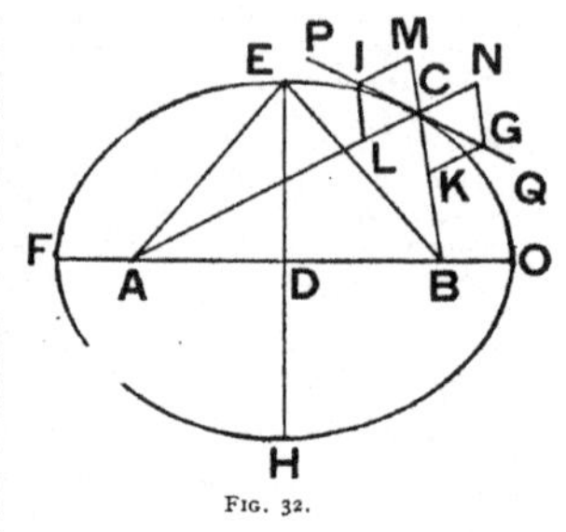

FIG. 32.

Analogia verticum binorum axium cum limitibus curvæ virium.

232. Quamobrem hic jam licebit contemplari in hac curva perimetro vicissitudinem limitum prorsus analogorum limitibus cohæsionis, & non cohæsionis, qui habentur in axe rectilineo curvæ primigeniæ figuræ 1. Erunt limites quidam in E, in F, in H, in O, in quibus nimirum vis erit nulla, cum in omnibus punctis C intermediis sit aliqua. Sed in E, & H crunt ejusmodi, ut si utravis ex parte punctum dimoveatur, per ipsam perimetrum, debeat redire versus ipsos ejusmodi limites, sicut ibi accidit in limitibus cohæsionis; at in F, & O crit ejusmodi, ut in utramvis partem, quantum libuerit, parvum inde punctum dimotum fuerit, sponte debeat inde magis usque recedere, prorsus ut ibi accidit in limitibus non cohæsionis.

Quando limites contrario modo positi: casus elegantissimi alternationis plurium limitum in perimetro ellipseos.

233. Contrarium accideret, si DO æquaretur distantiæ limitis non cohæsionis: tum enim distantia BC minor haberet attractionem CK, distantia major AC repulsionem CN, & vis composita per diagonalem CG rhombi CNGK haberet itidem directionem tangentis ellipseos; & in verticibus quidem axis utriusque haberetur limes quidam, sed punctum in perimetro collocatum tenderet versus vertices axis transversi, non versus vertices axis conjugati, & hi referrent limites cohæsionis, illi e contrario limites non cohæsionis. Sed adhuc major analogia in perimetro harum ellipsium habebitur cum axe curvæ primigeniæ figuræ 1; si fuerit DO æqualis distantiæ limitis cohæsionis AN illius, & DB in hac major, quam in fig. 1 amplitudo NL, NP; multo vero magis, si ipsa hujus DB superet plures ejusmodi amplitudines, ac arcuum æqualitas maneat hinc, & inde per totum ejusmodi spatium. Ubi enim AC hujus figuræ fiet æqualis abscissæ AP illius, etiam BC hujus fiet pariter æqualis AL illius. Quare in ejusmodi loco habebitur limes, & ante ejusmodi locum versus A distantia [108] longior AC habebit repulsionem, & BC brevior attractionem, ac rhombus erit KGNC, & vis dirigetur versus O. Quod si alicubi ante in loco adhuc propriore O distantiæ AC, BC æquarentur abscissis AR, AI figuræ 1; ibi iterum esset limes; sed ante eum locum rediret iterum repulsio pro minore distantia, attractio pro majore, & iterum rhombi diameter jaceret versus verticem axis conjugati E. Generaliter autem ubi semiaxis transversus æquatur distantiæ cujuspiam limitis cohæsionis, & distantia punctorum a centro ellipseos, sive ejus eccentricitas est major, quam intervallum dieti limitis a pluribus sibi proximis hinc, & inde, ac maneat æqualitas arcuum, habebuntur in singulis quadrantibus perimetri ellipeos tot limites, quot limites transibit eccentricitas hine translata in axem figuræ 1, a limite illo nominato, qui terminet in fig. 1 semiaxem transversum hujus ellipseos; ac prætererea habebuntur limites in verticibus amborum ellipseos axium; eritque incipiendo ab utrovis vertice axis conjugati in gyrum per ipsam perimetrum is limes primus cohæsionis, tum illi proximus esset non cohæsionis, deinde

from the foci of the ellipse, in which they were originally situated ; or, if they are forced to depart therefrom owing to the insufficient strength of the limit-point, they may be considered to be kept immovable in the same place by means of an external force, so that we may consider the relation of the third point to those two alone.

At remaining points of the perimeter the force directed along the perimeter is towards the vertices of the conjugate axis.

231. A point, then, which is situated at one of the vertices of the conjugate axis of the ellipse or at one of the vertices of the transverse axis remains motionless & under the action of no force. If it is placed at any point C in the perimeter of the ellipse, then, since AC, CB taken together are in the ellipse equal to the transverse axis, or double the semi-axis DO, AC will be as much longer than DO as BC is shorter. Hence, if in Fig. 1 A*u*, A*z* are equal to these lines AC, BC, we shall have in every case, in Fig. 1, *uy*, *zt* also equal to one another. Therefore, in Fig. 32, the attraction CL will be equal to the repulsion CM, & LIMC will be a rhombus, in which the inclination IC will bisect the angle LCM. Hence if it is produced on either side to P & Q, the angle ACP, which is the same as the angle LCI will be equal to the angle BCQ, which is vertically opposite to the angle ICM. Now this is a well-known property with respect to the tangent referred to the foci in the case of an ellipse ; & therefore PQ is the tangent. Hence the force on the point C is directed laterally along the tangent, i.e., in the direction of the arc of the ellipse ; & this is true, no matter where the point is situated on the perimeter, & the force is towards the nearest vertex of the conjugate axis ; if left to itself, the point will travel along the perimeter towards that vertex, except in so far as its motion is disturbed somewhat in addition, owing to centrifugal force.

Analogy between the vertices of the two axes & the limit-points of the curve of forces.

232. Hence we can consider in this curved perimeter the alternation of limit-points as being perfectly analogous to those of cohesion & non-cohesion, which were obtained in the rectilinear axis of the primary curve of Fig. 1. There will be certain limit-points at E, F, H, O, in which there is no force, whilst in all intermediate points such as C there will be some force. But at E & H they will be such that, if the point is moved towards either side along the perimeter, it must return towards such limit-points, just as it has to do in the case of limit-points of cohesion in Fig. 1. But at F & O, the limit-point would be such that, if the point is moved therefrom to either side by any amount, no matter how small, it must of its own accord depart still further from it ; exactly as it fell out in Fig. 1 for the limit-points of non-cohesion.

When the limit points are disposed in the opposite way ; most elegant instances of alternation of several limit-points in the perimeter of the ellipse.

233. Just the contrary would happen, if DO were equal to the distance corresponding to a limit-point of non-cohesion. For then the smaller distance BC would have an attraction CK, & the greater distance AC a repulsion CN ; the resultant force along the diagonal CG of the rhombus CNGK would in the same way have its direction along the tangent to the ellipse, & at the vertices of either axis there would be certain limit-points ; but a point situated in the perimeter would tend towards the vertices of the transverse axis, & not towards the vertices of the conjugate axis ; & the latter are of the nature of limit-points of cohesion & the former of non-cohesion. However, a still greater analogy in the case of the perimeter of these ellipses with the axis of the primary curve of Fig. 1 would be obtained, if DO were taken equal to the distance corresponding to the limit-point of cohesion AN in that figure, & in the present figure DB were taken greater than the width of NL, NP in Fig. 1 ; much more so, if DB were greater than several of these widths, & the equality between the areas on one side & the other held good throughout the whole of the space taken. For where AC in the present figure becomes equal to the abscissa AP of the former, BC in the present figure will likewise become equal to AL in the former. Hence at a position of this kind there will be a limit-point ; & before a position of this kind, towards O, the longer distance AC will have a repulsion & the shorter distance BC an attraction, KGNC will be a rhombus, & the force will be directed towards O. But if at some position, on the side of O, & still nearer to O, the distances AC, BC were equal to the abscissæ AR, AI of Fig. 1, then again there would be a limit-point ; but before that position there would return once more a repulsion for the smaller distance & an attraction for the greater, & once more the diagonal of the rhombus would lie in the direction of E, the vertex of the conjugate axis. Moreover, in general, whenever the transverse semiaxis is equal to the distance corresponding to any limit-point of cohesion, & the distance of the points from the centre of the ellipse, i.e., its eccentricity, is greater than the interval between the said limit-point & several successive limit-points on either side of it, & the equality of the arcs holds good, then for each quadrant of the perimeter of the ellipse there will be as many limit-points as the number of limit-points in the axis of Fig. 1 that the eccentricity will cover when transferred to it from the present figure, measured from that limit-point mentioned as terminating in Fig. 1 the transverse semiaxis of the ellipse of the present figure ; in addition there will be limit-points at the vertices of both axes of the ellipse. Beginning at either vertex of the conjugate axis, & going round the perimeter, the first limit-point will be one of cohesion, then the next to it one of non-cohesion, then

alter cohæsionis, & ita porro, donec redeatur ad primum, ex quo incœptus fuerit gyrus, vi in transitu per quemvis ex ejusmodi limitibus mutante directionem in oppositam. Quod si semiaxis hujus ellipseos æquetur distantiæ limitis non cohæsionis figuræ 1; res eodem ordine pergit cum hoc solo discrimine, quod primus limes, qui habetur in vertice semiaxis conjugati sit limes non cohæsionis, tum eundo in gyrum ipsi proximus sit cohæsionis limes, deinde iterum non cohæsionis, & ita porro.

Perimetri plurium ellipsium æquivalentes limitibus.

234. Verum est adhuc alia quædam analogia cum iis limitibus; si considerentur plures ellipses iisdem illis focis, quarum semiaxes ordine suo æquentur distantiis, in altera cujuspiam e limitibus cohæsionis figuræ 1, in altera limitis non cohæsionis ipsi proximi, & ita porro alternatim, communis autem illa eccentricitas sit adhuc etiam minor quavis amplitudine arcuum interceptorum limitibus illis figuræ 1, ut nimirum singulæ ellipsium perimetri habeant quaternos tantummodo limites in quatuor verticibus axium. Ipsæ ejusmodi perimetri totæ erunt quidam veluti limites relate ad accessum, & recessum a centro. Punctum collocatum in quavis perimetro habebit determinationem ad motum secundum directionem perimetri ejusdem; at collocatum inter binas perimetros diriget semper viam suam ita, ut tendat versus perimetrum definitam per limitem cohæsionis figuræ 1, & recedat a perimetro definita per limitem non cohæsionis; ac proinde punctum a perimetro primi generis dimotum conabitur ad illam redire; & dimotum a perimetro secundi generis, sponte illam adhuc magis fugiet, ac recedet.

Demonstratio.

235. Sint enim in fig. 33. ellipsium FEOH, F'E'O'H', F''E''O''H'' semiaxes DO, D'O', D''O'' æquales primus di-[109]-stantiæ AL limitis non cohæsionis figuræ 1; secundus distantiæ AN limitis cohæsionis; tertius distantiæ AP limitis iterum non cohæsionis, & primo quidem collocetur C aliquanto ultra perimetrum mediam F'E'O'H' : erunt AC, BC majores, quam si essent in perimetro, adeoque in fig. 1 factis A*u*, A*z* majoribus, quam essent prius, decrescet repulsio *zt*, crescet attractio *uy*; ac proinde hic in parallelogrammo LCMI erit attractio CL major, quam repulsio CM, & idcirco accedet directio diagonalis CI magis ad CL, quam ad CM, & inflectetur introrsum versus perimetrum mediam. Contra vero si C' sit intra perimetrum mediam, factis BC', AC' minoribus, quam si essent in perimetro media; crescet repulsio C'M', & decrescet attractio C'L', adeoque directio C'I' accedet magis ad priorem C'M', quam ad posteriorem C'L', & vis dirigetur extrorsum versus eandem mediam perimetrum. Contrarium autem accideret ob rationem omnino similem in vicinia primæ vel tertiæ perimetri : atque inde patet, quod fuerat propositum.

Alias curvas ellipsibus substituendas : ampla problematum seges, sed minus utilis immensa combinationum varietas.

236. Quoniam arcus hinc, & inde a quovis limite non sunt prorsus æquales; quanquam, ut supra observavimus num. 184, exigui arcus ordinatas ad sensum æquales hinc, & inde habere debeant; curva, per cujus tangentem perpetuo dirigatur vis, licet in exigua eccentricitate debeat esse ad sensum ellipsis, tamen nec in iis erit ellipsis accurate, nec in eccentricitatibus majoribus ad ellipses multum accedet. Erunt tamen semper aliquæ curvæ, quæ determinent continuam directionem virium, & curvæ etiam, quæ trajectoriam describendam definiant, habita quoque ratione vis centifugæ : atque hic quidem uberrima seges succrescit problematum Geometriæ, & Analysi exercendæ aptissimorum; sed omnem ego quidem ejusmodi perquisitionem omittam, cujus nimirum ad Theoriæ applicationem usus mihi idoneus occurrit nullus; & quæ huc usque vidimus, abunde sunt ad ostendendam elegantem sane analogiam alternationis in directione virium agentium in latus, cum virium primigeniis simplicibus, ac harum limitum cum illarum limitibus, & ad ingerendam animo semper magis casuum, & combinationum diversarum ubertatem tantam in solo etiam trium punctorum systemate simplicissimo; unde conjectare liceat, quid futurum sit, ubi immensus quidam punctorum numerus coalescat in massulas constituentes omnem hanc usque adeo inter se diversorum corporum multitudinem sane immensam.

Conversio totius systematis illæsi : impulsu per perimetrum ellipseos oscillatio : idea liquationis, & conglaciationis.

237. At præterea est & alius insignis, ac magis determinatus fructus, quem ex ejusmodi contemplationibus capere possumus, usui futurus etiam in applicatione Theoriæ ad Physicam. Si nimirum duo puncta A, & B sint in distantia limitis cohæsionis satis validi, & punctum tertium collocatum in vertice axis conjugati in E distantiam a reliquis habeat, quam habet limes itidem cohæsionis satis validus; poterit sane [110] vis, qua ipsum retinetur in eo vertice, esse admodum ingens pro utcunque exigua dimotione ab eo loco,

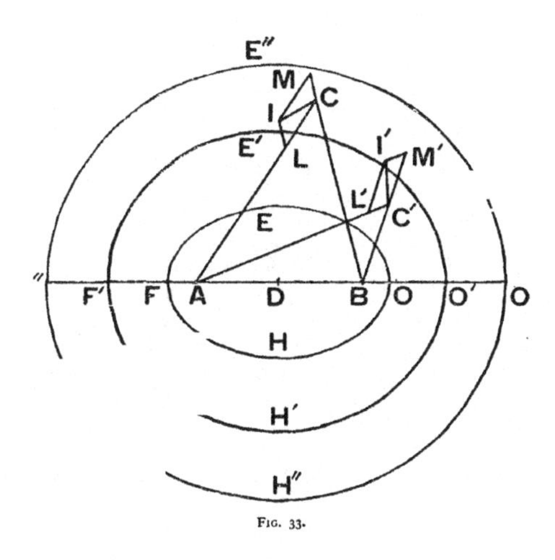

Fig. 33.

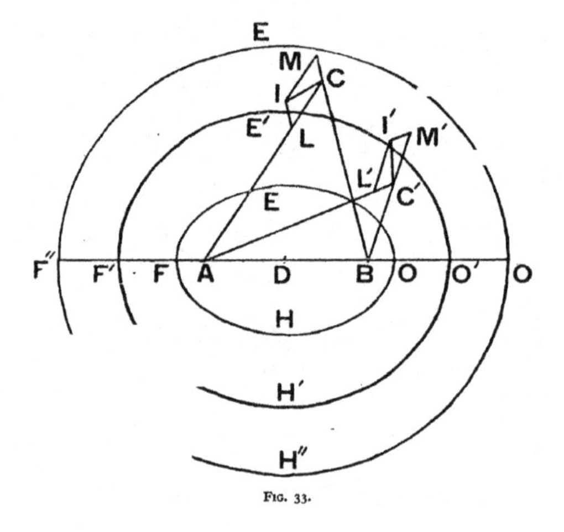

Fig. 33.

another of cohesion, & so on, until we arrive at the first of them, from which the circuit was commenced; & the force changes direction as we pass through each of the limit-points of this kind to the exactly opposite direction. But if the semiaxis of this ellipse is equal to the distance corresponding to a limit-point of non-cohesion in Fig. 1, the whole matter goes on as before, with this difference only, namely, that the first limit-point at the vertex of the conjugate semiaxis becomes one of non-cohesion; then, as we go round, the next to it is one of cohesion, then again one of non-cohesion, & so on.

The perimeters of several ellipses equivalent to limit-points.

234. Now there is yet another analogy with these limit-points. Let us consider a number of ellipses having the same foci, of which the semiaxes are in order equal to the distances corresponding to limit-points in Fig. 1, namely to one of cohesion for one, to that of non-cohesion next to it for the second, & so on alternately; also suppose that the eccentricity is still smaller than any width of the arcs between the limit-points of Fig. 1, so that each of the elliptic perimeters has only four limit-points, one at each of the four vertices of the axes. The whole set of such perimeters will be somewhat of the nature of limit-points as regards approach to, or recession from the centre. A point situated in any one of the perimeters will have a propensity for motion along that perimeter. If it is situated between two perimeters, it will always direct its force in such a way that it will tend towards a perimeter corresponding to a limit-point of cohesion in Fig. 1, & will recede from a perimeter corresponding to a limit-point of non-cohesion. Hence, if a point is disturbed out of a position on a perimeter of the first kind, it will endeavour to return to it; but if disturbed from a position on a perimeter of the second kind, it will of its own accord try to get away from it still further, & will recede from it.

Demonstration.

235. In Fig. 33, of the semiaxes DO, DO′, DO″ of the ellipses FEOH, F′E′O′H′, F″E″O″H″, let the first be equal to the distance corresponding to AL, a limit-point of non-cohesion in Fig. 1, the second to AN, one of cohesion, the third to AP, one of non-cohesion. In the first place, let the point C be situated somewhere outside the middle perimeter F′E′O′H′; then AC, BC will be greater than if they were drawn to the perimeter. Hence, in Fig. 1, since A*u*, A*z* would be made greater than they were formerly, the repulsion *zt* would decrease, & the attraction *uy* would increase. Therefore, in Fig. 33, in the parallelogram LCMI, the attraction CL will be greater than the repulsion CM, & so the direction of the diagonal CI will approach more nearly to CL than to CM, & will be turned inwards towards the middle perimeter. On the other hand, if C′ is within the middle perimeter, BC′, AC′ are made smaller than if they were drawn to the middle perimeter; the repulsion C′M′ will increase, & the attraction C′L′ will decrease, & thus the direction of C′I′ will approach more nearly to the former, C′M′, than to the latter, C′L′; & the force will be directed outwards towards the middle perimeter. Exactly the opposite would happen in the neighbourhood of the first or third perimeter, & the reasoning would be similar. Hence, the theorem enunciated is evidently true.

Other curves to be substituted for ellipses; an ample crop of theorems, but not of much use; great variety of combinations.

236. Now, since the arcs on either side of any chosen limit-point are not exactly equal, & yet, as has been mentioned above in Art. 184, very small arcs on either side are bound to have approximately equal ordinates; the curve, along the tangent to which the force is continually directed, although for small eccentricity it must be practically an ellipse, yet will neither be an ellipse accurately in this case, nor approach very much to the form of an ellipse for larger eccentricity. Nevertheless, there will always be certain curves determining the continuous direction of the force, & also curves determining the path described when account is taken of the centrifugal force. Here indeed there will spring up a most bountiful crop of problems well-adapted for the employment of geometry & analysis. But I am going to omit all discussion of that kind; for I can find no fit use for them in the application of my Theory. Also those which we have already seen are quite suitable enough to exhibit the truly elegant analogy between the alternation in direction of forces acting in a lateral direction & the simple primary forces, between the limit-points of the former & those of the latter; also for impressing on the mind more & more the great wealth of cases & different combinations to be met with even in the single very simple system of three points. From this it may be conjectured what will happen when an immeasurable number of points coalesce into small masses, from which are formed all that truly immense multitude of bodies so far differing from one another.

Rotation of the whole system intact; oscillation along the perimeter of the ellipse due to an impulse; the idea of liquefaction & congelation.

237. In addition to the above, there is another noteworthy & more determinate result to be derived from considerations of this kind, & one that will be of service in the application of the Theory to Physics. For instance, if the two points A & B are at a distance corresponding to a limit-point of cohesion that is sufficiently strong, & the third point situated at the vertex E of the conjugate axis is at a distance from the other two which corresponds to a limit-point of cohesion that is also sufficiently strong, then the force retaining the point at that vertex might be great enough, for any slight disturbance from that position, to prevent it from being moved any further, unless through the action of a huge external

ut sine ingenti externa vi inde magis dimoveri non possit. Tum quidem si quis impediat motum puncti B, & circa ipsum circumducat punctum A, ut in fig. 34 abeat in A'; abibit utique & E versus E', ut servetur forma trianguli AEB, quam necessario requirit conversatio distantiarum, sive laterum inducta a limitum validitate, & in qua sola poterit respective quiescere systema, ac habebitur idea quædam soliditatis cujus & supra injecta est mentio. At si stantibus in fig. 32 punctis A, B per quaspiam vires externas, quæ eorum motum impediant, vis aliqua exerceatur in E ad ipsum a sua positione deturbandum; donec ea fuerit mediocris, dimovebit illud non nihil; tum, illa cessante, ipsum se restituet, & oscillabit hinc, & inde ab illo vertice per perimetrum curvæ cujusdam proximæ arcui elliptico. Quo major fuerit vis externa dimovens, eo major oscillatio fiet; sed si non fuerit tanta, ut punctum a vertice axis conjugati recedens deveniat ad verticem axis transversi; semper retro cursus reflectetur, & describetur minus, quam semiellipsis. Verum si vis externa coegerit percurrere totum quadrantem, & transilire ultra verticem axis transversi; tum vero gyrabit punctum circumquaque per totam perimetrum motu continuo, quem a vertice axis conjugati ad verticem transversi retardabit, tum ab hoc ad verticem conjugati accelerabit, & ita porro, nec sistetur periodicus conversionis motus, nisi exteriorum punctorum impedimentis occurrentibus, quæ sensim celeritatem imminuant, & post ipsos ejusmodi motus periodicos per totum ambitum reducant meras oscillationes, quas contrahant, & pristinam debitam positionem restituant, in qua una haberi potest quies respectiva. An non ejusmodi aliquid accidit, ubi solida corpora, quorum partes certam positionem servant ad se invicem, ingenti agitatione accepta ab igneis particulis liquescunt, tum iterum refrigescentes, agitatione sensim cessante per vires, quibus igneæ particulæ emittuntur, & evolant, positionem priorem recuperant, ac tenacissime iterum servant, & tuentur? Sed hæc de trium punctorum systemate hucusque dicta sint satis.

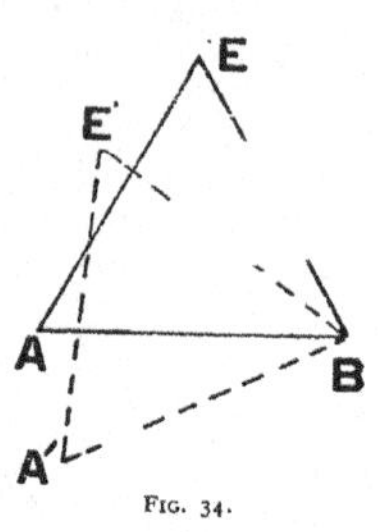

Fig. 34.

Systema punctorum quatuor, in eodem plano cum distantiis limitum, suæ formæ tenax.

238. Quatuor, & multo magis plurium, punctorum systemata multo plures nobis variationes objicerent; si rite ad examen vocarentur; sed de iis id unum innuam. Ea quidem in plano eodem possunt positionem mutuam tueri tenacissime; si singulorum distantiæ a reliquis æquentur distantiis limitum satis validorum figuræ 1: neque enim in eodem plano positionem respectivam mutare possunt, aut aliquod ex iis exire e plano dueto per reliqua tria, nisi mutet distantiam ab aliquo e reliquis, cum datis trium punctorum distantiis mutuis detur triangulum, quod constituere debent, tum datis distantiis quarti a duobus detur itidem ejus positio respectu eorum in eodem plano, & detur distantia ab eorum tertio, quæ, si id punctum exeat e [111] priore plano, sed retineat ab iis duobus distantiam priorem, mutari utique debet, ut facili negotio demonstrari potest.

Alia ratio systematis punctorum quatuor in eodem plano cum idea virgæ rigidæ, & flexilis: systema eorundem formæ pyramidalis: ordines varii particularum pyramidalium.

239. Quin immo in ipsa ellipsi considerari possunt puncta quatuor, duo in focis, & alia duo hinc, & inde a vertice axis conjugati in ea distantia a se invicem, ut vi mutua repulsiva sibi invicem elidant vim, qua juxta præcedentem Theoriam urgentur in ipsum verticem; quo quidem pacto rectangulum quoddam terminabunt, ut exhibet fig. 35, in punctis A, B, C, D. Atque inde si supra angulos quadratæ basis assurgant series ejusmodi punctorum exhibentium series continuas rectangulorum, habebitur quædam adhuc magis præcisa idea virgæ solidæ, in qua si basis ima inclinetur; statim omnia superiora puncta movebuntur in latus, ut rectangulorum illorum positionem retineant, & celeritas conversionis erit major, vel minor, prout major fuerit, vel minor vis illa in latus, quæ ubi fuerit aliquanto languidior, multo serius progredietur vertex, quam fundum, & inflectetur virga, quæ inflexio in omni virgarum genere apparet adhuc multo magis manifesta, si celeritas conversionis fuerit ingens. Sed extra idem planum possunt quatuor puncta collocata ita, ut positionem suam validissime tueantur, etiam ope unicæ distantiæ limitis unici satis validi. Potest enim fieri pyramis regularis, cujus latera singula triangularia habeant ejusmodi distantiam. Tum ea pyramis constituet particulam quandam suæ figuræ tenacissimam, quæ in puncta, vel pyramides ejusmodi aliquanto remotiores ita poterit agere, ut ejus puncta respectivum situm nihil ad sensum mutent. Ex quatuor ejusmodi particulis in aliam majorem pyramidem dispositis fieri poterit particula secundi ordinis aliquanto minus tenax ob majorem distantiam particularum primi eam componen-

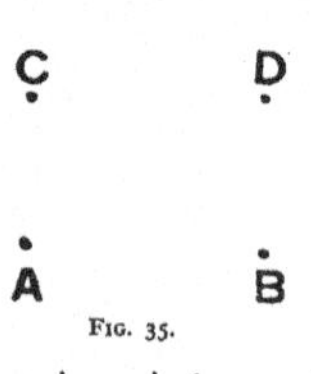

Fig. 35.

force. In that case, if the motion of the point B were prevented, & the point A were set in motion round B, so that in Fig. 34 it moved to A′, then the point E would move off to E′ as well, so as to conserve the form of the triangle AEB, as is required by the conservation of the sides or distances which is induced by the strength of the limits; & the system can be relatively at rest in this form only; thus we get an idea of a certain solidity, of which casual mention has already been made above. But if, in Fig. 2, whilst the points A,B, are kept stationary by means of an external force preventing their motion, some force is exerted on the point at E to disturb it from its position, then, as long as the force is only moderate, it will move the point a little; & afterwards, when the force ceases, the point will recover its position, & will then oscillate on each side of the vertex along a perimeter of the curve that closely approximates to an elliptic arc. The greater the external force producing the motion, the greater the oscillation will be; but if it is not so great as to make the point recede from the vertex of the conjugate axis until it reaches the vertex of the transverse axis, its path will always be retraced, & the arc described will be less than a semi-ellipse. But if the external force should compel the point to traverse a whole quadrant & pass through the vertex of the transverse axis, then indeed the point will make a complete circuit of the whole perimeter with a continuous motion; this will be retarded from the vertex of the conjugate axis to that of the transverse axis, then accelerated from there onwards to the vertex of the conjugate axis, & so on; there will not be any periodic reversal of motion, unless there are impediments met with from external points that appreciably diminish the speed; in which case, following on such periodic motions round the whole circuit, there will be a return to mere oscillations; & these will be shortened, & the original position restored, the only one in which there can possibly be relative rest. Probably something of this sort takes place, when solid bodies whose parts maintain a definite position with regard to one another, if subjected to the enormous agitation produced by fiery particles, liquefy; & once more freezing, as the agitation practically ceases on account of forces due to the action of which the fiery particles are driven out & fly off, recover their initial position & again keep & preserve it most tenaciously. But let us be content with what has been said above with regard to a system of three points for the present.

A system of four points, in one plane at distances corresponding to limit-points, which conserves its form.

238. Systems of four, & much more so for more, points would yield us many more variations, if they were examined carefully one after the other; but I will only mention one thing about such systems. It is possible that such systems, in one plane, may conserve their relative positions very tenaciously, if the distances of each from the rest are equal to the distances in Fig. 1 corresponding to limit-points of sufficient strength. For neither can they change their relative position in the plane, nor can any one of them leave the plane drawn through the other three; since, if the distances of three points from one another is given, then we are given the triangle which they must form; & then being given the distances of the fourth point from two of these, we are also given the position of this fourth point from them, & therefore also the distance from the third of them. If the point should depart from the plane mentioned, & yet preserve its former distances from the two points the distance from the third point must be changed in any case, as can be easily proved.

A further consideration of a system of four points in connection with the idea of rigid & flexible rods; a system of four points in the form of a pyramid; different arrangements of particular pyramids.

239. Again, we may consider the case of four points in the ellipse, two being at the foci, & the other two on either side of a vertex of the conjugate axis at such a distance from one another, that the mutual repulsive force between them will cancel the force with which they are urged towards that vertex, according to the preceding theorem. Thus, they are at the vertices of a rectangle, as is shown in Fig. 35, where they occupy the points A,B,C,D. Hence, if we have a series of points of this kind to stand above the four angles of the quadratic base, so as to represent continuous series of rectangles, we shall obtain from this supposition a more precise idea than hitherto has been possible of a solid rod, in which, if the lowest set of points is inclined, all the points above are immediately moved sideways, but so that they retain the positions in their rectangles; & the speed of rotation will be greater or less according as the force sideways was greater or less; even where this force is somewhat feeble, the top will move considerably later than the base & the rod will be bent; & the amount of bending in every kind of rod will be still more apparent if the speed of rotation is very great. Again, four points not in the same plane can be so situated that they preserve their relative position very tenaciously; & that too, when we make use of but a single distance corresponding to a limit-point of sufficient strength. For they can form a regular pyramid, of which each of the sides of the triangles is of a length equal to this distance. Then this pyramid will constitute a particle that is most tenacious as regards its form; & this will be able to act upon points, or pyramids of the same kind, that are more remote, in such a manner that its points do not alter their relative position in the slightest degree for all practical purposes. From four particles of this kind, arranged to form a larger pyramid, we can obtain a particle of the second order, somewhat less tenacious of form on account of the greater distance between the particles of the first order that compose it;

tium, qua fit, ut vires in easdem ab externis punctis impressæ multo magis inæquales inter se sint, quam fuerint in punctis constituentibus particulas ordinis primi; ac eodem pacto ex his secundi ordinis particulis fieri possunt particulæ ordinis tertii adhuc minus tenaces figuræ suæ, atque ita porro, donec ad eas deventum sit multo majores, sed adhuc multo magis mobiles, atque variabiles, ex quibus pendent chemicæ operationes, & ex quibus hæc ipsa crassiora corpora componuntur, ubi id ipsum accideret, quod Newtonus in postrema Opticæ questione proposuit de particulis suis primigeneis, & elementaribus, alias diversorum ordinum particulas efformantibus. Sed de particularibus hisce systematis determinati punctorum numeri jam satis, ac ad massas potius generaliter considerandas faciemus gradum.

Transitus ad massas: quid centrum gravitatis: theoremata hic de eo demonstrando.

240. In massis primum nobis se offerunt considerandæ elegantissimæ sane, ac & fœcundissimæ, & utilissimæ proprietates centri gravitatis, quæ quidem e nostra Theoria sponte propemodum fluunt, aut saltem ejus ope evidentissime demonstrantur. Porro centrum gravitatis a gravium æquilibrio nomen accepit suum, a quo etiam ejus consideratio ortum duxit; sed id quidem a gravi-[112]-tate non pendet, sed ad massam potius pertinet. Quamobrem ejus definitionem proferam ab ipsa gravitate nihil omnino pendentem, quanquam & nomen retinebo, & innuam, unde originem duxerit; tum demonstrabo accuratissime, in quavis massa haberi aliquod gravitatis centrum, idque unicum, quod quidem passim omittere solent, & perperam; deinde ad ejus proprietatem præcipuam exponendam gradum faciam, demonstrando celeberrimum theorema a Newtono propositum, centrum gravitatis commune massarum, sive mihi punctorum quotcunque, & utcunque dispositorum, quorum singula moveantur sola inertiæ vi motibus quibuscunque, qui in singulis punctis uniformes sint, in diversis utcunque diversi, vel quiescere, vel moveri uniformiter in directum: tum vero mutuas actiones quascunque inter puncta quælibet, vel omnia simul, nihil omnino turbare centri communis gravitatis statum quiescendi vel movendi uniformiter in directum, unde nobis & actionis, ac reactionis æqualitas in massis quibusque, & principia collisiones corporum definientia, & alia plurima sponte provenient. Sed aggrediamur ad rem ipsam.

Definitio centri gravitatis non pendens ab idea gravitatis: ejus congruentia cum idea communi.

241. Centrum igitur commune gravitatis punctorum quotcunque, & utcunque dispositorum, appellabo id punctum, per quod si ducatur planum quodcunque; summa distantiarum perpendicularium ab eo plano punctorum omnium jacentium ex altera ejusdem parte, æquatur summa distantiarum ex altera. Id quidem extenditur ad quascunque, & quotcunque massas; nam corum singulæ punctis utique constant, & omnes simul sunt quædam punctorum diversorum congeries. Nomen traxit ab æquilibrio gravium, & natura vectis, de quibus agemus infra: ex iis habetur illud, singula pondera ita connexa per virgas inflexiles, ut moveri non possint, nisi motu circa aliquem horizontalem axem, exerere ad conversionem vim proportionalem sibi, & distantiæ perpendiculari a plano verticali ducto per axem ipsum; unde fit, ut ubi ejusmodi vires, vel, ut ea vocant, momenta virium hinc, & inde æqualia fuerint, habeatur æquilibrium. Porro ipsa pondera in nostris gravibus, in quibus gravitatem concipimus, ac etiam ad sensum experimur, proportionalem in singulis quantitati materiæ, & agentem directionibus inter se parallelis, proportionalia sunt massis, adeoque punctorum eas constituentium numero; quam ob rem idem est, ea pondera in distantias ducere, ac assumere summam omnium distantiarum omnium punctorum ab eodem plano. Quod si igitur respectu aggregati cujuscunque punctorum materiæ quotcunque, & quomodocunque dispositorum sit aliquod punctum spatii ejusmodi, ut, ducto per ipsum quovis plano, summa distantiarum ab illo punctorum jacentium ex parte altera æquetur summæ distantiarum jacentium ex altera; concipiantur autem singula ea puncta animata viribus æqualibus, & parallelis, cujusmodi sunt vires, quas in nostris gravibus concipimus; illud utique consequitur, [113] suspenso utcunque ex ejusmodi puncto, quale definivimus gravitatis centrum, omni eo systemate, cujus systematis puncta viribus quibuscunque, vel conceptis virgis inflexibilus, & gravitate carentibus, positionem mutuam, & respectivum statum, ac distantias omnino servent, id systema fore in æquilibrio; atque illud ipsum requiri, ut in æquilibrio sit. Si enim vel unicum planum ductum per id punctum sit ejusmodi, ut summæ illæ distantiarum non sint æquales hinc, & inde; converso systemate omni ita, ut illud planum evadat verticale, jam non essent æquales inter se summæ momentorum hinc, & inde, & altera pars alteri præponderaret. Verum hæc quidem, uti supra monui, fuit occasio quædam nominis imponendi; at ipsum punctum ea lege determinatum longe ulterius extenditur, quam

for from this fact it comes about that the forces impressed upon these from external points are much more unequal to one another than they would be for the points constituting particles of the first order. In the same manner, from these particles of the second order we might obtain particles of the third order, still less tenacious of form, & so on; until at last we reach those which are much greater, still more mobile, & variable particles, which are concerned in chemical operations; & to those from which are formed the denser bodies, with regard to which we get the very thing set forth by Newton, in his last question in Optics, with respect to his primary elemental particles, that form other particles of different orders. We have now, however, said enough concerning these systems of a definite number of points, & we will proceed to consider masses rather more generally.

Passing on to masses; what is a centre of gravity? Theorems to be proved concerning it at this point.

240. In dealing with masses, the first matters that present themselves for our consideration are certain really very elegant, as well as most fertile & useful properties of the centre of gravity. These indeed come forth almost spontaneously from my Theory, or at least are demonstrated most clearly by means of it. Further, the centre of gravity derived its name from the equilibrium of heavy (*gravis*) bodies, & the first results in connection with the former were developed by means of the latter; but in reality it does not depend on gravity, but rather is related to masses. On this account, I give a definition of it, which in no way depends on gravity, although I will retain the name, & will mention whence it derived its origin. Then I will prove with the utmost rigour that in every body there is a centre of gravity, & one only (a thing which is usually omitted by everybody, quite unjustifiably). Then I will proceed to expound its chief property, by proving the well-known theorem enunciated by Newton; that the centre of gravity of masses, or, in my view, of any number of points in any positions, each of which is moved in any manner by the force of inertia alone, this force being uniform for the separate points but maybe non-uniform to any extent for different points, will be either at rest or will move uniformly in a straight line. Finally, I will show that any mutual action whatever between the points, or all of them taken together, will in no way disturb the state of rest or of uniform motion in a straight line of the centre of gravity. From which the equality of action & reaction in all bodies, & the principles governing the collision of solids, & very many other things will arise of themselves. However let us set to work on the matter itself.

Definition of the centre of gravity independent of any idea of gravitation; the agreement of this definition with the usual idea.

241. Accordingly, I will call the common centre of gravity of any number of points, situated in any positions whatever, that point which is such that, if through it any plane is drawn, the sum of the perpendicular distances from the plane of all the points lying on one side of it is equal to the sum of the distances of all the points on the other side of it. The definition applies also to masses, of any sort or number whatever; for each of the latter is made up of points, & all of them taken together are certain groups of different points. The name is taken from the equilibrium of weights (*gravis*), & from the principle of the lever, with which we shall deal later. Hence we obtain the principle that each of the weights, connected together by rigid rods in such a manner that the only motion possible to them is one round a horizontal axis, will exert a turning force proportional to itself & to its perpendicular distance from a vertical plane drawn through this axis. From which it comes about that, when the forces of this sort (or, as they are called, the moments of the forces) are equal to one another on this side & on that, then there is equilibrium. Further, the weights in our heavy bodies, in which we conceive the existence of gravity (& indeed find by experience that there is such a thing) proportional in each to the quantity of matter, & acting in directions parallel to one another, are proportional to the masses, & thus to the number of points that go to form them. Therefore, the product of the weights into the distances comes to the same thing as the sum of all the distances of all the points from the plane. If then, for an aggregate of points of matter, of any sort & number whatever, situated in any way, there is a point of space of such a nature that, for any plane drawn through it, the sum of the distances from it of all points lying on one side of it is equal to the sum of the distances of all the points lying on the other side of it; if moreover each of the points is supposed to be endowed with a force, & these forces are all equal & parallel to one another, & of such a kind as we conceive the forces in our weights to be; then it follows directly that, if the whole of this system is suspended in any way from a point of the sort we have defined the centre of gravity to be, the points of the system, on account of certain assumed forces or rigid weightless rods, preserving their mutual position, their relative state & their distances absolutely unchanged, then the system will be in equilibrium. Such a point is to be found, in order that the system may be in equilibrium. For, if any one plane can be drawn through the point, such that the sum of the distances on the one side are not equal to those on the other side, & the whole system is turned so that this plane becomes vertical, then the sums of the moments will not be equal to one another on each side, but one part will outweigh the other part. This indeed, as I said above, was the idea that gave rise to the term centre of gravity; but the point determined by this rule has

ad solas massas animatas viribus æqualibus, & parallelis, cujusmodi concipiuntur a nobis in nostris gravibus, licet ne in ipsis quidem accurate sint tales. Quamobrem assumpta superiore definitione, quæ a gravitatis, & æquilibrii natura non pendet, progrediar ad deducenda inde corollaria quaædam, quæ nos ad ejus proprietates demonstrandas deducant.

Corollarium generale pertinens ad summas distantiarum omnium punctorum massæ a plano transeunte per centrum gravitatis æquales utrinque.

242. Primo quidem si aliquod fuerit ejusmodi planum, ut binæ summæ distantiarum perpendicularium punctorum omnium hinc & inde acceptorum æquenter inter se : æquabuntur & summæ distantiarum acceptarum secundum quancunque aliam directionem datam, & communem pro omnibus. Erit enim quævis distantia perpendicularis ad quanvis in dato angulo inclinatam semper in eadem ratione, ut patet. Quare & summæ illarum ad harum summas erunt in eadem ratione, ac æqualitas summarum alterius binarii utriuslibet secum trahet æqualitatem alterius. Quare in sequentibus, ubi distantias nominavero, nisi exprimam perpendiculares, intelligam generaliter distantias acceptas in quavis directione data.

Bina theoremata pertinentia ad planum parallelum plano distantiarum æqualium cum eorum demonstrationibus.

243. Quod si assumatur planum aliud quodcunque parallelum plano habenti æquales hinc, & inde distantiarum summas ; summa distantiarum omnium punctorum jacentium ex parte altera superabit summam jacentium ex altera, excessu æquali distantiæ planorum acceptæ secundum directionem eandem ductæ in numerum punctorum : & vice versa si duo plana parallela sint, ac is excessus alterius summæ supra summam alterius in altero ex iis æquetur eorum distantiæ duetæ in numerum punctorum ; planum alterum habebit oppositarum distantiarum summas æquales. Id quidem facile concipitur ; si concipiatur, planum distantiarum æqualium moveri versus illud alterum planum motu parallelo secundum eam directionem, secundum quam sumuntur distantiæ. In eo motu distantiæ singulæ ex altera parte crescunt, ex altera decrescunt continuo tantum, quantum promovetur planum, & si aliqua distantia evanescit interea ; jam eadem deinde incipit tantundem ex parte contraria crescere. Quare patet excessum omnium citeriorum [114] distantiarum supra omnes ulteriores æquari progressui plani toties sumpto, quot puncta habentur, & in regressu destruitur e contrario, quidquid in ejusmodi progressu est factum, atque idcirco ad æqualitatem reditur. Verum ut demonstratio quam accuratissima evadat, exprimat in fig. 36 recta AB planum distantiarum æqualium, & CD planum ipsi parallelum, ac omnia puncti distribui poterunt in classes tres, in quorum prima sint omnia jacentia citra utrumque planum, ut punctum E ; in secunda omnia puncta jacentia inter utrumque, ut F, in tertia omnia puncta adhuc jacentia ultra utrumque, ut G. Rectæ autem per ipsa ductæ in directione data quacunque, occurrant rectæ AB in M, H, K, & rectæ CD in N, I, L ; ac sit quædam reacta directionis ejusdem ipsis AB, CD occurrens in O, P. Patet, ipsam OP fore æqualem ipsis MN, HI, KL. Dicatur jam summa omnium punctorum E primæ classis E, & distantiarum omnium EM summa e ; punctorum F secundæ classis F, & distantiarum f ; punctorum G tertiæ classis summa G, & distantiarum earundem g ; distantia vero OP dicatur O. Patet, summam omnium MN fore $E \times O$; summam omnium HI fore $F \times O$; summam omnium KL fore $G \times O$; erit autem quævis $EN = EM + MN$; quævis $FI = HI - FH$; quævis $GL = KG - KL$. Quare summa omnium EN erit $e + E \times O$; summa omnium $FI = F \times O - f$, & summa omnium $GL = g - G \times O$; adeoque summa omnium distantiarum punctorum jacentium citra planum CD, primæ nimirum, ac secundæ classis, erit $e + E \times O + F \times O - f$, & summa omnium jacentium ultra, nimirum classis tertiæ, erit $g - G \times O$. Quare excessus prioris summæ supra secundam erit $e + E \times O + F \times O - f - g + G \times O$; adeoque si prius fuerit $e = f + g$; deleto $e - f - g$, totus excessus erit $E \times O + F \times O + G \times O$, sive $(E + F + G) \times O$, summa omnium punctorum dueta in distantiam planorum ; & vice versa si is excessus respectu secundi plani CD fuerit æqualis huic summæ duetæ in distantiam O, oportebit $e - f - g$ æquetur nihilo, adeoque sit $e = f + g$, nimirum respectu primi plani AB summas distantiarum hinc, & inde æquales esse.

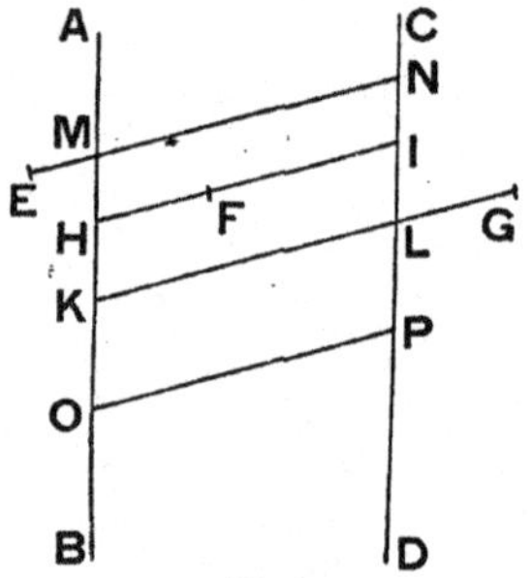

Fig. 36.

Complementum demonstrationis ut extendatur ad omnes casus.

244. Si aliqua puncta sint in altero ex iis planis, ea superioribus formulis contineri possunt, concepta *zero* singulorum distantia a plano, in quo jacent ; sed & ii casus involvi facile possent, concipiendo alias binas punctorum classes ; quorum priora sint in priore plano AB, posteriora in posteriore CD, quæ quidem nihil rem turbant : nam prioris classis

a far wider application than to the single cases of mass endowed with equal & parallel forces such as we have assumed to exist in our heavy bodies; & indeed such do not exist accurately even in the latter. Hence, taking the definition given above, which is independent of gravity & the nature of equilibrium of weights, I will proceed to deduce from it certain corollaries, which will enable us to demonstrate the properties of the centre of gravity.

General corollary relating to the sums of the distances of all the points of a mass, from a plane passing through the centre of gravity, being equal on either side of it.

242. First of all, then, if there should be any plane such that the two sums of the perpendicular distances of all the points on either side of it taken together are equal to one another, then the sums of the distances taken together in any other given direction, that is the same for all of them, will also be equal to one another. For, any perpendicular distance will evidently be in the same ratio to the corresponding distance inclined at a given angle. Hence the sums of the former distances will bear the same ratio to the sums of the latter distances; & therefore the equality of the sums in either of the two cases will involve the equality of the sums for the other also. Consequently, in what follows, whenever I speak of distances, I intend in general distances in any given direction, unless I expressly say that they are perpendicular distances.

Two theorems relating to a plane parallel to the plane of equal distances; & their demonstrations.

243. If now we take any other plane parallel to the plane for which the sums of the distances on either side are equal, then the sum of the distances of all the points lying on the one side of it will exceed the sum for those lying on the other side by an amount equal to the distance between the two planes measured in the like direction multiplied by the number of all the points. Conversely, if there are two parallel planes, & if the excess of the sum of the distances from one of them over the sum of the distances from the other is equal to the distance between the planes multiplied by the number of the points, then the second plane will have the sums of the opposite distances equal to one another. This is easily seen to be true; for, if the plane of equal distances is assumed to be moved towards the other plane by a parallel motion in the direction in which the distances the measured, then as the plane is moved each of the distances on the one side increase, & those on the other side decrease by just the amount through which the plane is moved; & should any distance vanish in the meantime, there will be an increase on the other side of just the same amount. Thus, it is evident that the excess of all the distances on the near side above the sum of all the distances on the far side will be equal to the distance through which the plane has been moved, taken as many times as there are points. On the other hand, when the plane is moved back again, this excess is destroyed, namely exactly the amount that was produced as the plane moved forward, & consequently equality will be restored. But to give a more rigorous demonstration, let the straight line AB, in Fig. 36, represent the plane of equal distances, & let CD represent a plane parallel to it. Then all the points can be grouped into three classes; let the first of these be that in which we have every point that lies on the near side of both the planes, as E; let the second be that in which every point lies between the two planes, as F; & the third, every point lying on the far side of both planes, as G. Let straight lines, drawn in any given direction whatever, through the points meet AB in M, H, K, & the straight line CD in N, I, L; also let any straight line, drawn in the same direction, meet AB, CD in O & P. Then it is clear that OP will be equal to MN, HI, or KL. Now, let us denote the sum of all the points of the first class, like E, by the letter E, & the sum of all the distances like EM by the letter e; & those of the second class by the letters F & f; those of the third class by G & g; & the distance OP by O. Then it is evident that the sum of all the MN's will be $E \times O$; the sum of all the HI's will be $F \times O$; the sum of all the KL's will be $G \times O$; also in every case, $EN = EM + MN$, $FI = HI - FH$, & $GL = KG - KL$. Hence the sum of the EN's will be $e + E \times O$, the sum of the FI's will be $F \times O - f$, & the sum of the GL's will be $g - G \times O$. Hence, the sum of all the distances of the points lying on the near side of the plane CD, that is to say, those belonging to the first & second classes, will be equal to $e + E \times O + F \times O - f$; & the sum of all those lying on the far side, that is, of the third class, will be equal to $g - G \times O$. Hence, the excess of the former over the latter will be equal to $e + E \times O + F \times O - f - g + G \times O$. Therefore, if at first we had $e = f + g$, then, on omitting $e - f - g$, we have the total excess equal to $E \times O + F \times O + G \times O$, or $(E + F + G) \times O$, i.e., the sum of all the points multiplied by the distance between the planes. Conversely, if the excess with respect to the second plane CD were equal to this sum multiplied by the distance O, it must be that $e - f - g$ is equal to nothing, & thus $e = f + g$; in other words the sum of the distances with respect to the first plane AB must be equal on one side & the other.

Completion of the proof, so as to include all possible cases.

244. If any of the points should be in one or other of the two planes, these may also be included in the foregoing formulæ, if we suppose that the distance for each of them is zero distance from the plane in which they lie. Then these cases may also be included by considering that there are two fresh classes of points; namely, first those lying in the first plane AB, & secondly those lying in the second plane CD; & these classes will in

distantiæ a priore plano crunt omnes simul *zero*, & a posteriore æquabuntur distantiæ O duetæ in eorum numerum, quæ summa accedit priori summæ punctorum jacentium citra; posterioris autem classis distantiæ a priore erant prius simul æquales summæ ipsorum duetæ itidem in O, & deinde fiunt nihil; adeoque [115] summæ distantiarum punctorum jacentium ultra, demitur horum posteriorum punctorum summa itidem dueta in O, & proinde excessui summæ citeriorum supra summam ulteriorum accedit summa omnium punctorum harum duarum classium dueta in eandem O.

Theoremata pro plano posito ultra omnia puncta: eorum extensio ad quævis plana.

245. Quod si planum parallelum plano distantiarum æqualium jaceat ultra omnia puncta; jam habebitur hoc theorema: *Summa omnium distantiarum punctorum omnium ab eo plano æquabitur distantiæ planorum ductæ in omnium punctorum summam, & si fuerint duo plana parallela ejusmodi, ut alterum jaceat ultra omnia puncta, & summa omnium distantiarum ab ipso æquetur distantiæ planorum ductæ in omnium punctorum numerum; alterum illud planum erit planum distantiarum æqualium.* Id sane patet ex eo, quod jam secunda summa pertinens ad puncta ulteriora, quæ nulla sunt, evanescat, & excessus totus sit sola prior summa. Quin immo idem theorema habebit locum pro quovis plano habente etiam ulteriora puncta, si citeriorum distantiæ habeantur pro positivis, & ulteriorum pro negativis; cum nimirum summa constans positivis, & negativis sit ipse excessus positivorum supra negativa; quo quidem pacto licebit considerare planum distantiarum æqualium, ut planum, in quo summa omnium distantiarum sit nulla, negativis nimirum distantiis elidentibus positivas.

Cuivis plano inveniri posse parallelum planum distantiarum æqualium.

246. Hinc autem facile jam patet, *dato cuivis plano haberi aliquod planum parallelum, quod sit planum distantiarum æqualium; quin immo data positione punctorum, & plano illo ipso, facile id alterum definitur.* Satis est ducere a singulis punctis datis rectas in data directione ad planum datum, quæ dabuntur; tum a summa omnium, quæ jacent ex parte altera, demere summam omnium, si quæ sunt, jacentium ex opposita, ac residuum dividere per numerum punctorum. Ad eam distantiam dueto plano priori parallelo, id erit planum quæsitum distantiarum æqualium. Patet autem admodum facile & illud ex eadem demonstratione, & ex solutione superioris problematis, dato cuivis plano non nisi unicum esse posse planum distantiarum æqualium, quod quidem per se satis patet.

Theorema præcipuum si tria plana distantiarum æqualium habeant unicum punctum commune; reliqua omnia per id transeuntia erunt ejusmodi.

247. Hisce accuratissime demonstratis, atque explicatis, progrediar ad demonstrandum haberi aliquod gravitatis centrum in quavis punctorum congerie, utcunque dispersorum, & in quotcunque massas ubicunque sitas coalescentium. Id fiet ope sequentis theorematis; *si per quoddam punctum transeant tria plana distantiarum æqualium se non in eadem communi aliqua recta secantia; omnia alia plana transeuntia per illud idem punctum erunt itidem distantiarum æqualium plana.* Sit enim in fig. 37, ejusmodi punctum C, per quod transeant tria plana GABH, XABY, ECDF, quæ omnia sint plana distantiarum æqualium, ac sit quodvis aliud planum KICL tran-[116]-siens itidem per C, ac secans primum ex iis recta CI quacunque; oportet ostendere, hoc quoque fore planum distantiarum æqualium, si illa priora ejusmodi sint. Concipiatur quodcunque punctum P; & per ipsum P concipiatur tria plana parallela planis DCEF, ABYX, GABH, quorum sibi priora duo mutuo occurrant in recta PM, postrema duo in recta PV, primum cum tertio in recta PO: ac primum occurrat plano GABH in MN, secundum vero eidem in MS, plano DCEF in QR, ac plano CIKL in SV, ducaturque ST parallela rectis QR, MP, quas, utpote parallelorum planorum intersectiones, patet fore itidem parallelas inter se, uti & MN, PO, DC inter se, ac MS, PTV, BA inter se.

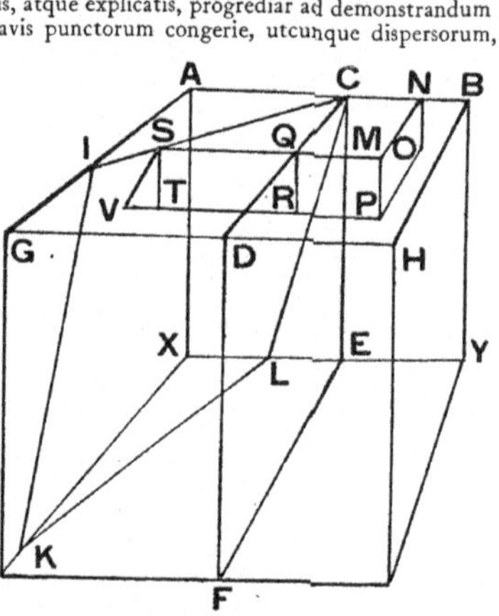

Fig. 37.

Demonstratio ejusdem.

248. Jam vero summa omnium dis antiarum a plano KICL secundum datam directionem BA erit summa omnium PV, quæ resolvitur in tres summas, omnium PR, omnium RT, omnium TV, sive eæ, ut figura exhibet in unam colligendæ sint, sive, quod in aliis plani novi inclinationibus posset accidere, una ex iis demenda a reliquis binis, ut habeatur omnium PV summa. Porro quævis PR est distantia a plano DCEF secundum candem eam directionem; quævis RT est æqualis QS sibi respondenti, quæ ob datas directiones laterum trianguli SCQ est ad CQ, æqualem MN, sive PO, distantiæ a plano XABY secundum

no way cause any difficulty. For the distances of the points of the first class from the first plane, all together, will be zero, & their distances from the second plane will, all together, be equal to the distance O multiplied by the number of them ; & this sum is to be added to the former sum for the points lying on the near side. Again, the distances of the points of the second class from the first plane were, all together, at first equal to the distance O multiplied by their number, & then are nothing for the second plane. Hence from the sum of the distances of the points lying on the far side, we have to take away the sum of these last points also multiplied by the distance O ; & thus, to the excess of the sum of the points on the near side over the sum of the points on the far side we have to add the sum of all the points in these two classes multiplied by the same distance O.

Theorems for a plane lying beyond all the points; extension of these theorems to any plane whatever.

245. Now, if the plane parallel to the plane of equal distances should lie on the far side of all the points then the following theorem is obtained. *The sum of all the distances of all the points from this plane will be equal to the distance between the planes multiplied by the sum of all the points ; & if there were two parallel planes, such that one of them lies beyond all the points, & if the sum of all the distances from this plane is equal to the distance between the planes multiplied by the number of points, then the other plane will be the plane of equal distances.* This is perfectly clear from the fact that in this case the second sum relating to the points that lie beyond the planes vanishes, for there are no such points, & the whole excess corresponds to the first sum alone. Further, the same theorem holds good for any plane even if there are points beyond it, if the distances of points on the near side of it are reckoned as positive & those on the far side as negative ; for the sum formed from the positives & the negatives is nothing else but the excess of the positives over the negatives. In precisely the same manner, we may consider the plane of equal distances to be a plane for which the sum of all the distances is nothing, that is to say, the positive distances cancel the negative distances.

Given any plane, there can be found a plane of equal distances, parallel to it.

246. From the foregoing theorem it is now clear that *for any given plane there exists another plane parallel to it, which is a plane of equal distances ; further, if we are given the position of the points, & also the plane is given, then the parallel plane is easily determined.* It is sufficient to draw from each of the points straight lines in a given direction to the given plane, & then these are all given ; then from the sum of all of them that lie on the one side to take away the sum of all those that lie on the other side, if any such there are ; & lastly to divide the remainder by the number of the points. If a plane is drawn parallel to the first plane, & at a distance from it equal to the result thus found, then this plane will be a plane of equal distances, as was required. Moreover it can be seen quite clearly, & that too from the very demonstration just given, that to any given plane there can correspond but one single plane of equal distances ; indeed this is sufficiently self-evident without proof.

The important theorem, that, if three planes of equal distances have a common point, then any other plane through the point will be of the same nature.

247. Now that the foregoing theorems have received rigorous demonstrations & explanation, I will proceed to prove that there is a centre of gravity for any set of points, no matter how they are dispersed or what the number of masses may be into which they coalesce, or where these masses may be situated. The proof follows from the theorem :— *If through any point there pass three planes of equal distances that do not all cut one another in some common line then all other planes passing through this same point will also be planes of equal distances.* In Fig. 37, let C be a point of this sort, & through it suppose that three planes, GABH, XABY, ECDF, pass ; also suppose that all the planes are planes of equal distances. Let KICL be any other plane passing through C also, & cutting the first of the three planes in any straight line CI ; we have to prove that this latter plane is a plane of equal distances, if the first three are such planes. Take any point P ; & through P suppose three planes to be drawn parallel to the planes DCEF, ABYX, GABH ; let the first two of these meet one another in the straight line PM, the last two in the straight line PV, & the first & third in the straight line PO. Also let the first meet the plane GABH in the straight line MN, the second meet this same plane in MS, & the plane DCEF in QR, the plane CIKL in SV, & let ST be drawn parallel to the straight lines QR & MP, which, since they are intersections with parallel planes, are parallel to one another ; similarly MN, PO, DC are parallel to one another, as also are MS, PTV & BA parallel to one another.

Proof of the theorem.

248. Now, the sum of all the distances from the plane KICL, in the given direction BA, will be equal to the sum of all the PV's ; & this can be resolved into the three sums, that of all the PR's, that of all the RT's, & that of all the TV's ; whether these, as are shown in the figure, have to be all collected into one whole, or, as may happen for other inclinations of a fresh plane, whether one of the sums has to be taken away from the other two, to give the sum of all the PV's. Now each PR is the distance of a point P from the plane DCEF, measured in the given direction ; & each RT is equal to the QS that corresponds to it, which, on account of the given directions of the sides of the triangle SCQ bears a given ratio to CQ , the latter being equal to MN or PO, the distance of P from the plane

datam directionem DC, in ratione data; & quævis VT est itidem in ratione data ad TS æqualem PM, distantiæ a plano GABH secundum datam directionem EC; ac idcirco etiam nulla ex ipsis PR, RT, TV poterit evanescere, vel directione mutata abire e positiva in negativam, aut vice versa, mutato situ puncti P, nisi sua sibi respondens ipsius puncti P distantia ex iis PR, PO, PM evanescat simul, aut directionem mutet. Quamobrem & summa omnium positivarum vel PR, vel RT, vel TV ad summam omnium positivarum vel PR, vel PO, vel PM, & summa omnium negativarum prioris directionis ad summam omnium negativarum posterioris sibi respondentis, crit itidem in ratione data; ac proinde si omnes positivæ directionum PR, PO, PM a suis negativis destruuntur in illis tribus æqualium distantiarum planis, etiam omnes positivæ PR, RT, TV a suis negativis destruentur, adeoque & omnes PV positivæ a suis negativis. Quamobrem planum LCIK erit planum distantiarum æqualium. Q.E.D.

[Haberi semper aliquod gravitatis centrum, atque id esse unicum.]

249. Demonstrato hoc theoremate jam sponte illud consequitur, *in quavis punctorum congerie, adeoque massarum utcunque dispersarum summa, haberi semper aliquod gravitatis centrum, atque id esse unicum, quod quidem data omnium punctorum positione facile determinabitur.* Nam assumpto puncto quovis ad arbitrium ubicunque, ut puncto P, poterunt duci per ipsum tria plana quæcunque, ut OPM, RPM, RPO. Tum singulis poterunt per num. 246 inveniri plana parallela, [117] quæ sint plana distantiarum æqualium, quorum priora duo si sint DCEF, XABY, se secabunt in aliqua recta CE parallela illorum intersectioni MP; tertium autem GABH ipsam CE debebit alicubi secare in C; cum planum RPO secet PM in P: nam ex hac sectione constat, hanc reetam non esse parallelam huic plano, adeoque nec illa illi crit, sed in ipsum alicubi incurret. Transibunt igitur per punctum C tria plana distantiarum æqualium, adeoque per num. 247 & aliud quodvis planum transiens per punctum idem C crit planum æqualium distantiarum pro quavis directione, & idcirco etiam pro distantiis perpendicularibus; ac ipsum punctum C juxta definitionem num. 241, erit commune gravitatis centrum omnium massarum, sive omnis congeriei punctorum, quod quidem esse unicum, facile deducitur ex definitione, & hac ipsa demonstratione; nam si duo essent, possent utique per ipsa duci duo plana parallela directionis cujusvis, & corum utrumque esset planum distantiarum æqualium, quod est contra id, quod num. 246 demonstravimus.

Necessitas demonstrandi haberi semper centrum gravitatis.

250. Demonstrandum necessario fuit, haberi aliquod gravitatis centrum, atque id esse unicum; & perperam id quidem a Mechanicis passim omittitur; si enim id non ubique adesset, & non esset unicum, in paralogismum incurrerent quamplurimæ Mechanicorum ipsorum demonstrationes, qui ubi in plano duas invenerunt rectas, & in solidis tria plana determinantia æquilibrium, in ipsa intersectione constituunt gravitatis centrum, & supponunt omnes alias rectas, vel omnia alia plana, quæ per id punctum ducantur, eandem æquilibrii proprietatem habere, quod utique fuerat non supponendum, sed demonstrandum. Et quidem facile est similis paralogismi exemplum præbere in alio quodam, quod magnitudinis centrum appellare liceret, per quod nimirum figura sectione quavis secaretur in duas partes æquales inter se, sicut per centrum gravitatis secta, secatur in binas partes æquilibratas in hypothesi gravitatis constantis, & certam directionem habentis plano secanti parallelam.

Centrum enim magnitudinis non semper haberi.

251. Erraret sane, qui ita definiret centrum magnitudinis, tum determinaret id ipsum in datis figuris eadem illa methodo, quæ pro centri gravitatis adhibetur. Is ex. gr. pro triangulo ABG in fig. 38 sic ratiocinationem institueret. Secetur AG bifariam in D, ducaturque BD, quæ utique ipsum triangulum secabit in duas partes æquales. Deinde, secta AB itidem bifariam in E, ducatur GE, quam itidem constat, debere secare triangulum in partes æquales duas. In earum igitur concursu C habebitur centrum magnitudinis. Hoc invento si progrederetur ulterius, & haberet pro æqualibus partes, quæ alia sectione quacunque facta per C obtinentur; erraret pessime. Nam dueta ED, jam constat, fore ipsam ED parallelam BG, & ejus dimidiam; adeoque similia fore triangula [118] ECD, BCG, & CD dimidiam CB. Quare si per C ducatur FH parallela AG; triangulum FBH, crit ad ABG, ut quadratum BC ad quadratum BD, seu ut 4 ad 9, adeoque segmentum FBH ad residuum FAGH est ut 4 ad 5, & non in ratione æqualitatis.

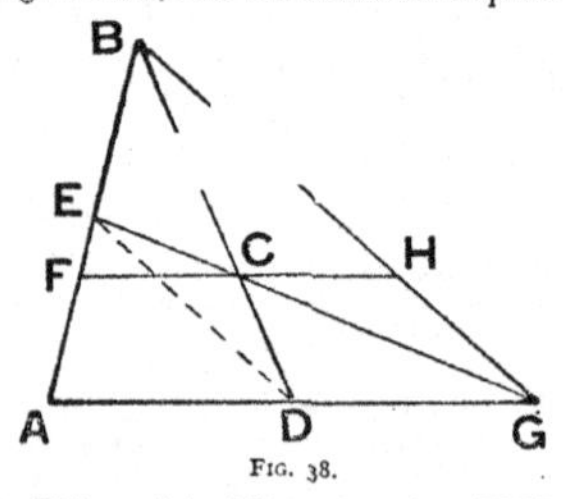

FIG. 38.

Ubi hæc primo demonstrata.

252. Nimirum quæcunque punctorum, & massarum congeries, adeoque & figura quævis, in qua concipiatur punctorum numerus auctus in infinitum, donec figura ipsa evadat continua, habet suum gravitatis centrum; centrum magnitudinis infinitæ earum non habent; & illud primum, quod hic accuratissime demonstravi, demonstraveram jam

XABY, measured in the given direction DC; lastly, VT is also in a given ratio to TS, the latter being equal to PM, the distance of the point P from the plane GABH, measured in the given direction EC. Hence, none of the distances PR, RT, TV can vanish or, having changed their directions, pass from positive to negative, or vice versa, by a change in the position of the point P, unless that one of the distances PR, PO, PM, of the point P, which corresponds to it vanishes or changes its direction at the same time. Therefore also the sum of all the positives, whether PR, or RT, or TV to the sum of all the positives, PR, or PO, or PM, & the sum of all the negatives for the first direction to the sum of all the negatives for the second direction which corresponds to it, will also be in a given ratio. Thus, finally, if all the positives out of the direction PR, PO, PM are cancelled by the corresponding negatives in the case of the three planes of equal distances; then also all the positive PR's, RT's, TV's are cancelled by their corresponding negatives, & therefore also all the positive PV's are cancelled by their corresponding negatives. Consequently, the plane LCIK will be a plane of equal distances. Q.E.D.

There is always one centre of gravity; and only one.

249. Now that we have demonstrated the above theorem, it follows immediately from it that, *for any group of points, & therefore also for a set of masses scattered in any manner, there exists a centre of gravity, & there is but one; & this can be easily determined when the position of each of the points is given.* For if a point is taken at random anywhere, like the point P there could be drawn through it any three planes, OPM, RPM, RPO. Then corresponding to each of these there could be found, by Art. 245, a parallel plane, such that these planes were planes of equal distances. If the first two of these are DCEF & XABY, they will cut one another in some straight line CE parallel to their intersection MP; also the third plane GABH must cut this straight line CE somewhere in C; for the plane RPO will cut PM in P, & from this fact it follows that the latter line is not parallel to the latter plane, & therefore the former line is not parallel to the former plane, but will cut it somewhere. Hence three planes of equal distances will pass through the point C, & therefore, by Art. 247, any other plane passing through this point C will also be a plane of equal distances for any direction, & thus also for perpendicular distances. Hence, according to the definition of Art. 241, the point C will be the common centre of gravity of all the masses, or of the whole group of points. That there is only one can be easily derived from the definition & the demonstration given; for, if there were two, there could in every case be drawn through them two parallel planes in any direction, & each of these would be a plane of equal distances; which is contrary to what we have proved in Art. 246.

The need for proving that there is a centre of gravity in every case.

250. It was absolutely necessary to prove that there always exists a centre of gravity, & that there is only one in every case; & this proof is everywhere omitted by Mechanicians, quite unjustifiably. For, if there were not one in every case, or if it were not unique, very many of the proofs given by these Mechanicians would result in fallacious argument. Where, for instance, they find two straight lines, in the case of a plane, & in the case of solids three planes, determining equilibrium, & suppose that all other lines, & all other planes, which are drawn through the point to have the same property of equilibrium; this in every case ought not to be a matter of supposition, but of proof. Indeed it is easy to give a similar example of fallacious argument in the case of something else, which we may call the centre of magnitude; for instance, where a figure is cut, by any section, into two parts equal to one another; just as when the section passes through the centre of gravity it is cut into two parts that balance one another, on the hypothesis of uniform gravitation acting in a fixed direction parallel to the cutting plane.

For there is not always a centre of magnitude.

251. He would indeed be much at fault, who would so define the centre of magnitude & then proceed to determine it in given figures by the same method as that used for the centre of gravity. For example, the reasoning he would use for the triangle ABG, in Fig. 38, would be as follows. Let AB be bisected in D, & through D draw BD; this will certainly divide the triangle into two equal parts. Then, having bisected AB also in E, draw GE; it is true that this also divides the triangle into two equal parts. Hence their point of intersection C will be the centre of magnitude. If then, having found this, he proceeded further, & said that those parts were equal, which were obtained by any other section made through C; he would be very much in error. For, if ED is drawn, it is well known that we now have ED parallel to BG & equal to half of it; & therefore the triangles ECD, BCG would be similar, & CD half of CB. Hence, if FH is drawn through C parallel to AG, the triangle FBH will be to the triangle ABG, as the square on BC is to the square on BD, or as 4 is to 9; & thus the segment FBH is to the remainder FAGH as 4 is to 5, & not in a ratio of equality.

Where the first proof of this was given.

252. Thus, any group of points or masses, & therefore any figure in which the number of points is supposed to be indefinitely increased until the figure becomes continuous, possesses a centre of gravity; but there are an infinite number of them which have not got a centre of magnitude. The first of these, of which I have here given a rigorous

olim methodo aliquanto contractiore in dissertatione *De Centro Gravitatis*; hujus vero secundi exemplum hic patet, ac in dissertatione *De Centro Magnitudinis*, priori illi addita in secunda ejusdem impressione, determinavi generaliter, in quibus figuris centrum magnitudinis habeatur, in quis desit; sed ea ad rem præsentem non pertinent.

Inde ubi sit centrum commune massarum duarum.

253. Ex hac generali determinatione centri gravitatis facile colligitur illud, centrum commune binarum massarum jacere in directum cum centris gravitatis singularum, & horum distantias ab eodem esse reciproce, ut ipsas massas. Sint enim binæ massæ, quarum centra gravitatis sint in fig. 39 in A, & B. Si per reetam AB ducatur planum quodvis, id debet esse planum distantiarum æqualium respectu utriuslibet. Quare etiam respectu summæ omnium punctorum ad utrumque simul pertinentium distantiæ omnes hinc, & inde acceptæ æquantur inter se; ac proinde id etiam respectu summæ debet esse planum distantiarum æqualium, & centrum commune debet esse in quovis ex ejusmodi planis, adeoque in intersectione duorum quorumcunque ex iis, nimirum in ipsa recta AB. Sit id in C, & si jam concipiatur per C planum quodvis secans ipsam AB; crit summa omnium distantiarum ab eo plano secundum directionem AB punctorum pertinentium ad massam A, si a positivis demantur negativæ, æqualis per num. 243 numero punctorum massæ A dueto in AC, & summa pertinentium ad B numero punctorum in B dueto in BC; quæ producta æquari debent inter se, cum omnium distantiarum summæ positivæ a negativis elidi debeant respectu centri gravitatis C. Erit igitur AC ad CB, ut numerus punctorum in B ad numerum in A, nimirum in ratione massarum reciproca.

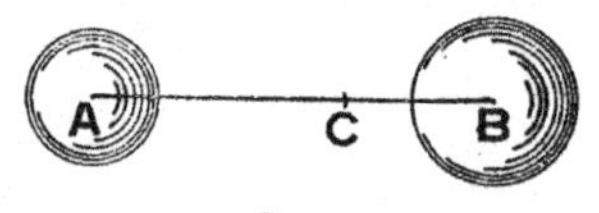

Fig. 39.

Inde & communis methodus pro quotcunque massis.

254. Hine autem facile deducitur *communis methodus inveniendi centrum gravitatis commune plurium massarum. Conjunguntur prius centra duarum, & eorum distantia dividitur in ratione reciproca ipsarum. Tum harum commune centrum sic inventum conjungitur cum centro tertiæ, & dividitur distantia in ratione reciproca summæ massarum priorum ad massam tertiam, & ita porro.* Quin immo *possunt seorsum inveniri centra gravitatis binarum quarumvis, ternarum, denarum quocunque* [119] *ordine, tum binaria conjungi cum ternariis, denariis, aliisque, ordine itidem quocunque, & semper eadem methodo devenitur ad centrum commune gravitatis massæ totius.* Id patet, quia quotcunque massæ considerari possunt pro massa unica, cum agatur de numero punctorum massæ tantummodo, & de summa distantiarum punctorum omnium; summæ massarum constituunt massam, & summæ distantiarum summam per solam conjunctionem ipsarum. Quoniam autem ex generali demonstratione superius facta devenitur semper ad centrum gravitatis, atque id centrum est unicum; quocunque ordine res peragatur, ad illud utique unicum devenitur.

Inde & theorema, ope cujus investigatur id in figuris continuis.

255. Inde vero illud consequitur, quod est itidem commune, *si plurium massarum centra gravitatis sint in eadem aliqua recta, fore etiam in eadem centrum gravitatis summæ omnium*; quod viam sternit ad investiganda gravitatis centra etiam in pluribus figuris continuis. Sic in fig. 38 centrum commune gravitatis totius trianguli est in illo puncto, quod a recta dueta a vertice anguli cujusvis ad mediam basim oppositam relinquit trientem versus basim ipsam. Nam omnium rectarum basi parallelarum, quæ omnes a recta BD secantur bifariam, ut FH, centra gravitatis sunt in eadem recta, adeoque & areæ ab iis contextæ centrum gravitatis est tam in recta BD, quam in recta GE ob eandem rationem, nempe in illo puncto C. Eadem methodus applicatur aliis figuris solidis, ut pyramidibus; at id, ut & reliqua omnia pertinentia ad inventionem centri gravitatis in diversis curvis lineis, superficiebus, solidis, hinc profluentia, sed meæ Theoriæ communia jam cum vulgaribus clementis, hic omittam, & solum illud iterum innuam, ea rite procedere, ubi jam semel demonstratum fuerit, haberi in massis omnibus aliquod gravitatis centrum, & esse unicum, ex quo nimirum hic & illud fluit, areas FAGH, FBH licet inæquales, habere tamen æquales summas distantiarum omnium suorum punctorum ab eadem recta FH.

Difficultas demonstrationis in communi methodo.

256. In communi methodo alio modo se res habet. Posteaquam inventum est in fig. 40 centrum gravitatis commune massis A, & B, juncta pro tertia massa DC, & secta in F in ratione massarum D, & A + B reciproca, habetur F pro centro communi omnium trium. Si prius inventum esset centrum commune E massarum D, B, & juncta AE, ea secta fuisset in F in ratione reciproca massarum A, & B + D; haberetur itidem illud

demonstration, I proved some time ago in a somewhat shorter manner in my dissertation *De Centro Gravitatis*; & a case of the second is here clearly shown; & in the dissertation *De Centro Magnitudinis*, which was added as a supplement to the former in the second edition, I determined in general the figures in which there existed a centre of magnitude & those in which there was none; but such things have no bearing on the matter now in question.

Hence to determine the position of the common centre of two masses.

253. From this general determination of the centre of gravity it is readily deduced that the common centre of two masses lies in the straight line joining the centres of each of the masses, & that the distances of the masses from this point will be reciprocally proportional to the masses themselves. For suppose we have two masses, & that their centres of gravity are, in Fig. 39, at A & B. If through the straight line AB any plane is drawn, it must be a plane of equal distances for either of the masses. Therefore also, with regard to the sum of the points of both masses taken together, all the distances taken on one side & on the other side will be equal to one another. Hence also with regard to this sum it must be a plane of equal distances; the common centre must lie in any one of these planes, & therefore in the line of intersection of any two of them, that is to say, in the straight line AB. Suppose it is at C; & suppose that any plane is drawn through C to cut AB. Then the sum of all the distances from this plane in the direction AB of all the points belonging to the mass A, the negatives being taken from the positives, will by Art. 243 be equal to the number of points in the mass A multiplied by AC; & the sum of those belonging to the mass B to the number of points in the mass B multiplied by BC. These products must be equal to each other, since the positives in the sum of all the distances must be cancelled by the negatives with regard to the centre of gravity C. Hence AC is to CB as the number in B is to the number of points in A, i.e., in the reciprocal ratio of the masses.

Hence, the usual method for any number of masses.

254. Further, from the foregoing theorem can be readily deduced *the usual method of finding the common centre of gravity of several masses. First of all the centres of two of them are joined, & the distance between them is divided in the reciprocal ratio of the masses. Then the common centre of these two masses, thus found, is joined to the centre of a third, & the distance is divided in the reciprocal ratio of the sum of the first two masses to the third mass; & so on.* Indeed, *we may find the centres of gravities of any groups of two, three, or ten, in any order, & then the groups of two may be joined to the threes, the tens, or what not, also in any order whatever; & in every case, in precisely the same manner, we shall arrive at the common centre of gravity of the whole mass.* This is evidently the case, for the reason that any number of masses can be reckoned as a single mass, since it is only a question of the number of points in the mass & the sum of the distances of all the points; the sum of the masses constitute a mass, & the sums of the distances a sum of distances, merely by taking them as a whole. Moreover, since, by the general demonstration given above, a centre of gravity is always obtained, & since this centre is unique, it follows that, no matter in what order the operations are performed, the same centre is arrived at in every case.

Hence, a theorem, by the help of which the centre of gravity for continuous figures may be investigated.

255. From the above we have a theorem, which is also well known, namely:—*If the centres of gravity of several masses all lie in one & the same straight line, then the centre of gravity of the whole set will also lie in the same straight line.* This indicates a method for investigating the centres of gravity also in the case of many continuous figures. Thus, in Fig. 38, the centre of gravity of the whole triangle is at that point, which cuts off, from the straight line drawn through the vertex of any angle to the middle point of the base opposite to it, one-third of its length on the side nearest to the base. For, the centre of gravity of every line drawn parallel to the base, such as FH, since each of them is bisected by BD, lies in this latter straight line. Hence the centre of gravity of the area formed from them lies in this straight line BD; as it also does in GE for a similar reason; that is to say, it is at the point C. The same method can be applied to some solid figures, such as pyramids. But I omit all this here, just as I do all the other matters relating to the finding of the centre of gravity for diverse curved lines, surfaces & solids, to be derived from what has been proved, but in which my theory is in agreement with the usual fundamental principles; I will only remark once again that these all will follow in due course when once it has been shown that for all masses there exists a centre of gravity, & that there is only one; and from this indeed there follows also the theorem that, although the areas FAGH, FBH are unequal, yet the sums of the distances from the straight line FH of all the points forming them are equal to one another.

The difficulty of proof in the ordinary method.

256. In the ordinary method it is quite another thing. Afterthat, in Fig. 40, the common centre of gravity of the masses A & B has been found, for the third mass, whose centre is D, join DC and divide it at F in the reciprocal ratio of D to A + B, then F is obtained as the common centre for all three masses. If, first of all, the common centre E of the masses D & B had been found, & AE were joined, & the latter divided at F in the reciprocal ratio of the masses A & B + D; then the point of section,

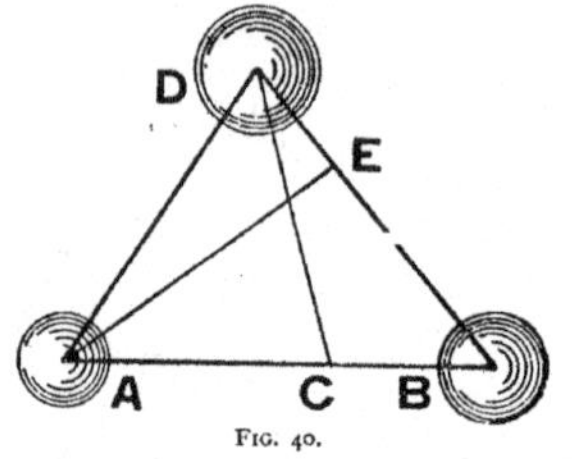

Fig. 40.

sectionis punctum pro centro gravitatis. Nisi generaliter demonstratum fuisset, haberi semper aliquod, & esse unicum gravitatis centrum; oporteret hic iterum demonstrare id novum sectionis punctum fore idem, ac illud prius: sed per singulos casus ire, res infinita esset, cum diversæ rationes conjungendi massas eodem redeant, quo diversi ordines litterarum conjungendarum in voces, de quarum multitudine immensa in exiguo etiam terminorum numero mentionem fecimus num. 114.

Similis difficultas in summa, & multiplicatione plurium numerorum, & in vi composita ex pluribus: methodus componendi simul omnes.

[120] 257. Atque hic illud quidem accidit, quod in numerorum summa, & multiplicatione experimur, ut nimirum quocunque ordine accipiantur numeri, vel singuli, ut addantur numero jam invento, vel ipsum multiplicent, vel plurium aggregata seorsum addita, vel multiplicata; semper ad eundem demum deveniatur numerum post omnes, qui dati fuerant, adhibitos semel singulos; ac in summa patet facile deveniri eodem, & in multiplicatione potest res itidem demonstrari etiam generaliter, sed ea huc non pertinent. Pertinet autem huc magis aliud ejusmodi exemplum petitum a compositione virium, in qua itidem si multæ vires componantur communi methodo componendo inter se duas per diagonalem parallelogrammi, cujus latera eas exprimant, tum hanc diagonalem cum tertia, & ita porro; quocunque ordine res procedat, semper ad eandem demum post omnes adhibitas devenitur. Hujusmodi compositione plurimarum virium generali jam indigebimus, & ad absolutam demonstrationem requiritur generalis expressio compositionis virium quotcunque, qua uti soleo. Compono nimirum generaliter motus, qui sunt virium effectus, & ex effectu composito metior vim, ut e spatiolo, quod dato tempusculo vi aliqua percurreretur, solet ipsa vis simplex quælibet æstimari. Assumo illud, quod & rationi est consentaneum, & experimentis constat, & facile etiam demonstratur consentire cum communi methodo componendi vires, ac motus per parallelogramma, nimirum punctum solicitatum simul initio cujusvis tempusculi actione conjuncta virium quarumcunque, quarum directio, & magnitudo toto tempusculo perseveret eadem, fore in fine ejus tempusculi in eo loci puncto, in quo esset, si singulæ eadem intensitate, & directione egissent aliæ post alias totidem tempusculis, quot sunt vires, cessante omni nova solicitatione, & omni velocitate jam producta a vi qualibet post suum tempusculum: tum reetam, quæ conjungit primum illud punctum cum hoc postremo, assumo pro mensura vis ex omnibus compositæ, quæ cum eadem perseveret per totum tempusculum; punctum mobile utique per unicam illam candem rectam abiret. Quod si & velocitatem aliquam habuerit initio illius tempusculi jam acquisitam ante; assumo itidem, fore in eo puncto loci, in quo esset, si altero tempusculo percurreret spatiolum, ad quod determinatur ab illa velocitate, altero spatiolum, ad quod determinatur a vi, sive aliis totidem tempusculis percurreret spatiola, ad quorum singula determinatur a viribus singulis.

Consensus ejus methodi cum communi per parallelogramma.

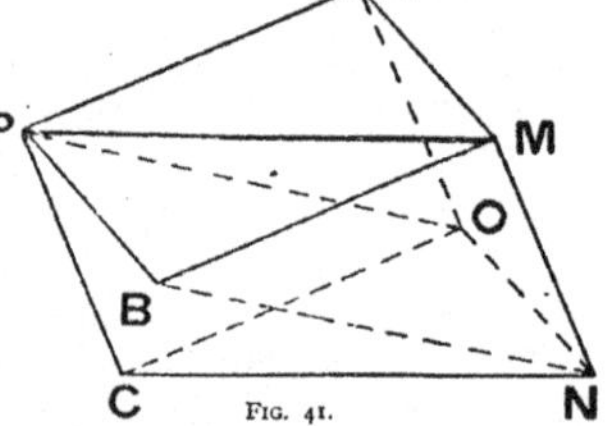

Fig. 41.

258. Huc recidere methodum componendi per parallelogramma facile constat; si enim in fig. 41 componendi sint plures motus, vel vires expressæ a rectis PA, PB, PC, &c, & incipiendo a binis quibusque PA, PB, eæ componantur per parallelogrammum PAMB, tum vis composita PM cum tertia PC per parallelogrammum PMNC, & ita porro; [121] patet, ad idem loci punctum N per hæc parallelogramma definitum debere devenire punctum mobile, quod prius percurrat PA, tum AM parallelam, & æqualem PB; tum MN parallelam, & æqualem PC, atque ita porro additis quotcunque aliis motibus, vel viribus, quæ per nova parallela, & æqualia parallelogrammorum latera debeant componi.

Demonstratio generalis methodi.

259. Deveniretur quidem ad idem punctum N, si alio etiam ordine componerentur ii motus, vel vires, ut compositis viribus PA, PC per parallelogrammum PAOC, tum vi PO cum vi PB per novum parallelogrammum, quod itidem haberet cuspidem in N; sed eo deveniretur alia via PAON. Hoc autem ipsum, quod tam multis viis, quam multas diversæ plurium compositiones motuum, ac virium exhibere possunt, eodem semper deveniri debeat, sic generaliter demonstro. Si assumantur ultra omnia puncta, ad quæ per ejusmodi compositiones deveniri potest, planum quodcunque; ubi punctum mobile percurrit lineolam pertinentem ad quencunque determinatum motum, habet eundem perpendicularem accessum ad id planum, vel recessum ab eo, quocunque tempusculo id fiat, sive aliquo e prioribus, sive

F, would again be obtained as the centre of gravity. Now, unless it had been already proved in general that there always was one centre of gravity, & only one, it would be necessary here to demonstrate afresh that the new point of section was the same as the first one. But to do this for every single instance would be an endless task; for diverse ways of joining the masses come to the same thing as diverse orders of joining up letters to form words; & I have already, in Art. 114, remarked upon the immense number of these even with a small number of letters.

A similar difficulty exists in the case of a sum or a product of several numbers; & also in a force compounded from several forces; the method of compounding them all at one time.

257. Indeed the same thing happens in the case of addition & multiplication; for we find, for instance, no matter what the order is in which the numbers are taken, whether they are taken singly, & added to the number already obtained, or multiplied, or whether the addition or multiplication is made with a group of several of them; the same number is arrived at finally after all those that have been given have been used each once. Now in addition it is easily seen that the result obtained is the same; & for multiplication also the matter can be easily demonstrated; but we are not concerned with these proofs here. Moreover, there is another example of this sort that is far more suitable for the present occasion, derived from the composition of forces. In this, if several forces are compounded in the ordinary manner, by compounding two of them together by means of the diagonal of the parallelogram whose sides represent the forces, & then this diagonal with a third force, & so on. In whatever order the operations are performed we always arrive at the same force finally, after all the given forces have been used. We shall now need a general composition of very many forces, & for rigorous proof we must have a general representation for the composition of any number of forces, such as the one I usually employ. Thus, I in general compound the motions, which are the effects of the forces, & measure the force from the resultant of the effects; so that any simple force is usually estimated by the small interval of space through which the force moves its point of application in a given short interval of time. I make an assumption, which is not only a reasonable one, but is also verified by experiment, & further one which can be easily shown to agree with the usual method for the composition of forces & motions by means of the parallelogram. Thus, I assume that a point, which is influenced simultaneously, at the beginning of any short interval of time by the joint action of any forces whatever, whose directions & magnitudes continue unchanged during the whole of the interval, will be at the end of the interval in the same position in space, as if each of the forces had acted independently, one after another, with the same intensity & in the same direction, during as many intervals of time as there are forces; where each fresh influence & the velocity already produced by any one of the forces ceases at the end of the interval that corresponds to it. Then I take the straight line which joins the initial point to the final point as the measure of the force that is the resultant of them all, & that this force will be represented by this same straight line during the whole of the interval of time, & that the moving point will traverse in every case that straight line & that one only. But if, moreover, at the beginning of the interval of time, the point should have a velocity previously acquired, then I also assume that it would occupy that position in space that it would have occupied if during another interval of time it had passed over an interval of space, determined by this other velocity, which is itself determined by the force; or if it had passed over as many intervals of spaces in as many intervals of time as there are forces determining the initial velocity.

Agreement of this method with the usual one by means of the parallelogram.

258. It is easily seen that the method of composition by means of the parallelogram comes to the same thing. For, if, in Fig. 41, the several motions or forces to be compounded are represented by PA, PB, PC, &c.; &, beginning with any two of them, PA & PB, these are compounded by means of the parallelogram PAMB, then the resultant force PM is compounded with a third PC by means of the parallelogram PMNC, & so on; it is clear that the moving point must reach the same point of space, N, determined by these parallelograms, as it would have done if it had traversed PA, then AM parallel & equal to PB, & then MN parallel & equal to PC; & so on, for any number of additional motions or forces, which have to be compounded by fresh straight lines equal & parallel to the sides of the parallelograms.

General proof of the method.

259. Now the same point N would be reached also, if these motions or forces were compounded in another order, say, by first compounding PA & PC by means of the parallelogram PAOC, then the force PO with the force PB by another parallelogram, which has its fourth vertex at N, although the point is reached by another path PAON. The fact that the same point is bound to be reached, by each of the many paths that correspond to the many different orders of compounding several motions or forces, I prove in general as follows. Imagine a plane drawn beyond any point that could be reached owing to compositions of this kind; then, when a moving point traverses a short path corresponding to any given motion, there is the same perpendicular approach towards the plane, or recession from it, in whichever of the short intervals of time it takes place, whether one of those at

aliquo e postremis, vel mediis. Nam ea lineola ex quocunque puncto discedat, ad quod deventum jam sit, habet semper eandem & longitudinem, & directionem, cum eidem e componentibus parallela esse debeat, & æqualis. Quare summa ejusmodi accessuum, ac summa recessuum erit eadem in fine omnium tempusculorum, quocunque ordine disponantur lineolæ hæ parallelæ, & æquales lineolis componentibus, adeoque etiam id, quod prodit demendo recessuum summam a summa accessuum, vel vice versa, crit idem, & distantia puncti postremi, ad quod deventum est ab illo eodem plano, crit eadem. Inde autem sponte jam fluit id, quod demonstrandum erat, nimirum punctum illud esse idem semper. Si enim ad duo puncta duabus diversis viis deveniretur, assumpto plano perpendiculari ad reetam, quæ illa duo puncta jungeret, distantia perpendicularis ab ipso non esset utique eadem pro utroque, cum altera distantia deberet alterius esse pars.

Theorema de statu centri gravitatis manente etiam ubi agant utcunque vires mutuæ, ac ejus demonstrationis initium.

260. Porro similis admodum est etiam methodus, qua utor ad demonstrandum præclarissimum Newtoni theorema, in quod coalescunt simul duo, quæ superius innui, & huc reducuntur. *Si quotcunque materiæ puncta utcunque disposita, & in quotcunque utcunque disjunctas massas coalescentia habeant velocitates quascunque cum directionibus quibuscunque, & præterea urgeantur viribus mutuis quibuscunque, quæ in binis quibusque punctis æqualiter agant in plagas oppositas ; centrum commune gravitatis omnium vel quiescet, vel movebitur uniformiter in directum eodem motu, quem haberet, si nulla adesset mutua punctorum actio in se invicem.* Hoc autem theorema sic generaliter, & admodum facile, ac luculenter demonstratur. [122] Concipiamus vires singulas per quodvis determinatum tempusculum servare directiones suas, & magnitudines : in fine ejus tempusculi punctum materiæ quodvis crit in eo loci puncto, in quo esset, si singularum virium effectus, vel effectus velocitatis ipsius illi tempusculo debitus, haberentur cum eadem sua directione, & magnitudine alii post alios totidem tempusculis, quot vires agunt. Assumantur jam totidem tempuscula, quot sunt punctorum binaria diversa in ea omni congerie, & præterea unum, ac primo tempusculo habeant omnia puncta motus debitos velocitatibus illis suis, quas habent initio ipsius, singula singulos ; tum assignato quovis e sequentibus tempusculis cuivis binario, habeat binarium quodvis tempusculo sibi respondente motum debitum vi mutuæ, quæ agit inter bina ejus puncta, ceteris omnibus quiescentibus. In fine postremi tempusculi omnia puncta materiæ erunt in hac hypothesi in iis punctis loci, in quibus revera esse debent in fine unici primi tempusculi ex actione conjuncta virium omnium cum singulis singulorum velocitatibus.

Progressus ejusdem demonstrationis

261. Concipiatur jam ultra omnia ejusmodi puncta planum quodcunque. Primo ex illis tot assumptis tempusculis alia puncta accedent, alia recedent ab eo plano, & summa omnium accessuum punctorum omnium demptis omnibus recessibus, si qua superest, vel vice versa summa recessuum demptis accessibus, divisa per numerum omnium punctorum, æquabitur accessui perpendiculari ad idem planum, vel recessui centri gravitatis communis ; cum summa distantiarum perpendicularium tam initio tempusculi, quam in fine, divisa per eundem numerum exhibeat ipsius communis centri gravitatis distantiam juxta num. 246. Sequentibus autem tempusculis manebit utique eadem distantia centri gravitatis communis ab eodem plano nunquam mutata ; quia ob æquales & contrarios punctorum motus, alterius accessus ab alterius recessu æquali eliditur. Quamobrem in fine omnium tempusculorum ejus distantia erit eadem, & accessus ad planum erit idem, qui esset, si solæ adfuissent ejusmodi velocitates, quæ habebantur initio ; adeoque etiam cum omnes vires simul agunt, in fine illius unici tempusculi habebitur distantia, quæ haberetur, si vires illæ mutuæ non egissent, & accessus æquabitur summæ accessuum, qui haberentur ex solis velocitatibus, demptis recessibus. Si jam consideretur secundum tempusculum in quo simul agant vires mutuæ, & velocitates ; debebunt considerari tria genera motuum : primum eorum, qui proveniunt a velocitatibus, quæ habebantur initio primi tempusculi ; secundum eorum, qui proveniunt a velocitatibus acquisitis actione virium durante per primum tempusculum ; tertium eorum, qui proveniunt a novis actionibus virium mutuarum, quæ ob mutatas jam positiones concipiantur aliis directionibus agere per totum secundum tempusculum. Porro quoniam hi posteriorum duorum generum motus [123] sunt in singulis punctorum binariis contrarii, & æquales ; illi itidem distantiam centri gravitatis ab eodem plano, & accessum, vel recessum debitum secundo tempusculo non mutant ;

the commencement, or one of those at the end, or one in the middle. For the short line, whatever point it has for its beginning & whatever point it finally reaches, must always have the same length & direction; for it is bound to be parallel & equal to the same one of the components. Hence the sum of these approaches, & the sum of these recessions, will be the same at the end of the whole set of intervals of time, no matter in what order these little lines, which are parallel & equal to the component lines, are disposed. Hence also, the result obtained by taking away the sum of the recessions from the sum of the approaches, or conversely, will be the same; & the distance of the ultimate point reached from the plane will be the same. Thus there follows immediately what was required to be proved, namely, that the point is the same point in every case. For, if two points could be reached by any two different paths, & a plane is taken perpendicular to the line joining those two points, then it is impossible for the perpendicular distance from this plane to be exactly the same for both points, since the one distance must be a part of the other.

Theorem relating to the permanent state of the centre of gravity even when there are mutual forces acting; the first steps of the proof.

260. Further, the method, which I make use of to prove a most elegant theorem of Newton, is exactly similar; in it the two noted above are combined, & come to the same thing. *If any number of points of matter, disposed in any manner, & coalescing to form any number of separate masses in any manner, have any velocities in any direction; & if, in addition, the points are under the influence of any mutual forces whatever, these forces acting on each pair of points equally in opposite directions; then the common centre of gravity of the whole is either at rest, or moves uniformly in a straight line with the same motion as it would have if there were no mutual action of the points upon one another.* Now this theorem is quite easily & clearly proved in all generality as follows. Suppose that each force maintains its direction & magnitude during any given short intervals of time; at the end of the interval any point of matter will occupy that point of space, which it would occupy if the effects for each of the forces (i.e., the effect of each velocity corresponding to that interval of time) were obtained, one after another, in as many intervals of time as there are forces acting, whilst each maintains its own direction & magnitude the same as before. Now take as many small intervals of time as there are different pairs of points in the whole group, & one interval in addition; & in the first interval of time let all the points have the motions due to the velocities that they have at the beginning of the interval of time respectively. Then, any one of the subsequent intervals of time being assigned to any chosen pair of points, let any pair have, in the interval of time proper to it, that motion which is due to the mutual force that acts between the two points of that pair, whilst all the others remain at rest. Then at the end of the last of these intervals of time, each point of matter will be, according to this hypothesis, at that point of space which it is bound to occupy at the end of a single first interval of time, under the conjoint action of all the mutual forces, each having its corresponding velocity.

Continuation of the demonstration.

261. Now imagine a plane situated beyond all points of this kind. Then, in the first place, for these little intervals of time of which we have assumed the number stated, some of the points will approach towards, & some recede from the plane; & the sum of all these approaches less the sum of all the recessions, if the former is the greater, & conversely, the sum of the recessions less the sum of the approaches, divided by the number of all the points, will be equal to the perpendicular approach of the common centre of gravity to the plane, or the recession from it. For, by Art. 246, the sum of the perpendicular distances, both at the beginning & at the end of the interval of time will represent the distance of the common centre of gravity itself. Further, in subsequent intervals, this distance of the common centre of gravity from the plane will remain in every case quite unchanged; because, on account of the equal & opposite motions of pairs of points, the approach of the one will be cancelled by the equal recession of the other. Hence, at the end of all the intervals the distance of the centre of gravity will be the same, & its approach towards the plane will be the same, as it would have been if there had existed no velocities except those which it had at the beginning of the interval; thus, too, when all the forces act together, at the end of the single interval of time there will be obtained that distance, which would have been obtained if the mutual forces had not been acting; & the approach will be equal to the sum of the approaches, less the recessions, acquired from the velocities alone. If now we would consider a second interval of time, in which we have acting the mutual forces, & the velocities; we shall have to consider three kinds of motions. Firstly, those that come from the velocities which exist at the beginning of the interval; secondly, those which arise from the velocities acquired through the action of forces lasting throughout the first interval; & thirdly, those which arise from the new actions of the mutual forces, which may be assumed to be acting in fresh directions, due to the change in the positions of the points during the whole of this second interval. Further, since the latter of the last two kinds, of motion are equal & opposite for each pair of points, these two kinds also will not change the distance of the centre of gravity from the plane & the approach towards it or recession

sed ea habentur, sicuti haberentur, si semper durarent solæ illæ velocitates, quæ habebantur initio primi tempusculi; & idem redit argumentum pro tempusculo quocunque: singulis advenientibus tempusculis accedet novum motuum genus durantibus cum sua directione, & magnitudine velocitatibus omnibus inductis per singula præcedentia tempuscula, ex quibus omnibus, & ex nova actione vis mutuæ, componitur quovis tempusculo motus puncti cujusvis: sed omnia ista inducunt motus contrarios, & æquales, adeoque summa accessuum, vel recessuum ortam ab illis solis initialibus velocitatibus non mutant.

Progressus ulterior.

262. Quod si jam tempusculorum magnitudo minuatur in infinitum, aucto itidem in infinitum intra quodvis finitum tempus eorundem numero, donec evadat continuum tempus, & continua positionum, ac virium mutatio; adhuc centrum gravitatis in fine continui temporis cujuscunque, adeoque & in fine partium quarumcunque ejusdem temporis, habebit ab eodem plano distantiam perpendicularem, quam haberet ex solis velocitatibus habitis initio ejus temporis, si nullæ deinde egissent mutuæ vires; & accessus ad illud planum, vel recessus ab eo, æquabitur summæ omnium accessuum pertinentium ad omnia puncta demptis omnibus recessibus, vel vice versa. Is vero accessus, vel recessus assumptis binis ejus temporis partibus quibuscunque, erit proportionalis ipsis temporibus. Nam singulorum punctorum accessus, vel recessus orti ab illis velocitatibus initialibus perseverantibus, adeoque ab motu æquabili, sunt in ratione eadem earundem temporis partium; ac proinde & eorum summæ in eadem ratione sunt.

Demonstrationis finis.

263. Inde vero prona jam est theorematis demonstratio. Ponamus enim, centrum gravitatis quiescere quodam tempore, tum moveri per aliquod aliud tempus. Debebit utique aliquo momento temporis esse in alio loci puncto, diverso ab eo, in quo erat initio motus. Sumatur pro prima e duabus partibus temporis continui pars ejus temporis, quo punctum quiescebat, & pro secunda tempus ab initio motus usque ad quodvis momentum, quo centrum illud gravitatis devenit ad aliud aliquod punctum loci. Ducta recta ab initio ad finem hujusce motus, tum accepto plano aliquo perpendiculari ipsi productæ ultra omnia puncta, centrum gravitatis ad id planum accederet secunda continui ejus temporis parte per intervallum æquale illi rectæ, & nihil accessisset primo tempore, adeoque accessus non fuissent proportionales illis partibus continui temporis. Quamobrem ipsum commune gravitatis centrum vel semper quiescit, vel movetur semper. Si autem movetur, debet moveri in directum. Si enim omnia puncta loci, per quæ transit, non jacent in directum, sumantur tria in dire-[124]-etum non jacentia, & ducatur recta per prima duo, quæ per tertium non transibit, adeoque per ipsam duci poterit planum, quod non transeat per tertium, tum ultra omnem punctorum congeriem planum ipsi parallelum. Ad id secundum nihil accessisset illo tempore, quo a primo loci puncto devenisset ad secundum, & eo tempore, quo ivisset a secundo ad tertium, accessisset per intervallum æquale distantiæ a priore plano, adeoque accessus iterum proportionales temporibus non fuissent. Demum motus erit æquabilis. Si enim ultra omnia puncta concipiatur planum perpendiculare rectæ, per quam movetur ipsum centrum commune gravitatis, jacens ad eam partem, in quam id progreditur, accessus ad ipsum planum erit totus integer motus ejusdem centri; adeoque cum ii accessus debeant esse proportionales temporibus; erunt ipsis temporibus proportionales motus integri; & idcirco non tantum rectilineus, sed & uniformis erit motus; unde jam evidentissime patet theorema totum.

Corollarium de quantitate motus in eandem plagam conservata in Mundo.

264. Ex eodem fonte, ex quo profluxit hoc generale theorema, sponte fluit hoc aliud ut consectarium: *quantitas motus in Mundo conservatur semper eadem, si ea computetur secundum directionem quacunque ita, ut motus secundum directionem oppositam consideretur ut negativus, ejusmodi motuum contrariorum summa subtracta a summa directorum.* Si enim consideretur eidem directioni perpendiculare planum ultra omnia materiæ puncta, quantitas motus in ea directione est summa omnium accessuum, demptis omnibus recessibus, quæ summa tempusculis æqualibus manet eadem, cum mutuæ vires inducant accessus, & recessus se mutuo destruentes; nec ejusmodi conservationi obsunt liberi motus ab anima nostra producti, cum nec ipsa vires ullas possit exerere, nisi quæ agant in partes oppositas æqualiter juxta num. 74.

Æqualitas actionis & reactionis in massis inde orta.

265. Porro ex illo Newtoniano theoremate statim jam profluit lex actionis, & reactionis æqualium pro massis omnibus. Nimirum si duæ massæ quæcunque in se invicem agant viribus quibuscunque mutuis, & inter singula punctorum binaria æqualibus; binæ illæ

from it corresponding to the second interval. Hence, these will be the same as they would have been, if those velocities that existed at the beginning of the first interval had persisted throughout; & the same argument applies to any interval whatever. Each interval as it occurs will yield a fresh kind of motions, all the velocities induced during each of the preceding intervals remaining the same in direction & magnitude; & from all of these, & the fresh action of the mutual force, there is compounded for any interval the motion of any point. But all the latter induce equal & opposite motions in pairs of points; & thus the sum of the approaches or recessions arising from the velocities alone are unchanged by the mutual forces.

Further steps in the demonstration.

262. Now if the length of the interval of time is indefinitely diminished, the number of intervals in any given finite time being thus indefinitely increased, until we acquire continuous time, & continuous change of position & forces; still the centre of gravity at the end of any continuous time, & thus also at the end of any parts of that time, will have that perpendicular distance from the plane, which it would have had, due to the velocities that existed at the beginning of the time, if no mutual forces had been acting. The approach towards the plane, or the recession from it, will be equal to the sum of all the approaches corresponding to all the points less the sum of all the recessions, or vice versa. Indeed, any two parts of the time being taken, this approach or recession will be proportional to these parts of the time. For the approach or recession, for each of the points, arising from the velocities that persist throughout & thus also from uniform motion, is proportional for all parts of the time; & hence also, their sums are proportional.

Conclusion of the demonstration.

263. The complete proof now follows immediately from what has been said above. For, let us suppose that the centre of gravity is at rest for a certain time, & then moves for some other time. Then at some instant of time it is bound to be at some other point of space different from that in which it was at the beginning of the motion. Of two parts of continuous time, let us take as the first part of the time, that in which the point is at rest; & for the second part, the time between the beginning of the motion & the instant when the centre of gravity reaches some other point of space. Draw a straight line from the beginning to the end of this motion, & take any plane perpendicular to this line produced beyond all the points; then the centre of gravity would approach towards the plane, in the second part of the continuous time, through an interval equal to the straight line, but in the first part of the time there would have been no approach at all; hence the approaches would not have been proportional to those parts of the continuous time. Hence the centre of gravity is always at rest, or is always in motion. Further, if it is in motion, it must move in a straight line. For, if all points of space, through which it passes, do not lie in a straight line, take three of them which are not collinear; & draw a straight line through the first two, which does not pass through the third; then it will be possible to draw through this straight line a plane which will not pass through the third point; & consequently, a plane parallel to it beyond the whole group of points. To this second plane there will be no approach at all for the time, during which the centre of gravity would travel from the first point of space to the second; & for that time, during which it would go from the second point to the third, there would be an approach through an interval equal to its distance from the first plane; & thus, once again, the approaches would not be proportional to the times. Lastly, the motion will be uniform. For, if we imagine a plane drawn beyond all the points, perpendicular to the straight line along which the centre of gravity moves, & on that side to which there is approach, then the approach to that plane will be the whole of the entire motion of the centre; hence, since these approaches must be proportional to the times, the whole motions must be proportional to the times; & therefore the motion must not only be rectilinear, but also uniform. Thus, the whole theorem is now perfectly clear.

Corollary with regard to the conservation of the quantity of motion in the Universe in a given direction.

264. From the same source as that from which we have drawn the above general theorem, there is obtained immediately the following also, as a corollary. *The quantity of motion in the Universe is maintained always the same, so long as it is computed in some given direction in such a way that motion in the opposite direction is considered negative, & the sum of the contrary motions is subtracted from the sum of the direct motions.* For, if we consider a plane perpendicular to this direction lying beyond all points of matter, the quantity of motion in this direction is the sum of all the approaches with the sum of the recessions subtracted; this sum remains the same for equal times, since the mutual forces induce approaches & recessions that cancel one another. Nor is such conservation affected by free motions that are the result of our will; since it cannot exert any forces either, except such as act equally in opposite directions, as was proved in Art. 74.

Equality of action & reaction for masses the result of this theorem.

265. Further, from the Newtonian theorem, we have immediately the law of equal action & reaction for all masses. Thus, if any two masses act upon one another with any mutual forces, which are also equal for each pair of points, the two masses will acquire,

massæ acquirent ab actionibus mutuis summas motuum æquales in partes contrarias, & celeritates acquisitæ ab earum centris gravitatis in partes oppositas, componendæ cum antecedentibus ipsarum celeritatibus, erunt in ratione reciproca massarum. Nam centrum commune gravitatis omnium a mutuis actionibus nihil turbabitur per hoc theorema, & sive ejusmodi vires agant, sive non agant, sed solius inertiæ effectus habeantur; semper ab eodem communi gravitatis centro distabunt ea bina gravitatis centra hinc, & inde in directum ad distantias reciproce proportionales massis ipsis per num. 253. Quare si præter priores motus ex vi inertiæ uniformes, ob actionem mutuam adhuc magis ad hoc commune centrum accedet alterum ex iis, vel ab eo recedet; accedet & alterum, [125] vel recedet, accessibus, vel recessibus reciproce proportionalibus ipsis massis. Nam accessus ipsi, vel recessus, sunt differentiæ distantiarum habitarum cum actione mutuarum virium a distantiis habendis sine iis, adeoque erunt & ipsi in ratione reciproca massarum, in qua sunt totæ distantiæ. Quod si per centrum commune gravitatis concipiatur planum quodcumque, cui quæpiam data directio non sit parallela; summa accessuum, vel recessuum punctorum omnium massæ utriuslibet ad ipsum secundum eam directionem demptis oppositis, quæ est summa motuum secundum directionem candem, æquabitur accessui, vel recessui centri gravitatis ejus massæ dueto in punctorum numerum; accessus vero, vel recessus alterius centri ad accessum, vel recessum alterius in directione eadem, erit ut secundus numerus ad primum; nam accessus, & recessus in quavis directione data sunt inter se, ut accessus, vel recessus in quavis alia itidem data; & accessus, ac recessus in directione, quæ jungit centra massarum, sunt in ratione reciproca ipsarum massarum. Quare productum accessus, vel recessus centri primæ massæ per numerum punctorum, quæ habentur in ipsa, æquatur producto accessus, vel recessus secundæ per numerum punctorum, quæ in ipsa continentur; nimirum ipsæ motuum summæ in illa directione computatorum æquales sunt inter se, in quo ipsa actionis, & reactionis æqualitas est sita.

Inde leges collisionum: discrimen virium in corporibus elasticis, & mollibus.

266. Ex hac actionum, & reactionum æqualitate sponte profluunt leges collisionis corporum, quas ex hoc ipso principio Wrennus olim, Hugenius, & Wallisius invenerunt simul, ut in hac ipsa lege Naturæ exponenda Newtonus etiam memorat Principiorum lib. I. Ostendam autem, quo pacto generales formulæ inde deducantur tam pro directis collisionibus corporum mollium, quam pro perfecte, vel pro imperfecte elasticorum. Corpora mollia dicuntur ea, quæ resistunt mutationi figuræ, seu compressioni, sed compressa nullam exercent vim ad figuram recuperandam, ut est cera, vel sebum: corpora elastica, quæ figuram amissam recuperare nituntur; & si vis ad recuperandam sit æqualis vi ad non amittendam; dicuntur perfecte elastica, quæ quidem, ut & perfecte mollia, nulla, ut arbitror, sunt in Natura; si autem imperfecte elastica sunt, vis, quæ in amittenda, ad vim, quæ in recuperanda figura exercetur, datam aliquam rationem habet. Addi solet & tertium corporum genus, quæ dura dicunt, quæ nimirum figuram prorsus non mutent; sed ea itidem in Natura nusquam sunt juxta communem sententiam, & multo magis nulla usquam in hac mea Theoria. Adhuc qui ipsa velit agnoscere, is mollia consideret, quæ minus, ac minus comprimantur, donec compressio evadat nulla; & ita quæ de mollibus dicentur, aptari poterunt duris multo meliore jure, quam alii elasticorum leges ad ipsa transferant, considerando elasticitatem infinitam ita, ut figura nec mutetur, nec se restituat; [126] nam si figura non mutetur, adhuc concipi poterit, impenetrabilitatis vi amissus motus, ut amitteretur in compressione; sed ad supplendam vim, quæ exeritur ab elasticis in recuperanda figura, non est, quod concipi possit, ubi figura recuperari non debet. Porro unde corpora mollia sint, vel elastica hic non quæro; id pertinet ad tertiam partem, quanquam id ipsum innui superius num. 199; sed leges quæ in eorum collisionibus observari debent, & ex superiore theoremate fluunt, expono. Ut autem simplicior evadat res, considerabo globos, atque hos ipsos circumquaque circa centrum, in eadem saltem ab ipso centro distantia, homogeneos, qui primo quidem concurrant directe; nam deinde ad obliquas etiam collisiones faciemus gradum.

Præparatio pro collisionibus globorum, planorum, circulorum.

267. Porro ubi globus in globum agit, & ambo paribus a centro distantiis homogenei sunt, facile constat, vim mutuam, quæ est summa omnium virium, qua singula alterius puncta agunt in singula puncta alterius, habituram semper directionem, quæ jungit centra;

as a result of the mutual actions, sums of motions that are equal in opposite directions; & the velocities acquired by their centres of gravity in opposite directions, being compounded of the foregoing velocities, will be in the inverse ratio of the masses. For, by the theorem, the common centre of gravity of the whole will not be disturbed in the slightest degree by the mutual actions, whether such forces act or whether they do not, but only the effects of inertia will be obtained; hence the two centres of gravity will always be distant from this common centre of gravity, one on each side of it, in a straight line with it, at distances that are reciprocally proportional to the masses, as was proved in Art. 253. Hence, if in addition to the former uniform motions due to the force of inertia, one of the two masses, on account of the mutual action, should approach still nearer to the common centre, or recede still further from it; then the other will either approach towards it or recede from it, the approaches or recessions being reciprocally proportional to the masses. For these approaches or recessions are the differences between the distances that are obtained when there is action of mutual forces & the distances when there is not; & thus, they too will be in the inverse ratio of the masses, such as the whole distances are. But if we imagine a plane drawn through the common centre of gravity, & that some given direction is not parallel to it, then the sum of the approaches or recessions of all the points of either of the masses with respect to this plane, the opposites being subtracted (which is the same thing as the sum of the motions in this direction) will be equal to the approach or the recession of the centre of gravity of that mass multiplied by the number of points in it. But the approach or recessions of the centre of the one is to the approach or recession of the centre of the other, in the same direction, as the second number is to the first; for the approaches or recessions in any given direction are to one another as the approaches or recessions in any other given direction; & the approaches or recessions along the line joining the two masses are inversely proportional to the masses. Therefore the product of the approach or recession of the centre of the first mass, multiplied by the number of points in it, is equal to the approach or recession of the centre of the second mass, multiplied by the number of points that are contained in it. Thus the sums of the motions in the direction under consideration are equal to one another; & in this is involved the equality of action & reaction.

Hence the laws for collision; the distinction between the forces for elastic bodies & soft bodies.

266. From this equality of action & reaction there immediately follow the laws for collision of bodies, which some time ago Wren, Huygens & Wallis derived from this very principle at about the same time, as Newton also mentioned in the first book of the Principia, when expounding this law of Nature. Now I will show how general formulæ may be derived from it, both for the direct collision of soft bodies, & also for perfectly or imperfectly elastic bodies. By soft bodies are to be understood those, which resist deformation of their shapes, or compression; but which, when compressed, exert no force tending to restore shape; such as wax or tallow. Elastic bodies are those that endeavour to recover the shape they have lost; & if the force tending to restore shape is equal to that tending to prevent loss of shape, the bodies are termed perfectly elastic; &, just as there are no perfectly soft bodies, there are none that are perfectly elastic, according to my thinking, in Nature. Lastly, they are imperfectly elastic, if the force exerted against losing shape bears to the force exerted to restore it some given ratio. It is usual to add a third class of bodies, namely, such as are called hard; & these never alter their shape at all; but these also, even according to general opinion, never occur in Nature; still less can they exist in my Theory. Yet, if anyone wishes to take account of such bodies, they could consider them as soft bodies which are compressed less & less, until the compression finally becomes evanescent; in this way, whatever is said about soft bodies could be adapted to hard bodies with far more justification than there is for applying some of the laws of elastic bodies to them, by considering that there is infinite elasticity of such a nature that the figure neither suffers change nor seeks to restore itself. For, if the figure remains unchanged, it is yet possible to consider the motion lost due to the force of impenetrability, & that thus it would be lost in compression; but to supply the force which in elastic bodies is exerted for the recovery of shape, there is nothing that can be imagined, when there is necessarily no recovery of shape. Further, what are the causes of soft or elastic bodies, I do not investigate at present; this relates to the third part, although I have indeed mentioned it above, in Art. 199. But I set forth the laws which have to be observed in collisions between them, these laws coming out immediately from the theorem given above. Moreover to make the matter easier, I consider spheres, & these too homogeneous round about the centre, at any rate for the same distance from that centre; & these indeed will in the first place collide directly; for from direct collision we can proceed to oblique impact also.

Preparation for the consideration of collisions of spheres, Planes & circles.

267. Now, where one sphere acts upon another, & both of them are homogeneous at equal distances from their centres, it is readily shown that the mutual force, which is the sum of all the forces with which each of the points of the one acts on each of the points of the other, must always be in the direction of the line joining the two centres. For,

nam in ea recta jacent centra ipsorum globorum, quæ in eo homogeneitatis casu facile constat, esse centra itidem gravitatis globorum ipsorum; & in eadem jacet centrum commune gravitatis utriusque, ad quod viribus illis mutuis, quas alter globis exercet in alterum, debent ad se invicem accedere, vel a se invicem recedere; unde fit, ut motus, quos acquirunt globorum centra ex actione mutua alterius in alterum, debeant esse in directione, quæ jungit centra. Id autem generaliter extendi potest etiam ad casum, in quo concipiatur, massam immensam terminatam superficie plana, sive quoddam immensum planum agere in globum finitum, vel in punctum unicum, ac vice versa: nam alterius globi radio in infinitum aucto superficies in planum desinit; & radio alterius in infinitum imminuto, globus abit in punctum. Quin etiam si massa quævis teres, sive circa axem quendam rotunda, & in quovis plano perpendiculari axi homogenea, vel etiam circulus simplex, agat, vel concipiatur agens in globum, vel punctum in ipso axe constitutum; res eodem redit.

Formulæ pro corpore molli incurrente in mollius progrediens in eandem plagam.

268. Præcurrat jam globus mollis cum velocitate minore, quem alius itidem mollis consequatur cum majore ita, ut centra ferantur in eadem recta, quæ illa conjungit, & hic demum incurrat in ilium, quæ dicitur collisio directa. Is incursus mihi quidem non fiet per immediatum contactum, sed antequam ad contactum deveniant, vi mutua repulsiva comprimentur partes posteriores præcedentis, & anteriores sequentis, quæ compressio fiet semper major, donec ad æquales celeritates devenerint; tum enim accessus ulterior desinet, adeoque & ulterior compressio; & quoniam corpora sunt mollia, nullam aliam exercent vim mutuam post ejusmodi compressionem, sed cum æquali illa velocitate pergunt moveri porro. Hæc æqualitas velocitatis, ad quam reducuntur ii duo glo-[127]-bi, una cum æqualiate actionis, & reactionis æqualium, rem totam perficient. Sit enim massa, sive quantitas materiæ, globi præcurrentis $= q$, insequentis $= Q$; celeritas illius $= c$, hujus $= C$: quantitas motus illius ante collisionem erit cq, hujus CQ; nam celeritas dueta per numerum punctorum exhibet summam motuum punctorum omnium, sive quantitatem motus; unde etiam fit, ut quantitas motus per massam divisa exhibeat celeritatem. Ob actionem, & reactionem æquales, hæc quantitas erit eadem etiam post collisionem, post quam motus totus utriusque massæ, crit $CQ + cq$. Quoniam autem progrediuntur cum æquali celeritate; celeritas illa habebitur; si quantitas motus dividatur per totam quantitatem materiæ; quæ idcirco erit $\frac{CQ + cq}{Q + q}$. Nimirum ad habendam velocitatem communem post collisionem, oportebit ducere singulas massas in suas celeritates, & productorum summam dividere per summam massarum.

Ejus extensio ad omnes casus: celeritas amissa, vel acquisita.

269. Si alter globus q quiescat; satis erit illius celeritatem c considerare $= 0$: & si moveatur motu contrario motui prioris globi; satis erit illi valorem negativum tribuere; ut adeo & hic, & in sequentibus formula inventa pro illo primo casu globorum in eandem progredientium plagam, omnes casus contineat. In eo autem si libeat invenire celeritatem amissam a globo Q, & celeritatem acquisitam a globo q, satis erit reducere singulas formulas

$$C - \frac{CQ + cq}{Q + q}, \ \& \ \frac{CQ + cq}{Q + q} - c$$

ad eundem denominatorem, ac habebitur

$$\frac{Cq - cq}{Q + q}, \ \& \ \frac{CQ - cQ}{Q + q},$$

ex quibus deducitur hujusmodi theorema: *ut summa massarum ad massam alteram, ita differentia celeritatum ad celeritatem ab altera acquisitam*, quæ in eo casu accelerabit motum præcurrentis & retardabit motum consequentis.

Transitus ad elasticorum collisiones.

270. Ex hisce, quæ pertinent ad corpora mollia, facile est progredi ad perfecte elastica. In iis post compressionem maximam, & mutationem figuræ inductam ab ipsa, quæ habetur, ubi ad æquales velocitates est ventum, agent adhuc in se invicem bini globi, donec deveniant ad figuram priorem, & hæc actio duplicabit effectum priorem. Ubi ad sphæricam figuram deventum fuerit, quod fit recessu mutuo oppositarum superficierum, quæ in compressione ad se invicem accesserant, pergent utique a se invicem recedere aliquanto magis eædem superficies, & figura producetur, sed opposita jam vi mutua inter partes ejusdem globi incipient retrahi, & productio perget fieri, sed usque lentius, donec ad maximam quandam productionem de-[128]-ventum fuerit, quæ deinde incipiet minui, & globus ad sphæricam accedet iterum, ac iterum comprimetur quodam oscillatorio, ac partium trepidatione hinc, & inde a figura sphærica, ut supra vidimus etiam duo puncta circa distantiam limitis

in that straight line lie the centres of the two spheres ; & these in the case of homogeneity are easily shown to be also the centres of gravity of the spheres. Also in this straight line lies the common centre of gravity of both spheres ; & to, or from, it the spheres must approach or recede mutually, owing to the action of the mutual forces with which one sphere acts upon the other. Hence it follows that the motion, which the centres of the spheres acquire through the mutual action of one upon the other, is bound to be along the line which joins the centres. The argument can also be extended generally, even to include the case in which it is supposed that an immense mass bounded by a plane surface, or an immense plane acts upon a finite sphere, or on a single point, or vice versa ; for, if the radius of either of the spheres is increased indefinitely, the surface ultimately becomes a plane, & if the radius of either becomes indefinitely diminished, the sphere degenerates into a point. Moreover, if any round mass, or one contained by a surface of rotation round an axis and homogeneous in any plane perpendicular to that axis, or even a simple circle, act, or is supposed to act upon a sphere or point situated in the axis ; it comes to the same thing.

Formulae for a soft body impinging upon another soft body proceeding more slowly in the same direction.

268. Now suppose that a soft body proceeds with a less velocity than another soft body which is following it with a greater velocity, in such a manner that their centres are travelling in the same straight line, namely that which joins them ; & finally let the latter impinge upon the former ; this is termed direct impact. This impact, in my opinion indeed, does not come about by immediate contact, but, before they attain actual contact, the hinder parts of the first body & the foremost parts of the second body are compressed by a mutually repulsive force ; & this compression becomes greater & greater until finally the velocities become equal. Then further approach ceases, & therefore also further compression ; &, since the bodies are soft, they exercise no further mutual force after such compression, but continue to move forward with that equal velocity. This equality in the velocity, to which the two spheres are reduced, together with the equality of action & reaction, finishes off the whole matter. For, supposing that the mass or quantity of matter of the foremost sphere is equal to q, that of the latter to Q ; the velocity of the former equal to c, & that of the latter to C. Then the quantity of motion of the former before impact is cq, & that of the latter is CQ ; for the velocity multiplied by the number of points represents the sum of the motions of all the points, i.e., the quantity of motion, & in the same way the quantity of motion divided by the mass gives the velocity. Now, since the action & reaction are equal to one another, this quantity will be the same even after impact ; hence after impact the whole motion of both the masses together will be equal to $CQ + cq$. Further, since they are travelling with a common velocity, this velocity will be the result obtained on dividing the quantity of motion by the whole quantity of matter ; & it will therefore be equal to $(CQ + cq)/(Q + q)$. That is to say, to obtain the common velocity after impact, we must multiply each mass by its velocity, & divide the sum of these products by the sum of the masses.

Extension to all cases ; velocity lost or gained.

269. If one of the two spheres is at rest, all that need be done is to put its velocity c equal to zero ; also, if it is moving in a direction opposite to that of the first sphere, we need only take the value of c as negative. Thus, both here & subsequently, the formula found for the first case, in which the spheres are moving forward in the same direction, includes all cases. Again, if in this case, we wish to find the velocity lost by the sphere Q, & the velocity gained by the sphere q, we need only reduce the two formulae $C - (CQ + cq)/(Q + q)$ & $(CQ + cq)/(Q + q) - c$ to a common denominator, when we shall obtain the formulae $(Cq - cq)/(Q + q)$ & $(CQ - cQ)/(Q + q)$. From these there can be derived the theorem :—*The sum of the masses is to either of the masses as the difference between the velocities is to the velocity acquired by the other mass ;* in the present case there will be an increase of velocity for the foremost body & a decrease for the hindmost.

Transition to impact between elastic bodies.

270. From these theorems relating to soft bodies we can easily proceed to those that are perfectly elastic. For such bodies, after the maximum compression has taken place, & the alteration in shape consequent on this compression, which is attained when equality of the velocities is reached, the two spheres still continue to act upon one another, until the original shape is attained ; & this action will duplicate the effect of the first action. When the spherical shape is once more attained, as this takes place through a mutual recession of the opposite surfaces of the spheres, which during compression had approached one another, these same surfaces in each sphere will continue to recede from one another still somewhat further, & the shape will be elongated ; but the mutual force between the parts of each sphere is now changed in direction & the surfaces begin to be drawn together again. Hence elongation will continue, but more slowly, until a certain maximum elongation is attained ; this then begins to be diminished & the sphere once more returns to a spherical shape, once more is compressed with a sort of oscillatory motion & forward & backward vibration of its parts about the spherical shape ; exactly as was seen above in the case of two points oscillating to & fro about a distance equal to that corresponding to a limit-point

cohæsionis oscillare hinc, & inde; sed id ad collisionem, & motus centrorum gravitatis nihil pertinebit, quorum status a viribus mutuis nihil turbatur; actio autem unius globi in alterum statim cessabit post regressum ad figuram sphæricam, post quem superficies alterius postica & alterius antica in centra jam retractæ ulteriore centrorum discessu a se invicem incipient ita distare, ut vires in se invicem non exerant, quarum effectus sentiri possit; & hypothesis perfecte elasticorum est, ut tantus sit mutuæ actionis effectus in recuperanda, quantus fuit in amittenda figura.

Formulæ pro perfecte elasticis.

271. Duplicato igitur effectu, globus ammittet celeritatem $\frac{2Cq - 2cq}{Q + q}$, & globus q acquiret celeritatem $\frac{2CQ - 2cQ}{Q + q}$. Quare illius celeritas post collisionem erit $C - \frac{2Cq - 2cq}{Q + q}$ sive $\frac{CQ - Cq + 2cq}{Q + q}$; hujus vero erit $c + \frac{2CQ - 2cQ}{Q + q}, = \frac{cq - cQ + 2CQ}{Q + q}$, & motus fient in eandem plagam, vel globus alter quiescet, vel fient in plagas oppositas; prout determinatis valoribus Q, q, C, c, formulæ valor evaserit positivus, nullus, vel negativus.

Formulæ pro imperfecte elasticis.

272. Quod si elasticitas fuerit imperfecta, & vis in amittenda ad vim in recuperanda figura fuerit in aliqua ratione data, erit & effectus prioris ad effectum posterioris itidem in ratione data, nimirum in ratione subduplicata prioris. Nam ubi per idem spatium agunt vires, & velocitas oritur, vel extinguitur tota, ut hic respectiva velocitas extinguitur in compressione, oritur in restitutione figuræ, quadrata velocitatum sunt ut areæ, quas describunt ordinatæ viribus proportionales juxta num. 176, & hinc areæ erunt in ratione virium, si, viribus constantibus, sint constantes & ordinatæ, cum inde fiat, ut scalæ celeritatum ab iis descriptæ sint rectangula. Sit igitur rationis constantis illarum virium ratio subduplicata m ad n, & erit effectus in amittenda figura ad summam effectuum in tota collisione, ut m ad $m + n$, quæ ratio si ponantur esse 1 ad r, ut sit $r = \frac{m + n}{m}$ satis erit, effectus illos inventos pro globis mollibus, sive celeritatem ab altero amissam, ab altero acquisitam, non duplicare, ut in perfecte elasticis, sed multiplicare per r, ut habeantur velocitates acquisitæ in partes contrarias, & componendæ cum velocitatibus [129] prioribus. Erit nimirum illa quæ pertinet ad globum $Q = \frac{rCq - rcq}{Q + q}$, & quæ pertinet ad globum q, erit $- \frac{rCQ - rcQ}{Q + q}$, adeoque velocitas illius post congressum erit $C - \frac{rCq - rcq}{Q + q}$, & hujus $c + \frac{rCQ - rcQ}{Q + q}$; quæ formulæ itidem reducuntur ad eosdem denominatores; ac tum ex hisce formulis, tum e superioribus quam plurima elegantissima theoremata deducuntur, quæ quidem passim inveniuntur in elementaribus libris, & ego ipse aliquanto uberius persecutus sum in Supplementis Stayanis ad lib. 2, § 2; sed hic satis est, fundamenta ipsa, & primarias formulas derivasse ex eadem Theoria, & ex proprietatibus centri gravitatis, ac motuum oppositorum æqualium, deductis ex Theoria eadem; nec nisi binos, vel ternos evolvam casus usui futuros infra, antequam ad obliquam collisionem, ac reflexionem motuum gradum faciam.

Casus, in quo globus perfecte elasticus incurrit in alium.

273. Si globus perfecte elasticus incurrat in globum itidem quiescentem, erit, $c = 0$, adeoque velocitas contraria priori pertinens ad incurrentem, quæ erat $\frac{2Cq - 2cq}{Q + q}$, erit $\frac{2Cq}{Q + q}$; velocitas acquisita a quiescente, quæ erat $\frac{2CQ - 2cQ}{Q + q}$, erit $\frac{2CQ}{Q + q}$; unde habebitur hoc theorema: *ut summa massarum ad duplam massam quiescentis, vel incurrentis, ita celeritas incurrentis, ad celeritatem amissam a secundo, vel acquisitam a primo;* & si massæ æquales fuerint, fit ea ratio æqualitatis; ac proinde globus incurrens totam suam velocitatem amittit, acquirendo nimirum æqualem contrariam, a qua ea elidatur, & globus quiescens acquirit velocitatem, quam ante habuerat globus incurrens.

Casus triplex globi incurrentis in planum immobile.

274. Si globus imperfecte elasticus incurrat in globum quiescentem immensum, & qui habeatur pro absolute infinito, cujus idcirco superficies habetur pro plana, in formula velocitatis acquisitæ a globo quiescente $\frac{rCQ - rcQ}{Q + q}$, cum evanescat Q respectu q absolute infiniti, & proinde $\frac{Q}{Q + q}$ evadat $= 0$, tota formula evanescit, adeoque ipse haberi potest pro plano immobili. In formula vero velocitatis, quam in partem oppositam acquiret globus incurrens, $\frac{rCq - rcq}{Q + q}$, evadit $c = 0$, [130] & Q evanescit itidem respectu q. Hinc habetur $\frac{rCq}{q}$, sive rC, nimirum ob $r = \frac{m + n}{n}$ fit $\left(\frac{m + n}{n}\right) \times C$, cujus prima pars $\frac{n}{m} \times C$,

of cohesion. However, this has nothing to do with the impact or the motion of the centres of gravity, nor are their states affected in the slightest by the mutual forces. Again, the action of one sphere on the other will cease directly after return to the spherical shape; for after that the hindmost surface of the one & the foremost surface of the other, being already withdrawn in the direction of their centres, will through a further recession of the centres from one another begin to be so far distant from one another that they will not exert upon one another any forces of which the effects are appreciable. We are left with the hypothesis, for perfectly elastic solids, that the effect of their action on one another is exactly the same in amount during alteration of shape & recovery of it.

Formula for perfectly elastic bodies.

271. Hence, the effect being duplicated, the sphere Q will lose a velocity equal to $(2Cq - 2cq)/(Q + q)$, & the sphere q will gain a velocity equal to $(2CQ - 2Cq)/(Q + q)$. Hence, the velocity of the former after impact will be $C - (2Cq - 2cq)/(Q + q)$ or $(CQ - Cq + 2cq)/(Q + q)$, & the velocity of the latter will be $c + (2CQ - 2cQ)/(Q + q)$ or $(cq - cQ + 2CQ)/(Q + q)$. The motions will be in the original direction, or one of the spheres may come to rest, or the motions may be in opposite directions, according as formula, given by the values of Q, q, C, & c, turns out to be positive, zero, or negative.

Formulæ for imperfectly elastic bodies.

272. But if the elasticity were imperfect, & the force during loss of shape were in some given ratio to the force during recovery of shape, then the effect corresponding to the former would also be in a given ratio to the effect due to the latter, namely, in the subduplicate ratio of the first ratio. For, when forces act through the same interval of space, & velocity is generated, or is entirely destroyed, as here the relative velocity is destroyed during compression & generated during recovery of shape, the squares of the velocities are proportional to the areas described by the ordinates representing the forces, as was proved in Art. 176. Hence these areas are proportional to the forces, if, the forces being constant, the ordinates also are constant; for from that it is easily seen that the measures of the velocities described by them are rectangles. Suppose then that the subduplicate ratio of the constant ratio of the forces be $m : n$; then the ratio of the effect during loss of shape to the sum of the effects during the whole of the impact will be $m : m + n$. If we call this ratio $1 : r$, so that $r = (m + n)/m$, we need only, instead of doubling the effects found for soft bodies, or the velocity lost by one sphere or gained by the other, multiply these effects by r, in order to obtain the velocities acquired in opposite directions, which are to be compounded with the original velocities. Thus, that for the sphere Q will be equal to $(rCq - rcq)/(Q + q)$, & that for the sphere q will be $(rCQ - rcQ)/(Q + q)$. Hence, the velocity of the former after impact will be $C - (rCq - rcq)/(Q + q)$ & the velocity of the latter will be $c + (rCQ - rcQ)/(Q + q)$; & these formulæ also can be reduced to common denominators. From these formulæ, as well as from those proved above, a large number of very elegant theorems can be derived, such as are to be found indeed everywhere in elementary books. I myself have followed the matter up somewhat more profusely in the Supplements to Stay's Philosophy, in Book II, § 2. But here it is sufficient that I should have derived the fundamentals themselves, together with the primary formulæ, from one & the same Theory, & from the properties of the centre of gravity & of equal & opposite motions, which are also derived from the same theory. Except that I will consider below two or three cases that will come in useful in later work, before I pass on to oblique impact & reflected motions.

Case of a perfectly elastic sphere striking another.

273. If a perfectly elastic sphere strikes another, & the second sphere is at rest, then $c = 0$, & the velocity, in the direction opposite to the original velocity, for the striking body, which was $(2Cq - 2cq)/(Q + q)$, will in this case be $2Cq/(Q + q)$; whilst the velocity gained by the body that was at rest, which was shown to be $(2CQ - 2cQ)/(Q + q)$, will be $2CQ/(Q + q)$. Hence we have the following theorem. *As the sum of the masses is to twice the mass of the body at rest, or to the body that impinges upon it, so is the velocity of the impinging body to the velocity lost by the second body, or to that gained by the first.* If the masses were equal to one another, this ratio would be one of equality; hence in this case the impinging body loses the whole of its velocity, that is to say it acquires an equal opposite velocity which cancels the original velocity; & the sphere at rest acquires a velocity equal to that which the impinging sphere had at first.

Threefold case of a sphere impinging on an immovable plane.

274. If an imperfectly elastic sphere impinges on an immense sphere at rest, which may be considered as absolutely infinite, & therefore its surface may be taken to be a plane; then, in the formula for the velocity gained by a sphere at rest, $(rCQ - rcQ)/(Q + q)$, since Q vanishes in comparison with q which is absolutely infinite, & thus $Q/(Q + q) = 0$, the whole formula vanishes, & therefore the immense sphere can be taken to be an immovable plane. Now, in the formula for the velocity which the impinging sphere acquires in the opposite direction to its original motion, namely, $(rCq - rcq)/(Q + q)$, we have $c = 0$, & Q also vanishes in comparison with q. Hence we obtain rCq/q, or rC; that is to say, since $r = (m + n)/m$, we have $C \times (m + n)/m$, of which the first part, $C \times m/m$, or C,

sive C, est illa, quæ amittitur, sive acquiritur in partem oppositam in comprimenda figura, & $\frac{n}{m} \times C$ est illa, quæ acquiritur in recuperanda, ubi si fit $n = 0$, quod accidit nimirum in perfecte mollibus; habetur sola pars prima; si $m = n$, quod accidit in perfecte elasticis, est $\frac{n}{m} \times C = C$, secunda pars æqualis primæ; & in reliquis casibus est, ut m ad n, ita illa pars prima C, sive præcedens velocitas, quæ per primam partem acquisitam eliditur, ad partem secundam, quæ remanet in plagam oppositam. Quamobrem habetur ejusmodi theorema. *Si incurrat ad perpendiculum in planum immobile globus perfecte mollis, acquirit velocitatem contrariam æqualem suæ priori, & quiescat; si perfecte elasticus, acquirit duplam suæ, nimirum æqualem in compressione, qua motus omnis sistitur, & æqualem in recuperanda figura, cum qua resilit; si fuerit imperfecte elasticus in ratione* m *ad* n, *in illa eadem ratione erit velocitas priori suæ contraria acquisita, dum figura mutatur, quæ priorem ipsam velocitatem extinguit, ad velocitatem, quam acquirit, dum figura restituitur, & cum qua resilit.*

Summa quadratorum velocitatis ductorum in massas manens in perfecte elasticis.

275. Est & aliud theorema aliquanto operosius, sed generale, & elegans, ab Hugenio inventum pro perfecte elasticis, quod nimirum summa quadratorum velocitatis ductorum in massas post congressum remaneat eadem, quæ fuerat ante ipsum. Nam velocitates post congressum sunt $C - \frac{2q}{Q+q} \times (C - c)$, & $c + \frac{2Q}{Q+q} \times (C - c)$; quadrata dueta in massas continent singula ternos terminos: primi crunt $QCC + qcc$; secundi erunt $(-CC + Cc) \times \frac{4Qq}{Q+q}$ & $(cC - cc) \times \frac{4Qq}{Q+q}$, quorum summa evadit $(-CC + 2Cc - cc) \times \frac{4Qq}{Q+q}$; postremi crunt $\frac{4Qqq}{(Q+q)^2} \times (CC - 2Cc + cc)$, & $\frac{4qQQ}{(Q+q)^2} \times (CC - 2Cc + cc)$, sive simul $\frac{4(Q+q) \times Qq}{(Q+q)^2}$ [131] $\times (CC - 2Cc + cc)$, vel $\frac{4Qq}{Q+q} (CC - 2Cc + cc)$, quod destruit summam secundi terminorum binarii, remanente sola illa $QCC + qcc$, summa quadratorum velocitatum præcedentium ducta in massas. Sed hæc æqualitas nec habetur in mollibus, nec in imperfecte elasticis.

Collisionis obliquæ communis methodus per virium resolutionem.

276. Veniendo jam ad congressus obliquos, deveniant dato tempore bini globi A, C in fig. 42 per rectas quascunque AB, CD, quæ illorum velocitates metiantur, in B, & D ad physicum contactum, in quo jam sensibilem effectum edunt vires mutuæ. Communi methodo collisionis effectus sic definitur. Junctis eorum centris per rectam BD, ducantur, ad eam productam, qua opus est, perpendicula AF, CH, & completis rectangulis AFBE, CHDG resolvantur singull motus AB, CD in binos; ille quidem in AF, AE, sive EB, FB, hic vero in CH, CG, sive GD, HD. Primus utrobique manet illæsus; secundus FB, & HD collisionem facit directam. Inveniantur per legem collisionis directæ velocitates BI, DK, quæ juxta ejusmodi leges superius expositas haberentur post collisionem diversæ pro diversis corporum speciebus, & componantur cum velocitatibus expositis per rectas BL, DQ jacentes in directum cum EB, GD, & illis æquales. His peractis expriment BM, DP celeritates, ac directiones motuum post collisionem.

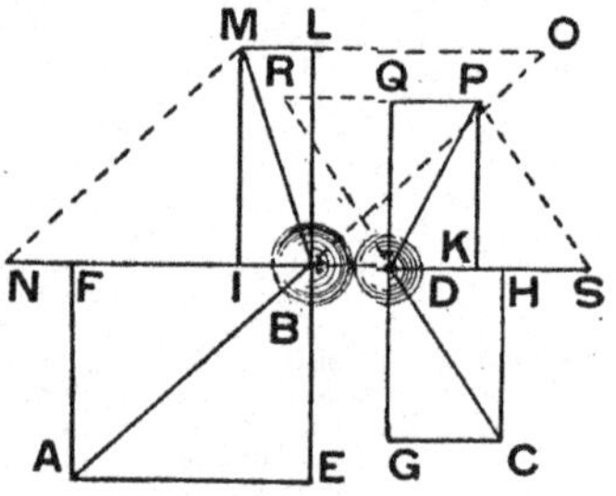

Fig. 42.

Compositio virium resolutioni substituta.

277. Hoc pacto consideratur resolutio motuum, ut vera quædam resolutio in duos, quorum alter illæsus perseveret, alter mutationem patiatur, ac in casu, quem figura exprimit, extinguatur penitus, tum iterum alius producatur. At sine ulla vera resolutione res vere accidit hoc pacto. Mutua vis, quæ agit in globos B, D, dat illis toto collisionis tempore velocitates contrarias BN, DS æquales in casu, quem figura exprimit, binis illis, quarum altera vulgo concipitur ut elisa, altera ut renascens. Eæ compositæ cum BO, DR jacentibus in directum cum AB, CD, & æqualibus iis ipsis, adeoque exprimentibus effectus integros præcedentium velocitatum, exhibent illas ipsas velocitates BM, DP. Facile enim patet, fore LO æqualem AE, sive FB, adeoque MO æqualem NB, & BNMO fore parallelogrammum; ac eadem demonstratione est itidem parallelogrammum DRPS. Quamobrem nulla ibi est vera resolutio, sed sola compositio motuum, perseverante nimirum velocitate priore per vim inertiæ, & ea composita cum nova velocitate, quam generant vires, quæ agunt in collisione.

is the part that is lost, or acquired in the opposite direction to the original velocity, during the compression, & $C \times n/m$ is the part that is acquired during the recovery of shape. In this, if $n = 0$, which is the case for perfectly soft bodies, there is only the first part; if $m = n$, which is the case for perfectly elastic bodies, then $C \times n/m$ will be equal to C, and the second part is equal to the first part; & in all other cases as m is to n, so is the first part C, or the original velocity, which is cancelled by the first part of the acquired velocity, to the second part, which is the final velocity in the opposite direction. Hence we have the following theorem. *If a perfectly soft sphere impinges perpendicularly upon an immovable plane, it will acquire a velocity equal & opposite to its original velocity, & will be brought to rest. If the body is perfectly elastic, it will acquire a velocity double of its original velocity but in the opposite direction, that is to say, an equal velocity during compression, by which the whole of the motion ceases, & an equal velocity during recovery of shape, with which it rebounds. If it were imperfectly elastic, the ratio being equal to that of m to n, the velocity acquired in the opposite direction to its original velocity whilst the shape is being changed, by which the original velocity is cancelled, will bear this same ratio to the velocity acquired whilst the shape is being restored, that is, the velocity with which it rebounds.*

The sum of the squares of the velocities, each multiplied by the corresponding mass, remains unaltered in the case of perfectly elastic bodies.

275. There is also another theorem, which is rather more laborious, but it is a general & elegant theorem, discovered by Huygens for perfectly elastic solids. Namely, that the sum of the squares of the velocities, each multiplied by the corresponding mass, remains the same after the impact as it was before it. Now, the velocities after impact are $C - \frac{2q}{Q+q} \times (C - c)$, & $c + \frac{2Q}{Q+q} \times (C - c)$; the squares of these, multiplied by the masses contain three terms each; the first are QCC & qcc: the second are $(-CC + Cc) \times \frac{4Qq}{Q+q}$ & $(cC - cc) \times \frac{4Qq}{Q+q}$, & the sum of these reduces to $(-CC + 2Cc - cc) \times \frac{4Qq}{Q+q}$: the last are $\frac{4Qqq}{(Q+q)^2}$ $(CC - 2Cc + cc)$, & $\frac{4qQQ}{(Q+q)^2} \times$ $(CC - 2Cc + cc)$, or added together $\frac{4(Q+q) \times Qq}{(Q+q)^2} \times (CC - 2Cc + cc)$, or $\frac{4Qq}{Q+q} \times (CC - 2Cc + cc)$, which will cancel the sum of the second terms; hence all that remains is $QCC + qcc$, the sum of the squares of the original velocities, each multiplied by the corresponding mass. This equality does not hold good for soft bodies, nor yet for imperfectly elastic bodies.

The usual method for oblique impact by means of resolution of forces.

276. Coming now to oblique impacts, suppose that, in Fig. 42, the two spheres A & C at some given time, moving along any straight lines AB, CD, which measure their velocities, come into physical contact in the positions B & D, where the mutual forces now produce a sensible effect. In the usual method the effect of the impact is usually determined as follows. Join their centres by the line BD, & to this line, produced if necessary, draw the perpendiculars AF, CH, & complete the rectangles AFBE, CHDG; resolve each of the motions AB, CD in two, the former into AF, AE, or EB, FB, & the latter into CH, CG, or GD, HD. In either pair, the first remains unaltered; the second, FB, & HD, give the effect of direct impact. The direct velocities BI, DK are found by the law of impact; & these, according to laws of the kind set forth above, will after impact be different for different kinds of bodies. They are compounded with velocities represented by the straight lines BL, DQ, which are in the same straight lines as EB, GD respectively, & equal to them. This being done, BM, DP will represent the velocities & the directions of motion after collision.

Composition of forces substituted for resolution.

277. In this method, there is considered to be a resolution of motions, as if there were a certain real resolution into two parts, of which the one part persisted unchanged, & the other part suffered alteration; & in the case, for which the figure has been drawn, the latter is altogether destroyed & a fresh motion is again produced. But the matter really proceeds without any real resolution in the following manner. The mutual force acting upon the spheres B, D, gives to them during the complete time of impact opposite velocities BN, DS, which are also equal, in the case for which the figure is drawn, to those two, of which the one is considered to be destroyed & the other to be produced. These motions, compounded with BO, DR, drawn in the directions of AB, CD & equal to them, & thus representing the whole effects of the original velocities, will represent the velocities BM, DP. For it is easily seen that LO is equal to AE, or FB; & thus MO is equal to NB, & BNMO will be a parallelogram; in the same manner it can be shown that DRPS is a parallelogram. Therefore, there is in reality no true resolution, but only a composition of motions, the original velocity persisting throughout on account of the force of inertia; & this is compounded with the new velocity generated by the forces which act during the impact.

Compositio resolutioni substituta etiam ubi globus incurrit in planum immobile.

278. Idem etiam mihi accidit, ubi oblique globus incurrit in planum, sive consideretur motus, qui haberi debet deinde, sive percussionis obliquæ energia respectu perpendicularis. Deveniat in fig. 43 globus A cum directione obliqua AB ad planum [132] CD consideratum ut immobile, quod contingat physice in N, & concipiatur planum GI parallelum priori ductum per centrum B, ad quod appellet ipsum centrum, & a quo resiliet, si resilit. Dueta AF perpendiculari ad GI, & completo parallelogrammo AFBE, in communi methodo resolvitur velocitas AB in duas AF, AE; sive FB, EB, primam dicunt manere illæsam, secundam destrui a resistentia plani: tum perseverare illam solam per BI æqualem ipsi FB; si corpus incurrens sit perfecte molle, vel componi cum alia in perfecte elasticis BE æquali priori EB, in imperfecte elasticis B*e*, quæ ad priorem EB habeat rationem datam, & percurrere in primo casu BI, in secundo BM, in tertio B*m*. At in mea Theoria globus a viribus in illa minima distantia agentibus, quæ ibi sunt repulsivæ, acquirit secundum dircetionem NE perpendicularem plano repellenti CD in primo casu velocitatem BE, æqualem illi, quam acquireret, si

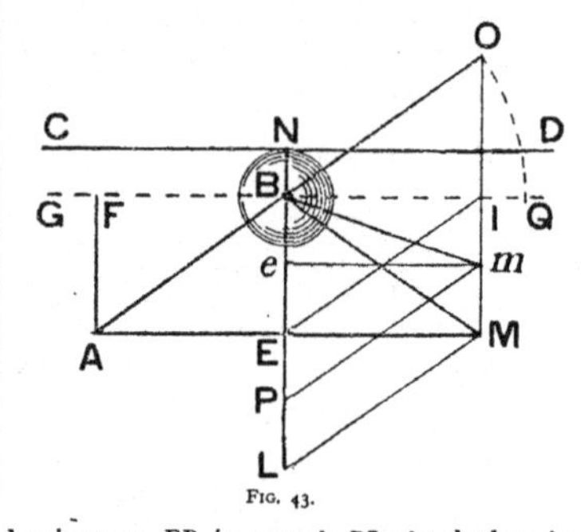

Fig. 43.

cum velocitate EB perpendiculariter advenisset per EB, in secundo BL ejus duplam, in tertio BP, quæ ad ipsam habeat illam rationem datam *r* ad 1, sive $m + n$ ad m, & habet deinde velocitatem compositam ex velocitate priore manente, ac expressa per BO æqualem AB, & positam ipsi in directum, ac ex altera BE, BL, BP, ex quibus constat, componi illas ipsas BI, BM, B*m*, quas prius; cum ob IO æqualem AF, sive EB, & IM, I*m* æquales BE, B*e*, sive EL, EP, totæ etiam BE, BP, BL totis OI, OM, O*m* sint æquales, & parallelæ.

Ubique in hac Theoria compositionem resolutioni substitui, easque sibi invicem æquivalere.

279. Res mihi per compositionem virium ubique eodem redit, quo in communi methodo per earum resolutionem. Resolutionem solent vulgo admittere in motibus, quos vocant impeditos, ubi vel planum subjectum, vel ripa ad latus procursum impediens, ut in fluviorum alveis, vel filum, aut virga sustentans, ut in pendulorum oscillationibus, impedit motum secundum eam directionem, qua agunt velocitates jam conceptæ, vel vires; ut & virium resolutionem agnoscunt, ubi binæ, vel plures etiam vires unius cujusdam vis alia directione agentis effectum impediunt, ut ubi grave a binis obliquis planis sustinetur, quorum utrumque premit directione ipsi plano perpendiculari, vel ubi a pluribus filis elasticis oblique sitis sustinetur. In omnibus istis casibus illi velocitatem, vel vim agnoscunt vere resolutam in duas, quarum utrique simul illa unica velocitas, vel vis æquivaleat, ex illis veluti partibus constituta, quarum si altera impediatur, debeat altera perseverare, vel si impediatur utraque, suum utraque effectum edat seorsum. At quoniam id impedimentum in mea Theoria nunquam habebitur ab immediato contactu plani rigidi subjecti, nec a virga vere rigida, & inflexili sustentante, sed semper a viribus mutuis repulsivis in primo casu, attractivis in secundo; semper habebitur nova velocitas, vel vis æqualis, & contraria illi, quam communis methodus elisam dicit, quæ cum [133] tota velocitate, vel vi obliqua composita eundem motum, vel idem æquilibrium restituet, ac idem omnino erit, in effectuum computatione considerare partes illas binas, & alteram, vel utramque impeditam, ac considerare priorem totam, aut velocitatem, aut vim, compositam cum iis novis contrariis, & æqualibus illi parti, vel illis partibus, quæ dicebantur elidi. In id autem, quod vel inferne, vel superne motum massæ cujuspiam impedit, vel vim, non aget pars illa prioris velocitatis, vel illius vis, quæ concipitur resoluta, sed velocitas orta a vi mutua, & contraria velocitati illi novæ genitæ in eadem massa, a vi mutua, vel ipsa vis mutua, quæ semper debet agere in partes contrarias, & cui occasionem præbet illa determinata distantia major, vel minor, quam sit, quæ limites, & æquilibrium constitueret.

Exemplum rei in ipso globo molli incurrente in planum immobile.

280. Id quidem abunde apparet in ipso superiore exemplo. Ibi in fig. 43 globus (quem concipamus mollem) advenit oblique per AB, & oblique impeditur a plano ejus progressus. Non est velocitas perpendicularis AF, vel EB, quæ extinguitur, durante AE, vel FB, uti diximus; nec illa ursit planum CD. Velocitas AB occasionem dedit globo accedendi ad planum CD usque ad eam exiguam distantiam, in qua vires variæ agerent;

Composition substituted also for resolution in the case of a sphere impinging on an immovable plane.

278. The same thing comes about in my theory, when a sphere impinges obliquely on a plane, whether the motion which it must have after impact is under consideration, or whether we are considering the energy of oblique percussion with regard to the perpendicular to the plane. Thus, in Fig. 43, suppose a sphere A to move along the oblique direction AB & to arrive at the plane CD, which is considered to be immovable, & with which the sphere makes physical contact at the point N. Now imagine a plane GI, parallel to the former, to be drawn through the centre B; to this plane the centre of the sphere will attain, & rebound from it, if there is any rebound. After drawing AF perpendicular to GI & completing the parallelogram AFBE, the usual method continues by resolving the velocity AB into the two velocities AF, AE, or FB, EB; of these, the first is stated to remain constant, whilst the second is destroyed by the resistance of the plane; & all that remains after impact is represented by BI, which is equal to FB, if the body is soft; or that this is compounded with another represented by BE, equal to the original velocity EB, in the case of perfectly elastic bodies; and in the case of imperfectly elastic bodies, it is compounded with B*e*, which bears a given ratio to the original EB. Then the sphere will move off, in the first case along BI, in the second case along BM, & in the third case along B*m*. But, according to my Theory the sphere, on account of the action of forces at those very small distances, which are in that case repulsive, acquires in the direction NE perpendicular to the repelling plane CD, in the first case a velocity BE equal to that which it would have acquired if it had travelled along EB with a velocity EB at right angles to the plane; in the second case, it acquires a velocity double of this, namely BL, & in the third a velocity BP, which bears to BE the given ratio r to 1, i.e., $m + n : m$. After impact it has a velocity compounded of the original velocity which persists, expressed by BO equal to AB, & drawn in the same direction as AB, with another velocity, either BE, BL, or BP; from which it is easily shown that there results either BI, BM, or B*m*, just as in the usual method. For, since IO, AF, or EB, & IM, I*m* are respectively equal to BE, B*e*, or EL, EP; hence the wholes BE, BP, BL are also respectively equal to the wholes OI, OM, O*m*, & are parallel to them.

In every case, in my Theory, composition is used instead of resolution; & these are equivalent to one another.

279. The matter, in my hands, comes to the same thing in every case with composition of forces, as in the usual method is obtained by resolution. In the usual method it is customary to admit resolution for motions which are termed impeded, for instance, when a bordering plane, or a bank, impedes progress to one side, as in the channels of rivers; a string, or a sustaining rod, as in the oscillations of pendulums hinders motion in the direction in which the velocities or forces are in that case supposed to be acting. In a similar manner, they recognize resolution of forces, when two, or even more forces impede the effect of some one force acting in another direction; for instance, when a heavy body is sustained by two inclined planes, each of which exerts a pressure on the body in a direction perpendicular to itself, or when such a body is suspended by several elastic strings in inclined positions. In all these cases, the velocity of force is taken to be really resolved into two; to both of these taken together the single velocity or force will be equivalent, being as it were compounded of these parts, of which if one is impeded, the other will still persist, or if both are impeded, they will each produce their own effect separately. Now, since in my Theory there never is such impediment, caused by an immediate contact with the bordering plane, nor by a truly rigid or inflexible sustaining rod, but always considered to be due to mutual forces, that are repulsive in the first case & attractive in the second case, a new velocity or force, equal & opposite to that which is in the usual theory supposed to be destroyed, is obtained. This velocity, or force, combined with the whole oblique velocity or force, will give the same motion or the same equilibrium; & it will come to exactly the same thing, when computing the effects, if we consider the two velocities, or forces, either one or the other, or both, to be impeded, as it would to consider the original velocity, or force, to be compounded with the new velocities, or forces, which are opposite in direction & equal to that part or parts which are said to be destroyed. Moreover, upon the object which hinders the motion, or force, of any mass upwards or downwards, it is not the part of the original velocity, or force, which is said to be resolved, that will act; but it is the velocity arising from the mutual force, opposite in direction to that velocity which is newly generated in the mass by the mutual force, or the mutual force itself. This must always act in opposite directions; & is governed by the given distance, greater or less than that which gives the limit-points & equilibrium.

A case in point where a soft sphere impinges on an immovable Plane.

280. This fact indeed is seen clearly enough in the example given above. There, in Fig. 43, the sphere, which we will suppose to be soft, travels obliquely along AB, & its progress is impeded, also obliquely, by the plane. It is not true that the perpendicular velocity AF, or EB is destroyed, whilst AE, or FB persists, as we have already proved; nor was there any direct pressure from it on the plane CD. The velocity AB made the sphere approach the plane CD to within a very small distance from it, at which various forces come into action;

donec ex omnibus actionibus conjunctis impediretur accessus ad ipsum planum, sive perpendicularis distantiæ ulterior diminutio. Illæ vires agent simul in directione perpendiculari ad ipsum planum juxta num. 266: debebunt autem, ut impediant ejusmodi ulteriorem accessum, producere in ipso globo velocitatem, quæ composita cum tota BO perseverante in eadem directione AB, exhibeat velocitatem per BI parallelam CD. Quoniam vero triangula rectangula AFB, BIO æqualia erunt necessario ob AB, BO æquales; erit BEIO parallelogrammum, adeoque velocitas perpendicularis, quae cum priore velocitate BO debeat componere velocitatem per rectam parallelam plano, debebit necessario esse contraria, & æqualis illi ipsi EB perpendiculari eidem plano, in quam resolvunt vulgo velocitatem AB. Interea vero vis, quæ semper agit in partes contrarias æqualiter, urserit planum tantundem, & omnes in eo produxerit effectus illos, qui vulgo tribuuntur globo advenienti cum velocitate ejusmodi, ut perpendicularis ejus pars sit EB.

Aliud globi descendentis per planum inclinatum.

281. Idem accidet etiam in reliquis omnibus casibus superius memoratis. Descendat globus gravis per planum inclinatum CD (fig. 44) oblique, quod in communi sententia continget hunc in modum. Resolvunt gravitatem BO in duas, alteram BR perpendicularem plano CD, qua urgetur ipsum planum, quod eum sustinet; alteram BI, parallelam eidem plano, quæ obliquum descensum accelerat. In mea Theoria gravitas cogit globum semper magis accedere ad planum CD; donec distantia ab eodem evadat ejusmodi; ut vires mutuæ [134] repulsivæ agant, & illa quidem, quæ agit in B, sit ejusmodi ut composita cum BO exhibeat BI parallelam plano ipsi, adeoque non inducentem ulteriorem accessum, sit autem perpendicularis plano ipsi. Porro ejusmodi est BE, jacens in directum cum RB, & ipsi æqualis, cum nimirum debeat esse parallela, & æqualis OI. Vis autem æqualis ipsi, & contraria, adeoque expressa per RB, urgebit planum.

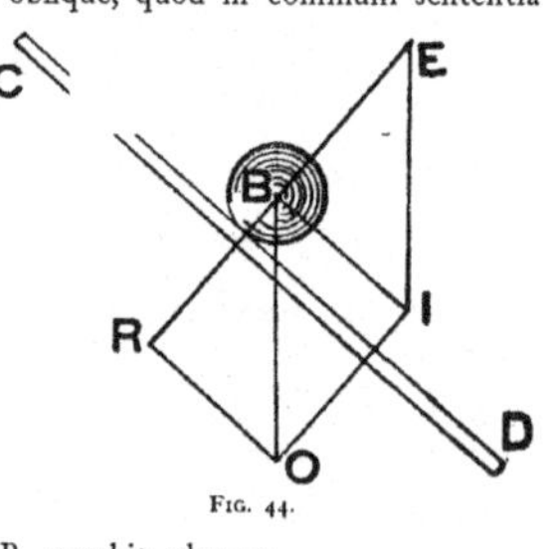

FIG. 44.

Aliud in pendulo.

282. Quod si grave suspensum in fig. 45 filo, vel virga BC debeat oblique descendere per arcum circuli BD; tum vero in communi methodo gravitatem BO itidem resolvunt in duas BR, BI, quarum prima filum, vel virgam tendat, & elidatur, secunda acceleret descensum obliquum, qui fieret ex velocitate concepta per rectam BA perpendicularem BC, ac præterea etiam tensionem fili agnoscunt ortum a vi centrifuga, quæ exprimitur per DA perpendicularem tangenti. At in mea Theoria res hoc pacto procedit. Globus ex B abit ad D per vires tres compositas simul cum velocitate præcedente; prima e viribus est vis gravitatis BO; secunda attractio versus C orta a tensione fili, vel virgæ, expressa per BE parallelam, & æqualem OI, adeoque RB, quæ solæ componerent vim BI; tertia est attractio in C expressa per BH æqualem AD orta itidem a tensione fili respondente vi centrifugæ, & incurvante motum. Adest præterca velocitas præcedens, quam exprimit BK æqualis IA, ut sit BI æqualis KA. His viribus cum ea velocitate simul agentibus crit globus in D in fine ejus tempusculi, cui ejusmodi effectus illarum virium respondent. Nam ibi debet esse, ubi esset, si aliæ ex illis causis agerent post alias: gravitate agente veniret per BO, vi BE abiret per OI, velocitate BK abiret per IA ipsi æqualem, vi BH abiret per AD. Quamobrem res tota itidem peragitur sola compositione virium, & motuum.

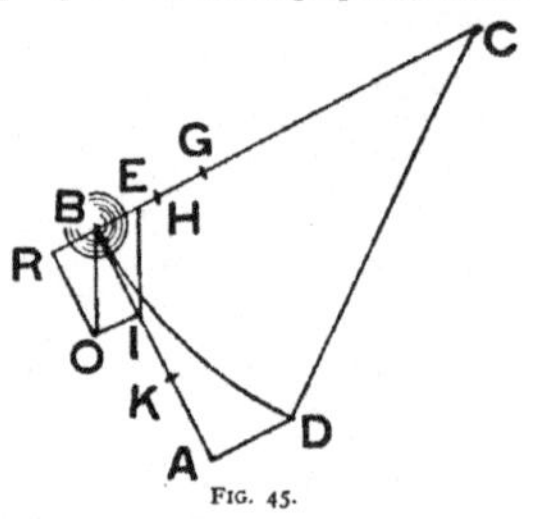

FIG. 45.

Alia ratio componendi vires in eodem casu.

283. Porro si sumatur EG æqualis BH; tum tota attractio orta a tensione fili crit BG, quæ prius considerata est tanquam e binis partibus in directum agentibus composita, ac res eodem redit; nam si prius componantur BH, & BE in BG (quo casu tota BG ut unica vis haberetur), tum BO, ac demum BK, ad idem punctum D rediretur juxta generalem demonstrationem, quam dedi num. 259. Jam vero vi expressa per totam BG attraheretur ad centrum suspensionis C ab integra tensione fili, ubi pars EG, vel BH ad partem BE habet proportionem pendentem a celeritate BK, ab angulo RBO, ac a radio CB; sed ista meæ Theoriæ cum omnium usitatis Mechanicæ elementis communia sunt, posteaquam compositionis hujus cum illa resolutione æquivalentia est demonstrata.

Aliud exemplum in globo sustentato a binis planis. Difficultas communis methodi in eodem.

284. Quæ de motu diximus facto vi oblique, sed non penitus impedita, eadem in æquilibrio habent locum, ubi omnis impeditur motus. Innitatur globus gravis B in fig. 46 binis planis AC, CD, quæ accurate, vel in mea Theoria [135] physice solum, contingat

then, under the combined actions of all the forces further approach toward the plane, or further diminution of the perpendicular distance from it, is impeded. The forces act together in the direction perpendicular to the plane, as was shown in Art. 266; & they must, in order to impede further approach of this kind, produce in the sphere itself a velocity which, compounded with the whole velocity that persists throughout, namely BP, in the direction of AB, will give a velocity represented by BI parallel to CD. But, since the right-angled triangles AFB, BIO are necessarily congruent on account of the equality of AB & BO, it follows that BEIO is a parallelogram. Hence, the perpendicular velocity, which has, when combined with the original velocity BO, to give a resultant represented by a straight line parallel to the plane, must of necessity be equal & opposite to that represented by EB, also perpendicular to the plane, into which commonly the velocity AB is resolved. Meanwhile, the force, which always acts equally in opposite directions, would act on the plane to precisely the same extent, & all those effects would be produced on it, which are commonly attributed to the sphere striking it with a velocity of such sort that its perpendicular part is EB.

Another case in point, that of a sphere descending along an inclined Plane.

281. The same thing happens also in the rest of the cases mentioned above. Let a heavy sphere descend along the inclined plane CD, in Fig. 44; the descent takes place, according to the customary idea, in the following manner. Gravity, represented by BO, is resolved into two parts, the one, BR, perpendicular to the plane CD, acts upon the plane & is resisted by it; the other, BI, parallel to the plane, accelerates the oblique descent. According to my Theory, gravity forces the sphere to approach the plane CD ever nearer & nearer, until the distance from it becomes such as that for which the repulsive forces come into action; that which acts on B is such that, when combined with BO, will give a velocity represented by BI parallel to the plane, & thus does not induce further approach; moreover it is perpendicular to the plane. Further, it is such as BE, lying in the same straight line as RB, & equal to it, because indeed it must be parallel & equal to OI. Lastly, a force that is equal & opposite, & so represented by BR, will act upon the plane.

The Pendulum is another case in point.

282. But if, in Fig. 45, a heavy body is suspended by a string or rod BC, it is bound to descend obliquely along the circular arc BD. Now, in the usual method, gravity, represented by BO, is again resolved into two parts, BR & BI; the first of these exerts a pull on the string or rod & is destroyed; the second accelerates the oblique descent, which would come about through a velocity supposed to act along BA perpendicular to BC; in addition to these, account is taken of the tension of the string arising from a centrifugal force, which is represented by DA perpendicular to the tangent. But, according to my Theory, the matter goes in this way. The sphere passes from B to D, under the action of three forces compounded with the original velocity. The first of these forces is gravity, BO; the second is the attraction towards C arising from the tension of the string or rod, & represented by BE, parallel & equal to OI, & thus also to RB, these two alone, taken together, give a force BI. The third is an attraction towards C, represented by BH, equal to AD, arising also from the tension of the string corresponding to the centrifugal force & incurving the motion. In addition to these, we have the original velocity, represented by BK, equal to IA, so that BI is equal to KA. If such forces as these act together with this velocity, the sphere will arrive at D at the end of the interval of time to which such effects of the forces correspond. For it must reach that point at which it would be, if all these causes acted one after the other; &, with gravity acting, it would travel along BO; with the force BE acting it would pass along OI; with the velocity BK, it would traverse IA, which is equal to BK; & with the force BH acting, it would go from A to D. Hence, in this case also, the whole matter is accomplished with composition alone, for forces & motions.

Another manner of compounding the forces in the case just considered.

283. Further, if EG is taken equal to BH; then the whole attraction arising from the tension of the string will be BG, which previously was considered only as being compounded of two parts acting in the same straight line; & it comes to the same thing as before. For, if BH & BE are first of all compounded into BG (in which case BG is reckoned as a single force), then BO is taken into account, & finally BK; we shall be led to the same point D, according to the general demonstration I gave in Art. 259. Now we have an attraction to the centre of suspension C due to a force expressed by the whole BG, where the part of it, EG, or BH, bears to the part BH a ratio that depends upon the velocity BK, the angle RBO, & the radius CB. The results of my Theory are in agreement with the elementary principles of Mechanics accepted by everyone else, as soon as the equivalence of my composition with their resolution has been demonstrated.

Another example; a sphere supported by two planes; difficulty with the usual method in this case.

284. The same things hold good in the case of equilibrium, where all motion is impeded, as those we have already spoken of with respect to motion derived from a force acting obliquely, but not altogether impeded. In Fig. 46, a heavy sphere is supported by two planes AC, CD, which actually, or in my Theory physically only, it touches at H & F;

in H, & F, & gravitatem referat recta verticalis BO, ac ex puncto O ad rectas BH, BF ducantur rectæ OR, OI parallelæ ipsis BF, BH, & producta sursum BK tantundem, ducantur ex K ipsis BF, BH parallelæ KE, KL usque ad easdem BH, BF; ac patet, fore rectas BE, BL æquales, & contrarias BR, BI. In communi methodo resolutionis virium concipitur gravitas BO resoluta in binas BR, BI, quarum prima urgeat planum AC, secunda DC; & quoniam si angulus HCF fuerit satis acutus; crit itidem satis acutus angulus R, qui ipsi æqualis esse debet, cum uterque sit complementum HBF ad duos rectos, alter ob parallelogrammum, alter ob angulos BHC, BFC rectos; fieri potest, ut singula latera BR, RO, sive BI, sint, quantum libuerit, longiora quam BO; vires singulæ, quæ urgent illa plana, possunt esse, quantum libuerit, majores, quam sola gravitas: mirantur multi, fieri posse, ut gravitas per solam ejusmodi applicationem tantum quodammodo supra se assurgat, & effectum tanto majorem edat.

Solutio in ipsa methodo communi: in hac Theoria nullum ipsi difficultati esse locum.

285. Difficultas ejusmodi in communi etiam sententia evitari facile potest exemplo vectis, de quo agemus infra, in quo sola applicatio vis in multo majore distantia collocatæ multo majorem effectum edit. Verum in mea Theoria ne ullus quidem difficultati est locus. Non resolvitur revera gravitas in duas vires BR, BI, quarum singulæ plana urgeant, sed gravitas inducit ejusmodi accessum ad ea plana, in quo vires repulsivæ perpendiculares ipsis planis agentes in globum componant vim BK æqualem, & contrariam gravitati BO, quam sustineat, & ulteriorem accessum impediat. Ad id præstandum requiruntur illæ vires BE, BL æquales, & contrariæ hisce BR, BI, quæ rem conficiunt. Sed quoniam vires sunt mutuæ, habebuntur repulsiones agentes in ipsa plana contrariæ, & æquales illis ipsis BE, BL, adeoque agent vires expressæ per illas ipsas BR, BI, in quas communis methodus gravitatem resolvit.

Aliud in globo suspenso filis obliquis.

286. Quod si globus gravis P in fig. 47 e filo BP pendeat, ac sustineatur ab obliquis filis AB, DB, exprimat autem BH gravitatem, & sit BK ipsi contraria, & æqualis, ac sint HI, KL parallelæ DB, & HR, KE parallelæ filo AB; communis methodus resolvit gravitatem BH in duas BR, BI, quæ a filis sustineantur, & illa tendant; sed ego compono vim BK gravitati contrariam, & æqualem e viribus BE, BL, quas exerunt attractivas puncta fili, quæ ob pondus P delatum deorsum sua gravitate ita distrahuntur a se invicem, donec habeantur vires attractivæ componentes ejusmodi vim contrariam, & æqualem gravitati.

Conclusio generalis, pro hac theoria quæ omnia præstat per solam compositionem.

287. Quamobrem per omnia casuum diversorum genera pervagati jam vidimus, nullam esse uspiam in mea Theoria veram aut virium, aut motuum resolutionem, sed omnia prorsus phænomena pendere a sola compositione virium, & motuum, adeo-[136]-que naturam eodem ubique modo simplicissimo agere, componendo tantummodo vires, & motus plures, sive edendo simul eum effectum, quem ederent illæ omnes causæ; si aliæ post alias effectus ederent suos æquales, & candem habentes directionem cum iis, quos singulæ, si solæ essent, producerent. Et quidem id generale esse Theoriæ meæ, patet vel ex eo, quod nulli possunt esse motus ex parte impediti, ubi nullus est immediatus contactus, sed in libero vacuo spatio punctum quodvis liberrime movetur parendo simul velocitati, quam habet jam acquisitam, & viribus omnibus, quæ ab aliis omnibus pendent materiæ punctis.

Resolutio tantum mente concepta sæpe utilis ad contrahendas solutiones.

288. Quanquam autem habeatur revera sola compositio virium; licebit adhuc vires imaginatione nostra resolvere in plures, quod sæpe demonstrationes theorematum, & solutionem problematum contrahet mirum in modum, ac expeditiores reddet, & elegantiores; nam licebit pro unica vi assumere vires illas, ex quibus ea componeretur. Quoniam enim idem omnino effectus oriri debet, sive adsit unica vis componens, sive reapse habeantur simul plures illæ vires componentes; manifestum est, substitutione harum pro illa nihil turbari conclusiones, quæ inde deducuntur: & si post resolutionem ejusmodi inveniatur vis contraria, & æqualis alicui e viribus, in quas vis illa data resolvitur; illa haberi potest pro nulla consideratis solis reliquis, si in plures resoluta fuit, vel sola altera reliqua, si resoluta fuit in duas. Nam componendo vim, quæ resolvitur, cum illa contraria uni ex iis, in quas resolvitur, eadem vis provenire debet omnino, quæ oritur componendo simul reliquas, quæ fuerant in resolutione sociæ illius elisæ, vel retinendo unicam illam alteram reliquam, si resolutio facta est in duas tantummodo; atque id ipsum constat pro resolutione in duas ipsis superioribus exemplis, & pro quacunque resolutione in vires quotcunque facile demonstratur.

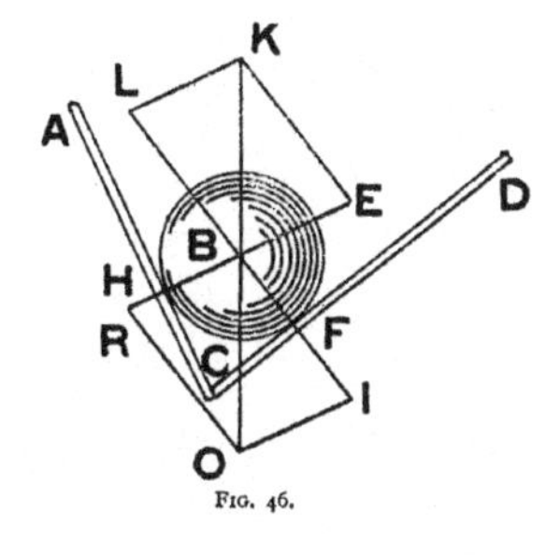

Fig. 46.

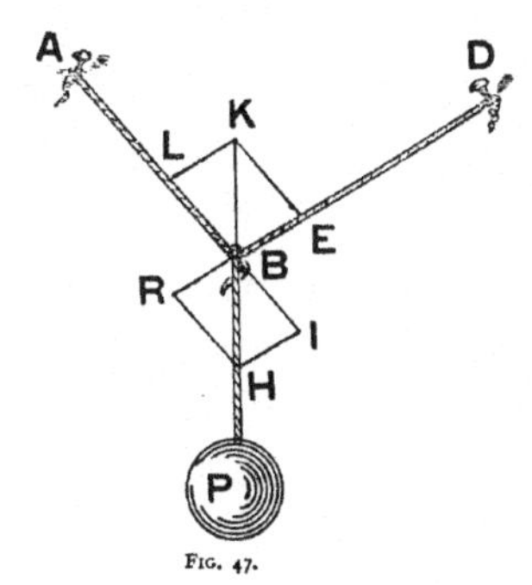

Fig. 47.

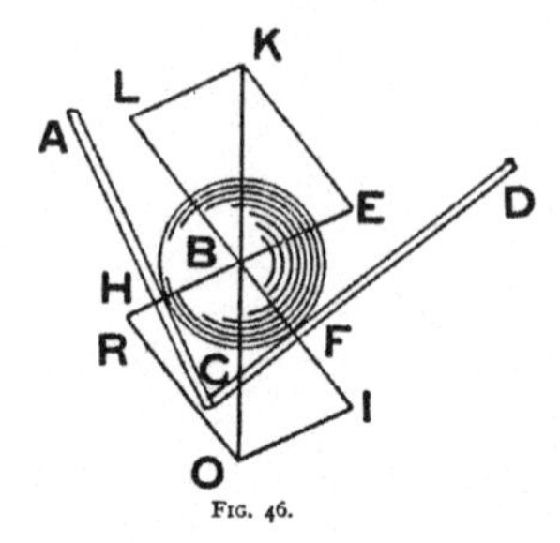

FIG. 46.

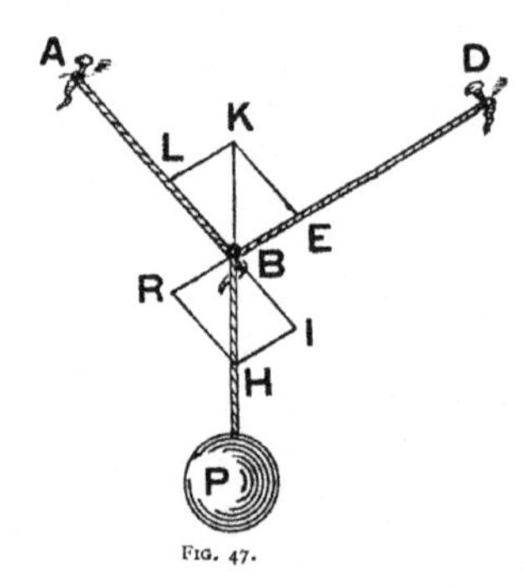

FIG. 47.

let the vertical line BO represent gravity, & draw from the point O, to meet the straight lines BH, BF, the straight lines OR, OI parallel to BF, BH; also producing BK upwards to the same extent, draw through K the straight line KE, KL, parallel to BF, BH to meet BH & BF. Then it is clear that BE, BL will be equal & opposite to BR, BI. Now, according to the usual method by means of resolution of forces, the gravity BO is supposed to be resolved into the two parts BR, BI, of which the first acts upon the plane AC & the second upon DC. Also if the angle HCF is sufficiently acute, then the angle at R is also sufficiently acute; for these angles must be equal to one another. For, each is the supplement of the angle HBF, the one in the parallelogram, the other on account of BHC & BFC being right angles. This being so, it may happen that each of the sides BR, RO, or BI, will be greater than BO, to any desired extent. Thus each of the forces, which act upon the planes, may be greater than gravity alone, to any desired extent. Many will wonder that it is possible that gravity, by a mere application of this kind, surpasses itself to so great an extent, & gives an effect that is so much greater.

Answer to the difficulty by the ordinary method; in my Theory there is no occasion for any difficulty.

285. A difficulty of this kind even according to the ordinary opinion is easily avoided by comparing the case of the lever, with which we will deal later; in it the mere application of a force situated at a much greater distance gives a far greater effect. But with my Theory there is no occasion for any difficulty of the sort. For there is no actual resolution of gravity into the two parts BR, BI, each acting on one of the planes; but gravity induces an approach to the planes, to within the distance at which repulsive forces acting perpendicular to the planes upon the sphere compound into a force BK, equal & opposite to the gravity BO; this force sustains the sphere & impedes further approach to the planes. To represent this, the forces BE, BL are required; these are equal & opposite to BR, BI; & that is all there need be said about the matter. Now, since the forces are mutual, there are repulsions acting upon the planes, & these repulsions are equal & opposite to BE, BL; & thus the forces acting are represented by BR, BI, which are those into which the ordinary method resolves gravity.

Explanation in the case of a sphere suspended by inclined strings.

286. But if, in Fig. 47, a heavy sphere P is suspended by a string BP, & this is held up by inclined strings AB, DB, & gravity is represented by BH; let BK be equal & opposite to it, & let HI, KL be parallel to the string DB, & HR, KE parallel to the string AB. The ordinary method resolves the gravity BH into the two parts BR, BI, which are sustained by the strings & tend to elongate them. On the other hand, I compound the force BK, equal & opposite to gravity from the two forces BE, BL; these attractive forces are put forth by the points of the string, which, owing to the heavy body P suspended beneath are drawn apart by its gravity to such a distance that attractive component forces are obtained such as will give a force that is equal & opposite to the gravity of P.

General summing up in favour of this Theory, which gives everything by composition alone.

287. Having thus considered all sorts of different cases, we now see that there is nowhere in my Theory any real resolution either of forces or of motions; but that all phenomena depend on composition of forces & motions alone. Thus, nature in all cases acts in the same most simple manner, by compounding many forces & motions only; that is to say, by producing at one time that effect, which all the causes would produce, if they acted one after the other, & each produced that effect which was equal & in the same direction as that which it would produce if it alone acted. That this is a general principle of my Theory is otherwise evident from the fact that no motions can be in part impeded, where there is no immediate contact; on the contrary, any point can move in a free empty space in the freest manner, subject to the combined action of the velocity it has already acquired, & to all the forces which come from all other points of matter.

Resolution, although only a mental fiction, is yet often useful in shortening solutions.

288. Now, although as a matter of fact we can only have compositions of forces, yet we may mentally resolve our forces into several; & this will often shorten the proofs of theorems & the solution of problems in a wonderful manner, & render them more elegant & less cumbrous; for we may assume instead of a single force the forces from which it is compounded. For, since the same effect must always be produced, whether a single component force is present, or whether in fact we have the several component forces taken all together, it is plain that the conclusions that are derived will in no way be disturbed by the substitution of the latter for the former. If after such resolution a force is found, equal & opposite to any one of the forces into which the given force is resolved, then these two can be taken together as giving no effect; & only the rest need be considered if the given force was resolved into several parts, or only the other force if the given force was resolved into two parts. For, by compounding the force which was resolved with that force which is equal & opposite to the one of the forces into which it was resolved, the same force must be obtained as would arise from compounding all the other forces which were partners of the cancelled force in the resolution, or from retaining the single remaining force when the resolution was into two parts only. This has been shown to be the case for resolution in the two examples given above, & can be easily proved for any sort of resolution into forces of any number whatever.

Methodus generalis resolvendi vim in alias quotcunque.

289. Porro quod pertinet ad resolutionem in plures vires, vel motus, facile est ex iis, quæ dicta sunt num. 257 definire legem, quæ ipsam resolutionem rite dirigat, ut habeantur vires, quæ datam aliquam componant. Sit in fig. 48, vis quæcunque, vel motus AP, & incipiendo ab A ducantur quotcunque, & cujuscunque longitudinis rectæ AB, BC, CD, DE, EF, FG, GP, continuo inter se connexæ ita, ut incipiant ex A, ac desinant in P; & si ipsis BC, CD, &c. ducantur parallelæ, & æquales A*c*, A*d*, &c.; vires omnes AB, A*c*, A*d*, A*e*, A*f*, A*g*, A*p* component vim AP; unde patet illud: ad componendam vim quamcunque posse assumi vires quotcunque, & quascunque, quibus assumptis determinari poterit una alia præterea, quæ compositionem perficiat; nam poterunt duci rectæ AB, BC, CD, &c. parallelæ, & æquales datis quibuscunque, & ubi postremo deventum fuerit ad aliquod punctum G, satis erit addere vim expressam per GP.

Evolutio resolutionis in duas tantum.

[137] 290. Eo autem generali casu continetur particularis casus resolutionis in vires tantummodo duas, quæ potest fieri per duo quævis latera trianguli cujuscunque, ut in fig. 49, si datur vis AP, & fiat quodcunque triangulum ABP; vis resolvi potest in duas AB, BP, & data illarum altera, datur & altera, quod quidem constat etiam ex ipsa compositione, seu resolutione per parallelogrammum ABPC, quod semper compleri potest, & in quo AC est parallela, & æqualis BP, ac binæ vires AB, AC componunt vim AP: atque idem dicendum de motibus.

Cur vis composita sit minor componentibus simul sumptis.

291. Ejusmodi resolutio illud etiam palam faciet; cur vis composita a viribus non in directum jacentibus, sit minor ipsis componentibus, quæ nimirum sunt ex parte sibi invicem contrariæ, & clisis mutuo contrariis, & æqualibus, remanet in vi composita summa virium conspirantium, vel differentia oppositarum pertinentium ad componentes. Si enim in fig. 50, 51, 52 vis AP componatur ex viribus AB, AC, quæ sint latera parallelogrammi ABPC, & ducantur in AP perpendicula BE, CF, cadentibus E, & F inter A, & P in fig. 50, in A, & P in fig. 51, extra in fig. 52; satis patet, fore in prima, & postrema æqualia triangula AEB, PFC, adeoque vires EB, FC contrarias, & æquales elidi; vim vero AP in primo casu esse summam binarum virium conspirantium AE, AF, æquari unicæ AF in secundo, & fore differentiam in tertio oppositarum AE, AF.

Cur ea crescere videatur in resolutione: nihil inde posse deduci pro viribus vivis.

292. In resolutione quidem vis crescit quodammodo; quia mente adjungimus alias oppositas, & æquales, quæ adjunctæ cum se invicem elidant, rem non turbant. Sic in fig. 52 resolvendo AP in binas AB, AC, adjicimus ipsi AP binas AE, PF contrarias, & præterea in directione perpendiculari binas EB, FC itidem contrarias, & æquales. Cum resolutio non sit realis, sed imaginaria tantummodo ad faciliorem problematum solutionem; nihil inde difficultatis afferri potest contra communem methodum concipiendi vires, quas huc usque consideravimus, & quæ momento temporis exercent solum nisum, sive pressionem; unde etiam fit, ut dicantur vires mortuæ, & idcirco solum continuo durantes tempore sine contraria aliqua vi, quæ illas elidat, velocitatem inducunt, ut causæ velocitatis ipsius inductæ: nec inde argumentum ullum desumi poterit pro admittendis illis, quas Leibnitius invexit primus, & vires vivas appellavit, quas hinc potissimum necessario saltem concipiendas esse arbitrantur nonnulli, ne nimirum in resolutione virium habeatur effectus non æqualis suæ causæ. Effectus quidem non æqualis, sed proportionalis esse debet, non causæ, sed actioni causæ, ubi ejusmodi actio contraria aliqua actione non impeditur vel tota, vel ex parte, quod accidit, uti vidimus, in obliqua compositione: ac utcunque & aliæ responsiones sint in communi etiam sententia pro casu resolutionis; [138] in mea Theoria, cum ipsa resolutio realis nulla sit, nulla itidem est, uti monui, difficultas.

Satis patere ex hac Theoria, *Vires Vivas in Natura nullas esse.*

293. Et quidem tam ex iis, quæ huc usque demonstrata sunt, quam ex iis, quæ consequentur, satis apparebit, nullum usquam esse ejusmodi virium vivarum indicium, nullam necessitatem; cum omnia Naturæ phænomena pendeant a motibus, & æquilibrio, adeoque a viribus mortuis, & velocitatibus inductis per earum actiones, quam ipsam ob causam in illa dissertatione *De Viribus Vivis*, quæ hujus ipsius Theoriæ occasionem mihi præbuit ante annos 13, affirmavi, *Vires Vivas in Natura nullas esse*, & multa, quæ ad eas probandas proferri solebant, satis luculenter exposui per solas velocitates a viribus non vivis inductas.

IG. 48.

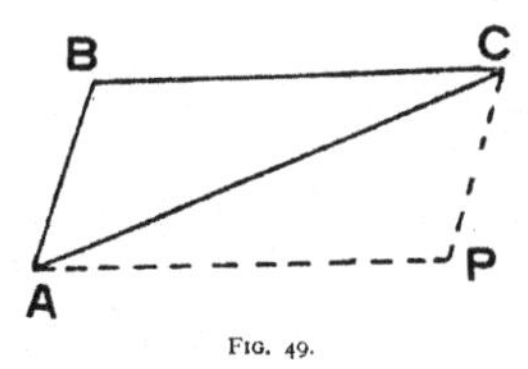

Fig. 49.

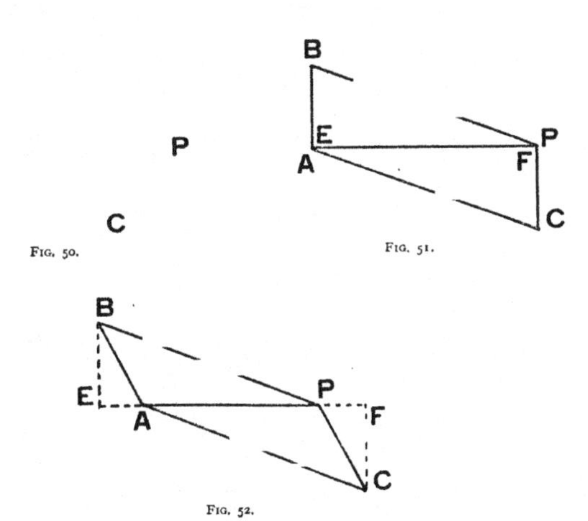

Fig. 50.

Fig. 51.

Fig. 52.

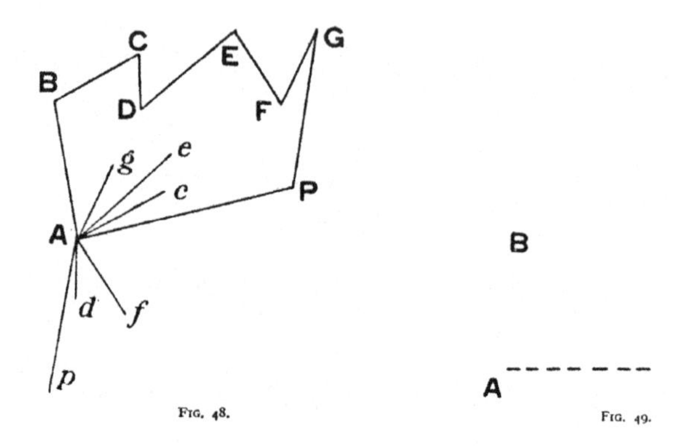

Fig. 48.

Fig. 49.

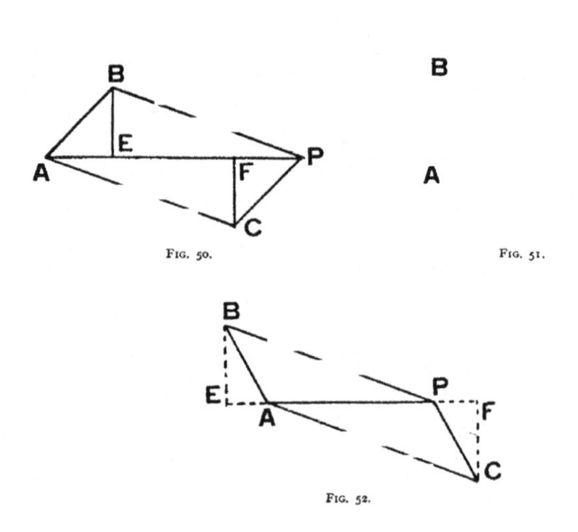

Fig. 50.

Fig. 51.

Fig. 52.

A general method for resolving a force into any number of other forces.

289. Further, as regards resolution into several forces or motions, it is easy, from what has been said in Art. 257, to determine a law which will rightly govern such resolution, so that the forces which compose any given force may be obtained. In Fig. 48, let AP be any force, or motion; starting from A, draw any number of straight lines of any length, AB, BC, CD, DE, EF, FG, GP, continuously joining one another so that they start from A & end up at P. Then, if to these lines AB, BC, &c., straight lines A*c*, A*d*, &c., are drawn, equal & parallel, all the forces represented by AB, A*c*, A*d*, A*e*, A*f*, A*g*, A*p*, will compound into a force AP. From this it is clear that, to make up any force, it is possible to assume any forces, & any number of them, & these being taken, it is possible to find one other force which will complete the composition. For, the straight lines AB, BC, CD, &c., can be drawn parallel & equal to any given lines whatever, & when finally they end up at some point G, it will be sufficient to add the force represented by GP.

Derivation of the principle of resolution in two directions only.

290. Moreover the particular case of resolution into two forces only is contained in the general case. This can be accomplished by means of any two sides of any triangle. In Fig. 49, if AP is the given force, & any triangle ABP is constructed, then the force AP can be resolved into the two parts AB, BP; & if one of these is given, the other also is given. This indeed is manifest even from the composition itself, or from resolution by means of the parallelogram ABPC, which can be completed in every case; in this AC is parallel & equal to BP, & the two forces AB, AC will compound into the force AP. The same may be said with regard to motions.

Why the resultant force is less than the two component forces taken together.

291. Such a resolution also brings out clearly the reason why a force compounded from forces that do not lie in the same straight line is less than the sum of these components. These are indeed partly opposite to one another; &, when the equals & opposites have cancelled one another, there remains in the force compounded of them the sum of the forces that agree in direction or the difference of the opposites which relate to the components. For, in Figs. 50, 61, 62, if the force AP is compounded from the forces AB, AC, which are sides of the parallelogram ABPC, & BE, CF are drawn perpendicular to AP, E & F falling between A & P in Fig. 50, at A & P in Fig. 51, & beyond them in Fig. 52; then it is plain that, in the first & last cases the triangles AEB, PFC are equal, & thus the forces EB & FC, which are equal & opposite, cancel one another. But the force AP in the first case is the sum of the two forces AE, AF acting in the same direction; it is equal to the single force AF in the second case; & in the third case it is equal to the difference of the opposite forces AE, AF.

The reason why the force seems to increase in resolution: no argument for living forces to be derived from this.

292. In resolution there is indeed some sort of increase of force. The reason for this is that mentally we add on other equal & opposite forces, which taken together cancel one another, & thus do not have any disturbing effect. Thus, in Fig. 52, by resolving the force AP into the two forces AB, AC, we really add to AP the two equal & opposite forces AE, PF, &, in addition, in a direction at right angles to AP, the two forces EB, FC, which are also equal & opposite. Now, since resolution is not real, but only imaginary, & merely used for the purpose of making the solution of problems easier; no exception can be taken on this account to the usual method of considering forces such as we have hitherto discussed, such as exert for an instant of time merely a stress or a pressure; for which reason they are termed dead forces, & because, whilst they last for a continuous time without any contrary force to cancel them, they yet only produce velocity, they are looked upon as the causes of the velocity produced. Nor from this can any argument be derived in favour of admitting the existence of those forces, which were first introduced by Leibniz, & called by him living forces. These forces some people consider must at least be supposed to exist, in order that in the resolution of forces, for instance, there should not be obtained an effect unequal to its cause. Now the effect must be proportional, & not equal; also it must be proportional, not to the cause, but to the action of the cause, where an action of this kind is not impeded, either wholly or in part, by some equal & opposite action, which happens, as we have seen, in oblique composition. But, whatever may be the various arguments, according to the usual opinion, to meet the difficulties in the case of resolution, since, in my Theory, there is no real resolution, there is no difficulty, as I have already said.

It is sufficient to prove from my Theory that there do not exist in Nature any living forces.

293. Indeed it will be sufficiently evident, both from what has already been proved, as well as from what will follow, that there is nowhere any sign of such living forces, nor is there any necessity. For all the phenomena of Nature depend upon motions & equilibrium, & thus from dead forces & the velocities induced by the action of such forces. For this reason, in the dissertation *De Viribus Vivis*, which was what led me to this Theory thirteen years ago, I asserted that *there are no living forces in Nature*, & that many things were usually brought forward to prove their existence, I explained clearly enough by velocities derived solely from forces that were not living forces.

Impactus obliquus globi elastici in quatuor globos, qui pro iis afferri solet

294. Unum hic proferam, quod pertinet ad collisionem globorum elasticorum obliquam, quæ compositionem resolutioni substitutam illustrat. Sint in fig. 53 triangula ADB. BHG, GML rectangula in D, H, M ita, ut latera BD, GH, LM, sint æqualia singula dimidiæ basi AB, ac sint BG, GL, LQ parallelæ AD, BH, GM. Globus A cum velocitate AB = 2 incurrat in B in globum C sibi æqualem jacentem in DB producta : ex collisione obliqua dabit illi velocitatem CE = 1, æqualem suæ BD, quam amittet, & progredietur per BG cum velocitate = AD = $\sqrt{3}$. Ibi eodem pacto si inveniat globum I, dabit ipsi velocitatem IK = 1, amissa sua GH, & progredietur per GL cum $\sqrt{2}$; tum ibi dabit, globo O velocitatem OP = 1, amissa sua LM, & abibit cum LQ = 1, quam globo R, directe in eum incurrens, communicabit. Quare, ajunt, illa vi, quam habebat cum velocitate = 2, communicabit quatuor globis sibi æqualibus vires, quæ junguntur cum velocitatibus singulis = 1 ; ubi si vires fuerint itidem singulæ = 1, erit summa virium = 4, quæ cum fuerit simul cum velocitate = 2, vires sunt, non ut simplices velocitates in massis æqualibus, sed ut quadrata velocitatum.

Ejus explicatio in hac Theoria sine viribus vivis per solam compositionem.

295. At in mea Theoria id argumentum nullam sane vim habet. Globus A non transfert in globum C partem DB suæ velocitatis AB resolutæ in duas DB, TB, & cum ea partem suæ vis. Agit in globos vis nova mutua in partes oppositas, quæ alteri imprimit velocitatem CE, alteri BD. Velocitas prior globi A expressa per BF positam in directum cum AB, & ipsi æqualem, componitur cum hac nova accepta BD, & oritur velocitas BG minor ipsa BF ob obliquitatem compositionis. Eodem pacto nova vis mutua agit in globos in G, & I, in L, & O, in Q, & R, & velocitates novas primi globi GL, LQ, *zero*, componunt velocitates GH, & GN ; LM, & LS ; LQ, & QL, sine ulla aut vera resolutione, aut translatione vis vivæ, Natura in omni omnino casu, & in omni corporum genere agente prorsus eodem pacto.

Quid notandum idcirco, quod nulli sunt globi continui, aut plana continua, aut mathematicus contactus.

296. Sed quod attinet ad collisiones corporum, & motus [139] reflexos, unde digressi eramus ; inprimis illud monendum duco ; cum nulli mihi sint continui globi, nulla plana continua ; pleraque ex illis, quæ dicta sunt, habebunt locum tantummodo ad sensum, & proxime tantummodo, non accurate ; nam intervalla, quæ habentur inter puncta, inducent inæqualitates sane multas. Sic etiam in fig. 43. ubi globus delatus ad B incurrit in CD, mutatio viæ directionis non fiet in unico puncto B, sed per continuam curvaturam ; ac ubi globus reflectetur, ipsa reflexio non fiet in unico puncto B, sed per curvam quandam. Recta AB, per quam globus adveniet, non erit accurate recta, sed proxime ; nam vires ad distantias omnes constanti lege se extendunt, sed in majoribus distantiis sunt insensibiles ; nisi massa, in quam tenditur, sit enormis, ut est totius Terræ massa in quam sensibili vi tendunt gravia. At ubi globus advenerit satis prope planum CD ; incipiet incurvari etiam via centri, quæ quidem, jam attracto, jam repulso globo, serpet etiam, donec alicubi repulsio satis prævaleat ad omnem ejus perpendicularem velocitatem extinguendam (utar enim imposterum etiam ego vocabulis communibus a virium resolutione petitis, uti & superius aliquando usu fueram, & nunc quidem potiore jure, posteaquam demonstravi æquipollentiam veræ compositionis virium cum imaginaria resolutione), & retro etiam motum reflectat.

Lex reflexionis perfecte, & imperfecte elasticorum.

297. Et quidem si vires in accessu ad planum, ac in recessu a plano fuerint prorsus æquales inter se ; dimidia curva ab initio sensibilis curvaturæ usque ad minimam distantiam a plano crit prorsus æqualis, & similis reliquæ dimidiæ curvæ, quæ habebitur inde usque ad finem curvaturæ sensibilis, ac angulus incidentiæ erit æqualis angulo reflexionis. Id in casu, quem exprimit fig. 43, curva ob insensibilem ejus tractum considerata pro unico puncto, pro perfecte elasticis patet ex eo, quod in triangulis rectangulis AFB, MIB latera æqualia circa angulos rectos secum trahant æqualitatem angulorum ABF, MBI, quorum alter dicitur angulus incidentiæ, & alter reflexionis, ubi in imperfecte elasticis non habetur ejusmodi æqualitas, sed tantummodo constans ratio inter tangentem anguli incidentiæ, & tangentem anguli reflexionis, quæ nimirum ad radios æquales BF, BI sunt FA, & I*m*, & sunt juxta denominationem, quam supra adhibuimus num. 272, & retinuimus huc usque, ut *m* ad *n*.

P

N

O

L

K

M

G F

H

D B C E

A T

Fig. 53.

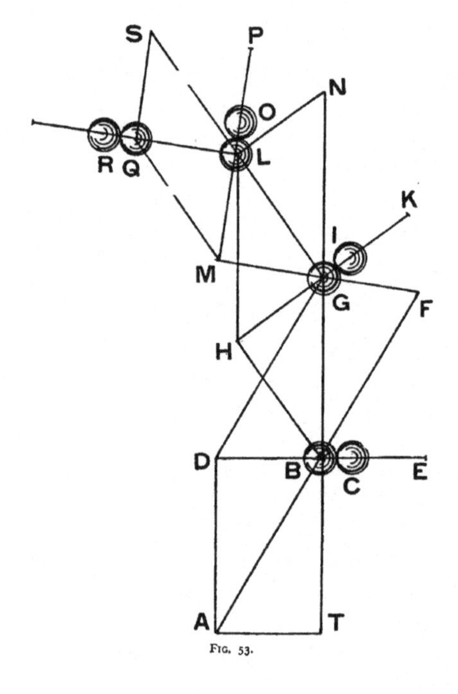

Fig. 53.

294. I will bring forward here one example, which deals with the oblique impact of elastic spheres; this will illustrate the substitution of composition for resolution. In Fig. 53, let ADB, BHG, GML, be right-angled triangles such that the sides BD, GH, LM are each equal to half the base AB, & let BG, GL, LQ be parallel to AD, BH, GM. Suppose the sphere A, moving with a velocity = 2, to impinge at B upon a sphere C, equal to itself, lying in DB produced. From the oblique impact, it will impart to C a velocity CE = 1, which is equal to its own velocity BD, which it loses; & it itself will go on along BG with a velocity equal to AD = $\sqrt{3}$. It will then come to the sphere I, will give to it a velocity IK = 1, losing its own velocity GH, & will go on along GL with a velocity equal to $\sqrt{2}$. Then it will give the sphere O a velocity OP = 1, losing its own velocity LM, & will go on with a velocity LQ = 1. This it will give up to the sphere R, on which it impinges directly. Wherefore, they contest, by means of the force which it had in connection with a velocity = 2, it will communicate to four spheres equal to itself forces, each of which is conjoined with a velocity = 1; hence, since, if each of the forces were also equal to 1, their sum would be equal to 4, & this sum was at the same time connected with a velocity = 2, it must be that the forces are not in the simple ratio of the velocities in equal masses but as their squares.

Oblique impact of a sphere on four spheres, an example usually brought forward in support of living forces.

295. But in my Theory this argument has no weight at all. The sphere A does not transfer to the sphere C that part DB of its velocity AB resolved into the two parts DB, TB; & with it part of its force. There acts on the spheres a new mutual force in opposite directions, which gives the velocity CE to the one sphere, & the velocity BD to the other. The previous velocity of the sphere A, represented by BF lying in the same direction as, and equal to, AB, is compounded with the newly received velocity BD, and the velocity BG, less than BF on account of the obliquity of the composition, is the result. In the same way, a new mutual force acts upon the spheres at G & I, at L & O, at Q & R, & the new velocities of the first sphere, GL, LQ & zero, are the resultants of the velocities GH & GN, LM & LS, & LQ & QL respectively; & there is not either any real resolution, or transference of living force. Nature in every case without exception, & for all classes of bodies acts in exactly the same manner.

Its explanation in my Theory without living forces by means of composition alone.

296. But we have digressed from the consideration of impact of bodies & reflected motions. Returning to them, I will first of all bring forward a point to be noted carefully. Since, to my idea, there are no such things as continuous spheres or continuous planes, many of the things that have been said are only true as far as we can observe, & only very approximately & not accurately; for the intervals, which exist between the points, induce a large number of inequalities. So also, in Fig. 43, where the sphere carried forward to B impinges upon the plane CD, the change in the direction of the path will not take place at the single point B, but by means of a continuous curvature. Also in the case where the sphere is reflected, the reflection will not occur at the single point B, but along a certain curve. The straight line AB, along which the sphere is approaching, will not accurately be a straight line, but approximately so; for the forces extend to all distances according to a fixed law, but at fairly great distances are insensible, unless the mass it is approaching is enormous, as in the case of the whole Earth, to which heavy bodies tend to approach with a sensible force. But as soon as the sphere comes sufficiently near to the plane CD, the path to the centre will begin to be curved, & indeed, as the sphere is first attracted & then repelled, the path will be winding, until it reaches a distance at which the repulsion will be strong enough to destroy all its perpendicular velocity (for in future I also will use the usual terms derived from resolution of forces, as I did once or twice in what has been given above; & this indeed I shall now do with greater justification seeing that I have proved the equivalence between true composition & imaginary resolution), & also will reflect the motion.

It is therefore to be noted that there are no continuous spheres or continuous planes, nor such a thing as mathematical contact.

297. Indeed, if the forces during the approach towards the plane & those during the recession from it were exactly equal to one another, then the half of the curve starting from the beginning of sensible curvature up to the least distance from the plane would be exactly equal & similar to the other half of the curve from this point to the end of sensible curvature, & the angle of incidence would be equal to the angle of reflection. This, in the case for which Fig. 43 is drawn, where on account of the insensible length of its arc the curve is considered as a single point, is evidently true for perfectly elastic bodies, from the fact that in the right-angled triangles AFB, MIB, the equal sides about the right angles involve the equality of the angles ABF, MBI, of which the first is called the angle of incidence & the second that of reflection; whereas, in imperfectly elastic bodies, there is no such equality, but only a constant ratio between the tangents of the angle of incidence & the tangent of the angle of reflection. For instance, these are, measured by the equal radii BF, BI, equal to FA, I*m*; & these latter are, according to the notation used above in Art. 272, & retained thus far, in the proportion of *m* to *n*.

Law of reflection for perfectly & imperfectly elastic bodies.

Eadem facta vi agente in aliqua distantia, considerata curvatura semitæ.

298. Curvaturam in reflexione exhibet figura 54, ubi via puncti mobilis repulsi a plano CO est ABQDM, quæ circa B, ubi vires incipiunt esse sensibiles, incipit ad sensum incurvari, & desinit in eadem distantia circa D. Ea quidem, si habeatur semper repulsio, incurvatur perpetuo in candem plagam, ut figura exhibet; si vero & attractio repulsionibus interferatur, serpit, uti monui; sed si paribus a plano distantiis vires æquales sunt; satis patet, & accuratissime demonstrari [140] etiam posset, ubi semel deventum sit alicubi, ut in Q, ad directionem parallelam plano, debere deinceps describi arcum QD prorsus æqualem, & similem arcui QB, & ita similiter positum respectu plani CO, ut ejus inclinationes ad ipsum planum in distantiis æqualibus ab eo, & a Q hinc, & inde sint prorsus æquales; quam ob causam tangentes BN, DP, quæ sunt quasi continuationes rectarum AB, MD, angulos faciunt ANC, MPO æquales, qui deinde habentur pro angulis incidentiæ, & reflexionis.

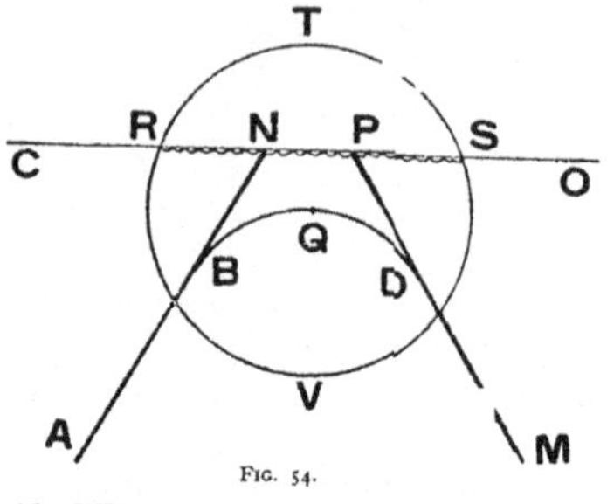

Fig. 54.

Quid, si planum sit asperum: applicatio ad reflexionem lucis.

299. Si planum sit asperum, ut Figura exhibet, & ut semper contingit in Natura; æqualitas illa virium utique non habetur. At si scabrities sit satis exigua respectu ejus distantiæ, ad quam vires sensibiles protenduntur; inæqualitas ejusmodi erit perquam exigua, & anguli incidentiæ, & reflexionis æquales crunt ad sensum. Si enim eo intervallo concipiatur sphæra VRTS habens centrum in puncto mobili, cujus segmentum RTS jaceat ultra planum; agent omnia puncta constituta intra illud segmentum, adeoque monticuli prominentes satis exigui respectu totius ejus massæ, satis exiguam inæqualitatem poterunt inducere; & proinde sensibilem æqualitatem angulorum incidentiæ, & reflexionis non turbabunt, sicut & nostri terrestres montes in globo oblique projecto, & ita ponderante, ut a resistentia aeris non multum patiatur, sensibiliter non turbant parabolicum motum ipsius, in quo bina crura ad idem horizontale planum eandem ad sensum inclinationem habent. Secus accideret, si illi monticuli ingentes essent respectu ejusdem sphæræ. Atque hæc quidem, qui diligentius perpenderit, videbit sane, & lucem a vitro satis lævigato resilire debere cum angulo reflexionis æquali ad sensum angulo incidentiæ; licet & ibi pulvisculus quo poliuntur vitra, relinquat sulcos, & monticulos, sed perquam exiguos etiam respectu distantiæ, ad quam extenditur sensibilis actio vitri in lucem; sed respectu superficierum, quæ ad sensum scabræ sunt, debere ipsam lucem irregulariter dispergi quaqua versus.

Quid in impactu obliquo globi mollis in planum: velocitas amissa, quæ manet illæsa in curvatura continua.

300. Pariter ubi globus non elasticus deveniat per AB in eadem illa fig. 43, & deinde debeat sine reflexione excurrere per BQ, non describet utique reetam lineam accurate, sed serpet, & saltitabit non nihil: crit tamen recta ad sensum: velocitas vero mutabitur ita; ut sit velocitas prior AB ad posteriorem BI, ut radius ad cosinum inclinationis OBI rectæ BO ad planum CD, ac ipsa velocitas prior ad velocitatum differentiam, sive ad partem velocitatis amissam, quam exprimit IQ determinata ab arcu OQ habente centrum in B, erit ut radius ad sinum versum ipsius inclinationis. Quoniam autem imminuto in infinitum angulo, sinus versus decrescit in infinitum etiam respectu ipsius arcus, adeoque summa omnium sinuum versorum pertinentium ad omnes inflexiones infinitesimas tempore finito factas adhuc in infinitum decrescit; ubi inflexio evadat [141] continua, uti fit in curvis continuis, ea summa evanescit, & nulla fit velocitas amissio ex inflexione continua orta, sed vis perpetua, quæ tantummodo ad habendam curvaturam requiritur perpendicularis ipsi curvæ, nihil turbat velocitatem, quam parit vis tangentialis, si qua est, quæ motum perpetuo acceleret, vel retardet; ac in curvilineis motibus quibuscunque, qui habeantur per quascunque directiones virium, semper resolvi potest vis illa, quæ agit, in duas, alteram perpendicularem curvæ, alteram secundum directionem tangentis, & motus in curva per hanc tangentialem vim augebitur, vel retardabitur eodem modo, quo si eædem vires agerent, & motus haberetur in eadem recta linea constanter. Sed hæc jam meæ Theoriæ communia sunt cum Theoria vulgari.

Theoremata pro vi accelerante descensum, vel retardante ascensum in planis inclinatis, & pendulis.

301. Communis est itidem in fig. 44, & 45 ratio gravitatis absolutæ BO ad vim BI, quæ obliquum descensum accelerat, vel ascensum retardat, quæ est, ut radius ad sinum anguli BOI, vel OBR, sive cosinum OBI. Angulum OBI est in fig. 44, quem continet directio BI, quæ est eadem, ac directio plani CD, cum linea verticali BO, adeoque angulus OBR est æqualis inclinationi plani ad horizontem, & angulus idem OBR in fig. 45 est is, quem continet cum verticali BO recta CB jungens punctum oscillans cum puncto suspensionis. Quare habentur hæc theoremata: *Vis accelerans descensum, vel retardans ascensum in planis*

298. Fig. 54 illustrates the curvature in reflection; here we have the path of a moving point repelled by a plane CO represented by ABQDM; this, near B, where the forces begin to be sensible, begins to be appreciably curved, & leaves off at the same distance from the plane, near the point D. The path, indeed, if there is always repulsion, will be continuously incurved towards the same parts, as is shown in the figure; but if attraction alternates with repulsion, the path will be winding, as I mentioned. However, if the forces at equal distances from the plane are equal to one another, it is sufficiently clear, & indeed it could be rigorously proved, that as soon as some point such as Q was reached where the direction of the path was parallel to the plane, it must thereafter describe an arc QD exactly equal & similar to the arc QB; & therefore similarly placed with respect to the plane CO; so that the inclinations of the parts at equal distances from the plane, & from Q on either side, are exactly equal. Hence, the tangents BN, DP, which are as it were continuations of the straight lines AB, MD, will make the angles ANC, MPO equal to one another; & these may then be looked upon as the angles of incidence & reflection.

The case of a force acting at a considerable distance; consideration of the curvature of the path.

299. If the plane is rough, as is shown in the figure, & such as always occurs in Nature, there will in no case be this equality of forces. But if the roughness is sufficiently slight in comparison with that distance, over which sensible forces are extended, such inequality will be very slight, & the angle of incidence will be practically equal to the angle of reflection. For if with a radius equal to that distance we suppose a sphere VRTS to be drawn, having its centre at the position of the moving point, & a segment RTS lying on the other side of the plane; then all the points contained within that segment exert forces; &, if therefore the little prominences are sufficiently small compared with the whole mass, they can only induce quite a slight inequality. Hence, they will not disturb the sensible equality of the angles of incidence & reflection; just as the mountains on our Earth, acting on a sphere projected in a direction inclined to the vertical, & of such a weight that it does not suffer much from the resistance of the air, do not sensibly disturb its parabolic motion, in which the two parts of the parabola have practically the same inclination to the same horizontal plane. It would be quite another matter, if the little prominences were of large size compared with the sphere. Anyone who will study these matters with considerable care will perceive clearly that light also must rebound from a sufficiently well polished piece of glass with the angle of reflection to all intents equal to the angle of incidence. Although it is true that the powder with which glasses are polished leaves little furrows & prominences; but these are always very slight compared with the distance over which the sensible action of glass on light extends. However, for surfaces that are sensibly rough, it will be perceived that light must be scattered irregularly in all directions.

What if the plane is rough; application to the reflection of light.

300. Similarly, when a non-elastic sphere travels along AB, in Fig. 43, & then without reflection has to continue along BQ, it will not describe a perfectly accurate straight line, but will wind irregularly to some extent; yet the line will be to all intents a straight line. Moreover, the velocity will be changed in such a way that the previous velocity AB will be to the new velocity BI, as the radius is to the cosine of OBI the inclination of the straight line BO to the plane CD; & the previous velocity is to the difference between the velocities, i.e., to the velocity that is lost, which is represented by IQ determined by the arc OQ having its centre at B, as the radius is to the versine of the same angle. Now, since, when the angle is indefinitely diminished, the versine decreases indefinitely with respect to the arc itself, & thus the sum of all the versines belonging to all the infinitesimal inflections made in a finite time still decreases indefinitely; it follows that, when the inflexion becomes continuous, as is the case with continuous curves, this sum vanishes, & therefore there is no loss of velocity arising from continuous inflection. There is a perpetual force, which is required for the purpose of keeping up the curvature, perpendicular to the curve itself, & therefore not disturbing the velocity at all; the velocity arises from a tangential force, if there is any, & this continuously accelerates or retards the motion. In curvilinear motions of all kinds, due to forces in all kinds of directions, it is always possible to resolve the force acting into two parts, one of them perpendicular to the curve, & the other along the tangent; the motion along the curve will be increased or retarded by the tangential force, in precisely the same manner as if these same forces acted & the motion was constantly in the same straight line. But all these matters are common to my theory and the usual theory.

What happens in the case of oblique impact of a soft sphere; the velocity lost, which remains unimpaired in continuous curvature.

301. In Fig. 44, 45, there is a common ratio between the absolute gravity BO & the force BI, which accelerates the descent or retards the ascent; & this ratio is equal to that of the radius to the sine of the angle BOI, or OBR, or the cosine of OBI. The angle OBI is, in Fig. 44, that which is contained by the direction BI, which is the same as the direction of the plane CD, with the vertical line BO; & thus the angle OBR is equal to the inclination of the plane to the horizon; & the same angle OBR, in Fig. 45, is that which is contained by the vertical BO with the straight line CB, which joins the point of oscillation with the point of suspension. Hence, we have the following theorems. *The force accelerating descent,*

Theorems with regard to the force accelerating descent or retarding ascent in the cases of the inclined plane & of the pendulum.

inclinatis, vel ubi oscillatio fit in arcu circulari, est ad gravitatem absolutam, ibi quidem ut sinus inclinationis ipsius plani, hic vero ut sinus anguli, quem cum verticali linea continet recta jungens punctum oscillans cum puncto suspensionis, ad radium. E quorum theorematum priore fluunt omnia, quæ Galilæus tradidit de descensu per plana inclinata ; ac e posteriore omnia, quæ pertinent ad oscillationes in circulo ; quia immo etiam ad oscillationes factas in curvis quibuscunque pondere per filum suspenso, & curvis evolutis applicato ; ac eodem utemur infra in definiendo centro oscillationis.

Applicatio Theoriæ ad refractionem : tres casus velocitatis normalis extinctæ, imminutæ, auctæ.

302. Hisce perspectis, applicanda est etiam Theoria ad motuum refractionem, ubi continentur elementa mechanica pro refractione luminis, & occurrit elegantissimum theorema a Newtono inventum huc pertinens. Sint in fig. 55 binæ superficies AB, CD parallelæ inter se, & punctum mobile quodpiam extra illa plana nullam sentiat vim, inter ipsa vero urgeatur viribus quibuscunque, quæ tamen & semper habeant directionem perpendicularem ad ipsa plana, & in æqualibus distantiis ab altero ex iis æquales sint ubique ; ac mobile deferatur ad alterum ex iis, ut AB, directione quacunque GE. Ante appulsum feretur motu rectilineo, & æquabili, cum nulla urgeatur vi : ejus velocitatem exprimat EH, quæ erecta ER, perpendiculari ad AB, resolvi poterit in duas, alteram perpendicularem ES, alteram parallelam HS. Post ingressum inter alia duo [142] plana incurvabitur motus illis viribus, sed ita, ut velocitas parallela ab iis nihil turbetur, velocitas autem perpendicularis vel minuatur, vel augeatur ; prout vires tendent versus planum citerius AB, vel versus ulterius CD. Jam vero tres casus haberi hinc possunt ; vel enim iis viribus tota velocitas perpendicularis ES extinguitur, antequam deveniatur ad planum ulterius CD ; vel perstat usque ad appulsum ad ipsum CD, sed imminuta, vi contraria prævalente viribus eadem directione agentibus ; vel perstat potius aucta.

Primo reflexionem induci.

303. In primo casu, ubi primum velocitas perpendicularis extincta fuerit alicubi in X, punctum mobile reflectet cursum retro per XI, & iisdem viribus agentibus in regressu, quæ egerant in progressu, acquiret velocitatem perpendicularem IL æqualem amissæ ES, quæ composita cum parallela LM, æquali priori HS, exhibebit obliquam IM in recta nova IK, quam describet post egressum, & erunt æquales anguli HIL, MES, adeoque & anguli KIB, GEA ; quod congruit cum iis, quæ in fig. 54. sunt exhibita, & pertinent ad reflexionem.

Secundo refractionem cum accessu ad superficiem refringentem, tertio itidem refractionem, sed cum recessu.

304. In secundo casu prodibit ultra superficiem ulteriorem CD, sed ob velocitatem perpendicularem OP minorem priore ES, parallelam vero PN æqualem priori HS, erit angulus ONP minor, quam EHS, adeoque inclinatio VOD ad superficiem in egressu minor inclinatione GEA in ingressu. Contra vero in tertio casu ob *op* majorem ES, angulus *uo*D crit major. In utroque autem hoc casu differentia quadratorum velocitatis ES, & OP vel *op*, crit constans, per num. 177 in adn. *m*, quæcunque fuerit inclinatio GE in ingressu, a qua inclinatione pendet velocitas perpendicularis SE.

Ratio constans sinus anguli incidentiæ, ad sinum anguli refracti.

305. Inde autem facile demonstratur, fore sinum anguli incidentiæ HES, ad sinum anguli refracti PON (& quidquid dicitur de iis, quæ designantur litteris PON, erunt communia iis, quæ exprimuntur litteris *pon*) in ratione constanti, quæcunque fuerit inclinatio rectæ incidentis GE. Sumatur enim HE constans, quæ exprimat velocitatem ante incidentiam : exprimet HS velocitatem parallelam, quæ erit æqualis rectæ PN exprimenti velocitatem parallelam post refractionem ; ac ES, OP exprimcnt velocitates perpendiculares ante, & post, quarum quadrata habebunt differentiam constantem. Sed ob HS, PN semper æquales, differentia quadratorum HE, ON æquatur differentiæ quadratorum ES, OP. Igitur etiam differentia quadratorum HE, ON erit constans ; cumque ob HE constantem debeat esse constans ejus quadratum ; erit constans etiam quadratum ON, adeoque constans etiam ipsa ON, & proinde constans erit & ratio HE ad ON ; quæ quidem ratio est eadem, ac sinus anguli NOP ad sinum HES : cum enim sit in quovis triangulo rectangulo radius ad latus utrumvis, ut basis ad sinum anguli oppositi ; in diversis triangulis rectangulis sunt sinus, ut latera opposita divisa per [143] bases, sive directe ut latera, & reciproce ut bases, & ubi latera sunt æqualia, ut hic HS, PN, crunt reciproce ut bases.

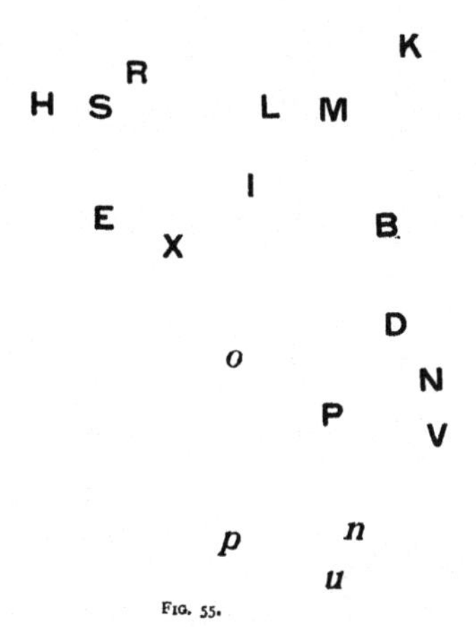

Fig. 55.

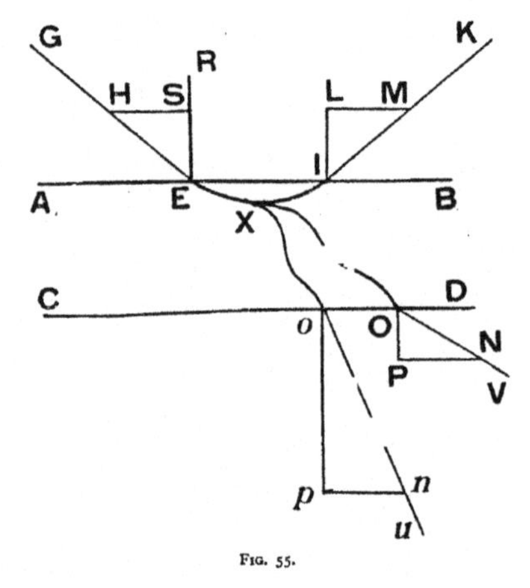

FIG. 55.

or retarding ascent, on inclined planes, or where there is oscillation in a circular arc, is to the absolute gravity, in the first case as the sine of the inclination of the plane to the radius, & in the second case as the sine of the angle between the vertical & the line joining the oscillating point to the point of suspension, is to the radius. From the first of these theorems there follow immediately all that Galileo published on the descent along inclined planes; & from the second, all matters relating to oscillations in a circle. Moreover, we have also all matters that relate to oscillations made in curves of all sorts by a weight suspended by a string wrapped round in volute curves; & we shall make use of the same idea later to define the centre of oscillation.

Application of the Theory to refraction; the three cases in which the normal velocity is extinguished, or diminished, or increased.

302. These matters being investigated, we now have to apply the Theory to the refraction of motions, in which are contained the mechanical principles of the refraction of light; here also we find a most elegant theorem discovered by Newton, referring to the subject. In Fig. 55, let AB, CD be two surfaces parallel to one another; & let a moving point feel the action of no force when outside those planes, but when between the two planes suppose it is subject to any forces, so long as these always have a direction perpendicular to the planes, & they are always equal at equal distances from either of them. Suppose the point to approach one of the planes, AB say, in any direction GE. Until it reaches AB it will travel with rectilinear & uniform motion, since it is acted upon by no force; let EH represent its velocity. Then, if ER is erected perpendicular to the plane AB, the velocity can be resolved into two parts, the one, ES, perpendicular to, & the other, HS, parallel to, the plane AB. After entry into the space between the two planes the motion will be incurved owing to the action of the forces; but in such a manner that the velocity parallel to the plane will not be affected by the forces; whereas the perpendicular velocity will be diminished or increased, according as the forces act towards the plane AB, or towards the plane CD. Now there are three cases possible; for, the whole of the perpendicular velocity may be destroyed before the point reaches the further plane CD, or it may persist right up to contact with the plane CD, but diminished in magnitude, owing to a force existing contrary to the forces in that direction, or it may continue still further increased.

In the first case reflection is induced.

303. In the first case, where the perpendicular velocity was first destroyed at a point X, the moving point will follow a return path along XI; & as the same forces act in the backward motion as in the forward motion, the point will acquire a perpendicular velocity IL, equal to ES, that which it lost; this, compounded with the parallel velocity LM, equal to the previous parallel velocity HS, will give a velocity IM, in an oblique direction along the new straight line IK, along which the point will move after egress. Now the angles HIL, MES will be equal, & therefore also the angles KIB, GEA; this agrees with what is represented in Fig. 54, & pertains to reflection.

In the second case we have refraction & nearer approach to the refracting surface; in the third, refraction & recession from the surface.

304. In the second case, the point will proceed beyond the further surface CD; but, since the perpendicular velocity OP is now less than the previous one ES, whilst the parallel velocity is the same as the previous one HS, the angle ONP will be less than the angle EHS, & therefore the inclination to the surface, VOD, on egress, will be less than the inclination, GEA, on ingress. On the other hand, in the third case, since *op* is greater than ES, the angle *uo*D will be greater than the angle GEA. But in either case, we here have the difference between the squares of the velocity ES, & that of OP, or *op*, constant, as was shown in Art. 177, note *m*, whatever may be the inclination on ingress, made by GE with the plane, upon which inclination depends the perpendicular velocity SE.

The constant ratio of the sine of the angle of incidence to the sine of the angle of refraction.

305. Further, from this it is easily shown that the sine of the angle of incidence HES is to the sine of the angle of refraction HON (& whatever is said with regard to these angles, denoted by the letters PON, will hold good for the angles denoted by the letters *pon*), in a constant ratio, whatever the inclination of the line of incidence, GE, may be. For, suppose HE, which represents the velocity before incidence, to be constant; then HS, representing the parallel velocity, will be equal to PN, which represents the parallel velocity after refraction. Now, if ES, OP represent the perpendicular velocities before & after refraction, they will have the difference between their squares constant. But, since HS, PN are equal, the difference between the squares of HE, ON will be equal to the difference between the squares of ES, OP. Hence the difference of the squares of HE, ON will be constant. But, since HE is constant, its square must also be constant; therefore the square of ON, & thus also ON itself, must be constant. Therefore also the ratio of HE to ON is constant; & this ratio is the same as that of the sine of the angle NOP to the sine of the angle HES. For, since in any right-angled triangle the ratio of the radius to either side is that of the base to the angle opposite, in different right-angled triangles, the sines vary as the sides opposite them divided by the bases, or directly as the sides & inversely as the bases; & where the sides are equal, as HS, PN are in this case, the sines vary as the bases.

Ratio sinuum constans, & ratio velocitatum reciproca rationis sinuum.

306. Quamobrem in refractionibus, quæ hoc modo fiant motu libero per intervallum inter duo plana parallela, in quo vires paribus distantiis ab altero eorum pares sint, ratio sinus anguli incidentiæ, sive anguli, quem facit via ante incursum cum recta perpendiculari plano, ad sinum anguli refracti, quem facit via post egressum itidem cum verticali, est constans, quæcunque fuerit inclinatio in ingressu. Præterea vero habetur & illud, fore celeritates absolutas ante, & post in ratione reciproca eorum sinuum. Sunt enim ejusmodi velocitates ut HE, ON, quæ sunt reciproce ut illi sinus.

Transitus ad Theorema, quod huic operi occasionem dedit.

307. Hæc quidem ad luminis refractiones explicandas viam sternunt, ac in Tertia Parte videbimus, quo pacto hypothesis hujusce theorematis applicetur particulis luminis. Sed interea considerabo vires mutuas, quibus in se invicem agant tres massæ, ubi habebuntur generalius ea, quæ pertinent etiam ad actiones trium punctorum, & quæ a num. 225, & 228 huc reservavimus. Porro si integræ vires alterius in alteram diriguntur ad ipsa centra gravitatis, referam hic ad se invicem vires ex integris compositas; sed etiam ubi vires aliam directionem habeant quancunque; si singulæ resolvantur in duas, alteram, quæ se dirigat a centro ad centrum; alteram, quæ sit ipsi perpendicularis, vel in quocunque dato angulo obliqua; omnia in prioribus habebunt itidem locum.

Consideratio directionis virium, quibus tres massæ in se mutuo agunt.

308. Agant in se invicem in fig. 56 tres massæ, quarum centra gravitatis sint A, B, C, viribus mutuis ad ipsa centra directis, & considerentur inprimis directiones virium. Vis puncti C ex utraque CV, C*d* attractiva erit C*e*; ex utraque repulsiva CY, C*a*, erit CZ, & utriusque directio saltem ad partes oppositas producta ingreditur triangulum, & secat illa angulum internum ACB, hæc ipsi ad verticem oppositum *a*CY. Vi CV attractiva in B, ac CY repulsiva ab A, habetur CX; & vi C*d* attractiva in A, ac C*a* repulsiva a B, habetur C*b*, quarum utraque abit extra triangulum, & secat angulos ipsius externos. Primæ C*e*, cum debeant respondere attractiones BP, AG, respondent cum attractionibus mutuis BN, AE, vires BO, AF, vel cum repulsionibus BR, AI, vires BQ, AH, ac tam priores binæ, quam posteriores, jacent ad candem partem lateris AB, & vel ambæ ingrediuntur triangulum tendentes versus ipsum, vel ambæ extra ipsum etiam productæ abeunt, & tendunt ad partes oppositas directionis C*e* respectu AB. Secundæ CZ debent respondere repulsiones BT, AL, quæ cum repulsionibus BR, AI, constituunt BS, AK, cum attractionibus BN, AE constituunt BM, AD, ac tam priores binæ, quam posteriores jacent ad eandem plagam respectu AB, & ambarum [144] directiones vel productæ ex parte posteriore ingrediuntur triangulum, sed tendunt ad partes ipsi contrarias, ut CZ, vel extra triangulum utrinque abeunt ad partes oppositas directioni CZ respectu AB. Quod si habeatur CX, quam exponunt CV, CY, tum illi respondent BP, & AL, ac si prima conjungitur cum BN, jam habetur BO ingrediens triangulum; si BR, tum habetur quidem BQ, cadens etiam ipsa extra triangulum, ut cadit ipsa CX; sed secunda AL jungetur cum AI, & habebitur AK, quæ producta ad partes A ingredietur triangulum. Eodem autem argumento cum vi C*b* vel conjungitur AF ingrediens triangulum, vel BS, quæ producta ad B triangulum itidem ingreditur. Quamobrem semper aliqua ingreditur, & tum de reliquis binis redeunt, quæ dicta sunt in casu virium C*e*, CZ.

Theorema pertinens ad directiones virium.

309. Habetur igitur hoc theorema. *Quando tres massæ in se invicem agunt viribus directis ad centra gravitatis, vis composita saltem unius habet directionem, quæ saltem producta ad partes oppositas secat angulum internum trianguli, & ipsum ingreditur: reliquæ autem duæ vel simul ingrediuntur, vel simul evitant, & semper diriguntur ad eandem plagam respectu lateris jungentis earum duarum massarum centra; ac in primo casu vel omnes tres tendunt ad interiora trianguli jacendo in angulis internis, vel omnes tres ad exteriora in partes triangulo oppositas jacendo in angulis ad verticem oppositis; in secundo vero casu respectu lateris jungentis eas binas massas tendunt in plagas oppositas ei, in quam tendit vis illa prioris massæ.*

Theorema elegantius ad eas pertinens cum ejus demonstratione.

310. Sed est adhuc elegantius theorema, quod ad directionem pertinet, nimirum: *Omnium trium compositarum virium directiones utrinque productæ transeunt per idem punctum: & si id jaceat intra triangulum; vel omnes simul tendunt ad ipsum, vel omnes simul ad partes ipsi contrarias: si vero jaceat extra triangulum; binæ, quarum directiones non ingrediuntur*

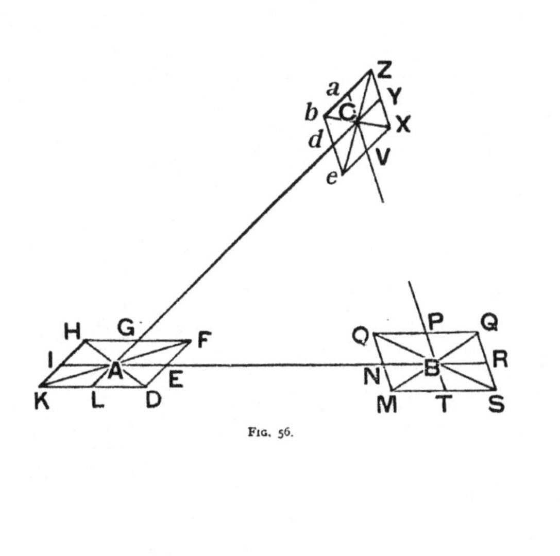

FIG. 56.

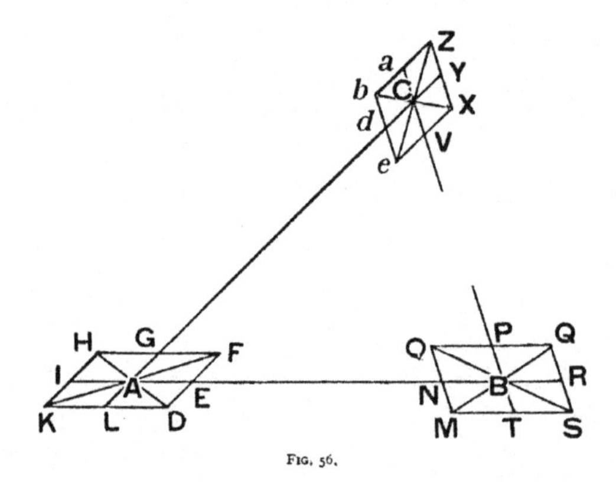

FIG. 56.

The ratio of the sines constant: the ratio of the velocities the reciprocal ratio of the sines.

306. Hence, in refractions, which arise in this way from a free motion between two parallel planes, where the forces at equal distances from one or the other of them are equal, the ratio of the sine of the angle of incidence, or the angle made by the path before refraction, with a straight line perpendicular to the plane, to the sine of the angle of refraction, or the angle made after refraction with the vertical also, is constant, whatever may be the inclination at ingress. We also obtain the theorem that the absolute velocities before and after refraction are in the inverse ratio of the sines. For such velocities are represented by HE, ON; & these are inversely as the sines in question.

Passing on to the theorem which gave rise to this work.

307. These facts suggest a method for explaining refraction of light; & in the Third Part we shall see the manner in which the hypothesis of the above theorem may be applied to particles of light. In the meanwhile, I will consider the mutual forces, with which three masses act upon one another; here we shall obtain more generally all those things that relate to the actions of three points also, such as I reserved from discussion in Art. 225, 228 until now. Further, if the total forces of the one or the other are directed towards their centres of gravity, I will here take account of the mutual forces compounded of these wholes. But, where the forces have any directions whatever, if each of them is resolved into two parts, of which one is directed from centre to centre & the other is perpendicular to this line, or makes some given inclination with it, then also all things that are true for the former hold good also in this case.

Investigation of the directions of the forces with which three masses act upon one another.

308. In Fig. 56, let three masses, whose centres of gravity are at A, B, C, act upon one another with mutual forces directed to their centres; & first of all let the directions of the forces be considered. The force on the point C, from the two attractive forces CV, C*d* will be C*e*; that from CY, C*a*, both repulsive, will be CZ; & the direction of both of these, produced backwards in one case, will fall within the triangle, the former dividing the angle ACB, & the latter the vertically opposite angle *a*CY, into two parts. But, from CV, attractive towards B, & CY, repulsive from A, we obtain CX; & from C*d*, attractive towards A, & C*a*, repulsive from B, we have C*b*; & the direction of each of these will fall without the triangle, & divide its exterior angles into two parts. To C*e*, the first of these, since we must have the corresponding attractions BP, AG, there correspond the forces BO, AF, from combination with the mutual attractions BN, AE; or the forces BQ, AH, from combination with the mutual repulsions BR, AI. Both the former of these pairs, & the latter, lie on the same side of AB; either both will fall within the triangle & tend in its direction, or both will, even if produced, fall without it; in each case, they will tend in the opposite direction to that of C*e* with respect to AB. To CZ, the second of the forces on C, there must correspond the repulsions BT, AL; these, combined with the repulsions BR, AI, give the forces BS, AK; & with the attractions BN, AE, the forces BM, AD. Both the former of these, & both the latter, lie on the same side of AB; & the directions of the two, either when produced backwards will fall within the triangle but tend in opposite directions to that of CZ with respect to it, or they will fall without the triangle & tend off on either side in directions opposite to that of CZ with respect to AB. Now if CX is obtained, given by CV, CY, then there will correspond to it BP & AL; &, if the first of these is compounded with BN, we shall then have BO falling within the triangle; or if compounded with BR, we shall have BQ, falling also without the triangle, just as CX does; but, in that case, the second action AL will be compounded with AI, & AK will be obtained, & this when produced in the direction of A will fall within the triangle. By the same argument, with the force C*b* there will be associated the force AF falling within the triangle, or the force BS, which when produced in the direction of B will also fall within the triangle. Hence, in all cases, some one of the forces falls within the triangle; & then what has been said in the case of C*e*, CZ will apply to the other two forces.

Theorem relating to the directions of the forces.

309. We therefore have the following theorem. *When three masses act upon one another with forces directed towards their centres of gravity, the resultant force, in at least one case, will have a direction which, produced backwards if necessary, will divide an internal angle of the triangle into two parts, & fall within the triangle. Also the remaining two forces will either both fall within, or both without, the triangle & will in all cases be directed towards the same side of the line joining the centres of the two masses. In the first case, all three forces either tend towards the interior of the triangle, falling within the interior angles, or outwards away from the triangle, falling within the angles that are vertically opposite to the interior angles. In the second case, on the other hand, they tend to opposite sides, of the line joining the two masses, to that towards which the force on the third mass tends.*

A still more elegant theorem with regard to the directions of the forces; & its demonstration.

310. But there is a still more elegant theorem with regard to the directions of the forces, namely:—*The directions of all three resultant forces, when produced each way, pass through the same point. If this point lies within the triangle, all three forces tend towards it, or all three away from it; but, if it lies without the triangle, those two forces, which do not*

triangulum, tendunt ad ipsum, ac tertia, cujus directio triangulum ingreditur, tendit ad partes ipsi contrariias ; vel illæ binæ ad partes ipsi contrarias, & tertia ad ipsum.

Prima pars, quod omnes transeant per idem punctum, sic demonstratur. In figura quavis a 57 ad 62, quæ omnes casus exhibent, vis pertinens ad C sit ea, quæ triangulum ingreditur, ac reliquæ binæ HA, QB concurrant in D : oportet demonstrare, vim etiam, quæ pertinet ad C, dirigi ad D. Sint CV, C*d* vires componentes, ac ducta CD, ducatur VT parallela CA, occurrens CD in T ; & si ostensum fuerit, ipsam fore æqualem C*d* ; res erit perfecta : ducta enim *d*T remanebit CVT*d* parallelogrammum, per cujus diagonalem CT dirigetur vis composita ex CV, C*d*. Ejusmodi autem æqualitas demonstrabitur considerando rationem CV ad C*d* compositam ex quinque intermediis, CV ad BP, BP ad PQ, PQ, sive BR ad AI, AI, sive HG ad AG, AG ad [145] C*d*. Prima vocando A, B, C massas, quarum ea puncta sunt centra gravitatum, est ex actione, & reactione æqualibus ratio massæ B ad C ; secunda *sin* PQB, sive ABD, ad *sin* PBQ, sive CBD ; tertia A ad B : quarta *sin* HAG, sive CAD, ad *sin* GHA, sive BAD : quinta C ad A. Tres rationes, in quibus habentur massæ, componunt rationem B × A × C ad C × B × A, quæ est 1 ad 1, & remanet ratio *sin* ABD × *sin* CAD ad *sin* CBD × *sin* BAD. Pro *sin* ABD, & *sin* BAD, ponantur AD, & BD ipsis proportionales ; ac pro *sinu* CAD, & *sin* CBD ponantur $\frac{sin\ ACD \times CD}{AD}$, & $\frac{sin\ BCD \times CD}{BD}$, ipsis æquales ex Trigonometria, & habebitur ratio *sin* ACD × CD ad *sin* BCD × CD sive *sin* ACD, vel CTV, qui ipsi æquatur ob VT, CA parallelas, ad *sin* BCD, sive VCT, nimirum ratio ejusdem illius CV ad VT. Quare VT æquatur C*d*, CVT*d* est parallelogrammum, & vis pertinens ad C, habet directionem itidem transeuntem per D.

Secunda pars patet ex iis, quæ demonstrata sunt de directione duarum virium, ubi tertia triangulum ingreditur, & sex casus, qui haberi possunt, exhibentur totidem figuris. In fig. 57, & 58 cadit D extra triangulum ultra basim AB, in 59, & 60 intra triangulum, in 61, & 62 extra triangulum citra verticem ad partes basi oppositas, ac in singulorum binariorum priore vis CT tendit versus basim, in posteriore ad partes ipsi oppositas. In iis omnibus demonstratio est communis juxta leges transformationis locorum geometricorum, quas diligenter exposui, & fusius persecutus sum in dissertatione adjecta meis *Sectionum Conicarum Elementis*, Elementorum tomo 3.

Corollarium pro casu directionum parallelarum.

311. Quoniam evadentibus binis HA, QB parallelis, punctum D abit in infinitum & tertia CT evadit parallela reliquis binis etiam ipsa juxta easdem leges ; patet illud : *Si binæ ex ejusmodi directionibus fuerint parallelæ inter se ; erit iisdem parallela & tertia : ac illa, quæ jacet inter directiones virium transeuntes per reliquas binas, quæ idcirco in eo casu appellari potest media, habebit directionem oppositam directionibus reliquarum conformibus inter se.*

Aliud generale tertiæ directionis datæ datis binis.

312. Patet autem, datis binis directionibus virium, dari semper & tertiam. Si enim illæ sint parallelæ ; crit illis parallela & tertia : si autem concurrant in aliquo puncto ; tertiam determinabit recta ad idem punctum dueta : sed oportet, habeant illam conditionem, ut tam binæ, quæ triangulum non ingrediantur, quam quæ ingrediantur, vel simul tendant ad illud punctum, vel simul ad partes ipsi contrarias.

Theorema præcipuum de magnitudine, quod toti Operi occasionem dedit. Ejus demonstratio expeditissima.

313. Hæc quidem pertinent ad directiones : nunc ipsas earum virium magnitudines inter se comparabimus, ubi statim occurret elegantissimum illud theorema, de quo mentionem feci num. 225 : *Vires acceleratrices binarum quarumvis e tribus massis in se mutuo agentibus sunt in ratione composita ex tribus,* [146] *nimirum ex directa sinuum angulorum quos continet recta jungens ipsarum centra gravitatis cum rectis ductis ab iisdem centris ad centrum tertiæ massæ ; reciproca sinuum angulorum, quos directiones ipsarum virium continent cum iisdem rectis illas jungentibus cum tertia ; & reciproca massarum.* Nam est BQ ad AH assumptis terminis mediis BR, AI in ratione composita ex rationibus BQ, ad BR, & BR ad AI, & AI ad AH. Prima ratio est sinus QRB, sive CBA ad sinum BQR, sive PBQ, vel CBD : secunda massæ A ad massam B : tertia sinus IHA, sive HAG, vel CAD, ad sinum HIA, sive CAB : eæ rationes, permutato solo ordine antecedentium, & consequentium, sunt rationes sinus CBA ad sinum CAB, quæ est illa prima e rationibus propositis directa ; sinus CAD ad sinum CBD, quæ est secunda reciproca : & massæ A ad massam B, quæ est tertia itidem reciproca. Eadem autem est prorsus demonstratio : si comparetur BQ, vel AH cum CT, ac in hac demonstratione, ut & alibi ubique, ubi de sinubus angulorum

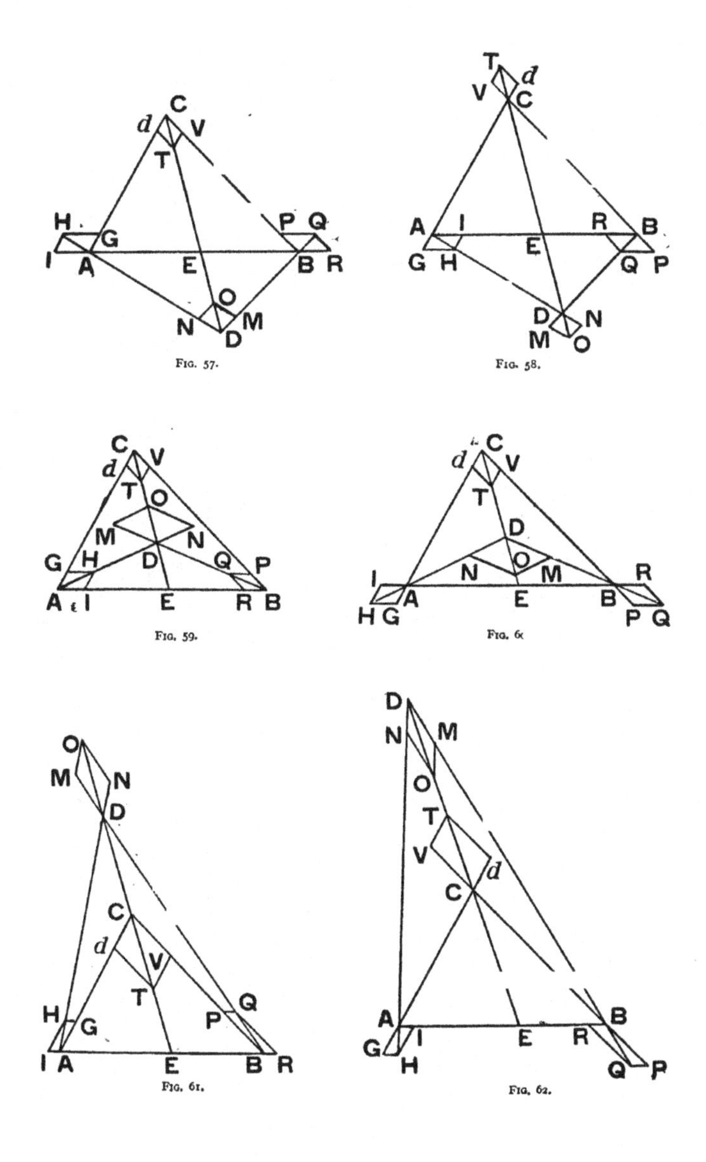

Fig. 57.

Fig. 58.

Fig. 59.

Fig. 60.

Fig. 61.

Fig. 62.

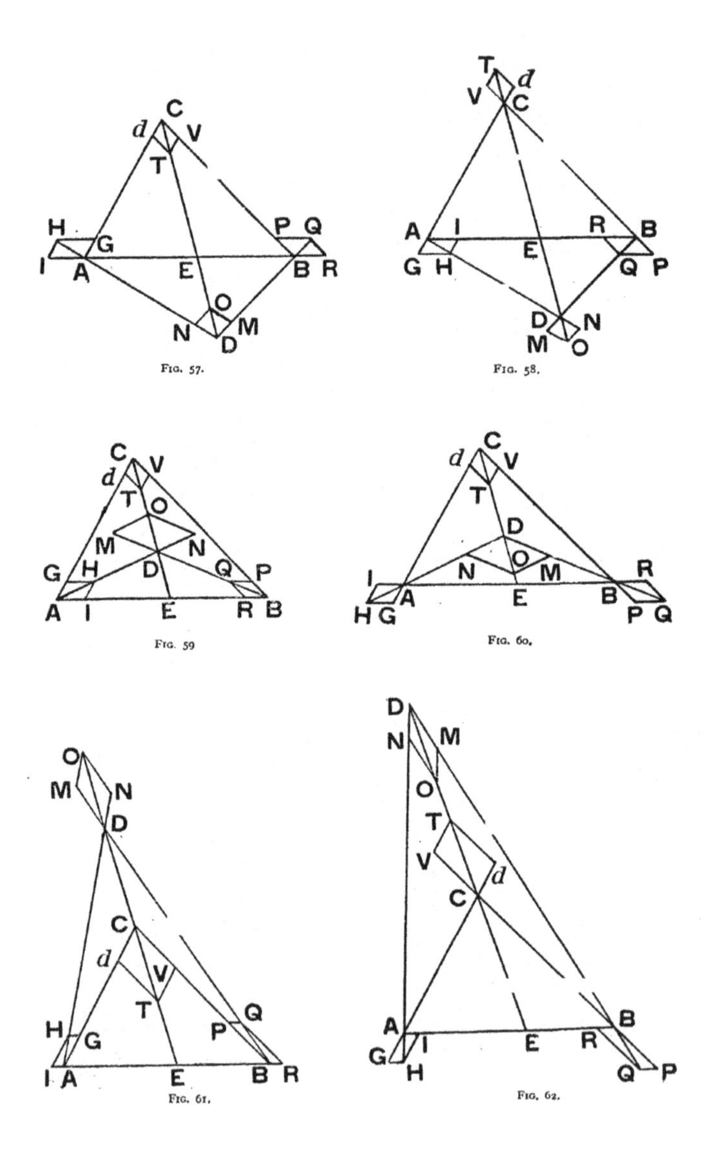

FIG. 57. FIG. 58.

FIG. 59 FIG. 60.

FIG. 61. FIG. 62.

fall within the triangle, tend towards it, & the third, whose direction does not fall within the triangle, tends away from it, or the former two tend away from the point & the third towards it. The proof of the first part of the theorem, that the forces all pass through the same point, is as follows. In any one of the diagrams from Fig. 57 to Fig. 62, which between them give all possible cases, let the force which acts on C be that which falls within the triangle; & let the other two, HA & QB, meet in the point D; then it has to be shown that the force which acts on C, also passes through D. Let CV, C*d* be the component forces; join CD & draw VT parallel to CA to meet CD in T; then, if it can be shown that VT is equal to C*d*, the proposition is proved; for, if *d*T is joined, CVT*d* will be a parallelogram, & the force compounded of CV & C*d* will be directed along its diagonal. Such equality will be proved by considering the ratio of CV to C*d*, compounded of the five intermediate ratios CV to BP; BP to PQ; PQ, or BR, to AI; AI, or HG, to AG; & AG to C*d*. The first of these, if we call the masses A, B, C, which have these points as their centres of gravity, will, on account of the equality of action & reaction, be the ratio of the mass B to the mass C; the second, the ratio of the sine of PQB, or ABD, to the sine of PBQ, or CBD; the third, that of the mass A to the mass B; the fourth, that of the sine of HAG, or CAD, to the sine of GHA, or BAD; the fifth, that of the mass C to the mass A. The three ratios, in which the masses appear, together give the ratio B × A × C to C × B × A, which is that of 1 to 1; & there remains the ratio of *sin*ABD × *sin*CAD to *sin*CBD × *sin*BAD. For *sin*ABD & *sin*BAD substitute AD & BD, which are proportional to them; & for *sin*CAD & *sin*CBD substitute *sin*ACD × CD/AD & *sin*BCD × CD/BD, which are equal to them by trigonometry. There will be obtained the ratio of *sin*ACD × CD to *sin*BCD × CD, or *sin*ACD to *sin*BCD; &, since VT & CA are parallel, this ratio is equal to that of *sin*CTV to *sin*VCT, that is, to the ratio of CV to VT. Therefore VT is equal to C*d*, CVT*d* is a parallelogram, & the force on C has also its direction passing through D. The second part is evident from what has already been proved with regard to the directions of two forces when the third falls within the triangle; & the six possible cases are shown in the six figures. In Fig. 57, 58, the point D falls without the triangle on the far side of the base AB; in Fig. 59, 60, it falls within the triangle; in Fig. 61, 62, outside the triangle on the side of the vertex remote from the base; & in the first of each pair of figures, the force CT tends towards the base, & in the latter away from it. In all of these the proof is the same, having regard to the laws of transformation of geometrical positions; these I have set forth carefully, & I investigated them more minutely in a dissertation added as a supplement to my *Sectionum Conicarum Elementa*, the third volume of my *Elementa Matheseos*.

Corollary for the case of parallel directions.

311. Now, since the point D will go off to infinity, when two of the forces, HA & QB, happen to be parallel, & the third also, according to the same laws, becomes parallel to the other two, we have this theorem. *If two of these forces are parallel to one another, the third also is parallel to them; & that force, which lies between the directions of the other two, & consequently in that case can be called the middle force, has its direction opposite to the directions of the other two, which are in agreement with one another.*

Another general theorem; the direction of the third force is given when the directions of the other two are given.

312. Further, it is clear that, when the directions of two of the forces are given, the direction also of the third force is given in all cases. For if the former are parallel, the third will be parallel to them; & if the former meet at a point, the straight line joining the third mass to this point will determine the third direction. But this condition holds; namely, that the two which do not fall within the triangle, or the pair which do fall within the triangle, either both tend towards the point D, or both tend away from it.

Fundamental theorem concerning magnitude which gave rise to the whole of this work.

313. So much with regard to directions; now we will go on to compare with one another the magnitudes of these forces. We immediately come to that most elegant theorem, which has already been mentioned in Art. 225. *The accelerating effects of any two masses out of three that mutually act upon one another are in a ratio compounded of three ratios; namely, the direct ratio of the sines of the angles made by the straight line joining the centres of gravity of these two with the straight lines joining each of these to the centre of gravity of the third mass: the inverse ratio of sines of the angles which the directions of the forces make with the straight lines joining the two masses to the third; & the inverse ratio of the masses.* For, if BR, AI are taken as intermediary terms, the ratio of BQ to AH is equal to the ratios compounded from the ratio of BQ to BR, that of BR to AI, & that of AI to AH. The first ratio is equal to that of the sine of QRB, or CBA, to the sine of BQR, or PBQ, or CBD; the second is that of the mass A to the mass B; & the third is equal to that of the sine of IHA, or HAG, or CAD to the sine of HIA, or CAB. These ratios are, by a simple permutation of the antecedents & consequents, as *sin*CBA is to *sin*CAB, which is the first direct ratio of those enunciated; as *sin*CAD to *sin*CBD, which is the second inverse ratio; & as the mass A to the mass B, which also is the third inverse ratio. Moreover the proof is precisely similar, if the ratio of BQ, or AH, to CT is considered; & in this proof, as also in all others,

agitur, angulis quibusvis substitui possunt, uti sæpe est factum, & fiet imposterum, eorum complementa ad duos rectos, quæ eosdem habent sinus.

Corollarium simplex pro viribus ipsis.

314. Inde consequitur, *esse ejusmodi vires reciproce, ut massas ductas in suas distantias a tertia massa, & reciproce, ut sinus, quos earum directiones continent cum iisdem rectis; adeoque ubi eæ ad ejusmodi rectas inclinentur in angulis æqualibus, esse tantummodo reciproce, ut producta massarum per distantias a massa tertia.* Nam ratio directa sinuum CBA, CAB est eadem, ac distantiarum AC, BC, sive reciproca distantiarum BC, AC, qua substituta pro illa, habentur tres rationes reciprocæ, quas exprimit ipsum theorema hic propositum. Porro ubi anguli æquales sunt, sinus itidem sunt æquales, adeoque corum sinuum ratio fit 1 ad 1.

Ratio virium motricium.

315. *Vires autem motrices sunt in ratione composita ex binis tantummodo, nimirum directa sinuum angulorum, quos continent distantiæ a tertia massa cum distantia a se invicem; & reciproca sinuum angulorum, quos continent cum iisdem distantiis directiones virium; vel in ratione composita ex reciproca illarum distantiarum, & reciproca horum posteriorum sinuum: ac si inclinationes ad distantias sint æquales, in sola ratione reciproca distantiarum.* Nam vires motrices sunt summæ omnium virium determinantium celeritatem in punctis omnibus secundum eam directionem, secundum quam movetur centrum gravitatis commune, quæ idcirco sunt præterea directe, ut massæ, sive ut numeri punctorum; adeoque ratio directa, & reciproca massarum mutuo eliduntur.

Ratio virium acceleratricium, ubi eæ diriguntur ad aliquod commune punctum.

316. Præterea *vires acceleratrices, si alicubi earum directiones concurrunt, sunt ad se invicem in ratione composita ex reciproca massarum, & reciproca sinuum angulorum, quibus inclinantur ad directionem tertiæ; & vires motrices in hac poste-*[147]*-riore tantum.* Nam ob latera proportionalia sinubus angulorum oppositorum, erit AC $\times$ *sin* CAD = CD $\times$ *sin* CDA; & pariter CB $\times$ *sin* CBA = CD $\times$ *sin* CDB. Quare ob CD communem, sola ratio sinuum ADC, BDC, quibus directiones AD, BD inclinantur ad CD, æquatur compositæ ex rationibus sinuum CAD, CBD, & distantiarum CA, CB, quæ ingrediebantur rationem virium B, & A; ac eodem pacto AC $\times$ *sin* ACD= AD $\times$ *sin* ADC, & AB $\times$ *sin* ABD = AD $\times$ *sin* ADB, adeoque AC $\times$ *sin* ACD ad AB $\times$ *sin* ABD, ut sinus ADC ad sinum ADB, quibus directiones CD, BD inclinantur ad AD; & eadem est demonstratio pro sinubus ADB, EDB assumpto communi latere BD.

Alia expressio tam virium motricium, quam acceleratricium in eodem casu.

317. *Si ducatur* MO *parallela* DA, *occurrens* BD, CD *in* M, O, & *compleatur parallelogrammum* DMON; *erunt vires motrices in* C, B, A *ad se invicem, ut rectæ* DO, DM, DN, & *vires acceleratrices præterea in ratione massarum reciproca.* Est enim ex præcedenti vis motrix in C ad vim in B, ut *sin* BDA ad *sin* CDA, vel ob AD, OM parallelas, ut *sin* DMO ad *sin* DOM, nimirum ut DO ad DM, & simili argumento vis in C ad vim in A, ut DO ad DN. Vires autem motrices divisæ per massas evadunt acceleratrices. *Quamobrem si, tres vires agerent in idem punctum cum directionibus, quas habent eæ vires motrices, & essent iis proportionales; binæ componerent vim oppositam, & æqualem tertiæ, ac essent in æquilibrio.* Id autem etiam directe patet: nam vires BQ, AH componuntur ex quatuor viribus BR, BP, AI, AG, quæ si ducantur in massas suas, ut fiant motrices; evadit prima æqualis, & contraria tertiæ, quam idcirco elidit, ubi deinde AH, BQ componantur simul, & in ejusmodi compositione remanent BP, AG, ex quarum oppositis, & æqualibus CV, C*d* componitur tertia CT.

Hic debere haberi ea, quæ habentur in compositione, & resolutione virium.

318. Hinc in hisce viribus motricibus habebuntur omnia, quæ habentur in compositione virium; dummodo capiatur [resolutio] compositæ contraria. Si nimirum resolvantur singulæ componentes in duas, alteram secundum directionem tertiæ, alteram ipsi perpendicularem, hæ posteriores elidentur, illæ priores conficient summam æqualem tertiæ, ubi ambæ candem directionem habent, uti sunt binæ, quæ simul ingrediantur, vel simul evitent triangulum; nam in iis, quarum altera ingreditur, altera evitat, tertia æquaretur differentiæ; & facile tam hic, quam in ratione composita, res traducitur ad resolutionem in aliam quamcunque directionem datam, præter directionem tertiæ, binis semper elisis, & reliquarum accepta summa; si rite habeatur ratio positivorum, & negativorum.

Alia expressio rationum earundem virium.

319. Est & illud utile: *tres vires motrices in* C, B, A *sunt inter se, ut* $\frac{AB \times ED}{AD \times BD}, \frac{AE}{AD}, \frac{BE}{BD}$, & *acceleratrices præterea* [148] *in ratione reciproca massarum.* Nam ex Trigonometria est $\frac{AB}{BD} = \frac{sin\,ADB}{sin\,BAD}$, & $\frac{AE}{ED} = \frac{sin\,ADE}{sin\,EAD}$. Quare cum divisor *sin* BAD, & *sin* EAD sit communis: erit *sin* ADB ad *sin* ADE, ut $\frac{AB}{AD}$ ad $\frac{AE}{ED}$, vel, ducendo utrunque terminum

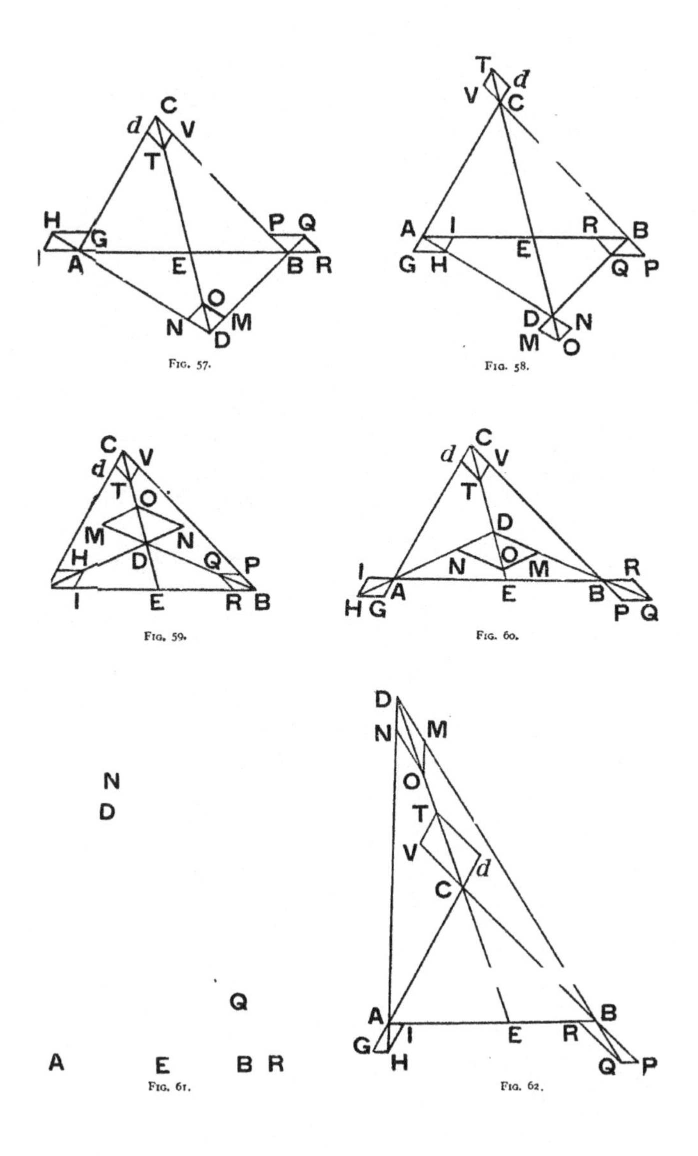

Fig. 57.

Fig. 58.

Fig. 59.

Fig. 60.

Fig. 61.

Fig. 62.

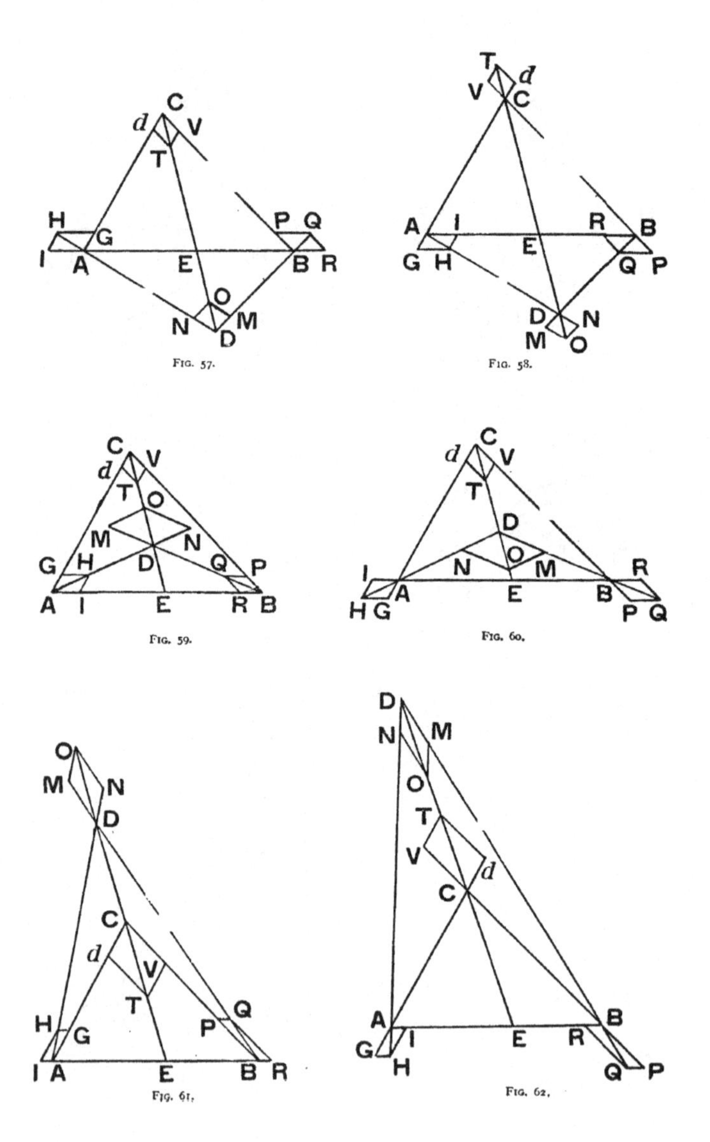

Fig. 57.

Fig. 58.

Fig. 59.

Fig. 60.

Fig. 61.

Fig. 62.

where sines of angles are considered, we can substitute for any of the angles, as often has been done, & as will be done hereafter, their supplements; for these have the same sines.

Simple corollary for the determination of the forces.

314. Hence we have the following corollary. *Such accelerating effects are inversely as the products of each of the two masses into its distance from the third mass, & inversely as the sines of the angles between their directions & these distances; & thus, if they are inclined at equal angles to these distances, the effects are inversely proportional to the products of the masses into the distances from the third mass only.* For the direct ratio of the sines of the angles CBA, CAB is the same as that of the distances AC, BC, or inversely as the distances BC, AC; & if the latter is substituted for the former, we have three inverse ratios, which are given in the enunciation of this corollary. Further, when the angles are equal, their sines are also equal, & their ratio is that of 1 to 1.

The ratio of the motive forces.

315. *The motive forces are in a ratio compounded of two ratios only, namely, the direct ratio of the sines of the angles the line joining each to the third mass & the line joining the two to one another; & the inverse ratio of the sines of the angles which their directions make with these distances; or the ratio compounded of the inverse ratio of these distances & the inverse ratio of the latter sines. Also, if the inclinations to the distances are equal to one another, the ratio is the simple inverse ratio of the distances.* For the motive forces are the sums of all the forces determining velocity for all points in the direction along which the common centre of gravity will move; & hence they are, other things apart, directly as the masses, or as the number of points; & thus the direct & the inverse ratio of the masses eliminate one another.

The ratio of the accelerations when they are directed towards some common point.

316. Further, *the accelerations, if their directions meet at a point, are to one another in the ratio compounded from the inverse ratio of the masses, & the inverse ratio of the sines of the angles between their directions & that of the third. The motive forces are in the latter ratio only.* For, since the sides of a triangle are proportional to the sines of the opposite angles, we have AC. *sin*CAD = CD. *sin*CDA, & similarly, CB. *sin*CBA = CD. *sin*CDB. Hence, since CD is common, the single ratio of the sines of ADC, BDC, the inclinations of AD, BD, to CD, is equal to that compounded from the ratios of the sines of CAD, CBD, & the distances CA, CB, which formed the ratio of the forces on B & A. In the same way, AC. *sin*ACD = AD. *sin*ADC, & AB. *sin*ABD = AD. *sin*ADB, & therefore AC. *sin*ACD is to AB. *sin*ABD as the sine of ADC is to the sine of ADB, the inclinations of CD, BD to AD. The proof is the same for the sines of the angles ADB, EDB, by using the common side DB.

Another expression for both the motive forces & the accelerations in the same case.

317. *If MO is drawn parallel to DA, meeting BD, CD in M, O respectively, & if the parallelogram DMON is completed, then the motive forces for C, B, A will be to one another as the straight lines DO, DM, DN; & for the accelerations, we have in addition the inverse ratio of the masses.* For, from the preceding article, the motive force for C is to the motive force for B as *sin*BDA is to *sin*CDA; that is to say, since AD, OM are parallel, as *sin*DMO is to *sin*DOM, or as DO is to DM. Similarly the force for C is to the force for B as DO is to DN. Now, the motive forces divided by the corresponding masses give the accelerations. *Hence, if three forces act at a point, having the same directions as the motive forces & proportional to them, the resultant compounded from any two of these will give a force equal & opposite to the third, & they will be in equilibrium.* This is immediately evident; for, the forces BQ, AH are compounded from the four forces BR, BP, AI, AG; & if these are multiplied by the corresponding masses, so as to give the motive forces, the first of them will come out equal & opposite to the third & will thus cancel it, when later AH, BQ are compounded together; & in such composition we are left with BP, AG; & from CV & C*d*, which are equal & opposite to these, the third force CT is compounded.

For these forces we must have all those things which hold good for composition & resolution of forces.

318. Hence for these motive forces, we have all those things which hold good in the composition of forces, so long as resolution is considered to be the inverse of composition. Thus, if each of the components is resolved into two parts, one in the direction of the third force, & the other perpendicular to it, the latter will cancel one another, & the former will give a sum equal to the third, when both have the same direction, as is the case when both of them either fall within the triangle or both of them are directed away from it; for those, in which one falls within the triangle & the other away from it, the third will be equal to the difference. The matter, both in this, & in the ratio compounded of these, is easily referred to a resolution in any chosen direction other than the direction of the third, the two at right angles always cancelling one another & the sum being taken of those that remain; provided due regard is had to positives & negatives.

Another expression for the ratio of the same forces.

319. Here is another useful theorem. *The three motive forces on C, B, & A are in the ratio of AB.ED/AD&BD, AE/AD, BE/BD, & the accelerations have, in addition, the inverse ratio of the masses.* For, by trigonometry, we have AB/BD = *sin*ADB/*sin*BAD, & AE/ED = *sin*ADE/*sin*EAD. Hence, since the divisors *sin*BAD, & *sin*EAD are equal, it follows that *sin*ADB is to *sin*ADE as AB/BD is to AE/ED; or, multiplying each term

in $\frac{ED}{AD}$, ut $\frac{AB \times ED}{AD \times BD}$ ad $\frac{AE}{AD}$. Simili autem argumento est itidem *sin* BDA. *sin* BDE : : $\frac{AB \times ED}{AD \times BD}$. $\frac{BE}{BD}$; ex quo patent omnia.

Expressio simplicior pro casu parallelismi.

320. Si punctum D abeat in infinitum, directionibus virium evadentibus parallelis; ratio rectarum ED, AD, BD, ad se invicem evadit ratio æqualitatis. Quare in eo casu illæ tres vires sunt ut AB, AE, EB, in quibus prima æquatur summæ reliquarum. Concipiantur rectæ parallelæ directioni virium ductæ per omnium trium massarum centra gravitatis, quarum massarum eam, quæ jacuerit inter reliquarum binarum parallelas diximus mediam: ac si ducantur in quavis alia directione data rectæ ab iis massis ad illas parallelas; erunt ejusmodi distantiæ ab iis parallelis, ut ipsæ AB, EB, ad quas erunt singulæ in ratione data, ob datas directiones. Quare pro viribus parallelis habetur hujusmodi theorema: *Vires parallelæ motrices binarum quarunvis ex tribus massis sunt inter se reciproce ut distantiæ a directione communi transeunte per tertiam: vires autem acceleratrices præterea in ratione reciproca massarum, & media est directionis contrariæ respectu reliquarum, ac vis media motrix æquatur reliquarum summæ, utralibet vero extrema differentiæ.*

Applicatio rationum superiorum ad centrum æquilibrii.

321. Hoc theorema primo quidem exhibet centrum æquilibrii, viribus utcunque divergentibus, vel convergentibus. Si nimirum sint tres massæ A, B, C (& nomine massarum etiam intelligi possunt singula puncta), quarum binæ, ut A, & B, solicitentur viribus motricibus externis; poterunt mutuis viribus illas elidere, ac esse in æquilibrio, & eas elident omnino, mutatis, quantum libuerit, parum mutuis distantiis; si fuerint ante applicationem earum virium externarum in satis validis limitibus cohæsionis, ac vis massæ C elidatur fulcro opposito in directione DC, vel suspensione contraria: dummodo binæ illæ vires ductæ in massas habeant conditiones requisitas in superioribus, ut nimirum ambæ tendant ad idem punctum, vel ab eodem, aut si fuerint parallelæ, ambæ candem directionem habeant, ubi simul ambæ ingrediantur, vel simul ambæ evitent triangulum ABC: ubi vero altera ingrediatur triangulum, altera evitet, tendat altera ad punctum concursus, altera ad partes illi oppositas: vel si fuerint parallelæ, habeant directiones [149] oppositas: & si parallelæ fuerint; sint inter se, ut distantiæ a directione virium transeunte per C; si fuerint convergentes, sint reciproce, ut sinus angulorum, quos earum directiones continent cum recta ex C tendente ad earum concursum, vel sint in ratione reciproca sinuum angulorum, quos continent cum rectis AC, BC, & ipsarum rectarum conjunctim.

Determinatio vis, quam fulcrum sustinet.

322. Determinabitur autem admodum facile per ipsa theoremata etiam vis, quam sustinebit fulcrum C, quæ in casu parallelismi æquabitur summæ, vel differentiæ reliquarum, prout ibi fuerit media, vel extrema: & in casibus reliquis omnibus æquabitur summæ pariter, vel differentiæ reliquarum ad suam directionem reductarum, reliquis binis in resolutione priorum sociis se per contrariam directionem, & æqualitatem elidentibus.

Consideratio massarum etiam intermediarum, quæ connectant massas viribus externis præditas, & positas in æquilibrio.

323. Habebitur igitur, quidquid pertinet ad æquilibrium virium agentium in eodem plano, & connexarum non per virgas inflexiles carentes omni vi præter cohæsionem, uti eas vulgo concipiunt, sed bisce viribus mutuis. Et Theoria quidem habebit locum tum hic, tum in sequentibus; licet massæ A, B, C non agant in se invicem immediate, sed sint aliæ massæ intermediæ, quæ ipsas jungant. Nam si inter massam B, & C sint aliæ massæ nullis externis viribus agitatæ, & positæ in æquilibrio cum hisce massis, & inter se, ac prima, quæ venit post B, agat in ipsam vi motrice æquali BP, aget & B in ipsam vi æquali: quare debebit illa ad servandum æquilibrium urgeri a secunda, quæ est post ipsam, vi æquali in partes contrarias. Hinc æquali contraria aget tertia in secundam, ut secunda in æquilibrio sit, & ita porro, donec deveniatur ad C, ubi habebitur vis motrix æqualis motrici, quæ erat in B, & erunt vires BP, CV acceleratrices in ratione reciproca massarum B, & C, cum vires illæ motrices æquales sint producta ex acceleratricibus ductis in massas. At si circumquaque sint massæ quotcunque cum vacuis quibuscunque, ac ubicunque interjectis, quæ connectantur cum punctis A, B, C, affectis illis tribus viribus externis, quarum una concipitur provenire a fulcro, una solet appellari potentia, & una resistentia, ac vires illæ externæ QB, HA concipiantur resolutæ singulæ in binas agentes secundum eas rectas,

of the ratio by ED/AD, as AB.ED/AD.BD is to AE/AD. By a similar argument we obtain also that *sin*BDA is to *sin*BDE as AB.ED/AD.BD is to BE/BD; from which the whole proposition is clear.

A more simple expression for the case of parallelism.

320. If the point D goes off to infinity, & the directions of the forces thus become parallel to one another, the ratios of the straight lines ED, AD, BD finally become ratios of equality. Hence, in that case, the three forces are to one another as AB to AE to EB; & the first of these is equal to the sum of the other two. Imagine straight lines drawn parallel to the directions of the forces, through the centres of gravity of all three masses, & let that one of the masses which lies between the parallels drawn through the other two be called the middle mass; then, if we draw in any given direction straight lines from the masses to meet the parallels, the distances from the parallels measured along these lines will be as AB, EB; for the distances bear the same given ratio to AB, EB, on account of the given directions. Hence for parallel forces we obtain the following theorem. *Parallel motive forces for any two out of three masses are to one another inversely as the distances from a common direction passing through the third; & the accelerations have in addition the inverse ratio of the masses. The middle acceleration is in an opposite direction to that of the others; & the middle motive force is equal to the sum of the other two, whilst either outside one is equal to the difference of the other two.*

Application of the above ratios to the centre of equilibrium.

321. The theorem of the preceding article will yield the centre of equilibrium for any forces, whether diverging or converging. For instance, if A, B, C are three masses (& in the term masses, single points can also be understood to be included), of which two, A & B say, are acted upon by external motive forces; then the mass will be able to eliminate these by means of mutual forces, & remain in equilibrium, & then to eliminate the mutual forces entirely by changing slightly their mutual distances, as required; provided that, before the application of those external forces, they were in positions corresponding to a sufficiently strong limit point of cohesion, & the force on the mass C was cancelled by a fulcrum opposite to the direction DC, or by a contrary suspension; & so long as the two forces multiplied each by its corresponding mass preserve the conditions stated as requisite in the above, namely, that both tend to the same point or both away from it, or if they are parallel both have the same direction, when they both together fall within the triangle ABC, or both tend away from it; or if, on the other hand, when one of them falls within the triangle & the other away from it, the one tends to the point of intersection & the other away from it, or if they are parallel have opposite directions. Further, if they are parallel, they are to one another as the distances from the direction of forces which passes through C; if they are convergent, they are inversely as the sines of the angles between their directions & the straight line through C to their point of intersection; or are in the inverse ratio of the sines of the angles between their directions & the straight lines AC, BC & the ratio of these straight lines jointly.

Determination of the force on the fulcrum.

322. It is moreover quite easy by means of the theorems to determine also the force on the fulcrum placed at C; this, in the case of parallelism, will be equal to the sum or the difference of the other two forces according as C is the middle or one of the outside masses. In all other cases, it will be equal to the sum or difference of the other forces, in a similar way, if these are reduced to the direction of the force on C, the remaining pairs of forces that are associated with the former in the resolution cancelling one another on account of their being equal & opposite.

Investigation of the case in which there are also intermediate masses connecting the masses upon which the external forces act, & placed in equilibrium.

323. Hence may be obtained all things that relate to the equilibrium of forces acting in one plane, & connected, not by inflexible rods lacking all force except cohesion, but by these mutual forces. The Theory holds good indeed, not only here, but also in what follows; that is to say, although the bodies A, B, C may not act upon one another directly, yet there are other intermediate masses which connect them. For, if between A & B there were other masses not influenced by any external forces, & placed in equilibrium with these masses & with one another, then the first mass which comes after B will act upon B with a motive force equal to BP, & B will act upon it with an equal force; hence, to preserve the equilibrium, this mass must be acted upon by the second, the one which comes next after it, with a force equal & opposite to this. Hence it follows that the third must act on this second with a force equal & opposite to that, in order that the second may be in equilibrium; & so on, until we come to C, where we have a motive force equal to that acting on B; & the accelerations BP, CV will be in the inverse ratio of the masses B & C, since the equal motive forces are proportional to the products of the accelerations into the masses. Moreover, if in any positions there are any number of masses, having any empty spaces interspersed anywhere, & these are in connection with three masses A, B, C, which are under the influence of those three forces, of which one is assumed to be produced by a fulcrum, one is usually termed the power, & the third the resistance; & if the external forces BQ, HA are considered to be resolved each into two parts acting along

quæ illa tria puncta conjungunt; poterit clisis mutuo reliquis omnibus æquilibrium constituentibus deveniri ad vires in punctis binis, ut A, & C, acceleratrices contrarias viribus BP, BR, & reciproce proportionales massis ipsarum respectu massæ B; licet ipsæ proveniant a massis quibusvis etiam non in eadem directione sitis, & agentibus in latus: nam per ejusmodi resolutionem, & ejusmodi virium considerationem adhuc habetur æquilibrium totius systematis affecti in illis tribus punctis per illas tres vires, cum assumantur in iis tantummodo vires motrices contrariæ, & æquales: unde fit, ut etiam illæ, quæ præterea ad has in illis considerandas assumuntur, & per quas connectuntur cum reliquis massis, se mutuo elidant.

Qui motus, ubi non habeatur æquilibrium.

[150] 324. Quod si vires ejusmodi non fuerint in ea ratione inter se; non poterunt puncta B, & A esse in æquilibrio, sed consequetur motus secundum directionem ejus, quæ prevalet: ac si omnis motus puncti C fuerit impeditus; habebitur conversio circa ipsum C.

Extensio ad æquilibrium quotcunque massarum, & inde principium generale pro machinis, & ratio momentorum.

325. Quod si non in tribus tantummodo massis habeantur vires externæ, sed in pluribus; licebit considerare quanvis aliam massam carentem omni externa vi, & eam concipere connexam cum singulis reliquarum massis, & massa C per vires mutuas, ac habebitur itidem Theoria pro æquilibrio omnium, cum positione omnium constanter servata etiam sine ulla figuræ mutatione, quæ sensu percipi possit. Quin immo si singulæ vires illæ externæ resolvantur in duas, quarum altera urgeat in directione rectæ transeuntis per C, ac elidatur vi proveniente a solo puncto C, & altera agat perpendiculariter ad ipsam, ut habeatur æquilibrium in singulis ternariis; oportebit esse singulas vires novæ massæ assumptæ ad vim ejus, cum qua conjungitur, in ratione reciproca distantiarum ipsarum massarum a C; cum jam sinus anguli recti ubique sit idem. Debebunt autem omnes vires, quæ in massam assumptam agunt directionibus contrariis, se mutuo elidere ad habendum æquilibrium. Quare debebit summa omnium productorum carum virium, quæ urgent conversione in unam plagam, per ipsarum distantias a centro conversionis, æquari summæ productorum carum, quæ urgent in plagam oppositam, per distantias ipsarum, ut habeatur æquilibrium; cumque arcus circulares in ea conversione descripti dato tempusculo sint illis distantiis proportionales, & proportionales sint ipsis arcubus velocitates; debebunt singularum virium agentium in unam plagam producta per velocitates, quas haberent puncta, quibus applicantur secundum suam directionem, si vincerentur, vel contra, si vincerent, simul sumpta æquari summæ ejusmodi productorum agentium in plagam oppositam. Atque inde habetur principium pro machinis & simplicibus, & compositis, ac notio illius, quod appellant momentum virium, deducta ex eadem Theoria.

Applicatio ad omnia vectium genera.

326. Casus trium tantummodo massarum exhibet vectem, cujus brachia sint utcunque inflexa. Quod si tres massæ jaceant in directum, efformabunt rectilineum vectem, qui quidem applicatis viribus inflectetur semper nonnihil, ut & in superioribus casibus semper non nihil a priore positione discedet systema novis viribus externis affectum; sed is discessus poterit esse utcunque exiguus, ut supra monui: si limites sint satis validi; adeoque poterit adhuc vectis esse ad sensum rectilineus. Tum vero vires externæ debebunt esse unius directionis, & contrariæ directioni vis mediæ, & binæ quævis ex iis erunt ad se invicem reciproce, ut distantiæ a tertia. Inde autem oriuntur tria genera vectium: si fulcrum, vel hypomochlium, sit in medio in E, vis in altero extremo A, [151] resistentia in altero B; vis ad resistentiam est, ut BE, distantia resistentiæ a fulcro, ad AE distantiam vis ab eodem: fulcrum autem sentiet summam virium. Et quod de hoc vectis genere dicitur, id omne ad libram pariter pertinet, quæ ad hoc ipsum vectis genus reducitur. Si fulcrum sit in altero extremo, ut in B, vis in altero, ut in A, & resistentia in medio, ut in E; vis ad resistentiam erit in ratione distantiæ EB ad distantiam majorem AB, cujus idcirco momentum, seu energia, augetur in ratione suæ distantiæ AB ad EB, ut nimirum possit tanto majori resistentiæ æquivalere. Si demum fuerit quidem fulcrum in altero extremo B, & resistentia in A, vis prior in E; tum e contrario erit resistentia ad vim in majore ratione AB ad EB, decrescente tantundem hujus energia, seu momento. In utroque autem casu fulcrum sentiet differentiam virium.

Consectaria doctrinæ de vectibus, & principium pro statera; cur totum pondus consideretur, ut collectum in centre gravitatis.

327. Quod si perticæ utcunque inclinatæ applicetur pondus in aliquo puncto E, & bini humeros supponant in A, & B, sentient ponderis partes inæquales in ratione reciproca distantiarum ab ipso; & si e contrario bina pondera suspendantur in A, & B utcunque inæqualia, assumpto autem puncto E, cujus distantiæ a punctis A, & B sint in ratione

the lines which join the three points; then it will be possible, all the other forces constituting the equilibrium cancelling one another, to arrive at accelerations for the two points A & C say, in opposite directions to the forces BP, BR, & inversely proportional to their masses with regard to the mass B. This will be the case, even although they may proceed from any masses not lying in the same direction, & acting to one side; for, by means of resolution of this kind, & a consideration of such forces, we yet have equilibrium of the whole system affected at the three points by the three forces, since here are assumed only motive forces such as are equal & opposite. Hence it follows that the former, which are assumed in addition for the consideration of the latter in such cases, & by which they are connected with the other masses, must also cancel one another.

The nature of the motion when equilibrium does not obtain.

324. But if such forces are not in this ratio to one another, the points B & A cannot be in equilibrium; but motion would follow in the direction of that which preponderated; also if all motion of the point C were prevented, then there would be rotation about C.

Extension to the equilibrium of any number of masses; & thence a general principle for machines & the ratio of moments.

325. Now if we have external forces acting, not on three masses only, but on several, we can consider any one mass to be without an external force, & suppose that this mass is connected to each of the others, & to the mass C, by mutual forces; & the Theory will hold good for the equilibrium of them all, with the position of them all constantly maintained without any change of figure so far as can be observed. Further, if all the external forces are resolved each into two parts, of which one acts along the straight line passing through C, & is cancelled by a force proceeding from C alone, & the other acts perpendicularly to this line, so that equilibrium is obtained for each set of three; then it will be necessary that each of the forces on the new mass chosen will be to the force of that to which it is joined in the inverse ratio of these masses from C, since now the sines of the right angles are everywhere the same. Also all the forces which act on the chosen mass in opposite directions, must cancel one another to maintain equilibrium. Hence the sum of all the forces which tend to produce rotation in one direction, each multiplied by its distance from the centre of rotation, must be equal to the sum of the products of the forces which tend to produce rotation in the opposite direction, multiplied by their distances, in order that equilibrium may be maintained. Since the circular arcs in this rotation which are described in any interval of time are proportional to the distances, & these are proportional to the velocities in the arcs, it follows that the products of each of the forces acting in one direction by the velocities which correspond to the points to which they are applied, in the direction of the forces if they are overcome, & in the opposite direction if they overcome, all together must be equal to the sum of the like products acting in the other direction. Hence is derived a principle for machines, both simple & complex; & also an idea of what is called the moment of forces; & these have been deduced from this same Theory.

Application to all kinds of levers.

326. The case of three masses only yields the case of the lever, whose arms are curved in any manner. But if the three masses lie in one straight line, they will form a rectilinear lever; now this, on the application of forces, will always be bent to some degree; just as, in the cases above, the system when affected by fresh external forces always departed from its original position to some extent. But this departure is exceedingly slight in every case, as I mentioned above, if only the limit-points are sufficiently strong; & thus the lever can still be considered as sensibly rectilinear. In this case, then, the external forces must be in the same direction, & in an opposite direction to that of the middle force, & any two of them must be to one another in the inverse ratio of their distances from the third. Now from this there arise three kinds of levers. If the fulcrum, or lever-support, is in the middle at E, the force acting on one end A & the resistance at the other end B; then the ratio of the force to the resistance is as BE, the distance of the resistance from the fulcrum, to AE the distance of the force from it; & the force on the fulcrum will be the sum of the two. What is said about this kind of lever applies equally well to the balance, which reduces to this kind of lever. If the fulcrum should be at one end, at B say, the force at the other, A, & the resistance in the middle, at E; then the force is to the resistance in the ratio of the distance EB to the greater distance AB; & therefore the moment, or energy, will increase in the ratio of the distance AB to EB, so that indeed it may be able to balance a much greater resistance in proportion. Finally, if the fulcrum were at one end, B, the resistance at A, & the former force at E; then, on the contrary, the resistance is to the force in the greater ratio of AB to EB, thus decreasing its energy or momentum in the same proportion. In both these latter cases the force on the fulcrum will be equal to the difference of the forces.

Consequences of this doctrine of levers, & the principle of the steelyard. The reason why the whole may be considered as if collected at the centre of gravity.

327. Now, if to a long pole, inclined at any angle to the horizontal, a weight is applied at any point E; & if two men place their shoulders under the pole at A & B; then they will support unequal parts of the weight, in the inverse ratio of their distances from it. Conversely, if two unequal weights of any sort are suspended from A & B, & a point E is taken whose distances from the points A & B are in the inverse ratio of the weights, & so

reciproca ipsorum ponderum, adeoque massarum, quibus pondera proportionalia sunt, quod idcirco erit centrum gravitatis; suspensa per id punctum pertica, vel supposito fulcro, habebitur æquilibrium, & in E habebitur vis æqualis summæ ponderum. Quin immo si pertica sit utcunque inflexa, & pendeant in A, & B pondera; suspendatur autem ipsa pertica per C ita, ut directio verticalis transeat per centrum gravitatis; habebitur æquilibrium, & ibi sentietur vis æqualis summæ ponderum, cum ob naturam centri gravitatis debeant esse singula pondera, seu massæ duetæ in suas perpendiculares distantias a linea verticali, quam etiam vocant lineam directionis, hinc, & inde æqualia. Nam vires ponderum sunt parallelæ, & in iis juxta num. 320 satis est ad æquilibrium, si vires motrices sint reciproce proportionales distantiis a directione virium transeunte per tertium punctum: sentietur autem in suspensione vis æqualis summæ ponderum. Atque inde fluit, quidquid vulgo traditur de æquilibrio solidorum, ubi linea directionis transit per basim, sive fulcrum, vel per punctum suspensionis, & simul illud apparet, cur in iis casibus haberi possit tota massa tanquam collecta in suo centro gravitatis, & habeatur æquilibrium impedito ejus descensu tantummodo. Gravitas omnium punctorum non applicatur ad centrum gravitatis, nec ibi ipsa agit per sese; sed ejusmodi esse debent distantiæ punctorum totius systematis, ut inter fulcrum, & punctum ipsi imminens habeatur vis quædam æqualis summæ virium omnium parallelarum, & directa ad partes oppositas directionibus illarum.

Theoriam exhibere egregie itidem centrum oscillationis. Quid ipsum sit.

[152] 328. At non minus feliciter ex eadem Theoria, & ex eodem illo theoremate, fluit determinatio centri oscillationis. Pendula breviora citius oscillant, remotiora lentius. Quare ubi connexa sunt inter se plura pondera, aliud propius axi oscillationis, aliud remotius ab ipso, oscillatio neque fiet tam cito, quam requirunt propiora, neque tam lente, quam remotiora, sed actio mutua debebit accelerare hæc, retardare illa. Erit autem aliquod punctum, quod nec accelerabitur, nec retardabitur, sed oscillabit, tanquam si esset solum. Illud dicitur centrum oscillationis. Determinatio illius ab Hugenio primum est facta, sed precario, & non demonstrato principio: tum alii alias itidem obliquas inierunt vias, ac præcipuas quasque methodos huc usque notas persecutus sum in Supplementis Stayanis § 4 lib. 3. En autem ejus determinationem simplicissimam ope ejusdem theorematis numeri 313.

Preparatio ad solutionem problematis quærentis ipsum centrum.

329. Sint plures massæ, quarum una A in fig. 63, mutuis viribus singulæ connexæ cum P, cujus motus sit impeditus suspensione, vel fulcro, & cum massa Q jacente in quavis recta PQ, cujus massæ Q motus a mutuo nexu nihil turbetur, quæ nimirum sit in centro oscillationis. Porro hic cum massas pono in punctis spatii A, P, Q, intelligo vel puncta singula, vel quævis aggregata punctorum, quæ concipiantur, ut compenetrata in iis punctis. Velocitati jam acquisitæ in descensu nihil obstabit is nexus, cum ea sit proportionalis distantiæ a puncto suspensionis P, nisi quatenus per eum nexum retrahentur omnes massæ a recta tangente ad arcum circuli, sustinente puncto ipso suspensionis justa num. 282 vim mutuam respondentem iis omnibus viribus centrifugis. Resoluta gravitate in duas partes, quarum altera agat secundum reetam, quæ jungit massam cum P, altera sit ipsi perpendicularis, idem punctum P sustinebit etiam priorem illam, posterior autem determinabit massas ad motus AN, QM, perpendiculares ipsis AP, QP, ac proportionales per num. 301 sinubus angulorum APR, QPR, existente PR verticali. Sed nexus coget describere arcus similes, adeoque proportionales distantiis a P. Quare si sit AO spatium, quod vi gravitatis obliquæ, sed ex parte impeditæ a nexu, revera percurrat massa A; quoniam Q non turbatur, adeoque percurrit totum suum spatium QM; erit QM ad AO, ut QP ad AP. Demum actio ex A in Q ad actionem ex Q in A proportionalem ON, crit ex theoremate numeri 314 ut est Q × QP ad A × AP, & omnes ejusmodi actiones ab omnibus massis in Q debebunt evanescere, positivis & negativis valoribus se mutuo elidentibus. Ex illis tribus proportionibus, & hac æqualitate res omnis sic facillime expeditur.

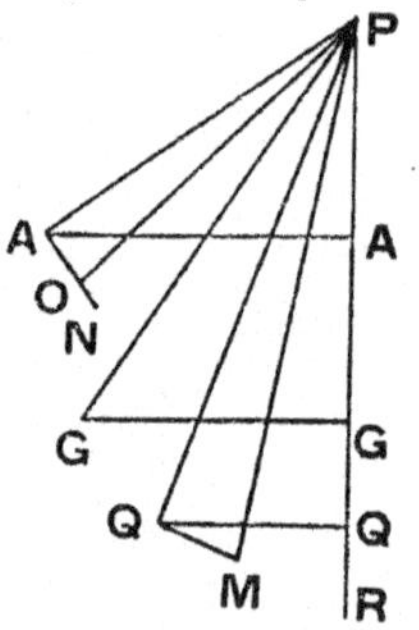

Fig. 63.

Solutio problematis, ac demonstratio.

330. Dicatur QM = V, sinus APR = a, sinus QPR = q. Erit ex prima proportione $q : a :: \text{QM} = \text{V} : \text{AN} = \frac{a}{q} \times \text{V}$. [153] Ex secunda $\text{QP}.\ \text{AP} :: \text{QM} = \text{V}.\ \text{AO} = \frac{\text{AP}}{\text{QP}} \times \text{V}$.

Quare $\text{ON} = \left(\frac{a}{q} - \frac{\text{AP}}{\text{QP}}\right) \times \text{V}$. Sed ex tertia

$$\text{Q} \times \text{QP}.\ \text{A} \times \text{AP} :: \text{ON} = \left(\frac{a}{q} - \frac{\text{AP}}{\text{QP}}\right) \times \text{V}.\ \left(\frac{a \times \text{A} \times \text{AP}}{q} - \frac{\text{A} \times \text{AP}^2}{\text{QP}}\right) \times \frac{\text{V}}{\text{Q} \times \text{QP}},$$

of the masses to which the weights are proportional, so that the point is their centre of gravity; then, if the pole is suspended by this point, or a fulcrum is placed beneath it, there will be equilibrium, & the force at E will be equal to the sum of the two weights. Further, if the pole were bent in any manner, & weights were suspended at A & B, & the pole itself were suspended at C, so that the vertical direction passes through the centre of gravity of the weights; then there would be equilibrium, & there would be a force at C equal to the sum of the weights. For, on account of the nature of the centre of gravity, each of the weights, or masses, multiplied by its perpendicular distance from the vertical line, which is also called the line of direction, must be equal on the one side & on the other. For the forces of the weights are parallel; & for such, according to Art. 320, it is sufficient for equilibrium, if the motive forces are proportional inversely to the distances from the direction of forces passing through the third point; moreover there will be experienced at the point of suspension a force equal to the sum of the weights. Hence is derived everything that is usually taught concerning the equilibrium of solids, where a line of direction passes through the base, or through the fulcrum, or through the point of suspension; at the same time we get a clear perception of the reason why in such cases the whole mass can be considered as if it were condensed at its centre of gravity, & equilibrium can be obtained by merely preventing the descent of this point. The gravity of all the points is not applied at the centre of gravity, nor does it act there of itself; but the distances of the points of the whole system must be such that between the fulcrum & the point hanging just over it there must be a certain force equal to the sum of all the parallel forces, & directed so as to be opposite to their direction.

The Theory affords an excellent explanation of the centre of oscillation as well.

328. In a no less happy manner there follows from this same Theory, & from the very same theorem, the determination of the centre of oscillation. Shorter pendulums oscillate more quickly, & longer ones more slowly. Hence when several weights are connected together, one nearer to the axis of oscillation, & another more remote from it, the oscillation is neither so fast as that required by the nearer, nor so slow as that required by the more remote; but a mutual action must accelerate the one & retard the other. Moreover there will be one point, which will be neither accelerated nor retarded, but will oscillate as if it were alone; that point is called the centre of oscillation. Its determination was first made by Huygens, but from a principle that was doubtful & unproved. After him, others came upon it indirectly, some in one way & some in another; & I investigated some of the best methods then known in the Supplements to Stay's Philosophy, § 4, Bk. 3. Now I present you with an exceedingly simple determination of it, derived from that same theorem of Art. 313.

Preparation for the solution of the problem of finding this centre.

329. Suppose there are several masses, of which in Fig. 63 one is at A, & that each of these is connected to P by mutual forces; & let the motion of P be prevented by suspension, or by a fulcrum; also let A be connected with a mass Q lying in a straight line PQ, & let the motion of this mass Q be in no way affected by the mutual connection, as will happen if Q is at the centre of oscillation. Now, when I place masses at the points of space A, P, Q, I intend single points of matter, or any aggregates of such points, which may be considered as condensed at those points of space. The connection will not oppose in any way the velocity already acquired in descent, since it is proportional to the distance from the point of suspension P; except in so far as all the masses are pulled out of the tangent line into a circular arc by the connection, the point of suspension itself being under the influence of a mutual force corresponding to all the centrifugal forces. If gravity is resolved into two parts, one of which acts along the straight line joining the mass to P, & the other perpendicular to it; then the point P will sustain the former of these as well, but the latter will give to the masses the motions AN, QM, respectively perpendicular to AP, QP, & proportional, by Art. 301, to the sines of the angles APR, QPR, where PR is the vertical. But the connection forces them to describe arcs that are similar, & therefore proportional to the distances from P. Hence, if AO is the space, which under the oblique force of gravity, but partly hindered by the connection, the mass A would really pass over; then, since Q is not affected, & will thus pass over the whole of its course QM, we shall have QM to AO as QP to AP. Lastly, the action of A on Q is to the action of Q on A, (which is proportional to ON), as $Q \times QP$ is to $A \times AP$, by the theorem of Art. 314; & all such actions from all the masses upon Q must vanish, the positive & negative values cancelling one another. From the three proportions & this equality the whole question is worked out in the easiest possible way.

Solution of the problem and its demonstration.

330. Suppose $QM = V$, the sine of $APR = a$, the sine of $QPR = q$. Then, since from the first proportion, $q : a = QM : AN$, therefore $AN = a.V/q$; &, since from the second proportion, $QP : AP = QM : AO$, therefore $AO = AP.V/QP$. Hence $ON = (a/q - AP/QP).V$. But, from the third proportion, $Q \times QP$ is to $A \times AP$ as ON is to the action of A on Q. Therefore the action on Q due to the connection with A

quæ crit actio in Q ex nexu cum A. At eodem pacto si esset alibi alia massa B itidem connexa cum P, & Q, actio in Q inde orta haberetur, positis B, b loco A, a; & ita porro in quibusquis massis C, D, &c. Omnes autem isti valores positi $= 0$, dividi possent per $\frac{V}{Q \times QP}$, utique commune omnibus, & deberent e valoribus conclusis intra parentheses ii, qui sunt positivi, æquales esse negativis. Quare habebitur

$$\frac{a \times A \times AP + b \times B \times BP}{q} = \frac{A \times AP^2 + B \times BP^2 \ \&c.}{QP},$$

& inde $QP = q \times \frac{A \times AP^2 + B \times BP^2 \ \&c.}{a \times A \times AP + b \times B \times BP \ \&c.}$

Evolutio casus ponderum jacentium in eadem recta cum puncto suspensionis.

331. Sint jam primo omnes massæ in eadem recta linea cum puncto suspensionis P, & cum centro oscillationis Q; & angulus QPR æquabitur cuivis ex angulis APR, ac ejus sinus q singulis sinubus a, b &c. Quare pro eo casu formula evadit $\frac{A \times AP^2 + B \times BP^2 \ \&c.}{A \times AP + B \times BP \ \&c.}$, quæ est ipsa formula Hugeniana pro ponderibus jacentibus recta transeunte per centrum suspensionis.

Et casus jacentium extra.

332. Quod si jaceant extra ejusmodi rectam in plano POR perpendiculari ad axem rotationis transeuntem per P; sit G centrum commune gravitatis omnium massarum, ducanturque perpendicula AA′, GG′, QQ′ ad PR, & crit ut radius $= 1$ ad a, ita AP ad $AA' = a \times AP$; & eodem pacto $QQ' = q \times QP$, $GG' = g \times GP$. Substitutis AA′ pro $a \times AP$ & eodem pacto BB′ (quam Figura non exprimit) pro $b \times BP$ &c. evadat $QP = q \times \frac{A \times AP^2 + B \times BP^2 \ \&c}{A \times AA' + B \times BB' \&c.}$. Sed si summa massarum dicatur M, est per num. 245 ex natura centri gravitatis, $A \times AA' + B \times BB' \ \&c. = M \times GG' = M \times g \times GP$. Habebitur igitur valor QP radii nihil turbati in ea inclinatione

$$\frac{q}{g} \times \frac{A \times AP^2 + B \times BP^2 \ \&c.}{M \times GP}.$$

Initium applicationis ad oscillationes in latus ponderum jacentium in eodem plano.

[154] 333. Is valor crit variabilis pro varia inclinatione ob valores sinuum q, & g variatos, nisi QP transeat per G, quo casu sit $q = g$; & quidem ubi G accedit in infinitum ad PR, decrescente g in infinitum, si PQ non transeat per G, manente finito q, valor $\frac{q}{g}$ excrescit in infinitum; contra vero appellente QP ad PR, evadit $q = 0$, & g remanet aliquid, adeoque $\frac{q}{g}$ evanescit. Id vero accidit, quia in appulsu G ad verticalem totum systema vim acceleratricem in infinitum imminuit, & lentissime acceleratur; adeoque ut radius PQ adhuc obliquus sit ipsi in ea particula oscillationis infinitesima isochronus, nimirum æque parum acceleratus, debet in infinitum produci. Contra vero appellente PQ ad PR ipsius acceleratio minima esse debet, dum adhuc acceleratio radii PG obliqui est in immensum major, quam ipsa; adeoque brevitate sua ipse radius compensare debet accelerationis imminutionem.

Finis ejusdem cum formula generali.

334. Quare ut habeatur pendulum simplex constantis longitudinis, & in quacunque inclinatione isochronum composito, debet radius PQ ita assumi, ut transeat per centrum gravitatis G, quo unico casu fit constanter $q = g$, & formula evadit constans $QP = \frac{A \times AP^2 + B \times BP^2 \ \&c.}{M \times GP}$, quæ est formula generalis pro oscillationibus in latus massarum quotcumque, & quomodocunque collocatarum in eodem plano perpendiculari ad axem rotationis, qui casus generaliter continet casum massarum jacentium in eadem recta transeunte per punctum suspensionis, quem prius eruimus.

Corollarium pro positione centri oscillationis, & gravitatis ex eadem parte a puncto suspensionis.

335. Inde autem pro hujusmodi casibus plura corollaria deducuntur. Inprimis patet: *gravitas centrum debere jacere in recta, quæ a centro suspensionis ducitur per centrum oscillationis*, uti demonstratum est num. 334. Sed & *debet jacere ad eandem partem cum ipso centro oscillationis.* Nam utcunque mutetur situs massarum per illud planum, manentibus puncto suspensionis P, & centro gravitatis G, signum valoris quadrati cujusvis AP, BP manebit semper idem. Quare formula valoris sui signum mutare non poterit;

will be $\left(\frac{a \times A \times AP}{q} - \frac{A \times AP^2}{QP}\right) \times \frac{V}{Q \times QP}$. In the same manner, if there were another mass somewhere else, also connected with P & Q, the action on Q arising from its presence would be obtained, if B & b were substituted for A & a; & so on for any masses C, D, &c. Now, putting all these values together equal to zero, they can be divided through by V/(Q × QP), which is common to every one of them; & those of the values included in the brackets that are positive must be equal to those that are negative. Hence we have

$$(a \times A \times AP + b \times B \times BP + \&c.)/q = (A \times AP^2 + B \times BP^2 + \&c.)/QP;$$

and hence $QP = q. \frac{A \times AP^2 + B \times BP^2 + \&c.}{a \times A \times AP + b \times B \times BP + \&c.}$

331. Suppose now, first of all, that all the masses lie in one straight line with the point of suspension P, & so with the point of oscillation Q; then the angle QPR will be equal to any one of the angles like APR, & its sine q will be equal to any one of the sines a, b, &c. Hence for this case the formula reduces to

Derivation of the case of weights hanging in the same straight line with the point of suspension.

$$\frac{A \times AP^2 + B\, BP^2 + \&c.}{A \times AP + B \times BP + \&c.};$$

& this is the selfsame formula found by Huygens for weights lying in the straight line passing through the centre of suspension.

332. But if the masses lie outside of any such line, in the plane POR, perpendicular to the axis of rotation passing through P, suppose that G is the common centre of gravity of all the masses, & let perpendiculars AA′, GG′, QQ′ be drawn to PR. Then, since the radius (= 1) : a = AP : AA′, therefore AA′ = a × AP : & in a similar manner, QQ′ = q × QP, & GG′ = g × GP. Now, if AA′ is substituted for a × AP, & similarly BB′ (not shown in the figure) for b × BP, & so on; the formula will become

The case of when the masses are not on this line.

$$QP = q. \frac{A \times AP^2 + B \times BP^2 + \&c.}{A \times AA' + B \times BB' + \&c.}$$

But, if the sum of the masses is denoted by M, then, by Art. 245, from the nature of the centre of gravity, we have A × AA′ + B × BB′ + &c. = M × GG′ = M × g × GP; & therefore we obtain the value of the radius QP, in a form that is independent of the inclination, namely,

$$\frac{q}{g} \times \frac{A \times AP^2 + B \times BP^2 + \&c.}{M \times GP}$$

333. The value obtained will vary with various inclinations, owing to the varying values of the sines q & g, unless QP passes through G; in which case $q = g$. Indeed, when G approaches indefinitely near to PR, & g thus decrease indefinitely, if PQ does not pass through G, thus leaving q finite, the value of q/g will increase indefinitely. On the other hand, when QP coincides with PR, q = O, & g will remain finite; & thus q/g will vanish. This indeed is just what does happen; for, when G approaches the vertical the whole system diminishes the accelerating force indefinitely, & it is accelerated exceedingly slowly; thus, in order that the radius PQ whilst still oblique may be isochronous during that infinitesimally small part of the oscillation, that is to say, may be accelerated by an equally small amount, it must be prolonged indefinitely. On the other hand, as PQ approaches PR, its acceleration must be very small, whilst the acceleration of the radius PG which is still oblique is immensely greater in comparison with it; & thus the radius PQ must by its shortness compensate for the diminution of the acceleration.

Commencement of the application to oscillations to one side of bodies lying in the same plane.

334. Hence, in order to obtain a simple pendulum of constant length, isochronous at any inclination with the composite pendulum, the radius PQ must be so taken that it passes through the centre of gravity G, in which case alone $q = g$, & the formula reduces to a constant value for QP, which

Conclusion of the same, with a general formula.

$$= \frac{A \times AP^2 + B \times BP^2 + \&c.}{M \times GP}.$$

This is a general formula for oscillations to one side of any number of masses, disposed in any way whatever in the same plane, the plane being perpendicular to the axis of rotation; & this case contains in general the case of masses lying in the same straight line through the point of suspension, which we have already solved.

335. Now for cases of this sort many corollaries can be derived from the theorem proved above. First of all, it is clear that :—*The centre of gravity must lie in the straight line joining the centres of oscillation & suspension*; this has been proved in Art. 335. But also *it must lie on the same side of the point of suspension as does the centre of oscillation.* For however the positions of the masses are changed in the plane, so long as the positions of the points of suspension P & of the centre of gravity G remain unaltered, the sign of the value of any square, such as AP, BP, will remain the same. Hence the formula cannot

Corollary with respect to the centres of oscillation & gravity on the same side of the point of suspension.

adeoque si in uno aliquo casu jaceat Q respectu P ad eandem plagam, ad quam jacet G; debebit jacere semper. Jacet autem ad eandem plagam in casu, in quo concipiatur, omnes massas abire in ipsum centrum gravitatis, quo casu pendulum evadit simplex, & centrum oscillationis cadit in ipsum centrum gravitatis, in quo sunt massæ. Jacebit igitur semper ad eandem partem cum G.

Centrum gravitatis debere esse inter bina reliqua ex iis punctis.

[155] 336. Deinde *debet centrum gravitatis jacere inter punctum suspensionis, & centrum oscillationis.* Sint enim in fig. 64 puncta A, P, G, Q eadem, ac in fig. 63, ducanturque AG, AQ, & Aa perpendicularis ad PQ; summa autem omnium massarum ductarum in suas distantias a recta quapiam, vel plano, vel in earum quadrata, designetur præfixa litera ∫ soli termino pertinente ad massam A, ut contractiores evadant demonstrationes. Erit ex formula inventa $PQ = \frac{\int.A \times AP^2}{M \times GP}$.
Porro est $AG^2 = AP^2 + GP^2 - 2\ GP \times Pa$, adeoque
$AP^2 = AG^2 - GP^2 + 2\ GP \times Pa$,
& $\int.A \times GP^2$ est $M \times GP^2$,
ob GP constantem; ac $\int.A \times Pa = M \times GP$, cum P$a$ sit æqualis distantiæ massæ a plano perpendiculari rectæ QP transeunte per P, & eorum productorum summa æquetur distantiæ centri gravitatis ductæ in summam massarum; adeoque $\int.A \times 2\ GP \times Pa$ erit $= 2\ M \times GP^2$.

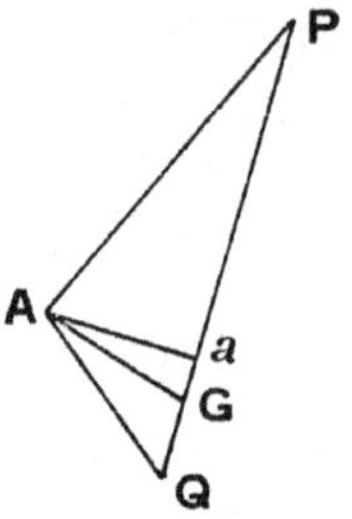

Fig. 64.

$$\text{Quare } \frac{\int.A \times AP^2}{M \times GP} = \frac{\int.A \times AG^2 - M \times GP^2 + 2\ M \times GP^2}{M \times GP} = \frac{\int.A \times AG^2}{M \times G} + GP.$$

Erit igitur PQ major, quam PG, excessu $GQ = \frac{\int.A \times AG^2}{M \times GP}$.

Valor constans producti ex binis distantiis centri gravitatis ab iisdem.

337. Ex illo excessu facile constat, mutato utcunque puncto suspensionis, rectangulum sub binis distantiis centri gravitatis ab ipso, & a centro oscillationis fore constans. Cum enim sit $QG = \frac{\int.A \times AG^2}{M \times GP}$, erit $GQ \times GP = \frac{\int.A \times AG^2}{M}$, quod productum est constans, & habetur hujusmodi elegans theorema: *singulæ massæ ducantur in quadrata suarum distantiarum a centro gravitatis communi, & dividatur omnium ejusmodi productorum summa per summam massarum, ac habebitur productum sub binis distantiis centri gravitatis a centro suspensionis & a centro oscillationis.*

Manente puncto suspensionis & centro gravitatis, manere centrum oscillationis.

338. Inde autem primo eruitur illud; *manente puncto suspensionis, & centro gravitatis, debere etiam centrum oscillationis manere nihil mutatum; utcunque totum systema, servata respectiva omnium massarum distantia, & positione ad se invicem convertatur intra idem planum circa ipsum gravitatis centrum;* nam illa GP inventa eo pacto pendet tantummodo a distantiis, quas singulæ massæ habent a centro gravitatis.

Centrum oscillationis, & punctum suspensionis reciprocari.

339. Sed & illud sponte consequitur: *Centrum oscillationis, & centrum suspensionis reciprocari ita, ut, si fiat suspensio per id punctum, quod fuerat centrum oscillationis; evadat oscillationis* [156] *centrum illud, quod fierat punctum suspensionis; & alterius distantia a centro gravitatis mutata, mutetur & alterius distantia in eadem ratione reciproca.* Cum enim earum distantiarum rectangulum debeat esse constans; si pro secunda ponatur valor, quem habuerat prima; debet pro prima obvenire valor, quem habuerat secunda, & altera debet æquari quantitati constanti divisæ per alteram.

Altera ex iis distantiis evanescente, abire alteram in infinitum.

340. Consequitur etiam illud: *Altera ex iis binis distantiis evanescente, abibit altera in infinitum, nisi omnes massæ in unico puncto sint simul compenetratæ.* Nam sine ejusmodi compenetratione summa omnium productorum ex massis, & quadratis distantiarum a centro gravitatis, remanet semper finita quantitas: adeoque remanet finita etiam, si dividatur per summam massarum, & quotus, manente diviso finito, crescit in infinitum; si divisor in infinitum decrescat.

Suspensione facta per centrum gravitatis, nullum haberi motum.

341. Hine vero iterum deducitur: *Suspensione facta per ipsum centrum gravitatis nullum motum consequi.* Evanescit enim in eo casu distantia centri gravitatis a puncto suspensionis, adeoque distantia centri oscillationis crescit in infinitum, & celeritas oscillationis evadit nulla.

Quæ distantia centri oscillationis omnium minima pro data positione mutua massarum datarum: maximam haberi nullam.

342. Quoniam utraque distantia simul evanescere non potest, potest autem centrum oscillationis abire in infinitum; nulla erit maxima e longitudinibus penduli simplicis isochroni pendulo facto per suspensionem dati systematis; sed aliqua debet esse minima, suspensione quadam inducente omnium celerrimam dati systematis oscillationem. Ea vero minima debet esse, ubi illæ binæ distantiæ æquantur inter se: ibi enim evadit minima earum summa, ubi altera crescente, & altera decrescente, incrementa prius minora decrementis, incipiunt esse majora, adeoque ubi ea æquantur inter se. Quoniam autem illæ binæ distantiæ mutantur in eadem ratione, utut reciproca; incrementum alterius

change the sign of its value; & thus, if in any one case, Q lies on the same side of P as G does, it must always lie on the same side. Now they lie on the same side for the case in which it is supposed that all the masses go to their common centre of gravity; for in this case the pendulum becomes a simple pendulum, & the centre of oscillation coincides with the centre of gravity, at which all the masses are placed. Hence it will always fall on the same side of the centre of suspension as G does.

Of the three points, the centre of gravity must lie between the other two.

336. Next, *the centre of gravity must lie intermediate between the centre of suspension & the centre of oscillation.* For, in Fig. 64, let the points A, P, G, O be the same points as in Fig. 63; & let AG, AQ, & Aa be drawn perpendicular to PQ. Then, the sum of all the masses, each multiplied into its distance from some chosen straight line or plane, or into their squares, may be designated by the letter ſ prefixed to the term involving the mass A alone, so as to make the proofs shorter. If this is done, the formula found will become $PQ = \int.A \times AP^2/M \times GP$. Now $AG^2 = AP^2 + GP^2 - 2GP \times Pa$, & therefore $AP^2 = AG^2 - GP^2 + 2GP \times Pa$; & $\int.A \times GP^2 = M \times GP^2$, since GP is constant; also $\int.A \times Pa = M \times GP$, since P$a$ is equal to the distance of the mass A from the plane perpendicular to the straight line QP, passing through P, & thus the sum of these products will be equal to the distance of the centre of gravity multiplied by the sum of the masses; hence $\int.A \times 2GP \times Pa = 2M \times GP^2$. Therefore

$$\int.A \times AP^2/M \times GP = \frac{\int.(A \times AG^2 - M \times GP^2 + 2M \times GP^2)}{M \times GP} = \frac{\int.A \times AG^2}{M \times GP} + GP.$$

Hence PQ will be greater than PG; & the excess GQ will be equal to $\int.A \times AG^2/M \times GP$.

The value of the product of the distances of the centre of gravity from the other two centres is constant.

337. From the value of this excess, it is readily seen that, however the point of suspension may be changed, the rectangle contained by the two distances of the centre of gravity from it & from the centre of oscillation, will be constant. For, since $QG = \int.A \times AG^2/M \times GP$, it follows that $GQ \times GP = \int.A \times AG^2/M$; & this product is constant. Hence we have the following elegant theorem:—*If each of the masses is multiplied by the square of its distance from the common centre of gravity, & the sum of all these products is divided by the sum of the masses, then the result obtained will be the product of the two distances of the centre of gravity from the centres of suspension & oscillation.*

If the centre of suspension & the centre of gravity remain unchanged, so also will the centre of oscillation.

338. Now, from this theorem, we can derive first of all the following theorem. *If the centre of gravity & the centre of suspension remain unchanged, then also the centre of oscillation must remain quite unchanged; no matter how the whole system is rotated about the centre of gravity, in the same plane, so long as the mutual distances of all the masses & their position with regard to one another are preserved.* For, the value of GP found in the manner above depends solely on the distances of the several masses from their centre of gravity.

The centre of oscillation & the centre of suspension are reversible.

339. But there is another theorem that also follows immediately. *The centre of oscillation & the centre of suspension are mutually related to one another in such a fashion that, if the suspension is made from the point which formerly was the centre of oscillation, then the new centre of oscillation will prove to be that point which was formerly the centre of suspension; & if the distance of either of them from the centre of gravity is changed the distance of the other will be also changed in the same ratio inversely.* For, since the rectangle contained by their distances remains constant, if for the second there is substituted that which the first had, then for the first there must be obtained the value which the second formerly had; & either of the two is equal to the constant quantity divided by the other.

If one of the distances vanishes, the other will become infinite.

340. It also follows that, *if either of the distances vanishes, the other must become infinite, unless all the masses are condensed at a single point.* For, unless there is condensation of this kind, the sum of all the products formed from the masses & the squares of their distances from their centre of gravity will always remain a finite quantity; & thus it will still remain finite if it is divided by the sum of the masses, & the quotient, still left finite after division, will increase indefinitely, if its divisor decreases indefinitely.

If the suspension is made from the centre of gravity, there is no motion.

341. Hence, again, it can be deduced that *if the suspension is made from the centre of gravity, no motion will ensue.* For, in this case, the distance of the centre of gravity from the centre of suspension vanishes and so the distance of the centre of oscillation increases indefinitely, & therefore the speed of the oscillation becomes zero.

To find the least distance of the centre of oscillation for a given position of the masses with regard to one another; there is no maximum distance.

342. Since both distances cannot vanish together, but the centre of oscillation can go off to infinity, there cannot be a maximum among the lengths of a simple pendulum isochronous with the pendulum made by the suspension of the given system; but there must be a minimum, since there must be one suspension of the given system which will give the greatest speed of oscillation. Indeed, this least value must occur, when the two distances are equal to one another; for their sum will be least when, as the one increases & the other decreases, the increments, which were before less than the decrements, now begin to be greater than the latter; & thus, at the time when they are equal to one another. Moreover since the two distances change in the same ratio, although inversely, the infinitesimal

infinitesimum crit ad alterius decrementum in ratione ipsarum, nec ea æquari poterunt inter se, nisi ubi ipsæ distantiæ inter se æquales fiant. Tum vero illarum productum evadit utriuslibet quadratum, & longitudo penduli simplicis isochroni æquatur eorum summæ; ac proinde habetur hujusmodi theorema: *Singulæ massæ ducantur in quadrata suarum distantiarum a centro gravitatis, ac productorum summa dividatur per summam massarum: & dupla radix quadrata quoti exhibebit minimam penduli simplicis isochroni longitudinem.* Vel Geometrice sic: *Pro quavis massa capiatur recta, quæ ad distantiam cujusvis massæ a centro gravitatis sit in ratione subduplicata ejusdem massæ ad massarum summam: inveniatur recta, cujus quadratum æquetur quadratis omnium ejusmodi rectarum simul: & ipsius duplum dabit quæsitum longitudinem mediam, quæ brevissimam præstet oscillationem.*

Superiora habere locum tantummodo, ubi omnes massæ sint in eodem plano perpendiculari ad axem rotationis: transitus ad centrum percussionis.

343. Hæc quidem omnia locum habent, ubi omnes massæ sint in unico plano perpendiculari ad axem rotationis, ut ni-[157]-mirum singulæ massæ possint connecti cum centro suspensionis, & centro oscillationis. At ubi in diversis sunt planis, vel in plano non perpendiculari ad axem rotationis, oportet singulas massas connectere cum binis punctis axis, & cum centro oscillationis, ubi jam occurrit systema quatuor massarum in se mutuo agentium (q); & relatio virium, quæ in latus agant extra planum, in quo tres e massis jaceant, quæ perquisitio est operosior, sed multo fœcundior, & ad problemata plurima rite solvenda magni usus; sed quæ hucusque protuli, speciminis loco abunde sunt; mirum enim, quo in hujusmodi Theoria promovenda, & ad Mechanicam applicanda progredi liceat. Sic etiam in determinando centro percussionis, virgam tantummodo rectilineam considerabo, speciminis loco futuram, sive massas in eadem recta linea sitas, & mutuis actionibus inter se connexas.

Præparatio ad inveniendum centrum percussionis massarum jacentium in eadem recta.

344. Sint in fig. 65 massæ A, B, C, D connexæ inter se in recta quadam, quæ concipiatur revoluta circa punctum P in ea situm, & quæratur in eadem recta punctum quoddam Q, cujus motu impedito debeat impediri omnis motus earumdem massarum per mutuas actiones; quod punctum appellatur *centrum percussionis.* Quoniam systema totum gyrat circa P, singulæ massæ habebunt velocitates A*a*, B*b* &c. proportionales distantiis a puncto P, adeoque singularum motus, qui per mutuas vires motrices extingui debent, poterunt exprimi per $A \times AP$, $B \times BP$ &c. Quare vires motrices in iis debebunt esse proportionales iis motibus. Concipiantur singulæ connexæ cum punctis P, & Q, & quoniam velocitas puncti P erat nulla; ibi omnium actionum summa debebit esse $= 0$: summa autem earum, quæ habentur in Q, elidetur a vi externa percussionem sustinente.

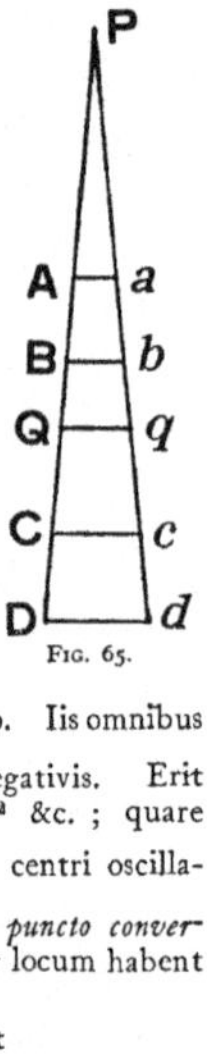

Fig. 65.

Calculus cum ejus determinatione.

345. Quoniam actiones debent esse perpendiculares eidem rectæ jungenti massas, erit per theorema numeri 314, ut PQ ad AQ, ita actio in $A = A \times AP$, ad actionem in $P = \frac{A \times AP \times AQ}{PQ}$, sive ob $AQ = PQ - AP$, erit ea actio [158] $\frac{A \times AP \times PQ - A \times AP^2}{PQ}$. Eodem pacto actio in P ex nexu cum B erit $\frac{B \times BP \times BQ - B \times PB^2}{PQ}$, & ita porro. Iis omnibus positis$=0$, divisor communis PQ abit, & omnia positiva æquantur negativis. Erit igitur $A \times AP \times PQ + B \times BP \times PQ$ &c. $= A \times AP^2 + B \times BP^2$ &c.; quare $PQ = \frac{A \times AP^2 + B \times BP^2 \text{ \&c.}}{A \times AP + B \times BP \text{ \&c.}}$, quæ formula est eadem, ac formula centri oscillationis, ac habetur hujusmodi theorema: *Distantia centri percussionis a puncto conversionis æquatur distantiæ centri oscillationis a puncto suspensionis;* adeoque hic locum habent in hoc casu, quæcunque de centro oscillationis superius dicta sunt.

Determinatio vis percussionis in ipso centro.

346. Quod si quis quærat vim percussionis in Q, hic habebit $QP . AP :: A \times AP . \frac{A \times AP^2}{PQ}$, quæ erit vis in Q ex nexu cum A. Eodem pacto invenientur vires ex reliquis: adeoque summa virium erit $\frac{A \times AP^2 + B \times BP^2}{PQ}$ &c.,

(q) *Systema binarum massarum cum binis punctis connexarum, & inter se, sed adhuc in eodem plano jacentium, persecutus fueram ante aliquot annos; quod sibi a me communicatum exhibuit in sua* Synopsi Physicæ Generalis *P. Benvenutus, ut ibidem ipse innuit. Id inde excerptum habetur hic in Supplementis* § 5.

Habetur autem post idem supplementum & Epistola, quam delatus Florentiam scripsi ad P. Scherfferum, dum hoc ipsum opus relictum Viennæ ante tres menses jam ibidem imprimeretur, quæ quidem adjecta est in ipsa prima editione in fine operis. Ibi & theoriam trium massarum extendi ad casum massarum quatuor ita; ut inde generaliter deduci possit & æquilibrium, & centrum oscillationis, & centrum percussionis, pro massis quotcunque, & utcunque dispositis.

increment of the one will be to the infinitesimal decrement of the other in the ratio of the distances themselves; & the former cannot be equal to one another, unless the distances themselves are equal to one another. In this case their product becomes the square of either of them, & the length of the simple isochronous pendulum will be equal to their sum. Hence we have the following theorem :—*If each mass is multiplied by the square of the distance from the centre of gravity, & the sum of all such products is divided by the sum of the masses; then, twice the square root of the quotient will give the least length of a simple isochronous pendulum.* This may be expressed geometrically as follows :—*For each mass, take a straight line, which is to the distance of that mass from the centre of gravity in the subduplicate ratio of the mass to the sum of all the masses; find a straight line whose square is equal to the sum of the squares on all the straight lines so found; then the double of this straight line will give the required mean length, which will afford the quickest oscillation.*

The theorems given above only hold good when all the masses are in the same plane perpendicular to the axis of rotation; let us pass on to the centre of percussion.

343. These theorems hold good when all the masses are in a single plane perpendicular to the axis of rotation, so that each of the masses can be connected with the point of suspension & the centre of oscillation. But, when they are in different planes, or all in a plane that is not perpendicular to the axis of rotation, it is necessary to connect each of the masses with a pair of points on the axis & with the centre of oscillation : & we thus have the case of a system of four masses acting upon one another (*q*), & the relation between the forces which act to one side, out of the plane in which three of the masses lie. This investigation is much more laborious, but also far more fertile, & of great use for the correct solution of a large number of problems. However, I have already given enough as examples; for it is wonderful how far one can go in developing a Theory of this kind, & in applying it to Mechanics. So also in determining the centre of percussion, I shall only consider a rectilinear rod, which will serve as an example, or masses in the same straight line, connected together by mutual actions.

Preparation for finding the centre of percussion for masses lying in the same straight line.

344. In Fig. 65, let A,B,C,D be masses connected together, lying in one straight line, which is supposed to be rotated about a point P situated in it; it is required to find in this straight line a point Q such that, if its motion is prevented, then the whole motion of the masses is also prevented through the mutual actions. This point is called the centre of percussion. Now, since the whole system rotates round P, each of the masses will have velocities, such as A*a*, B*b*, &c., proportional to their distances from the point P; & thus the motions of each, which have to be destroyed by the mutual motive forces,* can be represented by A × AP, B × BP, &c. Hence, the motive forces on them must be proportional to these motions. Suppose each of the masses to be connected with P & Q; then, since the velocity of the point P is zero, at P the sum of all the actions must be equal to zero; moreover, the sum of those that act at Q is cancelled by the external force sustaining the percussion.

The calculation giving the determination of this centre.

345. Since the actions must be perpendicular to the straight line joining the masses, we shall have, by Art. 314, PQ to AQ as the action on A, which is equal to A × AP, is to the action on P; hence the latter is equal to A × AP × AQ/PQ, or, since AQ = PQ − AP, this action will be equal to (A × AP × PQ−A × AP²)/PQ. In the same way, the action on P due to the connection with B is equal to (B × BP × PQ − B × BP²)/PQ, & so on. If all these together are put equal to zero, the common divisor PQ goes out, & all the positives will be equal to the negatives. Therefore A × AP × PQ + B × BP × PQ + &c. = A × AP² + B × BP² + &c. Hence

$$PQ = \frac{A \times AP^2 + B \times BP^2 + \&c.}{A \times AP + B \times BP + \&c.},$$

which is the same formula as the formula for the centre of oscillation. Thus we have the following theorem :—*The distance of the centre of percussion from the point of rotation is equal to the distance of the centre of oscillation from the centre of suspension.* Hence all that has been said above concerning the centre of oscillation holds good also for the centre of percussion.

Determination of the force of percussion at the centre of percussion.

346. Now, if the force of percussion at Q is required, we have QP is to AP as A × AP is to the force on Q due to the connection with A; hence this latter is equal to A × AP²/PQ. In the same way we can find the forces due to the rest; and thus the sum of all the forces will be (A × AP² + B × BP² + &c.)/PQ. Now, since PQ is equal to

(q) *I investigated the system of two masses connected with two points & with one another, yet all lying in the same plane, several years ago: &, when I had communicated the matter to Father Benvenuto, he expounded it in his* Synopsis Physicæ Generalis, *mentioning that he had obtained it from me. It is also included in this work, abstracted from the above, as Supplement 5.*

Moreover, after this supplement, it is also contained in a letter, which I wrote to Father Scherffer when I reached Florence, whilst this work, which I had left in his hands at Vienna three months before, was in the press there; & it was added to the first edition at the end of the work. In it I have also extended the theory of three masses to the case of four masses, in such a manner that from it it is possible to deduce, in a perfectly general way, the equilibrium, the centre of oscillation, & the centre of percussion for any number of masses disposed in any manner whatever.

sive ob $PQ = \frac{A \times AP^2 + B \times BP^2 \&c.}{A \times AP + B \times BP \&c.}$, summa illa erit $A \times AP + B \times BP$ &c.; nimirum ejusmodi vis erit æqualis summæ virium, quæ requiruntur ad sistendos omnes motus massarum A, B, &c., cum illis diversis velocitatibus progredientium, videlicet ejusmodi, quæ in massa percussionem excipiente possit producere quantitatem motus æqualem toti motui, qui sistitur in massis omnibus, quod congruit cum lege actionis, & reactionis æqualium, & cum conservatione ejusdem quantitatis motus in candem plagam, de quibus egimus num. 265, & 264.

Omitti hic multa quæ adhanc Theoriam pertinerent, ad quam pertinet universa Mechanica.

347. Haberent hic locum alia sane multa, quæ pertinent ad summas virium, quibus agunt massæ, compositarum e viribus, quibus agunt puncta, vel a Newtono, vel ab aliis demonstrata, & magni usus in Mechanica, & Physica: hujusmodi sunt ea omnia, quæ Newtonus habet sectione 12, & 13 libri I Princip. de attractionibus corporum sphæricorum, & non sphæricorum, quæ componantur ex attractionibus particularum; ubi habentur præclarissima theoremata tam pro viribus quibuscunque generaliter, quam pro certis virium legibus, ut illud, quod pertinet ad rationem reciprocam duplicatam distantiarum, in qua globus globum trahit, tanquam si omnis materia esset compenetrata in centris eorundem; punctum intra [159] orbem sphæricum, vel ellipticum vacuum nullas vires sentit, elisis contrariis; intra globos plenos punctum habet vim directe proportionalem distantiæ a centro; unde fit, ut in particulis exiguis ejusmodi vires fere evanescant, & ad hoc, ut vires adhuc etiam in iis sint admodum sensibiles, debeant decrescere in ratione multo majore, quam reciproca duplicata distantiarum. Hujusmodi etiam sunt, quæ Mac-Laurinus tradit de sphæroide elliptico potissimum, quæ Clairautius de attractionibus pro tubulis capillaribus, quæ D'Alembertus, Eulerus, aliique pluribus in locis persecuti sunt; quin omnis Mechanica, quæ agit vel de æquilibrio, vel de motibus, seclusa omni impulsione, huc pertinet, & ad diversos arcus reduci potest curvæ nostræ, qui possunt esse quantumlibet multi, habere quascunque amplitudines, sive distantias limitum, & areas quæ sint inter se in ratione quacunque, ac ad curvas quascunque ibi accedere, quantum libuerit; sed res in immensum abiret, & satis est, ea omnia innuisse.

Pressio fluidorum si puncta sint in recta verticali.

348. Addam nonnulla tantummodo, quæ generaliter pertinent ad pressionem, & velocitatem fluidorum. Tendant directione quacunque AB puncta disposita in eadem recta in fig. 66 vi quadam externa respectu systematis eorum punctorum, cujus actionem mutuis viribus elidant ea puncta, & sint in æquilibrio. Inter primum punctum A, & secundum ipsi proximum debebit esse vis repulsiva, quæ æquetur vi externæ puncti A. Quare urgebitur punctum secundum hac vi repulsiva, & præterea vi externa sua. Hinc vis repulsiva inter secundum, & tertium punctum debebit æquari vi huic utrique, adeoque erit æqualis summæ virium externarum puncti primi, & secundi. Adjecta igitur sua vi externa tendet deorsum cum vi æquali summæ virium externarum omnium trium; & ita porro progrediendo usque ad B. quodvis punctum urgebitur deorsum vi æquali summæ virium externarum omnium superiorum punctorum.

A

B

FIG. 66.

Eadem punctis utcunque dispersis, & cum omnibus directionibus agens.

349. Quod si non in directum disposita sint, sed utcunque dispersa per parallelepipedum, cujus basim perpendicularem directioni vis externæ exprimat recta FH in fig. 67, & FEGH faciem ipsi parallelam; adhuc facile demonstrari potest componendo, vel resolvendo vires; sed & per se patet, vires repulsivas, quas debebit ipsa basis exercere in particulas sibi propinquas, & ad quas vis ejus mutua pertinebit, fore æquales summæ omnium superiorum virium externarum; atque id erit commune tam solidis, quam fluidis. At quoniam in fluidis particulæ possunt ferri directione quacunque, quod unde proveniat, videbimus in tertia parte; quævis particula, ut ibidem videbimus, in omnem plagam urgebitur viribus æqualibus, & urgebit sibi proximas, quæ pressionem in alias propagabunt ita, ut, quæ sint in eodem plano LI, parallelo FH, in cujus directione [160] nulla vis externa agit, vires ubique eædem sint. Quamobrem quævis particula sita ubicunque in ea recta in N, habebit candem vim tam versus planum EF, quam versus planum EG, & versus FH, quam habet particula collocata in eadem linea in MK etiam, ubi addantur parietes AM, CK paralleli FE, cum planis LM, KI, parallelis FH, nimirum vi, quæ respondet altitudini MA: ac particula sita in O prope basim FH urgebitur, ut quaquaversum, ita & versus ipsam, iisdem viribus, quibus particula sita in BD sub AC. Ipsam urgebunt

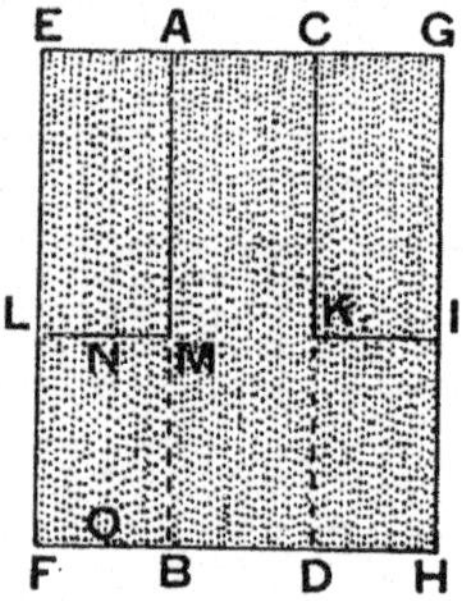

E A C G

L K I

N M

O

F B D H

FIG. 67.

$(A \times AP^2 + B \times BP^2 + \&c.)/(A \times AP + B \times BP + \&c.)$, this sum will be equal to $A \times AP + B \times BP + \&c.$ That is, the whole force will be equal to the sum of the forces, which are required to stop all the motions of the masses A, B, &c., which are proceeding with their several different velocities; in other words, a force which, acting on the mass receiving percussion, can produce a quantity of motion equal to the whole motion existing in all the masses; and this agrees with the law of equal action & reaction, & with the conservation of the same quantity of motion for the same direction, with which I dealt in Art. 265, & 264.

Many things pertaining to the Theory must here be omitted; for the whole of Mechanics pertains to this Theory.

347. Many other things indeed should find a place here, such as relate to the sums of forces, with which masses act, these being compounded from the forces with which points act; such as have been proved by Newton & others; & things that are of great use in Mechanics & Physics. Of this kind are all those which Newton has in the 12th & 13th sections of The First Book of the *Principia* concerning the attractions of spherical bodies, & non-spherical bodies, such as are compounded from the attractions of their particles. Here we have some most wonderful theorems, not only for forces in general, but also for certain laws of forces like that relating to the inverse square of the distances, where a sphere attracts another sphere as if the whole of its matter were condensed at the centre of each of them: the theorem that a point within a spherical or elliptic hollow shell is under the action of no force, equal & opposite forces cancelling one another; the theorem that within solid spheres a point is under the action of a force proportional to the distance from the centre directly. From this it follows that in exceedingly small particles of this kind the forces must almost vanish; & in order that the forces even then may be quite sensible, they must decrease in a much greater ratio than that of the inverse square of the distances. Also we have theorems such as Maclaurin enunciated with regard to the elliptic spheroid especially, & those which Clairaut gave with regard to attractions in the case of capillary tubes, & those which D'Alembert, Euler, & others have investigated in many places. Nay, the whole of Mechanics, which deals with equilibrium, or motions, impulse being excluded, belongs here: the whole of it can be reduced to different arcs of our curve; & these may be as many in number as you please, they can have any amplitudes, or distances between the limit-points, any areas, which may be in any ratio whatever to one another, & can approach as nearly as you please to any given curves. But the matter would become endless, & it is quite sufficient for me to have given all those that I have given.

Pressure of fluids when the points are all in a vertical line.

348. I will add a few things only that in general deal with pressure & velocity of fluids. Suppose we have a set of points, in Fig. 66, lying in a straight line, extended in any direction AB, under the action of some force external to the system of points; & suppose that the action of this external force is cancelled by the mutual forces between the points, & that the latter are in equilibrium. Then between the first point A & the next to it there must be a repulsive force which is equal to the external force on the point A. Then the second point will be under the action of this repulsive force in addition to the external force on it. Hence the repulsive force between the second & third points must be equal to both of these; &, further, it will be equal to the sum of the external forces on the first & second points. Hence, adding the external force on the third point, it will tend downwards with a force equal to the sum of the external forces on all three; & so on, until we reach B, any point will be under the action of a force equal to the sum of the external forces on all the points lying above it.

The same for points disposed in any manner, & acting in all directions.

349. Now if the points are not all situated in a straight line, but dispersed anyhow throughout a parallelepiped, & if, in Fig. 67, FH denotes the base of the parallelepiped, which is perpendicular to the direction of the external force, & FEGH is a face parallel thereto; then, it can yet easily be proved, either by composition or by resolution of forces, indeed it is self-evident, that the repulsive forces, which the base exerts on the particles next to it, & to which its mutual force will pertain, must be equal to the sum of the external forces on all points above it: & this will hold good for solids as well as for fluids. But, since in fluids the particles can move in any direction (we will leave the cause of this to be seen in the third part), any particle (as we shall also see there) will be urged in any direction with equal forces: & each will act on the next to it & propagate the pressure to the others in such a manner that the forces on those points which lie in the same plane LI, parallel to the base FH, in which direction there is no external force acting, will be everywhere the same. Hence, every particle situated anywhere in the straight line, at N say, will have the same force towards the plane EF as towards the plane EG, & towards FH; the same also as there is on a particle situated in the same straight line in MK also, where the partitions AM, CK are added parallel to FE, together with the planes LM, KI parallel to FH, namely, one equal to a force corresponding to the altitude MA. And a particle situated close to the base FH, at O say, will be urged in all directions & towards FH with the same forces as a particle situated in BD which is below AC. All the particles lying in the same horizontal

particulæ in eodem plano horizontali jacentes, & accedet ad omnes fluidi, & bascos particulas, donec vi contraria elidatur vis ejus tota ab ejusmodi pressione derivata. Quamobrem basis FH a fluido tanto minore FLMACKIH sentiet pressionem, quam sentiret a toto fluido FEGH : superficies autem LM sentiet a particulis N vim æqualem vi massæ LEAM, accedentibus ad ipsam particulis, donec vis mutua repulsiva ei vi æquetur.

Inde, cur exiguo fluidi pondere fieri possit ingens pressio.

350. Hinc autem patet, cur in fluidis nostris gravitate præditis basis FH sentiat pressionem tanto majorem massæ fluidæ incumbentis pondere, & cur pondere perquam exiguo fluidi AMKC elevetur pondus collocatum supra LM etiam immane, ubi repagulum LM sit ejusmodi, ut pressioni fluidi parere possit, quemadmodum sunt coriacea. At totum vas FLMACKIH bilanci impositum habebit pondus æquale ponderi suo, & fluidi contenti tantummodo : nam superficies vasis LM, KI horizontalis vi repulsiva mutua urgebit sursum, quantum urget deorsum puncta omnia N versus O, & illa pressio tantundem imminuit vim, quam in bilancem exercet vas, ac tota vis ipsius habebitur dempta pressione sursum superficiei LM, KI a pressione fundi FH facta deorsum : & pariter se mutuo elident vires exercitæ in parietes oppositos. Atque hæc Theoria poterit applicari facile aliis etiam figuris quibuscunque. Respondebit semper pressio superficiei, & toti ponderi fluidi, quod habeat basim illi superficiei æqualem, & altitudinem ejusmodi, quæ usque ad supremam superficiem pertinet inde accepta in directione illius externæ vis.

Pressio fluidorum cum compressione sensibili unde proveniat in hac Theoria.

351. Quod si vires particularum repulsivæ sint ejusmodi, ut ad eas multum augendas requiratur mutatio distantiæ, quæ ad distantiam totam habeat rationem sensibilem ; tum vero compressio massæ erit sensibilis, & densitas in diversis altitudinibus admodum diversa : sed in iisdem horizontalibus planis eadem. Si vero mutatio sufficiat, quæ rationem habet prorsus insensibilem ad totam distantiam ; tum vero compressio sensibilis nulla erit, & massa in fundo candem habebit ad sensum densitatem, quam prope superficiem supremam. Id pendet a lege virium mutua inter partienlas, & a curva, quæ illam expri-[161]-mit. Exprimat in fig. 68 AD distantiam quandam, & assumpta BD ad AB in quacunque ratione utcunque parva, vel utcunque sensibili, capiantur rectæ perpendiculares DE, BF itidem in quacunque ratione minoris inæqualitatis utcunque magna : poterit utique arcus MN curvæ exprimentis mutuas particularum vires transire per illa puncta F, F, & exhibere quodcunque pressionis incrementum cum quacunque pressione utcunque magna, vel utcunque insensibili.

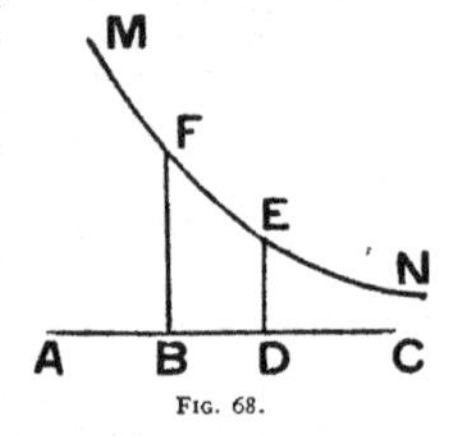

Fig. 68.

Compressio aeris a qua vi proveniat : aquæ compressio cur ad sensum nulla : unde mutatio in vapores tam elasticos.

352. Compressionem ingentem experimur in aere, quæ in eo est proportionalis vi comprimenti. Pro eo casu demonstravit Newtonus Princ. Lib. 3. prop. 23, vim particularum repulsivam mutuam debere esse in ratione reciproca simplici distantiarum. Quare in iis distantiis, quas habere possunt particulæ aeris perseverantis cum ejusmodi proprietate, & formam aliam non inducentis (nam & aerem posse e volatili fieri fixum, Newtonus innuit, ac Halesius inprimis uberrime demonstravit), oportet, arcus MN accedat ad formam arcus hyperbolæ conicæ Apollonianæ. At in aqua compressio sensibilis habetur nulla, utcunque magnis ponderibus comprimatur. Inde aliqui inferunt, ipsam elastica vi carere, sed perperam ; quin immo vires habere debet ingentes distantiis utcunque parum imminutis ; quanquam eædem particulæ debent esse prope limites, nam & distractioni resistit aqua. Infinita sunt curvarum genera, quæ possunt rei satisfacere, & satis est, si arcus EF directionem habeat fere perpendicularem axi AC. Si curvam cognitam adhibere libeat ; satis est, ut arcus EF accedat plurimum ad logisticam, cujus subtangens sit perquam exigua respectu distantiæ AD. Demonstratur passim, subtangentem logisticæ ad intervallum ordinatarum exhibens rationem duplam esse proxime ut 14 ad 10 ; & eadem subtangens ad intervallum, quod exhibeat ordinatas in quacunque magna ratione inæqualitatis, habet in omnibus logisticis rationem eandem. Si igitur minuatur subtangens logisticæ, quantum libuerit ; minuetur utique in eadem ratione intervallum BD respondens cuicunque rationi ordinatarum BF, DE, & accedet ad æqualitatem, quantum libuerit, ratio AB ad AD, a qua pendet compressio ; & cujus ratio reciproca triplicata est ratio densitatum, cum spatia similia sint in ratione triplicata laterum homologorum, & massa compressa possit cum eadem nova densitate redigi ad formam similem. Quare poterit haberi incrementum vis comprimentis

plane will act upon it & it will approach all the particles of the fluid & the base, until the whole of its force is cancelled by a contrary force derived from pressure of this kind. Hence the base FH would be subject, from the much smaller amount of fluid FLMACKIH, to the same pressure as it would be subject to from the whole fluid FEGH; & the surface LM would be subject to a force from the particles like N equal to the force of the mass LEAM, these particles tending to approach LM, until the mutual repulsive force is equal to this pressure.

Hence the reason why in a very small amount of fluid there can exist a very great pressure.

350. Further, from this the reason is evident, why the base FH should be subject, in our fluids possessed of gravity, to a pressure so much greater than the weight of the superincumbent fluid; & why by a very small weight of fluid, like AMKC, the weight collected above LM can be upheld, even though this is immensely great, when the restraint LM is of such a nature that it can submit to the pressure of the fluid; leather for example. But if the whole vessel FLMACKIH is placed on a balance it will only have a weight equal to its own weight plus that of the fluid contained. For, the horizontal surface LM, KI of the vessel will urge it upwards with its mutual repulsive force, just the same amount as all the points N will urge it downwards towards O, & this pressure will to the same extent diminish the force which the vessel exerts upon the balance; & the whole force will be obtained by taking away the pressure upwards on the surface LM, KI from the pressure produced downwards on the base FH. In the same way the forces exerted on the partitions will mutually cancel one another. The Theory can also easily be applied to any other figures whatever. The pressure on the surface will always correspond to the whole weight of the fluid having for its base an area equal to the surface, & for its height that which belongs to the highest surface from it measured in the direction of the external force.

The source of pressure for fluids with sensible compression according to this Theory.

351. Now if the repulsive forces of the particles are of such a kind that, in order to increase them to any sensible extent, a change of distance is required, which bears a sensible ratio to the whole distance; then the compression of the mass will also be sensible, & the density at different heights will be quite different; nevertheless, they will still be the same throughout the same horizontal planes. However, if a change, which bears to the whole distance a ratio that is quite insensible, is sufficient, then the mass at the bottom will have approximately the same density as near the top surface. This depends on the mutual law of forces between the particles, & on the curve which represents this law. In Fig. 68, let AD be any distance, & suppose that BD is taken in AB produced, bearing to AB any ratio however small, or however sensible; take the perpendicular straight lines DE, BF, also in any ratio of less inequality however great. In all cases, it will be possible for the arc MN of the curve representing the mutual forces of the particles to pass through the points E & F, & to represent any increment of pressure, together with any pressure however great, or however insensible, it may be.

The force that gives rise to the compression of air; the reason for the incompressibility of water; the source of the change in highly elastic vapours.

352. We find that in air there is great compression, & that this is proportional to the compressing force. For this case, Newton proved, in prop. 3, of the Third Book of his *Principia*, that the mutual repulsive force between the particles must be inversely proportional to the first power of the distance. Hence, for these distances, which the particles of air can have as it persists with a property of this kind, & does not induce another form (for Newton remarked that an air could from being volatile become fixed, & Hales especially gave a very full proof of this), the arc MN must approach the form of an arc of the rectangular hyperbola. But in water there is no sensible compression, however great the compressing weights may be. Hence some infer that it lacks elastic force; but that is not the case; nay rather, there are bound to be immense forces if the distances are diminished ever so slightly; although the particles must be near limit-points, for water also resists separation. There are infinitely many classes of curves which would satisfy the conditions; & it is sufficient if the arc EF has a direction that is nearly perpendicular to the axis AC. If it is desired to employ some known curve, it is sufficient to know that the arc EF approximates closely to the logistic curve whose subtangent is very small compared with the distance AD. Now it is proved that the subtangent of the logistic curve is to the interval corresponding to a double ratio between the ordinates very nearly as 14 is to 10; & the subtangent is to the interval, corresponding to a ratio of inequality between the ordinates of any magnitude, in the same ratio for all logistic curves. If therefore the subtangent of the logistic curve is diminished indefinitely, in every case there is a diminution in the same ratio of the interval BD corresponding to any ratio of the ordinates BF, DE, & the ratio of AB to AD, upon which depends the compression, will approach indefinitely near to equality. Now the ratio of the densities is the inverse triplicate of this ratio: for similar parts of space are in the triplicate ratio of homologous lengths, & the mass when compressed can be reduced to similar form having the same new density. Thus, we can have the increment of the compressing force, increased in

in quacunque ingenti ratione anetæ cum compressione utcunque exigua, & ratione densitatum utcunque accedente ad æqualitatem. Verum ubi ordinata ED jam satis exigua fuerit, debet curva recedere plurimum ab arcu logisticæ, ad quem accesserat, & qui in infinitum protenditur ex parte eadem, ac debet accedere ad axem AC, & ipsum secare, ut habeantur deinde vires attractivæ, quæ ingentes etiam esse possunt; tum post exiguum intervallum debet haberi alius arcus [162] repulsivus, recedens plurimum ab axe, qui exhibeat vires illas repulsivas ingentes, quas habent particulæ aqueæ, ubi in vapores abierunt per fermentationem, vel calorem.

Ubi pressio proportionalis altitudini, & unde.

353. In casu densitatis non immutatæ ad sensum, & virium illarum parallelarum æqualium, uti eas in gravitate nostra concipimus, pressiones erunt ut bases, & altitudines; nam numerus particularum paribus altitudinibus respondens erît æqualis, adeoque in diversis altitudinibus erit in earum ratione; virium autem æqualium summæ crunt ut particularum numeri. Atque id experimur in omnibus homogeneis fluidis, ut in Mercurio, & aqua.

Quomodo fiat acceleratio in effluxu.

354. Ubi facto foramine liber exitus relinquitur ejusmodi massæ particulis, erumpent ipsæ velocitatibus, quas acquirent, & quæ respondebunt viribus, quibus urgentur, & spatio, quo indigent, ut recedant a particulis se insequentibus; donec vis mutua repulsiva jam nulla sit. Prima particula relicta libera statim incipit moveri vi illa repulsiva, qua premebatur a particulis proximis : utcunque parum illa recesserit, jam secunda illi proxima magis distat ab ea, quam a tertia, adeoque movetur in eandem plagam, differentia virium accelerante motum; & eodem pacto aliæ post alias ita, ut tempusculo utcunque exiguo omnes aliquem motum habeant, sed initio eo minorem, quo posteriores sunt. Eo pacto discedunt a se invicem, & semper minuitur vis accelerans motum, donec ea evadat nulla; quin immo etiam aliquanto plus æquo a se invicem deinde recedunt particulæ, & jam attractivis viribus retrahuntur, accedentes iterum, non quod retro redeant, sed quod anteriores moveantur jam aliquanto minus velociter, quam posteriores; tum iterum aucta vi repulsiva incipiunt accelerari magis, & recedere, ubi & oscillationes habentur quædam hinc, & inde.

Unde velocitas aquæ effluentis subduplicata altitudinis.

355. Velocitates, quæ remanent post exiguum quoddam determinatum spatium, in quo vires mutuæ, vel nullæ jam sunt, vel æque augentur, & minuuntur, pendent ab area curvæ, cujus axis partes exprimant non distantias a proxima particula, sed tota spatia ab initio motus percursa, & ordinatæ in singulis punctis axis exprimant vires, quas in iis habebat particula. Velocitates in effluxu aquæ experimur in ratione subduplicata altitudinum, adeoque subduplicata virium comprimentium. Id haberi debet, si id spatium sit ejusdem longitudinis, & vires in singulis punctis respondentibus ejus spatii sint in ratione primæ illius vis. Tum enim areæ totæ erunt ut ipsæ vires initiales, & proinde velocitatum quadrata, ut ipsæ vires. Infinita sunt curvarum genera, quæ rem exhibere possunt; verum id ipsum ad sensum exhibere potest etiam arcus alterius logisticæ cujuspiam amplioris illa, quæ exhibuit distantias singularum particularum. Sit ea in fig. 69 MFIN. Tota ejus area infinita ad partes CN asymptotica a quavis ordinata [163] æquatur producto sub ipsa ordinata, & subtangente constanti. Quare ubi ordinata ED jam est perquam exigua respectu ordinatarum BF, HI tota area CDEN respectu CBFN insensibilis erit, & areæ CBFN, CHIN integræ accipi poterunt pro areis FBDE, IHDE, quæ idcirco erunt, ut vires initiales BF, HI.

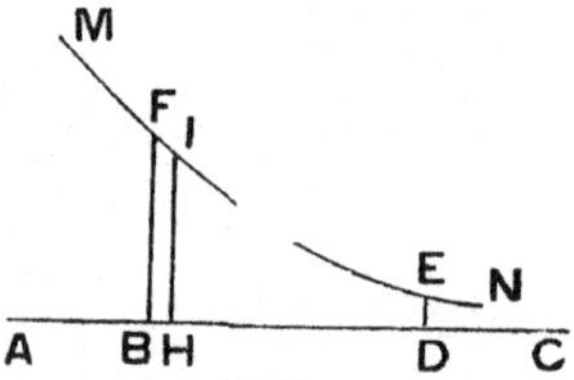

Fig. 69.

Quid requiritur, ut velocitas sit æqualis illi, quæ habetur cadendo per altitudinem.

356. Inde quidem habebuntur quadrata celeritatum proportionalia pressionibus, sive altitudinibus. Ut autem velocitas absoluta sit æqualis illi, quam particula acquireret cadendo a superficie suprema, quod in aqua experimur ad sensum; debet præterea tota ejusmodi area æquari rectangulo facto sub recta exprimente vim gravitatis unius particulæ, sive vis repulsivæ, quam in se mutuo exercent binæ particulæ, quæ se primo repellunt, sustinente inferiore gravitatem superioris, & sub tota altitudine. Deberet eo casu esse totum pondus BF ad illam vim, ut est altitudo tota fluidi ad subtangentem logisticæ, si FE est ipsius logisticæ arcus. Est autem pondus BF ad gravitatem primæ particulæ, ut numerus particularum in ea altitudine ad unitatem, adeoque ut eadem illa tota altitudo ad distantiam primarum particularum. Quare subtangens illius logisticæ deberet æquari

any very great ratio in conjunction with a compression that is small to any extent, & a ratio of densities which approaches indefinitely near to equality. But when the ordinate ED is sufficiently small, the curve must depart considerably from an arc of the logistic curve, to which it formerly approximated, & which proceeded to infinity in the same direction; it must approach the axis AC, & cut it, in order that attractive forces may be obtained, which may also become very great. Then, after a small interval, we must have another repulsive arc, receding far from the axis, to represent those very great repulsive forces, which the particles of water have, when they pass into vapour through fermentation or heat.

Where the pressure is proportional to the altitude, & the reason for this.

353. In the case of the density not being sensibly changed, & of those equal parallel forces, such as we suppose our gravity to be, the pressures will be proportional to the bases & the altitudes. For, the number of particles corresponding to equal altitudes will be equal, & therefore, in different altitudes, the numbers will be proportional to the altitudes; moreover the sums of the equal forces will be proportional to the numbers of particles. We find this to be the case in all homogeneous fluids, such as mercury & water.

How acceleration in efflux arises.

354. When, on making an opening, a free exit is left for the particles of a mass, they burst forth with the velocities which they acquire & which correspond to the forces urging them, & to the space to which it is necessary for them to recede from those particles that follow, before the mutual repulsive force becomes zero. The first particle, when left free, immediately begins to move under the action of the repulsive force by which it is pressed by the particles next to it. As soon as it has moved ever so little, the second particle next to it becomes more distant from it than from the third, & thus moves in the same direction as the difference of the forces accelerates the motion. Similarly, one after the other they acquire motion in such a manner that in any little interval of time, no matter how brief, all of them will have some motion; this motion at the commencement is so much the less, the farther back the particles are. In this way they separate from one another, & the force accelerating the motion ever becomes less until finally it vanishes. Nay rather, to speak more correctly, the particles still recede from one another, & come under the action of attractive forces, & approach one another; not indeed that they retrace their paths, but because the more forward particles are now moving with somewhat less velocity than those behind; then once more the repulsive force is increased & they begin to be accelerated more than those behind & to recede from them; & so oscillations to & fro are obtained.

Why the velocity of effluent water is the sub-duplicate of the height.

355. The velocities that are left after any determinate interval of space, in which the mutual forces are either nothing or are equally increased & diminished, depend on the area of the curve, of which parts of the axis represent not the distances from the next particle, but the whole spaces travelled from the beginning of the motion, & the ordinates at each point of the axis represent the forces which the particle had at those points. It is found that the velocities of effluent water are in the subduplicate ratio of the altitudes, & thus in the subduplicate ratio of the compressing forces. Now this is what must be obtained, if the space is of the same length, & the forces at each corresponding point of that space are in the ratio of that first force. For, then the total areas will be as the initial forces, & hence the squares of the velocities will be as the forces. There are an infinite number of classes of curves which will serve to represent the case; but this also can be represented by the arc of another logistic curve more ample than that which represented the distances of the single particles. Let MFIN be such a curve, in Fig. 69. The whole area, indefinitely produced in the direction of C & N, which are asymptotic, measured from any ordinate, will be equal to the product of that ordinate & the constant subtangent. Therefore when the ordinate ED is now very small with respect to the ordinates BF, HI, the whole area CDEN will be insensible with respect to the area CBFN; & thus the whole areas CBFN, CHIN can be taken instead of the areas FBDE, IHDE; & therefore these are to one another as the initial forces BF, HI.

What is required so that the velocity shall be equal to that acquired in falling from the given altitude.

356. From this, then, we have that the squares of the velocities are proportional to the pressures, or the altitudes. Now, in order that the absolute velocity may be equal to that which the particle would acquire in falling from the upper surface, as is found to be approximately the case for water, we must have, in addition, that the whole of such area must be equal to the rectangle formed by multiplying the straight line representing the force of gravity on one particle (or the repulsive force which a pair of particles mutually exert upon one another, when they first repel one another, the lower sustaining the gravity of the one above) by the whole altitude. In this case, the whole weight BF would be bound to be to the force as the whole altitude of the fluid is to the subtangent of the logistic curve, if FE is an arc of the logistic curve. Moreover, the weight BF is to the gravity of the first particle as the number of particles in the altitude is to unity; & thus in the ratio of the altitude to the distance between the primary particles. Hence the subtangent of the logistic curve would have to be equal to the distance between

illi distantiæ primarum particularum, quæ quidem subtangens erit itidem idcirco perquam exigua.

Tentandum an in omnibus fluidis id accidat. Transitus ad partem tertiam.

357. An in omnibus fluidis habeatur ejusmodi absoluta velocitas & an quadrata velocitatum in effluxu respondeant altitudinibus; per experimenta videndum est, ut constet, an curvæ virium in omnibus sequantur superiores leges, an diversas. Sed ego jam ab applicatione ad Mechanicam ad applicationem ad Physicam gradum feci, quam uberius in tertia Parte persequar. Hæc interea speciminis loco sint satis ad immensam quandam hujusce campi fœcunditatem indicandam utcunque.

xists in all fluids, & whether the squares of nd to the altitudes, must be investigated whether the curves of forces follow the laws pass on from the application to Mechanics ow out more fully in the third part. These some sort to indicate an immense fertility

It must be tested whether this happens in all fluids. We will now pass on to the third part.

[164] PARS III

Applicatio Theoriæ ad Physicam

Agendum hic primo de generalibus proprietatibus corporum, tum de discrimine inter varias species.

358. In secunda hujusce Operis parte, dum Theoriam meam applicarem ad Mechanicam, multa identidem immiscui, quæ applicationi ad Physicam sternerent viam, & vero etiam ad eandem pertinerent; at hic, quæ pertinent ad ipsam Physicam, ordinatius persequar; & primo quidem de generalibus agam proprietatibus corporum, quas omnes omnino exhibet illa lex virium, quam initio primæ partis exposui; tum ex eadem præcipua discrimina deducam, quæ inter diversas observamus corporum species, & mutationes, quæ ipsis accidunt, alterationes, atque transformationes evolvam.

Enumeratio earum, de quibus agetur, & ordo.

359. Primum igitur agam de Impenetrabilitate, de Extensione, de Figurabilitate, de Mole, Massa, & Densitate, de Inertia, de Mobilitate, de Continuitate motuum, de Æqualitate Actionis & Reactionis, de Divisibilitate, & Componibilitate, quam ego divisibilitati in infinitum substituo, de Immutabilitate primorum materiæ elementorum, de Gravitate, de Cohæsione, quæ quidem generalia sunt. Tum agam de Varietate Naturæ, & particularibus proprietatibus corporum, nimirum de varietate particularum, & massarum multiplici, de Solidis, & Fluidis, de Elasticis, & Mollibus, de Principiis Chemicarum Operationum, ubi de Dissolutione, Præcipitatione, Adhæsione, & Coalescentia, de Fermentatione, & emissione Vaporum, de Igne, & emissione Luminis; ac ipsis præcipuis Luminis proprietatibus, de Odore, de Sapore, de Sono, de Electricitate, de Magnetismo itidem aliquid innuam sub finem; ac demum ad generaliora regressus, quid Alterationes, Corruptiones, Transformationes mihi sint, explicabo. Verum in horum pluribus rem a mea Theoria deducam tantummodo ad communia principia, ex quibus peculiares singulorum tractatus pendent; ac alicubi methodum indicabo tantummodo, quæ ad rei perquisitionem aptissima mihi videatur.

Impenetrabilitas unde in hac Theoria.

360. Impenetrabilitas corporum a mea Theoria omnino sponte fluit; si enim in minimis distantiis agunt vires repulsivæ, quæ iis in infinitum imminutis crescant in infinitum ita, ut pares sint extinguendæ cuilibet velocitati utcunque magnæ, utique non potest ulla finita vis, aut velocitas efficere, ut distantia duorum punctorum evanescat, quod requiritur ad compenetrationem; sed ad id præstandum infinita Divina virtus, quæ infinitam vim exerceat, vel infinitam producat velocitatem, sola sufficit.

Aliud impenetrabilitatis genus: proprium huic Theoriæ.

[165] 361. Præter hoc impenetrabilitatis genus, quod a viribus repulsivis oritur, est & aliud, quod provenit ab inextensione punctorum, & quod evolvi in dissertationibus *De Spatio, & Tempore*, quas ex Stayanis Supplementis huc transtuli, & habetur hic in fine Supplementorum § 1, & 2. Ibi enim ex eo, quod in spatio continuo numerus punctorum loci sit infinities infinitus, & numerus punctorum materiæ finitus, erui illud: nullum punctum materiæ occupare unquam punctum loci, non solum illud, quod tunc occupat aliud materiæ punctum, sed nec illud, quod vel ipsum, vel ullum aliud materiæ punctum occupavit unquam. Probatio inde petitur, quod si ex casibus ejusdem generis una classis infinities plures contineat, quam altera, infinities improbabilius sit, casum aliquem, de quo ignoremus, ad utram classem pertineat, pertinere ad secundam, quam ad primam. Ex hoc autem principio id etiam immediate consequitur; si enim una massa projiciatur contra alteram, & ab omnibus viribus repulsivis abstrahamus animum; numerus projectionum, quæ aliquod punctum massæ projectæ dirigant per rectam transeuntem per aliquod punctum massæ, contra quam projicitur, est utique finitus; cum numerus punctorum in utraque massa finitus sit; at numerus projectionum, quæ dirigant puncta omnia per rectas nulli secundæ massæ puncto occurrentes, est infinities infinitus, ob puncta spatii in quovis plano infinities infinita. Quamobrem, habita etiam ratione infinitorum continui temporis momentorum, est infinities improbabilior primus casus secundo; & in quacunque projectione massæ contra massam nullus habebitur immediatus occursus puncti materiæ cum altero puncto materiæ, adeoque nulla compenetratio, etiam independenter a viribus repulsivis.

PART III

Application of the Theory to Physics

358. In the second part of this work, in applying my Theory to Mechanics, I brought in also at the same time many things which opened the road for an application to Physics, & really even belonged to the latter. In this part I will investigate in a more ordered manner those things that belong to Physics. First of all, I will deal with general properties of bodies ; & these will be given by that same law of forces that I enunciated at the beginning of the first part. After that, from the same law I will derive the most important of the distinctions that we observe between the different species of bodies, & I will discuss the changes, alterations & transformations that happen to them.

We will first of all here deal with the general properties of bodies, & then with the difference between the several species.

359. First, therefore, I will deal with Impenetrability, Extension, Figurability, Volume, Mass, Density, Inertia, Mobility, Continuity of Motions, the Equality of Action & Reaction, Divisibility, & Componibility (for which I substitute infinite divisibility), the Immutability of the primary elements of matter, gravity, & Cohesion ; all these are general properties. Then I will consider the Variety of Nature, & special properties of bodies ; such, for instance, as the manifold variety of particles & masses, Solids & Fluids, Elastic, & Soft bodies ; the principles of chemical operations, such as Solution, Precipitation, Adhesion & Coalescence, Fermentation, & emission of Vapours, Fire & the emission of Light ; also about the principal properties of Light, Smell, Taste, Sound, Electricity & Magnetism, I will say a few words towards the end. Finally, coming back to more general matters, I will explain my idea of the nature of alterations, corruptions & transformations. Now in most of these, I shall derive the whole matter from my Theory alone, & reduce it to those common principles, upon which depends the special treatment for each ; in certain cases I shall only indicate the method, which seems to me to be the most fit for a further investigation of the matter.

Enumeration of the matters to be dealt with, & the order in which they will be taken.

360. The Impenetrability of bodies comes naturally from my Theory. For, if repulsive forces act at very small distances, & these forces increase indefinitely as the distances decrease, so that they are capable of destroying any velocity however large ; then there never can be any finite force, or velocity, that can make the distance between two points vanish, as is required for compenetration. To do this, an infinite Divine virtue, exercising an infinite force, or creating an infinite velocity, would alone suffice.

The origin of Impenetrability, according to this Theory.

361. Besides this kind of Impenetrability, which arises from repulsive forces, there is also another kind, which comes from the inextension of the points ; this I discussed in the dissertations *De Spatio, & Tempore*, which I have abstracted from the Supplement to Stay's Philosophy, & set at the end of this work as Supplements, §§ 1, 2. From the fact that the number of points of position in a continuous space may be infinitely infinite, whilst the number of points of matter may be finite, I derive the following principle ; namely, that no point of matter can ever occupy either a point of position which is at the time occupied by another point of matter, or one which any other point of matter has ever occupied before. The proof is derived from the argument that, if of cases of the same nature one class of them contains infinitely more than another, then it is infinitely more improbable that a certain case, concerning which we are in doubt as to which class it belongs, belongs to the second class rather than to the first. It also follows immediately from this principle ; if one mass is projected towards another, & we disallow a directive mind in all repulsive forces, the number of the ways of projection, which direct any point of the projected mass along a straight line passing through any point of the mass against which it is projected, is finite ; for the number of points in each of the masses is finite. But the number of ways of projection, which direct all points along straight lines that pass through no point of the second mass, is infinitely infinite because the number of points of space in any plane is infinitely infinite. Therefore, even when the infinite number of moments in continuous time is taken into account, the first case is infinitely more improbable than the second. Hence, in any projection whatever of mass against mass there is no direct encounter of one point of matter with another point of matter ; & thus there can be no compenetration, even apart from the idea of repulsive forces.

Another kind of Impenetrability, peculiar to this Theory.

Sine viribus repulsivis debere haberi compenetrationem apparentem. Quid eæ præstent in particulis, & velo quodam, potissimum si habeantur asymptoti.

362. Si vires repulsivæ non adessent; omnis massa libere transiret per aliam quanvis massam, ut lux per vitra, & gemmas transit, ut oleum per marmora insinuatur; atque id semper fieret sine ulla vera compenetratione. Vires, quæ ad aliquod intervallum extenduntur satis magnæ, impediunt ejusmodi liberum commeatum. Porro hic duo casus distinguendi sunt; alter, in quo curva virium non babeat ullum arcum asymptoticum cum asymptoto perpendiculari ad axem, præter illum primum, quem exhibet figura 1, cujus asymptotus est in origine abscissarum; alter, in quo adsint alii ejusmodi arcus asymptotici. In hoc secundo casu si sit aliqua asymptotus ad aliquam distantiam ab origine abscissarum, quæ habeat arcum citra se attractivum, ultra repulsivum cum area infinita, ut juxta num. 188 puncta posita in minore distantia non possint acquirere distantiam majorem, nec, quæ in majore sunt, minorem; tum vero particula composita ex punctis in minore distantia positis, esset prorsus impenetrabills a particula posita in majore distantia ab ipsa, nec ulla finita velocitate posset cum illa commisceri, & in ejus locum irrumpere; & si duæ habeantur [166] asymptoti ejusmodi satis proximæ, quarum citerior habeat ulterius crus repulsivum, ulterior citerius attractivum cum areis infinitis, tum duo puncta collocata in distantia a se invicem intermedia inter distantias earum asymptotorum, nec possent ulla finita vi, aut velocitate acquirere distantiam minorem, quam sit distantia asymptoti citerioris, nec majorem, quam sit ulterioris; & cum eæ duæ asymptoti possint esse utcunque sibi invicem proximæ; illa puncta possent esse necessitata ad non mutandam distantiam intervallo utcunque parvo. Si jam in uno plano sit series continua triangulorum æquilaterorum habentium eas distantias pro lateribus, & in singulis angulis poneretur quicunque numerus punctorum ad distantiam inter se satis minorem ea, qua distent illæ duæ asymptoti, vel etiam puncta singula; fieret utique velum quoddam indissolubile, quod tamen esset plicatile in quavis e rectis continentibus triangulorum latera, & posset etiam plicari in gyrum more veterum voluminum.

Solidum indissolubile, & impermeabile.

363. Si autem sit solidum compositum ex ejusmodi velis, quorum alia ita essent aliis imposita, ut punctum quodlibet superioris veli terminaret pyramidem regularem habentem pro basi unum e triangulis veli inferioris, & in singulis angulis collocarentur puncta, vel massæ punctorum; id esset solidissimum, & ne plicatile quidem; etiamsi crassitudo unicam pyramidum seriem admitteret. Possent autem esse dispersa inter latera illins veli, vel hujus muri, puncta quotcunque, nec eorum ullum posset inde egredi ad distantiam a punctis positis in angulis veli, vel muri, majorem illa distantia ulterioris asymptoti. Quod si præterea ultra asymptotum ulteriorem haberetur area repulsiva infinita; nulla externa puncta possent perrumpere nec murum, nec velum ipsum, vel per vacua spatiola transire, utcunque magna cum velocitate advenirent; cum nullum in triangulo æquilatero sit punctum, quod ab aliquo ex angulis non distet minus, quam per latus ipsius trianguli.

Alia ratio acquirendi impenetrabilitatem, & nexum per asymptotos remotas ab origine abscissarum.

364. Quod si ejusmodi binæ asymptoti inter se proximæ sint in ingenti distantia a principio abscissarum, & in distantia media inter earum binas distantias ab ipso initio ponantur in cuspidibus trianguli æquilateri tria puncta materiæ, tum in cuspide pyramidis regularis habentis id triangulum æquilaterum pro basi ponantur quotcunque puncta, quæ inter se minus distent, quam pro distantia illarum asymptotorum; massula constans hisce punctis erit indissolubilis; cum nec ullum ex iis punctis possit acquirere distantiam a reliquis, nec reliqua inter se distantiam minorem distantia asymptoti citerioris, & majorem distantia ulterioris, & ipsa hæc particula impenetrabilis a quovis puncto externo materiæ, cum nullum ad reliqua illa tria puncta possit ita accedere, si distat magis, vel recedere, si minus, ut acquirat distantiam, quam hahent puncta ejus massæ. Ejusmodi massis ita cohibitis per terna puncta ad maximas distantias sita posset integer constare Mundus, qui ha-[167]-beret in suis illis massulis, seu primigeniis particulis impenetrabilitatem continuam prorsus insuperabilem, sine ulla extensione continua, & indissolubilitatem itidem insuperabilem etiam sine ullo mutuo nexu inter earum puncta, per solum nexum, quem haberent singula cum illis tribus punctis remotis.

In iis & aliis casibus resistentia continua sine continuo faciente vim, & absoluta impermeabilitas.

365. In omnibus hisce casibus habetur in massa non continua vis ita continua, ut nulla ne apparens quidem compenetratio, & permixtio haberi possit æque, ac in communi sententia de continua impenetrabilis materiæ extensione. Quod autem in illo velo, vel muro exhibuit triangulorum, & pyramidum series, idem obtineri potest per figuras alias

362. If there were no repulsive forces, every mass would pass freely through every other mass, as light passes through glass & crystals, & as oil insinuates itself into marble; but such a thing as this would always happen without any true compenetration. Forces, which extend to an interval that is sufficiently large for the purpose, prevent free passage of that kind. Further there are here two cases to be distinguished; one, in which the curve of forces has not any asymptotic arc with an asymptote perpendicular to the axis, except the first, as is shown in Fig. 1, where the asymptote occurs at the origin of abscissæ; the other, in which there are other such asymptotic arcs. In the second case, if there is an asymptote at some distance from the origin of abscissæ, which has an attractive arc on the near side of it, & on the far side a repulsive arc with an infinite area corresponding to it, so that, as was shown in Art. 188, points situated at a less distance cannot acquire a greater, & those at a greater distance cannot acquire a less; then particles that are made up of points situated at the less distance would be quite impenetrable by a particle situated at a greater distance from it; nor could any finite velocity force it to mingle with it or invade its position; and if there are two asymptotes of the kind sufficiently near together, of which the nearer to the origin has its further branch repulsive, & the further has its nearer branch attractive, the corresponding areas being infinite, then two points situated at a distance from one another that is intermediate between the distances of these asymptotes, cannot with any finite force or velocity acquire a distance less than that of the nearer asymptote or greater than that of the further asymptote. Now since these two asymptotes may be indefinitely near to one another, the two points may be forced to keep their distance unchanged within an interval of any smallness whatever. Suppose now that we have in a plane a continuous series of equilateral triangles having these distances as sides, & that at each of the angles there are placed any number of points at a distance from one another sufficiently less than that of the distance between the two asymptotes, or even single points; then, in every case, we should have a kind of unbreakable skin, which however could be folded along any of the straight lines containing sides of the triangles, or could even be folded in spirals after the manner of ancient manuscripts.

Without repulsive forces there must be apparent compenetration. What these forces may give us in particles, & a sort of skin, especially if there are asymptotes.

363. Moreover, if we have a solid composed of such skins, one imposed upon the other in such a manner that any point of an upper skin should terminate a regular pyramid having for its base one of the equilateral triangles of the skin beneath, & in each of them points were situated, or masses of points; then that would have very great solidity, & would not be even capable of being folded, even if its thickness only admitted of a single series of pyramids. Further, any number of points could be scattered between the sides of the former skin, or the wall of the latter, & none of these could get out of this position to a distance from the points situated at the angles of the skin, or of the wall, greater than the distance of the further asymptote. Now if, in addition to these, there happened to be beyond the further asymptote a corresponding infinite repulsive area, no external points could break into the skin or wall, nor could they pass through empty spaces in it, no matter how great the velocity with which they approached it. For, there is no point within an equilateral triangle that is at a less distance from the angular points than a side of the triangle.

An unbreakable & impermeable solid.

364. Again, if there are two asymptotes very near one another, at a great distance from the origin of abscissæ, & at a distance intermediate between their two distances from the origin there are placed three points of matter at the vertices of an equilateral triangle, & then at the vertex of a regular pyramid having for its base that equilateral triangle there are placed any number of points, which are at a less distance from one another than that between the two asymptotes, the little mass made up of these points will be unbreakable. For, none of these points can acquire from the rest, nor the rest from one another, a distance less than the distance of the nearer asymptote, nor greater than that of the further asymptote. This particle will also be impenetrable by any external point of matter; for no point can possibly approach those other three points so nearly, if the distance is greater, or recede from them so far, if the distance is less, as to acquire the same distance as that between the several points of the mass. The whole Universe may be made up of masses of this kind restrained by sets of three points situated at very great distances; & it would have in the little masses forming it, or in the primary particles, a continuous impenetrability that was quite insuperable, without any continuous extension; it would also have an insuperable unbreakableness without any mutual connection between the points forming it, simply owing to the connection existing between each of its points with the three remote points.

Another way in which impenetrability may be acquired, & the connection with asymptotes that are remote from the origin of abscissæ.

365. In all these cases there is obtained for a non-continuous mass a force that is continuous in such sort that there is not even apparent compenetration; & commingling can be had just as well as with the usual idea of continuous extension of impenetrable matter. Moreover, what has been represented by the skin or wall of a series of triangles or pyramids, can be obtained by means of very many other figures; & it can be obtained

In these & other cases, we have continuous resistance without imagining a continuous force, & also absolute impenetrability.

quamplurimas, & id multo pluribus abhuc modis obtineretur; si non in unica, sed in pluribus distantiis essent ejusmodi asymptotica repagula cum impenetrabilitate continua per non continuam punctorum dispersorum dispositionem.

Sine asymptoto omnes substantias permeabiles fore ab aliis si iis satis magnas velocitas imprimatur. Exemplum globuli ferrei inter magnetes transeuntis.

366. At in primo illo casu, in quo nulla habetur ejusmodi asymptotus præter primam, res longe alio modo se haberet. Patet in eo casu illud, si velocitas imprimi possit massæ cuipiam satis magna; fore, ut ea transeat per massam quancunque sine ulla perturbatione suarum partium, & sine ulla partium alterius; nam vires, ut agant, & motum aliquem finitum sensibilem gignant, indigent continuo tempore, quo imminuto in immensum, uti imminuitur, si velocitas in immensum augeatur, imminuitur itidem in immensum earum effectus. Rei ideam exhibebit globulus ferrens, qui debeat transire per planum, in quo dispersæ sint hac, illac plurimæ massæ magneticæ vim habentes validam satis. Si is globus cum velocitate non ita ingenti projiciatur per directionem etiam, quæ in nullam massam debeat incurrere; progredi ultra illas massas non poterit; sed ejus motus sistetur ab illarum attractionibus. At si velocitas sit satis magna, ut actiones virium magneticarum satis exiguo tempore durare possint, prætervolabit utique, nullo sensibili damno ejus velocitati illato.

Diversi effectus relate ad magnetes pro diversa velocitate ejus globuli.

367. Quin immo ibi considerandum & illud; si velocitas ejus fuerit exigua, ipsum globum facile sisti, exiguo motu a vi mutua æquali, seu reactione, impresso magnetibus, quo per solam plani fractionem, & mutuas eorum vires impedito, exigua in eorum positionibus mutatio fiat. Si velocitas impressa aliquantulum creverit; tum mutatio in positione magnetum major fiet, & adhuc sistetur globuli motus; sed si velocitas fuerit multo major, globulus autem transeat satis prope aliquas e massis magneticis; ab actione mutua inter ipsum, & eas massas communicabitur satis ingens motus iis ipsis massis, quo possint etiam ipsum non nihil retardatum, sed adhuc progredientem sequi, avulsæ, a cæteris, quæ ob actiones in majore distantia minores, & brevitatem temporis, remaneant ad sensum immotæ, & nihil turbatæ. Sed si velo-[168]-citas ipsa adhuc augeretur, quantum est opus, eo deveniri posset; ut massa utcunque proxima in globuli transitu nullum sensibilem motum auferret illi, & ipsa sibi acquireret.

Inde facilis explicatio phænomeni, quo globus sclopeto explosus perforat plana mobilia, nec movet: cur lumini data tanta velocitas.

368. Porro ejusmodi exemplum intueri licet, ubi globus aliquis contra obstaculum aliquod projicitur, quod, si satis magnam velocitatem habet, concuti totum, & diffringit ac eo majorem effectum edit, quo major est velocitas, ut in muris arcium accidit, qui tormentariis globis impetuntur. At ubi velocitas ad ingentem quandam magnitudinem devenerit; nisi satis solida sit compages obstaculi, sive vires cohæsionis satis validæ; jam non major effectus fit, sed potius minor, foramine tantum excavato, quod æquetur ipsi globo. Id experimur; si globus ferreus explodatur sclopeto contra portam ligneam, quæ licet semiaperta sit, & summam habeat super suis cardinibus mobilitatem; tamen nihil prorsus commovetur; sed excavatur tantummodo foramen æquale ad sensum diametro globi, quod in mea Theoria multo facilius utique intelligitur, quam si continuo nexu partes perfecte solidæ inter se complicarentur, & conjungerentur. Nimirum, ut in superiore magnetum casu, particulæ globi secum abripiunt particulas ligni, ad quas accesserunt magis, quam ipsæ ad sibi proximas accederent, & brevitas temporis non permisit viribus illis, a quibus distantium ligni punctorum nexus præstabatur, ut in iis motus sensibilis haberetur, qui nexum cum aliis sibi proximis a vi mutua ortum dissolveret, aut illis, & toti portæ satis sensibilem motum communicaret. Quod si velocitas satis adhuc augeri posset; ne iis quidem avulsis massa per massam transvolaret, nulla sensibili mutatione facta, & sine vera compenetratione haberetur illa apparens compenetratio, quam habet lumen, dum per homogeneum spatium liberrimo rectilineo motu progreditur; quam ipsam fortasse ob causam Divinus Naturæ Opifex tam immanem luci velocitatem voluit imprimi, quantam in ea nobis ostendunt eclipses Jovis satellitum, & annua fixarum aberratio, ex quibus Rœmerus, & Bradleyus deprehenderunt, lumen semiquadrante horæ percurrere distantiam æqualem distantiæ Solis a Terra, sive plura milliariorum millia singulis arteriæ pulsibus.

Cur in cinere remaneat illæsa forma plantæ avolante parte volatili per ignem.

369. Ac eodem pacto, ubi herbarum forma in cinere cum tenuissimis filamentis remanet intacta, avolantibus oleosis partibus omnibus sine ulla læsione structuræ illarum, id quidem admodum facile intelligitur, qui fiat: ibi nova vis excitata ingentem velocitatem parit brevi tempore, quæ omnem alium effectum impediat virium mutuarum inter olea, &

in a much greater number of ways as well, if not only at one, but at many distances, there were these asymptotic restraints, resulting in continuous impenetrability through a non-continuous disposition of scattered points.

366. Now, in the first case, where there is no such asymptote besides the first, there would be a far different result. In this case, it is evident that, if a sufficiently great velocity can be given to any mass, it would pass through any other mass without any perturbation of its own parts, or of the parts of the other. For, the forces have no continuous time in which to act & produce any finite sensible motion; since if this time is diminished immensely (as it will be diminished, if the velocity is immensely increased), the effect of the forces is also diminished immensely. We can illustrate the idea by the example of an iron ball, which is required to pass across a plane, in which lie scattered in all positions a great number of magnetic masses possessed of considerable force. If the ball is not projected with a certain very great velocity, even if its direction is such that it is not bound to meet any of the masses, yet it will not go beyond those masses; but its motion will be checked by their attractions. But if the velocity is great enough, so that the actions of the magnetic forces only last for a sufficiently short interval of time, then it will certainly get through & beyond them without suffering any sensible loss of velocity.

If there were no asymptotes, all substances would be permeable by one another, if sufficiently great velocity is given to them. Example, an iron globe passing between magnets.

367. Lastly, there is to be considered also this point; if the velocity of the ball were very small, the ball might easily be brought to rest, a slight motion due to an equal mutual force or reaction being communicated to the magnets; but this latter being prevented merely by the friction of the plane, the change in their positions would be very small. Then if the impressed velocity were increased somewhat, the change in the positions of the magnets would become greater, & still the ball might be brought to rest. But if the velocity was much greater, the ball may also pass near enough to some of the magnetic masses; & by the mutual action between it & the masses there will be communicated to the masses a sufficiently great motion, to enable them to follow it as it goes on with its velocity somewhat retarded; they will be torn from the rest, which owing to the smaller action corresponding to a greater distance, & the shortness of the time, remain approximately motionless, & in no wise disturbed. If the velocity is still further increased, to the necessary extent, it could become such that a mass, no matter how near it was to the path of the ball, would communicate no velocity to it, nor acquire any from it.

Relatively diverse effects with regard to the magnets, due to diverse velocities of the ball.

368. Further, an example of this sort of thing can be seen in the case where a ball is projected against an obstacle; if the velocity is sufficiently great, it agitates the whole & breaks it to pieces; & the effect produced is the greater, the greater the velocity, as is the case for the walls of forts bombarded with cannon-balls. But when the velocity reaches a certain very great magnitude, unless the fabric of the obstacle is sufficiently solid or the forces of cohesion sufficiently great, there will now be no greater effect, rather a less, a hole only being made, equal to the size of the ball. Let us consider this; suppose an iron ball is fired from a gun against a wooden door, & this door is partly open, & it has the utmost mobility to swing on its hinges; nevertheless, it will not be moved in the slightest. Merely a hole, approximately equal to the size of the ball, will be made. Now this is far more easily understood according to my Theory, than if we assume that there are perfectly solid parts united & joined together by a continuous connection. Indeed, as in the case of the magnets given above, the particles of the ball carry off with them particles of the wood, which they have approached more closely than these particles have approached to the particles of wood next to them; & the shortness of the time does not allow the forces, by which the connection between the distances of the points of the wood is maintained, to give to the particles a sensible motion in the latter, which would dissolve the connection with others next to them arising from the mutual force, or in the former, which would also communicate a sufficiently sensible motion in the whole door. But if the velocity is still further increased to a sufficient extent, not even the latter particles are torn away, & one mass will pass through the other, without any sensible change being made. Thus, without real compenetration, we should have that apparent compenetration that we have in the case of light, as it passes through a homogeneous space with a perfectly free rectilinear motion. Perchance that is the reason why the Divine Founder of Nature willed that so enormous a velocity should be given to light; how great this is we gather from the eclipses of Jupiter's satellites, & from the annual aberration of the fixed stars. From which Roemer & Bradley worked out the fact that light took an eighth of an hour to pass over the distance from the Sun to the Earth, or many thousands of miles in a single beat of the pulse.

Hence an easy explanation of the phenomenon in which a ball fired from a cannon will perforate a movable plane without moving it; & why such a great velocity is given to light.

369. In the same way, when the form of stalks remain intact in the ash with their finest fibres, after that the oleose parts have all been driven off without any breaking down of their structure, what happens can be quite easily understood. Here, a new force being excited produces in a brief space of time a mighty velocity, which prevents all that other effect arising from the mutual forces between the oily & the ashy parts; the oily particles

The reason why in the ash there remains unimpaired the form of the plant after that the volatile part has been driven off by the action of fire.

cineres, oleaginosis particulis inter terreas cum hac apparenti compenetratione liberrime avolantibus sine ullo immediato impactu, & incursu.

Compenetratio apparens, quæ haberetur, si possemus nobis imprimere velocitatem satis magnam.

370. Quod si ita res habet; liceret utique nobis per occlusas ingredi portas, & per durissima transvolare murorum se-[169]-pta sine ullo obstaculo, & sine ulla vera compenetratione, nimirum satis magnam velocitatem nobis ipsis possemus imprimere, quod si Natura nobis permisisset, & velocitates corporum, quæ habemus præ manibus, ac nostrorum digitorum celeritates solerent esse satis magnæ; apparentibus ejusmodi continuis compenetrationibus assueti, nullam impenetrabilitatis haberemus ideam, quam mediocritati nostrarum virium, & velocitatum, ac experimentis hujus generis a sinu materno, & prima infantia usque adeo frequentibus, & perpetuo repetitis debemus omnem.

Extensio necessario profluens a viribus repulsivis.

371. Ex impenetrabilitate oritur extensio. Ea sita est in eo, quod aliæ partes sint extra alias: id autem necessario haberi debet; si plura puncta idem spatii punctum simul occupare non possint. Et quidem si nihil aliunde sciremus de distributione punctorum materiæ; ex regulis probabilitatis constaret nobis, dispersa esse per spatium extensum in longum, latum, & profundum, atque ita constaret, ut de eo dubitare omnino non liceret, adeoque haberemus extensionem in longum, latum, & profundum ex eadem etiam sola Theoria deductam. Nam in quovis plano pro quavis recta linea infinita sunt curvarum genera, quæ eadem directione egressæ e dato puncto extenduntur in longum, & latum respectu ejusdem rectæ, & pro quavis ex ejusmodi curvis infinitæ sunt curvæ, quæ ex illo puncto egressæ habeant etiam tertiam dimensionem per distantiam ab ipso. Quare sunt infinities plures casus positionum cum tribus dimensionibus, quam cum duabus solis, vel unica, & idcirco infinities major est probabilitas pro uno ex iis, quam pro uno ex his, & probabilitas absolute infinita omnem eximit dubitationem de casu infinite improbabili, utut absolute possibili. Quin immo si res rite consideretur, & numeri casuum inter se conferantur; inveniemus, esse infinite improbabile, uspiam jacere prorsus accurate in directum plura, quam duo puncta, & accurate in eodem plano plura, quam tria.

Extensum ejusmodi esse physice, non mathematice continuum: realem esse: in quo id consistat.

372. Hæc quidem extensio non est mathematice, sed physice tantum continua: at de præjudicio, ex quo ideam omnino continuæ extensionis ab infantia nobis efformavimus, satis dictum est in prima Parte a num. 158; ubi etiam vidimus, contra meam Theoriam non posse afferri argumenta, quæ contra Zenonistas olim sunt facta, & nunc contra Leibnitianos militant, quibus probatur, extensum ab inextenso fieri non posse. Nam illi inextensa contigua ponunt, ut mathematicum continuum efforment, quod fieri non potest, cum inextensa contigua debeant compenetrari, dum ego inextensa admitto a se invicem disjuncta. Nec vero illud vim ullam contra me habet, quod nonnulli adhibent, dicentes, hujusmodi extensionem nullam esse, cum constet punctis penitus inexten-[170]-sis, & vacuo spatio, quod est purum nihil. Constat per me non solis punctis, sed punctis habentibus relationes distantiarum a se invicem: eæ relationes in mea Theoria non constituuntur a spatio vacuo intermedio, quod spatium nihil est actu existens, sed est aliquid solum possibile a nobis indefinite conceptum, nimirum est possibilitas realium modorum localium existendi cognita a nobis secludentibus mente omnem hiatum, uti exposui, in prima Parte num. 142, & fusius in ea dissertatione De Spatio & Tempore, quam hic ad calcem adjicio; constituuntur a realibus existendi modis, qui realem utique relationem inducunt realiter, & non imaginarie tantum diversam in diversis distantiis. Porro si quis dicat, puncta inextensa, & hosce existendi modos inextensos non posse constituere extensum aliquid; reponam facile, non posse constituere extensum mathematice continuum, sed posse extensum physice continuum, quale ego unicum admitto, & positivis argumentis evinco, nullo argumento favente alteri mathematice continuo extenso, quod potius etiam independenter a meis argumentis difficultates habet quamplurimas. Id extensum, quod admitto, est ejusmodi, ut puncta materiæ alia sint extra alia, ac distantias habeant aliquas inter se, nec omnia jaceant in eadem recta, nec in eodem plano omnia, sint vero multa ita proxima, ut eorum intervalla omnem sensum effugiant. In eo sita est extensio, quam admitto, quæ erit reale quidpiam, non imaginarium, & erit physice continua.

fly off between the earthy particles with this apparent compenetration, in the freest manner, without any immediate impulse or collision.

Apparent compenetration, such as would be obtained if we were able to give ourselves a velocity great enough.

370. But if this were the case, we could walk through shut doors, or pass through the hardest walled enclosures without any resistance, & without any real compenetration; that is to say, if we could impress upon ourselves a sufficiently great velocity. Now if Nature allowed us this, & the velocities of bodies which are around us, & the speed of our fingers were usually sufficiently great, we, being accustomed to such continuous apparent compenetration, should have no idea of impenetrability. We owe the whole idea of impenetrability to the mediocrity of our forces & velocities, & to experiences of this kind, which have happened to us from the time we were born, during infancy & up till the present time, frequently & continually repeated.

Extension necessarily arising from repulsive forces.

371. From impenetrability there arises extension. It is involved in the fact that some parts are outside other parts; & this of necessity must be the case, if several points cannot at the same time occupy the same point of space. Indeed, even if we knew nothing from any other source about the distribution of the points of matter, it would be manifest from the rules of probability that they were dispersed through a space extended in length, breadth & depth; & it would be so clear, that there could not be the slightest doubt about it; & thus we should obtain extension in length, breadth & depth as a consequence of my Theory alone. For, in any plane, for any straight line in it, there are an infinite number of kinds of curves, which starting in the same direction from a given point extend in length & breadth with respect to this same straight line; & for any one of these curves there are an infinite number of curves that, starting from that point, have also a third dimension through distance from the point. Hence, there are infinitely more cases of positions with three dimensions than with two alone or only one; & thus there is infinitely greater probability in favour of one of the former than for one of the latter; & as the probability is absolutely infinite, it removes any doubt about a case which is infinitely improbable, though absolutely possible. Indeed, if the matter is carefully considered, & the number of cases compared with one another, we shall find that it is infinitely improbable that more than two points will anywhere lie accurately in the same straight line, or more than three in the same plane.

Such extension is physically, not mathematically, continuous; it is real; of what it consists.

372. This extension is not mathematically, but only physically, continuous; & on the matter of the prejudgment, from which we have formed for ourselves the idea of absolutely continuous extension from infancy, enough has been said in the First Part, starting with Art. 158. There, too, we saw that there could not be brought forward against my Theory the arguments which of old were brought against the followers of Zeno, & which now are urged against the disciples of Leibniz, by which it is proved that extension cannot be produced from non-extension. For these disputants assume that their non-extended points are placed in contact with one another, so as to form a mathematical continuum; & this cannot happen, since things that are contiguous as well as non-extended must compenetrate; but I assume non-extended points that are separated from one another. Nor indeed have the arguments, which some others use, any validity in opposition to my Theory; when they say that there is no such extension, since it is founded on non-extended points & empty space, which is absolute nothing. According to my Theory, it is founded, not on points simply, but on points having distance relations with one another; these relations, in my Theory, are not founded upon an empty intermediate space; for this space has no actual existence. It is only something that is possible, indefinitely imagined by us; that is to say, it is the possibility of real local modes of existence, pictured by us after we have mentally excluded every gap, as I explained in the First Part in Art. 142, & more fully in the dissertation on Space & Time, which I give at the end of this work. The relations are founded on real modes of existence; & these in every case yield a real relation which is in reality, & not merely in supposition, different for different distances. Further, if anyone should argue that these non-extended points, or non-extended modes of existence, cannot constitute anything extended, the reply is easy. I say that they cannot constitute a mathematically extended continuum, but they can a physically extended continuum. The latter only I admit, & I prove its existence by positive arguments; none of these arguments being favourable to the other continuum, namely one mathematically extended. This latter, even apart from any arguments of mine, has very many difficulties. The extension, which I admit, is of such a nature that it has some points of matter that lie outside of others, & the points have some distance between them, nor do they all lie on the same straight line, nor all of them in the same plane; but many of them are so close to one another that the intervals between them are quite beyond the scope of the senses. In that is involved the extension which I admit; & it is something real, not imaginary, & it will be physically continuous.

Quomodo existat Geometria sublato continuo actu existente.

373. At crit fortasse, qui dicet, sublata extensione absolute mathematica tolli omnem Geometriam. Respondeo, Geometriam non tolli, quæ considerat relationes inter distantias, & inter intervalla distantiis intercepta, quæ mente concipimus, & per quam ex hypothesibus quibusdam conclusiones cum iis connexas ex primis quibusdam principiis deducimus. Tollitur Geometria actu existens, quatenus nulla linea, nulla superficies mathematice continua, nullum solidum mathematice continuum ego admitto inter ea, quæ existunt; an autem inter ea, quæ possunt existere, habeantur, omnino ignoro. Sed aliquid ejusmodi in communi etiam sententia accidit. Nulla existit revera in Natura recta linea, nullus circulus, nulla ellipsis, nec in ejusmodi lineis accurate talibus fit motus ullus, cum omnium Planetarum, & Terræ in communi sententia motus habeantur in curvis admodum complicatis, atque altissimis, &, ut est admodum probabile, transcendentibus. Nec vero in magnis corporibus ullam habemus superficiem accurate planam, & continuam, aut sphæricam, aut cujusvis e curvis, quas Geometræ contemplantur, & plerique ex iis ipsis, qui solida volunt elementa, simplices ejusmodi figuras ne in ipsis quidem elementis admittent.

Quid in ea imaginarium, quid reale: elegans analogia loci cum tempore in ordine ad æqualitatis mensuras.

374. Quamobrem Geometria tota imaginaria est, & idealis, sed propositiones hypotheticæ, quæ inde deducuntur, [171] sunt veræ, & si existant conditiones ab illa assumptæ, existent utique & conditionata inde eruta, ac relationes inter distantias punctorum imaginarias ope Geometriæ ex certis conditionibus deductæ, semper erunt reales, & tales, quales eas invenit Geometria, ubi illæ ipsæ conditiones in realibus punctorum distantiis existant. Ceterum ubi de realibus distantiis agitur, nec illud in sensu physico est verum, ubi punctum interiacet aliis binis in eadem recta positis, a quibus æque distet, binas illas distantias fore partes distantiæ punctorum extremorum juxta ea quæ diximus num. 67. Physice distantia puncti primi a secundo constituitur per puncta ipsa, & binos reales ipsorum existendi modos, ita & distantia secundi a tertio; quorum summa continet omnia tria puncta cum tribus existendi modis, dum distantia primi a tertio constituitur per sola duo puncta extrema, & duos ipsorum existendi modos, quæ ablato intermedio reali puncto manet prorsus eadem. Illæ duæ sunt partes illins tertiæ tantummodo in imaginario, & geometrico statu, qui concipit indefinite omnes possibiles intermedios existendi modos locales, & per eam cognitionem abstractam concipit continua intervalla, ac corum partes assignat, & ope ejusmodi conceptuum ratiocinationes instituit ab assumptis conditionibus petitas, quæ, ubi demum ad aliquod reale deducunt, non nisi ad verum possint deducere, sed quod verum sit tantummodo, si rite intelligantur termini, & explicentur. Sic quod aliqua distantia duorum punctorum sit æqualis distantiæ aliorum duorum, situm est in ipsa natura illorum modorum, quibus existunt, non in eo, quod illi modi, qui eam individuam distantiam constituunt, transferri possint, ut congruant. Eodem pacto relatio duplæ, vel triplæ distantiæ habetur immediate in ipsa essentia, & natura illorum modorum. Vel si potius velimus illam referre ad distantiam æqualem; dici poterit, eam esse duplam alterius, quæ talis sit, ut si alteri ex alterius punctis ponatur tertium novum ad æqualem distantiam ex parte altera; distantia nova hujus tertii a primo sit æqualis illi, quæ duplæ nomen habet, & sic de reliquis, ubi ad realem statum transitur. Neque enim in statu reali haberi potest usquam congruentia duarum magnitudinum in extensione, ut haberi nec in tempore potest unquam; adeoque nec æqualitas per congruentiam in statu reali haberi potest, nec ratio dupla per partium æqualitatem. Ubi decempeda transfertur ex uno loco in alium, succedunt alii, atque alii punctorum extremorum existendi modi, qui relationes inducunt distantiarum ad sensum æqualium : ea æqualitas a nobis supponitur ex cansis, nimirum ex mutuo nexu per vires mutuas, uti hora hodierna ope egregii horologii comparatur cum hesterna, itidem æqualitate supposita ex cansis, sed loco suo divelli, & ex uno die in alterum hora eadem traduci nequaquam potest. Verum hæc omnia ad Metaphysicam potius pertinent, & ea fusius cum omnibus [172] loci, ac temporis relationibus persecutus sum in memoratis dissertationibus, quas hic in fine subjicio.

Figurabilitas orta ab extensione: quid sit figura, & quam vaga, & incerta sit ejus idea etiam in communi sententia.

375. Ex extensione oritur figurabilitas, cum qua connectitur moles, & densitas supposita massa. Quoniam puncta disperguntur per spatium extensum in longum, latum, & profundum; spatium, per quod extenduntur, habet suos terminos, a quibus figura pendet. Porro figuram determinatam ab ipsa natura, & existentem in re, possunt agnoscere tantummodo in elementis ii, qui admittunt elementa ipsa solida, atque compacta, & continua,

How Geometry can stand, if an actually existing continuum is excluded.

373. But perhaps some one will say that, if absolutely continuous extension is barred, then the whole of Geometry is demolished. I reply, that Geometry is not demolished, since it deals with relations between distances, & between intervals intercepted in these distances; that these we mentally conceive, & by them we derive from certain hypotheses conclusions connected with them, by means of certain fundamental principles. Geometry, as actually existent, is demolished; in so far as there is no line, no surface, & no solid that is mathematically continuous, which I admit as being among things actually existing; whether they are to be numbered amongst things that might possibly exist, I do not know. But something of the sort does take place, according to the usual idea of things. As a matter of fact, there is in Nature no such thing as a straight line, or a circle, or an ellipse; nor is there motion in lines that are accurately such as these; for in the opinion of everybody, the motions of all the planets & the Earth take place in curves that are very complicated, having equations of a very high degree, or, as is quite possible, transcendent. Nor in large bodies do we have any surfaces that are quite plane, & continuous, or spherical, or shaped according to any of the curves which geometers investigate; & very many of these men, who accept solid elements, will not admit simple figures even in the very elements.

The imaginary & the real parts of Geometry; an elegant analogy between place & time as regards measures of equality.

374. Hence the whole of geometry is imaginary; but the hypothetical propositions that are deduced from it are true, if the conditions assumed by it exist, & also the conditional things deduced from them, in every case; & the relations between the imaginary distances of points, derived by the help of geometry from certain conditions, will always be real, & such as they are found to be by geometry, when those conditions exist for real distances of points. Besides, when we are dealing with real distances, it is not true in a physical sense, when a point lies between two others in the same straight line, equally distant from either, to say that the two distances are parts of the distance between the two outside points, according to what we have said in Art. 67. Physically speaking, the distance of the first point from the second is fixed by the two points & their two real modes of existence, & so also for the distance between the second & the third. The sum of these contains all three points & their three modes of existence; whilst the distance of the first from the third is fixed by the two end points only, together with their two modes of existence; & this remains unaltered if the intermediate real point is taken away. The two distances are parts of the third only in imagination, & in the geometrical condition, which in an indefinite manner conceives all the possible intermediate local modes of existence; & from that abstract conception forms a picture of continuous intervals, & assigns parts to them; then, by the aid of such imagery institutes chains of reasoning founded on assumed conditions; & these, when at last they lead to something real, will only do so, if it is possible for them to lead to something that is true, & something that is only true if the terms are correctly understood & explained. Thus, the fact, that the distance between two points is equal to the distance between two other points, rests upon the nature of their modes of existence, & not upon the idea that the modes, which constitute the individual distances, can be transferred, so as to agree with one another. In the same way, the idea of twice, or three times a distance, is obtained directly from the essential nature of those modes of existence. Or, if we prefer to refer it to the idea of equal distances, we can say that one distance is twice another when it is such that, if beyond the second point of the latter we place a new third point at a distance equal to that of the first point from the second, then the distance of this new third point from the first point will be equal to that to which the name double distance is given; & so on for other multiples, when the matter is reduced to a consideration of real state. For, in the real state, there never can be a congruence of two magnitudes in extension, just as there never can be such a congruence in time; & therefore there never can be an equality depending on congruence in the real state, nor a double ratio through equality of parts. When a length of ten feet is transferred from one place to another, there follow, one after the other, different modes of existence of the end points; & these modes introduce relations of practically equal distances. This equality is supposed by us to be due to causes; for instance, to the mutual connection in consequence of mutual forces; just as an hour of to-day may be compared with one of yesterday by the help of an accurate clock; but the same hour cannot be disjointed from its own position & transferred from one day to another in any way. But really, such matters have more to do with Metaphysics; & I have investigated them more fully, together with all the relations of space & time, in the dissertations I have mentioned, which I add at the end of this work.

Figurability arises from extension; the nature of shape, & how vague the idea of it is, even in the opinion usually held.

375. From extension arises the idea of figurability; with this is connected volume &, when we have conceived the idea of mass, density. Since points are scattered through extended space in length, breadth & depth, the space through which they are extended has its boundaries; & upon these boundaries depends shape. Further, it is in the elements alone that a shape, determinate by its very nature, & existing of itself, can be acknowledged by those who suppose the elements to be solid, compact & continuous; & by those who

& qui ab inextensis extensum continuum componi posse arbitrantur, ubi nimirum tota illa materia superficie continua quadam terminetur. Ceterum in corporibus hisce, quæ nobis sub sensum cadunt, idea figuræ, quæ videtur maxime distincta, est admodum vaga, & indefinita, quod quidem diligenter exposui agens superiore anno de figura Telluris in dissertatione inserta postremo Bononiensium Actorum tomo, in qua continetur Synopsis mei operis de *Expeditione Litteraria per Pontificiam ditionem*, ubi sic habeo ; *Inprimis hoc ipsum nomen figuræ terrestris, quod certam quandam, ac determinatam significationem videtur habere, habet illam quidem admodum incertam, & vagam. Superficies illa, quæ maria, & lacus, & fluvios, ac montes, & campos, vallesque terminat, est illa quidem admodum, nobis saltem, irregularis, & vero etiam instabilis : mutatur enim quovis utcunque minimo undarum, & glebarum motu, nec de hac Telluris figura agunt, qui in figuram Telluris inquirunt : aliam ipsi substituunt, quæ regularis quodammodo sit, sit autem illi priori proxima, quæ nimirum abrasis haberetur montibus, collibusque, vallibus vero oppletis. At hæc iterum terrestris figuræ notio vaga admodum est, & incerta. Uti enim infinita sunt curvarum regularium genera, quæ per datum datorum punctorum numerum transire possint, ita infinita sunt genera curvarum superficierum, quæ Tellurem ita ambire possint, atque concludere, ut vel omnes, vel datos contingant in datis punctis montes, collesque, vel si per medios transire colles, ac montes debeat superficies quædam ita, ut regularis sit, & tantundem materiæ concludat extra, quantum vacui aeris infra sese concludat usque ad veram hanc nobis irregularem Telluris superficiem, quam intuemur : infinitæ itidem, & a se invicem diversæ admodum superficies haberi possunt, quæ problemati satisfaciant, atque eæ ejusmodi etiam, ut nullam, quæ sensu percipi possit, præ se ferant gibbositatem, quæ ipsa vox non ita determinatam continet ideam.*

Quanto magis in hac Theoria.

376. Hæc ego ibi de Telluris figura, quæ omnino pertinent ad figuram corporis cujuscunque in communi etiam sententia de continua extensione materiæ : nam omnium fere corporum superficies hic apud nos utique multo magis scabræ sunt pro ratione suæ magnitudinis, quam Terra pro ratione magnitudinis suæ, & vacuitates internas habent quamplurimas. Ve-[173]-rum in mea Theoria res adhuc magis indefinita, & incerta est. Nam infinitæ sunt etiam superficies curvæ continuæ, in quibus tamen omnia jacent puncta massæ cujusvis : quin immo infinitæ numero curvæ sunt lineæ, quæ per omnia ejusmodi puncta transeant. Quamobrem mente tantummodo confingenda est quædam superficies, quæ omnia puncta includat, vel quæ pauciora, & a reliquorum coacervatione remotiora excludat, quod æstimatione quadam morali fiet, non accurata geometrica determinatione. Ea superficies figuram exhibebit corporis ; atque hic jam, quæ ad diversa figurarum genera pertinent ; id omne mihi commune est cum communi Theoria de continua extensione materiæ.

Moles a figura pendens : incerta ejus idea & in sententia communi, & multo magis in hac Theoria.

377. A figura pendet moles, quæ nihil est aliud, nisi totum spatium extensum in longum, latum, & profundum externa superficie conclusum. Porro nisi concipiamus superficiem illam, quam innui, quæ figuram determinet ; nulla certa habebitur molis idea : quin immo si superficiem concipiamus tortuosam illam, in qua jaceant puncta omnia ; jam moles triplici dimensione prædita erit nulla ; si lineam curvam concipimus per omnia transeuntem : nec duarum dimensionum habebitur ulla moles. Sed in eo itidem incerta æstimatione indiget sententia communis ob interstitia illa vacua, quæ habentur in omnibus corporibus, & scabritiem, juxta ea, quæ diximus, de indeterminatione figuræ. Hic autem itidem concepta superficie extima terminante figuram ipsam, quæ deinde de mole relata ad superficiem tradi solent, mihi communia sunt cum aliis omnibus, ut illud : posse eandem magnitudine molem terminari superficiebus admodum diversis, & forma, & magnitudine, ac omnium minimam esse sphæricæ figuræ superficiem respectu molis : in figuris autem similibus molem esse in ratione triplicata laterum homologorum, & superficiem in duplicata, ex quibus pendent phænomena sane multa, atque ea inprimis, quæ pertinent ad resistentiam tam fluidorum, quam solidorum.

Massa : quid in ejus idea incertum ob materiam exteram immixtam. Omnia corpora constare partibus diversæ naturæ.

378. Massa corporis est tota quantitas materiæ pertinentis ad id corpus, quæ quidem mihi erit ipse numerus punctorum pertinentium ad illud corpus. At hic jam oritur indeterminatio quædam, vel saltem summa difficultas determinandi massæ ideam, nec id tantum in mea, verum etiam in communi sententia, ob illud additum *punctorum pertinentium ad illud corpus*, quod heterogeneas substantias excludit. Ea de re sic ego quidem in Stayanis Supplementis § 10 Lib. 1 : *Nam admodum difficile est determinare, quæ sint illæ substantiæ heterogeneæ, quæ non pertinent ad corporis constitutionem. Si materiam spectemus ; ea & mihi, & aliis plurimis homogenea est, & solis ejus diversis combinationibus diversæ oriuntur*

think that an extended continuum can be formed out of non-extended points, when indeed the whole of the matter is bounded by a continuous surface. Besides, in those bodies that fall within the scope of our senses, the idea of figure, which seems to be very distinct, is however quite vague & indefinite; & I pointed this out fairly carefully, when dealing some time ago with the figure of the Earth, in a dissertation inserted in the last volume of the Acta Bononiensia; this contains the synopsis of my work, *Expeditio Litteraria per Pontificiam ditionem*, & there the following words occur. *Now, in the first place, this term, " the figure of the Earth," which seems to have a certain definite & determinate meaning, is really very vague & indefinite. The surface which bounds the seas, the lakes, the rivers, the mountains, the plains & the valleys, is really something quite irregular, at least to us; & moreover it is also unstable; for it changes with the slightest motion of the waves & the soil. But those who investigate " the figure of the Earth," do not deal with this figure of the Earth; they substitute for it another figure which, although to some extent regular, yet approximates closely to the former true figure; that is to say, it has the mountains & the hills levelled off, whilst the valleys are filled up. Now once more the idea of this figure of the Earth is vague & uncertain. For, just as there are infinite classes of regular curves that can be made to pass through a given number of given points; so also there are infinite classes of curved surfaces that can be made to go round the Earth & circumscribe it in such a manner that they touch all the mountains & hills, or at least certain given ones; or, if you like, some surface is bound to pass through the middle of the hills & mountains in such a way that it cuts off as much matter outside itself, as it encloses empty air-spaces within it & our true surface of the Earth, to our eyes, so irregular. Also, there can be an infinite number of surfaces, & these too quite different from one another, which satisfy the problem; & all of them, too, of such a kind that they have no manifest humps, as far as can be detected; & this term even contains no true definiteness.*

The vagueness is still greater in this Theory.

376. These are my words in that dissertation with regard to the figure of the Earth; & they apply in general to the figure of any body also, if considered according to the usual way with regard to the continual extension of matter. For, the surfaces of nearly all bodies here around us are in every case much rougher in comparison with their size than is the Earth in comparison with its magnitude; & they have many internal empty spaces. But, in my Theory, the matter is much more indefinite & uncertain still. For there are an infinite number of continuous curved surfaces, in which nevertheless all the points of any mass lie; nay, further, there are an infinite number of curved lines passing through all the points. Therefore we can only mentally conceive a certain surface which shall include all the points or exclude a few of them which are more remote by gathering the rest together; this can be done by a kind of moral assessment, but not by an accurate geometrical construction. This surface gives the shape of the body; & with that idea, all that relates to the different kinds of shapes of bodies is in agreement in my Theory with the usual theory of the continual extension of matter.

Volume depends on shape: the idea of this too is vague in the usual theory, & much more so in mine.

377. Volume depends upon shape; & volume is nothing else but the whole of the space, extended in length, breadth & depth, which is included by the external surface. Further, unless we picture that surface which I mentioned as determining the shape, there can be no definite idea of volume. Nay indeed, if we think of the tortuous surface in which all the points lie, we shall never have a volume possessed of a third dimension; whilst if we think of a curved line passing through all the points, no volume will be obtained that has even two dimensions. But in that the usual idea is also wanting, as regards indefinite assessment, owing to those empty interstices that are present in all bodies, & the roughness, as we have said, which arises from the indeterminateness of figure. Here again, if an outside surface is conceived as bounding the figure, all those things that are usually enunciated about volume in relation to figure agree in my theory with those of all others; for instance, that the same volume as regards magnitude can be bounded by surfaces that are quite different, both in shape & size, & that the least surface of all having the same volume is that of a sphere. Also that, in similar figures, the volumes are in the triplicate ratio of homologous sides, the surfaces in the duplicate ratio; & upon these depend a truly great number of phenomena, & especially those which are connected with the resistance both of fluids & of solids.

Mass: what there is in the idea of it that is indefinite owing to outside matter mingling with it. All bodies are composed of parts of different natures.

378. The mass of a body is the total quantity of matter pertaining to that body; & in my Theory this is precisely the same thing as the number of points that go to form the body. Here now we have a certain indefiniteness, or at least the greatest difficulty, in forming a definite idea of mass; & that, not only in my theory, but in the usual theory as well, on account of the addition of the words *points that go to form the body*; this excludes heterogeneous substances. On this point indeed, I made the following remarks in the Supplements to Stay's Philosophy:—*For it is very difficult to define what those heterogeneous substances may be, if they do not pertain to the constitution of a body. If we consider matter, it is in my opinion, & in that of very many others, homogeneous; & the different species of*

corporum species. Quare ab ipsa materia non potest desumi discrimen illud inter substantias pertinentes, & non pertinentes. Si autem & diversam [174] *illam combinationem spectemus, corpora omnia, quæ observamus, mixta sunt ex substantiis admodum dissimilibus, quæ tamen omnes ad ejus corporis constitutionem pertinent. Id in animalium corporibus, in plantis, in marmoribus plerisque, oculis etiam patet, in omnibus autem corporibus Chemia docet, quæ mixtionem illam dissolvit.*

Plures substantiæ non pertinentes ad substantiam corporis.

379. *Ex alia parte tenuissima ætherea materia, quæ omnino est aliqua nostro aere varior, ad constitutionem massæ nequaquam pertinere censetur, ut nec pro corporibus plerisque aer, qui meatibus internis interjacet. Sic aer inclusus spongiæ meatibus, ad ipsius constitutionem nequaquam censetur pertinere. Idem autem ad multorum corporum constitutionem pertinet: saltem ad fixam naturam redactus, ut Halesius demonstravit, plures & animalis regni, & vegetabilis substantias magna sui parte constare aere fixitatem adepto. Rursus substantiæ volatiles, aere ipso tenuiores multo, quæ in corporum dissolutione chemica in halitus, & fumos abeunt, & plures fortasse, quas nos nullo sensu percipimus, ad ipsa corpora pertinebant.*

Nec excludi omnia fluida, nec ea omnia includi posse, quæ translato corpore cum ipso transferuntur.

380. *Nec illud assumi potest, quidquid solidum, & fixum est, id tantummodo pertinere ad corporis massam; quis enim a corporis humani massa sanguinem omnem, & tot lymphas excludat, a lignis resectis succos nondum concretos? Præterquam quod massæ idea non ad solida solum corpora pertinet, sed etiam ad fluida, in quibus ipsis alia tenuiora aliorum densiorum meatibus interjacent. Nec vero dici potest, pertinere ad corporis constitutionem, quidquid materiæ translato corpore, simul cum ipso transfertur; nam aer, qui intra spongiam est, partim mutatur in ea translatione, is nimirum, qui orificio est propior, partim manet, qui nimirum intimior, & qui aliquandiu manet, mutatur deinde.*

Hinc indistinctam esse & massæ ideam. Quid densitas & raritas; utranque augeri, & minui posse in hac theoria in quacunque ratione.

381. *Hæc, & alia mihi diligentius perpendenti, illud videtur demum, ideam massæ non esse accurate determinatam, & distinctam, sed admodum vagam, arbitrariam, & confusam. Erit massa materia omnis ad corporis constitutionem pertinens; sed a crassa quadam, & arbitraria æstimatione pendebit illud, quod est pertinere ad ipsam ejus constitutionem.* Hæc ego ibi: tum ad molem transeo, de cujus indeterminatione jam hic superius egimus, ac deinde ad densitatem, quæ est relatio massæ, ad molem, eo major, quo pari mole est major massa, vel quo pari massa est minor moles. Hinc mensura densitatis est massa divisa per molem; & quæcunque vulgo proferuntur de comparationibus inter massam, molem, & densitatem, hæc omnia & mihi communia sunt. Massa est ut factum ex mole & densitate; moles ut massa divisa per densitatem. Raritas autem etiam mihi, ut & aliis, est densitatis inversa, ut nimirum idem sit dicere, corpus aliquod esse decuplo minus densum alio aliquo corpore, ac dicere, esse decuplo magis rarum. Verum quod ad densitatem & raritatem pertinet, in eo ego quidem a communi sententia discrepo, uti exposui num. 89, quod [175] ego nullum habeo limitem densitatis & raritatis, nec maximum, nec minimum; dum illi minimam debent aliquam raritatem agnoscere, & maximam densitatem possibilem, utut finitam, quæ illis idcirco per saltum quendam necessario abrumpitur; licet nullam agnoscant raritatem maximam, & minimam densitatem. Mihi enim materiæ puncta possunt & augere distantias a se invicem, & imminuere in quacunque ratione; cum data linea quavis, possit ex ipsis Euclideis clementis inveniri semper alia, quæ ad ipsam habeat rationem quancunque utcunque magnam, vel parvam; adeoque potest, stante eadem massa, augeri moles, & minui in quacunque ratione data; at illis potest quidem quævis massa dividi in quenvis numerum particularum, quæ dispersæ per molem utcunque magnam augeant raritatem, & minuant densitatem in immensum; sed ubi massa omnis ita ad contactus immediatos devenit, ut nihil jam supersit vacui spatii; tum vero densitas est maxima, & raritas minima omnium, quæ haberi possint, & tamen finita est, cum mensura prioris habeatur, massa finita per finitam molem divisa, & mensura posterioris, divisa mole per massam.

Inertia massarum orta ex inertia punctorum: ipsi respondens conservatio status centri gravitatis, & idea massæ unitæ in ipso.

382. Inertia corporum oritur ab inertia punctorum, & a viribus mutuis; nam illud demonstravimus num. 260, si puncta quæcunque vel quiescant, vel moveantur directionibus, & celeritatibus quibuscunque, sed singula æquabili motu; centrum commune gravitas vel quiescere, vel moveri uniformiter in directum, ac vires mutuas quascunque inter eadem puncta nihil turbare statum centri communis gravitatis sive quiescendi, sive movendi uniformiter in directum. Porro vis inertiæ in eo ipso est sita: nam vis inertiæ est

bodies arise solely from different combinations of it. Hence it is impossible to take away from matter the distinction between substances that pertain to a body & those that do not. Again if we consider the difference of combination, all bodies that come under our observation are mixtures of substances that are perfectly unlike one another ; & yet all of them are necessary to the constitution of the body. We have ocular evidence of this in the bodies of animals, in plants, in most of the marbles ; moreover, in all bodies, chemistry teaches us how to separate that mixture.

A large number of substances do not pertain to the substance of a body.

379. *In another respect, that very tenuous ethereal matter, which is something indeed much less dense than our air, can in no sense be considered to be a constituent part of a body ; nor indeed, in the case of most bodies can the air which is contained in its internal parts. Thus the air that is included in the passages of a sponge can in no sense be considered as being necessary to the constitution of the sponge. But the same thing pertains to the constitution of many bodies ; at least, when reduced to a fixed nature. For Hales has proved that many substances of the animal & vegetable kingdoms in a great part consist of air that has attained fixity. Again, volatile substances, more tenuous than air itself, which go off in vapours & fumes from bodies chemically decomposed, & perchance many which are not perceived by any of our senses, all pertained to these bodies.*

Nor can all fluids be excluded ; nor can all those things be included, which when the body is moved are carried with the body.

380. *Nor can it be assumed that only something solid & fixed can pertain to the mass of a body. For who would exclude from the mass of the human body the whole of the blood, & the large number of watery fluids, or from chips of wood the juices that are not yet congealed? Especially as the idea of mass pertains not only to solids alone but also to fluids ; & in these some of the more tenuous parts lie in the interstices of the more dense. On the other hand, it cannot be said that any kind of matter, which when the body is moved is carried with it, pertains of necessity to the constitution of the body. For the air which is within a sponge is partly moved by that translation, that is to say that part which is near an orifice ; whilst it partly remains, that is to say that part which is more internal, & remains for some length of time, & then is moved.*

Hence also the idea of mass is indefinite. The nature of density, & rarity ; either of them in this Theory can be increased or diminished to any extent.

381. *After carefully considering these & other matters, I have come to the conclusion that the idea of mass is not strictly definite & distinct, but that it is quite vague, arbitrary & confused. Mass will be the whole of the matter pertaining to the constitution of a body ; but what part of it actually does pertain to its constitution, will depend upon a non-scientific & arbitrary assessment.* These are my words ; & after that I pass on to volume, the indefiniteness of which I have already dealt with above, & after that to density, which is the relation of mass to volume ; being so much the greater as in equal volume there is so much the greater mass, or according as for equal mass there is so much the less volume. Hence the measure of density is mass divided by volume ; & whatever is usually said about comparisons between mass, volume & density, everything is in agreement with what I say. Mass is, so to speak, the product of volume & density ; & volume is mass divided by density. Rarity, with me, as well as with others, is the inverse of density ; thus it is the same thing to say that one body is ten times less dense than another body as to say that it is ten times more rare. But as regards the properties of rarity & density, here I indeed differ from the usual opinion. For, as I showed in Art. 89, I have no limiting value for either density or rarity, no maximum, no minimum ; whereas others must admit a minimum rarity, or a maximum density, as being possible ; &, since this must be something finite, it must of necessity involve a sudden break in continuity ; although they may not admit any maximum rarity or minimum density. For with me the points of matter can both increase & diminish their distances from one another in any ratio whatever ; since, given any line, it is possible, by the elementary principles of Euclid, to find another in every case, which shall bear to the given line any ratio however great or small. Thus, it is possible that, whilst the mass remains the same, the volume should be increased or diminished in any ratio whatever. But, in the case of other theories, it is indeed possible that a mass can be divided into any number of particles, which when dispersed throughout a volume of any size however great will increase the rarity or diminish the density to an indefinitely great extent ; but when the whole mass has been brought into a state of immediate contact of its particles in such a manner that there no longer exist any empty spaces between these particles, then indeed there is a maximum density or a minimum rarity obtainable, although this is finite ; for, a measure of the first may be obtained by dividing a finite mass by a finite volume, or of the second by dividing volume by mass.

The inertia of a mass arises from the inertia of its points ; the corresponding conservation of the state of the centre of gravity ; the idea of a mass being condensed at its centre of gravity.

382. The inertia of bodies arises from the inertia of their points & their mutual forces. For, in Art. 260, it was proved that, if any points are either at rest, or moving in any directions with any velocities, so long as each of the motions is uniform, then the centre of gravity of the set will either be at rest or move uniformly in a straight line ; & that, whatever mutual forces there may be between the points, these will in no way affect the state of the common centre of gravity, whether it is at rest or whether it is moving uniformly in a straight line. Further the force of inertia is involved in this ; for the force of inertia

determinatio perseverandi in eodem statu quiescendi, vel movendi uniformiter in directum : nisi externa vis cogat statum suum mutare : & cum ex mea Theoria demonstretur, eam proprietatem debere habere centrum gravitatis massæ cujuscunque compositæ punctis quotcunque, & utcunque dispositis ; patet, eam deduci pro corporibus omnibus : & hic illud etiam intelligitur, cur concipiantur corpora tanquam collecta, & compenetrata in ipso gravitatis centro.

Mobilitas : quiescibilitatem non haberi, exclusa prorsus quiete a Natura.

383. Mobilitas recenseri solet inter generales corporum proprietates, quæ quidem sponte consequitur vel ex ipsa curva virium : cum enim ipsa exprimat suarum ordinatarum ope determinationes ad accessum, vel recessum, requirit necessario mobilitatem, sive possibilitatem motuum, sine quibus accessus, & recessus ipsi haberi utique non possunt. Aliqui & quiescibilitatem adscribunt corporibus : at ego quidem corporum quietem saltem in Natura, uti constituta est, haberi non posse arbitror, uti exposui num. 86. Eam excludi oportere censeo etiam infinitæ improbabilitatis argumento, quo sum usus in ea dissertatione *De Spatio, & Tempore*, quam toties jam nominavi, & in Supplementis hic proferam § I, ubi [176] evinco, casum, quo punctum aliquod materiæ occupet quovis momento temporis punctum spatii, quod alio quopiam quocunque occuparit vel ipsum, vel aliud punctum quodcunque, esse infinities improbabilem, considerato nimirum numero punctorum materiæ finito, numero momentorum possibilium infinito ejus generis, cujus sunt infinita puncta in una recta, qui numerus momentorum bis sumitur, semel cum consideratur puncti dati materiæ cujuscunque momentum quodvis, & iterum cum consideratur momentum quodvis, quo aliud quodpiam materiæ punctum alicubi fuerit, ac iis collatis cum numero punctorum spatii habentis extensionem in longum, latum, & profundum, qui idcirco debet esse infinitus ordinis tertii respectu superiorum. Deinde ab omnium corporum motu circa centrum commune gravitatis, vel quiescens, vel uniformiter progrediens in recta linea, quies actualis itidem a Natura excluditur.

Quies exclusa etiam a continuitate omnium motuum : problema generale eo pertinens.

384. Verum ipsam quietem excludit alia mihi proprietas, quam omnibus itidem materiæ punctis, & omnium corporum centris gravitatis communem censeo, nimirum continuitas motuum, de qua egi num. 883, & alibi. Quodvis materiæ punctum seclusis motibus liberis, qui oriuntur ab imperio liberorum spirituum, debet describere curvam quandam lineam continuam, cujus determinatio reducitur ad hujusmodi problema generale : Dato numero punctorum materiæ, ac pro singulis dato puncto loci, quod occupent dato quopiam momento temporis, ac data directione, & velocitate motus initialis, si tum primo projiciuntur, vel tangentialis, si jam ante fuerunt in motu, ac data lege virium expressa per curvam aliquam continuam, cujusmodi est curva figuræ I, quæ meam hanc Theoriam continet, invenire singulorum punctorum trajectorias, lineas nimirum, per quas ea moventur singula. Id problema mechanicum quam sublime sit, quam omnem humanæ mentis excedat vim, ille satis intelliget, qui in Mechanica versatus non nihil noverit, trium etiam corporum motus, admodum simplici etiam vi præditorum, nondum esse generaliter definitos, uti monui num. 204, & consideret immensum punctorum numerum, ac altissimam curvæ virium tantis flexibus circa axem circumvolutæ elevationem.

Quid curvæ descriptæ a punctis non habeant. Problema inversum datis particulis descriptis tempusculo utcunque parvo.

385. Sed licet ejusmodi problema vires omnes humanæ mentis excedat ; adhuc tamen unusquisque Geometra videbit facile, problema esse prorsus determinatum, & curvas ejusmodi fore omnes continuas sine ullo saltu, si in lege virium nullus sit saltus. Quin immo & illud arbitror, in ejusmodi curvis nec ullas usquam cuspides occurrere ; nam nodos nullos esse consequitur ex eo, quod nullum materiæ punctum redeat ad idem punctum spatii, in quo ipsum aliquando fuerit, adeoque nullus habeatur regressus, qui tamen ad nodum est necessarius. Hujusmodi curvæ necessariæ essent omnes, & mens, [177] quæ tantum haberet vim, quanta requiritur ad ejusmodi problemata rite tranctanda, & intimius perspiciendas solutiones (quæ quidem mens posset etiam finita esse, si finitus sit punctorum numerus, & per finitam expressionem sit data notio curvæ exprimentis legem virium) posset ex arcu continuo descripto tempore etiam utcunque exiguo a punctis materiæ omnibus derivare ipsam virium legem, cum quidam finiti tantummodo positionum numeri finitos determinare possint numeros punctorum curvæ virium, & arcus continuus legem ipsam continuam : & fortasse solæ etiam positiones omnium punctorum cum dato arcu continuo percurso ab unico etiam puncto motu continuo, exiguo etiam aliquo tempusculo ad rem præstandam satis essent. Cognita autem lege virium & positione, ac velocitate, & directione punctorum omnium dato tempore, posset ejusmodi mens prævidere omnes futuros necessarios motus, ac status, & omnia Naturæ phænomena necessaria, ab iis utique pendentia, atque prædicere : & ex unico arcu descripto a quovis puncto, tempore continuo utcunque

consists in a propensity for staying in a state of rest or of maintaining a uniform state of motion in a straight line, unless some external force compels a change of this state. Now, since by my Theory it is proved that the centre of gravity of any mass, composed of any number of points disposed in any manner whatever, is bound to have this property, it is clear that the same property can be deduced for all bodies ; & by this it can also be understood why bodies can be conceived to be collected & condensed at their centres of gravity.

Mobility; quiescibility is impossible, since there is no such thing in Nature as absolute rest.

383. Mobility is usually considered as one of the general properties of bodies ; & indeed it follows immediately from the curve of forces. For, since this curve, by means of its ordinates, represents the propensity to approach or recede, it necessarily requires mobility, or the possibility of motion, without which approach or recession can certainly not be obtained. Now there are some, who ascribe quiescibility to bodies ; but I consider that absolute rest, at any rate in Nature as it is at present constituted, is impossible, as I explained in Art. 86. I think also that it must be excluded by the argument of infinite improbability, which I used in the dissertation *De Spatio, & Tempore*, which I have mentioned so many times already, & which I quote in this work as Supplement, § 1 ; in it I prove that the case in which any point of matter occupies at any instant of time a point of space, which at any other instant whatever either it or any other point whatever would occupy, is infinitely improbable ; this, by considering the finite number of points of matter, & the infinite number of instants of time possible, of that class for which there are an infinite number of points in the same straight line ; this number of instants is considered twice, once when any instant for any given point of matter is considered, & again when any instant is considered in which any other point of matter was somewhere else ; when these are compared with the number of points of a space which has extension in length, breadth & depth, the latter must be infinite of the third order with respect to those mentioned above. Finally, by the motion of all bodies about a common centre of gravity, whether this is at rest or travelling uniformly in a straight line, absolute rest is excluded from Nature.

Absolute rest is excluded also by the continuity of all motions ; general problem with respect to it.

384. In my opinion also, there is another property that excludes absolute rest, one which I consider is common also to all points of matter & to the centres of gravity of all bodies ; namely, continuity of motion, with which I dealt in Art. 88 & elsewhere. Any point of matter, setting aside free motions that arise from the action of arbitrary will, must describe some continuous curved line, the determination of which can be reduced to the following general problem. Given a number of points of matter, & given, for each of them, the point of space that it occupies at any given instant of time ; also given the direction & velocity of the initial motion if they were projected, or the tangential velocity if they are already in motion ; & given the law of forces expressed by some continuous curve, such as that of Fig. 1, which contains this Theory of mine ; it is required to find the path of each of the points, that is to say, the line along which each of them moves. How difficult this mechanical problem may become, how it may surpass all powers of the human mind, can be easily enough understood by anyone who is versed in Mechanics & is not quite unaware that the motions of even three bodies only, & these possessed of a perfectly simple law of force, have not yet been completely determined in general, & then will consider an immense number of points, & the extremely high degree of a curve of forces twisting round the axis with so many sinuosities.

What does not exist in curves described by the points. The inverse problem, given the curves described in any interval of time however small.

385. Now, although a problem of such a kind surpasses all the powers of the human intellect, yet any geometer can easily see thus far, that the problem is determinate, & that such curves will all be continuous without any break in them, so long as there is no discontinuity in the law of forces. Indeed, I think that, in such curves, there never occur any cusps ; for, it follows that there are no nodes, from the fact that no point of matter returns to the same point of space that it occupied at any time ; & thus there is none of that regression which is necessary for a node. All the curves must be of this kind ; & a mind which had the powers requisite to deal with such a problem in a proper manner & was brilliant enough to perceive the solutions of it (& such a mind might even be finite, provided the number of points were finite, & the notion of the curve representing the law of forces were given by a finite representation), such a mind, I say, could, from a continuous arc described in an interval of time, no matter how small, by all points of matter, derive the law of forces itself ; for, any merely finite number of positions can determine a finite number of points on the curve of forces, & a continuous arc the continuous law. Perhaps even the positions of all the points, together with a given continuous arc traversed with continuous motion by but a single one of them, & that too in an interval of time no matter how small, would be sufficient to obtain a solution of the problem. Now, if the law of forces were known, & the position, velocity & direction of all the points at any given instant, it would be possible for a mind of this type to foresee all the necessary subsequent motions & states, & to predict all the phenomena that necessarily followed from them. It would be possible from a single arc described by any point in an interval of continuous time, no matter how

parvo, quem aliqua mens satis comprehenderet, eadem determinare posset reliquum omnem ejusdem continuæ curvæ tractum utraque e parte in infinitum productum.

Cur ab humana mente solvi non possit. Quid officiat ei determinationi libertas: Harmoniæ præstabilitæ impugnatio.

386. Nos eo aspirare non possumus, tum ob nostræ mentis imbecillatatem, tum quia ignoramus numerum, & positionem, ac motum punctorum singulorum (nam nec motus absolutos intuemur, sed respectivos tantummodo respectu Telluris, vel ad summum respectu systematis planetarii, vel systematis fixarum omnium) tum etiam, quia curvas illas turbant liberi motus, quos producunt spirituales substantiæ. Harmonia præstabilita Leibnitianorum ejusmodi perturbationem tollit omnem, saltem respectu animæ nostræ, cum omne immediatum commercium demat inter corpus, & animam; & id, quod tantopere improbatum est in Theoria Cartesiana, quæ bruta redegerat ad automata, ad homines etiam ipsos transfert, quorum motus a machina provenire omnes, & necessarios esse in ea Theoria, facile constat: & quidem idcirco etiam mihi Theoria displicet plurimum, quam præterea si admitterem, nullam sane viderem, ne tenuissimam quidem rationem, quæ mihi suadere posset, præter animam meam, cujus ideæ per se, & sine ullo immediato nexu cum corpore evolvantur, me habere aliquod corpus, quod motus ullos habeat, & multo minus, ejusmodi motus esse conformes iis ideis, aut ullos alios esse homines, ullam naturam corpoream extra me; ad quæ omnia, & multo adhuc pejora, mentem suis omnia momentis librantem deducat omnino oportet ejusmodi sententia, quam promoveri passim, & vero etiam recipi, ac usque adeo gliscere, quin & omnino tolerari, semper miratus sum.

Motus liberos omnino ab anima progigni, sed non imprimi, nisi æqualiter in partes oppositas, & sine saltu.

387. Censeo igitur, & id intima vi, qua anima suarum [178] idearum naturam, & proprietates quasdam, atque originem novit, constare arbitror, motus liberos corporis ab anima provenire: ac quemadmodum virium lex necessaria, in ipsa fortasse materiæ natura sita, ejusmodi est; ut juxta eam bina materiæ puncta debeant ad se invicem accedere, vel a se invicem recedere, determinata & quantitate motus, & directione per distantias; ita esse alias leges virium liberas animæ, secundum quas debeant quædam puncta materiæ habentia ejusmodi dispositionem, quæ ad vivum, & sanum corpus organicum requiritur, ad ipsius animæ nutum moveri; sed hujusmodi leges itidem censeo requirere illud, ut nulli materiæ puncto imprimatur motus aliquis, nisi alicui alteri imprimatur alius contrarius, & æqualis, quod constat ex ipso nisu, quem semper exercemus in partes contrarias, juxta ea, quæ diximus num. 74: ac itidem arbitror, & id ipsum diligenti observatione, & reflexione facile colligitur, ejusmodi quoque motus imprimi non posse, nisi servata lege continuitatis sine ullo saltu, quod si ab omnibus spiritibus observari debeat; discedent quidem veri motus a curvis illis necessariis, & a libera voluntatis determinatione pendebunt curvæ descriptæ; sed motuum continuitas nequaquam turbabitur.

Conclusiones deductæ; potissimum exclusio quietis.

388. Porro inde constat, cur in motibus nullum uspiam deprehendamus saltum, cur nullum materiæ punctum ab uno loci puncto abeat ad aliud punctum loci sine transitu per intermedia, cur nulla densitas mutetur per saltum, cur & motus reflexi, & refracti fiant per curvaturam continuam, ac alia ejusmodi, quæ huc pertinent. Verum simul patebit & illud, in cujus gratiam hæc congessimus, nullam fore absolutam quietam, in qua nimirum continuatus ille curvæ descriptæ ductus abrumpatur ea continuitate læsa nihilo minus, quam læderetur, si curva continua desineret alicubi in rectam.

Aequalitas actionis, & reactionis, & ejus consectaria.

389. Jam vero ad actionis, & reactionis æqualitatem gradu facto, eam abunde deduximus a num. 265. pro binis quibusque corporibus ex actione, & reactione æqualibus in punctis quibuscunque. Cum nimirum mutuæ vires nihil turbent statum centri gravitatis communis, & centra gravitatis binarum massarum debeant cum ipso communi centro jacere in directum ad distantias hinc, & inde reciproce proportionales ipsis massis, ut ibidem demonstravimus; consequitur illud, motus quoscunque, quos ex mutua actione habebunt binarum massarum centra gravitatis, debere fieri in lineis similibus, & proportionalibus distantiæ singularum ab ipso gravitatis centro communi, adeoque reciproce proportionalibus ipsis massis; & quod inde consequitur, summam motuum computatorum secundum directionem quancunque, quam ex mutuis actionibus acquiret altera massa, fore semper æqualem summæ motuum computatorum secundum oppositam, quam massa altera acquiret simul, in quo ipso sita est actionis & reactionis æqualitas, ex qua corporum [179] collisiones deduximus in secunda parte, & ex qua multa phænomena pendent, in Astronomia inprimis.

small, which was sufficient for a mind to grasp, to determine the whole of the remainder of such a continuous curve, continued to infinity on either side.

Why the problem cannot be solved by the human intellect; what obstacle to its determination is due to freedom; argument against "pre-established harmony."

386. We cannot aspire to this, not only because our human intellect is not equal to the task, but also because we do not know the number, or the position & motion of each of these points (for we do not observe absolute motions, but merely relative motions with respect to the Earth, or at most those with respect to the planetary system or the system of all the fixed stars); & there is yet another reason, namely that the free motions produced by spiritual substances affect these curves. The "pre-established harmony" of the followers of Leibniz abrogates all such disturbing effect, at least as far as regards our will, since it does not admit any direct intercourse between body & spirit. What was so strongly condemned in the theory of Descartes, which reduced animals to automata, is transferred to men as well; & it is easily shown that all their motions arise from a mechanism, & that these are necessary upon that theory. For this reason, indeed, I am very much against the Cartesian theory; for, besides other things, if I admitted its principles, I should not be able to see any real reason, nay, not of the slightest kind, which would lead me to think that, in addition to my mind, ideas about which are evolved of itself & without any direct connection with the body, I had a body that had motions; much less, that these motions conformed to those ideas, or that there were any other men, or any corporeal nature outside myself. Such a philosophy must of necessity lead a mind that puts everything in the scales of its own impulses to such absurdities, & still worse; & I have always been astonished that this philosophy has gained ground & has even been accepted everywhere, & up to the present has been growing; I am amazed that it should have been tolerated at all.

Free motions are certainly produced by the mind, but are not impressed except equally in opposite directions, & without breach of continuity

387. I think, therefore, that the free motions of bodies arise from the mind; & that this is due to an inner force, by which the mind knows the nature, certain properties & the origin of its ideas, I think can be easily established. Just as we must have a law of forces, perhaps involved in the very nature of matter, of such a kind that according to it two points of matter must approach towards, or recede from, one another with a motion determined in magnitude & direction by the distance between the points; so there must be other free laws for the mind, according to which any points that have that disposition which a living & healthy body requires, must obey the command of the mind. But such laws, I also think, require the condition that a motion cannot be impressed on any point of matter, unless an equal & opposite motion is impressed on some other point of matter; this follows from the stress that we always exert in opposite directions, according to what has been said in Art. 74. Lastly, I consider, & the fact can be derived by diligent observation & reflection, that such motion can not be impressed, unless it follows a law of continuity without any break; & if this law is bound to be observed by all object-souls, the real motions will truly depart from the necessary curves, & the curves actually described will depend on a free determination of the will; but the continuity of the motions will not thereby be affected.

Conclusions deduced; especially the exclusion of absolute rest.

388. Further, it is hence evident why we nowhere get any discontinuity in motions, why no point of matter can ever pass from one position to another without passing through all intermediate positions, why density can in no case be suddenly changed, why reflected & refracted motions come about through continuous curvature, & other things of the sort relating to the matter in hand. But, in particular, there will at the same time be evident the fact, which is the purpose of all we have just done, namely, that there is no such thing as absolute rest; that is to say, such a thing as the sudden breaking off of the continuous drawing of the curve described, the continuity being destroyed just as much as it would be if a continuous curve finally became a straight line after reaching a certain point.

Equality of action & reaction; its consequences.

389. Passing on to the equality of action & reaction, we have already, in Art. 265, fully proved its truth for any two bodies from the equality of the action & reaction between any two points. For instance, since the mutual forces do not in any way affect the state of the common centre of gravity, & the centres of gravity of two masses must lie in a straight line with the common centre of the two, at distances on each side of the latter that are inversely proportional to the masses, as was also proved in the same article; it must follow that any motions, which owing to mutual action are possessed by the centres of gravity of the two masses, must take place along lines that are similar & proportional to the distances of each from the common centre of gravity, & thus inversely proportional to the masses. Also it then follows that the sum of the motions, reckoned in any direction, acquired by either of the masses on account of the mutual actions, must always be equal to the sum of the motions in the directly opposite direction, acquired simultaneously by the other mass; & in this is involved the equality of action & reaction; & from it we deduced the laws of the collisions of bodies in the second part; & upon it depend many phenomena, especially in Astronomy.

Inde an motus massæ proveniat a viribus internis, an ab externis.

390. Illud unum hic adnotandum censeo, per hanc ipsam legem comprobari plurimum ipsas vires mutuas inter materiæ particulas, & deveniri ad originem motuum plurimorum, quæ inde pendet; si nimirum particulæ massæ cujuslibet ingentem habeant motum reciprocum hac, illac, & interea centrum commune gravitatis iisdem iis motibus careat; id sane indicio est, eos motus provenire ab internis viribus mutuis inter puncta ejusdem massæ. Id vero accidit inprimis in fermentationibus, quæ habentur post quarundam substantiarum permixtionem, quarum particulæ non omnes simul jam in unam feruntur plagam, jam in aliam, sed singillatim motibus diversissimis, & inter se etiam contrariis, quos idcirco motus omnes illarum centra gravitatis habere non possunt; ii motus provenire omnino debent a mutuis viribus, & commune gravitatis centrum interea quiescet respectu ejus vasis, in quo fermentatio sit, & Terræ, respectu cujus quiescit vas.

Divisibilitas in infinitum spatii continui; & materiæ itidem si sit continua, & sine virtuali extensione.

391. Quod ad divisibilitatem pertinet, eam quidem in infinitum progredientem sine ullo limite in spatio continuo ille solus non agnoscet, qui Geometriæ etiam elementaris vim non sentiat, a qua pro ejusmodi divisibilitate in infinitum tam multa, & simplicia, & perspicua sane argumenta desumuntur. Ubi ad materiam sit transitus; si, ubi de ea agitur, quæ distinctas occupant loci partes, distincta etiam sunt; ab illa spatii continui divisibilitate in infinitum, materiæ quoque divisibilitas in infinitum consequitur evidentissime, & utcunque prima materiæ clementa atomos, sive Naturæ vi insectilia censeant multi, ut & Newtonus; adhuc tamen absolutam eorum divisibilitatem agnoscunt passim illi ipsi.

Virtualem extensionem non haberi.

392. Materiæ elementa extensa per spatium divisibile, sed omnino simplicia, & carentia partibus, admiserunt nonnulli e Peripateticis, & est etiam nunc, qui recentiorem Philosophiam professus admittat; at eam sententiam non ex præjudicio quodam, quanquam id etiam est ingens, & commune, sed ex inductionis principio, & analogia impugnavi in prima parte num. 83. Quamobrem arbitror, si quid corporeum extensionem habeat per totum quodpiam continuum spatium, id ipsum debere absolute habere partes, & esse divisibile in infinitum æque, ac illud ipsum est spatium.

Puncta esse indivisibilia; massas divisibiles usque ad certum limitem singulas.

393. At in mea Theoria, in qua prima elementa materiæ mihi sunt simplicia, ac inextensa, nullam, eorum divisibilitatem haberi constat. Massæ autem, quæcunque actu existant, sunt mihi congeries punctorum ejusmodi numero finitæ. Hinc eæ congeries dividi utique possunt in partes, sed non plures, quam sit ipse punctorum numerus massam constituentium, cum nulla pars minus continere possit, quam unum ex iis punctis. Nec Geometrica argumenta quidquam probant in mea Theo-[180]ria pro divisibilitate ultra eum limitem; posteaquam enim deventum fuerit ad intervalla minora, quam sit distantia duorum punctorum, sectiones ulteriores secabunt intervalla ipsa vacua, non materiam.

Componibilitas in infinitum.

394. Verum licet ego non habeam divisibilitatem in infinitum, habeo tamen componibilitatem, ut appellare soleo, in infinitum. In quovis dato spatio habebitur quidem semper certus quidam punctorum numerus, qui idcirco etiam finitus erit; neque enim ego admitto infinitum ullum in Natura, aut in extensione, neque infinite parvum in se determinatum, quod ego positiva demonstratione exclusi primum in mea Dissertatione *de Natura & usu infinitorum, & infinite parvorum;* tum & aliis in locis; quod tamen requireretur ad hoc, ut intra finitum spatium contineretur punctorum numerus indefinitus: at longe aliter se res habet; si consideremus, qui numerus punctorum in dato spatio possit existere: tum enim nullus est numerus finitus ita magnus, ut alius adhuc finitus ipso major haberi in eo spatio non possit. Nam inter duo puncta quæcunque potest in medio interseri aliud, quod quidem neutrum continget; aliter enim etiam ea duo se contingerent mutuo, & non distarent, sed compenetrarentur. Potest autem eadem ratione inter hoc novum, & priora illa interseri novum utrinque, & ita porro sine ullo limite: adeoque deveniri potest ad numerum punctorum quovis determinato utcunque magno majorem in unica etiam recta, & proinde multo magis in spatio extenso in longum, latum, & profundum. Hanc ego voco componibilitatem in infinitum. Numerus, qui in quavis data massa existit, finitus est; sed dum eum Naturæ Conditor determinare voluit, nullos habuit limites, quos non potuerit prætergredi, nullum ultimum habente terminum serie illa possibilium finitorum in infinitum crescentium.

Ejus æquivalentia cum divisibilitate in infinitum.

395. Hæc componibilitas in infinitum æquivalet divisibilitati in ordine ad explicanda Naturæ phænomena. Posita divisibilitate materiæ in infinitum, solvitur facile illud

Hence, the point as to whether the motion of a mass arises from internal or external forces.

390. I consider that in this connection it should be remarked that by means of this law especially the existence of these mutual forces between particles of matter is established, & that in it we attain to the source of most of the motions, which arises from it. For instance, considering that the particles of a mass may have an immense reciprocal motion, whilst the common centre of gravity is without any such motion, surely that is a token that these motions come from mutual internal forces between the particles of the mass. Now, this takes place, in particular, in fermentations, such as are obtained after making a mixture of certain substances; here the particles of the substances are not all at the same time moving first in one direction, then in another, but each of them separately in the most widely diverging directions, & even in opposite directions, to one another. Hence, as the centres of gravity cannot have all these motions, the motions must arise from mutual forces; &, besides, the common centre of gravity is at rest with regard to the vessel in which the fermentation takes place, & also with regard to the Earth, with respect to which the vessel is at rest.

Infinite divisibility of continuous space; the same of matter, if it is continuous, & without virtual extension.

391. Now, as concerning divisibility, that this can be carried on indefinitely without any limit in continuous space will be denied only by one who does not feel the force of the most elementary principles of geometry; for, from it may be derived so many simple & perfectly clear arguments in favour of such infinite divisibility. When we come to consider matter, if in dealing with it, we take it that what occupies a distinct part of space is itself distinct, then, from the infinite divisibility of continuous space, the infinite divisibility of matter also follows very clearly; &, although there are many who think that the primary elements of matter are atoms, that is to say, things that are incapable of further division by any Natural force, as Newton also thought, yet even they must still in all cases admit their absolute divisibility.

Virtual extension is non-existent.

392. Some of the Peripatetics admitted elements of matter extended through divisible space, but quite simple & without parts; & at the present day there is one professing a more modern philosophy who admits such elements. This idea, in Art. 83 of the first part of this work, I contradicted, not by the employment of any prejudgment, although there certainly exists one that is very forcible & generally acknowledged, but by the employment of the principle of induction & analogy. Hence, I think that, if anything has corporeal extension throughout the whole of any continuous space, it must also absolutely have parts & must be infinitely divisible, in exactly the same manner as the space is infinitely divisible.

Points are indivisible, whilst every mass is divisible up to a certain limit.

393. Now, in my Theory, in which the primary elements of matter are simple & non-extended, it is easily seen that there can be no divisibility of the elements. Also masses, in so far as they actually exist, are to me merely sets of such points finite in number. Hence these sets of points can at any rate be divided into parts, but not into a greater number of points than that given by the number of points constituting the mass, since no part can contain less than one of these points. Nor do geometrical arguments prove anything, as far as my Theory is concerned, in favour of divisibility beyond this limit; for, as soon as we reach intervals that are less than the distance between two points, further sections will cut these empty intervals & not matter.

Infinite componibility.

394. Now, although I do not hold with infinite divisibility, yet I do admit infinite componibility, as it is usually called. In any given space we can always have a certain number of points; & hence this number is finite. For, I do not admit anything infinite in Nature, or in extension, or a self-determined infinitely small. Such a thing I excluded by direct proof, for the first time in my dissertation *De Natura, & usu infinitorum, & infinite parvorum;* & later, in other writings; this, however, is required, if an indefinite number of points is to be included within a finite space. But the facts of the matter are quite different, if we consider how great a number of points can exist within a given space; for, then there is no finite number so great, but that a still greater finite number can be had within the space. For, between any two points it is possible to insert another midway, which will touch neither of the former; if this is not the case, then the two former points must touch one another, & not be at a distance from one another, but compenetrated. Further, in the same manner, between the new point & the first two points, we can insert a new one on either side; & so on without any limit. Thus we could arrive at a number of points greater than any given number, no matter how large, all of them even lying in a single straight line; much more then would this be the case in space extended in length, breadth & depth. This I call infinite componibility. The number of points present in any given mass is finite; but when the Creator of the Universe willed what that number was to be, he had no limits; for the series of possible finites increasing indefinitely has no last term.

The equivalence of componibility to infinite divisibility.

395. This infinite componibility is equivalent to divisibility for the purpose of explaining the phenomena of Nature. If we postulate infinite divisibility for matter, we have an easy

problema : *Datam massam utcunque parvam, ita distribuere per datum spatium utcunque magnum, ut in eo nullum sit spatiolum majus dato quocunque utcunque parvo penitus vacuum, & sine ulla ejus materiæ particula.* Concipitur enim numerus, quo illud magnum spatium datum continere possit hoc spatiolum exiguum, qui utique finitus est, & in se determinatus : concipitur in totidem particulas divisa massula, & singulæ particulæ destinantur singulis spatiolis ; quæ iterum dividi possunt, quantum libuerit, ut parietes spatioli sui convestiant, qui utique ad unam ejus transversam sectionem habent finitam rationem, adeoque continua sectione planis parallelis facta possunt ipsi parietes convestiri segmentis suæ particulæ, vel possunt ejus particulæ segmenta iterum per illud spatiolum utcunque dispergi. In [181] mea Theoria substituitur hujusmodi aliud problema : *Intra datum spatiolum collocare eum punctorum numerum, qui deinde distribui possit per spatium utcunque magnum ita, ut in eo nullum sit spatiolum cubicum majus dato quocunque utcunque parvo penitus vacuum, & quod in se non habeat numerum punctorum utcunque magnum.*

Demonstratur ea ipsa.

396. Quod in ordine ad explicanda phænomena hoc secundum problema æquivaleat illi primo, patet utique : nam solum deest convestitio parietum continua mathematice : sed illi succedit continuatio physica, cum in singulis parietibus collocari possit ejus ope quicunque numerus utcunque magnus, distantiis idcirco imminutis utcunque. Quod in mea Theoria secundum illud problema solvi possit ope expositæ componibilitatis in infinitum, patet : quia ut inveniatur numerus, qui ponendus est in spatiolo dato, satis est, ut numerus vicium, quo ingens spatium datum continet illud spatiolum posterius multuplicetur per numerum punctorum, quem velimus collocari in hoc ipso quovis posteriore spatiolo post dispersionem, & auctor Naturæ potuit utique intra illud spatiolum primum hunc punctorum numerum collocare.

Divisibilitas in Natura immanis;

397. Jam quod pertinet ad divisibilitatem immanem, quam nobis ostendunt Naturæ phænomena in coloratis quibusdam corporibus, immanem molem aquæ inficientibus eodem colore, in auro usque adeo ductili, in odoribus, & ante omnia in lumine, omnia mihi cum aliis communia crunt ; & quoniam nulla ex observationibus nobis potest ostendere divisibilitatem absolute infinitam, sed ingentem tantummodo respectu divisionum, quibus plerumque assuevimus ; res ex meo problemate æque bene explicabitur per componibilitatem ac in communi Theoria ex illo alio per divisibilitatem materiæ in infinitum.

Immutabilitas primorum elementorum materiæ: ordines diversi particularum minus ac minus immutabilium.

398. Prima materiæ elementa volunt plerunque immutabilia, & ejusmodi, ut atteri, atque confringi omnino non possint, ne nimirum phænomenorum ordo, & tota Naturæ facies commutetur. At elementa mea sunt sane ejusmodi, ut nec immutari ipsa, nec legem suam virium, ac agendi modum in compositionibus commutare ullo modo possint ; cum nimirum simplicia sint, indivisibilia, & inextensa. Ex iis autem juxta ea, quæ diximus num. 239 ad distantias perquam exiguas collocatis in limitibus virium admodum validis oriri possunt primæ particulæ minus jam tenaces suæ formæ, quam simplicia elementa, sed ob ingentem illam viciniam adhuc tenacissimæ idcirco, quod alia particula quævis ejusdem ordinis in omnia simul ejus puncta fere æqualiter agat, & vires mutuæ majores sint, quam sit discrimen virium, quibus diversa ejus puncta solicitantur ab illa particula. Ex hisce primi ordinis particulis possunt constare particulæ ordinis secundi ; adhuc minus tenaces, & ita porro ; quo enim plures compositiones sunt, & majores distantiæ, eo facilius fieri potest, ut inæqualitas [182] virium, quæ sola mutuam positionem turbat, incipiat esse major, quam sint vires mutuæ, quæ tendunt ad conservandam mutuam positionem, & formam particularum ; & tunc jam alterationes, & transformationes habebuntur, quas videmus in corporibus hisce nostris, & quæ habentur etiam in pluribus particulis postremorum ordinum, hæc ipsa nova corpora componentibus. Sed prima materiæ elementa crunt omnino immutabilia, & primorum etiam ordinum particulæ formas suas contra externas vires validissime tuebuntur.

Gravitas exhibita a postremo arcu curvæ accedens ad Newtonianam quam proxime : posse nostro concipiendi modo fieri absolute talem.

399. Gravitas etiam inter generales proprietates a Newtonianis inprimis numeratur, quibus assentior ; dummodo ea reipsa non habeat rationem reciprocam duplicatam distantiarum extensam ad omnes distantias, sed tantum ad distantias ejusmodi, cujusmodi sunt eæ, quæ interjacent inter distantiam nostrorum corporum a parte multo maxima

solution of the following problem. *Distribute a given mass, however small, within a given space, however large, in such a manner that there shall be no little space in it greater than any given one, no matter how small, that shall be quite empty, & without any particle of that matter.* For we assume a certain number to represent the number of times the large given space can contain the exceedingly small space, this number being in every case finite & self-determined; we assume the mass to be divided into the same number of particles, & one of the particles to be placed in each of the small spaces. The former can again be divided, as much as is desired, so that the new parts of each particle cover the boundary walls of the corresponding small space; & these in every case bear a finite ratio to one transverse section of it, so that, by making continuous sections with parallel planes, these boundary walls can be covered each with segments of the particle corresponding to it; or the segments of a particle can be scattered in any manner throughout the small space, repeating the above process. In my Theory another problem is substituted, such as the following:— *Place within a given small space such a number of points, that these can then be distributed throughout any space, however great, in such a manner that there shall be no little cubical space in it greater than any given one, however small, that shall be quite empty, & which does not contain in itself any number of points however great.*

Demonstration.

396. It is quite clear that, for the purpose of explaining the phenomena of Nature, the second problem is equivalent to the first; for, the only thing that is wanting in it is a continuous covering of the boundary walls, in a strictly mathematical sense; & instead of this we have a physical continuity, since in each of the walls there can be placed by means of it any number of particles, however great, & therefore at distances from one another which are indefinitely diminished. It is also clear that, in my Theory, the second problem can be solved by the employment of the infinite componibility that I have explained; for, in order to find the number to be placed in a given small space, it is sufficient that the number of times that the large given space contains the latter small space should be multiplied by the number of points which we desire to be placed in this latter small space after dispersion; & certainly the Author of Nature was able to place this number of points within that first small space.

The immense divisibility in Nature.

397. Now, as regards the immense divisibility, which the phenomena of Nature present to us in certain coloured bodies, when they stain an immense volume of water with the same colour, in the extremely great ductility of gold, in odours, & more than all in light, everything will be in agreement in my Theory with the theories of others. Moreover, since no observations can show us any divisibility that is absolutely infinite, but only such as is immensely great when compared with such divisions as we are for the most part accustomed to; it follows that the matter can be explained just as well from my problem by means of componibility, as in the usual theory it can be from the other problem by the infinite divisibility of matter.

Immutability of the primary elements of matter: different kinds of particles, less & less immutable.

398. The primary elements of matter are considered by most people to be immutable, & of such a kind that it is quite impossible for them to be subject to attrition or fracture, unless indeed the order of phenomena & the whole face of Nature were changed. Now, my elements are really such that neither themselves, nor the law of forces can be changed; & the mode of action when they are grouped together cannot be changed in any way; for they are simple, indivisible & non-extended. From these, by what I have said in Art. 239, when collected together at very small distances apart, in sufficiently strong limit-points on the curve of forces, there can be produced primary particles, less tenacious of form than the simple elements, but yet, on account of the extreme closeness of its parts, very tenacious in consequence of the fact that any other particle of the same order will act simultaneously on all the points forming it with almost the same strength, & because the mutual forces are greater than the difference between the forces with which the different points forming it are affected by the other particle. From such particles of the first order there can be formed particles of a second order, still less tenacious of form; & so on. For the greater the composition, & the larger the distances, the more readily can it come about that the inequality of forces, which alone will disturb the mutual position, begins to be greater than the mutual forces which endeavour to maintain that mutual position, i.e. the form of the particles. Then indeed we shall have changes & transformations, such as we see in these bodies of ours, & which are also obtained in most of the particles of the last orders, which compose these new bodies. But the primary elements of matter will be quite immutable, & particles of the first orders will preserve their forms in opposition to even very strong forces from without.

Gravity, as represented by the last arc of the curve, approximates to that given by the Newtonian law; possibility of its being exactly the same, according to my hypothesis.

399. Gravity also is counted as a general property, especially by followers of Newton; & I am of the same opinion, so long as it is not supposed to be in the inverse ratio of the squares of the distances for all distances, but merely for distances such as those that lie between the distance of our bodies from the far greatest part of the mass of the Earth,

massæ terrestris, & distantias a Sole apheliorum pertinentium ad cometas remotissimos, & dummodo in hoc ipso tractu sequatur non accuratissime, sed, quam libuerit, proxime, rationem ipsam reciprocam duplicatam, juxta ea, quæ diximus num. 121. Ejusmodi autem gravitas exhibetur ab arcu illo postremo meæ curvæ figuræ 1, qui, si gravitas extenditur cum eadem illa lege ad sensum, vel cum aliqua simili, in infinitum, erit asymptoticus. Posset quidem, ut monui num. 119, concipi gravitas etiam accurate talis, quæ extendatur ad quascunque distantias cum eadem lege, & præterea alia quædam vis exposita per aliam curvam, in quam vim, & in gravitatem accurate reciprocam quadratis distantiæ resolvatur lex virium figuræ 1; quæ quidem vis in illis distantiis, in quibus gravitas sequitur quam proxime ejusmodi legem, esset insensibilis; in aliis autem distantiis plurimis ingens esset: ac ubi figura 1 exhibet repulsiones, deberet esse vis hujus alterius conceptæ legis itidem repulsiva tanto major, quam vis legis primitivæ figuræ 1, quanta esset gravitas ibi concepta, quæ nimirum ab illo additamento vis repulsivæ elidi deberet. Sed hæc jam a nostro concipiendi modo penderent, ac in ipsa mea lege primitiva, & reali, gravitas utique est generalis materiæ, ac legem sequitur rationis reciprocæ duplicatæ distantiarum, quanquam non accurate, sed quamproxime, nec ad omnes extenditur distantias; sed illas, quas exposui.

Gravitatem generalem haberi in toto solari systemate, nec posse tribui pressioni fluidi.

400. Ceterum gravitatem generalem haberi in toto planetario systemate, ego quidem arbitror omnino evinci iisdem argumentis ex Astronomia petitis, quibus utuntur Newtoniani, quæ hic non repeto, cum ubique prostent, & quæ tum alibi ego quidem congessi pluribus in locis, tum in *Adnotationibus ad poema P. Noceti De Aurora Boreali.* Illud autem arbitror evidentissimum, illum accessum ad Solum cometarum, & planetarum primariorum, ac secundariorum ad primarios, quem videmus in descensu a recta tangente ad arcum curvæ, & multo magis alios motus a mutua gravitate pendentes haberi omnino [183] non posse per ullius fluidi pressionem; nam ut alia prætermittam sane multa, id fluidum, quod sola sua pressione tantum possit in ejusmodi globos, multo plus utique posset occursu suo contra illorum tangentialem velocitatem, quæ omnino deberet imminui per ejusmodi resistentiam, cum ingenti perturbatione arearum, & totius Astronomiæ Mechanicæ perversione; adeoque id fluidum vel resistentiam ingentem deberet parere planetæ, aut cometæ progredienti, vel ne pressione quidem ullum ipsi sensibilem imprimit motum.

Eam ex ipsa Theoria respondere massæ directe, & quadrato distantiæ reciproce.

401. Ejus autem præcipuæ leges sunt, ut directe respondeat massæ, & reciproce quadratis distantiarum a singulis punctis massæ ipsius, quod in mea Theoria est admodum manifestum ita esse debere; nam ubi ventum est ad arcum illum meæ curvæ, qui gravitatem refert, vires omnes jam sunt attractivæ, & eandem illam ad sensum sequuntur legem, adeoque aliæ alias non elidunt contrariis directionibus, sed summa earum respondet ad sensum summæ punctorum; nisi quatenus ob inæqualem punctorum distantiam, & positionem, ad habendam accurate ipsam summam, ubi moles sunt aliquanto majores, opus erit illa reductione, qua Mechanici utuntur passim, & cujus ope inveniuntur leges, secundum quas punctum in data distantia, & positione situm respectu massæ habentis datam figuram, ab ipsa attrahitur; ubi, quemadmodum indicavimus num. 347, globus in globum ita gravitat, ut gravitaret, si totæ eorum massæ essent compenetratæ in eorum centris: at in aliis figuris longe aliæ leges obveniunt.

Commendatio Theoriæ ex conformitate omnium corporum in ea, & discrimine in totis aliis.

402. Verum hic illud maxime Theoriam commendat meam, quod num. 212 notandum dixi, quod videamus tantam hanc conformitatem in vi gravitatis in omnibus massis; licet eædem in ordine ad alia phænomena, quæ a minoribus distantiis pendent, tantum discrimen habeant, quantum habent diversa corpora in duritie, colore, sapore, odore, sono. Nam diversa combinatio punctorum materiæ inducit summas virium admodum diversas pro iis distantiis, in quibus adhuc curva virium contorquetur circa axem; proinde exigua mutatio distantiæ vires attractivas mutat in repulsivas, ac vice versa summis differentias substituit; dum in distantiis illis, in quibus gravitas servat quamproxime leges, quas diximus, curva ordinatas omnes ejusdem directionis habet, & vero etiam distantia parum mutata, fere easdem; quod necessario inducit tanta priorum casuum discrimina, & tantam in hoc postremo conformitatem.

Omnia fere a gravitate pendentia sunt communia huic Theoriæ cum communi: nonnullorum in ea facilior deductio.

403. Distinctio gravitatis (quæ est ut massa, in quam tenditur, directe, & quadratum distantiæ reciproce) a pondere (quod est præteræ ut massa, quæ gravitat) est mihi eadem, ac Newtonianis, & omnibus Mechanicis; & illa vim acceleratricem exhibet, hoc vim

& the distances from the Sun of the aphelia of the most remote comets; & so long as in this region it is not assumed to follow the law of the inverse squares exactly, but only very approximately to any desired degree of closeness, as I said in Art. 121. Now gravity of this kind is represented by the last arc of my curve in Fig. 1; & this, if gravity goes on indefinitely according to this same or any similar law, will be an asymptotic branch. Indeed, it may be, as I remarked in Art. 119, assumed that gravity is even accurately as the inverse square, & that it extends to all distances according to the same law, but that in addition there is some other force represented by another curve; then the law of forces of Fig. 1 is to be resolved into this force & into gravity reckoned as being exactly as the inverse square of the distance. This force, then, at those distances, for which gravity follows very approximately such a law, will be an insensible force; but at most other distances it would be very great. Where Fig. 1 gives repulsions, the force that is assumed to follow this other law would also have to be repulsive, & greater than the force, given by the law of the primitive curve of Fig. 1, by an amount equal to the supposed value of gravity at that place; & this must be cancelled by the addition of this repulsive force. However, this would depend upon our manner of assumption; & in this my own primitive & actual law, I consider that gravity is indeed universal & follows the law of the inverse squares of the distances, although not exactly, but very closely; I consider that it does not extend to all distances, but only to those I have set forth.

Gravity exists throughout the whole solar system, & it cannot possibly be explained by fluid pressure.

400. For the rest, that gravity exists universally throughout the whole planetary system, I think is thoroughly demonstrated by those arguments derived from Astronomy which are used by the disciples of Newton; these I do not repeat here, since they are set forth everywhere; I too have discussed them in several places, besides including them in *Adnotationes ad poema P. Noceti De Aurora Boreali*. But I consider that it is most evident that the approach to the Sun of the comets & primary planets, & that of the secondaries to the primaries, such as we see in the descent from the rectilinear tangent to the arc of the curve, & to a far greater degree other motions depending on mutual gravitation cannot possibly be due to fluid pressure. For, to omit other reasons truly numerous, the fluid that can avail so much in its action on spheres of this kind merely by its pressure, would certainly have a much greater effect upon their tangential velocities, by its opposition; these would in every case be bound to be diminished by such resistance, with a huge perturbation of areas; & the perversion of the whole of astronomical mechanics. Thus the fluid would either be bound to set up a huge resistance to the progress of a planet or a comet, or else it does not even by its pressure impress any sensible motion upon it.

Gravitation, according to my Theory, varies as the mass directly & as the square of the distance inversely.

401. Now, the principal laws of gravitation are that it varies directly as the mass & inversely as the square of the distances from each of the points of that mass; & in my Theory it is quite clear that this must be the case. For, as soon as we reach the arc of my curve that represents gravitation, all the forces are attractive, & to all intents obey the same law; & so some of them do not cancel others in opposite directions, but their sum approximately corresponds to the number of points. Except in so far as, on account of the inequality between the distances of the points, & their relative positions, there will be need, in order to obtain the sum of the forces accurately when the volumes are somewhat large, to make use of the reduction usually employed by mechanicians; by the aid of which are found the laws according to which a point situated at a given distance & in a given position from a mass of given shape is attracted by that mass. Here, as I indicated in Art. 347, one sphere gravitates towards another sphere in the manner that it would if the whole of their masses were for each condensed at their respective centres; whilst for other figures we meet with altogether different laws.

Support to be derived for my Theory from the conformity of all bodies in having gravitation, whilst there are so many differences in other properties.

402. But the greatest support for my Theory lies in a statement in Art. 212, which I said ought to be noticed; namely, in the fact that we see so much uniformity in all masses with regard to the force of gravity; in spite of the fact that these same masses, for the purpose of other phenomena depending on the smaller distances apart, have differences so great as those possessed by different bodies as regards hardness, colour, taste, smell & sound. For, a different combination of the points of matter induces totally different sums for those distances up to which the curve of forces still twists about the axis; where a very slight change in the distances changes attractive forces into repulsive, & substitutes, vice versa, differences for sums. Whereas, at those distances for which gravity obeys the laws we have stated very approximately, the curve has its ordinates all in the same direction &, even if the distance is slightly altered, practically unaltered in length. This of necessity produces a huge difference in the former case, & a very great uniformity in the latter.

Nearly everything depending on gravity in my Theory is in agreement with the usual theory: but the deduction of some of them is easier in mine.

403. The distinction between gravitation (which is proportional to the mass on which it acts, directly, & as the square of the distance, inversely) & weight (which is, in addition, proportional to the mass causing the gravitation) is just the same in my Theory as in that of Newton & all mechanicians. The former gives the accelerating force, the latter the motive

motricem, cum illa determinet vim puncti gravitantis cujusvis, a qua pendet celeritas massæ; [184] hoc summam virium ad omnia ejusmodi puncta pertinentium. Pariter communia mihi sunt, quæcunque pertinet ad gravium motus a Galilæo, & Hugenio definitos, nisi quod gravitatis resolutionem in descensu per plana inclinata, & in gravibus sustentatis per bina obliqua plana, vel obliqua fila, reducam ad compositionem juxta num. 284, & 286, & centrum oscillationis, una cum centro æquilibrii, & vecte, & libra, & machinarum principiis deducam e consideratione systematis trium massarum in se mutuo agentium, ac potissimum a simplici theoremate ad id pertinente, quæ fuse persecutus sum a num. 307. Communia pariter mihi sunt, quæcunque habentur in cælesti Newtoniana Mechanica jam ubique recepta de planetarum, & cometarum motibus, de perturbationibus motuum potissimum Jovis, & Saturni in distantiis minoribus a se invicem, de aberrationibus Lunæ, de maris æstu, de figura Telluris, de præcessione æquinoctiorum, & nutatione axis; quin immo ad hæc postrema problemata rite solvenda, multo tutior, & expeditior mihi panditur via, quæ me eo deducet post considerationem systematis massarum quatuor jacentium etiam non in eodem plano communi, & connexarum invicem per vires mutuas, uti ad centrum oscillationis etiam in latus in eodem plano, & ad centrum percussionis in eadem recta tam facile me deduxit consideratio systematis massarum trium.

Immobilitas fixarum quomodo a Newtonianis explicetur.

404. Illud mihi præterea non est commune, quod pertinet ad immobilitatem stellarum fixarum, quam contra generalem Newtoni gravitatem vulgo solent objicere, quæ nimirum debeant ea attractione mutua ad se invicem accedere, & in unicam demum coire massam. Respondent alii, Mundum in infinitum protendi, & proinde quamvis fixam æque in omnes partes trahi. Sed in actu existentibus infinitum absolutum, ego quidem censeo, haberi omnino non posse. Recurrent alii ad immensam distantiam, quæ non sinat motum in fixis oriundum a vi gravitatis, ne post immanem quidem sæculorum seriem sensu percipi. Ii in eo verum omnino affirmant; si enim concipiamus fixas Soli nostro æquales & similes, vel saltem rationem luminum, quæ emittunt, non multum discedere a ratione massarum; quoniam & vis ipsis massis proportionalis est, ac præterea tam vis, quam lumen decrescit in ratione reciproca duplicata distantiarum; erit vis gravitatis nostri solaris systematis in omnes stellas, ad vim gravitatis nostræ in Solem, quæ multis vicibus est minor, quam vis gravitatis nostrorum gravium in Terram, ut est tota lux, quæ provenit a fixis omnibus, ad lucem, quæ provenit a Sole, quæ ratio est eadem, ac ratio noctis ad diem in genere lucis. Quam exiguus motus inde consequi possit eo tempore, cujus temporis ad nos devenire potuit notitia, nemo non videt. Si fixæ omnes ad eandem etiam jaceant plagam, is motus omnino haberi posset pro nullo.

Difficultas residua sublata ab hac Theoria.

405. Adhuc tamen, quoniam nostra vita, & memoria respectu immensi fortasse subsecuturi ævi est itidem fere nihil; [185] si gravitas generalis in infinitum protendatur cum eadem illa lege, & eodem asymptotico crure, utique non solum hoc systema nostrum solare, sed universa corporea natura ita, paullatim utique, sed tamen perpetuo ab eo statu recederet, in quo est condita, & universa ad interitum necessario rueret, ac omnis materia deberet demum in unicam informem massam conglobari, cum fixarum gravitas in se invicem, nullo obliquo, & curvilineo motu elidatur. Id quidem haud ita se habere, demonstrari omnino non potest; adhuc tamen Divinæ Providentiæ videtur melius consulere Theoria, quæ ejus etiam ruinæ universalis evitandæ viam aperiat, ut aperit sane mea. Fieri enim potest, uti notavimus n. 170, ut postremus ille curvæ meæ arcus, qui exhibet gravitatem, posteaquam recesserit ad distantias majores, quam sint cometarum omnium ad nostrum solare systema pertinentium distantiæ maximæ a Sole, incipiat recedere plurimum ab hyperbola habente ordinatas reciprocas quadratorum distantiæ, ac iterum axem secet, & contorqueatur. Eo pacto posset totum aggregatum fixarum cum Sole esse unica particula ordinis superioris ad eas, quæ hoc ipsum systema componunt, & pertinere ad systema adhuc in immensum majus & fieri posset ut plurimi sint ejus generis ordines particularum ejusmodi etiam, ut ejusdem ordinis particulæ sint penitus a se invicem segregatæ sine ullo possibili commeatu ex una in aliam per asymptoticos arcus plures meæ curvæ juxta ea, quæ exposui a num. 171.

Cohæsio: explicatio per quietem, vel motus conspirantes.

406. Hoc pacto difficultas quæ a necessario fixarum accessu repetebatur contra Newtonianam Theoriam, in mea penitus evanescit ac simul a gravitate jam gradum fecimus ad cohæsionem, quam ex generalibus materiæ proprietatibus posueram postremo loco.

force; since the former gives the force of any gravitating point, upon which depends the velocity of the mass, & the latter the sum of all the forces pertaining to all such points. Similarly, the agreement is the same in my Theory with regard to anything relating to the motions of heavy bodies stated by Galileo & Huygens; except that, in descent along inclined planes, or bodies supported by two inclined planes or inclined strings, I substitute for their resolution of gravity the principle of composition, as in Art. 284, 286; & I deduce the centre of oscillation, as well as the centre of equilibrium, the lever, the balance & the principles of machines from a consideration of three masses acting mutually upon one another; & this more especially by means of a simple theorem depending on that consideration, which I investigated fully in Art. 307. The agreement is just as close in my Theory with regard to anything occurring in the celestial mechanics of Newton, now universally accepted, with regard to the motions of planets & comets, particularly the perturbations of the motions of Jupiter & Saturn when at less than the average distances from one another, the aberrations of the Moon, the flow of the tides, the figure of the Earth, the precession of the equinoxes, & the nutation of the axis. Finally, for the correct solution of these latter problems, a much safer & more expeditious path is opened to me, such as will lead me to it after an investigation of the system of four masses, not even lying in the same common plane, connected together by mutual forces; just as the consideration of a system of three masses led me with such ease to the centre of oscillation even to one side in the same plane, & to the centre of percussion in the same straight line.

The manner in which the immobility of the fixed stars was explained by Newton.

404. In addition to these, there is one thing in which I do not agree, namely, in that which relates to the immobility of the fixed stars; it is usually objected to the universal gravitation of Newton, that in accordance with it the fixed stars should by their mutual attraction approach one another, & in time all cohere into one mass. Others reply to this, that the universe is indefinitely extended, & therefore that any one fixed star is equally drawn in all directions. But in things that actually exist, I consider that it is totally impossible that there can be any absolute infinity. Others fall back on the immense distance, which they say will not permit the motion arising in the fixed stars from the force of gravity to be perceived by the senses, even after an immense number of ages. In this they assert nothing but the truth; for if we consider the fixed stars equal & similar to our sun, or at any rate the amounts of light that they emit, as not being far different from the ratio of their masses; then since also the force is proportional to the masses, & in addition both force & light decrease in the inverse ratio of the squares of the distances, it must be that the force of gravity of our solar system on all the stars is to the force of our gravity on the Sun, which latter is many times less than the force of gravity of our heavy bodies on the Earth, as the total light which comes from all the stars is to the light which comes from the Sun; & this ratio is the same as the ratio of night to day in respect of light. How slight is the motion that can arise from this in the time (the comparatively short time available for observation) nobody can fail to see. Even if all the fixed stars were situated in the same direction, the motion could be considered as absolutely nothing.

The remaining difficulty taken away in this Theory.

405. However, since our period of life & memory, in comparison with the immense number of ages perchance to follow, is almost as nothing, if universal gravitation extends indefinitely with the same law, & the same asymptotic branch, not only this solar system of ours indeed, but the universe of corporeal nature, would, little by little in truth, but still continuously, recede from the state in which it was established, & the universe would necessarily fall to destruction; all matter would in time be conglomerated into one shapeless mass, since the gravity of the fixed stars on one another will not be cancelled by any oblique or curvilinear motion. That this is not the case cannot be absolutely proved; & yet a Theory which opens up a possible way to avoid this universal ruin, in the way that my Theory does, would seem to be more in agreement with the idea of Divine Providence. For it may be that, as I remarked in Art. 170, the last arc of my curve, which represents gravity, after it has reached distances greater than the greatest distances from the Sun of all the comets that belong to our solar system, will depart very considerably from the hyperbola having its ordinates the reciprocals of the squares of the distances, & once more will cut the axis & be twined about it. In this way, it may be that the whole aggregate of the fixed stars, together with the Sun, is a single particle of an order higher than those of which the system is composed; & that it belongs to a system immensely greater still. It may even be the case that there are very many such orders of particles, of such a kind that particles of the same class are completely separated from one another without any possible means of getting from one to the other, owing to several asymptotic arcs to my curve, as I explained in Art. 171.

Cohesion; explanation by means of rest, or of motions in the same direction.

406. In this way, the difficulty, which has been repeatedly brought against the Newtonian theory on account of this necessary mutual approach of the fixed stars, disappears altogether in my Theory. At the same time, we have now passed on from gravity to cohesion, which

Cohæsionem explicuerunt aliqui per puram quietam ut Cartesiani alii per motus conspirantes, ut Joannes Bernoullius, ac Leibnitius, quam explicationem illustrarunt exemplo illius veli aquæ, quod in fontibus quibusdam cernimus, quod velum sit tantummodo ex conspirante motu guttularum tenuissimarum, & tamen si quis digito velit perrumpere, eo majorem resistentiam sentit, quo velocitas aquæ effluentis est major, ut idcirco multo adhuc major conspirantis motus velocitas videatur nostrorum cohæsionem corporum exhibere, quæ non nisi immani vi confringimus, ac in partes dividimus. Utraque explicandi ratio eodem redit, si quietis nomine intelligatur non utique absoluta quies, quæ translata Tellure a Cartesianis nequaquam admittebatur, sed respectiva: nam etiam conspirantes motus nihil sunt aliud, nisi quies respectiva illarum partium, quæ conspirant in motibus.

Illas exponere effectum, causam non exhibere.

407. At neutra eam explicat, quam cohæsionem reipsa dicimus, sed cohæsionis quendam velut effectum. Ea, quæ cohærent, utique respective quiescunt, sive motus conspirantes habent, & id quidem ipsum in hac mea Theoria accidit [186] itidem, in qua cum singula puncta materiæ suam pergant semper eandem continuam curvam describere, ea, quæ cohærent inter se, toto eo tempore, quo cohærent, arcus habent curvarum suarum inter se proximos, & in arcubus ipsis conspirantes motus. Sed in iis, quæ cohærent, id ipsum, quod motus ibi sint conspirantes, non est sine causa pendente a mutuis eorum viribus, quæ causa impediat separationem alterius ab altero, ac in ea ipsa causa stat discrimen cohærentium a contiguis. Si duo lapides in plano horizontali jaceant, utique habent motum conspirantem, quem circa Solem habet Tellus; sed si tertius lapis in alterutrum incurrit, vel ego ipsum submoveo manu, statim sine ulla vi mutua, quæ separationem impediat, dividuntur, & motus desinit esse conspirans. Hanc ipsam quærimus causam, dum in cohæsionem inquirimus. Nec velocitas motus, & exemplum veli aquæ rem conficit. Motus conspirans duorum lapidum contiguorum cum tota Tellure est utique multo velocior, quam motus particularum aquæ proveniens a gravitate in illo velo, & tamen sine ullo, difficultate separantur. In aqua experimur difficultatem perrumpendi velum, quia illa motus conspirans non est communis etiam nobis & Telluri, ut est motus illorum lapidum; unde fit, ut vis, quam pro separatione applicamus singulis particulis, perquam exiguo tempore possit agere, & ejus effectus citissime cesset, iis decidentibus, & supervenientibus semper novis particulis, quæ cum tota sua ingenti respectiva velocitate incurrunt in digitum. At in corporibus, in quibus partes cohærentes cernimus, eæ partes nullam habent velocitatem respectivam respectu nostri, nec aliæ succedunt aliis fugientibus. Quamobrem longe aliter in iis se res habet, & oportet invenire causam longe aliam, præter ipsum solum conspirantem motum, ut explicetur difficultas, quam experimur in iis separandis, & in inducendo motu non conspirante.

Explicatio petita a pressione fluidi; cur adhiberi non possit.

408. Sunt, qui adducant pressionem fluidi cujuspiam tenuissimi, uti pressio atmosphæræ extracto aere ex hemisphæriis etiam vacuis ipsorum separationem impedit vi respondente ponderi ipsius atmosphæræ, quæ vis cum in vulgaribus cohæsionibus, & vero etiam in hemisphæriis bene ad se invicem adductis, sit multis vicibus major, quam pondus atmosphæræ ipsius, quod se prodit in suspensione mercurii in barometris; aliud auxilio advocant tenuius fluidum. At inprimis ejus fluidi hypothesis precaria est; deinde huc illud redit, quod supra etiam monui, ubi de gravitatis causa egimus, quod nimirum meo quidem judicio explicari nullo modo possit, cur illud fluidum, quod sola pressione tantum potest, nihil omnino ad sensum possit incursu suo contra celerrimos planetarum, & cometarum motus. Accedit etiam, quod distractio & compressio fibrarum, quæ habetur ante fractionem solidorum corporum, ubi franguntur appenso inferne, vel superne imposito [187] pondere ingenti, non ita bene cum ea sententia conciliari posse videatur.

Explicatio Newtoni ab attractione in minimis distantiis: cur admitti non possit.

409. Newtonus adhibuit ad eam rem attractionem diversam ab attractione gravitatis, quanquam is quidem videtur eam repetere itidem a tenuissimo aliquo fluido comprimente; repetit certe sub finem Opticæ a spiritu quodam intimas corporum substantias penetrante, cujus spiritus nomine quid intellexerit, ego quidem nunquam satis assequi potui; cujus

I had put in the last place amongst the general properties of matter. Some have explained cohesion from the idea of absolute rest, for instance, the Cartesians; others, like Johann Bernoulli, & Leibniz, by means of equal motions in the same direction. They illustrate the explanation by means of the film of water, which we see in certain fountains; this film is formed merely from the equal motions in the same direction of the tiniest little drops, & yet, if anyone tries to break it with his finger, he feels a resistance that is the greater, the greater the velocity of the effluent water; so that from this illustration it would seem that a far greater velocity of equal motion in the same direction would account for the cohesion of the bodies around us, which we cannot fracture & divide up into parts unless we use a huge force. Either of these methods of explaining the matter reduces to the same thing, if by the term 'rest' we understand not only absolute rest which, since the Earth is in motion, has in no sense been admitted by the Cartesians, but also relative rest. For, equal motions in the same direction are nothing else but the relative rest of the parts that have equal motions in the same direction.

But these methods only explain the effect & not the cause of cohesion.

407. Neither of these ideas explains that which we call cohesion in a real sense, but only an effect of cohesion. Things which cohere are certainly relatively at rest; or they have equal motions in the same direction. This is exactly what happens in my Theory also; for, in it, since each point of matter always keeps on describing the same continuous curve which is peculiar to itself, those points that cohere to one another, during the whole of the time in which they cohere, have the arcs of their respective curves very near to one another, & the motions in those arcs equal & in the same direction. But in points that cohere, the fact that their motions are then equal & in the same direction is not without a cause; & this depends on their mutual forces, which prevent separation of one point from another; & in this cause is involved the difference between cohering & contiguous points. If two stones lie in the same horizontal plane, they will in all cases have equal motions in the same direction as the Earth has round the Sun; but if a third stone strikes against either of them, or if I move this third stone up to the others with my hand, immediately, without any mutual force preventing separation, the two are divided, & the equal motion in the same direction comes to an end. This cause of cohesion is just what we want to find, when we seek to investigate cohesion; & velocity of motion, or the example of the film of water will not effect the solution. The equal motions in the same direction as the whole Earth, possessed by the two contiguous stones, is certainly much greater than the motions of the particles of water produced by gravity in the film; & yet the two stones can be separated without any difficulty. In the water we encounter a difficulty in breaking the film, because the equal motion in the one direction is not common to us & the Earth, as the motion of the stones is. Hence it comes about that the force, which we apply to separate the several particles, can only act for an exceedingly small interval of time; & the effect of this force ceases very quickly, as those particles continually fall away & fresh ones come up; & these strike the finger with the whole of their relatively huge velocity. But, in bodies in which we perceive coherent parts, those parts have no relative velocity with regard to ourselves, nor as one part flies off does another take its place. Therefore the matter has to be explained in a totally different manner; & we must find a totally different cause to the idea of mere equality of motion in the same direction, in order to solve the difficulty that is experienced in separating the parts & inducing in them motions that are not equal & in the same direction.

Explanation sought from fluid pressure; why it is impossible that this should be the case.

408. There are some who bring forward the pressure of some fluid of very small density as an explanation. Just as the pressure of the atmosphere, when the air has been abstracted from a pair of hollow hemispheres, prevents them from being separated with a force corresponding to the weight of the atmosphere; &, since this force in ordinary cohesions, & indeed also in the case of two hemispheres that fit one another very well, becomes many times greater than the weight of the atmosphere, as shown in the suspension of mercury in the barometer, they invoke the aid of another fluid of less density. But, first of all, the hypothesis of such a fluid is uncertain; next, there here arises the same objection that I remarked upon above, when discussing the cause of gravity. Namely, that, in my opinion no manner of reason could be given as to why this fluid, which by its mere pressure could produce so great an effect, had as far as observation could discern absolutely no effect on the swiftest motions of planets & comets, owing to impact with them. Also there is this point in addition, that the extension & compression of fibres, which takes place before fracture in solid bodies, when they are broken by hanging a weight beneath or by setting a weight on top of them, does not seem to be in much conformity with this idea.

The reason why it is impossible to admit the explanation from attraction at very small distances, as given by Newton.

409. Newton derived an explanation of the matter from an attraction of a different kind to gravitation; although he indeed seems to seek to obtain this attraction from some compressing fluid of very small density. In fact, he seeks to obtain it, at the end of his *Optics*, from a 'spirit' permeating the inmost substances of bodies; but I never was able to grasp clearly what he intended by the term 'spirit'; & even he confessed that the

quidem agendi modum & sibi incognitum esse profitetur. Is posuit ejusmodi attractionem imminutis distantiis crescentem ita, ut in contactu sit admodum ingens, & ubi primigeniæ particulæ se in planis continuis, adeoque in punctis numero infinitis contingant, sit infinities major, quam ubi particulæ primigeniæ particulas primigenias in certis punctis numero finitis contingant, ac eo minor sit, quo pauciores contactus sunt respectu numeri particularum primigeniarum, quibus constant particulæ majores, quæ se contingunt, quorum contactuum numerus cum eo sit minor, quo altius ascenditur in ordine particularum a minoribus particulis compositarum, donec deveniatur ad hæc nostra corpora; inde ipse deducit, particulas ordinum altiorum minus itidem tenaces esse, & minime omnium hæc ipsa corpora, quæ malleis, & cuneis dividimus. At mihi positiva argumenta sunt contra vires attractivas crescentes in infinitum, ubi in infinitum decrescant distantiæ, de quibus mentionem feci num. 126; & ipsa meæ Theoriæ probatio evincit, in minimis distantiis vires repulsivas esse, non attractivas, ac omnem immediatum contactum excludit: quamobrem alibi ego quidem cohæsionis rationem invenio, quam mea mihi Theoria sponte propemodum subministrat.

Cohæsionem repetendam a limitibus virium.

410. Cohæsio mihi est igitur juxta num. 165 in iis virium limitibus, in quibus transitur a vi repulsiva in minoribus distantiis, ad attractivam in majoribus; & hæc quidem est cohæsio inter duo puncta, qua fit, ut repulsio diminutionem distantiæ impediat, attractio incrementum, & puncta ipsa distantiam, quam habent, tueantur. At pro punctis pluribus cohæsio haberi potest, tum ubi singula binaria punctorum sunt inter se in distantiis limitum cohæsionum, tum ubi vires oppositæ eliduntur, cujusmodi exemplum dedi num. 223.

Cohæsio duorum punctorum: limites cohæsionis posse esse quotcunque, utcunque fortes, quocunque ordine positos.

411. Porro quod ad ejusmodi cohæsionem pertinet, multa ibi sunt notatu digna. Inprimis ubi agitur de binis punctis, tot diversæ haberi possunt distantiæ cum cohæsione, quot exprimit numerus intersectionum curvæ virium cum axe unitate auctus, si forte sit impar, ac divisus per duo. Nam primus quidem limes, in quo curva ab arcu asymptotico illo primo, sive a repulsionibus impenetrabilitatem exhibentibus transit ad primum attractivum arcum, est limes cohæsionis, & deinde alterni intersectionum limites sunt non cohæsionis, & cohæ-[188]sionis, juxta num. 179; unde fit, ut si intersectionum se consequentium assumatur numerus par; dimidium sit limitum cohæsionis. Hinc quoniam in solutione problematis expositi num. 117 ostensum est, curvam simplicem illam meam habere posse quemcunque demum intersectionum numerum; poterit utique etiam pro duobus tantummodo punctis haberi quicunque numerus distantiarum differentium a se invicem cum cohæsione. Poterunt autem ejusmodi cohæsiones ipsæ esse diversissimæ a se invicem soliditatis, ac nexus, limitibus vel validissimis, vel languidissimis utcunque, prout nimirum ibi curva secuerit axem fere ad perpendiculum, & longissime abierit, vel potins ad illum inclinetur plurimum, & parum admodum discedat; nam in priore corum casuum vires repulsivæ imminutis, & attractivæ auctis utcunque parum distantiis, ingentes erunt; in posteriore plurimum immutatis, perquam exiguæ. Poterunt autem etiam e remotioribus limitibus aliqui esse multo languidiores, & alii multo validiores aliquibus e propioribus; ut idcirco cohæsionis vis nihil omnino pendeat a densitate, sed cohæsio possit in densioribus corporibus esse vel multo magis, vel multo minus valida, quam in rarioribus, & id in ratione quacunque.

In massis numerus limitum multo major: problema pro iis inveniendis quomodo solvendum.

412. Quæ de binis punctis sunt dicta, multo magis de massis continentibus plurima, puncta, dicenda sunt. In iis numerus limitum est adhuc major in immensum, & discrimen utique majus. Inventio omnium positionum pro dato punctorum numero, in quibus tota massa haberet limitem quendam virium, esset problema molestum, & calculus ad id solvendum necessarius in immensum excresceret, existente aliquo majore punctorum numero. Sed tamen data virium lege solvi utique posset. Satis esset assumere positiones omnium punctorum respectu cujusdam puncti in quadam arbitraria recta ad arbitrium collocati, & substitutis singulorum binariorum distantiis a se invicem in æquatione curvæ primæ pro abscissa, ac valoribus itidem assumptis pro viribus singulorum punctorum pro ordinatis, eruere totidem æquationes, tum reducere vires singulas singulorum punctorum ad tres datas directiones, & summam omnium candem directionem habentium in quovis puncto ponere $= 0$: orirentur æquationes, quæ paullatim eliminatis valoribus incognitis assumptis, demum ad æquationes perducerent definientes punctorum distantias necessarias ad æquilibrium, & respectivam quietem, quæ altissimæ essent, & plurimas

mode of action was unknown to him. He supposed that there was such an attraction, which, as the distances were diminished, increased in such a manner that at contact it was exceedingly great; & when the primary particles touched one another along continuous planes, & thus in an infinite number of points, this attraction became infinitely greater than when primary particles touched primary particles in a definite finite number of points; & the less the number of contacts compared with the number of primary particles forming the larger particles which touch one another, the less the attraction becomes; & since the number of these contacts becomes smaller the higher we go in the orders of particles formed from smaller particles, he deduces from this that particles of higher orders are also of less tenacity, & the least tenacious of all are those bodies that we can divide with mallet & wedge. But in my opinion there are positive arguments against attractive forces increasing indefinitely, when the distances decrease indefinitely, as I remarked in Art. 126; the very demonstration of my Theory gives convincing proof that the forces at very small distances are repulsive, not attractive, & ezcludes all immediate contact. So that I find the cause of cohesion from other sources; & my Theory supplies me with this cause almost spontaneously.

Cohesion is to be ascribed to the limit-points on the curve of forces.

410. Cohesion, then, in my opinion is, as I have said in Art. 165, to be ascribed to the limit-points on the curve of forces, where there is a passage from a repulsive force at a smaller distance to an attractive force at a greater distance; that is to say, this is the cause of cohesion between two points, for here a repulsion prevents decrease, & attraction increase, of distance; & so the points preserve the distance at which they are. Cohesion for more than two points can be obtained, both when each of the pairs of points is at a distance corresponding to a limit-point of cohesion, & also when the opposite forces cancel one another, an example of which I gave in Art. 223.

Cohesion of two points; limit-points of cohesion can be anything whatever as regards number, strength & order of occurrence.

411. Further, with regard to such cohesion, there are many points that are worthy of remark. First of all, in connection with two points, we can have as many different distances corresponding with cohesion as is represented by the number of intersections of the curve of forces with the axis (increased by one if perchance the number is odd) divided by two. For the first limit-point, at which the curve passes from the first asymptotic arc, i.e., from repulsions that represent impenetrability, to the first attractive arc, is a limit-point of cohesion; & after that the points of intersection are alternately limit-points of non-cohesion & cohesion, as was shown in Art. 179. Hence it comes about that, if the number of intersections following one after the other are assumed to be even, half of them are limit-points of cohesion. Hence, since, in the solution of the problem given in Art. 117, it was shown that that simple curve of mine could have any number of intersections, it will be possible for two points only to have any number of different distances from one another that would correspond to limit-points of cohesion. Moreover these cohesions could be of very different kinds, as regards solidity & connection, the limit-points being either very strong or very weak; that is to say, according as the curve at these points was nearly perpendicular to the axis & departed far from it, or on the other hand was much inclined to the perpendicular & only went away from the axis by a very small amount. For, in the first case, the repulsive forces on diminishing the distances, or the attractive forces on increasing the distances, ever so slightly, will be very great; in the second case, even when the distances are altered a good deal, the forces are very slight. Again also, it is possible that some of the more remote limit-points would be much weaker, & others much stronger, than some of the nearer limit-points. Thus, with me, the force of cohesion is altogether independent of density; the strength of cohesion, in denser bodies, can be either much greater or much less than that in less dense bodies, & the ratio can be anything whatever.

In masses the number of limit-points is much greater; how the problem of finding them is to be solved.

412. What has been said concerning two points applies also & in a far greater degree to masses made up of a large number of points. In masses, the number of limit-points is immensely greater still, & the difference between them is greater in every case. The finding of all the positions for a given number of points, at which the whole mass has a limit-point of forces, would be a troublesome undertaking; & the calculation necessary for its solution would increase immensely in proportion to the greater number of points taken. However, it can certainly be solved, if the law of forces is given. It would be sufficient to assume the positions of all the points with respect to any one point in any arbitrary straight line in any arbitrary way, & having substituted the distances for each pair from one another for the abscissa in the equation of the primary curve, & taking the values of the forces for each of the points as ordinates, to make out as many equations; then to resolve each of the forces into three chosen directions, & to put the sum of all those in the same direction for any point equal to zero. We shall thus obtain equations which, as the unknown assumed values are one by one eliminated, will finally lead to equations determining the distances of the points necessary for equilibrium, & relative rest; but these would be of very high

haberent radices; nam æquationes, quo altiores sunt, eo plures radices habere possunt, ac singulis radicibus singuli limites exhiberentur, vel singulæ positiones exhibentes vim nullam. Inter ejusmodi positiones illæ, in quibus repulsioni in minoribus distantiis habitæ, succederent attractiones in majoribus, exhiberent limites cohæsionis, qui adhuc essent quam plurimi, & inter se magis diversi, quam limites ad duo tantummodo pun-[189]cta pertinentes; cum in compositione plurium semper utique crescat multitudo, & diversitas casuum. Sed hæc innuisse sit satis.

Cur partes solidi fracti ad se invicem appressæ non acquirant cohæsionem priorem, ratio in Theoria Newtoniana.

413. Ubi confringitur massa aliqua, & dividitur in duas partes, quæ prius tenacissime inter se cohærebant, si iterum illæ partes adducantur ad se invicem; cohæsio prior non redit, utcunque apprimantur. Ejus rei ratio apud Newtonianos est, quod in illa divisione non æque divellantur simul omnes particulæ, ut textus remaneat idem, qui prius: sed prominentibus jam multis, harum in restitutione contactus impediat, ne ad contactum deveniant tam multæ particulæ, quam multæ prius se mutuo contingebant, & quam multis opus esset ad hoc, ut cohæsio fieret iterum satis firma: at ubi satis lævigatæ binæ superficies ad se invicem apprimantur, sentiri primo resistentiam ingentem dicunt, donec apprimuntur; sed ubi semel satis appressæ sint, cohærere multis vicibus majore vi, quam sit pondus aeris comprimentis; quia antequam deveniatur ad eos contactus, haberi debet repulsiva vis ingens, quam in majoribus distantiis, sed adhuc exiguis, agnovit Newtonus ipse, cui cum deinde succedat in minoribus vis attractiva, quæ in contactu evadat maxima, & in lævigato marmore satis multi contactus obtineantur simul; idcirco deinde satis validam cohæsionem consequi.

Ejusdem ratio in mea Theoria.

414. Quidquid ipsi de contactibus dicunt, id in mea Theoria dicitur æque de satis validis cohæsionis limitibus. In scabra superficie satis multæ prominentes particulæ progressæ ultra limites, in quibus ante sibi cohærebant, repulsionem habent ejusmodi, quæ impediat accessum reliquarum ad limites illos ipsos, in quibus fuerant ante divulsionem. Inde fit, ut ibi nimis paucæ simul reduci possint ad cohæsionem particulæ, dum in lævigatis corporibus adducuntur simul satis multæ. Ubi autem duo marmora, vel duo quæcunque satis solida corpora, bene complanata, & lævigata sola appressione cohæserunt invicem, illa quidem admodum facile divelluntur; si una superficies per alteram excurrat motu ipsis superficiebus parallelo; licet motu ad ipsas superficies perpendiculari usque adeo difficulter distrahi possint: quia particulæ eo motu parallelo delatæ, quæ adhuc sunt procul a marginibus partium congruentium, vires sentiunt hine, & inde a particulis lateralibus, a quibus fere æquidistant, fere æquales, adeoque sentitur resistentia earum attractionum tantummodo, quas in se invicem exercent marginales particulæ, dum augent distantias limitum: nam mihi citra limitem quenvis cohæsionis est repulsio, ultra vero attractio; licet ipsi deinde adhuc aliæ & attractiones, & repulsiones possint succedere. Ubi autem perpendiculariter distrahuntur, debet omnium simul limitum resistentia vinci.

Discrimen massæ primigeniæ, a binis frustis etiam lævigatis ad se invicem appressis.

415. Nec vero idem accidit, ubi marmora integra, & nunquam adhuc divisa, inter se cohærent; tum enim fibræ possunt esse multæ, quarum particulæ adhuc in minori-[190]bus distantiis, & multo validioribus limitibus inter se cohæreant, ad quos sensim devenerint aliæ post alias iis viribus, quibus marmor induruit, ad quos nunc iterum reduci nequeant omnes simul, dum marmora apprimuntur, quæ ulteriorum limitum minus adhuc validorum, sed validorum satis repulsivas vires simul sentiunt, ob quas non possunt denticuli, qui adhuc supersunt perquam exigui post quamvis lævigationem, in foveolas se immittere, & ad ulteriores limites validiores devenire; præterquam quod attritione, & lævigatione illa plurimarum particularum ordinis proximi massis nobis sensibilibus inducitur discrimen satis amplum inter massam solidam primigeniam, & binas massas complanatas, lævigatasque ad se invicem appressas.

Distractio, & compressio fibrarum ante fractionem hic commode ex ea.

416. Inde autem in mea Theoria satis commode explicatur & distractio, & compressio fibrarum ante fractionem; cum nimirum nihil apud me pendeat ab immediato contactu, sed a limitibus, quorum distantia mutatur vi utcunque exigua: sed si satis validi sint, ad

degree & would have very many roots. For, the higher the degree, the more the roots given by the equations; & for each of the roots there would be a corresponding limit-point, or a position representing zero force. Amongst such positions, those, in which we have repulsion at a less distance followed by attraction at a greater distance, would yield limit-points of cohesion; & these would be as great in number & as different from one another as were the limit-points pertaining to two points only; for in a composition of several things there certainly is always an increasing multitude & diversity of cases. But let it suffice that I have called attention to these matters.

The reason given in the Newtonian theory to account for the fact that the parts of a broken solid, when brought closely together, do not attain their former cohesion.

413. When a mass is broken, & divided into two parts, which originally cohered most tenaciously, if the parts are again brought into contact with one another, the previous cohesion does not return, however much they are pressed together. The reason of this, according to the followers of Newton, is that in the division all the particles are not equally torn apart simultaneously, leaving the texture the same as before; but as many of them now jut out beyond the rest, the contact between these in restitution prevents as many particles coming into contact as there were touching one another originally, which number is necessary for the purpose of again establishing a sufficiently strong cohesion. But when two surfaces that are sufficiently well polished are brought closely together, they say that at first there is felt a resistance of very great amount, until they are pressed into contact; but when once the surfaces are pressed together sufficiently closely, they cohere with a force that is many times greater than that due to the weight of the air pressing upon them. The reason they give is that, before actual contact is reached, there must be obtained a very great repulsive force, such as Newton himself recognized as existing at comparatively large, but actually very small, distances; & after that, there followed an attractive force at still smaller distances, which became exceedingly great when contact was reached. Thus, in polished marble, a sufficiently great number of contacts was obtained simultaneously; & in consequence a comparatively great cohesion was obtained.

The reason for the same thing according to my Theory.

414. All that the Newtonians say with regard to contacts applies in my Theory equally well with regard to sufficiently strong limit-points of cohesion. In a rough surface, a sufficient number of jutting particles, pushed out beyond the distances corresponding to those of the limit-points, at which they previously cohered, give rise to a repulsion of such sort as prevents the other particles from approaching to the distances of the limit-points, at which they were before being torn apart. Thus it comes about that in this case too few of the particles can be brought into a state of cohesion; whilst in the case of polished bodies we have a sufficient number of particles brought together simultaneously. Moreover, when two pieces of marble, or any two bodies of comparatively great solidity, after being well smoothed & polished, cohere when they are merely pressed together, they can be forced apart perfectly easily. If, for instance, one surface traverses the other with a motion parallel to the surfaces; although they can with difficulty be torn apart with a motion perpendicular to the surfaces. For, particles carried along by this parallel motion, such as are still far from the marginal surfaces of the parts in contact, feel the effects of forces on one side & on the other, due to laterally situated particles from which they are nearly equidistant, that are nearly equal to one another; & thus resistance is only experienced from the attractions which the particles in the marginal surfaces exert upon one another, whilst they increase the distances of the limit-points. The reason is that with me there is repulsion on the near side of any limit-point of cohesion, & attraction on the far side; although thereafter still other attractions & repulsions may follow. But when the bodies are drawn apart perpendicularly, the resistance due to every limit-point must be overcome simultaneously.

Distinction between the cases of a primitive mass & two pieces that have been broken off, polished & pressed together.

415. The same arguments do not apply to the case of whole pieces of marble that have not as yet been broken at any time, when they cohere. For, in that case, there may be many filaments, the particles of which hitherto have been cohering at less distances & in much stronger limit-points; these limit-points they would gradually reach one after the other with the forces that have given the marble its hardness; but they cannot be reduced to them once more all at once, whilst the pieces of marble are being pressed together. At the same time they feel the effect of the repulsive forces due to further limit-points still less strong, but yet fairly powerful; & on account of these, the little teeth which still are left, though very small, after any polishing, cannot insert themselves into the little hollows, & so reach the strong limit-points beyond. Besides, by this attrition & polishing of the greater number of the particles of an order next to such masses as are sensible to us there is induced a sufficiently wide distinction between a primitive solid mass & two masses that have been smoothed & polished & then pressed together.

Distension & compression of fibres before fracture have a good explanation from my Theory.

416. Hence also, in my Theory, we can give a fairly satisfactory explanation of the distension & compression of fibres that precedes fracture; for, with me, everything depends not on immediate contact, but on the limit-points, the distance of which is changed by

vincendam satis magno accessu omnem repulsionem, vel recessu attractionem, requiritur satis magna vis : quæ quidem repulsio, & attractio in aliis limitibus longe mihi alia est, tam si vis ipsa consideretur quam si consideretur spatii, per quod ea agit, magnitudo, quæ omnia pendent a forma, & amplitudine arcuum, quibus hinc, & inde circa axem contorquetur mea virium curva. Hinc in aliis corporibus ante fractionem compressiones, & distractiones esse possunt longe majores, vel minores, & longe major, vel minor vis requiri potest ad fractionem ipsam, quæ vis, ubi distantiis immutatis, superaverit maximam arcus ulterioris repulsivam vim in recessu, superatis multo magis reliquis omnibus posterioribus viribus repulsivis ope celeritatis quoque jam acquisitæ per ipsam vincentem vim, & per attractivas intermixtas vires, quæ ipsam juvant, defert particulas massam constituentes ad illas distantias, in quibus jam nulla vis habetur sensibilis, sed ad tenuissimum gravitatis arcum acceditur.

Hinc cur solida corpora nimio pondere pressa confringantur.

417. Hinc autem etiam illud in mea Theoria commodius accidit, quam in communi, quod in mea statim apparet, cur pila quæcunque utcunque solidi corporis post certa imposita pondera confringatur, & confringatur etiam solidus globus utrinque compressus ; cum multo magis appareat, quo pacto textus, & dispositio particularum necessaria ad summam virium satis validam mutari possit, ubi omnia puncta a se invicem distant in vacuo libero, quam ubi continuæ compactæ partes se contingant, nec ulla mihi est possibilis solida pila, quæ Mundum totum, si vi gravitatis in certam plagam feratur totus, sustineat, ut in sententia de continua extensione materiæ pila perfecte solida utcunque tenuis ad eam rem abunde sufficeret.

Communia esse huic theoriæ tum communi multa, quæ pertinent ad explorandam cohæsionis vim, & resistentiam ad fractionem in diversis positionibus.

418. Hisce omnibus jam accurate expositis, communia mihi sunt ea omnia, quæ pertinent ad methodos explorandi per [191] experimenta diversam diversorum corporum cohæsionis vim, quod argumentum diligenter, ut solet, excoluit Musschenbroekius, & comparandi resistentiam ad fractionem, ubi divisio fieri debeat divulsione perpendiculari ad superficies divellendas, ut ubi trabi verticali ingens pondus appenditur inferne, cum resistentia, quæ habetur, ubi circa latus suum aliquod gyrare debeat superficies, quæ divellitur, quod accidit, ubi extremæ parti trabis horizontalis pondus appenditur ; quam perquisitionem a Galileo inchoatam, sed sine ulla consideratione flexionis & compressionis fibrarum, quæ habetur in ima parte, alii plures excoluerunt post ipsum ; & in quibus omnibus discrimina inveniuntur quamplurima. Illud unum hic addam : posse cohæsionem ingentem acquiri ab iis, quæ per se nullum haberent, nova materia interposita, ut ubi cineres, qui oleis actione ignis avolantibus inter se inertes remanserunt, oleis novis in massam cohærentem rediguntur iterum, ac in aliis ejusmodi casibus ; sed id jam pendet a discrimine inter diversas particulas, & massas, ac pertinet ad soliditatem explicandam inprimis, non generaliter ad cohæsionem, de quibus jam agam gradu facto a generalibus corporum proprietatibus ad multiplicem varietatem Naturæ, & proprietates corporum particulares.

Discrimen inter particulas diversas, a numero punctorum, a mole, a densitate, a figura, quæ potest esse quævis, cum quavis vi ad eam retinendam.

419. Et primo quidem se hic mihi offert ingens illud plurium generum discrimen, quod haberi potest inter diversas punctorum congeries, quæ constituunt diversa genera particularum corpora constituentium. Primum discrimen, quod se objicit, repeti potest ab ipso numero punctorum constituentium particulam, qui potest esse sub eadem etiam mole admodum diversus. Deinde moles ipsa diversa itidem esse potest, ac diversa densitas, ut nimirum duæ particulæ nec massam habeant, nec molem, nec densitatem æqualem. Deinde data etiam & massa, & mole, adeoque data densitate media particulæ ; potest haberi ingens discrimen in ipsa figura, sive in superficie omnia includente puncta & eorum sequente ductum. Possunt enim in una particula disponi puncta in sphæram, in alia in pyramidem, vel quadratum, vel triangulare prisma. Sumatur figura quæcunque, & in eam disponantur puncta utcunque : tot erunt ibi distantiæ, quot erunt punctorum binaria, qui numerus utique finitus erit. Curva virium potest habere limites cohæsionis quotcunque, & ubicunque. Fieri igitur potest, ut limites iis ipsis distantiis respondeant, & tum eam ipsam formam habebit particula, & ejus formæ poterit esse admodum tenax. Quin immo per unicam etiam distantiam cum repagulo infinitæ resistentiæ, orto a binis asymptotis parallelis, & sibi proximis, cum area hinc attractiva, & inde repulsiva infinita,

any force, however small this force may be. If these are sufficiently strong, then, to overcome all repulsion by a sufficient great approach, or all attraction by a similar recession, there will be required a force that is sufficiently great for the purpose. This repulsion & attraction, with me, varies considerably for different limit-points, both when the force itself is considered, & when the magnitude of the space through which it acts is taken into account; & all of these things depend on the form & size of the arcs with which my curve of forces is twined round the axis, first on one side & then on the other. Hence, in different bodies, there may occur, before fracture takes place, compressions & distensions that are far greater or far less, & a force may be required for that fracture that is far greater or far less; & this force, when the distances are changed, having overcome the maximum repulsive force of the further arc as it recedes, would (all the rest of the repulsive forces due to the first arcs having been overcome all the more by the help of the velocity already acquired through the overcoming force, assisted by the attractive forces that come in between) carry off the particles forming the mass to those distances, at which there is no sensible force, but the arc of exceedingly small amplitude corresponding to gravity is reached.

Hence the reason why solid bodies will be broken under the pressure of too great a weight.

417. Hence, more easily in my Theory than in the common theory, because in mine it follows immediately, we have an explanation as to the reason why any pillar whatever, made of a solid body, is broken when certain weights are imposed upon it; & also why a solid sphere is crushed when compressed on both sides. For, it is much clearer how the texture & disposition of the particles, necessary to give such a comparatively great sum of forces, can be changed, if all the points lie apart from one another in a free vacuum, than if we suppose continuous compact parts that touch one another; nor can I imagine as possible any solid pillar that would sustain the whole Universe, if by the force of gravity the whole of it were borne in a given direction; & yet in the common idea of continuous extension of matter a pillar that was perfectly solid, of no matter what thinness, would be quite sufficient to do this.

There are many points of agreement between my Theory & the usual one, relating to the investigation of the forces of cohesion & resistance to fracture in different positions.

418. These matters having now been accurately explained, I proceed in the ordinary manner in all things that relate to methods of experimental investigation of the different force of cohesion in different bodies, a mode of demonstration that Mussenbroeck assiduously practised with his usual care; & methods of comparing the resistance to fracture in the case when division must take place by a fracture perpendicular to the surfaces to be broken, such as occur when a great weight is hung beneath a vertical beam, with the resistance that is obtained in the case when the surface has to rotate about one of its sides, which is torn off, as happens when a weight is hung at the end of a horizontal beam. This investigation, first started by Galileo, but without considering bending or the compression of the fibres that takes place on the under side of the beam, was carried on by several others after him; & in all cases of these there are very great differences to be found. I will here add but this one thing; it is possible for a very great cohesion to be acquired by things, which of themselves have no cohesion, by the interposition of fresh matter. For instance in the case of ashes, which, after the oily constituents have been driven off by the action of fire, remained inert of themselves; but, as soon as fresh oily constituents have been added, become once more a coherent mass; & in other cases of like nature. But this really depends on the distinction between different kinds of particles & masses, & refers to the explanation of solidity in particular, & not to cohesion in general. With such things I will now deal, passing on from general properties of bodies to the multiplicity & variety of Nature, & to particular properties of bodies.

Distinction between different kinds of particles arising from the number of points in them, their volume, their density, their shape; for the latter anything is possible, & any corresponding force can be had for the purpose of maintaining this shape.

419. The first thing that presents itself is the huge difference, of many kinds, which there can be amongst different groups of points such as form the different kinds of particles of which bodies are formed. The first difference that calls our attention can be derived from the number of points that form the particle; this number can be quite different within the same volume. Then the volume itself may be different, as also may the density; for, of course, two particles need not have either equal masses, equal volumes, or equal densities. Then, even if the mass & the volume be given, that is to say, the mean density of the particle is given, there may be a huge difference in shape, that is to say, in the surface enclosing all the points, & conforming with them. For, the points in one particle may be disposed in a sphere, in another in a pyramid, or a square or triangular prism. Take any such figure, & suppose the points are disposed in any particular manner whatever; then there will be as many distances as there are pairs of points, & their number will be finite in every case. The curve of forces can have any number of limit-points of cohesion, & these can occur anywhere along it. Therefore it must be the case that limit-points can be found to correspond to those distances, & on account of these the particle will have that particular form, & can be extremely tenacious in keeping that form. Indeed, through a single distance, with a restraint of infinite resistance, arising from a pair of parallel asymptotes close to one another, having the area on one side

potest haberi in quavis massa cujuscunque figuræ soliditas etiam infinita, sive vis, quæ impediret dispositionis mutationem non minorem data quacunque. Nam intra illam figuram [192] posset inscribi continuata series pyramidum juxta num. 363 habentium pro lateribus illas distantias nunquam mutandas magis, quam pro distantia binarum illarum asymptotorum, & positis punctis ad singulos angulos, haberetur massa punctorum, quorum nullum jaceret extra ejusmodi figuram, nec ullum adesset intra illam figuram, vel in ejus superficie spatii punctum, a quo ad distantiam minorem illa distantia data non haberetur punctum materiæ aliquod. Possent autem intra massam haberi hiatus ubicunque, & quotcunque prorsus vacui, inscriptis in solo residuo spatio pyramidibus illis, & in angulis quibusvis posset haberi quivis numerus punctorum distantium a se invicem minus, quam distent illæ binæ asymptoti, & quivis eorum numerus collocari posset inter latera, & facies pyramidum. Quare posset variari densitas ad libitum. Sed absque eo, quod singulis distantiis respondeant in curva primigenia singuli limites, vel singula asymptotorum binaria, vel ullæ sint ejusmodi asymptoti præter illam primam, innumera sunt sane figurarum genera, in quibus pro dato punctorum numero haberi potest æquilibrium, & cohæsionis limes per elisionem contrariarum virium, ex solutione problematis indicati num. 412. Hoc discrimen est maxime notatu dignum.

Discrimen in punctorum distributione per figuram eandem.

420. Data etiam figura potest adhuc in diversis particulis haberi discrimen maximum ob diversam distributionem punctorum ipsorum. Sic in eadem sphæra possunt puncta esse admodum inæqualiter distributa ita, ut etiam paribus distantiis ex altera parte sint plurima, ex altera paucissima, vel in diversis locis superficiei ejusdem concentricæ esse congeries plurimæ punctorum conglobatorum, in aliis eorum raritas ingens, & hæc ipsa loca possunt in diversis a centro distantiis jacere ad plagas admodum diversas in eadem etiam particula, & in eadem a centro distantia esse in diversis particulis admodum diversis modis distributa. Verum etiam si particulæ habeant eandem figuram, ut sphæricam, & in singulis circumquaque in eadem a centro distantia puncta æqualiter distributa sint; ingens adhuc discrimen esse poterit in densitate diversis a centro distantiis respondente. Possunt enim in altera esse fere omnia versus centrum, in altera versus medium, in altera versus superficiem extimam: & in hisce ipsis discrimina, tam quod pertinet ad loca densitatum earundem, quam quod pertinet ad rationem inter diversas densitates, possunt in infinitum variari.

Discrimen in vi, qua figuram conentur retinere: posse esse talem, ut nulla finita vi dissolvi possit.

421. Hæc omnia discrimina pertinent ad numerum, & distributionem punctorum in diversis particulis: sed ex iis oriuntur alia discrimina præcipua, quæ maximam corporum, & phænomenorum varietatem inducunt, quæ nimirum pertinent ad vires, quibus puncta particulam constituentia agunt inter se, vel quibus tota una particula agit in totam alteram. Possunt inprimis, & in tanta dispositionum varietate debent, [193] puncta constituentia eandem particulam habere vires cohæsionis admodum inter se diversas, ut aliæ multo facilius, aliæ multo difficilius dispositionem mutent mutatione, quæ aliquam non ita parvam rationem habeat ad totum. Est autem casus, in quo possint puncta particulæ cohærere inter se ita, ut nulla finita vi nexus dissolvi possit, ut ubi adsint asymptotici arcus in curva primitiva, juxta ea, quæ persecutus sum num. 362.

Particulæ aliæ se attrahentes, aliæ repellentes, aliæ inertes inter se.

422. Discrimina autem virium, quas una particula exercet in aliam, debent esse adhuc plura. Inprimis ex num. 222 patet, fieri posse, ut una particula constans etiam duobus punctis tertium punctum in iisdem distantiis collocatum ab earum medio attrahat per totum quoddam intervallum, vel repellat per idem intervallum totum, vel nec usquam in eo repellat, nec attrahat, conspirantibus in primo casu binis attractionibus, in secundo binis repulsionibus itidem conspirantibus, & in tertio attractione, & repulsione æqualibus se mutuo elidentibus. Multo autem magis summa virium totius cujusdam particulæ in aliam totam in eadem etiam distantia sitam, si medium utriusque spectetur, erit pro diversa dispositione punctorum admodum inter se diversa, ut nimirum in una attractiones prævaleant, in alia repulsiones, in alia vires oppositæ se mutuo elidant. Inde habebuntur, particulæ in se invicem agentes viribus admodum diversis, pro diversa sua constitutione & particulæ ad sensum inertes inter se, quæ quidem persecutus sum ipso num. 222.

attractive & on the other side repulsive, there can be obtained in any mass of any form whatever a solidity that is also infinite, or a force that would prevent any change of disposition of the particles equal to or greater than any given change. For within that form there could be inscribed a continued series of pyramids, after the manner of Art. 363, having for sides those distances which are never to be altered by more than that corresponding to the distance between the pair of asymptotes. If the points are placed one at each of the angles, there would be obtained a mass consisting of points no one of which would lie outside a figure of this sort; & no other point could get within that figure or occupy a point of space on its surface, from which there would not be some point of matter at a less distance than the given distance. Further, within the figure, there may be any kind & any number of gaps quite empty of points, the pyramids being described only in the remainder of the space; & at the angles there may be any number of points distant from one another less than the distance between the asymptotes; & there may be any number of them situated along the sides & faces of the pyramids. Hence, the density can be varied to any extent. But, apart from the fact that to each distance there corresponds a limit-point in the primary curve, or that there are pairs of asymptotes, or any other asymptotes of the sort except the first, there are really an innumerable number of kinds of figures, in which with a given number of points there can be equilibrium, & a limit-point of cohesion due to the cancelling of equal & opposite forces, as can be seen from the solution of the problem indicated in Art. 412. The following distinction is especially worth remark.

Difference in the distribution of points within the same figure.

420. Even if the figure is given, there can still be obtained a great difference between different particles on account of the different disposition of the points that form it. Thus, in the same sphere, the points may be quite unequally distributed, in such a way that, even at equal distances, there may be very many in one part & very few in another; or in different places on the same concentric surface there may be very many groups of points condensed together, whilst in others there are very few of them; these very places may be at quite different distances in different places even within the same particle, & in different particles at the same distance from the centre they may be distributed in ways that are altogether different. Further, even if particles have the same figure, say spherical, & in each of them, round about, & at the same distance from, the centre the points are distributed uniformly; yet even then there may be a huge difference in the density corresponding to different distances from the centre. For, in the one, they may be all grouped near the centre, in another towards the middle surface, & in a third close to the outer surface. In these the differences, both as regards the positions of equal density, & also as regards the ratio of the different densities, can be varied indefinitely.

Difference in the force with which particles try to conserve their figures; this may be such that the particle can be broken up by no finite force.

421. All such differences pertain to the number & distribution of points in the different particles. From them arise the principal differences that are left for consideration; these lead to the greatest variety in bodies & in phenomena. Such as those that relate to the forces with which the points forming a particle act upon one another, or the forces with which the whole of one particle acts upon the whole of another particle. First of all, the points forming the same particle may, & in such a great variety of distribution must, have forces of cohesion that are quite different one from the other; so that some of them much more easily, & others with much more difficulty, change this distribution with a change that bears a ratio to the whole that is not altogether small. There is also the case, in which the points of a particle can cohere so strongly together that the connection between them cannot be broken by any finite force; this happens when we have asymptotic arcs in the primary curve, as I showed in Art. 362.

Some particles attract, others repel one another, & some have no action on one another.

422. Moreover we may have still more differences between the forces which one particle exerts upon another particle. First of all, it is evident from Art. 222, that it may happen that a particle consisting of even two points may attract a third point situated at the same distances from the middle point of the distance between the two points throughout the whole of a certain interval of space, or they may repel it throughout the whole of the same interval, or neither repel or attract it anywhere; in the first case we have a pair of attractions that are equal & in the same direction, in the second case a pair of repulsions that are also equal & in the same direction, & in the third case an attraction & a repulsion that are equal to one another cancelling one another. Also, to a far greater degree, the sum of the forces for the whole of any particle upon the whole of another particle even when situated at this same distance, if the mean for each is considered, will be altogether different from one another for a different distribution of the points. Thus, in one particle attractions will prevail, in another repulsions, & in a third equal & opposite forces will cancel one another. Hence there will be particles acting upon one another with forces that are altogether different, according to the different constitutions of the particles; & there will be particles that are approximately without any action upon one another, such as I investigated also in the above-mentioned Art. 222.

Particulæ quæ in certis punctis se repellant, in aliis attrahant: quæ se urgeant in latus, quæ circumquaque eandem vim exerceant.

423. Aliud discrimen admodum notabile inter ejusmodi particularum vires est illud, quod eadem particula ex altera parte poterit datam aliam particulam attrahere, ex altera repellere; quin immo possunt esse loca quotcunque in superficie particulæ etiam sphæricæ, quæ alteram particulam in eadem a centro distantia sitam attrahant, quæ repellant, quæ nihil agant; cum nimirum in iis locis possint vel plura, vel pauciora esse puncta, quam in aliis, & ea ad diversas a centro, & a se invicem distantias collocata. Inde autem & illud fieri poterit, ut, quemadmodum in iis, quæ vidimus a num. 231, unum punctum a duorum aliorum altero attractum, ab altero repulsum, vi composita urgetur in latus, ita etiam una particula ab una alterius parte attracta, & repulsa ab altera in altera directione sita, urgeatur itidem in latus, & certam assecuta positionem respectu ipsius, ad eam tuendam determinetur, nec consistere possit, nisi in ea unica positione respectu ipsius, vel in quibusdam determinatis positionibus, ad quas trudatur ab aliis rejecta. Quod si particula sphærica sit, & in omnibus concentricis superficiebus puncta æqualiter distributa sint, ad distantias a se invicem perquam exiguas; tum ejus, & alterius ejus similis particulæ vires mutuæ dirigentur ad sensum ad earum centra, & fieri poterit, ut in quibusdam distantiis se repellant mutuo, in aliis se attrahant, quo casu habebitur quidem diffi-[194]cultas in avellenda altera ab altera, sed nulla difficultas habebitur in altera circa alteram circumducenda in gyrum, sicut si Terræ superficies horizontalis ubique sit, & egregie lævigata; globus ponderis cujuscunque posset quavis minima vi rotari per superficiem ipsam, elevari non posset sine vi, quæ totum ipsius pondus excedat.

Quo minores particulæ, eo difficilius dissolubiles.

424. In hac actione unius particulæ in aliam generaliter, quo particulæ ipsæ minorem habuerint molem, eo minus ceteris paribus perturbabitur earum respectiva positio ab alia particula in data quavis distantia sita: nam diversitas directionis & intensitatis, quam habent vires agentes in diversas ejus partes, quæ sola positionem turbare nititur, viribus æqualibus & parallelis nullam mutuæ positionis mutationem inducentibus, eo erit minor, quo distantiarum, & directionum discrimen minus erit: atque idcirco, quemadmodum jam exposni num. 239, inferiorum ordinum particulæ difficilius dissolvi possunt, quam partieulæ ordinum superiorum.

Discrimina inter particulas oriri ex punctorum vicinia; quanto magis debeant differre corpora, quæ ex iis constant.

425. Hæc quidem præcipue notatu digna mihi sunt visa inter particularum ex homogeneis etiam punctis compositarum discrimina, quæ tamen, quod ad vires pertinet, intra admodum exiguos distantiarum limites sistunt: nam pro majoribus distantiis particularum omnium vires sunt prorsus uniformes, uti ostensum jam est num. 212, nimirum attractivæ in ratione reciproca duplicata distantiarum ad sensum. Porro hinc illud admodum evidenter consequitur, massas majores ex adeo diversis particulis compositas, nimirum hæc ipsa nostra majora corpora, quæ sub sensum cadunt, debere esse adhuc multo magis diversa inter se in iis, quæ ad eorum nexum pertinent, & ad phænomena exhibita a viribus se extendentibus ad distantias illas exiguas, licet omnia in lege gravitatis generalis, quæ ad illas pertinet majores distantias, conformia sint penitus, quod etiam supra num. 402 notandum proposui. De hoc autem discrimine, & de particularibus diversorum corporum proprietatibus ad diversas pertinentium classes jam agere incipiam.

Quæ natura solidorum, & fluidorum; quid in solidis rigida, quid virgæ elasticæ: in fluidis duid viscosa, quid humida.

426. Prima se mihi offerunt solida, & fluida, quorum discrimina quæ sint, & quomodo a mea Theoria ortum ducant, est exponendum. Solida ita inter se connexa sunt, ut quemlibet aliquot particularum motum sequantur reliquæ: promotæ, si illæ promoventur: retractæ, si illæ retrahuntur: conversæ in latus, si linea, in qua ipsæ jacent, directionem mutet: & in eo soliditas est sita: porro ea dicuntur rigida; si ingenti etiam adhibita vi positio, quam habet recta ducta per duas quasvis particulas massæ, respectu rectæ, quæ jungit alias quascunque, mutari ad sensum non possit, sed ad inclinandam unam partem oporteat inclinare totam massam, & basim, & quanvis ejus reetam eodem angulo; nam in iis, quæ flexilia sunt, ut elasticæ virgæ, pars una directionem positionis mutat, & [195] inclinatur, altera priorem positionem servante: & priora illa franguntur, alia majore, alia minore vi adhibita; hæc posteriora se restituunt. Fluida autem passim non utique carent vi mutua inter particulas, immo pleraque exercent, & aliqua satis magnam, repulsivam vim, ut aer, qui ad expansionem semper tendit, aliqua attractivam, & vel non exiguam, ut aqua, vel etiam admodum ingentem, ut mercurius, quorum liquorum particulæ se in globum etiam conformant mutua particularum suarum attractione, & tamen separantur admodum facile a se invicem majores eorum massæ, ac aliquot partibus motus facile ita imprimitur: ut eodem tempore ad remotas satis sensibilis non protendatur; unde fit

423. There is another difference that is well worth while mentioning amongst forces of this sort, namely, that the same particle in one part may exert attraction on another particle, & repulsion from another part; indeed, there may be any number of places in the surface of even a spherical particle, which attract another particle placed at the same distance from the centre, whilst others repel, & others have no action at all. For, at these places there may be a greater or less number of points than in other places, & these may be situated at different distances from the centre & from one another. Thus, just as we saw for the cases considered in Art. 231, that it may happen that a point is attracted by one of two points & repelled by the other, & be urged to one side by the force that is the resultant of these two, so also one particle may be attracted by one part of another particle, & repelled by another part situated in another direction, & also be urged to one side; & having gained a certain position with respect to it, is inclined to preserve that position; nor can it stay in any position with regard to the other except the one, or perhaps in several definite positions, to which it is forced when driven out from others. But if the particle is spherical, & the points are equally distributed in all concentric surfaces, at very small distances from one another; then the mutual forces of it & another similar particle are directed approximately to their centres; & it may happen that at certain distances they repel one another, & at other distances attract one another; & in the latter case there will be some difficulty in tearing them apart, but none in making them rotate round one another. Just as, if the Earth's surface was everywhere horizontal, & perfectly smooth, a ball of any weight whatever could be made to rotate along that surface by using any very small force, whereas it could not be lifted except by using a force which exceeded its own weight.

Particles which at certain points repel & at others attract; some which urge one another to one side, & which exert the same force to produce rotation.

424. In general, in this action of one particle on another, the smaller volume the particles have, the less, other things being equal, is their relative position affected by another particle situated at any given distance from it. For the differences in the directions & intensities of the forces acting on different parts of it (which alone try to alter their positions, since equal & parallel forces induce no alteration of mutual position) will be the less, the less the difference in the distances & directions. Hence, just as I explained in Art. 239, particles of lower orders will be broken up with more difficulty than particles of higher orders.

The smaller the particle, the greater difficulty there is in breaking it up.

425. The things given above seemed to me to be those especially worthy of remark amongst the differences between particles formed from even homogeneous points, which yet remained, as far as forces are concerned, within certain very narrow limits. For, as regards greater distances, the forces of all the particles are quite uniform; that is to say, they are attractive forces varying approximately as the inverse square of the distances. Further, from them it follows perfectly clearly that greater masses, formed from these already composite particles of different sorts, that is to say, the bodies that lie about us of considerable size, such as come within the scope of our senses, must be still much more different from one another in matters that have to do with the ties between them, & with the phenomena exhibited by forces extending over very small distances; although all of them are quite uniform as regards the law of universal gravitation, which pertains to greater distances, a point to which I also called attention in Art. 402. But I will now start to consider this difference & the particular properties of different bodies belonging to different classes.

Differences arise from the nearness of points to one another; how much more should bodies formed from them differ from one another.

426. The first matters that offer themselves to me for explanation are the differences that exist between solids & fluids & how these arise according to my Theory. Solids are so connected together that the motion of any number of the particles is followed by the remaining particles; if the former move forward, so do the latter; if they are retracted, so are the rest; if a line in which they lie changes its direction, they are moved to one side; & in these facts solidity is defined. Further, solids are said to be rigid, if the position of a straight line drawn through any two particles of the mass cannot be sensibly changed with regard to the straight line joining any other pair of particles by using even a very large force; but in order to incline any one part of the mass it is necessary to incline the whole mass, the base, & any straight line in the mass at the same angle. For, in those that are flexible, such as elastic rods, one part may change the direction of its position & be inclined, whilst the rest maintains its original position. The first are broken by using in some cases a greater, & in others a less, force; whereas the latter recover their form. Now fluids in every case do not lack mutual force between their particles throughout; indeed very many of them exert, & some of them a fairly great, repulsive force, such as air, which always tends to expand; whilst others exert an attractive force, that is either not very small, as in the case of water, or may even be very great, as in the case of mercury. Of these liquids, the particles even form themselves into balls by the mutual attraction of the particles forming them; & yet larger masses of them are quite easily separated, & motion is easily given to any number of parts in such a manner that the motion does not

The nature of solids & fluids; what in solids are rigid, & what elastic rods; what in fluids are viscous, & what are watery.

ut fluida cedant vi cuicunque impressæ, ac cedendo facile moveantur, solida vero nonnisi tota simul moveri possint, & viribus impressis idcirco resistant magis : quæ autem resistunt quidem multum, sed non ita multum, ut solida, dicuntur viscosa. Ipsa vero fluida dicuntur humida, si solido admoto adhærescant, & sicca, si non adhæreant.

Unde fluiditas : tria fluidorum genera.

427. Hæc omnia phænomena præstari possunt per illa sola discrimina, quæ in diverso particularum textu consideravimus. Ut enim a fluiditate incipiamus, inprimis in ipsis fluidis omnes, particulæ in æquilibrio esse debent, dum quiescunt, & si nulla externa vi comprimantur, vel in certam dirigantur plagam ; id æquilibrium debebit haberi a solis mutuis actionibus : sed ejusmodi casum non habemus hic in nostris fluidis, quæ incumbentis massæ premuntur pondere, & aliqua, ut aer, etiam continentis vasis parietibus comprimuntur, in quibus idcirco omnibus aliqua haberi debet repulsiva vis inter particulas proximas, licet inter remotiores haberi possit attractio, ut jam constabit. Tria autem genera fluidorum considerari poterunt : illud, in quo in majoribus ejus massulis nulla se prodit mutua particularum vis : illud, in quo se prodit vis repulsiva : illud, in quo vis attractiva se prodit. Primi generis fere sunt pulveres, & arenulæ, ut illæ, ex quibus etiam horologia clepsydris veterum similia construuntur, & ad fluidorum naturam accedunt maxime, si satis lævigatam habeant superficiem, quod in quibusdam granulis cernimus, ut in milio : nam plerumque scabritiem habent aliquam & inæqualitates, quæ motum difficiliorem reddunt. Secundi generis sunt fluida elastica, ut aer : tertii vero generis liquores, ut aqua, & mercurius. Porro in primis ostensum est num. 222, & 422, posse binas particulas eodem etiam punctorum numero constantes, sed diverso modo dispositas, ita diversas habere virium summas in iisdem etiam centrorum distantiis, ut aliæ se attrahant, aliæ se repellant, aliæ nihil in se invicem agant. Quamobrem ejusmodi discrimina exhibet abunde Theoria. Verum multa in singulis diligenter notanda sunt ; nam ibi etiam, ubi nulla se prodit vis attractiva, habetur inter proximas particulas repulsio, ut innui paullo ante, & jam patebit.

Unde facilis motus in fluidis primi generis.

[196] 428. Porro in primo casu statim apparet, unde facilis ille habeatur motus. Quoniam, aucta distantia, nulla sensibili vi se attrahunt particulæ ; altera non sequetur motum alterius ; nisi ubi illa versus hanc promota ita accesserit, ut vi repulsiva mutua, quemadmodum in corporum collisionibus accidit, cogatur illi loco cedere, quæ cessio, si satis lævigatæ superficies fuerint, ut prominentes monticuli in exiguos hiatus ingressi motum non impediant, & sit locus aliquis, versus quem possint vel in gyrum actæ particulæ, vel elevatæ, vel per apertum foramen erumpentes, loco cedere ; facile fiet, nec alia requiretur vis ad eum motum, nisi quæ ad inertiæ vim vincendam requiritur, vel si graves particulæ sint versus externam massam, ut hic versus Tellurem, & fluidum motu impresso debeat ascendere, vis, quæ requiritur ad vincendam gravitatem ipsam : verum ad vincendam solam vim inertiæ, satis est quæcunque activa vis utcunque exigua, & ad vincendam gravitatem, in hoc fluidorum genere, si perfecta sit lævigatio ; satis est vis utcunque paullo major pondere massæ fluidæ ascendentis : quanquam nisi excessus fuerit major ; lentissimus erit motus ; ipsum autem pondus coget particulas ad se invicem accedere nonnihil, donec obtineatur vis repulsiva ipsum elidens, uti supra ostendimus num. 348 ; adeoque in statu æquilibrii se particulæ, in hoc etiam casu, repellent, sed erunt citra, & propre ejusmodi limites, ultra quos vis attractiva sit ad sensum nulla. Quod si figura particularum præterea fuerit sphærica, multo facilior habebitur motus in omnem plagam ob ipsam circumquaque uniformem figuram.

Eadem ratio, & in reliquis binis : discrimen inter ipsa.

429. In secundo, ac tertio genere motus itidem habebitur facilis, si particulæ sphæricæ sint, & paribus a centro distantiis homogeneæ, ut nimirum vires dirigantur ad centra. In ejusmodi enim particulis motus quidem unius particulæ circa aliam omni difficultate carebit, & vires mutuæ solum accessum vel recessum impedient. Hine impresso motu particulis aliquot, poterunt ipsæ moveri in gyrum aliæ circa alias, & alia succedere poterit loco ab alia relicto, quin partes remotiores motum ejusmodi sentiant : quanquam fere semper fortuita quædam particularum dispositio hiatus, qui necessario relinqui debent inter globos, & directio impressionis varia inducent etiam accessus & recessus aliquos, quibus fiet, ut

spread simultaneously in any sensible degree to parts further off. Hence it comes about that fluids yield to any impressed force whatever, &, in doing so, are easily moved; but solids cannot be moved except all together as a whole, & thus offer greater resistance to an impressed force. Those fluids which offer a considerable resistance, but one that is not so great as it is in the case of solids, are called viscous; again, fluids are said to be moist when they adhere to a solid that is moved away from them, & dry if they do not do so.

The origin of fluidity; three kinds of fluids.

427. All these phenomena can be presented by means of the single difference, which I have already considered in the different texture of particles. For, to begin with fluidity, we have first of all that in fluids all the particles must be in equilibrium, whilst they are at rest; &, if they are not under the action of an external force, or driven in a certain direction, that equilibrium must be due to the mutual actions alone. But we do not have this sort of case here, when considering the fluids about us, which are under the action of the weight of a superincumbent mass, & some of them, like air, are also acted upon by the walls of the vessel in which they are enclosed; hence, in all of these, there must be some repulsive force between the particles next to one another, although, as will now be evident, there may also be an attraction between more remote particles. Now, three kinds of fluids can be considered; one kind, in which, amongst its greater parts, no mutual force between its particles is shown; another kind, in which a repulsive force appears; & a third kind, in which there is an attractive force. Of the first kind are nearly all powders & sands, such as those, from which are constructed clocks similar to the clepsydras of the ancients; & these approximate very closely to the nature of fluids, if they have sufficiently polished surfaces, such as we see in some grains, like millet; for, the greater part of them have some roughness, & inequalities, which render motion more difficult. To the second class belong the elastic fluids, such as the air; & of the third kind are such liquids as water & mercury. Further, it has been shown particularly in Art. 222, 422, that it is possible for two particles, made up even of the same number of points, though differently distributed, to have the sums of the forces corresponding to them so different, even at the same distances from the centre, that some of them attract, some repel, & some have no action at all upon one another: hence, my Theory furnishes such differences in abundance. However, there are many things to be carefully noted in each case; for even when no attractive force is in evidence, there is a repulsive force between adjacent particles, as I mentioned just above; & this will be evident without saying anything further.

The source of the easy movement of particles of fluids of the first kind.

428. Moreover, in the first case it is at once apparent why there is easy movement of the particles. For, since when the distance is increased the particles do not attract one another with any sensible force, the one does not follow the motion of the other; except when the former moves towards the latter & approaches it to such an extent that, just as happens in the cases of impact of bodies, it is forced to give way to it by a mutual repulsive force; & this giving way would easily take place, if the surfaces were sufficiently smooth, so that the projecting hillocks of one did not hinder the motion by sticking into the tiny gaps of another; & if there were some place, to which the particles could be forced in a curved path, or elevated, or could break through an orifice opened to them, they might give way. This may easily happen; no other force would be required for the motion except that necessary to overcome the force of inertia; or, if heavy particles are attracted towards an external mass, as with us towards the Earth, & the fluid has to ascend, then no other force is required save that necessary to overcome gravity. But to overcome the force of inertia alone any active force, however small, is sufficient; & to overcome gravity, in this kind of fluids, if there is perfect smoothness, any force that is a little greater than the weight of the ascending part of the fluid will suffice; although, unless the excess were considerable, the motion would be very slow. Moreover, the weight of the fluid will force the particles somewhat closer together, until a mutual repulsive force is produced which will cancel it, as I showed above in Art. 348. Thus, when in a state of equilibrium the particles, even in this case, will repel one another; but they will lie on the near side of, & close to such limit-points as have the attractive force on the far side of them practically zero. But if, in addition the shape of the particles should be spherical, there would be much easier movement in all directions due to the uniformity of shape all round.

The same argument also holds good for the other two kinds; differences between them.

429. In the second & third classes of fluids there is also easy movement, if the particles are spherical, & homogeneous at equal distances from their centres, that is to say, so that the forces are directed towards their centres. For, in the case of such particles, the motion of one particle round another lacks difficulty of any sort, & the mutual forces prevent approach or recession only. Hence, if a motion be impressed on any number of particles, they could move in curved paths round one another, & some could take the place left free by others, without the parts further off feeling the effects of such motion; although nearly always the accidental arrangement of the gaps empty of particles, which must of necessity be left between the spheres, & the varied direction of the pressures will lead also to approach

motus ad remotiores etiam particulas deveniat, sed eo minor, quo major fuerit earum distantia. Verum hic notandum erit discrimen ingenis inter duos casus, in quibus partes fluidi se repellunt, & casus, in quibus se attrahunt.

In elasticis fluidis particulas esse extra limites sub arcubus repulsivis latis.

430. In primo casu particulæ proximæ debebunt se omnino repellere, & vis ex parte altera elidet vim ex altera; sed si repente relinquatur libertas ex parte quavis, sine ulla externa vi, sed sola illa particularum actione mutua, recedent reipsa particulæ a se invicem, & fluidum dilatabitur; quin [197] immo externa vi opus est, ad continendam in eo statu massam ejusmodi, uti aerem gravitas superioris atmosphæræ continet, vel in vase occluso vasis ipsius parietes; & aucta illa externa vi comprimente augeri poterit compressio, imminuta imminui. Particulæ illæ inter se non crunt in limitibus quibusdam cohæsionis, sed crunt sub repulsivo arcu curvæ exprimentis vires compositas particularum ipsarum.

In fluidis humidis limitem validum cohæsionis fore proximum, & si abeat in vapores debere haberi prope validissimum arcum repulsivum.

431. At in tertio genere particulæ quidem proximæ se mutuo repellent, repulsione æquali illi vi, quæ necessaria est ad elidendam vim externam, & ad elidendam pressionem, quæ oritur a remotiorum attractionibus: verum si fluidum est parum admodum compressibile, vel etiam nihil ad sensum, ut aqua; debent esse citra, & admodum prope limitem, ultra quem vel immediate, vel potius, si id fluidum neque distrahitur (ut nimirum durante sua forma nequeat acquirere spatium multo majus, quod itidem in aqua accidit) habeat post limites alios satis inter se proximos arcum attractivum ad distantias aliquanto majores protensum, a quo attractio illa prodeat, quæ se in ejusmodi fluidorum massulis prodit; licet si iterum id fluidum majore vi abire possit in elasticos vapores, ut ipsa aqua post eum attractivum arcum; arcus repulsivus debeat succedere satis amplus, juxta ea, quæ diximus num. 195.

Motus non obstante vi mutua facilis, quod ad motum aliquot particularum non debeant moveri remotæ simul ut in solidis. Exemplum in quadam hypothesi globorum gravium.

432. In hoc fluidi genere illud mirum videri potest, quod illa attractiva vis, quæ in majoribus succedit distantiis, & ille validus cohæsionis limes, qui & compressionem & rarefactionem impedit, non impediat divisionem massæ, & separationem unius partis massæ ab alia. At quomodo id facile fieri ibi possit, & non possit in solidis, patchit hoc exemplo. Concipiatur Terræ superficies sphærica accurate, & bene lævigata, ac gravitas sit ejusmodi, ut in distantia perquam exigua fiat jam insensibilis, ut vis magnetica in exigua distantia sensum jam effugit. Sint autem globi multi itidem læves mutua attractiva vi præditi, quæ vim in totam Terram superet. Si quis unum ejusmodi globum apprehendat, & attollat; secundus ipsi adhærebit relicta Terra, & post ipsum ascendet, reliquis per superficiem Terræ progredientibus, donec alii post alios eleventur, vi in globum jam elevatum superante vim in Terram. Is, qui primum manu teneret globum, sentiret, & deberet vincere vim unius tantummodo globi in Terram, quem separat, cum nulla sit difficultas in progressu reliquorum per superficiem Terræ, quo distantia non augetur, & globorum jam altiorum vis in Terram ponatur insensibilis. Vinceret igitur aliorum vim post vim aliorum, & vis ab eo adhibita major tantummodo vi globi unici requireretur ad rem præstandam. At si illi globi deberent elevari simul, ut si simul omnes colligati essent per virgas rigidas; deberent utique omnes illæ vires omnium in Terram simul superari, & requireretur vis major omnibus simul. Res eodem redit, ac ubi fasciculus virgarum [198] debeat totus frangi simul, vel potins debeant aliæ post alias frangi virgæ.

Applicatio exempli ad fluida, & solida: successiva particularum separatio in fluidis.

433. Id ipsum est discrimen inter fluida hujus generis, & solida. In his motus particularum circa particulas liber ob earum uniformitatem permittit, ut separentur aliæ post alias; dum in solidis vis in latus, de qua egimus jam in pluribus locis, & anguli prominentes, ac figurarum irregularitas, impediunt ejusmodi liberum motum, qui fiat sine mutatione distantiarum, & cogunt divulsionem plurimarum particularum simul: unde oritur difficultas illa ingens dividendi a se invicem particulas solidas, quæ in divisione fluidorum est adeo tenuis, ac ad sensum nulla.

& recession of some kind ; & through these it will come about that the effect of the motion will reach the particles further off, although this will be the less, the greater the distance they are away. But here we have to notice the great difference between the two cases, the one, in which the parts of the fluid repel, & the other, in which they attract, one another.

In elastic fluids the particles are outside the limit-points, & under wide repulsive arcs.

430. In the first case adjacent particles must repel one another, in every instance, & the force from one part must cancel the force from another part. Moreover, if all at once freedom of movement is left in any one part, without any external force to prevent it, then by the mutual action of the particles alone, these particles will of themselves recede from one another & the fluid will expand. Indeed, what is more, there is need of an external force to maintain a mass of this kind in its original state, just as the gravity of the upper atmosphere constrains the air, or the walls of a vessel the air contained within it. When this compressing external force is increased the compression can be increased, & if diminished diminished. The particles themselves will not be at distances from one another corresponding to limit-points of cohesion of any sort ; but these will correspond to a repulsive arc of the curve that represents the resultant forces of the particles.

In watery fluids the nearest limit-point must be a very strong one, of cohesion ; & if the fluid goes off as a vapour, there must be close to it a very strong repulsive arc.

431. Again, in the third kind, adjacent particles must indeed repel one another, the repulsion being equal to that force that is necessary to cancel the external force, & also the pressure which arises from the attractions of points further off. But, if the fluid is only very slightly compressible, or not to any appreciable extent (like water, for example), then the particles must be on the near side, & quite close to, a limit-point ; & on the far side of this limit-point, either there must follow immediately a comparatively ample attractive arc ; or, more strictly speaking, if the fluid does not expand (that is to say, whilst it maintains its form, it cannot acquire much more space, which is also the case with water), then it has, after several other limit-points fairly close to one another, an attractive arc extending to somewhat greater distances, to which is due that attraction which is seen in small globules of fluids ; but if, with a greater force applied, the fluid can after that go off to still further distances in the form of elastic vapours (as water does), then, after the attractive arc we must have the above-mentioned comparatively ample repulsive arc ; as was shown in Art. 195.

Mutual force not hindering, movement is easy, because Particles further away need not move at the same time, when any number of particles are moved, as is the case for solids. Example in the hypothesis of heavy spheres.

432. In this kind of fluid it may appear strange that the attractive force which follows at greater distances, or the strong limit-point of cohesion, which prevents both compression & rarefaction, does not, either of them, prevent division of the mass or the separation of one part of it from the other. But the reason why this can take place here, & not in the case of solids, will become evident on considering the following example. Suppose the surface of the Earth to be perfectly spherical, & quite smooth ; & suppose gravity to be such, that when the distance becomes very small it becomes insensible, just as magnetic force practically vanishes at a very small distance. Then, suppose we have a number of smooth spheres endowed with an attractive force for one another, which exceeds the force each has for the whole Earth. If one of these spheres is taken & lifted, a second one will adhere to it & leave the ground, & ascend after it ; the rest will move along the surface of the Earth, until one after the other they are also lifted up, the attraction towards the sphere just lifted exceeding the attraction towards the Earth. The person, who took hold of the first sphere, would feel & would have to overcome the force of only the one sphere towards the Earth, namely, that of the one he takes away ; for there is no difficulty about the progress of the rest of the spheres along the surface of the Earth, supposing that the distance is not increased, & assuming that the force towards the Earth of spheres already lifted is quite insensible. Hence the force of one after that of another would be overcome, & the whole business would be accomplished by his using a force that was just greater than the force due to a single sphere. But if all the spheres had to be raised at once, as if they were all bound together by rigid rods, it would be necessary to overcome at one time all the forces of all the spheres upon the Earth, & there would be required a force greater than all these put together. It is just the same sort of thing as when a whole bundle of rods has to be broken at the same time, or rather the rods have to be broken one after another.

Application of the example to the case of fluids & solids ; successive separation of the particles in the case of fluids.

433. This is exactly what causes the difference between fluids of this kind & solids. With the former, the free motion of the particles about one another, due to their uniformity, allows them to be separated one after the other. Whilst, with solids, lateral force, with which we have already dealt in several places, projecting angles & irregularities of shape, prevent such freedom of motion, as (with fluids) takes place without any change in the mutual distances ; & they compel us to tear away a very great number of particles all at once. This is the cause of the very great difficulty in the way of dividing the particles of solids from one another ; & is the reason why the difficulty is very slight, or practically nothing, when dividing fluids.

Exemplum ipsius in aqua : resistentiam in fluidis ad separationem fieri eandem, ac in solidis, si velocitas debeat esse ingens.

434. Successivam hujusmodi separationem particularum aliarum post alias videmus utique in ipsis aquæ guttis pendentibus, quæ ubi ita excreverunt : ut pondus totius guttæ superet vim attractivam mutuam partium ipsius ; non divellitur tota simul ingens ejus aliqua massa, sed a superiore parte, utut brevissimo tempore, attenuatur per gradus ;. donec illud veluti filum jam tenuissimum penitus superetur. Fuerunt prius mille particulæ in superficie, quæ guttam pendentem connectebant cum superiore parte aquæ, quæ relinquitur adhærens corpori, ex quo pendebat gutta, fiunt paullo post ibi 900, 800, 700 : & ita porro imminuto carum numero per gradus, dum laterales accedunt ad se invicem, & attenuatur figura : quarum idcirco resistentia facile vincitur, ut ubi in illo virgarum fasciculo frangantur aliæ post alias. At ubi celerrimo motu in fluidum ejusmodi incurritur ita ; ut non possint tam brevi tempore aliæ aliis particulæ locum dare, & in gyrum agi ; tum vero fluida resistunt, ut solida. Id experimur in globis tormentariis, qui ex aqua resiliunt, in eam satis oblique projecti, ut manente satis magna horizontali velocitate collisio in perpendiculari fiat more solidorum : ac candem quoque resistentiam in aqua scindenda experiuntur, qui se ex editiore loco in eam demittunt.

Soliditatis causa in vi, & motu in latus : exemplum in parallelepipedis.

435. Hinc autem pronum est videre, unde soliditatis phænomena ottum ducant. Nimirum ubi particularum figura recedit plurimum a sphærica, vel distributio punctorum intra particulam inæqualis est, ibi nec habetur libertas illa motus circularis, & omnia, quæ ad soliditatem pertinent, consequi debent ex vi in latus. Cum enim una particula respectu alterius non distantiam tantummodo, sed & positionem servare debeat ; non solum, ea promota, vel retracta, alteram quoque promoveri, vel retrahi necesse est ; sed præterea, ea circa axem quencunque conversa, oportet & illam aliam loco cedere, ac eo abire, ubi positionem priorem respectivam acquirat ; quod cum & tertia respectu secundæ præstare debeat, & omnes reliquæ circunquaque circa illam positæ ; patet utique, non posse motum in eo casu imprimi parti cuipiam systematis ; quin & totius systematis motus consequatur respectivam po-[199]-sitionem servantis, quæ est ipsa superius indicata solidorum natura. Res autem multo adhuc magis manifesta fit, ubi figura multum abludat a sphærica, ut si sint bina parallelepipeda inter se constituta in quodam cohæsionis limite, alterum ex adverso alterius. Alterum ex iis moveri non poterit, nisi vel utrinque a lateribus accedat ad alterum, vel utrinque recedat, vel ex altero latere accedat, & recedat ex altero. In primo casu imminuta distantia habetur repulsiva vis, & illud alterum progreditur : in secundo, eadem aucta, habetur attractio, & illud secundum ad prioris motum consequitur ; in tertio casu, qui haberi non potest, nisi per inclinationem prioris parallelepipedi, altero latere attracto, & altero repulso inclinari necesse est etiam secundum ; quo pacto si ejusmodi parallelepipedorum sit series quædam continua, quæ fibram longiorem, vel virgam constituat ; inclinata basi, inclinatur illico series tota : & si ex ejusmodi particulis massa constet ; tota moveri debet ac inclinari, inclinato latere quocunque.

Idem in figuris omnibus : und discrimen inte flexilia, & rigida.

436. Quod de parallelepipedis est dictum, id ipsum ad figuras quascunque transferri potest inæquales utcunque, quæ ex altero latere possint accedere ad aliam particulam, ex altero recedere : habebitur semper motus in latus, & habebuntur soliditatis phænomena, nisi paribus a centro distantiis homogeneæ, & sphæricæ formæ particulæ sint. Verum ingens in eo motu discrimen crit inter diversa corpora. Si nimirum vires illæ hinc, & inde a limite, in quo particulæ constitutæ sunt, sint admodum validæ ; motus in latus fiet celerrime, & nulla flexio in virga, aut in massa apparebit ; quanquam crit utique semper aliqua. Si minores sint vires ; longiore tempore opus erit ad motum, & ad positionem debitam acquirendam, quo casu, inclinata parte ima virgæ, nondum pars summa obtinere potest positionem jacentem in directum cum ipsa, adeoque habebitur inflexio, quæ quidem eo erit major, quo major fuerit celeritas conversionis ipsius virgæ, uti omnino per experimenta deprehendimus.

Discrimen inter flexilia, & fragilia unde.

437. Nec vero minus facile intelligitur illud, quid intersit inter flexilia solida corpora, & fragilia. Si nimirum vires hinc, & inde ab illo limite, in quo sunt particulæ, extenduntur ad satis magnas distantias eædem, arcu utroque habente amplitudinem non ita exiguam ;

Example of this in the case of water; the resistance to separation in fluids becomes as great as that in solids, if the velocity has to be very great.

434. We certainly see an example of this kind of successive separation of particles, one after the other, in the case of drops of water hanging suspended; here, as soon as they have increased up to a point where the weight of the whole drop becomes greater than the mutual attractive force of its parts, any great part is not torn away as a whole; but by degrees, though in a time that is exceedingly short, the drop is attenuated at its upper part, until the neck, which has by now become exceedingly narrow, is finally broken altogether. There were, say, initially, a thousand particles in the surface connecting the hanging drop to the upper part of the water which is left adhering to the body from which the drop was suspended; these a little afterwards became 900, then 800, then 700, & so on, their number being gradually diminished as the sides of the neck approach one another, & its figure is narrowed. Hence, their resistance is easily overcome, just as when, in the bundle of rods, the rods are broken one after the other. But, when it is a case of an onset with high speed, so that the time is too short to allow the particles to give way one after the other, & move in curved paths round one another; then, indeed, fluids resist in just the same way as solids. This is to be observed in the case of cannon-balls, which rebound from the surface of water, when projected at sufficiently small inclination to it; so that, whilst the horizontal velocity remains sufficiently great, the vertical impact takes place in the manner of that between solids. Also, those who dive into water from a fairly great height will experience the same resistance in cleaving the surface.

The cause of solidity lies in lateral force & motion; example of this in parallelepipeds.

435. Further, from what has been said, it can be seen without difficulty whence the phenomena of solidity derive their origin. For instance, when the shape of the particles is very far from being spherical, or the distribution of the points within the particle is not uniform, then there is not that freedom of circular motion; & all things that pertain to solidity must follow from the presence of lateral force. For, since one particle must preserve not only its distance, but also its position with regard to another; not only, when the one is driven forwards or backwards, must the other also be driven forwards or backwards, but also if the one is turned about any axis, it is necessary that the other should give way & move off to the place in which it will acquire its original relative position. Since also the third must do the same thing with respect to the second, & all the rest of the particles round it in all directions, it is quite clear that in this case motion cannot be imparted to any part of the system, without a motion of the whole system following it, in which the mutual position is preserved; & this is the very nature of solids that was mentioned above. Moreover, the matter becomes even still more evident, when the shape differs considerably from the spherical; for instance, if we have a pair of parallelepipeds situated with regard to one another at a distance corresponding to a limit-point of cohesion, opposite one another. It will not be possible for one of them to be moved, unless either it approaches the other laterally at both ends, or recedes at both ends, or else approaches at one end & recedes at the other. In the first case, the distance being diminished, we have a repulsive force, & the second particle will move away; in the second case, the distance being increased, there will be an attraction, & the second particle will follow the motion of the first. In the third case, which cannot take place unless there is an inclination of the first parallelepiped, one end of the second being attracted, & the other repelled, it is necessary that the second particle should also be inclined. In this way, if there is a continuous series of such parallelepipeds, forming a fairly long fibre or rod, then, when the base is inclined, the whole rod must be inclined along with it; & if a mass is formed from such particles, then if any side of the mass is inclined, the whole of the mass must move along with it & be also inclined.

The same thing for all shapes; hence the difference between flexible & rigid bodies.

436. What has been said with regard to parallelepipeds can be said also about any figures whatever which are at all irregular, if they can approach another particle at one side & recede from it on the other side; there will in every case be motion to one side, & the phenomena of solidity will be obtained, unless the particles are homogeneous at equal distances from the centre & spherical in form. But in this motion there is a very great difference among different bodies. If, for instance, the forces on either side of the limit-point, in which the particles are situated, are quite strong, the lateral motion will be very swift, & no bending will be observed in the rod or in the mass; although there certainly will be some taking place. If the forces are not so great, there will be need of a longer time for it to acquire motion & the proper position; & in this case, if the bottom part of the rod is inclined, the top part of the rod cannot for a little while attain to a position lying in a straight line with the base, & thus there will be bending; & this indeed will be all the greater, the greater the speed with which the rod is turned; as is proved by experiment to be always the case.

The reason of the difference between flexible & fragile bodies.

437. Nor will it be less easy to understand the reason why there is a difference between flexible solids & fragile bodies. For instance, if the forces on each side of the limit-point, at which the particles are, are extended unaltered over sufficiently great distances from it, & the

tum vero, vi externa adhibita utrique extremo, vel majore velocitate impressa alteri, incurvabitur virga, atque inflectetur, sed sibi relicta ad positionem abibit suam, & in illo inflexionis violento statu vim exercebit perpetuam ad regressum, quod in elasticis virgis accidit. Si vires illæ non diu durent hinc, & inde eædem, vel per satis magnum intervallum sit ingens frequentia limitum; tum quidem inflexio habebitur sine conatu ad se restituendam, & sine fractione, tam vi adhibita utrique extremo, quam ingenti velocitate impressa alteri, ut videmus accidere in maxime ductilibus, [200] velut in plumbo. Si demum vires hinc, & inde per exiguum intervallum durent, post quod nulla sit actio, vel ingens repulsivus arcus consequatur, qui sequentes attractivos superet; habebitur virga rigida, & fractio, ac eo major erit soliditas, & illa, quæ vulgo appellatur durities, quo vires illæ hinc & inde statim post limites fuerint majores.

Quid, & unde viscositas.

438. Atque hic quidem jam etiam ad discrimen devenimus inter elastica, & mollia; verum antequam ad ea faciamus gradum, adnotabo non nulla, quæ adhuc pertinent ad solidorum, & fluidorum naturam, ac proprietates. Inprimis media inter solida, & fluida, sunt viscosa corpora, in quibus est aliqua vis in latus, sed exigua. Ea resistunt mutationi figuræ, sed eo majore, vel minore vi, quo majus, vel minus est in diversis particularum punctis virium discrimen, a quo oritur vis in latus. Viscosa autem præter tenacitatem, quam habent inter se, habent etiam vim, qua adhærent externis corporibus, sed non omnibus, in quo ad humidos liquores referuntur. Humiditas enim est itidem respectiva. Aqua, quæ digitis nostris adhæret illico, & per vitrum, ac lignum diffunditur admodum facile, oleaginosa, & resinosa corpora non humectat, in foliis herbarum pinguibus extat in guttulas eminens, & avium plurium plumas non inficit. Id pendet a vi inter particulas fluidi, & particulas externi corporis; & jam vidimus pro diversa punctorum distributione particulas easdem respectu aliarum debere habere in eadem directione vim attractivam, respectu aliarum repulsivam vim & respectu aliarum nullam.

Organicorum corporum efformatio per vires in latus versus certa superficiei puncta.

439. In particulis illis, quæ ad soliditatem requiruntur, invenitur admodum expedita ratio phænomeni ad solida corpora pertinentis, quod Physicos in summam admirationem rapit, nimirum dispositio quædam in peculiares quasdam figuras, quæ in salibus inprimis apparent admodum constantes, in glacie, & in nivium stellulis potissimum adeo sunt elegantes etiam, & ad certas quasdam leges accedunt, quas itidem cum constanti admodum figurarum forma in gemmarum succis simplicibus observamus, quæ vero nusquam magis se produnt, quam in organicis vegetabilium, & animalium corporibus. In hac mea Theoria in promptu est ratio. Si enim particulæ in certis suæ superficiei partibus quasdam alias particulas attrahunt, in aliis repellunt; facile concipitur, cur non nisi certo ordine sibi adhæreant, in illis nimirum locis tantummodo, in quibus se attrahunt, & satis firmos limites nancisci possunt, adeoque non nisi in certas tantummodo figuras possint coalescere. Quoniam vero præterea eadem particula, eadem sui parte, qua alteram attrahit, alteram pro ejus varia dispositione repellit; dum massa plurium particularum temere agitata prætervolat; eæ tantummodo sistentur, quæ attrahuntur, & ad ea se applicabunt puncta, ad quæ maxime attrahuntur, ac in illis hæbebunt, in quibus post accessum maxime tenaces limites [201] nanciscentur; unde & secretionis, & nutritionis, vegetationis, & certarum figurarum patet ratio admodum manifesta. Et hæc quidem ad nutritionem, & ad certas figuras pertinentia jam innueram num. 222, & 423.

Atomistarum systema posse deduci totum ex hac Theoria, & cum illa bene cohærere, explicata præterea cohæsione partium in atomis.

440. Quoniam ostensum est, qui fieri possit, ut certam figuram acquirant certa particularum genera, cujus admodum tenacia sint, si quis omnem veterum corpuscularium sententiam, quam Gassendus, ac e recentioribus alii secuti sunt, adhibentes variarum figurarum atomos, ut ad cohæsionem uncinatas, ab hac eadem Theoria velit deducere, is sane poterit, ut patet, & ejusmodi atomos adhibere ad explicationem eorum omnium phænomenorum, quæ pendent a sola cohæsione, & inertia, quæ tamen non ita multa sunt: poterunt autem haberi ejusmodi atomi cum infinita figuræ suæ tenacitate, & cohæsione mutua suarum partium per solas etiam binas asymptotos illas, de quibus num. 419, inter se satis proximas. Et si curva virium habeat tantummodo in minimis distantiis duas ejusmodi asymptotos, tum post crus repulsivum ulterioris statim consequatur arcus attractivus, primo quidem plurimum recedens ab axe cum exiguo recessu ab asymptoto, tum

arc on either side of it has an amplitude that is not altogether small; then, if an external force is applied at both ends of the rod, or a fairly great velocity is impressed upon one of the two ends, the rod will be curved, & bent; but if it is left to itself it will return to its original position; & whilst in that violent state of inflection, it will continuously exert a force of restoration, such as occurs in elastic rods. If the forces do not continue the same for such a distance on each side of the limit-point, or if in a sufficiently large interval there exist a considerable number of limit-points, then there will be bending without any endeavour towards restoration, & without fracture, both when we apply a force to each end, & when a great velocity is impressed upon one of them; we see this happen in solids that are extremely ductile, like lead. Finally, if the forces on either side of the limit-point only continue for a very short space, after which there is no action at all, or if a large repulsive arc follows, such as overcomes the attractive arcs that follow it; then the rod will be rigid, & there will be fracture; & the solidity, & what is commonly called the hardness, will be the greater the greater the forces on each side of the limit-points, & following immediately after them.

The nature & source of viscosity.

438. And now we come to the difference between elastic & soft bodies. But, before we pass on to them, I will mention a few matters that have to do with the nature & properties of solids & fluids. First of all, intermediate between solids & fluids come viscous bodies; in these there is indeed some force to one side, but it is very slight. They resist a change of shape; but, the force of resistance is the greater or the less, the greater or the less the difference of the forces on different points of the particles, from which arises the force to one side. Viscous bodies, in addition to the tenacity which they have within their own parts, have also another force with which they adhere to outside bodies, but not to all; & in this they are related to watery liquids. For humidity is also itself but relative. Water, which adheres immediately to our fingers, & is quite easily diffused over glass or wood, will not wet oily or resinous bodies; on the greasy leaves of plants it stands up in little droplets; nor does it make its way through the feathers of the greater number of the birds. This depends on the force between the particles of the fluid, & those of the external body; & we have already seen that, for a different distribution of their points, the same particles may have with respect to some, in the same direction, an attractive force, with respect to others a repulsive force, & with respect to others again no force at all.

The formation of organic bodies by means of transverse forces directed towards certain points of the surface.

439. In particles, such as are necessary for solidity, there is found quite easily the reason for a phenomenon pertaining to solid bodies, which is a source of the greatest wonder to physicists. That is, a disposition in certain special shapes, which in salts especially seem to be quite constant; in ice, & the star-like flakes of snow more especially, they are wonderfully beautiful; & they observe certain definite laws, such as we also see, together with a constant shape of figure, in the simple constituents of crystals. But these are nowhere to be found so frequently as in the organic bodies of the vegetable & animal kingdoms. The reason for this comes out directly in this Theory of mine. For, if particles, at certain parts of their surfaces, attract other particles, & at other parts repel other particles, it can easily be understood why they should adhere to one another only in a certain manner of arrangement; that is to say, in such places only as there is attraction, & where there can be produced limit-points of sufficient strength; & thus, they can only group themselves together in figures of certain shapes. But since, in addition to this, the same particle, at the same part of its surface, with which it attracts one particle, will repel another particle situated differently with respect to it; whilst the mass of the great number of particles, set in motion at random, will slip by, those only will stay, which are attracted; & they will attach themselves to the points to which they are most attracted, & will adhere to those points in which, after approach, limit-points of the greatest tenacity are produced. From this the reason for secretion, nutrition, the growth of plants, & fixity of shape, is perfectly evident. I have indeed already remarked on these matters, as far as they pertain to nutrition & fixity of shape, in Arts. 222 & 423.

The whole of the system formulated by the Atomists can be derived from this Theory, with which it agrees very well; in addition, the cohesion of the parts of their atoms is explained by it.

440. Since it has been shown how it may be possible for certain kinds of particles to acquire certain definite shapes, of which they are quite tenacious; if anyone should wish to derive from this same theory the whole idea of the ancient corpuscularians, such as Gassendi & others of the more modern philosophers have followed, employing atoms of various shapes, hooked together for cohesion; he will certainly be able, as is evident, to use atoms of this sort to explain all these phenomena that depend upon cohesion alone, & inertia; but the number of these is not very great. Moreover, atoms of this sort can be had with an infinite tenacity of shape, & mutual cohesion of their parts, by even the sole assumption of those pairs of asymptotes sufficiently close to one another, of which I spoke in Art. 419. Even if the curve of forces should have at very small distances two such asymptotes only, & then immediately after the repulsive arc of the far one of these there should follow an attractive arc, such as first of all recedes a great distance from the axis whilst it recedes only slightly from the asymptote, & then returns towards the axis & approximates immediately to the

ad axem regrediens, & accedens statim ad formam gravitati exhibendæ debitam; haberentur per ejusmodi curvam atomi habentes impenetrabilitatem, gravitatem, & figuræ suæ tenacitatem ejusmodi, ut ab ea discedere non possent discessu quantum libuerit parvo; cum enim possint illæ duæ asymptoti sibi invicem esse proximæ intervallo utcunque parvo, posset utique ita contrahi intervallum istud, ut figuræ mutatio æqualis datæ cuicunque utcunque parvæ mutationi eviteatur. Ubi enim cuicunque figuræ inscripta est series continua cubulorum, & puncta in singulis angulis posita sunt, mutari non potest figura externorum punctorum ductum sequens mutatione quadam data, per quam quædam puncta discedant a locis prioribus per quædam intervalla data, manentibus quibusdam, ut manente basi, nisi per quædam data intervalla a se invicem recedant, vel ad se invicem accedant saltem aliqua puncta, cum, data distantia puncti a tribus aliis, detur etiam ejus positio respectu illorum, quæ mutari non potest, nisi aliqua ex iisdem tribus distantiis mutetur, unde fit, ut possit data quævis positionis mutatio impediri, impedita mutatione distantiæ per intervallum ad eam mutationem necessarium. Quod si illæ binæ asymptoti essent tantillo remotiores a se invicem, tum vero & mutatio distantiæ haberi posset tantillo major, & idcirco singulis distantiis illata vi aliqua posset figura non nihil mutari, & quidem exigua mutatione distantiarum singularum posset in ingenti serie punctorum haberi inflexio figuræ satis magna orta ex pluribus exiguis flexibus. Sic & spirales atomi efformari possent, quarum spiris per vim contractis sentiretur ingens elastica vis, sive determinatio ad expansionem, ac per hujusmodi atomos possent iti-**[202]**-dem plurima explicari phænomena, ut & nexus massarum per uncos uncis, vel spiris insertos, quo pacto explicari itidem posset etiam illud, quomodo in duabus particulis, quarum altera ad alteram cum ingenti velocitate accesserit, oriatur ingens nexus novus, nimirum sine regressu a se invicem, unco nimirum alterius in alterius foramen injecto, & intra illud converso per virium inæqualitatem in diversas unci partes agentium, ut jam prodire non possit; nam unci cavitas, & foramen, seu porus alterius particulæ, posset esse multo amplior, quam pro exigua illa distantia insuperabili, ut idcirco inseri posset sine impedimento orto a viribus agentibus in minore distantia. Eædem autem atomi haberi possunt, etiam si curva habeat reliquos omnes flexus, quos habet mea, quo pacto ad alia multo plura, ut ad fermentationes inprimis, ac vaporum, & luminis emissionem multo aptiores erunt; & sine asymptoticis arcubus, qui vires exhibeant extra originem abscissarum in infinitum excrescentes, idem obtineri poterit per solos limites cohæsionis admodum validos cum tenacitate figuræ non quidem infinita, sed tamen maxima, ubi, quod illi veteres non explicarunt, cohæsio partium atomorum inter se, adeoque atomorum soliditas, ut & continuata impenetrabilitatis resistentia, & gravitas, ex eodem generali derivaretur principio, ex quo & reliqua universa Natura. Illud unum hic notandum superest, ejusmodi atomos habituras necessario ubique distantiam a se invicem majorem, quam pro illa insuperabili distantia, ad quam externa puncta devenire ibi non possunt.

Cur non omnia corpora sint fluida; licet omnia puncta sint circumquaque ejusdem vis.

441. Huc etiam pertinet solutio hujusmodi difficultatis, quæ sponte se objicit: si omnia materiæ puncta simplicia sunt, & vires in quavis directione circumquaque exercent easdem; omnia corpora ex iis utique composita erunt fluida multo potiore jure, quam fluida esse debeant, quæ globulis constent easdem in omni circum directione vires exercentibus. Huic difficultati hic facile occurritur: si particularum puncta possent vi adhibita mutare aliquanto magis distantias inter se, nam aliqua etiam ad circulationem exigua mutatio requiritur; posset autem imprimi exiguo numero punctorum constituentium unam e particulis primorum ordinum, quin imprimatur simul omnibus ejusmodi punctis, vel satis magno eorum numero, motus ad sensum idem; tum utique haberetur idem, quod habetur in fluidis, & separatis aliis punctis post alia, motus facilis per omnes omnium corporum massas obtineretur. At particulæ primi ordinis ab indivisibilibus punctis ortæ, ut & proximorum ordinum particulæ ortæ ab iis, sua ipsa parvitate molis tueri possunt juxta num. 424 formam suam, & positionem punctorum: nam differentia virium exercitarum in diversa earum puncta potest esse perquam exigua, summa virium prohibente tantum accessum unius particulæ ad alteram, quo tamen accessu inæqualitas virium, &

form proper to represent gravitation; by such a curve we should get atoms having impenetrability, gravitation, & tenacity of shape of such a kind that they would not be able to depart from this shape by any small amount we wish to assign. For, since the two asymptotes can be very close together, distant from one another by any interval no matter how small, this interval can in every case be contracted to such an extent, that the change of shape will be just less than any given change no matter how small. For, if within any figure there is inscribed a continuous series of little cubes, & points are situated at each of their corners, the figure cannot be changed, following the lead of external points, by any given change through which certain points depart from their original positions through certain given intervals, whilst others stay where they are, i.e., whilst the base, say, stays where it was; unless they recede from one another through a certain given interval, or approach one another, or some of the points do so at least. For, if the distances of a point from three other points are given, its position with regard to them is also given; & this cannot be changed without altering some one of the three distances; hence, any change of position can be prevented by preventing the change of distance through any interval that is necessary to such a change of position. But if the pair of asymptotes were just a little further away from one another, then in truth there would be possibility of getting a change of distance that was also just a little greater; & thus, a force being produced at each distance, the figure might suffer some change; & by a very slight change of each of the distances in a very long series of points there might be obtained a bending of the figure of comparatively large amount, due to a large number of these slight bendings. In such a way atoms might be formed like spirals; &, if these spirals were compressed by a force, there would be experienced a very great elastic force or propensity for expansion; also by means of atoms of this nature an explanation could be given of a very large number of phenomena, such as the connection of masses by means of hooks inserted into hooks or coils; & in this way also an explanation could be given of the reason why, in the case of two particles of which one has approached the other with a very great velocity, there arises a fresh connection of great strength, that is, one so strong that there is no rebound of the particles from one another. For instance, it may be said that the hook of the one is introduced into an opening in the other, & twisted within it by the inequality of the forces acting on different parts of the hook, so that it cannot get out again. For the concavity of the hook, & the opening or pore of the second particle, may be much wider than that corresponding to that very slight distance limiting nearer approach; & thus the hook can be inserted without hindrance due to forces acting at those very small distances. These same atoms might be obtained, even if the curve had all the inflected arcs that are present in mine; & then such atoms would be much more suitable to explain fermentations especially, as well as the emission of vapours & of light. If there were no asymptotic arcs representing indefinitely increasing forces beyond the origin of abscissæ, the same result could be obtained by means of limit-points of cohesion alone; with tenacity of figure, not indeed infinite, but still very great if these were very powerful. In this case, there could be derived from the same general principle, from which is derived the whole of Nature in general, an explanation of the cohesion of the parts of the atoms (which the ancients did not explain), & therefore of their solidity; & also the continued resistance of impenetrability, & gravitation too. There remains but one thing for me to mention; namely, that atoms of this kind will necessarily keep to a greater distance from one another than that corresponding to the distance limiting further approach, beyond which external points cannot come.

The reason why all bodies are not fluid, although all points in all directions round are under the same force.

441. Here also is the place to solve a difficulty that spontaneously presents itself. If all points of matter are simple, & if they exert the same forces in all directions round themselves; then it is far more natural to expect that all bodies that are composed of such points would be fluid than that those, which consist of little spheres exerting the same forces in all directions around, are bound to be fluid. The answer to this difficulty is easily given; if the points of particles can, by application of force, increase their mutual distances by a fair amount (for some slight change is necessary even for circulation), and if further it were possible to impress a practically equal motion on a very small number of points forming one of the particles of the first order, without at the same time giving this motion to all such points, or even to any considerable number of them; in that case we certainly should obtain the same effect as is obtained in the case of fluids; & the points being separated one after the other, an easy movement would be obtained throughout all masses of all bodies. But, particles of the first order, formed from indivisible points, as also those of the next orders formed from the first, can, owing to their very smallness of volume, preserve their form & the mutual arrangement of their points, as was shown in Art. 424. For, the difference between the forces acting on different points of them may be extremely small, since the sum of the forces prevents too close an approach of one particle to the other; & yet by this approach an inequality in the forces & an obliquity in their directions is obtained,

obliquitas directionum ha-[203]-beatur adhuc satis magna ad vincendas vires mutuas, mutandam positionem, qua positione manente, manet inæqualitas virium, quas diversa puncta ejus particulæ exercent in aliam particulam. Ea inæqualitas itidem potest non esse satis magna, ut possit illins mutuas vires vincere, & textum dissolvere, sed esse tanta, ut motum inducat in latus, ac ejus motus obliquitas, & virium inæqualitas eo deinde erit major, quo ad altiores ascenditur particularum ordines, donec deveniatur ad corpora, quæ a nobis sentiuntur.

Difficultas determinandi resistentiam fluidorum: methodi indirectæ id præstandi eædem in hac Theoria ac in communi.

442. Solida externum corpus ad ea delatum intra suam massam non recipiunt, ut vidimus: at fluida solidum intra se moveri permittunt, sed resistunt motui. Resistentiam ejusmodi accurate comparare, & ejus leges accurate definire, est res admodum ardua. Oporteret nosse ipsam virium legem determinate, & numerum, & dispositionem punctorum, ac habere satis promotam Geometriam, & Analysin ad rem præstandam. Sed in tanta particularum, & virium multitudine, quam debeat esse res ardua, & humano captu superior determinatio omnium motuum, satis constat ex ipso problemate trium corporum in se mutuo agentium, quod num. 204 diximus nondum satis generaliter solutum esse. Hine alii ad alias hypotheses confugiunt, ut rem perficiant, & omnes ejusmodi methodi æque cum mea, ac cum communi Theoria, consentire possunt.

Bini resistentiæ fontes, & utriusque lex.

443. Ut tamen aliquid innuam etiam de eo argumento, duplex est resistentiæ fons in fluidis; primo quidem oritur resistentia ex motu impresso particulis fluidi; nam juxta leges collisionis corporum, corpus imprimens motum alteri, tantundem amittit de suo. Deinde oritur resistentia a viribus, quas particulæ exercent, dum aliæ in alias incurrunt, quæ earum motum impediunt, quo casu comprimuntur non nihil particulæ ipsæ etiam in fluidis non elasticis egressæ e limitibus, & æquilibrio: acquirunt autem motus admodum diversos, gyrant, & alias impellunt, quæ a tergo urgent non nihil corpus progrediens, quod potissimum a fluidis elasticis a tergo impellitur, dilatato ibi fluido, dum a fronte a fluido ibi compresso impeditur: sed ea omnia, uti diximus, accurate comparare non licet. Illud generaliter notari potest: resistentia, quæ provenit a motu communicato particulis fluidi, & quæ dicitur orta ab inertia ipsius fluidi, est ut ejus densitas, & ut quadratum velocitatis conjunctim: ut densitas quia pari velocitate eo pluribus dato tempore particulis motus idem imprimitur, quo densitas est major, nimirum quo plures in dato spatio occurrunt particulæ: ut quadratum velocitatis, quia pari densitate eo plures particulæ dato tempore loco movendæ sunt, quo major est velocitas, nimirum quo plus spatii percurritur, & eo major singulis imprimitur motus, quo itidem velocitas est major. Resistentia autem, quæ oritur a viribus, quas in se exercent particulæ, si vis ea esset eadem in singulis, quacunque velocitate [204] moveatur corpus progrediens, esset in ratione temporis, sive constans: nam plures quidem eodem tempore particulæ eam vim exercent, sed breviore tempore durat singularum actio, adeoque summa evadit constans. Verum si velocitas corporis progredientis sit major; particulæ magis compinguntur, & ad se invicem accedunt magis, adeoque major est itidem vis. Quare ejusmodi resistentia est partim constans, sive, ut vocant, in ratione momentorum temporis, & partim in aliqua ratione itidem velocitatis.

Quam legem videantur innuere experimenta: in viscosis resistentiam esse majorem.

444. Porro ex experimentis nonnullis videtur erui, resistentiam in nonnullis fluidis esse partim in ratione duplicate velocitatum, partim in ratione earum simplici, & partim constantem, sive in ratione momentorum temporis, quanvis ubi velocitas est ingens, deprehendatur major: & ubi fluiditas est ingens, ut in aqua, ut secundum resistentiæ genus, quod est magis irregulare, & incertum, fit respectu prioris exiguum, satis accedit resistentia ad rationem duplicatam velocitatum. Sed & illud cum Theoria conspirat, quod viscosa fluida multo magis resistunt, quam pro ratione suæ densitatis, & velocitate corporis progredientis: nam in ejusmodi fluidis, quæ ad solida accedunt, illud secundum resistentiæ genus est multo majus, quod quidem in solidis usque adeo crescit: quanquam & in iis intrudi per ingentem vim intra massam potest corpus extraneum, ut clavus in murum, vel in metallum, quæ tamen, si fragilia sunt, & sensibilem compressionem non admittant, diffringuntur.

Problemata alia ad resistentiam pertinentia itidem communia huic Theoriæ.

445. Jam vero quæcunque a Newtono primum, tum ab aliis demonstrata sunt de motu corporum, quibus resistitur in variis rationibus velocitatum, ea omnia consentiunt itidem cum mea Theoria, & hujus sunt loci, ac ad illam pertinent Mechanicæ partem, quæ agit de motu solidorum per fluida. Sic etiam determinatio figuræ, cui minimum

which is sufficiently great to overcome the mutual forces & to alter their position ; & when this position stays as it was, so also does the inequality between the forces, which the different points of the particle exert upon another particle. Again, this inequality may not be great enough to overcome the mutual forces of that particle, & break up its formation ; but yet great enough to induce lateral motion ; the obliquity of this motion, & the inequality of forces will therefore be so much the greater, the further we ascend in the orders of the particles, until we finally reach such bodies as affect our senses.

The difficulty of determining the resistance of fluids; the indirect methods for accomplishing this are the same in my Theory as in the usual one.

442. As we see, solids do not receive within their mass an external body that is brought close up to them ; but fluids allow a solid to be moved within their mass, resisting however the motion. To find such resistance accurately, & to make out the laws which govern it, is a matter of great difficulty. It would be necessary to know the law of forces exactly, the number & arrangement of the points, & to be in possession of fairly advanced geometry & analysis to accomplish a solution. But, when dealing with such a great number of points & forces, how difficult the matter is bound to be can be fairly seen by reference to that problem of the three bodies acting upon one another, which I said, in Art. 204, had not yet been solved at all generally. Hence, others resort to other hypotheses for their purposes ; all such methods can be reconciled as well with my theory as with the common one.

Two sources of resistance, & the laws of each.

443. So that I may not leave the point altogether untouched, I will just remark that the source of resistance in fluids is twofold. First, we have resistance due to the motion impressed on the particles of the fluid ; for, according to the laws of the impact of bodies, the body which impresses the motion on the other will lose just as much of its own motion. Secondly, there is resistance due to the forces exerted by the particles, as they approach one another, which hinders their motion ; & in this case, the particles themselves are compressed to some extent, even in non-elastic fluids, as they go beyond the limit-points & equilibrium. Moreover they acquire different motions, they gyrate & drive off others that are driving the moving body to some extent from the back ; & especially in the case of elastic fluids we have this force at the back of the body, owing to the fluid being there dilated, whilst at the same time there is a hindering force at the front due to the fluid being compressed there. But all these things, as I have said, cannot be accurately determined. It can, however, be in general noted that the resistance due to the motion communicated to the particles of a fluid, which is said to arise from the inertia of the fluid, varies as its density & the squares of the velocities jointly. As the density, because in the same time, for equal velocities, the same motion is impressed upon a number of particles which is the greater, the greater the density, i.e., the greater the number of particles occupying the same space. As the squares of the velocities, because in the same time, for equal densities, the number of particles to be moved in position is the greater, the greater the velocity, that is to say, the greater the space to be traversed ; & the motion that is impressed on each point is the greater, the greater the velocity. Again, the resistance that is due to the forces which the particles exert on one another, if the force is the same for each of them, with whatever velocity the moving body proceeds, would be in proportion to the time, or constant. For, it is true that a large number of particles exert this force in the same time, but the action of each only lasts for a quite short time ; & thus the sum turns out to be constant. If the velocity of the moving body is greater, the particles are driven together more closely, & approach one another more nearly, & so also the force is greater. Hence this kind of resistance is partly constant, or, as it is usually termed, proportional to instants of time, & partly in some way proportional to the velocity as well.

The law that experiments seem to indicate : the resistance is greater in viscous fluids.

444. Further the results of some experiments seem to indicate that the resistance in some fluids is partly as the squares of the velocities, partly as the velocities simply, & partly constant, or as the instants of time, although where the velocity is very great, it is found to be greater. Also when the fluidity is great, as in the case of water, the second kind of resistance, which is the more irregular & uncertain of the two, becomes exceedingly small compared with that of the first kind, & the total resistance approaches fairly closely to a variation as the squares of the velocities. It is also in agreement with the Theory that the resistance for viscous fluids is much greater than that corresponding to the ratio of densities & the velocities of the moving bodies. For, in such fluids, which are a near approach to solids, the second kind of resistance is by far the greater, & indeed increases to so great an extent as in solids. Although, in solids also, an extraneous body can be introduced within their mass by means of a very great force, just as a nail may be driven into a wall, or into metal ; yet if these are fragile & do not admit of sensible compression, they are broken.

Other problems relating to resistance that are common also to this Theory.

445. But there are several other things, first demonstrated by Newton, & afterwards by others, concerning the motion of bodies, under a resistance varying as different powers of the velocity ; & all of these are also in agreement with my Theory, & come in in this connection ; they belong also to that part of Mechanics which deals with the motion of solids through fluids. So also the determination of the figure of least resistance, the

resistitur, determinatio vis fluidi solidum impellentis directionibus quibuscunque, mensura velocitatis inde oriundæ per corporum objectorum resistentiam observatione definitam, innatatio solidorum in fluidis, ac alia ejusmodi, & mihi communia sunt: sed oportet rite distinguere, quæ sunt hypothetica tantummodo, ab iis, quæ habentur reapse in Natura.

Alia pertinentia huc pertractata in parte secunda: discrimen inter elastica, & mollia.

446. Ad fluida & solida pertinent itidem, quæcunque in parte secunda demonstrata sunt de pressione fluidorum, & velocitate in effluxu, quæcumque de æquilibrio solidorum, de vecte, de centro oscillationis, & percussionis, quæ quidem in Mechanica pertractari solent. Illud unum addo, ex motu facili particularum fluidi aliarum circa alias, & irregulari earum congestione, facile deduci, debere pressionem propagari quaquaversus. Sed de his jam satis, quæ ad soliditatem, & fluiditatem pertinent: illud vero, quod pertinet ad discrimen inter elastica, & mollia, brevi expediam. Elastica sunt, quæ post mutationem [205] figuræ redeunt ad formam priorem; mollia, quæ in nova positione perseverant. Id discrimen Theoria exhibet per distantiam, vel propinquitatem limitum, juxta ea, quæ dicta num. 199. Si limites proximi illi, in quo particulæ cohærent, hinc, & inde plurimum ab eo distant, imminuta multum distantia, perstat semper repulsiva vis; aucta distantia, perstat vis attractiva. Quare sive comprimatur plus æquo, sive plus æquo distrahatur massa, ad figuram veterem redit; ubi rediit, excurrit ulterius, donec contraria vi elidatur velocitas concepta, ac oritur tremor, & oscillatio, quæ paullatim minuitur, & extinguitur demum, partim actione externorum corporum, ut per aeris resistentiam sistitur paullatim motus penduli, partim actione particularum minus elasticarum, quæ admiscentur, & quæ possunt tremorem illum paullatim interrumpere frictione, ac contrariis motibus, & sublapsu, quo suam ipsam dispositionem nonnihil immutent. Si autem limites sint satis proximi; causa externa, quæ massam comprimit, vel distrahit, posteaquam adduxit particulas ab uno cohæsionis limite ad alium, ibi eas itidem cogit subsistere, quæ ibidem semel constitutæ itidem in æquilibrio sunt, & habetur massa mollis.

Fluida elastica, quorum partes non sunt in limitibus cohæsionis; omnia & solida, & fluida elastica esse, sed non dici, quia sensibilem compressionem non patiuntur

447. Quædam elastica fluida non habent particulas positas inter se in limitibus cohæsionis, sed in distantiis repulsionum, & quidem ingentium, ut aer: sed vel incumbente pondere, vel parietibus quibusdam impeditur recessus ille, & sunt quodammodo ibidem in statu violento; licet semper puncta singula in æquilibrio sint, oppositis repulsionibus se mutuo elidentibus. Omnia autem & solida, & fluida, quæ videntur nec comprimi, nec ullas habere vires mutuas inter particulas, sed in limitibus esse, adhuc elastica sunt, sive vim repulsivam exercent inter particulas proximas, saltem quæ sensibili gravitate sunt prædita, quæ nimirum vis repulsiva vim gravitatis elidat. Verum ea distantias parum admodum mutant, mutatione, quæ idcirco sensum omnem effugiat; quod accidit in aqua, quæ in fundo putei, & prope superficiem supremam habet eandem ad sensum densitatem, & in metallis, & in marmoribus, & in solidis corporibus passim, quæ pondere majore imposito nihil ad sensum comprimuntur. Sed ea idcirco appellari non solent elastica, & ad ejusmodi appellationem non sufficit vis repulsiva etiam ingens inter particulas proximas: sed etiam requiritur mutatio sensibilis distantiæ respectu distantiæ totalis respondens sensibili mutationi virium.

Dura nulla esse: quæ dicantur: unde fragilitas, & ductilitas.

448. Dura corpora in eo sensu, in quo a Physicis duritiei nomen accipitur, ut nimirum figuram nihil prorsus immutent, nulla sunt in mea Theoria, ut & nulla compacta penitus, ac plane solida, quemadmodum diximus etiam num. 266; sed dura vocat vulgus, quæ satis magnam exercent vim, ne figuram mutent, sive elastica sint, sive fragilia, sive mollia. Fragilitas, unde ortum ducat, expositum est paullo su- [206] -perius num. 437, & inde etiam quid ductilitas, ac malleabilitas sit, facile intelligitur. Ductilia nimirum a mollibus non differunt, nisi in majore, vel minore vi, qua figuram tuentur suam: ut enim mollia pressione tenui, & ipsis digitis comprimuntur, vel saltem figuram mutant, sed mutatam retinent, ita ductilia ictu validiore mallei mutant itidem figuram suam veterem, & retinent novam, quam acquirunt.

Superiora omnia profluere ex Theoria: ejus fœcunditas: illa omnia a densitate non pendere.

449. Atque hoc demum pacto quæcunque pertinent ad fluidorum, & solidorum diversa genera, nam & elastica, mollia, ductilia, fragilia eodem referuntur, invenimus omnia in illo particularum discrimine orto ex sola diversa combinatione punctorum, quam nobis Theoria rite applicata exhibuit, in quibus omnibus immensa varietas itidem haberi poterit,

determination of the force of a fluid driving a solid in any directions, the measurement of the velocity arising thence by means of the observed resistance of bodies placed in the way, the flotation of bodies in fluids, & other things of the same kind, are all common to my Theory. But it is necessary to distinguish which of them are only hypothetical & which of them really occur in Nature.

Other matters that were discussed in the second part really pertain to this connection; distinction between elastic & soft bodies.

446. To fluids & solids are to be referred all those matters, which in the second part were demonstrated with regard to pressure of fluids, & velocity of efflux; & all matters relating to equilibrium of solids, the lever, the centre of oscillation, & the centre of percussion; all of which indeed are usually considered in connection with Mechanics. I will but add that, from the ease of movement of the particles of a fluid about one another, & from their irregular grouping, it readily follows that in them pressure must be propagated in every direction. But I have now said enough about those matters that refer to solidity & fluidity; however, I will make a few remarks on matters that relate to the distinction between elastic & soft bodies. Those bodies are elastic, which after change of shape return to their original form; & those are soft, which remain in their new state. This distinction my Theory shows to be consequent upon the distance or closeness of the limit-points; as I said in Art. 199. If the limit-points, that are next to the one in which the particles cohere, are far distant from it on either side, then, when the distance is much diminished, there will still be a repulsive force all the time; & if the distance is increased there will be a similar attractive force. Hence, whether the mass is compressed more than is natural, or expanded more than is natural, it will return to its original form. When it has returned to its original form, it will go beyond it, until the velocity attained is cancelled by the opposite force; and a tremor, or oscillation, will be produced, which will be gradually diminished and ultimately destroyed, partly by the action of external bodies, just as the motion of a pendulum is stopped by the resistance of the air, & partly by the action of less elastic particles which are interspersed, which can gradually break down the oscillation by their friction, & also by contrary motions, & a relapse by which they change their own distribution somewhat. But if these limit-points are fairly close, the external cause, which compresses or expands the mass, after that it has brought the particles from one limit-point of cohesion to another, will force them also to stay at the latter; & these, when once grouped in this position, will also be in equilibrium, & a soft mass will be the result.

Elastic fluids whose particles are not at limit-points of cohesion. All solids & fluids are really elastic, but are not called so, because they do not suffer sensible compression.

447. The particles of some elastic fluids are not at limit-points of cohesion with respect to one another, but are at distances corresponding to repulsions, & these too very great; for instance, air. But recession is prevented either by superincumbent weight, or by enclosing walls; these are in some sort of violent condition at these distances, although each point is always in equilibrium, due to the opposite repulsions cancelling one another. Moreover, all solids & fluids, which appear neither to suffer compression, nor to have any mutual forces between their particles, but to be at limit-points, are however elastic; that is to say, they exert a repulsive force between their adjacent particles; at least those do which are possessed of sensible gravitation, for it is this repulsive force that cancels the force of gravity. The distances are in fact changed very slightly, the change being therefore one that is beyond the scope of our senses. This is the case for water; with it, the density is practically the same at the bottom of a well as it is at the upper surface; the same thing happens in the case of metals & marbles & in all solid bodies, in which if a fairly large weight is superimposed there is no sensible compression. But such things are not usually termed elastic, for the reason that a repulsive force between adjacent particles, even if it is very great, is not sufficient for such an appellation; in addition, there is required to be a sensible change of distance, compared with the whole distance, to correspond with a sensible change in the forces.

There are no hard bodies; what bodies are called hard; hence fragility & ductility.

448. There are in my Theory none of those bodies, that are hard in the sense in which hardness is accepted by Physicists, namely such as do not suffer the slightest change of shape; & also there are none that are perfectly compact, or quite solid, as I said in Art. 266. But those are usually termed hard, which exert a fairly great force to prevent change of form; they may be either elastic, fragile or soft. The source of fragility has been explained just above, in Art. 437; & from this also the nature of ductility & malleability can be easily understood. For instance, ductile & malleable solids only differ from one another in the greater or less strength with which they preserve their form; for, just as soft bodies under slight pressure, even of the fingers, are compressed, or change their form, but retain the form thus changed; so ductile bodies under the stronger force of a blow with a mallet also change their original shape, & retain the new form that they acquire.

All the above properties are derived from my Theory; all of them do not depend upon density.

449. Finally, in this way, whatever properties there may be relating to different kinds of fluids & solids (for elastic, soft, ductile & fragile bodies all come to the same thing), we have made them all out from the difference between particles that is produced by the difference in the combination of the points alone; this will be shown by my Theory if

& debebit; si curva primigenia ingentem habeat numerum intersectionum cum axe, & particulæ primi ordinis, ac reliquæ ordinum superiorum dispositiones, quæ in infinitum variari possunt, habuerint plurimas, & admodum diversas inter se, ac eas inprimis, quæ ad hæc ipsa figurarum, & virium discrimina requiruntur. Illud unum hic diligenter notandum est, quod ipsam Theoriam itidem commendat plurimum, hasce proprietates omnes a densitate nihil omnino pendere. Fieri enim potest, uti num. 183 notavimus, ut curva virium primigenia limites, & arcus habeat quocunque ordine in diversis distantiis permixtos quocunque numero, ut validiores, & minus validi, ac ampliores, & minus ampli commisceantur inter se utcunque, adeoque phænomena eadem figurarum, & virium æque inveniri possint, ubi multo plura, & ubi multo pauciora puncta massam constituunt.

Communia quatuor elementa quid sint.

450. Jam vero illa, quæ vulgo elementa appellari solent, Terra, Aqua, Aer, Ignis, nihil aliud mihi sunt, nisi diversa solida, & fluida, ex iisdem homogeneis punctis composita diversimode dispositis, ex quibus deinde permixtis alia adhuc magis composita corpora oriuntur. Et quidem Terra ex particulis constat inter se nulla vi conjunctis, quæ soliditatem aliarum admixtione particularum acquirunt, ut cineres oleorum ope, vel etiam aliqua mutatione dispositionis internæ, ut in vitrificatione evenit, quæ transformationes quo pacto accidant, dicemus postremo loco. Aqua est fluidum liquidum elasticitate carens cadente sub sensum per compressionem sensibilem, licet ingentem exerceant repulsivam vim ejus particulæ, sustinentes vel externæ vis, vel sui ipsius ponderis pressionem sine sensibili distantiarum imminutione. Aer est fluidum elasticum, quem admodum probabile est constare particulis plurimorum generum, cum e plurimis etiam fixis corporibus generetur admodum diversis, ut videbimus, ubi de transformationibus agendum erit, ac propterea continet vapores, & exhalationes plurimas, & heterogenea corpuscula, quæ in eo innatant: sed ejus particulæ satis magna vi se repellunt, [207] & ea repulsiva particularum vis imminutis distantiis diu perdurat, ac pertinet ad spatium, quod habet ingentem rationem ad eam tanto minorem distantiam, ad quam compressione reduci potest, & in qua adhuc ipsa vis crescit, areu curvæ adhuc recedente ab axe: is vero arcus ad axem ipsum deinde debet ruere præceps, ut circa proximum limitem adhuc ingentes in eo residuo spatio variationes in arcubus, & limitibus haberi possint. Porro extensionem tantam arcus repulsivi evincit ipsa immanis compressio, ad quam ingenti vi aer compellitur, qui ut babeat compressiones viribus prementibus proportionales, debet, ut monuimus num. 352, habere vires repulsivas reciproce proportionales distantiis particularum a se invicem. Is autem etiam in fixum corpus, & solidum transire potest, quod qua ratione fieri possit, dicam itidem, ubi de transformationibus agemus in fine. Ignis etiam est fluidum maxime elasticum, quod violentissimo intestino motu agitatur, ac fermentationem excitat, vel etiam in ipsa fermentatione consistit, emittit vero lucem, de quo pariter agemus paullo inferius, ubi de fermentatione, & emissione vaporum egerimus inter ea, quæ ad Chemicas operationes pertinet, ad quas jam progredior.

Chemicarum operationum genera deduci facile ex illo particularum discrimine: singularium effectuum causas singulares non posse cognosci a mente humana.

451. Chemicarum operationum principia ex eodem deducuntur fonte, nimirum ex illo particularum discrimine, quarum aliæ inter se, & cum quibusdam aliis inertes, alias ad se attrahunt, alias repellunt constanter per satis magnum intervallum, ubi attractio ipsa cum aliis est major, cum aliis minor, aueta vero satis distantia, evadit ad sensum nulla; quarum itidem aliæ respectu aliarum habent ingentem virium alternationem, quam mutato nonnihil textu suo, vel conjunctæ, & permixtæ cum aliis mutare possunt, succedente pro particulis compositis alia virium lege ab ea, quæ in simplicibus observabatur. Hæc omnia si habeantur ob oculos; mihi sane persuasum est, facile inveniri posse in hac ipsa Theoria rationem generalem omnium Chemicarum operationum: nam singulares determinationes effectuum, qui a singulis permixtionibus diversorum corporum, per quas unice omnia præstantur in Chemia, sive resolvantur corpora, sive componantur, requirerent intimam cognitionem textus particularum singularum, & dispositionis, quam habent in massis singulis, ac præterea Geometriæ, & Analyseos vim, quæ humanæ mentis captum excedit longissime. Verum illud in genere omnino patet, nullam esse Chemiæ partem, in qua præter inertiam massæ, & specificam gravitatem, alia virium mutuarum genera inter particulas non ubique se prodant, & vel invitis incurrant in oculos, quod quidem vel in sola postrema quæstione Opticæ Newtoni abunde patet, ubi tam multa & attractionum,

properly applied, & in all such things also an immense variety can & must be produced. Provided that the primary curve has a number of intersections with the axis, & provided that particles of the first order, & the rest of the higher orders, have arrangements (which indeed can be infinitely varied) that are great in number & all different from one another; & those especially that are required for these differences in shape & forces. Now, one thing is at this point to be noted carefully, one that also supports the Theory itself very strongly, namely, that all these properties are totally independent of density. For it is possible that, as I mentioned in Art. 183, the primary curve of forces may have limit-points & arcs mixed together in any order at different distances, and there may be any number of either; so that stronger & weaker limit-points, more & less ample arcs may be intermingled in any manner amongst themselves; & thus the same phenomena of shapes & forces can be met with when the number of points constituting a mass is much larger or much smaller.

The nature of the four Elements, commonly so-called.

450. Now those things, which are commonly called the Elements, Earth, Water, Air & Fire, are nothing else in my Theory but different solids & fluids, formed of the same homogeneous points differently arranged; & from the admixture of these with others, other still more compound bodies are produced. Indeed Earth consists of particles that are not connected together by any force; & these particles acquire solidity when mixed with other particles, as ashes when mixed with oils; or even by some change in their internal arrangement, such as comes about in vitrification; we will leave the discussion of the manner in which these transformations take place till the end. Water is a liquid fluid devoid of elasticity such as comes within the scope of the senses through a sensible compression; although there is a strong repulsive force exerted between its particles, which is sufficient to sustain the pressure of an external force or of its own weight without sensible diminution of the distances. Air is an elastic fluid, which in all probability consists of particles of very many different sorts; for it is generated from very many totally different fixed bodies, as we shall see when we discuss transformations. For that reason, it contains a very large number of vapours & exhalations, & heterogeneous corpuscles that float in it. Its particles, however, repel one another with a fairly large force; & this repulsive force of the particles lasts for a long while as the distances are diminished, & pertains to a space that bears a very large ratio to the so much smaller distance, to which it can be reduced by compression; & at this distance too the force still increases, the arc of the curve corresponding to it still receding from the axis. But after that, the curve must return very steeply, so that in the neighbourhood of the next limit-point there may yet be had in the space that remains great variations in the arcs & the limit-points. Further such great extension of the repulsive arc is indicated by the great compression induced by the pressure due to a large force; & this, in order that the compression may be proportional to the impressed force, shows, as we pointed out in Art. 352, that there must be repulsive forces inversely proportional to the distances of the particles from one another. Moreover it can pass into & through a fixed & solid body; & the reason of this also I will state when I deal with transformations towards the end. Fire is also a highly elastic fluid, which is agitated by the most vigorous internal motions; it excites fermentations, or even consists of this very fermentation; it emits light, with which also we will deal a little later, when we discuss fermentation & emission of vapours amongst other things referring to chemical operations; to these we will now pass on.

The different kinds of chemical operations are readily derived from the differences between particles; the special causes of each of the effects are beyond the intelligence of the human mind.

451. The principles of chemical operations are derived from the same source, namely, from the distinctions between particles; some of these being inert with regard to themselves & in combination with certain others, some attract others to themselves, some repel others continuously through a fairly great interval; & the attraction itself with some is greater, & with others is less, until when the distance is sufficiently increased it becomes practically nothing. Further, some of them with respect to others have a very great alternation of forces; & this can vary if the structure is changed slightly, or if the particles are grouped & intermingled with others; in this case there follows another law of forces for the compound particles, which is different to that which we saw obeyed by the simple particles. If all these things are kept carefully in view, I really think that there can be found in this Theory the general theory for all chemical operations. For the special determination of effects that arise from each of the different mixtures of the different bodies, through which alone all effects in chemistry are produced, whether the bodies are resolved or compounded, would require an intimate knowledge of the structure of each kind of particle, & the arrangement of these in each of the masses; &, in addition, the whole power of geometry & analysis, such as exceeds by far the capacity of the human mind. But in general it is quite evident that there is no part of chemistry, in which, in addition to inertia of mass, & specific density, there are not everywhere produced other kinds of mutual forces between the particles; & these will meet our eyes without our looking for them, as is indeed abundantly evident in the single question that comes last at the end of Newton's

& vero etiam repulsionum indicia, atque argumenta proferuntur. Omnia etiam genera eorum, quæ ad Chemiam pertinent, singillatim persequi, infinitum esset: evolvam speciminis loco præcipua quædam.

Quid sint: dissolutio & præcipitatio: prima quomodo fiat, & quæ sit ejus causa.

[208] 452. Primo loco se mihi offerunt dissolutio, & ipsi contraria præcipitatio. Immissis in quædam fluida quibusdam solidis, cernimus, mutuum nexum, qui habebatur inter eorum particulas, dissolvi ita, ut ipsa jam nusquam appareant, quæ tamen ibidem adhuc manere in particulas perquam exiguas redacta, & dispersa, ostendit præcipitatio. Nam immisso alio corpore quodam, decidit ad fundum pulvisculus tenuissimus ejus substantiæ, & quodammodo depluit. Sic metalla in suis quæque menstruis dissolvuntur, tum ope aliarum substantiarum præcipitantur: aurum dissolvit aqua regia, quod immisso etiam communi sale præcipitatur. Rei ideam est admodum facile sibi efformare satis distinctam. Si particulæ solidi, quod dissolvitur, majorem habent attractionem cum particulis aquæ, quam inter se; utique avellentur a massa sua, & singulæ circumquaque acquirent fluidas particulas, quæ illas ambiant, uti limatura ferri adhæret magnetibus, ac fient quidam veluti globuli similes illi, quem referret Terra; si ei tanta aquarum copia affunderetur, ut posset totam alte submergere, vel illi, quem refert Terra submersa in aere versus eam gravitante. Si, ut reipsa debet accidere, illa vis attractiva in distantiis paullo majoribus sit insensibilis; ubi jam crit ad illam distantiam saturata eo fluido particula solidi, ulterius fluidum non attrahet, quod idcirco aliis eodem pacto particulis solidi immersi affundetur. Quare dissolvetur solidum ipsum, ac quidam veluti globuli terrulas suas cum ingenti affusa marium vi exhibebunt, quæ terrulæ ob exiguam molem effugient nostros sensus, nec vero decident sustentatæ a vi, quæ illas cum circumfuso mari conjungit: sed globuli illi ipsi constituent quandam veluti continui fluidi massam. Ea est dissolutionis idea.

Quomodo fiat præcipitatio, & quæ sit ejus causa.

453. Quod si jam in ejusmodi fluidum immittatur alia substantia, cujus particulæ particulas fluidi ad se magis attrahant, & fortasse ad majores etiam distantias, quam attrahuntur ab illis; dissolvetur utique hæc secunda substantia, & circa ipsius particulas affundentur particulæ fluidi, quæ prioris solidi particulis adhærebant, ab illis avulsæ, & ipsis ereptæ. Illæ igitur nativo pondere intra fluidum specifice levius depluent, & habebitur præcipitatio. Pulvisculus autem ille veterem particularum suarum nexum non acquiret ibi per sese, vel quia & gluten fortasse aliquod admodum tenue, quo connectebantur invicem, dissolutum simul jam deest in superficiebus illis, quarum separatio est facta, vel potius quia, ut ubi per limam, per tunsionem, vel aliis similibus modis solidum in pulverem redactum est, vel utcunque confractum, juxta ea, quæ diximus num. 413, non potest iterum solo accessu, & appressione deveniri ad illos eosdem limites, qui prius habebantur.

Pluviam fortasse esse quoddam præcipitationis genus: mira phænomena commixtionum quomodo explicentur.

454. Hoc pacto dissolutionis, & præcipitationis acquiritur idea admodum distincta; & fortasse etiam pluvia est quoddam præcipitationis genus, nec provenit e sola unione par-[209]-ticularum aquæ, quæ prius tantummodo dispersæ temere fuerint, & ob solam tenuitatem suam sustentatæ ac suspensæ innataverint. Apparet ibi etiam, qua ratione binæ substantiæ commisceri possint, & in unam massam coalescere. Id quidem in fluido commixto cum solido patet ex ipso superiore exemplo solutionis. In binis fluidis facile admodum fit, & si sint ejusdem ad sensum specificæ gravitatis, solo motu, & agitatione impressa fieri quotidie cernimus, ut in aqua, & vino, sed etiam si sint gravitatum admodum diversarum, attractione particularum unius in particulas alterius fieri potest unius dissolutio in altero, & commixtio. Fieri autem potest, ut ejusmodi commixtione e binis etiam fluidis oriatur solidum, cujusmodi exempla in coagulis cernimus: & in Physica illud quoque observatur quandoque, binas substantias commixtas coalescere in massam unicam minorem mole, quam fuerit prius, cujus phænomeni prima fronte admodum miri in promptu est causa in mea Theoria: cum particulæ, quæ nimirum se immediate non contingebant, aliis interpositis possint accedere ad se magis, quam prius accesserint. Sic si haberetur massa ingens elastrorum e ferro distractorum, quorum singulis inter cuspides adjungerentur globuli magnetici; hac nova accessione materiæ minueretur moles, vieta repulsione mutua

Optics, where there are many indications of both attractions & repulsions as well, & arguments are brought forward with regard to them. Further, to investigate separately all matters that relate to chemistry would be an endless task; so I will discuss certain of the more important, by way of example.

The nature of solution & precipitation; how the first comes about, & its cause.

452. In the first place there occur to me solution & its converse, precipitation. When certain solids are mixed with certain fluids, we see that the mutual connection which there used to be between the particles of each is dissolved in such a way that the solids are no longer visible; & yet that they are still there, reduced to extremely small particles & dispersed, is shown by precipitation. For, if a certain other body is introduced, there falls to the bottom an extremely fine powder of the original solid, as if it rained down. So metals, each in its own solvent, dissolve, & with the help of other substances are precipitated. "Aqua regia" dissolves gold; & this, on the addition of common salt, is precipitated. It is quite easy to get a clear idea of the matter. Suppose that the particles of the solid have a greater attraction for the particles of the water than for one another; then they will certainly be torn away from their own mass, & each of them will gather round itself fluid particles, which will surround it, in the same manner as iron filings adhere to a magnet; & each would become something in the nature of little spheres similar to what the Earth would resemble, if a sufficiency of water were to be poured over it to submerge it deeply, or to what the Earth does resemble, submerged as it is in the air gravitating towards it. If, as is bound to happen, the attractive force becomes insensible at distances a little greater, then, when a particle of a solid has become saturated to that distance with the fluid, it will no longer attract the fluid; & therefore the latter will surround other particles of the immersed solid in the same manner. Hence the solid will be dissolved, & each of the little spheres, so to speak, would represent a little earth with its great abundance of sea surrounding it; & these little earths, on account of their exceedingly small volume will escape our notice; & they cannot fall, sustained as they are by the force that connects them with the sea which surrounds them. Now these little globes themselves form a certain mass of as it were continuous fluid; hence we get an idea of the nature of solution.

The manner in which precipitation occurs; & its cause.

453. If now another substance is introduced into a fluid of this kind, the particles of which attract the particles of the fluid to themselves with a stronger force, & perhaps too at greater distances, than they are attracted by the particles of the first solid; then this second solid will be dissolved in every case, & its particles will be surrounded by the particles of the fluid, which formerly adhered to the particles of the first solid, being torn away from the latter & seized by the particles of the second solid. The particles of the first solid will then rain down on account of their own weight within the fluid which is specifically lighter, & there will be precipitation. Further, the fine powder will not of itself then acquire the former connection between its particles; this may be because a sort of very thin cement, by which the particles were connected together, has perhaps been at the same time dissolved, & this is now absent from the surfaces which have been separated; but more probably it is because, just as when, by means of a file or a hammer or the like, a solid has been reduced to powder, or broken up in any manner, it cannot by mere approach & pressing together get back once more to the same limit-points as before, as I said in Art. 413.

Perhaps rain is some sort of precipitation; how certain wonderful phenomena in connection with mixtures are explained.

454. In this way a perfectly clear idea of solution & precipitation is acquired. Perhaps also rain is some sort of precipitation, & does not merely come from the union of particles of water which previously had been only dispersed at random, & had floated, sustained & suspended in the air, owing to their extreme tenuity alone. Also, we can now see how two substances can be mixed together to coalesce into a single mass. This indeed, in the case of a fluid mixed with a solid, is evident from the example of solution given above. It takes place quite easily in the case of two fluids, &, if they are practically of the same specific gravity, we see it happening every day by mere motion & the agitation impressed; as in the case of water & wine. But even if their specific gravities are quite different, by the attraction of the particles of the one upon the particles of the other, there may be solution of the one in the other, & thus a mixture of the two. Further, it may happen that from a mixture of this kind, even of two fluids, there may be produced a solid; we see examples of such a thing in rennet. In Physics also, it is observed sometimes that two substances mixed together coalesce into a single mass having a smaller volume than before; the cause of this phenomenon, which at first sight appears wonderful, is to be found immediately with my Theory. For, the particles, which originally did not immediately touch one another, when others are interposed, may approach nearer to one another than they did before. Thus, if we have a large heap of springs made of iron, & to them we add a number of little magnetic spheres, placing one between the tips of each spring; then, with this fresh addition of matter, the whole volume is diminished, the mutual

per attractionem magneticam, qua cuspides elastrorum ad se invicem accederent.

Cur ad commixtionem solidorum requiratur contusio: quid ad eam præstet ignis: unde ars separandi metalla.

455. Ubi solidum cum solido commiscendum est, ut fiat unica massa, ibi quidem oportet solida ipsa prius contundere, vel etiam dissolvere, ut nimirum exiguæ particulæ seorsim possint ad exiguas alterius solidi accedere, & cum iis conjungi. Id autem fit potissimum per ignem, cujus vehementi agitatione, & vero etiam fortasse actione ingenti mutua inter ejus particulas, & inter quædam peculiaria substantiarum genera, ut olea, & sulphur, quæ ut gluten quoddam conjungebant inter se vel inertes particulas, vel etiam mutua repulsione præditas, dissolvit omnium corporum nexus mutuos, & massas omnes demum, si satis validus sit, cogit liquari, & ad naturam fluidorum accedere. Dissolutarum, ac liquescentium massarum particulæ commiscentur, & in unam massam coalescunt: ubi autem sic coaluerunt, possunt iterum sæpe dissimiles separari eadem actione ignis, qui aliquas prius, alias posterius, cogit minore vi abire per evaporationem, & maxime fixas majore vi reddit volatiles. Inæqualibus ejusmodi diversarum substantiarum attractionibus, & inæqualibus adhæsionibus inter earum particulas, omnis fere nititur ars separandi metalla a terris, cum quibus in fodinis commixta sunt, & alia aliorum ope prius uniendi, tum etiam a se invicem separandi, quæ omnia singillatim persequi infinitum foret. Generalis omnium explicatio facile repetitur ab illa, quam exposui, particularum diversa constitutione, quarum aliæ respectu aliarum inertes sunt, respectu aliarum activitatem habent, sed admodum diversam, tum [210] quod pertinet ad directionem, tum quod ad intensitatem virium.

Liquationem, & volatilizationem fieri posse per agitationem ingentem particularum. Prima quomodo fiat.

456 De Liquatione, & volatilizatione dicam illud tantummodo, eas fieri posse etiam sola ingenti agitatione particularum fluidi cujuspiam tenuissimi, cujus particulæ ad solidi, & fixi corporis particulas accedant satis, & inter ipsarum etiam intervalla irrumpant; qui motus intestinus, unde haberi possit, jam exponam, ubi de fermentatione egero, & effervescentia. Nam inprimis ea intestina agitatione induci potest in particulas corporis solidi, & fixi motus quidam circa axes quosdam, qui ubi semel inductus est, jam illæ particulæ vim exercent circunquaque circa illum axem ad sensum candem, succedentibus sibi invicem celerrime punctis, & directionibus, in quibus diversæ vires exercentur, qui etiam axes si celerrime mutentur, irregulari nimirum impulsu, habebitur in iis particulis id, quod æquivaleat sphæricitati & homogeneitati particularum, ex qua fluiditatem supra repetivimus, atque hujus ipsius rei exemplum habuimus num. 237 in motu puncti per peripheriam ellipseos, cujus focos bina alia puncta occupent. Hæc fluiditas erit violenta, & desinente tanta illa agitatione, ac cessante vi, quæ agitationem inducebat, cessabit, ac fluidum etiam sine admixtione novæ substantiæ poterit evadere solidum. Poterit autem paullatim cessare motus ille rotationis tam per inæqualitatem exiguam, quæ semper remanet inter vires in diversis locis particulæ diversas, & obsistit semper nonnihil rotationi, quam per ipsam expulsionem illius agitatæ substantiæ, ut igneæ, & per resistentiam circumjacentium.

Alia liquationis ratio per separationem partium heterogenearum.

457. Deinde haberi etiam poterit liquatio per subtractionem heterogenearum, & difformium particularum, quæ magis homogeneas, & ad sphæricitatem accedentes particulas alligabant quodammodo impedito motu in gyrum. Id sane videtur accidere in pluribus substantiis, quæ quo magis depurantur, & ad homogeneitatem reducuntur, eo minus tenaces evadunt, & viscosæ. Sic viscositas est minima in petroleo, major in naphtha, & adhuc major in asphalto, aut bitumine, in quibus substantiis Chemia ostendit, eo majorem haberi viscositatem, quo habetur major compositio.

Quomodo fiat volatilizatio: fixatio, & volatilizatio aeris

458. Quod si priore modo liquatio accidat, & in eo motu particulæ a limitibus cohæsionis, in quibus erant, abeant ad distantias paullo majores, in quibus habeatur ingens repulsivus arcus, se repente fugient, quo pacto corpus fixum evadet volatile. Eandem autem volatilitatem acquiret; si particulæ quæ fixum corpus componebant, erant quidem inter se in distantiis repulsionum validissimarum, sed per interjacentes particulas alterius substantiæ cohibebatur illa repulsiva vis superata ab attractione, quam exercebat in eas nova intrusa particula: si enim hæc agitatione illa excutiatur, vel ab alia, quæ ipsam attrahat magis, prætervolante ad exiguam distantiam abripia-[211]-tur; tum vero repulsiva vis particularum prioris substantiæ reviviscit quodammodo, & agit, ac ipsa substantia evadit volatilis, quæ iterum nova earundem particularum intrusione figitur. Id sane videtur accidere in aere, qui potest ad fixum redigi corpus, & Halesius

repulsion being overcome by the magnetic attraction, with which the tips of the springs would approach one another.

455. When a solid has to be mixed with a solid to form a single mass, it is necessary to first of all crush the solids, or even to dissolve them, so that the exceedingly small particles of the one can separately approach those of the other solid, & combine with them. Now this especially takes place in the case of fire; by its vigorous internal movement, & perhaps too through a very great mutual attraction between its particles & those of certain particular kinds of substance, like oils & sulphur, these two causes acting as a sort of cement to join together either inert particles, or even particles possessed of a mutual repulsion, fire dissolves the mutual connections of all bodies & finally forces, if it is sufficiently powerful, all masses to melt, & to approach fluids in their natures. The particles of the masses thus dissolved & in a molten condition mingle together & coalesce into one single mass. Moreover, after they have thus coalesced, the dissimilar substances can once more be separated by the same action of fire, which forces, some at first & others later, the particles to go off, with a smaller force through evaporation, & renders volatile the most refractory particles when the intensity is greater. Upon the unequal attractions of different substances of this kind, & upon the unequal adhesions between their particles, depends almost entirely the art of separating metals from the earths with which they are mixed in the ores; & some metals from others, by means of first uniting them & then separating them once more; but to investigate all these matters singly would be an endless task. The general explanation of them all is easily derived from that diverse constitution of the particles that I have expounded; namely, that some particles are inert with respect to others, & have activity with respect to yet others; where this activity is altogether varied, both as regards the directions, & as regards the intensities, of the forces.

Why crushing is necessary for a mixture of solids; the effect of fire in this respect; whence the art of separating metals.

456. With regard to liquefaction & volatilization, I will only say this: that these phenomena can take place simply through a violent agitation of some very tenuous fluid, whose particles approach sufficiently close to the particles of the solid fixed body, & push into the intervals between them. How this internal motion can happen I will explain, when I discuss fermentation & effervescence. First of all, owing to the internal agitation, there can be induced in the particles of the solid fixed body motions about certain axes; & when these motions have once been set up, the particles will exert a rotary force about the axis which is practically uniform, the points following one another extremely quickly, & also the directions in which the different forces are exerted; & if these axes are also changed very rapidly, due, say to an irregular impulse, we shall have in the particles what is equivalent to the sphericity & homogeneity of particles, from which we have derived fluidity in a preceding article; we had also an example of this kind of thing, in Art. 237, in the motion of a point along the perimeter of an ellipse, of which two other points occupied the foci. This fluidity will be very violent, &, as soon as the great agitation ends & the force which caused the agitation ceases, the agitation will cease as well, & the fluid will be able to become solid once more, without the admixture of any fresh substance. Further, this motion of rotation may gradually cease, owing not only to the slight inequality that will always remain between the different forces at different places of a particle, ever tending to hinder the rotation to some extent, but also to the expulsion of the substance in agitation (fire, say), & through the resistance of the particles lying in the neighbourhood.

Liquefaction & volatilization can take place owing to a very great agitation of the particles; the manner in which the first happens.

457. Secondly, there may be liquefaction through the subtraction of heterogeneous & non-uniform particles, which bound together the more homogeneous particles which approximate to sphericity, in such a way as to hinder their rotary motion. This is in fact seen to happen in several substances, which become less tenacious & viscous, the more they are purified & reduced to homogeneity. Thus the viscosity is very small in rock-oil, greater in naphtha, still greater in asphalt or bitumen; &, in these substances, chemistry shows that the viscosity is the greater, the more compound the substance.

Another reason for liquefaction is through the separation of heterogeneous parts.

458. But if liquation should take place in the first manner, & due to the motion the particles should go off from the limit-points at which they were to distances a little greater, & if for these distances there should be a very large repulsive arc, then the particles will fly off with great speed; & in this way a fixed body will become volatile. Moreover it will acquire the same volatility, if the particles which form the body were at such distances from one another as correspond to very strong repulsions, but are held together by intervening particles of another substance, the repulsive force being overcome by the attractions exerted upon them by the new particles that have been introduced between them. For, if these are displaced by the agitation, or are seized by others, which attract them more strongly, as they fly past at a slight distance, then the repulsive force of the first substance will revive, as it were, & come into action; & the substance will become volatile, & will once again become fixed on a fresh introduction of the same intervening particles. This in fact is seen to happen in the case of air, which can be reduced to a fixed body. Hales has proved

How volatilization takes place; fixation & volatilization of air.

demonstravit per experimenta, partem ingentem lapidum, qui in vesica oriuntur, & calculorum in renibus constare puro aere ad fixitatem reducto, qui deinde potest iterum statum volatilem recuperare: ac halitus inprimis sulphurei, & ipsa respiratio animalium ingentem acris copiam transfert a statu volatili ad fixum. Ibi non habetur aeris compressio sola facta per cellularum parietes ipsum concludentes; ii enim disrumperentur penitus, cum aer in ejusmodi fixis corporibus reducatur ad molem etiam millecuplo minorem, in quo statu, si integras haberet elasticas vires, omnia sane repagula illa diffringeret. Halesius putat, eum in illo statu amittere elasticitatem suam, quod fieret utique, si particulæ ipsius ad eam inter se distantiam devenirent, in qua jam vis repulsiva nulla sit, sed potius attractiva succedat: sed fieri itidem potest, ut vim quidem repulsivam adhuc ingentem habeant illæ particulæ, sed ab interposita sulphurei halitus particula attrahantur magis, ut paullo ante vidimus in elastris a globulo magnetico cohibitis, & constrictis. Tum quidem elasticitas in aere ad fixitatem redacto maneret tota, sed ejus effectus impediretur a prævalente vi. Atque id quidem animadverti, & monui ante aliquot annos in dissertatione *De Turbine*, in qua omnia turbinis ipsius phænomena ab hac acris fixatione repetii.

Causa agitationis particularum in igne, fermentationibus, effervescentiis repetita a contorsione curvæ circa axem.

459 Porro agitatio illa particularum in igne, ac in fermentationibus, & effervescentiis, unde oriatur, facile itidem est in mea Theoria exponere. Ut primum crus meæ curvæ mihi impenetrabilitatem exhibuit, postremum gravitatem, intersectiones autem varia cohæsionum genera; ita alternatio arcuum jam repulsivorum, jam attractivorum, fermentationes exhibet, & evaporationes variorum generum, ac subitas etiam deflagrationes, & explosiones, illas, quæ occurrunt in Chemia passim, & quam in pulvere pyrio quotidie intuemur. Quæ autem huc ex Mechanica pertinet, jam vidimus num. 199. Dum ad se invicem accedunt puncta cum velocitate aliqua, sub omni arcu attractivo velocitatem augent, sub omni repulsivo minuunt: contra vero dum a se invicem recedunt, sub omni repulsivo augent, sub omni attractivo minuunt, donec in accessu inveniant arcum repulsivum, vel in recessu attractivum satis validum ad omnem velocitatem extinguendam. Ubi eum invenerint, retro cursum reflectunt, & oscillant hinc, & inde, in quo itu, & reditu perturbato, ac celeri, fermentationis habemus ideam satis distinctam.

Oscillationes in accessu semper sisti a primo crure repulsivo, pro recessu bini casus. In primo cruris attractivi asymptotici semper sisti recessum etiam.

460. Et in accessu quidem semper devenitur ad arcum repulsivum aliquem parem extinguendæ velocitati cuilibet utcun-[212]-que magnæ; devenitur enim saltem ad primum asymptoticum crus, quod in infinitum protenditur: at pro recessu duo hic casus occurrunt potissimum considerandi. Vel enim etiam in recessu devenitur ad aliquod crus asymptoticum attractivum cum area infinita, de cujusmodi casibus egimus jam num. 195, vel devenitur ad arcum attractivum recedentem longissime, & continentem arcam admodum ingentem, sed finitam. In utroque casu actio punctorum, quæ extra massam sunt sita, aliorum punctorum massæ intestino illo motu agitatæ oscillationem augebit aliorum imminuet, & puncta alia post alia procurrent ulterius versus asymptotum, vel limitem terminantem attractivas vires: quin etiam actiones mutuæ punctorum non in directum jacentium in massa multis punctis constante, mutabunt sane singulorum punctorum maximos excursus hinc, & inde, & variabunt plurimum accessus mutuos, ac recessus, qui in duobus punctis solis motum habentibus in recta, quæ illa conjungit, deberent, uti monuimus num. 192, sine externis actionibus esse constantis semper magnitudinis. In accessu tamen in utroque casu ad compenetrationem sane nunquam deveniretur: in recessu vero in primo casu cruris asymptotici, & attractionis in infinitum crescentis cum area curvæ in infinitum aneta, itidem nunquam deveniretur ad distantiam illins asymptoti. Quare in eo primo casu utcunque vehemens esset interna massæ fermentatio, utcunque magnis viribus, ab externis punctis in majore distantia siris perturbaretur eadem massa, ipsius dissolutio per nullam finitam vim, aut velocitatem alteri parti impressam haberi unquam posset.

In secundo casu arcus attractivi ingentis, sed finiti egressus partis punctorum excursorum e fine oscillationis sine regressu.

461. At in secundo casu, in quo arcus attractivus ille ultimus ejus spatii ingens esset, sed finitus, posset utique quorundam punctorum in illa agitatione augeri excursus usque ad limitem, post quem limitem succedente repulsione, jam illud punctum a massa illa quodammodo velut avulsum avolaret, & motu accelerato recederet. Si post eum limitem summa arearum repulsivarum esset major, quam summa attractivarum, donec deveniatur ad arcum illum, qui gravitatem exprimit, in quo vis jam est perquam exigua, & area asymptotica ulterior in infinitum etiam producta, est finita, & exigua; tum vero puncti elapsi recessus ab illa massa nunquam cessaret actione massæ ipsius, sed ipsum punctum pergeret recedere, donec aliorum punctorum ad illam massam non pertinentium viribus sisteretur, vel detorqueretur utcunque. In fortuita autem agitatione interna, ut & in

by means of experiments that the great part of stones, that are produced in the bladder, & of the small ones in the kidneys, consists of pure air reduced to fixation; & that this can once again recover its volatile state. In this case the compression of the air is not obtained simply by the boundaries that enclose it; for these would be completely broken down, since the air in such fixed solids is reduced to a volume that is even a thousand times less; & in this state, if the elastic forces still were unimpaired, all restraints would be easily overcome. Hales thought that, when in this state, it loses its elasticity; & this would indeed happen if its particles attained that distance from one another, in which there is no repulsive force, but rather an attractive force succeeds the repulsive force. It might also happen that these particles still possess a very large repulsive force, but by the interposition of particles of a sulphurous vapour they are attracted to a greater extent than they are repelled; as just above we saw was the case for springs restrained & constricted by little magnetic spheres. Then, indeed, the elasticity in air reduced to fixity would remain unaltered, but its effect would be prevented by a superior force. I considered this point of view & mentioned it some years ago in my dissertation *De Turbine*, in which all the phenomena of the whirlwind are derived from this fixation of the air.

Cause of the agitation of the particles in fire, fermentations, & effervescence, derived from the contortions of the curve round the axis.

459. Further, the source of the agitation of the particles in fire, fermentation, & effervescence is also easily explained by my Theory. Just as the first branch of my curve gives me impenetrability, & the last branch gravitation, & the intersections with the axis the various kinds of cohesions; so also the alternation of the arcs, now repulsive, now attractive, represent fermentations & evaporations of various kinds, as well as sudden conflagrations & explosions; such things as occur everywhere in chemistry, & what we see every day in the case of gunpowder. Those things from Mechanics that belong here we have already seen in Art. 199. So long as points approach one another with any velocity, they increase the velocity under every attractive arc, & diminish it under every repulsive arc. On the other hand, so long as they recede from one another, they increase the velocity under every repulsive arc & increase it under every attractive arc; until, in approach, they come to a repulsive arc, or in recession, to an attractive arc, which is sufficiently strong to destroy the whole of the velocity. When they have reached this, they retrace their paths, & oscillate backwards & forwards; & in this, the backward & forward motion being perturbed & rapid, we have a sufficiently clear notion of what fermentation is.

Oscillations on approach are always stopped by that first repulsive branch, but for recession, there are two cases. In the first, where there is an asymptotic attractive branch, recession also is always stopped.

460. Now, on approach, there is always reached some repulsive arc or other, which is capable of destroying any velocity however great; for at least finally the first asymptotic branch, which goes off to infinity, is reached. But on recession, there are two cases met with, which have to be considered in this connection. For, on recession, either there is reached an asymptotic attractive branch having an infinite area, cases of which kind I dealt with in Art. 194; or else we come to an attractive arc receding very far from the axis, & containing an exceedingly great but finite area. In either case, the action of points situated outside the mass will increase the oscillation of some of the points of the mass that is agitated by the internal motion, & will diminish that of other points; & one point after another will go off beyond the mass towards the asymptote, or the limit-point bounding the attractive forces. Moreover, the mutual actions, of points not lying in the same straight line in a mass consisting of many points, will change considerably the largest oscillations of each of the points; especially will they alter their mutual approach & recession, which for two points only, having a motion in the straight line joining them, must be, except for external action, always of constant magnitude, as I remarked in Art. 192. On approach, however, in either case, the position corresponding to compenetration can never really be reached. But, on recession, in the first case, where there is an asymptotic branch, & an attraction indefinitely increased along with an area of the curve also increasing indefinitely, in this case also it can never attain the distance of that asymptote. Hence, in the first case, however fierce the internal fermentation of the mass may be, no matter with how great forces from external points situated further off the mass may be affected, its dissolution can never be effected by any finite force, or velocity impressed on any one part of it.

In the second case, where there is a very great but finite attractive arc, there will be separation of some of the points at the end of an oscillation, & these will fly off without returning.

461. Now, in the second case, in which the attractive arc at the end of the space is very large, but finite, it will indeed be possible for the motion of some points in the agitation to be increased right up to the limit-point; &, as repulsion follows the limit-point that point of the mass will now be as it were torn off, & it will fly away & leave the mass with accelerated motion. If after the limit-point, the sum of the repulsive areas should be greater than the sum of those that are attractive, that is, until that arc is reached which represents gravity, where the force then becomes exceedingly small, & the asymptotic area, when produced still further, is finite & very small; then indeed the recession of the point that has left the mass will never cease owing to any action of the mass itself, but the point will go on receding, until it is stopped by the forces from other points not belonging to that mass, or its path is contorted in some manner. Moreover, in irregular internal agitation,

externa perturbatione fortuita, illud accidet, quod in omnibus fortuitis combinationibus accidit, ut numerus casuum cujusdam dati generis in dato ingenti numero casuum æque possibilium dato tempore recurrat ad sensum idem, adeoque effluxus eorum punctorum, si massa perseveret ad sensum eadem, erit dato tempore ad sensum idem, vel, massa multum imminuta, imminuetur in aliqua ratione [213] massæ, cum a multitudine punctorum pendeat etiam casuum possibilium multitudo.

Inde pro diversa arcuum, forma evaporatio lenta.

462. Hic jam plurima considerari possent, & casuum differentium, ac combinationum numerus in immensum excrescit; sed pauca quædam adnotabimus. Ubi intervallum, quod massam claudit inter limites accessus, & recessus, est aliquanto majus, & posteriorum arearum repulsivarum summa non multum excedit summam attractivarum, fiet paullatim lenta quædam evaporatio: puncta quæ in fortuita agitatione ad eum finem deveniunt, erunt pauca respectu totius massæ, quæ tamen in ingenti massa, & eodem fermentationis statu erunt eodem tempore ad sensum æquali numero, ac, massa imminuta, imminuetur & is numerus, massa autem diu perseverabit ad sensum nihil mutata. Habebitur ibi quædam velut ebullitio, & vaporum quantitas, ac vis in egressu in diversis substantiis variari plurimum poterit, cum pendeat a situ, in quo illa puncta collocata sint intra curvam: nam possunt in aliis substantiis esse citra alios ingentes arcus attractivos, quorum posteriores vel sint prioribus minus validi, vel arcus repulsivos se subsequentes minus validos habeant.

Vel subita explosio, & deflagratio; ac transformationes variæ, avolante parte mixti.

463. Sed si intervallum, quod massam claudit inter limites accessus, & recessus, sit perquam exiguum, arcus attractivus postremus non sit ita validus, & succedat arcus repulsivus validissimus; fieri utique poterit, ut massa, quæ respective quiescebat, adveniente, exiguo motu a particulis externis satis proxime accedentibus, ut possint inæqualem motum imprimere punctis particularum massæ, agitatio ejusmodi in ipsa massa oriatur, qua brevissimo tempore puncta omnia transcendant limitem, & cum ingenti repulsiva vi, ac velocitate a se invicem discedant. Id videtur accidere in explosione subita pulveris pyrii, qui plerumque non accenditur contusione sola; sed exigua scintilla accedente dissilit fere momento temporis, & tanta vi repulsiva globum e tormento ejicit. Idem apparet in iis phosphoris, quæ deflagrant solo aeris contactu: ac nemo non videt, quanta in iis omnibus haberi possunt discrimina. Possunt nimirum alia facilius, alia difficilius deflagrare, alia serius, alia citius: potest sine lenta evaporatione solvi tota massa tempore brevissimo; potest, ubi massa fuerit heterogenea, avolare unum substantiæ genus aliis remanentibus. & interea possunt ex iis, quæ remanent, fieri alia mixta admodum diversa a præcedentibus, mutato etiam textu particularum altiorum ordinum per id, quod plures particulæ ordinum inferiorum, quæ pertinebant ad diversas particulas superiorum, coalescant in particulam ordinis superioris novi generis: hinc tam multæ compositiones, & transformationes in Natura, & in Chemia inprimis: hinc tam multa, tam diversa vaporum genera, & in aere elastico a tam diversis corporibus fixis genito tantum discrimen. Patet ubique immensus excursui campus: sed eo relicto [214] progredior ad alia nonnulla, quæ ad fermentationes, & evaporationes itidem pertinent.

Concretiones evaporato fluido, & certæ figuræ residui, ut in salibus.

464. Substantia, quæ fuerat dissoluta, non solum per præcipitationem colligitur iterum, ut ubi metalla cadunt suo pondere in tenuem pulvisculum redacta; sed etiam per evaporationem, ut diximus, in salibus, qui evaporato illo fluido, in quo fuerant dissoluti, remanent in fundo. Et quidem sales non remanent sub forma tenuis pulvisculi, particulis minutissimis prorsus inertibus, sed colliguntur in massulas grandiusculas habentes certas figuras quæ in aliis salibus aliæ sunt, & angulosæ in omnibus, ac in maxime corrosivis horrendum in modum cuspidatæ, ac serratæ, unde & sapores salium acutiores, & aliquorum ex iis, quæ corrosiva sunt, fibrillarum tenuium in animantibus proscissio, ac destructio organorum necessariorum ad vitam. Quo autem pacto eas potissimum figuras induere possint, id patet ex num. 439, ut & figuræ crystallorum & succorum, ex quibus gemmæ, & duri lapides fiunt ubi simplices sunt, & suam quique figuram affectant, ac aliorum ejusmodi, quæ post evaporationem concrescunt, haberi utique possunt, ut ibidem ostensum est, per hoc, quod in certis tantummodo lateribus, & punctis particulæ alias particulas positas ad certas distantias attrahant, adeoque sibi adjungant certo illo ordine, qui respondet illis punctis, vel lateribus.

just as also in irregular external perturbation, the same thing happens, as always does happen in irregular combinations; namely, out of a given very large number of cases of a given kind, all equally possible, the same number of cases will recur in any given interval of time. Hence, so long as the mass remains practically the same, there will be the same number of points going off; & when the mass is much diminished this number will also be diminished in some way proportional to the mass; for on the number of points depends also the number of possible cases.

Hence from a different form of the arcs comes a slow evaporation.

462. We may now consider a very large number of matters; & indeed the number of different cases & combinations increases immensely; but we will only mention just a few of them. When the interval, which encloses the mass between limits of approach & recession, is somewhat large, & the sum of the later repulsive areas does not greatly exceed that of the attraction, then a slow evaporation will take place. Points which, in the irregular agitation, arrive at the outside, will be few in comparison with the whole mass; & yet these, in a very large mass, in the same state of fermentation, will be practically of the same number in the same time; & this number will be diminished if the mass is diminished, but the mass itself will remain for a long time practically unaltered. Then there will be a sort of ebullition; & the amount of the vapour, & the force on egress may be very different in different substances; for it will depend on the position at which the points are situated within the curve. In some substances they may be on the near side of some, & in others of other, very great attractive arcs; & of these the later arcs may be either less powerful than those in front, or they may have less powerful repulsive arcs following them.

Or there may be sudden explosion & deflagration; & various transformations, as a part of the mixture flies off.

463. But if the interval, which encloses the mass between limits of approach & recession should be exceedingly small, the last attractive arc may not be so very strong, & a very strong repulsive arc may follow it. Then indeed, it may happen that, as the mass, which was in a state of relative rest, coming up to the limit with but a slight motion due to external points approaching close enough to it to be capable of impressing a non-uniform motion on the points of the particles, an agitation within the mass will be produced of such a kind that owing to it all the points in an extremely short time will cross the limit, & then they will fly off from one another with a huge repulsive force & a high velocity. This kind of thing is seen to take place in the sudden explosion of gunpowder, which commonly is not set on fire by a blow alone; but on contact with the smallest spark goes off almost at once, & with a very great repulsive force drives out the ball from the cannon. The same thing is seen in phosphorous substances, which go on fire merely on contact with the air; & nobody can fail to see the differences that may exist in all these things. Thus, some of them go on fire comparatively easily, others with greater difficulty, some slowly & others more suddenly; the whole of the mass may be broken up without any slow evaporation in an exceedingly short time. If the mass was originally heterogeneous, one part may fly off while the rest remains; & while this happens, the parts that remain may form fresh mixtures altogether different from the original, the structure of the particles of the higher orders even being altered; owing to the fact that several particles of lower orders, which originally belonged to different particles of higher orders, now coalesce into a particle of a higher order of a fresh kind. From this we get such a large number of compositions & transformations in Nature, & more especially in chemistry; hence we get such a large number of different kinds of vapours, & the great differences in elastic air, which is formed from such different fixed bodies. An immense field for inquiry is laid open; but I must leave it & go on to some other matters, which also refer to fermentations & evaporations.

Concretions, after evaporating the solvent; definite shapes in the residues, as for instance in salts.

464. A substance, which has been dissolved, can be once more obtained, not only by precipitation, as when metals fall by their own weight reduced to the form of an impalpable powder, but also by evaporation, as we have said, in the case of salts, which, on the fluid in which they were dissolved being evaporated, remain behind at the bottom. Nor indeed do salts remain behind in the form of a fine powder, with their minutest particles quite inert; but they are grouped together in fairly large masses having definite shapes, which differ for different salts; these are angular in all salts, & fearfully pointed & jagged in those salts of a particularly corrosive nature. In consequence, the salts are rather sharp to the taste; & with some of them, which are corrosive, there is a power of cutting the slender fibres of living things, & of destroying the organs that are necessary to life. The manner in which they can acquire these shapes especially is clear from Art. 439; as also the shapes of crystals & those jellies from which are formed gems & hard stones, when they are simple, & each adheres to its own shape; & also of some of the same kind, which take form after evaporation; & in every case this possibility is explained, as was also shown in the same article, from the fact that particles attract other particles situated at certain distances only at certain of their sides & points; & thus they will only attach them to themselves in a certain definite manner that corresponds to the particular points, or sides.

Quomodo possit fermentatio cessar.

465. Fermentatio paullatim minuitur, & demum cessat, cujus imminuti motus causas attigi pluribus locis, ut num. 197. Eodem autem pertinet illud etiam, quod innui num. 440. Irregularitas particularum, ex quibus corpora constant, & inæqualitas virium, plurimum confert ad imminuendum, & demum sistendum motum. Ubi nimirum aliquæ particulæ, vel totæ irruerunt in majorum cavitates, vel ubi suos uncos quosdam aliarum uncis, vel foraminibus inseruerunt, explicari non possunt, & sublapsus quidam, & compressiones particularum accidunt in massa temere agitata, quæ motum imminuunt & ad sensum extinguunt, quo & in mollibus sisti motus potest post amissam figuram. Multum itidem potest ad minuendum, ac demum sistendum motum sola asperitas ipsa particularum, ut motus in scabro corpore sistitur per frictionem; multum incursus in externa puncta, ut aer pendulum sistit: multum particulæ, quæ emittuntur in omnes plagas, ut in evaporatione, vel ubi corpus refrigescit, excussis pluribus igneis particulis, quæ dum evolant actione paticularum massæ, ipsis massæ particulis procurrentibus motum in partes contrarias imprimunt, & dum illæ, quæ oscillationem auxerant, aliæ post alias aufugiunt, illæ, quæ remanent, sunt, quæ oscillationes ipsas internis, & externis actionibus minuebant.

Cur quædam substantiæ fermentent cum quibusdam, & non cum aliis; cur quædam, ut fermentent, debeant contundi.

466. Porro non omnes substantiæ cum omnibus fermentant, sed cum quibusdam tantummodo: acida cum alcalinis; & [215] quod quibusdam videtur mirum, sunt quædam, quæ apparent acida respectu unius substantiæ, & alcalina respectu alterius. Ea omnia in mea Theoria facilem admodum explicationem habent: nam vidimus, particulas quasdam respectu quarundam inertes esse, cum quibus commixtæ idcirco non fermentant, respectu aliarum exercere vires varias: adeoque si respectu quarundam habeant pro variis distantiis diversas vires, & alternationem satis magnam attractionum, ac repulsionum; statim, ac satis prope ad ipsas accesserint, fermentant. Sic si limatura ferri cum sulphure commisceatur, & inspergatur aqua, oritur aliquanto post ingens fermentatio, quæ & inflammationem parit, ac terræmotuum exhibet imaginem quandam, & vulcanorum. Oportuit ferrum in tenues particulas discerpere, ac ad majorem mixtionem adhuc adhibere aquam.

Ignem esse fermentationis genus: quomodo excitetur tanta fermentatio ab exigua scintilla.

467. Ignem ego itidem arbitror esse quoddam fermentationis genus, quod acquirat vel potissimum, vel etiam sola sulphurea substantia, cum qua fermentat materia lucis vehementissime, si in satis magna copia collecta sit. Ignem autem voco eum, qui non tantum rarefacit motu suo, sed & calefacit, & lucet, quæ omnia habentur, quando materia illa sulphurea satis fermentescit. Porro ignis comburit, quia in substantiis combustibilibus multum adest substantiæ cujusdam, quæ sulphure abundat plurimum, & quæ idcirco sulphurea appellari potest, quæ vel per lucem in satis magna copia collectam, vel per ipsam jam fermentescentem sulphuream substantiam satis prægnantem ipsa lucida materia sibi admotam fermentescit itidem, & dissolvitur, ac avolat. Is ingens motus intestinus particularum excurrentium fit utique per vires mutuas inter particulas, quæ erant in æquilibrio: sed mutatis parum admodum distantiis exigui etiam punctorum numeri per exiguum unius scintillæ, vel tenuissimorum radiorum accessum, jam aliæ vires succedunt, & per earum reciprocationem perturbatur punctorum motus, qui cito per totam massam propagatur.

Exemplum aviculæ dimota arenula in summo monte dejicientis lapillos, saxa, rupes, & excitantis in mari subjecto undas immanes.

468. Imaginem rei admodum vividam habere possumus in sola etiam gravitate. Emergat e mari satis editus mons, per cujus latera dispositæ sint versus fundum ingentes lapidum prægrandium moles, tum quo magis ascenditur, eo minores; donec versus apicem lapilli sint, & in summo monte arenulæ: sint autem omnia fere in æquilibrio pendentia ita, ut vi respectu molis exigua devolvi possint. Si avicula in summo monte commoveat arenulam pede; hæc decidit, & lapillos secum dejicit, qui, dum ruunt, majores lapides secum trahunt, & hi demum ingentes illas moles: fit ruina immanis, & ingens motus, qui, decidentibus in mare omnibus, mare ipsum commovet, ac in eo agitationem ingentem, & undas immanes ciet, motu aquarum vehementissimo diutissime perdurante. Avi-[216]-cula æquilibrium arenulæ sustulit vi perquam exigua: reliquos motus gravitas edidit, quæ occasionem agendi est nacta ex illo exiguo motu aviculæ. Hæc imago quædam est virium intestinarum agentium, ubi cum vires crescere possint in immensum, mutata utcunque parum distantia; multo adhuc major effectus haberi potest, quam in casu gravitatis, quæ

465. The fermentation diminishes gradually, & at length ceases; I have touched upon the causes of this diminished motion in several places, for instance, in Art. 197. The remarks I made in Art. 440 also refer to the same thing. The irregularity of the particles, from which the bodies are formed, & the inequality of the forces, especially contribute to the diminution & final stoppage of the motion. Thus, when certain particles, or the whole of them enter cavities in larger particles, or when they insert their hooks into the hooks or openings of others, these cannot be disentangled, & certain relapses & compressions of the particles happen in a mass irregularly agitated, which diminish the motion & practically destroy it altogether; & due to this the motion even in soft bodies can be stopped after a loss of shape. Also the roughness of the particles alone may do much toward diminishing & finally stopping the motion; just as motion in a rough body is stopped by friction. Impact with external bodies has a great effect, e.g., the air stops a pendulum. Much may be due to the emission of particles in all directions, as in evaporation; or when a body freezes, many igneous particles being driven off in the process; & as these particles fly off by the action of the particles of the mass, impress a motion in the opposite direction on those particles as they move; & while those that had increased the oscillation, one after the other fly off, those that are left are such as were diminishing these oscillations by internal & external actions.

The manner in which fermentation may cease.

466. Further, all substances do not ferment with every substance, but with some of them only. Thus, acids ferment only with alkalies; &, what to some seems to be wonderful, there are some substances that appear to be acid with respect to one substance, & alkaline with respect to another. Now, all these things have a perfectly easy explanation in my Theory. For, we have seen that certain particles are inert with regard to certain other particles, & therefore when these are mixed together there will be no fermentation. With regard to ethers, again, they exert various forces; hence, if with respect to certain of them they have different forces for different distances, & a sufficiently great alternation of attractions & repulsions, they will immediately ferment on being brought into sufficiently close contact with them. Thus, if iron-filings are mixed with sulphur, & moistened with water, there will be produced in a little time a great fermentation; & this also produces inflammation, & exhibits phenomena akin to earthquakes & volcanoes. It is necessary, however, that the iron should be powdered very finely, & that water should be used to give a still closer mingling of the particles.

The reason why some substances ferment with certain substances & not with others; why some must be powdered before they will ferment.

467. I believe also that fire itself is some kind of fermentation, which is acquired, either more especially, or even solely by some sulphurous substance, with which the matter forming light ferments very vigorously, if it is concentrated in sufficiently great amount. Moreover I apply the term fire to that which not only rarefies through its own motion, but also produces heat & light; & all these conditions are present when the sulphurous substance ferments sufficiently. Further, fire burns, because in combustible substances there is present much of a substance largely consisting of something like sulphur, for which reason it may be termed a sulphurous substance. Such a substance, either by contact with light concentrated in sufficiently great amount, or by contact with the already fermenting sulphurous substance which is charged with the matter of light to a sufficient degree, will also ferment, & be broken up, & fly off. The very great internal motion of the particles flying off is in every case due to the mutual forces between the particles, which originally were in equilibrium; but, the distances of even a very small number of points being changed ever so little, by the slightest accession of a spark, or of its feeblest rays, other forces then take their place, the motion of the points is also disturbed by their oscillations, & this is quickly propagated throughout the whole of the mass.

Fire is some sort of fermentation; the manner in which so great a fermentation can be excited by the slightest of sparks.

468. We can obtain a really vivid picture of the matter, even in the case of gravity alone. Suppose that from the sea there rises a mountain of considerable height, & that along the sides of it there lie immense masses of huge stones, & the higher one goes, the smaller the stones are; until towards the top the stones are quite small, & at the very summit they are mere grains of sand. Also suppose that all of these are just in equilibrium, so that they can be rolled down by a very slight force compared with their whole volume. If, now, a little bird on the top of the mountain moves with his foot just one grain of the sand, this will fall, & bring down with it the small stones; these, as they fall, will drag with them the larger stones, & these in their turn will move the huge boulders. There will be an immense collapse & a huge motion; &, as all these stones fall into the sea, the motion will communicate itself to the sea & cause in it a huge agitation & immense waves, & this vigorous motion of the water will last for a very considerable time. The little bird disturbed the equilibrium of the grain of sand with a very slight force; gravity produced the remaining motions, & it obtained its opportunity for acting through the slight motion of the little bird. This is a kind of picture of the internal forces that act, when, owing to the possibility of the forces increasing indefinitely, on the distance being changed ever so slightly, a much

As example, in the case of a little bird, by moving a single grain of sand on the top of a mountain, hurling down stones, rocks, boulders, & exciting huge waves in the sea that lies at the foot of the mountain.

quidem perseverat eadem, aneta tantummodo velocitate descensus per novas accelerationes.

Quæ careant penitus materia sulphuræ, ab igne non debere lædi: hinc fortasse in ipso Sole posse manere substantias illæsas.

469. Quod si ignis excitatur tantummodo per sulphureæ substantiæ fermentationem; ubi nihil adsit ejus substantiæ, nullus erit metus ab igne. Videmus utique, quo minus ejusmodi substantiæ corpora habeant, eo minus igni obnoxia esse, ut ex amianto & telæ fiant, quæ igne moderato purgantur, non comburuntur. Censeo autem idcirco nostras basce terrestres substantias ab igne satis intenso dissolvi omnes, & inflammari, quod omnes ejusmodi substantiæ aliquid admixtum habeant, quod nectat etiam inter se plurimas inertes particulas. At si corpora haberentur aliqua, quæ nihil ex ejusmodi substantia haberent admixtum; ea in medio igne vehementissimo illæsa perstarent, nec ullum motum acquirerent, quem nimirum nostra hæc corpora acquirunt ab igne non per incursum, sed per fermentationem ab internis viribus excitatam. Hine in ipso Sole, & fixis, ubi nostra corpora momento fere temporis conflagrarent, & in vapores abirent tenuissimos, possunt esse corpora ea substantia destituta, quæ vegetent, & vivant sine ulla organici sui textus læsione minima. Videmus certe maculas superficiei Solis proximas durantes aliquando per menses etiam plures, ubi nostræ nubes, quibus eæ videntur satis analogæ, brevissimo tempore dissiparentur.

Exemplum fermentationis, quam cum aceto habent aliquæ terræ, aliis illæsis.

470. Id mirum videbitur homini præjudiciis præoccupato; nec intelliget, qui fieri possit, ut vivat aliquid in Sole ipso, in quo tanto major esse debet vis ustoria, dum hic exiguus radiorum solarium numerus majoribus cavis speculis, vel lentibus collectus dissolvit omnia. At ut evidenter pateat, cujusmodi præjudicium id sit: fingamus nostra corpora compacta esse ex illis terris, quas bolos vocant, quæ a diversis aquis mineralibus deponuntur, quæ cum acidis fermentant, ac omnia corpora, quæ habemus præ manibus, vel ex eadem esse terra, vel plurimum ex ea habere admixtum. Acetum nobis haberetur loco ignis: quæcunque corpora in acetum deciderent, ingenti motu excitato dissolverentur citissime, & si manum immitteremus in acetum: ea ipsa per fermentationem exortam amissa, protinus horrore concuteremur ad solam aceti viciniam, & eodem modo videretur nobis absurdum quoddam, ubi audiremus, esse substantias, quæ acetum non metuant, & in eo diu perstare possint sine minimo motu, atque sui textus læsione, quo vulgus rem prorsus absurdam censebit, si audiat, in medio igne, in ipso Sole, posse haberi corpora, quæ **[217]** nullam inde læsionem accipiant, sed pacatissime quiescant, & vegetent, ac vivant.

De lumine: sententiam de emissione luminis præferendam omnino undis fluidi elastici.

471. Hæc quidem de igne; jam aliquid de luce, quam ignis emittit, & quæ satis collecta ipsum excitat. Ipsa lux potest esse effluvium quoddam tenuissimum, & quasi vapor fermentatione ignea vehementi excussus. Et sane validissima, meo quidem judicio, argumenta sunt, contra omnes alias hypotheses, ut contra undas, per quas olim phænomena lucis explicare conatus est Hugenius, quam sententiam diu consepultam iterum excitare conati sunt nuper summi nostri ævi Geometræ, sed meo quidem judicio sine successu (r): nam explicarunt illi quidem, & satis ægre, paucas admodum luminis proprietates, aliis intactis prorsus, quas sane per eam hypothesim nullo pacto explicari posse censeo, & quarum aliquas ipsi arbitror omnino opponi: sed eam sententiam impugnare non est hujus loci, quod quidem alibi jam præstiti non semel. Mirum sane, quam egregie in effluviorum emanantium sententia ex mea Theoria profluant omnes tam variæ lucis proprietates, quam explicationem fuse persecutus sum in secunda parte dissertationis *De Lumine:* præcipua capita hic attingam; interea illud innuam, videri admodum rationi consentaneam ejusmodi sententiam materiæ effluentis, vel ex eo, quod in ingenti agitatione, quam habet ignis, debet utique juxta id, quod vidimus num. 195, evolare copia quædam particularum, ut in ebullitionibus, effervescentiis, fermentationibus passim evaporationes habentur.

Proprietates luminis, quarum reddenda est ratio.

472. Præcipuæ proprietates luminis sunt ejus emissio constans, & ab æquali massa, ut ab eodem Sole, ab ejusdem candelæ flamma, ad sensum eadem intensitate: immanis velocitas, nam semidiametrorum terrestrium 20 millia, quanta est circiter Solis a Terra

(r) *Cum hæc scriberem, nondum prodierant Opera Taurinensis Academiæ; nec vero huc usque, dum hoc Opus reimprimitur, adhuc videre potui, quæ Geometra maximus La Grange hoc in genere protulit.*

greater effect can be obtained, than is the case for gravity; for, this remains the same, the velocity of descent being only increased by fresh accelerations.

Substances, that are quite without sulphurous matter, are not necessarily impaired by fire; hence, perhaps in the Sun itself there may remain substances uninjured.

469. But if fire is excited only by the fermentation of sulphurous matter; then, when none of this matter is present, there will be no danger from fire. We see indeed, the less of this substance the bodies have, the less liable they are to be injured by fire; thus, a material is woven from asbestos, & this is only purified, but not burned, by moderate fire. Further, I consider that all our earthy substances are broken up by fire, provided it is sufficiently intense, & are set on fire, just because all substances of this kind have something mixed with them, which connects a large number of inert particles together. However, if there were any bodies which had nothing at all of such a substance mixed with them, these would be unaltered in the heart of the most vigorous fire, & would not acquire any motion, that is to say, such motion as the bodies about us acquire from fire, not through the entrance of fiery particles, but through fermentation excited by internal forces. Hence, in the Sun itself, & in the stars, in which our terrestrial bodies would burn up in an instant of time & go off into the thinnest of vapours, there may exist bodies altogether lacking in such a substance; & these may grow & live without the slightest injury of any kind to their organic structure. Indeed we see spots very close to the Sun lasting sometimes for several months even; whereas our clouds, to which these spots seem to bear a considerable analogy, would be dissipated in a very short time.

Example, in the case of fermentation which some earths have with vinegar, while others are unaffected.

470. Now this will appear wonderful to a man who is obsessed by prejudices; nor will he be able to understand why it is that anything can live in the Sun, in which there is bound to be ever so much greater burning force, while on earth an exceedingly small number of solar rays, collected by fairly large concave mirrors or by lenses, will break up all substances. However, in order to make plain how such a prejudice arises, let us suppose that our substances are formed from those earths, which are termed boluses, such as are deposited by certain minerals of different kinds & ferment with acids; & that all bodies around us either are formed out of this earth or are largely impregnated with it. Let vinegar be taken to represent fire; then if any of these bodies fell into the vinegar, they would be very quickly broken up by the huge motion induced; & if we placed our hands in the vinegar, they too being lost by the fermentation produced, we should be forthwith struck with horror at the mere vicinity of vinegar. It would seem to us that it was something ridiculous if we were told that there were substances which were in no fear of vinegar, but could last in it for a long time without slightest motion or injury to their structure; in exactly the same way as an ordinary man would think it ridiculous, if he were told that in the heart of fire, or in the Sun itself, there might exist bodies which received no injury from it, but remained at rest in the most calm fashion, & grew & lived.

Light; the theory of emission of light to be preferred altogether before that of waves in an elastic fluid.

471. So much on the subject of fire; now I will make a few remarks about light, which is given off by fire, & which, when present in sufficient quantity, excites fire. It is possible that light may be a sort of very tenuous effluvium, or a kind of vapour forced out by the vigorous igneous fermentation. Indeed, in my judgment, there are very strong arguments in favour of this hypothesis, as opposed to all other hypotheses, such as that of waves. On the hypothesis of waves, Huygens once tried to explain all the phenomena of light; & the most noted of the geometers of our age have tried to revive this theory, which had been buried with Huygens; but, as I think, unsuccessfully [r]. For, they have explained, & even then poorly enough, a very few of the properties of light, leaving the rest untouched; & indeed I consider that such properties can not be explained in any way by this hypothesis of waves, & my opinion is that some of them are altogether contrary to it. But this is not the right place to impugn this theory; indeed I have already, more than once, presented my view in other places. It is really marvellous how excellently, on the hypothesis of emanating effluvia, all the different properties of light are derived from my Theory in a straightforward way. I gave a very full explanation of this in the second part of my dissertation, *De Lumine*; & the principal points of this work I will touch upon here. Meanwhile, I will just mention that the idea of effluent matter seems to be altogether reasonable; more especially from the fact that, in a very great agitation amongst particles, such as there is in the case of fire, there is always bound to be, in accordance with what we have seen in Art. 195, an abundance of particles flying off, just as we have evaporations in ebullition, effervescence & fermentation.

Those properties of light for which we have to find the reason.

472. The principal properties of light are:—its constant emission, & the fact that the intensity is always the same from the same mass, such as from the Sun, or from the flame of the same candle; its huge velocity, for it traverses a distance equal to twenty thousand times the semidiameter of the Earth, which is about the distance of the Sun

(r) *When I wrote this, the Transactions of the Academy of Turin had not been published; and even now, at the time of this reprint of my work, I have so far been unable to see what that excellent geometer La Grange has published on the subject.*

distantia, percurrit semiquadrante horæ; velocitatum discrimen exiguum in diversis radiis, nam celeritatis discrimen in radiis homogeneis vix ullum esse, si quod est, colligitur pluribus indiciis: propagatio rectilinea per medium diaphanum ejusdem densitatis ubique cum impedimento progressus per media opaca, sine ullo impedimento sensibili ex impactu in se invicem radiorum tot diversas directiones habentium, aut in partes internas diaphanorum corporum utcunque densorum: reflexio partis luminis ad angulos æquales in mutatione medii, parte, quæ reflectitur, eo majore respectu luminis, quo obliquitas incidentiæ est major; refractio alterius partis eadem mutatione cum lege constantis rationis inter sinum incidentiæ, & sinum anguli refracti; quæ ratio [218] in diversis coloratis radiis diversa est, in quo stat diversa diversorum coloratorum radiorum refrangibilitas: dispersio & in reflexione, & in refractione exiguæ partis luminis cum directionibus quibuscunque quaquaversus: alternatio binarum dispositionum in quovis radio, in quarum altera facilius reflectatur, & in altera facilius transmittatur lux delata ad superficiem dirimentem duo media heterogenea, quas Newtonus vocat vices facilioris reflexionis, & facilioris transmissus, cum intervallis vicium, post quæ nimirum dispositiones maxime faventes reflexioni, vel refractioni redeunt, æquabilis in eodem radio ingresso in idem medium, & diversis coloratis radiis, in diversis mediorum densitatibus, & in diversis inclinationibus, in quibus radius ingreditur, ex quibus vicibus, & earum intervallis diversis in diversis coloratis radiis pendent omnia phænomena laminarum tenuium, & naturalium colorum tam permanentium, quam variabilium, uti & crassarum laminarum colores, quæ omnia satis luculenter exposuit in celebri dissertatione *De Lumine* P. Carolus Benvenuti e Soc. nostra Scriptor accuratissimus: ac demum illa, quam vocant diffractionem, qua radii in transitu prope corporum acies inflectuntur, & qui diversum colorem, ac diversam refrangibilitatem habent, in angulis diversis.

Emissio quomodo fiat: qui fiat, ut quædam simul citissime dissolvantur dum lumen emittunt, ut ignis subitus, quædam, ut Sol, diutissime persistent sine sensibili jactura.

473. Quod pertinet ad emissionem jam est expositum num. 199, & num 461; ubi etiam ostensum est illud, manente eadem massa quæ emittit effluvia, ipsorum multitudinem dato tempore esse ad sensum eandem. Porro fieri potest, ut massa, quæ lumen emittit, penitus dissolvatur, ut in ignibus subitis accidit, & fieri potest, ut perseveret diutissime, Id potissimum pendet a magnitudine intervalli, in quo fit oscillatio fermentationis, & a natura arcus attractivi terminantis id intervallum juxta num. 195. Quin immo si Auctor Naturæ voluit massam vehementissima etiam fermentatione agitatam prorsus indissolubilem quacunque finita velocitate, potuit facile id præstare juxta num. 460 per alios asymptoticos arcus cum areis infinitis, intra quorum limites sit massa fermentescens; quorum ope ea colligari potest ita, ut dissolvi omnino nequeat, ponendo deinde materiam luminis emittendi, ultra intervallum carum asymptotorum respectu particularum ejus massæ, & citra arcum attractivum ingentis areæ, sed non infinitæ, ex quo aliæ lucidæ particulæ evolare possint post alias. Nec illud, quod vulgo objici solet, tanta luminis effusione debere multum imminui massam Solis, habet ullam difficultatem, posita illa componibilitate in infinitum & illa solutione problematis quæ habetur num. 395. Potest enim in spatiolo utcunque exiguo haberi numerus utcunque ingens punctorum, & omnis massa luminis, quæ diffusa tam immanem molem occupat, potest in Sole, vel prope Solem occupavisse spatiolum, quantum libuerit, parvum, ut idcirco Sol post quotcunque sæ-[219]-culorum millia ne latum quidem unguem decrescat. Id pendet a ratione densitatis luminis ad densitatem Solis, quæ ratio potest esse utcunque parva; & quidem pro immensa luminis tenuitate sunt argumenta admodum valida, quorum aliqua proferam infra.

Unde tanta velocitas; cum velocitatis discrimen exiguum, & in radiis homogeneis multo minus.

474. Celeritas utcunque magna haberi potest ab arcubus repulsivis satis validis, qui occurrant post extremum limitem oscillationis terminatæ ab arcu ingenti attractivo juxta num. 194: nam si inde evadat particula cum velocitate nulla; quadratum velocitatis totius definitur ab excessu arearum omnium repulsivarum supra omnes attractivas juxta num. 178, qui excessus cum possit esse utcunque magnus; ejusmodi celeritas potest itidem esse utcunque magna. Verum celeritatis discrimen in particulis homogeneis erit prorsus insensibile, qui a particulæ luminis ejusdem generis ad finem oscillationis advenient cum velocitatibus fere nullis: nam eæ, quæ juxta Theoriam expositam num. 195, paullatim augent oscillationem suam, demum adveniunt ad limitem cohibentem massam, & avolant;

from the Earth, in an eighth of an hour ; the slight differences of velocity that exist in different rays, for it is proved from several indications that there is scarcely any difference for homogeneous light, if there is any at all ; its rectilinear propagation through a transparent medium everywhere equally dense, along with hindrance to progression through opaque media ; & this without any sensible hindrance due to impact with one another of rays having so many different directions, or any that prevents passage into the inner parts of transparent bodies, no matter how dense they may be ; reflection of part of the light at equal angles at the surface of separation of two media, the part that is reflected being greater with regard to the whole amount of light, according as the obliquity of incidence is greater ; refraction of the other part at the same surface of separation, with the law of a constant ratio between the sines of the angle of incidence & the angle of refraction, the ratio being different for differently coloured rays, upon which depends the different refrangibility of the differently coloured rays ; dispersion, both in reflection & in refraction, of a very small part of the light in directions of every description whatever ; the alternation of propensity in any one ray, in one of which the light falling upon the surface of separation between two media of different nature is the more easily reflected & in the other is the more easily transmitted, which Newton calls 'fits' of easier reflection & easier transmission, with intervals between these fits, after which the propensities mostly favouring reflection or refraction return, these intervals being equal in the same ray entering the same medium, & different for differently coloured rays, for different densities of the medium, & for the different inclinations at which the ray enters the medium ; upon these fits & the different intervals between them for differently coloured rays depend all the phenomena of thin plates, & of natural colours, both variable & permanent, as well as the colours of thick plates, all of which have been discussed with considerable clearness by Fr. C. Benvenuti, a most careful writer of our Society, in his well-known dissertation, *De Lumine*. Last of all, we have that property, which is called diffraction, in which rays, passing near the edge of a body, are bent inwards, having a different colour & different refrangibility for different angles.

How emission takes place ; how it happens that some bodies are very quickly broken up at the time they emit light, like a sudden flash of fire, while others, like the Sun persist for a very long time without any apparent loss.

473. What pertains to emission has been already explained in Art. 199 & Art. 461 ; there also it was shown that, if the mass emitting the effluvia remained the same, then the amount emitted is practically the same in any given time. Further, it may happen that the mass emitting the light is completely broken up, as takes place in sudden flashes of fire ; or it may be that this mass persists for a very long time. This to a very great extent depends on the size of the interval in which the oscillation due to fermentation takes place, & on the nature of the attractive arc at the end of that interval, by Art. 195. Nay, if the Author of Nature had wished that a mass, agitated by the most vigorous fermentation even, should be quite irreducible by any finite force whatever, he could easily have accomplished this, as shown in Art. 460, by other asymptotic arcs with infinite areas, between the confines of which the fermenting mass would be situated. By the aid of these arcs the mass could be so bound together, that it would not admit of the slightest dissolution ; & then by placing the material for emitting light further from the particles of the mass than the interval between those asymptotes, & within the distance corresponding to an attractive arc of huge but finite area ; from which we should have particles, one after the other, of light flying off. Nor is there any difficulty from the usual argument that is raised in objection to this, that the mass of the Sun must be much diminished by such a large emission of light ; if we suppose indefinitely great componibility, & the solution of the problem, given in Art. 395. For in any exceedingly small space there may be any huge number of points whatever ; & the whole mass of the light, which is diffused throughout & occupies such an immense volume, may, in the Sun or near the Sun, have occupied a space as small as ever one likes to assign ; so that the Sun, after the lapse of any number of thousands of centuries, will not therefore have decreased by even a finger's breadth. It all depends on the ratio of the density of light to the density of the Sun, & this ratio can be any small ratio whatever. Indeed there are perfectly valid arguments for the immense tenuity of light, some of which I will give below.

Whence comes the great velocity, notwithstanding the slight differences in velocity, & the still less differences in homogeneous rays.

474. Any velocity, no matter how great, can be obtained from sufficiently powerful repulsive arcs, if these occur after the last limit of oscillation within the confines of a very great attractive arc, as shown in Art. 194. For if a particle goes off from here with no velocity, the square of the whole velocity is defined by the excess of all the repulsive areas over all the attractive, as was shown in Art. 178 ; &, as this excess can be of any amount whatever, the velocity can also be of any magnitude whatever. Again, the difference of velocity for homogeneous particles is quite insensible, because particles of light of the same kind come to the end of their oscillation with velocities that are almost zero ; for those which, according to the Theory set forth in Art. 195, increase their oscillation gradually, arrive at the boundary limiting the mass at last, & then fly off. Now, if, at the time they

quo si tum, cum avolant, advenirent cum ingenti velocitate, advenissent utique eodem, & effugissent in oscillatione præcedenti. Demonstravimus autem ibidem, exiguum discrimen velocitatis in ingressu spatii, in quo datæ vires perpetuo accelerant motum, & generant velocitatem ingentem, inducere discrimen velocitatis genitæ perquam exiguum etiam respectu illius exigui discriminis velocitatis initialis, quod demonstravimus ibi ratione petita a natura quadrati quantitatis ingentis conjuncti cum quadrato quantitatis multo minoris, quod quantitatem exhibet a priore illa differentem multo minus, quam sit quantitas illa parva, cujus quadratum conjungitur. Discrimen aliquod sensibile haberi poterit; siqua effugiunt, non sint puncta simplicia, sed particulæ non nihil inter se diversæ: nam curva virium, qua massa tota agit in ejusmodi particulas, potest esse nonnihil diversa pro illis diversis particulis, adeoque excessus summæ arearum repulsivarum supra summam attractivarum potest esse nonnihil diversus & quadratum velocitatis ipsi respondens nonnihil itidem diversum. Hoc pacto particulæ luminis homogeneæ habebunt velocitatem ad sensum prorsus æqualem; particulæ heterogeneæ poterunt habere nonnihil diversam, uti ex observatione phænomenorum videtur omnino colligi. Illud unum hac in re notandum superest, quod curva virium, qua massa tota agit in particulam positam jam ultra terminum oscillationum, mutatis per oscillationem ipsam punctis massæ, mutabitur nonnihil: sed quoniam in fortuita ingenti agitatione massæ totius celerrime succedunt omnes diversæ positiones punctorum; summa omnium erit ad sensum eadem, potissimum pro particula diutius hærente in illo initio suæ fugæ, ad quod advenit, uti diximus, cum velocitate perquam exigua, ut idcirco homogenearum velocitas, [220] ubi jam deventum fuerit ad arcum gravitatis, & vires exiguas, debeat esse ad sensum eadem, & discrimen aliquod haberi possit tantummodo in heterogeneis particulis a diverso earum textu. Patet igitur, unde celeritas ingens provenire possit, & si quod est celeritatis discrimen exiguum.

Unde propagatio rectilinea: incursum immediatum punctorum lucis, in puncta medii nullum haberi: virium in medio homogeneo exiguam inæqualitatem eludi a tenuitate, & celeritate luminis.

475. Quod pertinet ad propagationem rectilineam per medium homogeneum diaphanum, & ad motum liberum sine ullo impedimento a particulis ipsius luminis, vel medii diaphani, id in mea Theoria admodum facile exponitur, quod in aliis ingentem difficultatem parit. Et quidem quod pertinet ad impedimenta, si curva virium nullum habeat arcum asymptoticum perpendicularem axi præter primum; ostensum est num. 362, sola satis magna velocitate obtineri posse apparentem compenetrationem duarum substantiarum, quam tenuitas, & homogeneitas spatii, per quod transitur, plurimum juvat. Quoniam respectu punctorum materiæ prorsus indivisibilium, & inextensorum infinities infinita sunt puncta spatii existentia in eodem plano; infinities infinite est improbabilis pro quovis momento temporis directio motus puncti materiæ cujusvis accurate versus aliud punctum materiæ, ac improbabilitas pro summa momentorum omnium contentorum dato quovis tempore utcunque longo evadit adhuc infinita. Ingens quidem est numerus punctorum lucis, & propemodum immensus, sed in mea Theoria utique finitus. Ea puncta quovis momento temporis directiones motuum habent numero propemodum immenso, sed in mea Theoria finito. Verum quidem est, ubicunque oculus collocetur in immensa propemodum superficie sphæræ circa unam fixam remotissimam descripta, immo intra ipsam sphæram, videri fixam, & proinde aliquam luminis particulam afficere nostrum oculum: sed id fit in mea Theoria non quia accurate in omnibus absolute infinitis directionibus adveniant radii, sed quod pupilla, & fibræ oculorum non unicum punctum sunt, & vires punctorum particulæ luminis agunt ad aliquod intervallum. Hinc quovis utcunque longo tempore nullus debet accidere casus in mea Theoria, in quo punctum aliquod luminis directe tendat contra aliquod aliud punctum vel luminis, vel substantiæ cujusvis, ut in ipsum debeat incurrere. Quamobrem per incursum, & immediatum impactum nullum punctum luminis aut sistet motum suum, aut deflectet.

Si satis magnam velocitatem habeant; quævis, solida etiam, transitura trans alia solida sine ulla motuum perturbatione.

476. Id quidem commune est omnibus corporibus, quæ corpora inter se congrediuntur. Ea nullum habent in mea Theoria punctum immediatum incurrens in aliud punctum; quam ob causam & illud ibidem dixi, si nullæ vires mutuæ adessent, debere utique haberi apparentem quandam compenetrationem omnium massarum; sed adhuc vel ex hoc solo capite veram compenetrationem haberi nunquam omnino posse. Vires igitur quæ ad aliquam distantiam protenduntur, im- [221] -pediunt progressum. Eæ vires si circumquaque essent semper æquales; nullum impedimentum haberet motus, qui vi inertiæ deberet

fly off, they should reach this boundary with a very great velocity, then it is certain that they would have reached it & flown off in a previous oscillation. Further, in the same article, we have proved that a slight difference of velocity on entering a space, in which given forces continually accelerate the motion & generate a huge velocity, also induces a difference in the velocity generated that is very small even when compared with the slight difference in the initial velocity. This we there prove from an argument derived from the nature of the square of a very large quantity compounded with the square of a quantity much less than it; this gives a quantity differing from the first quantity by something much less than the small quantity of which the square was added. A sensible difference may be obtained, if what fly off are not simple points, but particles somewhat different from one another. For the curve of forces, with which the mass acts upon such particles, can be somewhat different for those different particles; & thus, the excess of the sum of the repulsive areas over the sum of the attractive may be somewhat different, & therefore the square of the velocity corresponding to this excess may be somewhat different. In this way particles of homogeneous light will have velocities that are practically equal; but particles of heterogeneous light may have velocities that are somewhat different; as seems to be conclusively shown from observations of phenomena. One thing remains to be noted in this connection, namely, that the curve of forces, with which the whole mass acts upon a particle placed already beyond the limit of the oscillation, when the points of the mass are changed on account of the oscillation, will be somewhat altered. But since in a very large irregular agitation of the entire mass all the different positions of the points follow on after one another very quickly, the sum of all the forces will be practically the same, especially in the case of a particle stopping for some time at the beginning of its flight; which point it has reached, as we have said, with a velocity that is exceedingly small. Thus, the velocity of homogeneous particles must on that account be practically the same, when they have reached the arc representing gravitation; & a difference can only be obtained in heterogeneous particles owing to their structure. It is therefore clear from what source the very great velocity can come, & also the slight differences, if there are any.

The reason for rectilinear propagation; there can be no immediate collision between the points of light & the points of the medium; slight inequality of the forces in a homogeneous medium are eluded by the tenuity & the velocity of light.

475. That which relates to the rectilinear propagation through a transparent homogeneous medium, & the free motion, without hindrance, by particles either of the light or of the transparent medium, is quite easily explained in my Theory, whereas in other theories it begets a very great difficulty. Also as regards hindrance to this motion, so long as the curve of forces has no asymptotic arc perpendicular to the axis besides the first, it has been shown, in Art. 362, that merely with a sufficiently great velocity there can be obtained an apparent compenetration of two substances; & tenuity & homogeneity of space traversed will assist this to a very great extent. Now, since, compared with perfectly indivisible & non-extended points of matter, there are an infinitely infinite number of points of space existing in the same plane, there is an infinitely infinite improbability that, for any instant of time chosen, the direction of motion of any one point of matter should be accurately directed towards any other point of matter; & this improbability, when we consider the sum of all the instants contained in any given time, however long, still comes out simply infinite. The number of points of light is indeed very large, not to say enormous, but in my Theory it is at least finite. These points at any chosen instant of time have an almost immeasurable number of directions of motion, but this number is finite in my Theory. It is indeed true that, no matter where an eye is situated upon the well-nigh immeasurable surface of a sphere described about one of the remotest stars as centre, nay, or within that sphere, the star will be seen; & thus, it is true that some particle of light must affect our eye. But in my Theory, that does not come about because rays of light come to it accurately in every one of an absolute infinity of directions; but because the pupil & the nerves of the eye do not form a single point, & the forces due to the points of a particle of light act at some distance away. Hence, in any chosen time, no matter how long, there need not happen in my Theory any case, in which any point of light is directed exactly towards any other point either of light, or of any substance, so that it is bound to collide with it. Hence, no point of light stays its motion, or deflects it, through collision or immediate impact.

If they possess great enough velocity, any bodies, even solids, will pass through other solids, without any disturbance of their motion.

476. This is indeed a common property of all bodies, that is, of bodies that approach one another. In my Theory, they have no point directly colliding with any other point. For this reason I also stated, in the above-mentioned article, that, if no mutual forces were present, there is always bound to be an apparent compenetration of all bodies. Yet, from this article alone, it is utterly impossible that there ever can be real compenetration. Hence, forces extending over some distance will hinder the progressive motion. If these forces are always equal in all directions, there would be no impediment to the motion, & it would necessarily be rectilinear owing to the force of inertia. Hence, nothing but a difference in the

esse rectilineus. Quare sola differentia virium agentium in punctum mobile obstare potest. At si nulla occurrat infinita vis arcus asymptotici cujuspiam post primum; vires omnes finitæ sunt, adeoque & differentia virium secundum diversas directiones agentium finita est semper. Igitur utcunque ea sit magna, ipsam finita quædam velocitas elidere potest, quin permittat ullam retardationem, accelerationem, deviationem, quæ ad datam quampiam utcunque parvam magnitudinem assurgat: nam vires indigent tempore ad producendam novam velocitatem, quæ semper proportionalis est tempori, & vi. Hinc si satis magna velocitas haberetur; quævis substantia trans aliam quanvis libere permearet sine ullo sensibili obstaculo, & sine ulla sensibili mutatione dispositionis propriorum punctorum, & sine ulla jactura nexus mutui inter ipsa puncta, & cohæsionis, quod ibidem illustravi exemplo ferrei globuli inter magnetes dispersos cum satis magna velocitate libere permeantis, ubi etiam illud vidimus, in hoc casu virium ubique finitarum impenetrabilitatis ideam, quam habemus, nos debere soli mediocritati nostrarum velocitatum, & virium, quarum ope non possumus imprimere satis magnam velocitatem, & libere trans murorum septa, & trans occlusas portas pervadere.

Si per asymptoticos arcus particulæ essent prorsus impermeabiles, tum recurrendum ad molem imminutam quantum oportet.

477. Id quidem ita se habet, si nullæ præter primam asymptoti habeantur, quæ vires absolute infinitas inducant: nam si per ejusmodi asymptoticos arcus particulæ fiant & indissolubiles, & prorsus impenetrabiles juxta num. 362; tum vero nulla utcunque magna velocitate posset una particula alteram transvolare, & res eodem recideret, quo in communi sententia de continua extensione materiæ. Tum nimirum oporteret lucis particulas minuere, non quidem in infinitum (quod ego absolute impossibile arbitror, quemadmodum & quantitates, quæ revera infinite parvæ sint in se ipsis tales, ac independenter ab omni nostro cogitandi modo determinatæ: nec vero carum usquam habetur necessitas in Natura) sed ita, ut adhuc incursus unius particulæ in aliam pro quovis finito tempore sit, quantum libuerit, improbabilis, quod per finitas utique magnitudines præstari potest. Si enim concipiatur planum per lucis particulam quancunque ductum, & cum ea progrediens; corum planorum numerus dato quovis finito tempore utcunque longo erit utique finitus; si particulæ inter se distent quovis utcunque exiguo intervallo, quarum idcirco finito quovis tempore non nisi finitum numerum emittet massa utcunque lucida. Porro quodvis ex ejusmodi planis ad medias, qua latissimæ sunt, alias particulas luminis inter se distantes finito numero vicium appellet utique intra finitum quodvis tempus, cum id per intervalla finita tantummodo debeat accidere, [222] & summa ejusmodi accessuum pertinentium ad omnia plana particularum numero finitarum finita crit itidem, utcunque magna. Licebit autem ita particularum diametros maximas imminuere, ut spatium plani ad datam quamvis distantiam protensi circunquaque etiam exiguam, habeat ad sectionem maximam particulæ rationem, quantum libuerit, majorem illa, quam exprimit ille ingens, sed finitus accessuum numerus: ac idcirco numerus directionum, per quas possint transire omnia illa plana ad omnes particulas pertinentia sine incursu in ullam particulam, erit numero carum, per quas fieri possit incursus, major in ratione ingenti, quantum libuerit; etiam si cum ea lege progredi deberent, ut altera non deberet transire in majore distantia ab altera, quam sit intervallum illud determinans exiguum illud spatium, ad quod assumpta est particularum sectio minor in ratione, quantum libuerit, magna. Infinito nusquam opus erit in Natura, & series finitorum, quæ in infinitum progreditur, semper aliquod finitum nobis offert ita magnum, vel parvum, ut ad physicos usus quoscunque sufficiat.

Asymptoticis iis cruribus nullum esse opus: ea potius excludenda: quam bene omnia explicentur sine ipsis.

478. Quod de particulis inter se collatis est dictum, idem locum habet & in particulis respectu corporum quoruncunque, potissimum si corpora juxta meam Theoriam constituta sint particulis distantibus a se invicem, & non continuo nexu colligatis, sive extensionis vere continuæ illins veli, aut muri continuam infinitam objicientis resistentiam, de quo egimus num. 362, & 363. Verum ejusmodi asymptoticorum arcuum nulla mihi est necessitas in mea Theoria, & hic itidem per nexus, ac vires limitum ingentis, quantum libuerit, quanquam non etiam infiniti valoris, omnia præstari possunt in Natura: & si principio inductionis inhærere libeat; debemus potius arbitrari, nullos esse alios ejusmodi asymptoticos arcus in curva, quam Natura adhibet: cum in ingenti intervallo a fixis ad particulas minimas, quas intueri per microscopia possumus, nullus ejusmodi nexus occurrat, quod indicat motus continuus particularum luminis per omnes ejusmodi tractus; nisi forte primus ille repulsivus, & postremus ejus naturæ arcus, ad gravitatem pertinens, indicio sint, esse & alios alibi in distantiis, quæ citra microscopiorum, vel ultra telescopiorum potestatem

forces acting on a moving point can hinder it. But if no infinite force occurs corresponding to any asymptotic arc after the first, all the forces are finite; & so also the difference between the forces acting in different directions will be always finite. Therefore, no matter how great the force may be, there is some finite velocity capable of overcoming it, without suffering any retardation, acceleration, or deviation amounting to any given magnitude, no matter how small. For, the forces require time to produce a new velocity, this being always proportional to the force & the time. Hence, if there were a sufficiently great velocity, any substance would pass freely through any other substance, without any sensible hindrance, & without any sensible change in the situation of the points belonging to either substance, & without any destruction of the mutual connection between the points, or of cohesion. There also I gave an illustration of an iron ball making its way freely through a group of magnets with a sufficiently great velocity; & here also we saw that we owe what idea we have of impenetrability, in the case of forces that are everywhere finite, merely to the moderate nature of our velocities & forces; for by their help alone we cannot impress a sufficiently great velocity, & freely pass through barrier-walls, or shut doors.

If, owing to the presence of asymptotic arcs, particles become impermeable, then we should have to fall back upon diminution of volume, as far as was necessary.

477. Now, this is the case, so long as there are no asymptotic arcs besides the first, to induce absolutely infinite forces; but if, owing to such asymptotic arcs, the particles become incapable both of dissolution & penetration, as in Art. 362, then indeed by no velocity, however great, could one particle pass through another; & the matter would be reduced to the same idea, as is held generally about the continuous extension of matter. Thus, in that case it would be necessary to diminish the size of the particles of light; not indeed infinitely—for I consider that that would be altogether impossible, just as also I think that there are no quantities infinitely small in themselves, and so determined without reference to any process of human thought; nor is there anywhere in Nature any necessity for such quantities. But they must be so diminished that the direct collision of one particle with another in any chosen finite time will still be improbable, to any extent desired; & this can be secured in every case by finite magnitudes. For suppose a plane area circumscribing each particle of light, & that this plane moves with the particle; then the number of these planes in any given finite time, however long, will in every case be finite, so long as the particles are distant from one another by any interval at all, no matter how small; & thus, in any given finite time the mass, however luminous, can only emit a finite number of these particles. Further, any one of these planes will impinge, at their broadest parts, upon the middle of other particles of light distant from one another by a finite number of fits, in every case in a finite time; for, this can only take place through a finite interval. The sum of such approaches pertaining to all the planes of the particles, finite in number, will also be finite, no matter how great the number may be. But we may so diminish the greatest diameters of the particles that the area of the plane, extended in all directions round to any given distance, however small, may bear to the greatest section of the particle a ratio greater, to any arbitrary extent, than that which is expressed by the huge but finite number of the approaches. Hence, the number of directions, by which all the planes pertaining to all the particles may pass without colliding with any particle, will be greater than the number of directions in which there may be collision, the ratio being one that is as immense as we please. And this will even be the case, if they should have to move in accordance with the law that one must not pass at a greater distance from the other than that interval which determines the very small space, to which it is supposed that the section of the particle bears a ratio of less inequality, no matter what the magnitude. There will nowhere be any need of the infinite in Nature; a series of finites, extended indefinitely, will always give us something finite, which is large enough or small enough to satisfy any physical needs.

There is no need for the asymptotic branches; rather, they should be excluded; how well all things can be explained without them.

478. All that has been said with regard to particles referred to one another, the same will hold good for particles in reference to any bodies; & especially if the bodies are formed, in accordance with my Theory, of particles distant from one another, & not bound together by a continuous connection, or possessing the truly continuous extension of the skin or wall offering a continuous infinite resistance, with which we dealt in Art. 362, 363. But really there is no necessity for such asymptotic arcs in my Theory; in it also, by means of connections & forces of limits of any value however great, though not actually infinite, everything in Nature can be accomplished. If we are to adhere to the principle of induction, we are bound rather to think that there are no other asymptotic arcs in the curve which Nature follows. For, in the mighty interval between the stars & the smallest particles that are visible under the microscope, no connections of this kind occur, as is indicated by the continuous motion of the particles of light throughout the whole of these regions. Unless, perhaps, that first repulsive branch, & that last arc of the nature that pertains to gravity, are to be taken as a sign that there are also somewhere others like them, at distances which are less than microscopical, or greater than those within the range of the

contrahuntur, vel protenduntur. Ceterum si vires omnes finitæ sint, & puncta materiæ juxta meam Theoriam simplicia penitus, & inextensa: multo sane facilius concipitur, qui fiat, ut habeatur hæc apparens compenetratio sine ullo incursu, & sine ulla dissolutione particularum cum transitu aliarum per alias.

Quomodo rem conficiant velocitas satis magna, & æqualitas sensibilis virium circumquaque. Quomodo hæc in homogeneo medio habeatur.

479. Porro duo sunt, quorum singula rem præstare possunt, velocitas satis magna, quæ nimirum utcunque magnam virium inæqualitatem potest eludere, & virium circumquaque positarum æqualitas, quæ differentiam relinquat omnino nullam. Differentia nunquam sane habebitur omnino nulla, ubi [223] punctum materiæ prætervolet per quandam punctorum veluti silvam, quorum alia ab aliis distent: necessario enim mutabit distantiam ab iis, a quibus minimum distat, jam accedens nonnihil, jam recedens. Verum ubi distributio particularum ad æqualitatem quandam multum accesserit, inæqualitas virium erit perquam exigua; si omnium virium habeatur ratio, quas exercent omnia puncta disposita circa id punctum ad intervallum, ad quod satis sensibiles meæ curvæ vires protenduntur. Concipiamus enim sphæram quandam, quæ habeat pro semidiametro illam distantiam, ad quam protenduntur flexus curvæ virium primigeniæ, sive ad quam vires singulorum punctorum satis sensibiles pertingunt. Si medium satis ad homogeneitatem accedat; secta illa sphæra in duas partes utcunque per centrum, in utraque numerus punctorum materiæ erit quamproxime idem, & summa virium quam proxime eadem, se compensantibus omnibus exiguis inæqualitatibus in tanta multitudine, quod in omnibus fit satis numerosis fortuitis combinationibus: adeoque sine ullo sensibili impedimento, sine ingenti flexione progredietur punctum quodcunque motu vel rectilineo, vel tremulo quidem nonnihil, sed parum admodum, & ad sensum æque in omnem plagam.

Quomodo ingens velocitas exiguam inæqualitatem eludat: exemplum in turbine ligneo non cadente.

480. Quod si accedat ingens velocitas; multo adhuc minor crit inæqualitatum effectus, tum quod multo minus habebunt temporis vires ut agant, tum quod in ipso continuato progressu inæqualitates jam in unam plagam prævalebunt, jam in aliam, quibus sibi mutuo celerrime succedentibus, magis adhuc uniformis, & rectilineus erit progressus. Sic ubi turbo ligneus gyrat celerrime circa verticalem axem cuspide tenuissima innixum solo, stat utique, inæqualitate ponderis, quæ ad casum determinat, jam ad aliam plagam jacente, & totam inclinante molem, jam ad aliam, qui, celeritate motus circularis imminuta, decidit inclinatus, quo exigit præponderantia.

Accedere nexum inter puncta particulæ: quid is præstet.

481. Quod autem homogeneitas medii, & velocitas præstant simul, id adhuc auget multo magis is nexus, qui est inter materiæ puncta particulam componentia, & æquali ad sensum velocitate delata, qui mutuis viribus cum accessum ad se invicem punctorum particulam componentium, & recessum impediat, cogit totam particulam simul trepidare eo solo motu, quem inducit summa inæqualitatum pertinentium ad puncta omnia, quæ summa adhuc magis ad æqualitatem accedit: nam in fortuitis, & temere hac, illac dispersis, vel concurrentibus casu circumstantiis, quo major numerus accipitur, eo inæqualitatum irregularium summa decrescit magis.

Raritatem plurimum prodesse: omnes eas quatuor causas habere locum in lumine non turbato a radiis alia directione delatis quaquaversum: priores tres in mediis densioribus pellucidis.

482. Demum raritas medii ad id ipsum confert adhuc magis: quo enim major est raritas, eo minor occurrit punctorum numerus intra illam sphæram, adeoque eo minor virium componendarum multitudo, & inæqualitas adhuc multo mi-[224]-nor. Porro omnes hæ quatuor causæ æqualitatis concurrunt, ubi agitur de radiis collatis cum aliis radiis: homogeneitas, nam lumen a dato puncto progrediens suam densitatem imminuit in ratione reciproca duplicata distantiarum a puncto radiante, adeoque in tam exiguo circunquaque circa quodvis punctum intervallo, quantum est id, ad quod virium actio sensibilis protenditur, ad homogeneitatem accedit in immensum: celeritas, quæ tanta est, ut singulis arteriæ pulsibus quævis luminis particula fere bis centum millia Romanorum milliariorum percurrat: nexus particularum mutuus, nam ipsæ luminis particulæ ad diversos coloratos radios pertinentes habent perennes proprietates suas, quas constanter servant, ut certum refrangibilitatis gradum, & potentiam certo impulsu agitandi oculorum fibras, per quam certam certi coloris sensationem eliciant: ac demum tenuitas immanis, qua opus est ad tantam diffusionem, & tam perennem effluxum sine ulla sensibili imminutione solaris massæ, & cujus indicium aliquod proferam paullo inferius. Ubi vero agitur de lumine comparato cum substantiis pellucidis, per quas pervadit, priora illa tria tantummodo locum habent respectu particularum luminis, & omnia quatuor respectu particularum pellucidi corporis, quarum nexus non dissolvitur, nec positio turbatur quidquam ab intervolantibus radiorum particulis. Quamobrem errat qui putat, mea

telescope. Besides, if all the forces are finite, and points of matter, in accordance with my Theory, are perfectly simple & non-extended, it is far more easily understood why there can be this apparent compenetration, without any collision, & without any dissolution of the particles as they pass through one another.

How the matter can be accomplished by a sufficiently large velocity & a sensible equality of the forces in all directions. How these are to be had in a homogeneous medium.

479. Further, there are two things, each of which can accomplish the matter; namely, a sufficiently great velocity, such as will foil the inequality of the forces, however great that may be; & an equality of the forces in all directions, such as will leave the difference absolutely zero. Now the difference can never really be altogether zero, when a point of matter passes through, so to speak, a forest of points, which are separated from one another. For, of necessity, it will change its distance from those points, from which it is least distant, at one time approaching & at another receding. But when the distribution of the particles approaches very closely to an equality, the inequality of the forces will be exceedingly small, so long as account is taken of all the forces exerted by all the points situated about that point at an interval equal to that over which the forces of my curve extend while still fairly sensible. For, imagine a sphere, that has for its semi-diameter the distance over which the windings of the primary curve extend, that is, the distance up to which the forces of each of the points are fairly sensible. If the medium approximates sufficiently closely to homogeneity, & the sphere is divided into any two parts by a plane through the centre, the number of points of matter in each part will be nearly the same; & the sum of the forces will be very approximately the same, as the slight differences taken as a whole compensate one another in so great a multitude; for this is always the case in sufficiently numerous fortuitous combinations. Thus, without any impediment, without any very great flexure, any point will proceed with a motion that is rectilinear, or maybe somewhat but very slightly wavy, & practically uniformly so in every direction.

How a very great velocity foils a slight inequality; example, from a wooden spinning-top not falling.

480. But if the velocity is very great, the effect of inequalities will be still less; both because the forces will have much less time in which to act, & because in such a continued progress the inequalities will prevail first on one side & then on the other; & as these follow one another very quickly, the progress will be still more uniform & nearer to rectilinear motion. Thus, when a wooden spinning-top spins very quickly about a vertical axis with a very fine point resting on the ground, it stays perfectly upright; for, the inequality of its weight, which disposes it to fall, lies first on one side & inclines the whole mass that way, & then on the other side; while, as soon as the circular motion decreases, it becomes inclined to the side to which the preponderance forces it.

In addition there is a connection between the points of a particle; the effect of this.

481. Again the effect produced by the homogeneity of the medium & the great velocity together is still further increased by the connection that exists between the points of matter forming the particle & moving together with practically the same velocity. This connection, since, through the mutual forces, it prevents the mutual approach or recession of the points forming the particle, will force the entire particle to move as a whole with the single motion that is induced by the sum of the inequalities pertaining to all its points; & this sum will still further approximate to equality. For, in circumstances that are fortuitous, distributed here & there at random, or concurring by chance, the greater the number taken, the more the sum of the irregular inequalities decreases.

Great effect of small density; all four of these causes hold for light undisturbed by rays proceeding in any other directions, & the first three of them in the more dense transparent media.

482. Lastly, rarity of the medium is of still further assistance; for, the greater the rarity, the smaller the number of points that occur within the sphere imagined above, & therefore the smaller the number of forces to be compounded, & much smaller still the inequality. Further, all four of these causes of inequality occur together, when we are dealing with rays of light in regard to other rays. Homogeneity we have, because light proceeding from a given point diminishes its density in the inverse ratio of the squares of the distances from the radiant point; & thus, in the exceedingly small interval round about any point, whatever the distance may be over which a sensible action of the forces extend, the approach to homogeneity is exceedingly great. Velocity also we have, so great that in a single beat of the pulse a particle of light travels a distance of nearly two hundred thousand Roman miles. Mutual connection of the particles also, for the particles of light pertaining to differently coloured rays have all their special lasting properties, which they keep to unaltered, such as a definite refrangibility & the power of affecting the nerves of the eye with a definite impulse, through which they give it a definite sensation of a definite colour. Lastly, an extremely great tenuity, such as is necessitated by the greatness of the diffusion & the endurance of the efflux without sensible diminution of mass in the case of the Sun; & of this I will bring forward some evidence a little further on. But when we are dealing with light in regard to transparent substances, through which the light passes, the first three only hold good with regard to the particles of light, but all four with regard to the particles of the transparent body; the connections between the particles of the body are not broken, nor is their relative position affected to any extent by the particles of the rays of light passing through them. Therefore he will be mistaken, who thinks

indivisibilia puncta prædita insuperabili potentia repulsiva pertingente ad finitam distantiam esse tam subjecta collisionibus, quam sunt particulæ finitæ magnitudinis, & idcirco nulli adminiculo esse pro comprehendenda mutua lucis penetratione; nam sine cruribus illis asymptoticis posterioribus meæ vires repulsivæ non sunt insuperabiles, nisi ubi puncta congredi debeant in recta, quæ illa jungit, qui casus in Natura nusquam occurrit.

Pelluciditatem oriri a sola homogeneitate: solam heterogeneitatem impedire posse progressum per inæqualitatem virium.

483. Et vero sola homogeneitas pelluciditatem parit, uti jam olim notavit Newtonus, nec opacitas oritur ab impactu in partes corporum solidas, & a defectu pororum jacentium in directum, uti alii ante ipsum plures censuerant, sed ab inæquali textu particularum heterogenearum, quarum aliæ aliis minus densis, vel etiam penitus vacuis amplioribus spatiolis intermixtæ satis magnam inducunt inæqualitatem virium, qua lumen in omnes partes detorquent, ac distrahunt, flexu multiplici, & ambagibus per internos meatus continuis, quibus fit, ut si paullo crassior occurrat massa corporis ex heterogeneis particulis coalescentis, nullus radius rectilineo motu totam pervadat massam ipsam, quod nimirum ad pelluciditatem requiritur. Indicia rei habemus quamplurima præter ipsam omnem superiorem Theoriam, quæ rem sola evinceret; cum nimirum sine inæqualitate virium nullum haberi possit libero rectilineo progressui impedimentum. Id sane colligitur ex eo, quod omnium corporum tenuiores laminæ pellucidæ sunt, uti norunt, qui microscopiis tractandis assueverunt: id [225] evincunt illæ substantiæ, quæ aliarum poris injectæ easdem ex opacis pellucidas reddunt, ut charta oleo imbuta fit pellucida, supplente aerem ipso oleo, cum quo multo minus inæqualiter in lumen agunt particulæ chartæ, quam agerent soli aeri, vel vacuo spatio intermixtæ. Rem autem oculis subjicit vitrum contusum in minores particulas, quod sola irregularitate figuræ particularum temere ex contusione nascentium, & acris intermixti inæqualitate fit opacum per multiplicationem reflexionum, & refractionum irregularium: nec aliam ob causam aqua in glaciem bullis continuis interruptam abiens pelluciditatem amittit, ut & alia corpora sane multa, quæ, dum concrescunt vacuolis interrupta, illico opaca fiunt.

Reflexionem non oriri ab impactu, sed ab inæqualitate virium in mutatione medii; ubi pro refractionis explicatione præmissa principia.

484. Quamobrem nec reflexio inde ortum ducit, sed habetur etiam in pellucidis corporibus ex inæqualitate virium seu repellentium, seu attrahentium, uti in Optica sua Newtonus tam multis notissimis argumentis demonstravit, quorum unum est illud ipsum ex asperitate superficiei cujuscunque cujusvis corporis, utcunque nobis, nudo potissimum inspectantibus oculo, lævis appareat, & perpolita, quod num. 299 exposuimus; & ex eadem causa oritur etiam refractio. Si velocitas luminis esset satis magna; impediret etiam hujusce inæqualitatis effectum, qui provenit a diversa mediorum constitutione: sed ex ipsis reflexionibus, & refractionibus in mutatione medii, conjunctis cum propagatione rectilinea per medium homogeneum, patet, celeritatem illam tantam luminis satis esse magnam ad eludendam illam inæqualitatem tanto minorem, quæ habetur in mediis homogeneis, non illam tanto majorem, quæ oritur a mediorum discrimine. Quod vero ad refractionis explicationem ex Mechanica requiritur, exposuimus a num. 302, ubi adhibuimus principium illud virium inter duo plana parallela agentium æque in distantiis æqualibus ab eorum utroque, cujus explicationem ad luminis particulas jam expediemus.

Consideratio sphæræ, ad quam extenditur vis sensibilis agens in lumen: inde vis inter bina plana parallela superficiei dirimenti media; inter quæ vis agit.

485. Concipiatur (f) illa sphærula, cujus semidiameter [226] æquatur distantiæ illi, ad quam agunt actione satis sensibili particulæ corporum in lucis particulam, quæ cum

---

(f) *Refert* MN *in fig.* 70 *superficiem dirimentem duo media,* GE *viam radii advenientis,* H *particulam luminis;* HE *celeritatem, ejus absolutam,* HS *parallelam,* SE *perpendicularem, quæ est eo minor, quo radius incidit magis obliquus:* abc *est sphæra, intra quam habetur actio sensibilis in particulam* H, *quæ est adhuc tota in priore medio:* X, X′, X″ *sunt loca plura particulæ progredientis inter plana* AB, CD *parallela superficiei* MN, *sita ad distantiam ab ea æqualem semidiametro sphæræ* Hc. *Particula sita inter illa plana ubicunque, ut in* X, *ea sphæra habebit suum segmentum* FRL *ultra superficiem* MN: *sit ejus axis* RT, & *eodem axe segmentum* QTZ *priori æquale, ac* mn *planum per centrum parallelum* MN. *Segmenta* mFLn, mQZn *ejusdem medii agent æqualiter. Segmenta* FRL, QTZ *inæqualiter, sed eorum vires dirigentur per axem* TR *in alteram e binis plagis oppositis: adeoque* & *differentia virium dirigetur per eundem, qui quidem perpendicularis est utique planis* AB, CD. *Ea actione via incurva radii sinuatur per* XX′X″. *Prout vis dirigetur versus* CD, *vel versus* AB, *curva erit cava versus easdem,* & *in mutatione directionis vis ipsius mutabitur flexus curvæ. Si autem curva evaserit alicubi parallela plano* AB; *flectet cursum retro; nisi id accidat accurate in situ vis* = 0, *qui*

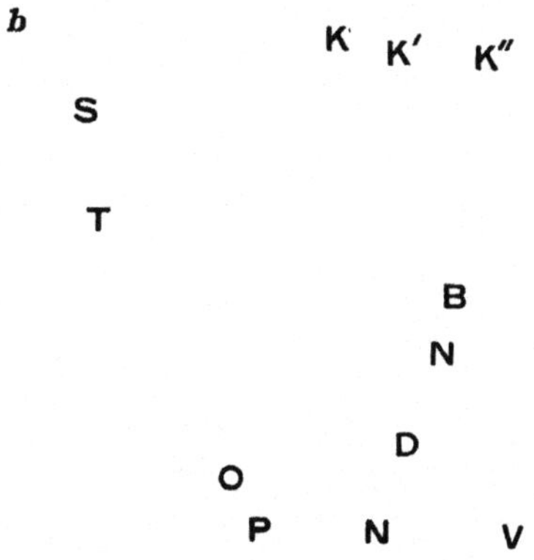

b

K K′ K″

S

T

B

N

D

O

P N V

Fig. 70.

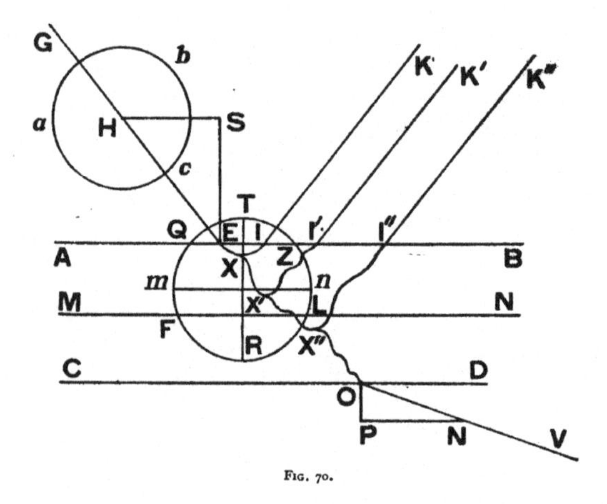

FIG. 70.

that my indivisible points, endowed with an insuperable repulsive force extending to a finite distance, are just as subject to collisions as particles of finite magnitude; & therefore that there is no assistance to be derived from them in understanding the mutual penetration of light; for, unless there are those asymptotic branches after the first, my repulsive forces are not insuperable, except when points are bound to move together in one straight line joining them, a circumstance which never occurs in Nature.

Transparence arises from homogeneity alone; & heterogeneity alone is capable of preventing progressive motion through inequality of forces

483. Indeed homogeneity by itself creates transparence, as was long ago stated by Newton; & opacity does not arise from impact with the solid parts of bodies, or through a lack of pores lying in a straight line, as many others before Newton thought, but from the unequal structure of heterogeneous particles; of which some are interspersed amongst others of less density, or even in perfectly empty little spaces, of considerable size, and thus induce an inequality great enough to distort the light in all directions, & to harass it with manifold windings & continuous meandering through internal channels; from which it comes about that, if a somewhat thick mass occurs of a body formed from heterogeneous particles, no ray with rectilinear motion will pass through the whole of that mass; which is the requirement for transparence. We have very many pieces of evidence on the subject, in addition to the whole of the Theory given above, which of itself is sufficient to prove it. For, indeed, without inequality of forces there can be no impediment to free rectilinear progressive motion. This can truly be deduced from the fact that fairly thin plates of all bodies are transparent, as is known to those who have been accustomed to microscopical work. Evidence is also afforded by such substances as, on injection into the pores of other substances, turn the latter from opaque to transparent; thus, paper soaked with oil becomes transparent, the oil taking the place of the air; for, with it the particles of paper act far less unequally upon the light than they would act, if merely air, or an empty space were interspersed. Moreover, glass broken up into fine particles brings the matter right before our eyes; for, from the mere irregularity of the shape of the particles randomly produced by the powdering, & the inequality of the interspersed air, it becomes opaque on account of the multiplication of reflections & refractions occurring irregularly. From no other cause does water, turning into ice interrupted by continuous bubbles, lose its transparence; it is just the same also with many other bodies, which, as they grow, are interspersed with little empty spaces, & from this cause alone become opaque.

Reflection does not take place through impact, but owing to inequality of forces on the medium being changed: where the principles for the explanation of refraction have been premised.

484. Therefore also reflection does not arise from impact; but it is even found in transparent bodies due to the inequality of forces, whether repulsive or attractive. This was proved by Newton in his Optics by a large number of arguments that are well known; one of these is that very reason that was stated in Art. 299, derived from the roughness of any surface of any body, no matter how smooth & polished it appears to us, especially when viewed with the naked eye. Refraction also arises from the same cause. If the velocity of light were great enough, it would prevent even the effect of this inequality that arises from the different constitution of the media. But, from the fact that there are these reflections & refractions on a change of medium, taken in conjunction with the fact of rectilinear propagation through a homogeneous medium, it is clear that the great velocity of light is enough to foil the comparatively small inequality that is found in homogeneous media, but is not enough for the comparatively greater inequality that arises from a difference in the media traversed. But that which is necessary for the mechanical explanation of refraction has been stated in Art. 302 onwards; where we employed the idea of forces acting between two parallel planes, the forces being equal for equal distances from either of the planes; we will now apply this idea to particles of light.

Consideration of the sphere whose radius is the distance to which the sensible force of light extends: thence the force between two planes parallel to the surface of separation of the media, between which the force acts.

485. Imagine (*f*) a sphere, of which the semidiameter is equal to the distance up to which the particles of a body act upon a particle of light with a fairly sensible action; &

(f) *In Fig.* 70, MN *is the surface of separation between the two media,* GE *the path of an approaching ray,* H *a particle of light,* HE *its absolute velocity,* HS *the parallel,* SE *the perpendicular component, which latter is the less, the more oblique the incidence of the ray.* abc *is the small sphere, within which there is sensible action on the particle* H, *which is as yet altogether in the first medium.* X,X′,X″ *are positions of the particle as it passes between the planes* AB, CD, *parallel to the surface* MN, *and situated at a distance from it equal to the semidiameter of the sphere* Hc. *If the particle is situated anywhere between the two planes, as at* X, *the sphere will have its segment* FRL *on the far side of the surface* MN. *Let the axis of the segment be* RT, *and let* QTZ *be a segment having the same axis and equal to the former segment, and let* mn *be a plane through the centre parallel to* MN. *Then the segments* mFLn, mQZn, *lying in the same medium, will act equally; but the segments* FRL, QTZ *will act unequally; yet their forces will be directed along the axis* TR *in one or other of the two opposite directions, and thus also the difference between these forces will act along the same straight line, which is perpendicular to the planes* AB, CD *in every case. Owing to this action the curved path of the ray will wind along through* X,X′,X″. *According as the force is directed towards* CD *or towards* AB, *the curve will be concave with respect to these same planes, and when the force changes its direction the flexure of the curve will also change. Moreover, if the curve should anywhere happen to become parallel to the plane* AB, *the path will be reflected; unless it should fall out that exactly in that position the force was zero, a case that is infinitely*

lucis particula progrediatur simul. Donec ipsa sphærula est in aliquo homogeneo medio tota, vires in particulam circunquaque æquales erunt ad sensum, & cum nullus habeatur immediatus incursus, motus inertiæ vi factus erit ad sensum rectilineus, & uniformis. Ubi illa sphærula aliquod aliud ingressa fuerit diversæ naturæ medium, cujus eadem moles exerceat in particulas luminis vim diversam a prioris medii vi; jam illa pars novi medii, quæ intra sphærulam immersa erit, non exercebit in ipsam particulam vim æqualem illi, quam exeret pars sphærulæ ipsi respondens ex altera centri parte, & facile patet, differentiam virium debere dirigi per axem perpendicularem illis segmentis sphærulæ, per quem singulæ utriusque segmenti vires diriguntur, nimirum perpendiculariter ad superficiem dirimentem duo media, quæ illud prius segmentum terminat: & quoniam ubicunque particula sit in æquali distantia a superficie, illud segmentum erit magnitudinis ejusdem; vis motum perturbans in iisdem a superficie illa distantiis eadem erit. Durabit autem ejusmodi vis, donec ipsa sphærula tota intra novum medium immergatur. Incipiet autem immergi ipsa sphærula in novum medium, ubi particula advenerit ad distantiam ab ipsius superficie æqualem radio sphærulæ, & immergetur tota, ubi ipsa particula jam immersa fuerit, ac ad distantiam eandem processerit. Quare si concipiantur duo plana parallela ipsi superficiei dirimenti media, quæ superficies in exiguo tractu habetur pro plana, ad distantias citra, & ultra ipsam æquales radio illius sphærulæ, sive intervallo actionis sensibilis; particula constituta inter illa plana habebit vim secundum directionem perpendicularem ipsis planis, quæ in data distantia ab eorum altero utrovis æqualis erit.

Tres casus, qui exhibent reflexionem, vel refractionem cum recessu a perpendiculo, vel ipsam refractionem cum accessu.

486. Porro id ipsum est id, quod assumpsimus num. 302, & unde derivavimus reflexionis, ac refractionis legem: nimirum si concipiatur ejusmodi vis resoluta in duas, alteram parallelam iis planis, alteram perpendicularem: illa vis pot-[227]-est perpendicularem velocitatem vel extinguere totam ante, quam deveniatur ad planum ulterius, vel imminuere, vel augere. In primo casu debet particula retro regredi, & describere curvam similem illi, quam descripsit usque ad ejusmodi extinctionem, recuperando iisdem viribus in regressu, quod amiserat in progressu, adeoque debet egredi in angulo reflexionis æquali angulo incidentiæ: in secundo casu habetur refractio cum recessu a perpendiculo, in tertio refractio cum accessu ad ipsum, & in utrolibet casu, quæcunque fuerit inclinatio in ingressu, debet differentia quadratorum velocitatis perpendicularis in ingressu, & egressu esse constantis cujusdam magnitudinis ex principio mechanico demonstrato num. 176 in adn. & inde num. 305 est erutum illud, sinum anguli incidentiæ ad sinum anguli refracti debere esse in constanti ratione, quæ est celeberrima lucis proprietas, cui tota innititur Dioptrica & præterea illud num. 306 velocitatem in medio præcedente ad velocitatem in medio sequente esse in ratione reciproca sinuum eorundem.

Lumen debere in corpora reagere æqualiter: hinc immensa lucis tenuitas: qui effectus ipsi falso tribuantur a nonnullis.

487. Hoc pacto ex uniformi Theoria deductæ sunt notissimæ, ac vulgares leges reflexionis, ac refractionis, ex quibus plura consectaria deduci possunt. Imprimis quoniam debet actio semper esse mutua, dum corpora agunt in lumen ipsum reflectendo, & refringendo; debet ipsum lumen agere in corpora, ac debet esse velocitas amissa a lumine ad velocitatem acquisitam a centro gravitatis corporis sistentis lumen, ut est massa corporis ad massam luminis. Inde deducitur immensa luminis tenuitas: nam massa tenuissima levissimæ plumulæ suspensæ filo tenui, si impetatur a radio repente immisso, nullum progressivum acquirit motum, qui sensu percipi possit. Cum tam immanis sit velocitas amissa a lumine; facile patet, quam immensa sit tenuitas luminis. Newtonus etiam radiorum impulsioni tribuit progressum vaporum cometicorum in candam; sed eam ego sententiam satis valido, ut arbitror, argumento rejeci in mea dissertatione *De Cometis*. Sunt, qui auroras boreales tribuant halitibus tenuissimis impulsis a radiis solaribus, quod miror fieri etiam ab aliquo, qui radios putat esse undas tantummodo, nam undæ progressivum

---

casus est in infinitum improbabilis. Id accidet in aliis radiis citius, in aliis radiis serius, pro diversa absoluta celeritate radii, pro diversa inclinatione incidentiæ, & pro diversa natura, vel constitutione particulæ, abeuntibus aliis particulis per QXIK, *aliis per* QXX'I'K', *aliis per* QXX'X"I"K". *Porro perquam exiguum discrimen in vi, vel celeritate, potest curvam uno aliquo in loco a positione proxima parallelismo ad ipsum parallelismum traducere, quo loco superato adhuc summa actionum usque ad* O *potest esse ad sensum eadem. Reliqua sunt hic, ut num.* 306.

suppose that this sphere moves along with the light particle. So long as the little sphere is altogether in a homogeneous medium, the forces on the particle all round it are practically equal; &, since no immediate impact can take place, the motion will be kept practically rectilinear & uniform by the force of inertia. When the little sphere enters some other medium of a different nature, the same volume of which exerts on the particles of light a force different from the force due to the first medium, then, that part of the new medium which is intercepted within the little sphere will not exert on the particle a force equal to that which the corresponding part on the other side of the centre exerts; & it is easily seen that the difference of the forces must be directed along the axis perpendicular to these segments of the sphere, for the forces due to each segment separately are so directed; that is to say, perpendicular to the surface of separation between the two media, which is the bounding surface of the first of the two segments. Now, since that segment will be of the same magnitude whenever the distance of the particle from the surface of separation is the same, the force determining the change of motion will be the same at equal distances from that surface. Further, such force will continue unchanged so long as the little sphere is altogether immersed in the new medium. Now, the little sphere will commence to be immersed in the new medium as soon as the particle reaches a distance from the surface of separation equal to the radius of the little sphere; & it will become altogether immersed in it as soon as the particle itself, after entering it, has gone forward a further distance equal to the radius. Hence, if two planes are imagined to be drawn parallel to the surface of separation of the media, & this surface is supposed to be plane, for the very small region extending on every side to a distance equal to the radius of the little sphere, or the interval corresponding to sensible action; then, a particle situated between those planes will be under the influence of a force in the direction perpendicular to the planes, which will be the same for equal distances from either of them.

Three cases, exhibiting respectively reflection, refraction with recession from the normal, & refraction with approach to the normal.

486. Now, this reduces to that very same supposition that we made in Art. 302, from which we derived the laws of reflection & refraction. Thus, if such a force is supposed to be resolved into two parts, one parallel & the other perpendicular to the planes, the latter force may either destroy the whole of the perpendicular velocity before the further plane is reached, or it may reduce it, or it may increase it. In the first case the particle must turn back in its path & describe a curve similar to that which it has already described up to the point at which its perpendicular velocity was described; & on its return it will recover the velocity it lost during its advance, with the same forces; & thus, it must leave the second medium with an angle of reflection equal to its angle of incidence. In the second case there will be refraction with recession from the normal; & in the third case, refraction with approach to the normal. In either of these cases, whatever the inclination was on entering the second medium, the difference between the squares of the velocities on entering & leaving must be of some constant magnitude, from the mechanical principle demonstrated in the note to Art. 176. From which, in Art. 305, I have deduced that the sine of the angle of incidence must bear a constant ratio to the sine of the angle of refraction; & this is the very well known property of light, upon which is established the whole theory of dioptrics. Also, in addition, in Art. 306, I deduced that the velocity in the first medium is to the velocity in the second in the inverse ratio of the sines of these angles.

Light must have an equal reaction on the bodies; hence the extreme tenuity of light; these effects are falsely attributed to light itself by some people.

487. In this way, from a uniform theory, all the principal well-known laws of reflection & refraction have been derived; & from these a large number of corollaries can be deduced. First of all, because the action must always be mutual, so long as bodies act upon light, reflecting or refracting it, the light must react on the bodies; & the velocity lost by the light must bear a ratio to the velocity gained by the centre of gravity of the body resisting the motion of the light, which is equal to the ratio of the mass of the body to the mass of the light. From this we deduce the extreme tenuity of light. For, the tiniest mass of the lightest feather suspended by the finest of strings, if it should be struck by a ray of light suddenly falling upon it, still would acquire no progressive motion, such as could be perceived. Since the velocity lost by the light is so huge, it can be clearly seen how exceedingly small must be the density of light. Newton even attributed to the impact of light rays the progressive motion tail first of the vapours of comets; but I overthrew this idea, by an argument which I consider to be perfectly sound, in my dissertation *De Cometis*. Some people attribute the aurora borealis to exhalations of extremely small density impelled by solar light-rays; & I am astonished that this should be put forward by anyone who considers

improbable. This reflection will take place sooner in some rays than in others, according to different velocities of the rays, different angles of incidence, different natures and constitutions of the particle; some of the particles will pass along a path QXIK, *others along* QXX'I'K', *and others again along* QXX'X''I''K''. *Further, a very slight difference in the force or velocity will be enough to turn the curve in some one position of the particle from being very nearly parallel to being exactly parallel; if this position is once passed, the sum of the actions thereafter as far as* O *may be practically the same. The rest is now similar to that which has been stated in Art.* 306.

motum per se se non imprimunt: qui autem censent, & fluvios retardari orienti Soli contrarios, & Terræ motus fieri ex impulsu radiorum Solis, ii sane nunquam per legitima Mechanicæ principia inquisiverunt in luminis tenuitatem.

Tenuissimum motum imprimi a lumine particulis corporum: calorem, & ustionem provenire ab earum viribus internis, quod ipsum probatur hic.

488. Solis particulis tenuissimis corporum imprimunt motum radii, ex quo per internas vires aueto oritur calor, & quidem in opacis corporibus multo facilius, ubi tantæ sunt reflexionum, & refractionum internæ vicissitudines: exiguo motu impresso paucis particulis, reliqua internæ mutuæ vires agunt juxta ea, quæ diximus num. 467. Sic ubi radiis solaribus speculo collectis comburuntur aliqua, alia calcinantur [228] etiam; omnes illi motus ab internis utique viribus oriuntur, non ab impulsione radiorum. Regulus antimonii ita calcinatus auget aliquando pondus decima sui parte. Sunt, qui id tribuant massæ radiorum ibi collectæ. Si ad ita esset; debuisset citissime abire illa substantia cum parte decima velocitatis amissæ a lumine, sive citius, quam binis arteriæ pulsibus ultra Lunam fugere. Quamobrem alia debet esse ejus phænomeni causa, qua de re fusius egi in mea dissertatione *De Luminis Tenuitate*.

Densiora agere in lumen fortius: sed sulphurosa, & oleosa pari densitate plus: cur id ipsum.

489. Quoniam lumen in sulphuris particulas agit validissime, nam sulphurosæ, & oleosæ substantiæ facillime accenduntur; eæ contra in lumen validissime agunt. Substantiæ generaliter eo magis agunt in lumen, quo densiores sunt, & attractionum summa prævalet, ubi radius utrumque illud planum transgressus refringitur: & idcirco generaliter ubi sit transitus a medio rariore ad densius, refractio fit per accessum ad perpendiculum, & ubi a medio densiore ad rarius, per recessum. Sed sulphurosa, & oleosa corpora multo plus agunt in lucem, quam pro ratione suæ densitatis. Ego sane arbitror, uti monui num. 467, ipsum ignem nihil esse aliud, nisi fermentationem ingentem lucis cum sulphurea substantia.

Lumen in progressu nullam pati resistentiam, positive probatur.

490. Lumen per media homogenea progredi motu liberrimo, & sine ulla resistentia medii, per quod propagetur, eruitur etiam ex illo, quod velocitas parallela maneat constans, uti assumpsimus num. 302, quod assumptum si non sit verum, manentibus ceteris; ratio sinus incidentiæ ad sinum anguli refracti non esset constans: sed idem eruitur etiam ex eo, quod ubi radius ex aere abivit in vitrum, tum e vitro in aerem progressus est, si iterum ad vitrum deveniat; candem habeat refractionem, quam habuit prima vice. Porro si resistentiam aliquam pateretur, ubi secundo advenit ad vitrum; haberet refractionem majorem: nam velocitatem haberet minorem, quæ semel amissa non recuperatur per hoc, quod resistentia minuatur, & eadem vis mobile minori velocitate motum magis detorquet a directione sui motus.

Unde lux in phosphoris quibusdam.

491. Posteaquam lux intra opaca corpora tam multis, tam variis erravit ambagibus aliqua saltem sui parte deveniet iterum ad superficiales particulas, & avolabit. Inde omnino ortum habchit lux illa tam multorum phosphororum, quæ deprehendimus, e Sole retracta in tenebras lucere per aliquot secunda, & a numero secundorum licet conjicere longitudinem itineris confecti per tot itus, ac reditus intra meatus internos. Sed progrediamur jam ad reliqua, quæ num. 472 proposuimus.

Cur in majore obliquitate plus luminis reflectatur.

492. Primo quidem illud facile perspicitur, ex Theoria, quam exposuimus, cur, ubi radius incidit cum majore inclinatione ad superficiem, major luminis pars reflectatur. Et quidem in dissertatione, quam superiore anno die 12 Novembris legit [229] Bouguerius in Academiæ Parisiensis conventu publico, uti habetur in *Mercurio Gallico* hujus anni ad mensem Januarii, profitetur, se invenisse in aqua in inclinatione admodum ingenti reflexionem esse æque fortem, ac in Mercurio ut nimirum reflectantur duo trientes, dum in incidentia perpendiculari vix quinquagesima quinta pars reflectatur. Porro ratio in promptu est. Quo magis inclinatur radius incidens ad superficiem novi medii, eo minor est perpendicularis velocitas, uti patet: quare vires, quæ agunt intra illa duo plana, eo facilius, & in pluribus particulis totam velocitatem perpendicularem elident, & reflexionem determinabunt.

Diversam refrangibilitatem non pendere a sola diversa celeritate particularum luminis, sed etiam a diverso earum textu inducente vim diversam.

493. Verum id quidem jam supponit, non in omnes lucis particulas eandem exerceri vim, sed in iis discrimen haberi aliquod. Ejusmodi discrimina diligenter evolvam. Inprimis discrimen aliquod haberi debet ex ipso textu particularum luminis, ex quo pendeat constans discrimen proprietatum quarundam, ut illud imprimis diversæ radiorum refrangibilitatis. Quod idem radius refringatur ab una substantia magis, ab alia minus in eadem

that light-rays are only waves; for, waves do not give any progressive motion of themselves. Further there are some who consider that rivers running in a direction opposite to the rising Sun are retarded, & that the motion of the Earth is due to impulse of solar rays; but really such people can never have investigated the tenuity of light by means of legitimate mechanical principles.

There is a very slight motion given to the particles of bodies by light; heat & combustion arise from their internal forces, as is here proved.

488. The rays of the Sun impress a motion on the exceedingly small particles of bodies; & from this, when increased by internal forces, arises heat, & this all the more easily in the case of opaque bodies, where there are such a number of internal alternations of reflections & refractions. If a slight motion is impressed on but a few particles, the internal mutual forces do all the rest, as we stated in Art. 467. Thus, when some substances are set on fire by solar rays collected by a mirror, while some are even reduced to powder, all the motions arise in every case from internal forces, & not from the impulse of the light-rays. Regulus of antimony (stibnite), thus calcined, will sometimes increase its weight by a tenth part of it; & there are some who attribute this fact to the mass of the rays so collected. But if this were the case, the substance would have to fly off very quickly with a velocity equal to a tenth part of the velocity lost by the light, or more quickly than would be necessary to get beyond the Moon in two beats of the pulse. Hence there must be other causes to account for this phenomenon, with which I have dealt fairly fully in my dissertation *De Luminis Tenuitate.*

Denser substances act more strongly on light; but sulphurous & oily substances more so than others of equal density; the reason for this.

489. Since light acts very strongly on the particles of sulphur, for sulphurous & oily substances are very easily set on fire, these on the other hand act very strongly on light. In general, substances have the greater action on light, the denser they are; & the sum of the attractions will be stronger when the ray is refracted as it passes through each of the planes. For this reason, in general, when a ray passes from a less dense to a more dense medium, refraction takes place with approach to the normal, & when from a more dense to a less dense, with recession from the normal. But sulphurous & oily bodies act much more vigorously upon light than in proportion to their density. I am firmly convinced that fire is nothing else but an exceedingly great fermentation of light with some sulphurous substance, as I stated in Art. 467.

Positive demonstration that light does not suffer any resistance in its progressive motion.

490. That light progresses through homogeneous media with a perfectly free motion, without suffering any resistance from the medium through which it is propagated, is proved by the fact that the parallel component of the velocity remains unaltered. We made this assumption in Art. 302; & if the assumption is not true, other things being unaltered, the ratio of the sine of incidence to the sine of refraction cannot be constant. Now the same thing is also proved by the fact that when a light-ray goes from air into glass, & then proceeds from the glass into air, then, if once more it should come to glass, it will have the same refraction as it had in the first instance. Moreover, if it suffered any resistance, when for the second time it came to glass, it would have a greater refraction; for, the velocity would be less, & once having lost this velocity, the particle could not regain it simply because the resistance was diminished; & the same force will cause a body moving with a smaller velocity to deviate from the direction of its motion to a greater degree.

The source of the light in certain phosphorous bodies.

491. After light has wandered through so many & various paths within opaque bodies, at some part at least it will once more arrive at the superficial particles of the bodies & fly off. This alone will give rise to the light that we perceive with so many phosphorous bodies, which on being withdrawn from the Sun into the shade shine for some seconds; & from the number of seconds one may conjecture the length of the path described by so many backward & forward journeys within the internal channels. But let us now go on to the rest of those things that we set forth in Art. 472.

Why at greater obliquity there is more of the light reflected.

492. In the first place, then, it is easily seen from the Theory which I have expounded, why the proportion of light reflected is greater, when the ray falls on the surface with greater inclination to it. Indeed, in a dissertation, read on November 12th of last year by Bouguer before a public convention of the Paris Academy, as is reported in the French *Mercury* for January of this year the author professed to have found for water at a very great inclination a reflection equal to that with mercury; that is to say, two-thirds of the light was reflected, while at perpendicular incidence barely a fifty-fifth part is reflected. Now, the reason for this is not far to seek. The more inclined the incident ray is to the surface of the new medium, the less is its perpendicular velocity, as is quite clear; hence, the forces that act between the two planes will the more easily, & for a larger number of particles, destroy the whole of the perpendicular velocity, & thus determine reflection.

Different refrangibility does not depend on different velocities of the particles of light alone, but also on their different structure which induces different forces.

493. But this supposes that the same force is not exerted on all particles of light, but that even for them there is some difference. I will carefully discuss these differences. First of all, there is bound to be some difference owing to the structure of the particles of light; & upon this will depend a constant difference in some of its properties, such as that of the different refrangibilities of rays, in particular. The fact that the same ray is refracted by

etiam inclinatione incidentiæ, id quidem provenit a diversa natura substantiæ refringentis, uti vidimus : ac eodem pacto e contrario, quod e diversis radiis ab eodem medio, & cum eadem inclinatione, alius refringatur magis, alius minus, id provenire debet a diversa constitutione particularum pertinentium ad illos radios. Debet autem id provenire vel a diversa celeritate in particulis radiorum, vel a diversa vi. Porro demonstrari potest, a sola diversitate celeritatis non provenire, atque id præstiti in secunda parte meæ dissertationis *De Lumine :* quanquam etiam radii diversæ refrangibilitatis debeant habere omnino diversam quoque celeritatem ; nam si ante ingressum in medium refringens habuissent æqualem ; jam in illo inæqualem haberent, cum velocitas præcedens ad velocitatem sequentem sit in ratione reciproca sinus incidentiæ ad sinum anguli refracti : & hæc ratio in radiis diversæ refrangibilitatis sit omnino diversa. Quare provenit etiam a vi diversa, quæ cum constanter diversa sit, ob constantem in eodem radio, utcunque reflexo, & refracto, refrangibilitatis gradum, debet oriri a diversa constitutione particularum, ex qua sola potest provenire diversa summa virium pertinentium ad omnia puncta. Cum vero diversa constanter sit harum particularum constitutio : nihil mirum, si diversam in oculo impressionem faciant, & diversam ideam excitent.

Ex eadem refractione radiorum ejusdem coloris emissorum ab omnibus lucidis corporibus evinci eandem in iis celeritatem, & textum.

494. At quoniam experimentis constat, radios ejusdem coloris eandem refractionem pati ab eodem corpore, sive a stellis fixis provenerint, sive a Sole, sive a nostris ignibus, sive etiam a naturalibus, vel artificialibus phosphoris, nam ea omnia eodem telescopio æque distincta videntur : manifesto patet, omnes radios ejusdem coloris pertinentes ad omnia ejusmodi lucida corpora eadem velocitate esse præditos, & eadem [230] dispositione punctorum : neque enim probabile est, (& fortasse nec fieri id potest), celeritatem diversam a diversa vi compensari ubique accurate ita, ut semper eadem habeatur refractio per ejusmodi compensationem.

Vices facilioris reflexionis &c., oriri a contractione, & expansione particularum in progressu inducente discrimen.

495. Sed oportet invenire aliud discrimen inter diversas constitutiones particularum pertinentium ad radios ejusdem refrangibilitatis ad explicandas vices faciliores reflexionis, & facilioris transmissus ; ac inde mihi prodibit etiam ratio phænomeni radiorum, qui in reflexione, & refractione irregularitur disperguntur, & ratio discriminis inter eos, qui reflectuntur potius, quam refringantur, ex quo etiam fit, ut in majore inclinatione reflectantur plures. Newtonus plures innuit in Optica sua hypotheses ad rem utcunque adumbrandam, quarum tamen nullam absolute amplectitur : ego utar hic causa, quam adhibui in illa dissertatione *De Lumine* parte secunda, quæ causa & existit & rei explicandæ est idonea : quamobrem admitti debet juxta legem communem philosophandi. Ubi particula luminis a corpore lucido excutitur fieri utique non potest, ut omnia ejus puncta eandem acquisierint velocitatem, cum a punctis repellentibus diversas distantias habuerint. Debuerunt igitur aliqua celerius progredi, quæ sociis relictis processissent, nisi mutuæ vires, acceleratis lentioribus, ea retardassent, unde necessario oriri debuit particulæ progredientis oscillatio quædam, in qua oscillatione particula ipsa debuit jam produci non nihil, jam contrahi : & quoniam dum per medium homogeneum particula progreditur, inæqualitas summæ actionum in punctis singulis debet esse ad sensum nulla ; durabit eadem per ipsum medium homogeneum reciprocatio contractionis, ac productionis particulæ, quæ quidem productio, & contractio poterit esse satis exigua ; si nimirum nexus punctorum sit satis validus : sed semper erit aliqua, & potest itidem esse non ita parva, nec vero debet esse eadem in particulis diversi textus.

In limitibus ejus velut oscillationis diutius perstare formam : in diversa parte ejusdem virium summam esse diversam.

496. Porro in ea reciprocatione figuræ habebuntur limites quidam productionis maximæ, & maximæ contractionis, in quibus juxta communem admodum indolem maximorum, & minimorum diutissime perdurabitur, motu reliquo, ubi jam inde discessum fuerit ad distantiam sensibilem cum ingenti celeritate peracto, uti in pendulorum oscillationibus videmus, pondus in extremis oscillationum limitibus quasi hærere diutius, in reliquis vero locis celerrime prætervolare : ac in alio virium genere diverso a gravitate constanti, illa mora in extremis limitibus potest esse adhuc multo diuturnior, & excursus in distantiis sensibilibus ab utrovis maximo multo magis celer. Deveniet autem particula ad medium extremarum illarum duarum dispositionum diutius perseverantium post æqualia temporum intervalla, ut æquales pendulorum oscillationes sunt æque diuturnæ, ac idcirco dum particula progreditur per medium homogeneum, recurrent illæ ipsæ binæ dispositiones post æqua-[231]-lia intervalla spatiorum pendentia a constanti velocitate particulæ, &

one substance more, & by another substance less, even for the same inclination of incidence, is due to the different nature of the refracting substance, as we have seen; & in the same way, on the other hand, the fact that, of different rays, & with the same inclination, one ray is refracted & another less, by the same medium, is due to the different constitution of the particles pertaining to those rays. Further, it is bound to be due either to a different velocity in the particles of the rays, or to a different force. Lastly, it can be proved that it is not due to the difference of velocity alone; & this I showed in the second part of my dissertation *De Lumine*; although indeed rays of different refrangibilities are bound to have altogether different velocities also. For, if before entering the refracting substance they had equal velocities, then after entering they would have unequal velocities; since the first velocity is to the second in the inverse ratio of the sines of the angles of incidence & refraction; & this ratio for rays of different refrangibilities is altogether different. Hence it must also be due to a difference of force; & since this must be constantly different, on account of the constant degree of refrangibility in the same ray, however it may be reflected or refracted, it must be due to a difference in the constitution of the particles, from which alone there can arise a difference in the sum of the forces pertaining to all points forming them. Now, since the constitution of these particles is constantly different, it is no wonder that they make a different impression on the eye, & incite a different sensation.

From the equality of refraction of rays of the colour coming from all bodies emitting light it is proved that for such rays there is the same velocity & structure.

494. Now, since it is proved by experiment that rays of the same colour suffer the same refraction by the same body, whether they come from the fixed stars, or from the Sun, or from our fires, or even from natural or artificial phosphorous substances, for they all appear equally distinct when viewed with the same telescope; it is clearly evident that all rays of the same colour pertaining to such light-giving bodies are endowed with the same velocities, & the same distribution of their points. For, it is very improbable, not to say impossible, that a difference in velocity should be everywhere exactly balanced by a difference in force to such a degree that by means of such a balance there should always be the same refraction obtained.

Fits of easier reflection, &c., are due to contraction & expansion of the particles, which induce a difference in the progressive motion.

495. But another difference must be found amongst the different constitutions of the particles belonging to rays of the same refrangibility, to account for the fits of easier reflection & easier transmission. From it I shall obtain also the reason for the phenomenon of rays that are irregularly scattered in reflection & refraction; & the reason for the difference between those that are reflected in preference to being refracted, from which also it comes about that the greater the angle the more numerous the rays reflected. Newton suggests several hypotheses, in his Optics, to give a rough idea of the matter; but he does not adhere absolutely to any one of them. I will use in this connection the reason that I employed in the dissertation *De Lumine*, in the second part; this reason both really exists & is fitted for explaining the matter; & therefore, according to the usual rule in philosophizing, this reason should be admitted. When a particle of light is driven off from a light-giving body, it cannot in any case happen that all the points forming it have acquired the same velocity; for, they will have been at different distances from the repelling points of the body. Therefore some of them are bound to progress more quickly than others, & the former would have left their fellows behind in their advance, unless the mutual forces had retarded them, while the slower ones were accelerated. Owing to this, there must necessarily have arisen a certain oscillation of the particle as it goes along, & due to this oscillation the particle itself must have been alternately extended & contracted to some extent. Now, since during the progress of a particle through a homogeneous medium inequality of the sum of the actions at all points of it must be practically zero, the same alternation of extension & contraction of the particle will continue right through the homogeneous medium, although the contraction & expansion will indeed be but slight, if the connections between the points are fairly strong. But there will always be some oscillation, & it may also not be so very small, nor need it be the same for particles of different structure.

At the boundaries of this oscillation, so to speak, the particle will preserve its shape longer; & the sum of the forces at different parts will be different.

496. Further, in this alternation of figure there will be certain bounding forms, corresponding to maximum extension & maximum contraction; & in these, according to a universal property of all maxima & minima, there will be quite a long pause; whereas, the rest of the motion, after a departure from them has taken place to a sensible distance, is accomplished with a great velocity. Thus, we see in the oscillations of pendulums that the weight at the extreme ends of the oscillations seems to pause for a considerable time, whereas in other positions it flies past very quickly. In another kind of forces different from constant gravitation, this delay at the extreme ends may be still more prolonged, & the motion at sensible distances from either maximum much more swift. Moreover the particle will reach the mean, between the two extreme dispositions that last for some considerable time, after equal intervals of time; just as equal oscillations of pendulums are of equal duration. Hence, as a particle proceeds through a homogeneous medium, those two dispositions recur after equal intervals of space, depending on the constant velocity

a constanti tempore, quo particulæ cujusvis oscillatio durat. Demum summa virium, quam novum medium, ad quod accedit particula, exercet in omnia particulæ puncta, non erit sane eadem in diversis illis oscillantis particulæ dispositionibus.

Inde binæ dispositiones constantes vicium in maxima particularum parte appellente in iis limitibus: in parte exigua appellente inter eos dispersio.

497. Hisce omnibus rite consideratis, concipiatur jam ille fere continuus affluxus particularum etiam homogenearum ad superficiem duo heterogenea media dirimentem. Multo maximus numerus adveniet in altera ex binis illis oppositis dispositionibus, non quidem in medio ipsius, sed prope ipsam, & admodum exiguus erit numerus earum, quæ adveniunt cum dispositione satis remota ab illis extremis. Quæ in hisce intermediis adveniunt, mutabunt utique dispositiones suas in progressu inter illa duo plana, inter quæ agit vis motum particulæ perturbans, ita, ut in datis ab utrovis plano distantiis vires ad diversas particulas pertinentes, sint admodum diversæ inter se. Quare illæ, quæ retro regredientur, non candem ad sensum recuperabunt in regressu velocitatem perpendicularem, quam habuerunt in accessu, adeoque non reflectentur in angulo reflexionis æquali ad sensum angulo incidentiæ, & illæ, quæ superabunt intervallum illud omne, in appulsu ad planum ulterius, aliæ aliam summam virium expertæ, habebunt admodum diversa inter se incrementa, vel decrementa velocitatum perpendicularium, & proinde in admodum diversis angulis egredientur dispersæ. At quæ advenient cum binis illis dispositionibus contrariis, habebunt duo genera virium, quarum singula pertinebunt constanter ad classes singulas, cum quarum uno idcirco facilius in illo continuo curvaturæ flexu devenietur ad positionem illis planis parallelam, sive ad extinctionem velocitatis perpendicularis cum altero difficilius: adeoque habebuntur in binis illis dispositionibus oppositis binæ vices, altera facilioris, altera difficilioris reflexionis, adeoque facilioris transitus, quæ quidem regredientur post æqualia spatiorum intervalla, quanquam ita, ut summa facilitas in media dispositione sita sit, a qua quæ minus, vel magis in appulsu discedunt, magis e contrario, vel minus de illa facilitate participent. Is ipse accessus major, vel minor ad summam illam facilitatem in media dispositione sitam in Benvenutiana dissertatione superius memorata exhibetur per curvam quandam continuam hinc, & inde æque inflexam circa suum axem, & inde reliqua omnia, quæ ad vices, & carum consectaria pertinent, luculentissime explicantur.

Unde discrimen rationis luminis reflexi ad transmissum.

498. Porro hinc & illud patet, qui fieri possit, ut e radiis homogeneis ad eandem superficiem advenientibus alii transmittantur, & alii reflectantur, prout nimirum advenerint in altera e binis dispositionibus: & quoniam non omnes, qui cum altera ex extremis illis dispositionibus adveniunt, adve-[232]-niunt prorsus in media dispositione, fieri utique poterit, ut ratio reflexorum ad transmissos sit admodum diversa in diversis circumstantiis, nimirum diversi mediorum discriminis, vel diversæ inclinationis in accessu: ubi enim inæqualitas virium est minor, vel major perpendicularis velocitas per illam extinguenda ad habendam reflexionem, non reflectentur, nisi illæ particulæ, quæ advenerint in dispositione illi mediæ quamproxima, adeoque multo pauciores quam ubi vel inæqualitas virium est major, vel velocitas perpendicularis est minor, unde fiet, ut quemadmodum experimur, quo minus est mediorum discrimen, vel major incidentiæ angulus, eo minor radiorum copia reflectetur: ubi & illud notandum maxime, quod ubi in continuo flexo curvaturæ viæ particulæ cujusvis, quæ via jam in alteram plagam est cava, jam in alteram, prout prævalent attractiones densioris medii, vel repulsiones, devenitur identidem ad positionem fere parallelam superficiei dirimenti media, velocitate perpendiculari fere extincta, exiguum discrimen virium potest determinare parallelismum ipsum, sive illius perpendicularis velocitatis extinctionem totalem: quanquam eo veluti anfractu superato, ubi demum reditur ad planum citerius in reflexione, vel ulterius in refractione, summa omnium actionum quæ determinat velocitatem perpendicularem totalem, debeat esse ad sensum eadem, nimirum nihil mutata ad sensum ab exigua illa differentia virium, quam peperit exiguum dispositionis discrimen a media dispositione.

Unde discrimen in intervallis vicium.

499. Atque hoc pacto satis luculenter jam explicatum est discrimen inter binas vices, sed superest exponendum, unde discrimen intervalli vicium, quod proposuimus num. 472. Quod diversi colorati radii diversa habeant intervalla, nil mirum est: nam & diversæ

of the particle, & on the constant time for which any oscillation of the particle lasts. Lastly, the sum of the forces, which the new medium, approached by the particle, exerts upon all the points of the particle, will not really be the same for the different dispositions of the oscillating particle.

Hence, we have the two constant dispositions yielding fits, with the greater proportion of particles, which are striking in those limiting states; & for the few that strike in states intermediate between them, we have dispersion.

497. All such things being duly considered, a conception can be now formed of the almost continuous flow of even homogeneous particles towards the surface of separation of two unlike media. By far the greater number of them will arrive at the surface in one or other of those two opposite dispositions; not indeed exactly so, but very nearly so. A very few of them will reach the surface with a disposition considerably removed from those extremes. Those that do arrive in these intermediate states, will in all cases change their dispositions in their passage between the two planes, between which the force disturbing the motion of the particle acts; & in such a manner that at any given distance from either plane the forces pertaining to different particles will be altogether different. Therefore, those which return on their path, will not recover a velocity on the return, that is practically equal to that perpendicular velocity that it had on approach; & thus, it will not be reflected at an angle of reflection practically equal to the angle of incidence. Those, which manage to pass over the whole of the interval between the two planes, on moving away from the further plane, will, under the influence of different sums of forces for different particles, have quite different increments or decrements of the perpendicular velocities; & they will emerge at quite different angles from one another, in all directions. But, those that reach the surface with either of those two opposite dispositions will have but two kinds of forces; & each of these will remain constant for its corresponding class of particles. Hence, with one of these classes there will be more easy approach in its continually curving path to a position parallel to the planes, corresponding to the extinction of the perpendicular velocity; & with the other, this will be more difficult. Therefore there will be produced, in consequence of the two opposite dispositions, two fits, the one of more easy, & the other of more difficult reflection, or more easy transmission; these fits recur at equal intervals of space. However, these will take place in such a manner that the greatest facility of reflection will correspond to the mean disposition; & the less or more the particles depart from this mean on striking the surface, the more or the less, respectively, will they participate in that facility. This greater or less approach to the maximum facility, corresponding to the mean disposition, has been represented in the dissertation by Benvenuti mentioned above by a continuous curve, which is equally inflected on each side of its axis; & from this curve all the other points that relate to fits & their consequences are explained in a most excellent manner.

The cause of the difference in the ratio of the amount of light reflected to that which is transmitted.

498. Further, from this also it is clear how it comes about that, out of a number of homogeneous rays reaching the same surface, some are transmitted & others are reflected, according as they reach it in one or other of two dispositions. Since, of those particles which do [not] reach the surface with one of the two extreme dispositions, not all reach it in the mean disposition exactly; it may happen that the ratio of reflections to transmissions will be altogether different in different circumstances of, say, various differences between the media, or different inclinations of approach. For when the inequality of the forces is less or the perpendicular velocity, which has to be destroyed by the inequality to produce reflection, is greater, only those particles are reflected which reach the surface in dispositions very near to that mean disposition; & so, much fewer are reflected than is the case when the inequality of forces is greater or the perpendicular velocity is less. Hence, it comes about that the less the difference between the media, or the greater the angle of incidence, the smaller the proportion of rays reflected; which is in agreement with experience. In this connection also it is especially to be observed that when in the continuous winding of the curved path of any particle, the path being at one time concave on one side & at another time on the other, according as the attractions or the repulsions of the denser medium are more powerful, a position nearly parallel to the surface of separation between the media is attained several times in succession, as the perpendicular velocity is nearly destroyed, a very slight difference of the forces will be sufficient to produce exact parallelism, or the total extinction of that perpendicular velocity. Although, when these, so to speak, tortuosities are ended as the particle at length reaches the nearer plane in reflection & the further plane in refraction, the sum of all the actions, which determines the total perpendicular velocity, must be practically the same; that is to say, in no wise changed to any sensible extent by the slight difference of forces, such as produced the slight difference of disposition from the mean disposition.

The cause of the difference in the intervals between successive fits.

499. In this way we have a sufficient explanation of the difference between the two fits; but we have still to explain the source of the difference in the intervals between the fits, which we propounded in Art. 472. There is nothing wonderful in the fact that differently coloured rays should have different intervals. For, different velocities require

velocitates diversa requirunt intervalla spatii inter vices oppositas, quando etiam eæ vices redeant æqualibus temporis intervallis, & diversus particularum heterogenearum textus requirit diversa oscillationum tempora. Quod in diversis mediis particulæ ejusdem generis habeant diversa intervalla, itidem facile colligitur ex diversa velocitate, quam in iis haberi post refractionem ostendimus num. 493; sed præterea in ipsa mediorum mutatione inæqualis actio inter puncta particulam componentia potest utique, & vero videtur etiam debere oscillationis magnitudinem, & fortasse etiam ordinem mutare, adeoque celeritatem oscillationis ipsius. Demum ejusmodi mutatio pro diversa inclinatione viæ particulæ advenientis ad superficiem, diversa utique esse debet, ob diversam positionem motuum punctorum ad superficiem ipsam, & ad massam agentem in ipsa puncta. Quamobrem patet, eas omnes tres causas debere discrimen aliquod exhibere inter diversa intervalla, uti reapse ex observatione colligitur.

Discrimen id non posse definiri, nisi per observationes: non pendere a sola velocitate.

500. Si possemus nosse peculiares constitutiones particula-[233]-rum ad diversos coloratos radios pertinentium, ordinem, & numerum, ac vires, & velocitates punctorum singulorum; tum mediorum constitutionem suam in singulis, ac satis Geometriæ, satis imaginationis haberemus, & mentis ad omnia ejusmodi solvenda problemata; liceret a priori determinare intervallorum longitudines varias, & eorundem mutationes pro tribus illis diversis circumstantiis exhibere. Sed quoniam longe citra eum locum consistimus debemus illas tantummodo colligere per observationes, quod summa dexteritate Newtonus, præstitit, qui determinatis per observationem singulis, mira inde consectaria deduxit, & Naturæ phænomena explicavit, uti multo luculentius videre est in illa ipsa Benvenutiana dissertatione. Illud unum ex proportionibus a Newtono inventis haud difficulter colligitur, ea discrimina non pendere a sola particularum celeritate, nam celeritatum proportiones, novimus per sinuum rationem: & facile itidem deducitur ex Theoria, quod etiam multo facilius infertur partim ex Theoria, & partim ex observatione, radium, qui post quotcunque vel reflexiones, vel refractiones regulares devenit ad idem medium, eandem in eo velocitatem habere semper; nam velocitates in reflexione manent, & in mutatione mediorum sunt in ratione reciproca sinus incidentiæ ad sinum anguli refracti: ac tam Theoria, quam observatio facile ostendit, ubi planis parallelis dirimantur media quotcunque, & radius in data inclinatione ingressus e primo abeat ad ultimum, eundem fore refractionis angulum in ultimo medio, qui esset, si a primo immediate in ultimum transivisset. Sed hæc innuisse sit satis.

Quod de crystallo Islandica Newtonus prodidit, id in hac Theoria nullam habere difficultatem.

501. Illud etiam innuam tantummodo, quod Newtonus in Opticis Quæstionibus exponit, esse miram quandam crystalli Islandicæ proprietatem, quæ radium quemvis, dum refringit, discerpit in duos, & alium usitato modo refringit, alium inusitato quodam, ubi & certæ quædam observantur leges, quarum explicationes ipse ibidem insinuat haberi posse per vires diversas in diversis lateribus particularum luminis, ac solum adnotabo illud, ex num. 423 patere, in mea Theoria nullam esse difficultatem agnoscendi in diversis lateribus ejusdem particulæ diversas dispositiones punctorum, & vires, qua ipsa diversitate usi sumus superius ad explicandam solidorum cohæsionem, & organicam formam, ac certas figuras tot corporum, quæ illas vel affectant constanter, vel etiam acquirunt.

Diffractionem esse inchoatam reflexionem, vel refractionem.

502. Remanet demum diffractio luminis explicanda, quam itidem num. 472 proposueramus. Ea est quædam velut inchoata reflexio, & refractio. Dum radius advenit ad eam distantiam a corpore diversæ naturæ ab eo, per quod progreditur, quæ virium inæqualitatem inducit, incurvat viam vel accedendo, vel recedendo, & directionem mutat. Si corporis superficies ibi esset satis ampla, vel reflecteretur ad angulos æquales, vel immergeretur intra novum illud medium, & refrin-[234]-geretur; at quoniam acies ibidem progressum superficiei interrumpit; progreditur quidem radius aciem ipsam evitans & circa illam prætervolat; sed egressus ex illa distantia directionem conservat postremo loco acquisitam, & cum ea, diversa utique a priore, moveri pergit: ut adeo tota luminis Theoria sibi ubique admodum conformis sit, & cum generali Theoria mea apprime consentiens, cujus rami quidam sunt bina Newtoni præclarissima comperta virium, quibus cælestia corpora motus peragunt suos & quibus particulæ luminis reflectuntur, refringuntur, diffringuntur. Sed de luce, & coloribus jam satis.

De sapore, & odore: multorum error in ratione densitatis odoris propagati.

503. Post ipsam lucem, quæ oculos percellit, & visionem parit, ac ideam colorum excitat, pronum est delabi ad sensus ceteros, in quibus multo minus immorabimur, cum circa eos multo minora habeamus comperta, quæ determinatam physicam explicationem ferant. Saporis sensus excitatur in palato a salibus. De angulosa illorum forma jam

different intervals of space between opposite fits, when those fits recur also at equal intervals of time; & a difference in the structure of heterogeneous particles requires a difference in the periods of oscillation. It is also easily seen that particles of the same kind have different intervals in different media, owing to that difference in velocity, which, in Art. 493, was proved to exist after refraction. But, in addition, on changing the medium, an unequal action between the points composing the particle certainly can and, apparently indeed, is bound to alter the magnitude of the oscillation also, & perhaps even the order; & thus the velocity of that oscillation must alter. Further, such a change, for a difference in the inclination of the path of the particle approaching the surface, is in every case bound to be different, on account of the difference in situation of the motions of the points with respect to the surface & the mass acting upon the points. Hence, it is clear that all three of these causes must stand for some difference between diverse intervals; & indeed we can deduce as much from observation.

This difference cannot be definitely given, unless by observation; it does not depend on velocity alone.

500. If we could know the particular constitutions of particles for differently coloured rays, the order, number, forces & velocities of each point, & the constitution of each medium for each ray, and if we had a sufficiency of geometry, imagination & intelligence to solve all problems of this kind, we could determine from first principles the various lengths of the intervals, & could give the changes due to each of the three different circumstances. But since this is far beyond us, we are bound to deduce them from observation alone. This Newton accomplished with the greatest dexterity; having determined each by observation, he deduced from them wonderful consequences; & explained the phenomena of Nature; as also it is to be seen much better in the dissertation by Benvenuti. There is one thing that can be without much difficulty derived from the proportions discovered by Newton, namely, that the differences do not solely depend upon the velocities of the particles; for we know the proportions of the velocities by the ratio of the sines. It can also easily be deduced from the Theory, & indeed much more easily can it be inferred partly from the Theory & partly from observation, that a ray which, after any number of regular reflections & refractions, comes to the same medium will always have the same velocity in it as at first. For the velocities remain unaltered in reflection, & on a change of medium they are in the inverse ratio of the sine of the angle of incidence to the sine of the angle of refraction. Both the Theory, & observation, clearly show that, when any number of media are separated by parallel planes, & a ray, entering at a given inclination, leaves the first & reaches the last, there will be the same angle of refraction in the last medium as there would have been, if it had passed directly from the first medium into the last. But a mere mention of these things is enough.

That which Newton recorded concerning Iceland Spar presents no difficulty in my Theory.

501. I will also merely mention that, as was stated by Newton in his Questions at the end of his Optics, there is a wonderful property of Iceland Spar; namely, that when it refracts a ray of light it divides it into two, refracting one part according to the normal manner, & the other in an unusual way; with the latter also definite laws are observed. Newton himself suggested that the explanation of these laws could be attributed to different forces on different sides of the particles of light; & I will only remark that, according to Art. 423, it is evident that in my Theory there is no difficulty over admitting for different sides of the same particle different dispositions of the points, & different forces; we have already employed this sort of difference to explain cohesion of solids, & organic form, & all those shapes of bodies, such as they always endeavour to acquire, & indeed do acquire.

Diffraction is incomplete reflection or refraction.

502. Finally, we have to explain diffraction, which we also enunciated in Art. 472. This is, so to speak, an incomplete reflection or refraction. When a ray of light attains the distance, from a body of a different nature from one through which it passes, which induces an inequality of forces, its path becomes curved, either by approach or recession, & the direction is altered. If the surface of the body at the point in question is sufficiently wide, the ray will either be reflected at equal angles, or it will enter the new medium & be reflected. But when a sharp edge terminates the run of the surface, the ray will pass on, slipping by the edge, & flying past & round it. But, on emergence from that distance, the ray will preserve the direction acquired in the last position, & with this direction, which will be altogether different from that which it had originally, it will continue its motion. Thus the whole theory of light will be quite consistent, & in close agreement with my Theory. Of this Theory, the two most noted discoveries of Newton with respect to forces are just branches; namely, the forces with which the heavenly bodies keep up their motions, & those by which particles of light are reflected, refracted & diffracted. But I have now said sufficient about light & colour.

Concerning taste & smell; the error of many people with regard to the ratio of the density of a propagated odour.

503. After light, which affects the eyes, begets vision, & excites the idea of colours, we naturally come to the other senses; over these I will spend far less time, since we have far less knowledge of them, such as will help us to give a definite physical explanation. The sense of taste is excited in the palate by salts. I have already spoken of the

diximus num. 464, quæ ad diversum excitandum motum in papillis palati abunde sufficit; licet etiam dum dissolvuntur, vires varias pro varia punctorum dispositione exercere debeant, quæ saporum discrimen inducant. Odor est quidam tenuis vapor ex odoriferis corporibus emissus, cujus rei indicia sunt sane multa, nec omnino assentiri possum illi, qui odorem etiam, ut sonum, in tremore medii cujusdam interpositi censet consistere. Porro quæ evaporationum sit causa, explicavimus abunde num. 462. Illud unum hic innuam, errare illos, uti pluribus ostendi in prima parte meæ dissertationis *De Lumine*, qui multi sane sunt, & præstantes Physici, qui odoribus etiam tribuunt proprietatem lumini debitam, ut nimirum eorum densitas minuatur in ratione reciproca duplicata distantiarum a corpore odorifero. Ea proprietas non convenit omnibus iis, quæ a dato puncto diffunduntur in sphæram, sed quæ diffunduntur cum uniformi celeritate, ut lumen. Si enim concipiantur orbes concentrici tenuissimi datæ crassitudinis; ii erunt ut superficies, adeoque ut quadrata distantiarum a communi centro, ac densitas materiæ erit in ratione ipsorum reciproca: si massa sit eadem: ut ea in ulterioribus orbibus sit eadem, ac in citerioribus; oportet sane, tota materia, quæ erat in citerioribus ipsis, progrediatur ad ulteriores orbes motu uniformi, quo fiet, ut, appellente ad citeriorem superficiem orbis ulterioris particula, quæ ad citeriorem citerioris appulerat, appellat simul ad ulteriorem ulterioris quæ appulerat simul ad ulteriorem citerioris, materia tota ex orbe citeriore in ulteriorem accurate translata: quod nisi fiat, vel nisi loco uniformis progressus habeatur accurata compensatio velocitatis imminutæ, & impeditæ a progressu partis vaporum, quæ compensatio accurata est admodum improbabilis; non habebitur densitas reciproce proportionalis orbibus, sive eorum superficiebus, vel distantiarum quadratis.

De sono difficultas in determinandis undis excitatis in fluido elastico.

[235] 504. Sonus geometricas determinationes admittit plures, & quod pertinet ad vibrationes chordæ elasticæ, vel campani æris, vel motum impressum aeri per tibias, & tubas, id quidem in Mechanica locum habet, & mihi commune est cum communibus theoriis. Quod autem pertinet ad progressum soni per aerem usque ad aures, ubi delatus ad tympanum excitat eum motum, a quo ad cerebrum propagato idea soni excitatur, res est multo operosior, & pendet plurimum ab ipsa medii constitutione: ac si accurate solvi debeat problema, quo quæratur ex data medii fluidi elasticitate propagatio undarum, & ratio inter oscillationum celeritates, a qua multipliciter variata pendent omnes toni, & consonantiæ, ac dissonantiæ, & omnis ars musica, ac tempus, quo unda ex dato loco ad datam distantiam propagatur; res est admodum ardua; si sine subsidiariis principiis, & gratuitis hypothesibus tractari debeat, & determinationi resistentiæ fluidorum est admodum affinis, cum qua motum in fluido propagatum communem habet. Exhibebo hic tantummodo simplicissimi casus undas, ut appareat, qua via ineundam censeam in mea Theoria ejusmodi investigationem.

Quo pacto oriantur undæ in serie continua punctorum se invicem repellentium.

505. Sit in recta linea disposita series punctorum ad data intervalla æqualia a se invicem distantium, quorum bina quæque sibi proxima se repellant viribus, quæ crescant imminutis distantiis, & dentur ipsæ. Concipiatur autem ea series utraque parte in infinitum producta, & uni ex ejus punctis concipiatur externa vi celerrime agente in ipsum multo magis, quam agant puncta in se invicem, brevissimo tempusculo impressa velocitas quædam finita in ejusdem rectæ directione versus alteram plagam, ut dexteram, ac reliquorum punctorum motus consideretur. Utcunque exiguum accipiatur tempusculum post primam systematis perturbationem, debent illo tempusculo habuisse motum omnia puncta. Nam in momento quovis ejus tempusculi punctum illud debet accessisse ad punctum secundum post se dexterum, & recessisse a sinistro, velocitate nimirum in eo genita majore, quam generent vires mutuæ, quæ statim agent in utrumque proximum punctum, aucta distantia a sinistro, & imminuta a dextero, qua fiet, ut sinistrum urgeatur minus ab ipso, quam a sibi proximo secundo ex illa parte, & dexterum ab ipso magis, quam a posteriore ipsi proximo, & differentia virium producet illico motum aliquem, qui quidem initio, ob differentiam virium tempusculo infinitesimo infinitesimam, erit infinities minor motu puncti impulsi, sed erit aliquis: eodem pacto tertium punctum utraque ex parte debet illo tempusculo infinitesimo habere motum aliquem, qui erit infinitesimus respectu secundi, & ita porro.

angular forms of salts, in Art. 464; these are quite sufficient for the excitement of different motions in the papillæ of the palate; although, even when they are dissolved, they must exert different forces for different dispositions of the points, which induce differences in taste. Smell is a sort of tenuous vapour emitted by odoriferous bodies; of this there are really many points in evidence. I cannot agree altogether with one who thinks that smell, like sound, consists of a sort of vibration of some intervening medium. Moreover, I have fully explained, in Art. 462, what is the cause of evaporations. I will but mention here this one thing, namely, that, as I showed in several places in the first part of my dissertation *De Lumine*, those many and distinguished physicists are mistaken who attribute to smell the same property as that proper to light, namely, that the density diminishes in the inverse ratio of the squares of the distances from the odoriferous body. That is a property that does not apply to all things that are diffused throughout a sphere from a given point; but only with those that are thus diffused with uniform velocity, as light is. For if we imagine a set of concentric spherical shells of given very small thickness, they will be like surfaces. Hence, they will be in the same ratio as the squares of the distances from the common centre; &, the density of matter will be inversely proportional to them, if the mass is the same. Now, in order that it may be the same in the outer shells as it is in the inner, it is necessary that the whole of the matter which was in the inner shells should proceed to the outer shells with a uniform motion; then, it would come about that two particles, which have reached simultaneously the inner & outer surfaces of the inner shell respectively, will reach simultaneously the inner & outer surfaces of the outer shell; & the whole of the matter will be transferred accurately from the inner shell to the outer. If this is not the case, or, failing uniform progression, if instead there is not an accurate compensation of the velocity thus diminished & hindered by the advance of part of the vapours (& such an accurate compensation is in the highest degree improbable), then the density cannot be inversely proportional to the shells, i.e., to their surfaces, or the squares of the distances.

Sound; difficulty in determining the waves excited in an elastic fluid.

504. Sound admits of several geometrical determinations; & matters pertaining to vibrations of an elastic cord or bell-metal, or the motion given to the air by flutes & trumpets, all belong to the science of Mechanics; & for them my Theory is in agreement with the ordinary theories. But, with respect to the progression of sound through the air to the ears, where it is carried to the ear-drum & excites the motion by means of which, when propagated to the brain, the idea of sound is produced, the matter is much more laborious, & depends to a very large extent on the constitution of the medium itself. If it is necessary to solve the problem, in which it is desired to find the propagation of waves from a given elasticity of a fluid medium, & the ratio between the velocities of the oscillations upon which, in its manifold variations, depend all musical sounds, harmonious or discordant, the whole art of music, & the time in which a wave is propagated from a given point to a given distance; then, the matter is very hard, especially if it has to be treated without the help of subsidiary principles or unfounded hypotheses. It is closely allied to the determination of the resistance of fluids, with which subject it has common ground in the motion propagated in a fluid. I will explain here merely waves of the very simplest kind; so that the manner in which I consider in my Theory such an investigation should be undertaken will be seen.

The manner in which waves may arise in a continuous series of points repelling one another.

505. Suppose we have a series of points situated in one straight line at given equal intervals of distance from one another; & of these let any two consecutive points repel one another with forces, which increase as the distance decreases, & suppose that the magnitudes of these forces are also given. Also suppose that this series is continued on either side to infinity; & suppose that, by means of an external force acting very quickly on one of the points of the series much more than the points act upon one another, there is impressed upon it in a very short time a certain finite velocity in the direction of the straight line towards either end of it, say towards the right; then we have to consider the motion of all the other points. No matter how small the interval of time taken, after the initial disturbance of the system, in that interval all points must have had motion. For, in any instant of that interval of time, that point must have approached the next point to it on the right, & have receded from the one on the left; a velocity being generated in it greater than that which the mutual forces would give. These forces immediately act on the points next to it on either side, the distance on the left being increased, & on the right diminished. Thus, the point on the left will be impelled by that point less than by the next one to it on its left, & the one on the right more than by the next one to the right of it. The difference of forces will immediately produce some motion; this motion indeed at first, owing to the difference of forces in an infinitesimal time being itself infinitesimal, will be infinitely less than the motion of the point under the action of the external force; but there will be some motion. In the same way, a third point on either side must in that infinitesimally small time have some motion, which will be infinitesimal with respect to that of the second;

Post tempusculum utcunque exiguum omnia puncta æquilibrium amittent, & motum habebunt aliquem. Interea cessante actione vis impellentis punctum primum incipiet ipsum retar-[236]-dari vi repulsiva secundi dexteri prævalente supra vim secundi sinistri, sed adhuc progredietur, & accedet ad secundum, ac ipsum accelerabit : verum post aliquod tempus retardatio continua puncti impulsi, & acceleratio secundi reducent illa ad velocitatem eandem : tum vero non ultra accedent ad se invicem, sed recedent, quo recessu incipiet retardari etiam punctum primum dexterum, ac paullo post extinguetur tota velocitas puncti impulsi, quod incipiet regredi : aliquanto post incipiet regredi & punctum secundum dexterum, & aliquanto post tertium, ac ita porro aliud. Sed interea punctum impulsum, dum regreditur, incipiet urgeri magis a primo sinistro, & acceleratio minuetur : tum habebitur retardatio, tum motus iterum reflexus. Dum id punctum iterum incipit regredi versus dexteram, crit aliquod e dexteris, quod tunc primo incipiet regredi versus sinistram, & dum per easdem vices punctum impulsum iterum reflexit motum versus sinistram, aliud dexterum remotius incipiet regredi versus ipsam sinistram, ac ita porro motus semper progreditur ad dexteram major, & incipient regredi nova puncta alia post alia. Undæ amplitudinem determinabit distantia duorum punctorum, quæ simul cunt & simul redeunt, ac celeritatem propagationis soni tempus, quod requiritur ad unam oscillationem puncti impulsi, & distantia a se invicem punctorum, quæ simul cum eo eunt, & redeunt ; & quod ad dexteram accidit ad sinistram. Sed & ea perquisitio est longe altioris indaginis, quam ut hic institui debeat ; & ad veras soni undas elasticas referendas non sufficit una series punctorum jacentium in directum, sed congeries punctorum, vel particularum circumquaque dispersarum, & se repellentium.

Solutio difficultatis pertinentis ad propagationem rectilineam diversorum sonorum admodum facilis in hac Theoria.

506. Interea illud unum adjiciam, in mea Theoria admodum facile solvi difficultatem, quam Eulerus objecit Mairanio, explicanti propagationem diversorum sonorum, a quibus diversi toni pendent, per diversa genera particularum elasticarum, quæ habentur in aere, quorum singula singulis sonis inserviant, ut diversi sunt colorati radii cum diverso constanti refrangibilitatis gradu, & colore. Eulerus illud objicit, uti tam multa sunt sonorum genera, quæ ad nostras, & aliorum aures simul possint deferri, ita debere haberi continuam seriem particularum omnium generum ad ea deferenda, quod haberi omnino non possit, cum circa globum quenvis in eodem plano non nisi sex tantummodo alii globi in gyrum possint consistere. Difficultas in mea Theoria nulla est, cum particulæ aliæ in alias non agant per immediatum contactum, sed in aliqua distantia, quæ diametro globorum potest esse major in ratione quacunque utcunque magna. Cum igitur certi globuli in iisdem distantiis possint esse inertes respectu certorum, & activi respectu aliorum ; patet, posse multos diversorum generum globulos esse permixtos ita, ut actionem aliorum sentiant alii. Quin [237] immo licet activi sint globuli, fieri debet, ut alii habeant motus conformes tum eos, qui pendent a viribus mutuis inter duos globulos, a quibus proveniunt undæ, tum eos qui pendent ab interna distributione punctorum, a qua proveniunt singularum particularum interni vibratorii motus, & qui itidem ad diversum sonorum genus plurimum conferre possint, & dissimilium globorum oscillationes se mutuo turbent, similium perpetuo post primas actiones actionibus aliis conformibus augeantur, quemadmodum in consonantibus instrumentorum chordis cernimus, quarum una percussa sonant & reliquæ. Ubique libertas motuum, & dispositionis, quæ sublato immediato impulsu, & accurata continuitate in corporum textu, acquiritur ad explicandam naturam, est perquam idonea, & opportuna.

De calore & frigore : materiæ cientis calorem expansio orta ab elasticitate : fixatio ejusdem, & velocitas ut torrentis cujusdam.

507. Quod pertinet ad tactiles proprietates, quid sit solidum, fluidum, rigidum, molle elasticum, flexile, fragile, grave, abunde explicavimus : quid lævigatum, quid asperum, per se patet. Caloris causam repono in motu vehementi intestino particularum igneæ, vel sulphureæ substantiæ fermentescentis potissimum cum particulis luminis, & qua ratione id fieri possit, exposuimus. Frigus haberi potest per ipsum defectum ejusmodi substantiæ, vel defectum motus in ipsa. Haberi possunt etiam particulæ, quæ frigus cieant actione sua, ut nitrosæ, per hoc, quod ejusmodi particularum motum sistant, & eas, attractione

& so on. Thus, after the lapse of any short interval of time, however small, all points will lose their equilibrium & have some motion. Further, the action of the force acting upon the first point will itself begin to be retarded by the repulsive force of the next point on the right prevailing over the force from the next on the left; but it will still progress, approach the second & accelerate it. However, after some time, the continuous retardation of the first point, & the acceleration of the second, will reduce them to the same velocity; & then they will no longer approach one another, but will recede from one another. When this recession starts, the first point on the right will also begin to be retarded, & a little while afterwards the whole of the velocity of the point impelled by the external force will be destroyed, & it will commence to go backwards; shortly afterwards, the second point on the right will also commence to go backwards; shortly after that, the third point; & so on, one after the other. But meanwhile, as it returns, the point, that was impelled by the external force, will be more under the action of the first point on the left, & its acceleration will be diminished; there will follow first a retardation, & then once more a reversal of motion. When the point once more begins to move towards the right, there will be some one of the points on the right, which then for the first time is beginning to move backwards to the left; & when, after the same changes, the point impelled once more reverses its motion & moves towards the left, there will be another point on the right, further off, which will begin to move backwards towards the left. In this way, the motion will always proceed further to the right, & fresh points, one after the other, will begin to reverse their motion. The distance between two points, which go forward & backward simultaneously, will determine the amplitude of the wave; the velocity of propagation of sound will be found from the time that is required for one oscillation of the impelled point, & the distance between points, whose motion backwards & forwards is simultaneous; & what happens on the right will also happen on the left. But the investigation is one of far too great difficulty to be properly treated here; to render an account of the true elastic waves of sound, one series of points lying in a straight line is insufficient; we must have groups of points or of particles, scattered in all directions round about, & repelling one another.

The solution of the difficulty with respect to the rectilinear propagation of different sounds comes quite easily from my Theory.

506. I will add just one other thing; in my Theory, it is quite easy to give a solution of the difficulty, which Euler brought forward in opposition to Mairan; the latter tried to explain the propagation of the different sounds, upon which different musical tones depend, by the presence of different kinds of elastic particles in the air; each kind of particle was of service to the corresponding sound, just as there are differently coloured rays of light, having a constant different degree of refrangibility, & a different colour. Euler's objection was that there are so many kinds of sounds, which can be borne simultaneously to our ears & to those of others, that there must be a continuous series of particles of all the different kinds to carry these sounds; & that this was quite impossible, since only six spheres could lie in a circle in the same plane round a sphere. There is no such difficulty in my Theory, since particles do not act upon one another by immediate contact, but at some distance, such as can bear to the diameter of the spheres any ratio whatever, however large. Since, then, certain little spheres can be inert, when placed at the same distances, with regard to some & active with regard to others, it is clear that a large number of little spheres of different kinds can be so intermingled that some of them feel the action of others. Nay indeed, even if the little spheres are active, there are bound to be some that have congruent motions; not only those motions which depend upon the mutual forces between two little spheres by which waves are produced, but also those which depend on the internal distribution of the points forming them from which arise the internal vibratory motions of the several particles. These, too, may contribute towards a different class of sounds to a very great extent; & they will disturb the mutual oscillations of unlike spheres, &, after the first actions, the oscillations of like spheres will be increased by congruent actions; just as in the consonant strings of instruments we see that, when one of them is struck, all the others sound as well. The freedom of motion everywhere, & of arrangement, which is acquired by the removal of the ideas of immediate impact & accurate continuity in the structure of bodies, is most suitable & convenient for the purpose of explaining the nature of sound.

Heat & cold; the expansion of the matter producing heat arises from elasticity; fixation of the same, & a velocity as of a torrent.

507. With respect to tactile properties, we have had full explanations of solid, fluid, rigid, soft, elastic, flexible, fragile & heavy bodies; what a smooth, or a rough, body is, is self-evident. I consider the cause of heat to consist of a vigorous internal motion of the particles of fire, or of a sulphurous substance fermenting more especially with particles of light; & I have shown the mode in which this may take place. Cold may be produced by a lack of this substance, or by a lack of motion in it. Also there may be particles which produce cold by their own action, such as nitrous substances, through something which stops the motion of such particles, &, as their attraction overcomes their

mutuas ipsarum vires vincente, ad se rapiant, ac sibi affundant quodammodo, veluti alligatas. Potest autem generari frigus admodum intensum in corpore calido per solum etiam accessum corporis frigefacti ob solum ejusmodi substantiæ defectum. Ea enim, dum fermentat, & in suo naturali volatilizationis statu permanet, nititur elasticitate sua ipsa ad expansionem, per quam, si in aliquo medio conclusa sit, utcunque inerte respectu ipsius, ad æqualitatem per ipsum diffunditur, unde fit, ut si uno in loco dematur aliqna ejus pars, statim illuc ex aliis tantum devolet, quantum ad illam æqualitatem requiritur. Hinc nimirum, si in aere libero cesset fermentantis ejusmodi substantiæ quantitas, vel per imminutam continuationem impulsuum ad continuandum motum, ut imminuta radiorum Solis copia per hyemem, ac in locis remotioribus ab Æquatore, vel per accessum ingentis copiæ particularum sistentium ejusdem substantiæ motum, unde fit, ut in climatis etiam non multum ab Æquatore distantibus ingentia pluribus in locis habeantur frigora, & glacies per nitrosorum, effluviorum copiam ; e corporibus omnibus expositis aeri perpetuo crumpet magna copia ejusdem fermentescentis ibi adhuc, & elasticæ materiæ igneæ ; & ea corpora remanebunt admodum frigida per solam imminutionem ejus materiæ, quibus si manum admoveamus, ingens illico ex ipsa manu particularum earundem multitudo avolabit transfusa illuc, ut res ad æqualitatem redu-**[238]**-catur, & tam ipsa cessatio illius intestini motus, qua immutabitur status fibrarum organici corporis, quam ipse rapidus ejus substantiæ in aliam irrumpentis torrens, eam poterit, quam adeo molestam experimur, frigoris sensationem, excitare.

Imago in aeris fixatione, & affluxu.

508. Torrentis ejusmodi ideam habemus in ipso velocissimo aeris motu, qui si in aliqua spatii parte repente ad fixitatem reducatur in magna copia, ex aliis omnibus advolat celerrime, & horrendos aliquando celeritate sua effectus parit. Sic ubi turbo vorticosus, & aerem inferne exsugens prope domum conclusam transeat, aer internus expansiva sua vi omnia evertit : avolant tecta, diffringuntur fenestræ, & tabulata, ac omnes portæ, quæ cubiculorum mutuam communicationem impediunt, repente dissiliunt, & ipsi parietes nonnunquam evertuntur, ac corruunt, quemadmodum Romæ ante aliquot observavimus annos, & in dissertatione *De Turbine* superius memorata, quam tum edidi, pluribus exposui.

Attractio, quæ potest intestinum motum sistere, & fixare : communicatio ad æqualem saturitatem post partem fixatam saturitatis varia discrimina.

509. Verum hæc sola substantiæ hujusce fermentantis expansiva vis non est satis ad rem explicandam, sed requiritur etiam certa vis mutua, qua ejusmodi substantia in alias quasdam attrahatur magis, in alias minus, quod qui fieri possit, vidimus, ubi de dissolutione, & præcipitatione egimus : & ejusmodi attractio potest esse ita valida, ut motum ipsum intestinum prorsus impediat appressione ipsa, ac fixationem ejus substantiæ inducat, quæ si minor sit, permittet quidem motus fermentatorii continuationem, sed a se totam massam divelli non permittet, nisi accedente corpore, quod majorem exerceat vim, & ipsam sibi rapiat. Hic autem raptus fieri potest ob duplicem causam ; primo quidem, quod alia substantia majorem absolutam vim habeat in ejusmodi substantiam igneam, quam alia, pari etiam particularum numero : deinde, quod licet ea æque, vel etiam minus trabat, adhuc tamen cum utraque in minoribus distantiis trahat plus, in majoribus minus, illa habeat ejus substantiæ multo minus etiam pro ratione attractionis suæ, quam altera ; nam in hoc secundo casu, adhuc ab hac posteriore avellerentur particulæ affusæ ipsius particulis ad distantias aliquanto majores, & affunderentur particulis prioris substantiæ, donec in utravis substantia haberetur æqualis saturitas, si ejus partes inter se conferantur, & æqualis itidem attractiva vis particularum substantiæ igneæ maxime remotarum a particulis utriusque substantiæ, quibus ea affunditur : sed copia ipsius substantiæ igneæ possit adhuc esse in iis binis substantiis in quacunque ratione diversa inter se ; cum possit in altera ob vim longius pertinentem certa vis haberi in distantia majore, quam in altera, adeoque altitudo ejusmodi veluti marium in altera esse major, minor in altera, & in iisdem distantiis possit in altera haberi ob vim majorem densitas major substantiæ ipsius igneæ affusæ, quam in altera. Ex bisce quidem principiis, ac diversis combinationibus, mirum sa-**[239]**-ne, quam multa deduci possint ad explicationem Naturæ perquam idoneis.

Quæ a diffusione ad æqualitatem consequantur potissimum respectu refrigerationis, & conglaciationis.

510. Sic etiam ex hac diffusione ad ejusmodi æqualitatem eandem inter diversas ejusdem substantiæ partes, sed admodum diversam inter substantias diversas, facile intelligitur, qui fiat, ut manus in hyeme exposita libero aeri minus sentiat frigoris, quam solido cuipiam satis denso corpori, quod ante ipsi aeri frigido diu fuerit expositum, ut marmori, & inter ipsa corpora solida, multo majus frigus ab altero sentiat, quam ab altero, ac ab aere humido multo plus, quam a sicco, rapta nimirum in diversis ejusmodi circumstantiis

mutual forces, these substances draw these particles towards themselves & surround themselves with them as if the particles were bound to them. Moreover, a very intense cold can be produced in a warm body merely by the approach of a body made cold by a mere defect of such a substance. For, the substance, while it ferments, & remains in its natural state of volatilization, avails itself of its own elasticity to expand; & thereby, if it is enclosed in any medium, however inert it may be with respect to the medium, the substance diffuses through the medium equally. Hence, it comes about that, if from any one place there is taken away some part of the substance, immediately there flies to it from other places just that quantity which is required for equality. Thus, for instance, if in the open air a quantity of such fermenting substance is lacking, whether through a diminution in the continued impulses necessary for the continued motion, such as the diminished supply of rays from the Sun in winter, or in places more remote from the equator, or whether through the presence of a large supply of particles that stop such motion of the substance, due to which there is, even in regions not far distant from the equator, great coldness in several places, & ice, through an abundance of nitrous exhalations; then, from all bodies exposed to such air there will rush forth a great abundance of the substance still fermenting in them, & of the elastic matter of fire. The bodies themselves will remain quite cold, merely by the diminution of this matter; & if we touch them with the hand, immediately a large number of these particles will fly out of the hand & be transfused into the bodies, so as to bring about equality; & not only the cessation of that internal motion by which the state of the nerves of the organic body is altered, but also the rapid rush of the substance entering into the other, will give rise to that feeling of cold which we experience so keenly.

An illustration from the fixation & inflow of air.

508. We have an idea of such a rush in the very swift motion of the air; if the air in some part of space is suddenly reduced to fixation in large quantities, air will rush in violently from all other places, & sometimes produces dreadful effects by its velocity. Thus, when a whirlwind, sucking out the air below, passes near to a house that is shut up, the air inside the house overcomes everything by its expansive force; roofs fly off, windows are broken, the floors & all the doors that prevent mutual communication between the rooms are suddenly burst apart, & the very walls are sometimes overthrown & fall down; just as was seen at Rome some years ago, & as I fully explained in the dissertation ***De Turbine*** already mentioned, which I published at the time.

The attraction which can stop & fix internal motion; motion shared so as to give equality of saturation after a part is fixed; different kinds of saturation.

509. But the mere expansive force of such a fermenting substance is insufficient to explain thoroughly what happens; we require also a certain mutual force, due to which the substance is attracted more by some bodies & less by others; & the manner in which this can happen was explained when we dealt with solution & precipitation. Such an attraction may be so powerful as to prevent that internal motion altogether by its pressure, & lead to fixation of the substance; but if this is fairly small, it will indeed allow some fermentatory motion to go on, but will not allow the whole mass to be broken up, unless a body approaches which exerts a greater force & draws the substance to itself. Now this attraction can take place in two ways. In the first, because one substance has a greater absolute force on this fiery substance than another, for the same number of particles; in the second, because although the one attracts the substance equally or even less than the other, yet, since either of them attracts it more at smaller distances & less at greater distances, the one has much less of the substance in proportion to its attraction than the other. In this second case, particles will still be torn away from the latter body, intermingled with particles of the substance, to distances somewhat greater, & will be surrounded with particles of the former, until in both there will be an equal saturation when parts of it are compared with one another; & also an equal attractive force for particles of the fiery substance that are remote from particles of either of the substances by which it is surrounded. But there still may be an abundance of the fiery substance in each of the two substances, in any ratio, different for each. For, in the one, due to a more extended continuation of the force, there may be had a given force at a greater distance than in the other; & thus the depth, so to speak, of the oceans surrounding the one may be greater than for the other; & for the same distances, for the one there may be, on account of the greater force, a greater density of the affused fiery substance, than for the other. From these principles, & different combinations of them, it is truly wonderful how many things can be derived extremely suitable to explain the phenomena of Nature.

The consequences of this diffusion tending to establish equality; especially in the matter of refrigeration & congelation.

510. Thus, from the principle of such diffusion tending to establish the same equality between different parts of the same substance, but an equality that is quite different for different substances, it is easily seen how it comes about that in winter the hand when exposed to the open air, feels the cold less than when exposed to a solid body of sufficient density, such as marble, which has previously been exposed to the same cold air for a long time; & amongst solids, feels far more cold from some than from others, from damp air much more than from dry. For, in different circumstances of the same kind, in the same time,

eodem tempore admodum diversa copia igneæ substantiæ, quæ calorem in manu fovebat. Atque hic quidem & analogiæ sunt quædam cum iis, quæ de refractione diximus: nam plerumque corpora, quæ plus habent materiæ, nisi oleosa, & sulphorosa sint, majorem habent vim refractivam, pro ratione densitatis suæ, & corpora itidem communiter, quo densiora sunt, eo citius manum admotam calore spoliant, quæ idcirco si lineam telam libero expositam aeri contingat in hyeme, multo minus frigescit, quam si lignum, si marmora si metalla. Fieri itidem potest, ut aliqua substantia ejusmodi substantiam igneam repellat etiam, sed ob aliam substantiam admixtam sibi magis attrahentem, adhuc aliquid surripiat magis, vel minus, prout ejus admixtæ substantiæ plus habet, vel minus. Sic fieri posset, ut aer ejusmodi substantiam igneam respueret, sed ob heterogenea corpora, quæ sustinet, inter quæ inprimis est aqua in vapores elevata, surripiat nonnihil; ubi autem in ipso volitantes particulæ, quæ ad fixitatem adducunt, vel expellunt ejusmodi substantiam igneam, accedant ad alias, ut aqueas, fieri potest, ut repente habeantur & concretiones, atque congelationes, ac inde nives, & grandines. A diffusione vero ad æqualitatem intra idem corpus fieri utique debet, ut ubi altius infra Terræ superficiem descensum sit, permanens habeatur caloris gradus, ut in fodinis, ad exiguam profunditatem pertinente effectu vicissitudinum, quas habemus in superficie ex tot substantiarum permixtionibus continuis, & accessu, ac recessu solarium radiorum, quæ omnia se mutuo compensant saltem intra annum, antequam sensibilis differentia haberi possit in profundioribus locis: ac ex diversa vi, quam diversæ substantiæ exercent in ejusmodi substantiam igneam, provenire debet & illud, quod experimenta evincunt, ut nimirum nec eodem tempore æque frigescant diversæ substantiæ aeri libero expositæ, nec caloris imminutio certam densitatum rationem sectetur, sed varietur admodum independenter ab ipsa. Eodem autem pacto & alia innumera ex iisdem principiis, ubique sane conformibus admodum facile explicantur.

Eodem pacto explicari & electricitatem: Principia Franklinianæ theoriæ Electricitatis.

511. Patet autem ex iisdem principiis repeti posse explica-[240]-tionem etiam præcipuorum omnium ex Electricitatis phænomenis, quorum Theoriam a Franklino mira sane sagacitate inventam in America & exornavit plurimum, & confirmavit, ac promovit Taurini P. Beccaria vir doctissimus opere egregio ea de re edito ante hos aliquot annos. Juxta ejusmodi Theoriam huc omnia reducuntur: esse quoddam fluidum electricum, quod in aliis substantiis & per superficiem, & per interna ipsarum viscera possit pervadere, per alias motum non babeat, licet saltem harum aliquæ ingentem contineant ejusdem substantiæ copiam sibi firmissime adhærentem, nec sine frictione, & motu intestino effundendam, quarum priora sint per communicationem electrica, posteriora vero electrica natura sua: in prioribus illis diffundi statim id fluidum ad æqualitatem in singulis; licet alia majorem, alia minorem ceteris paribus copiam ejusdem poscant ad quandam sibi veluti connaturalem saturitatem: hinc e duobus ejusmodi corporibus, quæ respectu naturæ suæ non eundem habeant saturitatis gradum, esse alterum respectu alterius electricum per excessum, & alterum per defectum, quæ ubi admoveantur ad eam distantiam, in qua particulæ circa ipsa corpora diffusæ, & iis utcunque adhærentes ad modum atmosphærarum quarundam, possint agere aliæ in alias, e corpore electrico per excessum fluere illico ejusmodi fluidum in corpus electricum per defectum, donec ad respectivam æqualitatem deventum sit, in quo effluxu & substantiæ ipsæ, quæ fluidum dant, & recipiunt, simul ad se invicem accedant, si satis leves sint, vel libere pendeant, & si motus coacervatæ materiæ sit vehemens, explosiones habeantur, & scintillæ, & vero etiam fulgurationes, tonitrua, & fulmina. Hinc nimirum facile repetuntur omnia consueta electricitatis phænomena, præter Batavicum experimentum phialæ, quod multo generalius est, & in Frankliniano plano æque habet locum. Id enim phænomenum ad aliud principium reducitur: nimirum ubi corpora natura sua electrica exiguam habent crassitudinem, ut tenuis vitrea lamella, posse in altera superficie congeri multo majorem ejus fluidi copiam, dummodo ex altera ipsi ex adverso respondente æqualis copia fluidi ejusdem extrahatur recepta in alterum corpus per communicationem electricum, quod ut per satis amplam superficiei partem fieri possit, non excurrente fluido per ejusmodi superficies; aqua affunditur superficiei alteri, & ad alteram manus tota apprimitur, vel auro inducitur superficies utraque, quod sit tanquam vehiculum, per quod ipsum fluidum possit inferri, & efferri, quod tamen non debet usque ad marginem deduci, ut citerior inauratio cum ulteriore conjungatur, vel ad illam satis accedat: si enim id fiat, transfuso statim fluido ex altera superficie in alteram, obtinetur æqualitas, & omnia cessant electrica signa.

a different quantity of the fiery substance is seized, & this originally kept the hand warm. Here, too, there are certain analogies with what we have said about refraction. For, very many bodies possessing a considerable amount of matter, unless they are oily or sulphurous, have a greater refractive force in proportion to their density; & commonly, too, the denser they are, the more quickly they withdraw heat from the hand that touches them; & thus, if the hand touches a linen cloth exposed to the open air in winter, it is made cold to a far less degree than it would be in the case of wood, marble, or metal. Further it may be that some substance of this sort even repels the fiery substance; but, owing to the fact that another substance mixed with it has a stronger attraction, it will still carry off some of the fiery substance, more or less in amount according as there is more or less of the second substance mixed with it. Thus, it might be the case that air would reject a fiery substance of this sort; but, owing to the presence of heterogeneous bodies in it, amongst which there is in particular water uplifted in the form of vapour, it seizes some portion of it. Also, when particles hovering in it, which either induce fixity, or repel such fiery substance, approach others, like those of water-vapour, it may happen that sudden concretions & congelations take place; & thus cause snow & hail. But from a diffusion tending to produce equality within the same body it must come about that, when one goes deeper down beneath the surface of the Earth, there is a permanent degree of warmth. Thus, in mines, the effect of the vicissitudes which take place on the surface owing to the continual mingling of so many substances, & the accession & recession of the solar rays, only continues for a very small depth; for these all compensate one another in the course of a year at any rate, before any sensible difference can be produced in places of fair depth. Because of this, and also on account of the different force exerted by different substances on this fiery substance, it must come about, as is proved experimentally, that different bodies are not cooled equally in the same time when exposed to the open air, nor is the diminution of heat in a fixed ratio to the density, but varies altogether independently of it. In the same way, innumerable other things can be quite readily derived from these same principles, which agree with one another perfectly.

Electricity can also be explained in the same way; Franklin's principles of the theory of electricity.

511. Further, it is clear that from these principles there can be derived an explanation of all the chief phenomena in electricity; the theory of these, discovered by Franklin in America with truly marvellous sagacity, has been greatly embellished & confirmed, & even further developed at Turin by Fr. Beccaria, a most learned man, in his excellent work on this subject, published some years ago. According to such theory, all things reduce to this; there is a certain electric fluid, which can in some substances move along the surface & also through their inward parts; but has no motion through others, although some of these at any rate hold an abundance of the substance very firmly adherent to themselves, & not to be loosened without friction & internal motion. Of these, the former are electric by communication, the latter electric by nature. In the former, the fluid is immediately diffused to produce equality on each of them; although some of them require more, others less, of the fluid to produce, so to speak, an intrinsic saturation, other things being the same. Thus, of two of these bodies, of which the saturation corresponding to their natures is not the same, one will be electric by excess, & the other by defect, with respect to one another. If these bodies approach one another to within that distance, for which the particles surrounding the bodies, & adhering to them like atmospheres, can act upon one another; then, from the body that is electric by excess this fluid will immediately flow towards the one that is electric by defect, until equality is reached. During this flow, the substances which respectively yield & receive the fluid will simultaneously approach one another, if they are light enough, or if they are freely suspended; & if the motion of the concentrated matter is vigorous, there will be explosions, & sparks, & even lightning, thunder, & thunderbolts. Hence, forsooth, can be derived all the customary phenomena of electricity, besides the experiment of the Leyden Jar, which is much more general, & the same holds equally good for Franklin's plate. For this phenomenon reduces to another principle; namely, that when bodies that are naturally electric have a very small thickness, such as a thin glass plate, there can be collected on one of the surfaces a much greater amount of the fluid, & at the same time from the other surface exactly opposite to it there can be withdrawn an equal amount of the fluid, & this may be passed into another body by electric communication. In order that this can take place over a sufficiently ample part of the surface, as the fluid does not run away from such surfaces, water is brought into contact with one surface, & the other is pressed with the whole hand; or each of the surfaces is overlaid with gold, which forms as it were a medium through which the fluid can be borne either in or out. The gold, however, must not be brought right up to the edge, so that the inner gilding touches the outer, or even approaches it too closely; for if this happens, the fluid is immediately transfused from one surface to the other, equality is obtained, & all signs of electricity cease.

Eorum explicatio in hac Theoria.

512. Hujusmodi Theoriæ ea pars, quæ continet respectivam [241] illam saturitatem, conspirat cum iis, quæ diximus de ignea substantia, ubi ipsam respectivam saturitatem abunde explicavimus. Dum autem fluidum vi mutua agente abit ex altera substantia in alteram : facile patet, debere ipsa etiam ea corpora, quorum particulæ ipsum fluidum, quanquam viribus inæqualibus, ad se trahunt, ad se invicem accedere, ac facile itidem patet, cur aer humidus, in quo ob admixtas aquæ particulas vidimus citius manum frigescere, electricis phænomenis contrarius sit, vaporibus abripientibus illico, quod in catena a globi sibi proximi frictione in ipso excitatum, & avulsum congeritur. Secunda pars, ex qua Batavicum experimentum pendet, & successus plani Frankliniani, aliquanto difficilior, explicatione tamen sua non caret. Fieri utique potest, ut in certis corporibus ingens sit ejus substantiæ copia ob attractionem ingentem, & ad exiguas distantias pertinentem, congesta, quæ in aliquanto majore distantia in repulsionem transeat, sed attractioni non prævalentem. Hæc repulsio cum illa copia materiæ potest esse in causa, ne per ejusmodi substantias transire possit is vapor, & ne per ipsam superficiem excurrat, nec vero ad eam accedat satis ; nisi alterius substantiæ adjunctæ actio simul superveniat, & adjuvet. Tum vero ubi lamina sit tenuis, potest repulsio, quam exercent particulæ fluidi prope alteram superficiem siti, agere in particulas sitas circa superficiem alteram : sed adhuc fieri potest, ut ea non possit satis ad vincendam attractionem, qua hærent particulis sibi proximis : verum si ea adjuvetur ex una parte ab attractione corporis admoti per communicationem electrici, & ex altera crescat accessu novi fluidi advecti ad superficiem oppositam, quod vim ipsam repulsivam intendat : tum vero ipsa prævaleat. Ipsa autem prævalente, effluet ex ulteriore superficie ejus fluidi pars novum illud corpus admotum ingressa, ac ex ejus partis remotione, cessante parte vis repulsivæ, quam nimirum id, quod effluit, exercebat in particulas citerioris superficiei, ipsi citeriori superficiei adhæreat jam idcirco major copia fluidi electrici admota per aquam, vel aurum, donec tamen, communicatione extrorsum restituta per seriem corporum sola communicatione electricorum, defluxus ex altera superficie pateat ad alteram. Porro explicationem hujusmodi & illud confirmat, quod experimentum in lamina nimis crassa non succedit. Quod autem per substantiam natura sua electricam non permeet, ut æqualitatem acquirat, id ipsum provenire posset ab exigua distantia, ad quam extendatur ingens ejus attractiva vis in illam substantiam fluidam, & aliquanto majore distantia suarum particularum a se invicem : nam in eo casu altera particula substantiæ per se electricæ, utut spoliata magna parte sui fluidi, non poterit rapere partem satis magnam fluidi alteri parti affusi, & appressi.

Quod videatur esse discrimen inter materiam electricam, & igneam.

513. Hæc quidem an eo modo se habeant, definire non licet [242] nisi & illud ostendatur simul, rem aliter se habere non posse. Sed illud jam patet, Theoriam meam, servato semper eodem agendi modo, suggerere ideam earum etiam dispositionum materiæ, quæ possint maxime omnium ardua, & composita explicare Naturæ phænomena, ac corporum discrimina. Illud unum hic addam ; quoniam & ingens inter igneam substantiam, & electricum fluidum analogia deprehenditur, & habetur itidem discrimen aliquod ; fieri etiam posse, ut inter se in eo tantummodo discrepent, quod altera sit cum actuali fermentatione, & intestino motu, quamobrem etiam comburat, & calefaciat, & dilatet, ac rarefaciat substantias, altera ad fermentescendum apta sit, sed sine ulla, saltem tanta agitatione, quantam fermentatio, inducit orta ex collisione ingenti mutua, vel ex aliarum admixtione substantiarum, quæ sint ad fermentandum idoneæ.

De magnetica vi : directionem, & ejus variationem pendere ab attractione, & mutatione massarum ingentium attrahentium.

514. Quod ad magneticam vim pertinet, adnotabo illud tantummodo, ejus phænomena omnia reduci ad solam attractionem certarum substantiarum ad se invicem. Nam directio, ad quam & inclinatio, & declinatio reducitur, repeti utique potest ab attractione ipsa sola. Videmus acum magneticam inclinari statim prope fodinas ferri, intra quas idcirco nullus est pyxidis magneticæ usus. Si ingens adesset in ipsis polis, & in iis solis, massa ferrea ; omnes acus magneticæ dirigerentur ad polos ipsos : sed quoniam ubique terrarum fodinæ ferreæ habentur, si circa polos eædem sint in multo majore copia, quam alibi ; dirigentur utique acus polos versus, sed cum aliqua deviatione in reliquas massas per totam Tellurem dispersas, quæ nunquam poterit certum superare graduum numerum ; nisi plus æquo ad fodinam aliquam accedatur. Declinatio ejusmodi diversa erit in diversis locis, ob diversam

Explanation of these matters in my Theory.

512. That part of this theory, which deals with the relative saturation, agrees with what we have said with respect to the fiery substance, when we gave a full explanation of its relative saturation. Moreover, when the fluid, under the action of a mutual force, passes from one substance to another, it is readily seen that those bodies, of which the particles attract the fluids to themselves although with unequal forces, must also attract one another. It is also quite clear why moist air, in which, on account of the admixture of water particles, we see that the hand is cooled more rapidly, works in an exactly opposite manner with electric phenomena, the vapour immediately carrying off the fluid, that is accumulated in a chain, after it has been excited in a sphere very close to it by friction & expelled from it into the chain. The second part, upon which the Leyden jar experiment depends, as also the Franklin plate, is somewhat more difficult, yet does not altogether lack an explanation. For, it may indeed be the case that in certain bodies there may be concentrated a huge amount of the substance, due to a huge attraction, which however only lasts for exceedingly small distances; & this attraction for a somewhat greater distance may pass into a repulsion, without however overcoming the attraction. This repulsion taken in conjunction with the large amount of matter may be for the purpose of preventing the possibility of this vapour from passing through such bodies, or of running along its surface, or even of approaching very near to it; unless the action of some other substance adjoined simultaneously supervenes & assists it. Then, indeed, when the plate is thin, there can be a repulsion, exerted by the particles of the fluid situated on one of the surfaces, acting on particles situated near the other surface. Still, it may be that this is not sufficient to overcome the attraction by which the particles adhere to those that are next to them. But, if this is assisted on the one side by the attraction of a body, which is electric by communication, moving towards it, & on the other side it is increased by a fresh accession of fluid brought up to the opposite surface, because this will augment the repulsive force also; then, the repulsive force will overcome the attraction. Now, when this is the case, part of the fluid will flow off from the further surface & enter the new body that has been brought close to it; & since part of the repulsive force ceases owing to the removal of this part of the fluid (namely, that repulsive force that was exerted on the particles of the nearer surface by the part of the fluid that flowed off), in consequence, there will adhere to the nearer surface a greater amount of the electric fluid brought to it by the water or the gold; until, however, communication being restored from without by means of a series of bodies that are merely electric by communication, the flow of the fluid from one surface to the other will be unhindered. Moreover, this explanation is confirmed by the fact that, if the experiment is tried with a plate that is too thick, it will not succeed. Further, the fact that the fluid will not pass through a substance that is naturally electric, so that equality is produced, can be produced by the very small distance over which the huge attractive force on the fluid substance extends, & the somewhat greater distance of its particles from one another. For, in this case, one particle of the naturally electric substance, when it has lost the greater part of its fluid, will not seize upon any great part of the fluid surrounding another part, & in close contact with it.

The manner in which electric matter seems to differ from fire.

513. Whether these things are indeed as stated cannot be determined, unless it can be shown at the same time that it is impossible for them to be otherwise. But this fact is clear, that my Theory, always maintaining the same mode of action, suggests also the idea of these dispositions of matter, such as are most of all capable of explaining the difficult & compound phenomena of Nature, & the differences between bodies. I will add but one thing further: since we can detect a very great analogy between the fiery substance & the electric fluid, & also some difference, it may possibly be that they only differ from one another in the fact that the one occurs in conjunction with actual fermentation & internal motion, due to which it burns, heats, dilates & rarefies substances; while the other is suitable to the setting up of fermentation, but without that agitation, or at least without an agitation so great as that produced by fermentation arising from a very great mutual collision, or from admixture of other substances that are liable to fermentation.

Magnetic force; its direction & variation depends upon the attraction & alteration in the large masses attracting.

514. With regard to magnetic force, I will make but the one observation, that all phenomena with regard to it reduce to a mere attraction of certain substances for one another. For direction, to which both inclination & declination can be reduced, can always be derived from attraction alone. We notice that a magnetic needle is immediately inclined near iron mines; & therefore within these a magnetic compass-box is of no service. If there were present at the poles, & there only, immense masses of iron, every magnetic needle would be directed towards those poles. But, since there are iron mines in all lands, if about the poles there were the same in much greater abundance than in other places, then, in every case needles would be directed towards the poles, but with some deviation towards the other masses scattered over the whole Earth; this deviation could never exceed a certain number of degrees, unless it was taken too near some one mine. Declination of

eorum locorum positionem ad omnes ejusmodi massas, & vero etiam variabitur, cum fodinæ ferri & destruantur in dies novæ, & generentur, ac augeantur, & minuantur in horas. Variatio intra unum diem exigua erit, cum eæ mutationes in fodinis intra unum diem exiguæ sint : procedente tempore evadet major, eritque omnino irregularis ; si mutationes, quæ in fodinis accidunt, sint etiam ipsæ irregulares.

Attractionem, & polos cohærere cum hac Theoria : difficultas de distantia ad quam vis ea extenditur : conjectura de solutione ipsius.

515. Quod autem ad attractionem pertinet eam in particulis haberi posse patet, & ab earum textu debere pendere : plurima autem sunt magnetismi phænomena, quæ ostendant, mutata dispositione particularum generari magneticam vim, vel destrui, & multo frequentius intendi, vel remitti, cujus rei exempla passim occurrunt apud eos, qui de magneticis agunt. Poli autem ex altera parte attractivi, ex altera repulsivi, qui habentur in magnetismo itidem, cohærent cum Theoria ; cum virium summa ex altera parte possit esse major, quam ex altera. Difficultatem aliquam majorem parit distantia ingens, ad quam ejusmodi vis extenditur : at fieri utique id ipsum potest per aliquod effluviorum intermedium genus, quod tenui-[243]-tate sua effugerit huc usque observantium oculos, & quod per intermedias vires suas connectat etiam massas remotas, si forte ex sola diversa combinatione punctorum habentium vires ab eadem illa mea curva expressas id etiam phænomenon provenire non possit. Sed ad hæc omnia rite evolvenda, & illustranda singulares tractatus, & longæ perquisitiones requirerentur ; hic mihi satis est indicasse ingentem Theoriæ meæ fœcunditatem, & usum in difficillimis quibuscunque Physicæ etiam particularis partibus pertractandis.

Quid materia, & unde ejus vires : tria diversa principia, a quibus provenire possunt.

516. Superest, ut postremo loco dicamus hic aliquid de alterationibus, & transformationibus corporum. Pro materia mihi sunt puncta indivisibilia, inextensa, prædita vi inertiæ, & viribus mutuis expressis per simplicem continuam curvam habentem determinatas illas proprietates, quas expressi a num. 117, & quæ per æquationem quoque algebraicam definiri potest. An hæc virium lex sit intrinseca, & essentialis ipsis indivisibilibus punctis ; an sit quiddam substantiale, vel accidentale ipsis superadditum, quemadmodum sunt Peripateticorum formæ substantiales, vel accidentales ; an sit libera lex Auctoris Naturæ, qui motus ipsos secundum legem a se pro arbitrio constitutam dirigat : illud non quæro, nec vero inveniri potest per phænomena, quæ eadem sunt in omnibus iis sententiis. Tertia est causarum occasionalium ad gustum Cartesianorum, secunda Peripateticis inservire potest, qui in quovis puncto possunt agnoscere materiam, tum formam substantialem exigentem accidens, quod sit formalis lex virium, ut etiam, si velint, destructa substantia, remanere eadem accidentia in individuo, possint conservare individuum istud accidens, unde sensibilitas remanebit prorsus eadem, & quæ pro diversa combinatione ejusmodi accidentium pertinentium ad diversa puncta, erit diversa. Prima sententia videtur esse plurimorum e Recentioribus, qui impenetrabilitatem, & activas vires, quas admittunt Leibnitiani, & Newtoniani passim, videntur agnoscere pro primariis materiæ proprietatibus in ipsa ejus essentia sitis. Potest utique hæc mea Theoria adhiberi in omnibus hisce philosophandi generibus, & suo cujusque peculiari cogitandi modo aptari potest.

Homogeneitas elementorum. Si ea non admittatur, quanto plures combinationes per diversas leges virium : formam substantialem, & accidentia posse Peripateticos, si velint, agnoscere in ipsis punctis.

517. Hæc materia mihi est prorsus homogenea, quod pertinet ad legem virium, & argumenta, quæ habeo pro homogeneitate, exposui num. 92. Siqua occurrent Naturæ phænomena, quæ per unicum materiæ genus explicari non possint ; poterunt adhiberi plura genera punctorum cum pluribus legibus inter se diversis, atque id ita, ut tot leges sint, quot sunt binaria generum, & præterea, quot sunt ipsa genera, ut illarum singulæ exprimant vires mutuas inter puncta pertinentia ad bina singulorum binariorum genera, & harum singulæ vires mutuas inter puncta pertinentia ad idem genus, singulæ pro generibus singulis. Porro inde mirum sane, quanto major [244] combinationum numerus orietur, & quanto facilius explicarentur omnia phænomena. Possent autem illæ leges exponi per curvas quasdam, quarum aliquæ haberent aliquid commune, ut asymptoticum impenetrabilitatis arcum, & arcum gravitatis, ac aliæ ab aliis possent distare magis, ut habeantur quædam genera, & quædam differentiæ, quæ corporum elementa in certas classes distribuerent ; & hic Peripateticis, si velint, occasio daretur admittendi materiam ubique homogeneam, ac formas substantiales diversas, quæ accidentalem virium formam diversam exigant, & vero etiam plures accidentales formas, quæ diversas determinent vires, ex quibus componatur vis totalis unius elementi respectu sui similium, vel respectu aliorum.

this kind will be different in different places, on account of the different situation of these places with respect to all such masses; & it will vary, since mines of iron are destroyed & generated every day, & are increased & diminished hourly. The variation within a day will be very slight, since the daily change in mines is very small; as time goes on it becomes greater, & it will be quite irregular, if the changes that take place in mines are themselves also irregular.

Attraction, & the poles, are consistent with my Theory; difficulty over the distance to which the force extends; conjecture as to the solution of this problem.

515. With regard to attraction, it is clear that this can be had in the particles, & that it must depend upon their structure. Moreover, there are very many phenomena of magnetism, which will show that magnetic force is generated by changing the disposition of the particles, or is destroyed, or more frequently is augmented or abated; examples of this everywhere come under the observation of those who study magnets. Moreover, poles that are attractive on one side & repulsive on the other, which are also had in magnetism, agree with my Theory; for, the sum of the forces on one side may be greater than the sum of the forces on the other. A somewhat greater difficulty arises from the huge distance to which this kind of force extends. But even this can take place through some intermediate kind of exhalation, which owing to its extreme tenuity has hitherto escaped the notice of observers, & such as by means of intermediate forces of its own connects also remote masses; if perchance this phenomenon cannot be derived from merely a different combination of points having forces represented by that same curve of mine. But to explain all these things properly, & to furnish them with illustrations would require separate treatment & long investigations. It is enough for me that I have pointed out the extreme fertility of my Theory, & its use in any of the most difficult & special problems of physics.

The nature of matter, & the source of its forces; three different principles from which they may arise.

516. It remains for me here to say a few words finally about alterations & transformations of bodies. To me, matter is nothing but indivisible points, that are non-extended, endowed with a force of inertia, & also mutual forces represented by a simple continuous curve having those definite properties which I stated in Art. 117; these can also be defined by an algebraical equation. Whether this law of forces is an intrinsic property of indivisible points; whether it is something substantial or accidental superadded to them, like the substantial or accidental shapes of the Peripatetics; whether it is an arbitrary law of the Author of Nature, who directs those motions by a law made according to His Will; this I do not seek to find, nor indeed can it be found from the phenomena, which are the same in all these theories. The third is that of occasional causes, suited to the taste of followers of Descartes; the second will serve the Peripatetics, who can thus admit the existence of matter at any point; & then a substantial form producing a circumstance (accidens) which becomes a formal law of forces; so that, if they wish, having destroyed the substance, that the same circumstances shall remain in the individual, they can preserve that individual circumstance. Hence the sensibility will remain the same exactly, & such as will be different for a different combination of such circumstances pertaining to different points. The first theory seems to be that of most of the modern philosophers, who seem to admit impenetrability & active forces, such as the followers of Leibniz & Newton all admit, as the primary properties of matter founded on its very essence. This Theory of mine can indeed be used in all these kinds of philosophising, & can be adapted to the mode of thought peculiar to any one of them.

Homogeneity of the elements. If this is not admitted, there will be all the more combinations through different laws of forces; & the Peripatetics can, if they wish, admit substantial form & circumstances into these points.

517. Matter, in my opinion, is perfectly homogeneous; what pertains to the law of forces, & the arguments which I have in favour of homogeneity, I have stated in Art. 92. If there are any phenomena of Nature, which cannot be explained by a single kind of matter, then we should have to make use of many different kinds of points, with many laws that differ from one another; & this, too, in such a manner that there are as many laws as there are pairs of kinds of points; &, in addition, as many more as there are kinds of points. For, each of the former express the mutual forces between the points belonging to two kinds of each pair, & each of the latter the mutual forces between points belonging to the same kind, one for each kind. Further, from this it is truly marvellous how much greater the number of combinations will become, & how much more easily all phenomena can be explained. Moreover, the laws can be expressed by curves, some of which would have something in common, such as the asymptotic arc of impenetrability, or the arc of gravitation; while some might be considerably different from others, so that certain classes & certain differences could be obtained, such as would distribute the elements of bodies into certain classes. This would give the Peripatetics an opportunity, if they so wished, of admitting matter that was everywhere homogeneous, as well as substantial forms of different kinds such as would necessitate a different accidental form of forces; & also many accidental forms, which determine different laws, from which is compounded the total force of one element upon others similar to it, or upon others that are not.

Mira variatas consectariorum: possibilitas quotlibuerit Mundorum in eodem spatio cum apparenti compenetratione, sine ulla notitia unius cujusvis in aliis.

518. Posset autem admitti vis in quibusdam generibus nulla, & tunc substantia unius ex iis generibus liberrime permearet per substantiam alterius sine ullo occursu, qui in numero finito punctorum indivisibilium nullus haberetur, adeoque transiret cum impenetrabilitate reali, & compenetratione apparente: ac posset unum genus esse colligatum cum alio per legem virium, quam habeant cum tertio, sine ulla lege virium mutua inter ipsa, vel possent ea duo genera nullum habere nexum cum ullo tertio: atque in hoc posteriore casu haberi possent plurimi Mundi materiales, & sensibiles in eodem spatio ita inter se disparati, ut nullum alter cum altero haberet commercium, nec alter ullam alterius notitiam posset unquam acquirere. Mirum sane, quam multæ aliæ in casibus illius nexus cujuspiam duorum generum cum tertio combinationes haberi possint ad explicanda Naturæ phænomena: sed argumenta, quæ pro homogeneitate protuli, locum habent pro omnibus punctis, cum quibus nos commercium aliquod habere possumus, pro quibus solis inductio locum habere potest. An autem sint alia punctorum genera vel hic in nostro spatio, vel alibi in distantia quavis, vel si id ipsum non repugnat, in aliquo alio spatii genere, quod nullam habeat relationem cum nostro spatio, in quo possint esse puncta sine ulla relatione distantiæ a punctis in nostro existentibus, nos prorsus ignoramus, nihil enim eo pertinens omnino ex Naturæ phænomenis colligere possumus, & nimis est audax, qui eorum omnium, quæ condidit Divinus Naturæ Fabricator limitem ponat suam sentiendi, & vero etiam cogitandi vim.

Formam in homogeneitatis suppositione esse numerum, & dispositionem punctorum, quæ sunt radix omnium proprietatum: quæ dici possint formæ specificæ: unde alterationes & transformationes.

519. Sed redeundo ad meam homogeneorum elementorum Theoriam, singulares corporum formæ erunt combinatio punctorum homogeneorum, quæ habetur a distantiis & positionibus, ac præter solam combinationem velocitas, & directio motus punctorum singulorum; pro individuis vero corporum massis accedit punctorum numerus. Dato numero & dispositione punctorum in data massa, datur radix omnium proprietatum, quas habet eadem massa in se, & omnium relationum, [245] quas eadem habere debet cum aliis massis, quas nimirum determinabunt numeri, & combinationes, ac motus earum, & datur radix omnium mutationum, quæ ipsi possunt accidere. Quoniam vero sunt quædam combinationes peculiares, quæ exhibent quasdam peculiares proprietates constantes, quas determinavimus, & exposuimus, nimirum suæ pro cohæsione, & variis soliditatum gradibus, suæ pro fluiditate, suæ pro elasticitate, suæ pro mollitie, suæ pro certis acquirendis figuris, suæ pro certis habendis oscillationibus, quæ & per se, & per vires sibi affixas diversos sapores pariant, & diversos ordores, & colorum diversas constantes proprietates exhibeant, sunt autem aliæ combinationes, quæ inducunt motus, & mutationes non permanentes, uti est omne fermentationum genus; possunt a primis illis constantium proprietatum combinationibus desumi specificæ corporum formæ, & differentiæ, & per hasce posteriores habebuntur alterationes, & transformationes.

Discrimen inter transformationem, & alterationem.

520. Inter illas autem proprietates constantes possunt seligi quædam, quæ magis constantes sint, & quæ non pendeant a permixtione aliarum particularum, vel etiam, quæ si amittantur, facile, & prompte acquirantur, & illæ haberi pro essentialibus illi speciei, quibus constanter mutatis habeatur transformatio, iisdem vero manentibus, habeatur tantummodo alteratio. Sic si fluidi particulæ alligentur per alias, ut motum circa se invicem habere non possint, sed illarum textus, & virium genus maneat idem; conglaciatum illud fluidum dicetur tantummodo alteratum, non vero etiam mutatum specifice. Ita alterabitur etiam, & non specifice mutabitur corpus, aucta quantitate materia igneæ, quam in poris continet, vel aucta quantitate materiæ igneæ, quam in poris continet, vel aueto motu ejusdem, vel etiam aucta aliqua suarum partium oscillatione, ac dicetur calefactione nova alteratum tantummodo: & aquæ massa, quæ post ebullitionem redit ad priorem formam, erit per ipsam ebullitionem alterata, non transformata: figuræ itidem mutatio, ubi ex cera, vel metallo diversa fiunt opera, alterationem quandam inducet. At ubi mutatur ille textus, qui habebatur in particulis, atque id mutatione constanti, & quæ longe alia phænomena præbeat; tum vero dicetur corrumpi, & transformari corpus. Sic ubi e solidis corporibus generetur permanens aer elasticus, & vapores elastici ex aqua, ubi aqua in terram concrescat, ubi commixtis substantiis pluribus arete inter se cohæreant novo nexu carum particulæ, & novum mixtum efforment, ubi mixti particulæ separatæ per solutionem nexus ipsius, quod accidit in putrefactione, & in fermentationibus plurimis, novam singulæ constitutionem acquirant, habebitur transformatio.

518. Also, in some of these classes, the absence of any force may be admitted ; & then the substance of one of these classes will pass perfectly freely through the substance of another without any collisions ; for, with a finite number of indivisible points, there would not be any ; & thus the substance would pass through with real impenetrability & apparent compenetration. Also it would be possible for one kind to be bound up with another by means of a law of forces, which they have with a third, without any mutual law of forces between themselves, or these two kinds might have no connection with any third. In this latter case there might be a large number of material & sensible universes existing in the same space, separated one from the other in such a way that one was perfectly independent of the other, & the one could never acquire any indication of the existence of the other. It is truly wonderful how many other combinations in cases of any such connection of two kinds with a third could be obtained for the purpose of explaining the phenomena of Nature. But the arguments, which I brought forward in favour of homogeneity, hold good for all points, with which we can have any relation ; & for these alone the principle of induction can hold good. Further, whether there may be other kinds of points, either here in the space around us, or somewhere else at a distance from us, or, if the idea of such a thing is not opposed to our reason, in some other kind of space having no relation with our space, in which there may be points that have no distance-relation with points existing in our space ; of this we can know nothing. For, nothing relating to it in the slightest degree can be gathered from the phenomena of Nature ; & it would be great presumption for any one to fix as a limit his own power of perception, or even of imagination, of all the things that the Divine Author of Nature has founded.

Wonderful variety of consequences ; the possibility of any number of Universes, occupying the same space with apparent compenetration, without any indication of the presence of any one of them in the others.

519. But, to return to my Theory of homogeneous elements, the several forms of bodies will consist of a combination of homogeneous points, which comes from their distances & positions, &, in addition to combination alone, the velocity & direction of the motion of each of the points ; also for individual masses of bodies there is to be added the number of points that form them. Given the number & disposition of the points in a given mass, the basis of all its properties, which are inherent in the mass, is given ; & also that of all the relations that the same mass must have with other masses ; that is to say, those determined by their numbers, combinations & motions ; moreover, the basis of all changes that can happen to it is also given. Now, since there are certain special combinations, representing certain special constant properties, which we have determined & explained, namely, those corresponding to cohesion, & various degrees of solidity, those for fluidity, for elasticity, for softness, for the acquisition of certain shapes, for the existence of certain oscillations, which combinations, both of themselves & through forces connected with them, produce different tastes & different smells, & exhibit the different constant properties of colours ; & also there are other combinations which induce motions & changes that are not permanent, like all sorts of fermentations ; there can be derived from the primary combinations of constant properties the specific forms of bodies & their differences, & from the latter also can be obtained alterations & transformations in these forms.

Form in the hypothesis of homogeneity is the number & disposition of the points ; these constitute the basis of all properties ; what may be said about specific form ; hence, alterations & transformations.

520. Now, amongst these constant properties there may be chosen some that are more constant than others ; such as do not depend upon admixture with other particles, & also such as, if they should be lost, would be easily & quickly acquired. These properties could be considered to be essential to the species ; & if such properties suffered a permanent change, we should have a transformation ; whereas, if they persisted, there would only be an alteration. Thus, if the particles of a fluid were bound together by other particles, so that they could have no motion about one another, but their structure & the kind of forces corresponding to them remained the same, the fluid thus congealed would be said to have been merely altered, & not to have been specifically changed as well. Thus also, a body would be said to be altered, but not specifically changed, if the quantity of fiery matter which it contains in its pores is increased ; or if there is an increase in its motion, or even in some oscillation of its parts ; similarly, it would be said to be merely altered by a fresh accession of heat. A mass of water, which after ebullition returns to its original form, will be altered by that ebullition, but not transformed ; & a change of shape, as when different things are made from wax & metal, gives some sort of alteration. But when the structure in the particles is changed, & the change is such as will give far different phenomena, then the body would be said to have been broken down & transformed. Thus, when from solid bodies there is generated a permanent elastic gas, & elastic vapour from water, when water is congealed into earth, when several substances are intimately mixed with one another & in consequence adhere with some fresh connection between their particles, & form a new mixture, when the mixed particles, separated by the breaking of this connection, as happens in the case of putrefaction & in most fermentations, severally acquire fresh constitutions ; then a transformation takes place.

Distinction between transformation & alteration.

Quid requireretur ad inspiciendam formam intimam, unde liceret a priori reducere massas ad genera, & species: quid præstandum, cum id non liceat.

521. Si possemus inspicere intimam particularum constitutionem, & textum, ac distinguere a se invicem particulas ordinum gradatim altiorum a punctis elementaribus ad hæc nostra corpora; fortasse inveniremus aliqua particularum genera [246] ita suæ formæ tenacia, ut in omnibus permutationibus ea nunquam corrumpantur, sed mutentur quorundam altiorum ordinum particulæ per solam mutationem compositionis, quam habent a diversa dispositione particularum constantium ordinis inferioris; liceret multo certius dividere corpora in suas species, & distinguere clementa quædam, quæ haberi possent pro simplicibus, & inalterabilibus vi Naturæ, tum compositiones mixtorum specificas, & essentiales ab accidentalibus proprietatibus discernere. Sed quoniam in intimum ejusmodi textum penetrare nondum licet; eas proprietates debemus diligenter notare, quæ ab illo intimo textu proveniunt, & nostris sensibus sunt perviæ, quæ quidem omnes consistunt in viribus, motu, & mutatione dispositionis massularum grandiuscularum, quæ sensibus se nostris objiciunt, & constanter habitas, vel facile, & brevi recuperatas distinguere a transitoriis, vel facile, & constanter amissas, & ex illarum aggregato distinguere species, hasce vero habere pro accidentalibus.

Videri, nos nunquam posse devenire ad cognoscendam intimam substantiam, & essentiam, ac discrimina specifica.

522. Verum quod ad omne hoc argumentum pertinet, non crit abs re, si postremo loco huc transferam ex Stayana *Recentiore Philosophia*, ac meis in eam adnotationibus, illud, quod habeo ad versum 547 libri i: "Quamvis intrinsecam corporum naturam intueri non liceat, non esse adjiciendum, affirmat, Naturæ investigandæ studium: posse ex externis illis proprietatibus plures detegi in dies: ad ipsum summæ laudi esse: ideam sane, quam habemus confusam substantiæ eas habentis proprietates, proprietatibus ipsis auctis extendimus. Rem illustrat aptissimo exemplo ejus substantiæ, quam aurum appellamus, ac seriem proprietatum eo ordine proponit, quo ipsas detectas esse verosimiliter arbitratur; colorem fulvum, pondus gravissimum, ductilitatem, fusilitatem, quod in fusione nihil amittat, quod rubiginem non contrahat. Diu his tantummodo proprietatibus auri substantiam contineri est creditum, sero additum, solvi per illam, quam dicunt aquam regiam, & præcipitari immisso sale. Porro & aliæ supererunt plurimæ ejusmodi proprietates olim fortasse detegendæ: quo plures detegimus eo plus ad confusam illam naturæ auri cognitionem accedimus: a clara, atque intima ipsius naturæ contemplatione adhuc absumus. Idem, quod in hoc vidimus peculiari corpore, de corporis in genere natura affirmat. Investigandas proprietates, quibus detectis illum intimum proprietatum fontem attingi nunquam posse: nil nisi inania proferri vocabula, ubi intimæ proprietates investigantur."

Quid tamen præstari possit circa generales proprietates, & generalia principia: id esse hic præstitum.

523. Hæc ego quidem ex illo: tum meam hanc ipsam Theoriam respiciens, quam & ipse libro 10 exposuit nondum edito, sic persequor: "Quid autem, si partim observatione partim ratiocinatione adhibita, constaret demum, materiam homogeneam esse, ac omne discrimen inter corpora prove-[247]-nire a forma, nexu, viribus, & motibus particularum, quæ sint intima origo sensibilium omnium proprietatum. Ea nostros sensus non alia effugiunt ratione, nisi ob nimis exiguam particularum molem: nec nostræ mentis vim, nisi ob ingentem ipsarum multitudinem, & sublimissimam, utut communem, virium legem, quibus fit, ut ad intimam singularum specierum compositionem cognoscendam aspirare non possimus. At generalium corporis proprietatum, & generalium discriminum explicationem libro 10 ex intimis iis principiis petitam, exhibebimus fortasse non infeliciter: peculiarium corporum textum olim cognosci, difficillimum quidem esse, arbitror, prorsus impossibile, affirmare non ausim."

Quo pacto interea species distinguamus.

524. Demum ibidem illud addo, quod pertinet ad genera, & species: "Interea specificas naturas æstimamus, & distinguimus a collectione illa externarum proprietatum, in quo plurimum confert ordo, quo deteguntur. Si quædam collectio, quæ sola innotuerat, inveniatur simul cum nova quadam proprietate conjuncta, in aliis fere æquali numero cum alia diversa; eam, quam pro specie infima habebamus, pro genere quodam habemus continente sub se illas species, & nomen, quod prius habuerant, pro utraque retinemus. Si diu invenimus cunjunctam ubique cum aliqua nova, deinde vero alicubi multo posterius inveniatur sine illa nova: tum, nova illa jam in naturæ ideam admissa, hanc substantiam ea carentem ab ejusmodi natura arcemus, nec ipsi id nomen tribuimus. Si nunc inveniretur massa, quæ ceteras omnes enumeratas auri proprietates haberet, sed aqua regia non solveretur,

521. If we could inspect the innermost constitution of particles & their structure, & distinguish particles from one another & separate them into classes, step by step of higher orders, from elementary points up to our own bodies; then, perhaps, we should find some classes of particles to be so tenacious of their form that in all changes they would never be broken down; but the particles of higher orders would be changed by mere change of the composition that they have owing to a different disposition of the particles of a lower order from which they are formed. It would then be possible to divide with far greater certainty bodies into their species, & to distinguish certain elements which could be taken as the simple elements, unalterable by any force in Nature; & then to distinguish the specific & essential compositions of mixtures from accidental properties. But, since we cannot as yet penetrate into the innermost structure of this sort, we must carefully observe those properties, that arise from this innermost structure, & are accessible to our senses; these indeed all consist of the forces, motion & change of disposition of those comparatively large, though really small, masses that meet our senses; & we must distinguish between those properties that are constantly possessed, or easily & quickly recovered, & those that are transitory, or easily lost & lost for good; & from the aggregate of the former to distinguish the species, while considering the latter as accidental properties.

What is required to enable us to look into the innermost constitution, in order that from it we might be able from first principles to reduce matter to classes & species; what is to be done, since such a thing is impossible.

522. But, with respect to all this argument, it will not be out of place if, in the last place, I here quote from Stay's *Recentior Philosophia*, & my notes thereon, that which I have written on verse 547 of Book I. "Although we cannot peer into the intrinsic nature of bodies, the endeavour to investigate Nature, he states, must not be abandoned. Many things can be detected daily from those external properties. This is worthy of all praise; we truly extend the idea, which we have in a confused form of a substance possessing these properties, if the properties are increased. He illustrates the matter with a very fitting example of the substance, which we call gold, & enunciates the series of properties in the order in which he considers that in all probability they were detected:—yellow colour, very heavy weight, ductility, fusibility, that nothing is lost in fusion, that it does not rust. For a long time it was believed that the substance of gold was comprised in these properties only; later, there was added, that it was dissolved by what is called aqua regia, & precipitated from the solution by salt. Moreover, there will be in addition very many other properties of this kind, perhaps to be detected in the future; & the more of these we find out, the nearer we shall approach to that hazy knowledge of the nature of gold; but we are still far from obtaining a clear & intimate view of this nature. He asserts the same thing about the nature of a body in general, as we have seen in the case of this particular body. He states that the properties should be investigated, although from their detection the inmost source of the properties can never be reached; that nothing except empty words can be produced, when fundamental properties are investigated."

It is thus to be seen that we can never arrive at a full knowledge of the innermost & essential substance, or the distinction between species.

523. These were my words in that book; then considering my own Theory, which he also explained in Book 10, not yet published, I went on thus:—"But what if, partly by observation & partly by using deduction, it should finally be established that matter is homogeneous, & that all distinction between bodies comes from form, connection, forces, & motions of the particles, such as may be the fundamental origin of all sensible properties? These escape our senses for no other reason than the exceedingly small volume of the particles; nor are they beyond the powers of our intelligence, except on account of their huge number, & the very complicated, though general, law of forces. Owing to these, we cannot hope to obtain an intimate knowledge of the composition of each species. But we will present, perhaps not unsuccessfully, in book 10, an explanation of the general properties of a body & the general distinctions between them, derived from such fundamental principles. I consider that the attainment of a knowledge of the structure of particular bodies in the future will be very difficult; that it will be altogether impossible, I will not dare to assert."

What may, however, be accomplished with regard to general properties & general principles, has been done in this work.

524. Lastly, I add this in the same connection, relating to classes & species:—"Amongst other things, we estimate specific natures, & distinguish them from the collection of external properties; & in this the order in which they are detected is of special assistance. If any collection, which had alone been observed, should be discovered conjoined with some fresh property, & in others of nearly equal number conjoined with something different; then that, which we had considered as a fundamental species, we should now consider as a class containing within it both these species; & the name that they had originally, we should retain for both species. If for some time we found it conjoined with some fresh property, & then at another time much later it is found without that fresh property; then, this fresh property being admitted into the idea of nature, we should exclude the substance lacking this property from a nature of this kind, & should not give it that name. If now a mass should be found, which had all the other enumerated properties of gold, but was not dissolved by aqua regia, we should say that it was not gold. If at the beginning it was

The manner, amongst other things, in which we shall distinguish between species.

eam non esse aurum diceremus. Si initio compertum esset, alias ejusmodi massas solvi, alias non solvi per aquam regiam, sed per alium liquorem, & utrumque in æquali fere earum massarum numero notatum esset, putatum fuisset, binas esse auri species, quarum altera alterius liquoris ope solveretur."

Hæc ego ibi; unde adhuc magis patet, quid specificæ formæ sint, & inde, quid sit transformatio. Sed de his omnibus jam satis.

e same sort were dissolved by aqua regia, but that others
another liquid; & each of the two phenomena was
l number of masses; then, it would be considered that
at one sort was dissolved by one liquid, & the other by

them it can be easily seen what specific forms are;
is. But I have now said sufficient on the point.

[248] APPENDIX

AD METAPHYSICAM PERTINENS

DE ANIMA ET DEO

Argumentum hujus Appendicis, & cur sit addita.

525. Quæ pertinent ad discrimen animæ a materia, & ad modum, quo anima in corpus agit, rejecta Leibnitianorum harmonia præstabilita, persecutus jam sum in parte prima a num. 153. Hic primum & id ipsum discrimen evolvam magis, & addam de ipsius animæ, & ejus actuum vi, ac natura, nonnulla, quæ cum eodem operis argumento arctissime connectuntur : tum ad eum colligendum, qui semper maximus esse debet omnium philosophicarum meditationum fructus, nimirum ad ipsum potentissimum, ac sapientissium Auctorem Naturæ conscendam.

Discrimen inter animam & corpus : in hoc omnia peragi per distantias locales, motus, ac vires inducentes motum localem.

526. Imprimis hic iterum patet, quantum discrimen sit inter corpus, & animam, ac inter ea, quæ corporeæ materiæ tribuimus, & quæ in nostra spirituali substantia experimur. Ibi omnia perfecimus tantummodo per distantias locales, & motus, ac per vires, quæ nihil aliud sunt, nisi determinationes ad motus locales, sive ad mutandas, vel conservandas locales distantias certa lege necessaria, & a nulla materiæ ipsius libera determinatione pendentes. Nec vero ullas ego repræsentativas vires in ipsa materia agnosco, quarum nomine haud scio, an ii ipsi, qui utuntur, satis norint, quid intelligant, nec ullum aliud genus virium, aut actionum ipsi tribuo, præter illud unum, quod respicit localem motum, & accessus mutuos, ac recessus.

In anima nos experiri sensationes, & cogitationes, ac volitiones : Vim esse in nobis innatam, qua videamus harum discrimina, & relationem quam habent ad substantias, a quibus procedunt essentialiter diversas.

527. At in ea nostra substantia, qua vivimus, nos quidem intimo sensu, & reflexione, duplex aliud operationum genus experimur, & agnoscimus, quarum alterum dicimus sensationem, alterum cogitationem, & volitionem. Profecto idea, quam de illis habemus intimam, & prorsus experimentalem, est longe diversa ab idea, quam habemus, localis distantiæ, & motus. Et quidem illud mihi, ut in prima parte innui, omnino persuasum est, inesse animis nostris vim quandam, qua ipsas nostras ideas, & illos, non locales, sed animasticos motus, quos in nobis ipsis inspicimus, intime cognoscamus, & non solum similes a dissimilibus possimus discernere, quod omnino facimus, cum post equi visi ideam, se nobis idea piscis objicit, & hunc dicimus non esse equum ; vel cum in [249] primis principiis ideas conformes affirmando conjungimus, difformes vero separamus negando ; verum etiam ipsorum non localium motuum, & idearum naturam immediate videamus, atque originem ; ut idcirco nobis evidenter constet per sese, alias oriri in nobis a substantia aliqua externa ipsi animo, & admodum discrepante ab ipso, utut etiam ipsi conjuncta, quam corpus dicimus, alias earum occasione in ipso animo exurgere, atque enasci per longe aliam vim : ac primi generis esse sensationes ipsas, & directas ideas, posterioris autem omne reflexionum genus, judicia, discursus, ac voluntatis actus tam varios : qua interna evidentia, & conscientia sua illi etiam, qui de corporum, de aliorum extra se objectorum existentia dubitare vellent, ac idealismum, & egoismum affectant, coguntur vel inviti internum ejusmodi ineptissimis dubitationibus assensum negare, & quotiescunque directe, & vero etiam reflexe, ac serio cogitant, & loquuntur, aut agunt, ita agere, loqui, cogitare, ut alia etiam extra se posita sibi similia, & spiritualia, & materialia entia agnoscant : neque enim libros conscriberent, & ederent, & suam rationibus confirmare sententiam niterentur ; nisi illis omnino persuasum esset, existere extra ipsos, qui, quæ scripserint, & typis vulgaverint, perlegant, qui eorum rationes voce expressas aure excipiant, & victi demum se dedant

Duo genera actuum vitalium, quae in nobis perspicimus, sensationes, & cogitationes ac volitiones, quas possumus etiam sine corpore exercere.

528. Et vero ex motibus quibusdam localibus in nostro corpore factis per impulsum ab externis corporibus, vel per se etiam eo modo, quo ab externis fierent, ac delatis ad cerebrum (in eo enim alicubi videtur debere esse saltem præcipua sedes animæ, ad quam nimirum tot nervorum fibræ pertingunt idcirco, ut impulsiones propagatæ, vel per succum

APPENDIX

RELATING TO METAPHYSICS

THE MIND AND GOD

525. What relates to the distinction between the mind & matter, & the manner in which the mind acts on the body, I have already investigated in the First part, from Art. 153 on, after rejecting the pre-established harmony of the followers of Leibniz. Here I will first of all consider more fully this distinction; & I will add something with regard to the mind itself, the force of its actions, & its nature; these are closely connected with the very theme of this work. After that, I will proceed to consider that which always ought to be the most profitable of all philosophical meditations, namely, the power & wisdom of the Author of Nature.

The theme of this appendix; & the reason for adding it.

526. Here, in the first place, it is clear how great a distinction there is between the body & the mind, & between those things that we term corporeal matter & those which we feel in our spiritual substance. In Art. 153, we did everything by the sole means of local distances & motions, & by forces that are nothing else but propensities to local motions, or propensities to change, or preserve, local distances in accordance with a certain necessary law; & these do not depend on any free determination of the matter itself. But I do not recognize any representative forces in matter itself—I do not know whether those, who use the term, are really sure of what they mean by it—nor do I attribute to it any other type of forces or actions besides that one which has to do with local motions & mutual approach & recession.

Distinction between the mind & the body: in this everything is accomplished by local distances & motions, & forces inducing local motions.

527. But in this substance of ours, by which we live, we feel & recognize, by an inner sense & thought, another twofold class of operations; one of which we call sensation, & the other thought or will. Without any doubt, the idea which we have within us, which is altogether the result of experience, of the former, is far different to that which we have of local distance & motion. Indeed I am quite of the opinion, as I remarked in the First Part, that there is in our minds a certain force, by means of which we obtain full cognition of our very ideas & those non-local, but mental, motions that we observe in our own selves; & we can distinguish between like & unlike, as we assuredly do, when after the idea of a horse that has been seen there presents itself the idea of a fish, & we say that this is not a horse; or when, in elementary principles, we join together affirmatively like ideas, & separate unlike ideas with a negation. Indeed, we also see immediately the nature & origin of these non-local motions & ideas. Hence, it is self-evident to us that some of them arise through a substance external to the mind, & altogether different from it, but yet in connection with it, which we call the body; & that others take rise from direct encounter with the mind itself, & spring from a far different force. We see that to the first class belong sensations & direct ideas, & to the second all kinds of reflections, decisions, trains of reasoning, & the numerous different acts of the will. By this internal evidence, & their own consciousness, even those, who would like to doubt the existence of bodies, & other objects external to themselves, & affect idealism & egoism, are forced to refuse, though unwillingly, their inward assent to such very absurd doubts. As often as directly, or even reflectively & seriously, they think, speak, or act, they are forced so to act, speak, or think, that they recognize other entities situated external to themselves, which are like to themselves, both spiritual & material. For, they would not write & publish books, or try to corroborate their theory with arguments; unless they were fully persuaded that, external to themselves, there exist those who will read what they have written & published in printed form, & those who will hear the reasons they have spoken, & at length acknowledge themselves convinced.

In the mind we feel sensations, thoughts & purpose; force is innate within us, by which we see the differences between these things, & the relation that they bear to essentially different substances, from which they proceed.

528. Now, certain local motions in our body are engendered by impulse from external bodies, or even self-produced by the manner in which they come from without, & these are carried to the brain. For in the brain, somewhere, it seems that the seat of the mind must be situated; & that is why so many nerve-fibres extend to it, so that the impulses can be carried to it, propagated either by a volatile juice or by rigid fibres in all directions,

Two kinds of vital acts which we perceive in ourselves, sensations, & thought or will, which we can exercise even without the body

volatilem, vel per rigidas fibras quaquaversus deferri possint, & inde imperium in universum exerceri corpus) exurgunt motus quidam non locales in animo, nec vero liberi, & ideæ coloris, saporis, odoris, soni, & vero etiam doloris, qui oriuntur quidem ex motibus illis localibus ; sed intima conscientia teste, qua ipsorum naturam, & originem intuemur, longe aliud sunt, quam motus ipsi locales : sunt nimirum vitales actus, utut non liberi. Præter hos autem in nobis ipsis illud aliud etiam operationum genus perspicimus cogitandi, ac volendi, quod alii & brutis itidem attribuunt, cum quibus illud primum operationum genus commune nobis esse censent jam omnes, præter Cartesianos paucos, Philosophi : nam & Leibnitiani brutis ipsis animam tribuunt, quanquam non immediate agentem in corpus : sed ex iis, qui ipsam cogitandi, & volendi vim brutis attribuunt, in iis agnoscunt passim omnes, qui sapiunt, nostra inferiorem longe, & ita a materia pendentem, ut sine illa nec vivere possint, nec agere ; dum nostras animas etiam a corpore separatas credimus posse eosdem æque cogitationis, & volitionis actus exercere.

Si ea brutis conveniant, quanto imperfectiora in iis esse debeant, & quid de voce *spiritus*

[250] 529. Porro ex his, qui cogitationem, & voluntatem brutis attribuunt, alii utrique generi applicant nomen spiritus, sed distinguunt diversa spirituum genera, alii vocem spiritualis substantiæ tribuunt illis solis, quæ cogitare, & velle possint etiam sine ullo nexu cum corpore & sine ulla materiæ organica dispositione, & motu, qui necessarius est brutis, ut vivant. Atque id quidem admodum facile revocari potest ad litem de nomine, & ad ideam, quæ affigatur huic voci *spiritus*, vel *spiritualis*, cujus vocis latina vis originaria non nisi tenuem flatum significat : nec magna erit in vocum usurpatione difficultas ; dummodo bene distinguantur a se invicem materia expers omni & sentiendi, & cogitandi, ac volendi vi, a viventibus sensu præditis ; & in viventibus ipsis anima immortalis, ac per se ipsam etiam extra omne organicum corpus capax cogitationis, & voluntatis, a brutis longe imperfectioribus, vel quia solum sentiendi vim habeant omnis cogitationis, & voluntatis expertia, vel quia, si cogitent, & velint, longe imperfectiores habeant ejusmodi operationes, ac dissoluto per organici corporis corruptionem nexu cum ipso corpore, prorsus dispereant.

Discrimen inter motus, a quibus idea excitatur, & ideam ipsam : quatuor acceptiones vocis *color*

530. Ceterum longe aliud profecto est & tenuitas lamellæ, quæ determinat hunc potius, quam illum coloratum radium ad reflexionem, ut ad oculos nostros deveniat, in quo sensu adhibet coloris nomen vulgus, & opifices ; & dispositio punctorum componentium particulam luminis, quæ certum ipsi conciliat refrangibilitatis gradum, certum in certis circumstantiis intervallum vicium facilioris reflexionis, & facilioris transmissus, unde fit, ut certam in oculi fibris impressionem faciat, in quo sensu nomen coloris adhibent Optici ; & impressio ipsa facta in oculo, & propagata ad cerebrum, in quo sensu coloris nomen Anatomici usurpare possunt ; & longe aliud quid, & diversum ab iis omnibus, ac ne analogum quidem illis, saltem satis arcto analogiæ, & omnimodæ similitudinis genere, est idea illa, quæ nobis excitatur in animo, & quam demum a prioribus illis localibus motibus determinatam intuemur in nobis ipsis, ac intima nostra conscientia, & animi vis, de cujus vera in nobis ipsis existentia dubitare omnino non possumus, evidentissima voce admonent ea de re, & certos nos reddunt.

Commercium animæ cum corpore continere tria legum genera : quae sint priora duo.

531. Porro commercium illud inter animam, & corpus, quod unionem appellamus, tria habet inter se diversa legum genera, quarum bina sunt prorsus diversa ab ea etiam, quæ habetur inter materiæ puncta, tertium in aliquo genere convenit cum ipsa, sed ita longe in allis plurimis ab ea distat, ut a materiali mechanismo penitus remotum sit. Priores sunt in ordine ad motus locales organici nostri corporis, vel potius ejus partis, sive ea sit fluidum quoddam tenuissimum, sive sint solidæ fibræ ; & ad motus non locales, sed animasticos nostri a-[251]-nimi, nimirum ad excitationem idearum, & ad voluntatis actus. Utroque legum genere ad quosdam motus corporis excitantur quidam animi actus, & vice versa, & utrumque requirit inter cetera positionem certam in partibus corporis ad se invicem, & certam animæ positionem ad ipsas : ubi enim læsione quadam satis magna organici corporis ea mutua positio partium turbatur, ejusmodi legum observantia cessat : nec vero ea locum habere potest, si anima procul distet a corpore extra ipsum sita.

In altero ex iis nexus inter animam, & corpus necessarius, in altero liber : exponuntur ambo.

532. Sunt autem ejusmodi legum duo genera : alterum genus est illud, cujus nexus est necessarius, alterum, cujus nexus est liber : habemus enim & liberos, & necessarios motus, & sæpe fit, ut aliquis apoplexia ictus amittat omnem, saltem respectu aliquorum membrorum, facultatem liberi motus ; at necessarios, non eos tantum, qui ad nutritionem pertinent, & a sola machina pendent, sed & eos, quibus excitantur sensationes, retineat.

& from it control can be exercised over the whole body. From these local motions there arise certain non-local motions in the mind, that are not indeed free motions, such as the ideas of colour, taste, smell, sound, & even grief, all of which indeed arise from such local motions. But, on the evidence of our inner consciousness, by means of which we observe their nature & origin, they are something far different to these local motions; that is to say, they are vital actions, although not voluntary. Besides these we also perceive in our own selves that other kind of operations, those of thinking & willing. This kind some people also attribute to brutes as well; & all philosophers, except a few of the Cartesians, already believe that the first kind of operations is common to the brutes & ourselves. The followers of Leibniz attribute a mind even to the brutes, although one that does not act directly on the body. But of those who attribute to the brutes the power of thinking & willing, all those that have any understanding admit that in the brutes it is far inferior to our own; & so dependent on matter, that without it they cannot live or act; while they believe that our minds, even if separated from the body, are capable of exercising the same acts of thought & will just as well.

If these powers are to be credited to the brutes, they must be much more imperfect in them; the term "spirit."

529. Again, of those who attribute to brutes the power of thought & will, some apply to either class the term "spirit," but distinguish between two different kinds of spirits; others attribute the name of spiritual substance to those only that can think & will without any connection with the body, & without any organic disposition of matter, & the motion that is necessary to the brutes in order that they may live. This may quite easily be reduced to a quarrel over a mere term, & the idea that is assigned to the word *spirit*, or *spiritual*, of which the original Latin signification is merely "a tenuous breath." There will not be any great difficulty over the use of the terms, so long as matter (which is devoid of all power of feeling, thinking & willing) & living things possessed of feeling are carefully distinguished from one another; & also amongst living things, the immortal mind, &, on account of it, in addition also every organic body capable of thinking & willing, from the far more imperfect brutes; either, because they have the power of feeling only, & are unable to think or will; or because, if they do think & will, they have these powers far more imperfectly, &, if the connection with the body is destroyed by some corruption of the organic body, they perish altogether.

Distinction between the motion by which an idea is excited & the idea itself; four acceptations of the term *colour*.

530. Besides, there is certainly a very great difference between thinness of the plate, which determines one coloured ray of light rather than another to be reflected, so that it comes to the eyes, in which sense ordinary people & craftsmen use the term colour; & the disposition of the points forming a particle of light, to which corresponds a definite degree of refrangibility, & in certain circumstances a definite interval between the fits of easier reflection & easier transmission, whence there arises the fact that it makes a definite impression upon the nerves of the eyes, in which sense the term colour is used by investigators in Optics; & the impression itself that is made upon the eyes, & propagated to the brain, in which sense anatomists may employ the term; & something far different, & of a diverse nature to all the foregoing, being not even analogous to them, or only with a kind of analogy, & total similitude that is sufficiently close, is the idea itself, which is excited in our minds, & which, determined at length by the former local motions, we perceive within ourselves; & our inner consciousness, & the force of the mind, concerning the existence of which within us there cannot be the slightest doubt, warn us with no uncertain voice about the matter, & make us acquainted with it.

The intercourse of the mind with the body contains three kinds of laws; the nature of the first two.

531. Now, the intercourse between the mind & the body, which we term union, has three kinds of laws different from one another; & of these, two are also quite different also from that which obtains between points of matter; while the third in some sort agrees with it, but is so far different from it in very many other ways that it is altogether remote from any material mechanism. The two former are especially applicable to local motions, of our organic bodies, or rather of part of them, whether that part consists of a very tenuous fluid, or of solid fibres; & to motions that are not local motions, but to mental motions of our minds, such as the excitation of ideas, & acts of the will. According to each of these laws, certain acts of the mind are transmitted to certain motions of the body, & vice versa; & each kind demands, amongst other things, a certain relative situation of parts of the body, & a certain situation of the mind with regard to these parts. For, when this mutual situation between the parts is sufficiently disturbed by a sufficiently great lesion of the organic body, observance of these laws ceases; nor indeed does it hold, if the mind is far away from the body situated outside it.

In one of these, the connection between the mind & the body is of a necessary nature, in the other it is free; explanation of each of them.

532. Moreover, of such laws there are two kinds; the one kind is that in which the connection is necessary, while in the other the connection is free. For, we have both necessary & free motions; & it often happens that one who is stricken with apoplexy loses all power of free motions, at least with respect to some of his limbs; while he retains the necessary motions, not only those which relate to nutrition, & depend solely upon a mechanism,

Unde apparet & illud, diversa esse instrumenta, quibus ad ea duo diversa motuum genera utimur. Quanquam & in hoc secundo legum genere fieri posset, ut nexus ibi quidem aliquis necessarius habeatur, sed non mutuus. Ut nimirum tota libertas nostra consistat in excitandis actibus voluntatis, & eorum ope etiam ideis mentis, quibus semel libero animastico motu intrinseco excitatis, per legem hujus secundi generis debeant illico certi locales motus exoriri in ea corporis nostri parte, quæ est primum instrumentum liberorum motuum, nulli autem sint motus locales partis ullius nostri corporis, nullæ ideæ nostræ mentis, quæ animum certa lege determinent ad hunc potius, quam illum voluntatis liberum actum ; licet fieri possit, ut certa lege ad id inclinent, & actus alios aliis faciliores reddant, manente tamen semper in animo, in ipsa illa ejus facultate, quam dicimus voluntatem, potestate liberrima eligendi illud etiam, contra quod inclinatur, & efficiendi, ut ex mera sua determinatione præponderet etiam illud, quod independenter ab ea minorem habet vim. In eodem autem genere nexus quidam necessarii erunt itidem inter motus locales corporis, ac ideas mentis, cum quibusdam indeliberatis animi affectionibus, quæ leges, quam multæ sint, quam variæ, & an singula genera ad unicam aliquam satis generalem reduci possint, id vero nobis quidem saltem huc usque est penitus inaccessum.

Tertium genus in quo conveniat cum nexu mutuo inter puncta materiæ, & in quo ab eo plurimum differat.

533. Tertium legum genus convenit cum lege mutua punctorum in hoc genere, quod ad motum localem pertinet animæ ipsius, ac certam ejus positionem ad corpus, & ad certam organorum dispositionem. Durante nimirum dispositione, a qua pendet vita, anima necessario debet mutare locum, dum locum mutat corpus, atque id ipsum quodam necessario nexu, non libero : si enim præceps gravitate sua corpus ruit, si ab alio repente impellitur, si vehitur navi, si ex ipsius ani-[252]-mæ voluntate progreditur, moveri utique cum ipso debet necessario & anima, ac illam candem respectivam sedem tenere, & corpus comitari ubique. Dissoluto autem eo nexu organicorum instrumentorum, abit illico, & a corpore, jam suis inepto usibus, discedit. At in eo hæc virium lex localem motum animæ respiciens plurimum differt a viribus materiæ, quod nec in infinitum protenditur, sed ad certam quandam satis exiguam distantiam, nec illam habet tantam reciprocationem determinationis ad accessum, & recessum cum tot illis limitibus, vel saltem nullum earum terum habemus indicium. Fortasse nec in minimis distantiis a quovis materiæ puncto determinationem ullam habet ad recessum, cum potius ipsa compenetrari cum materia posse videatur : nam ex phænomenis nec illud certo colligi posse arbitror, an cum ullo materiæ puncto compenetretur. Deinde nec hujusmodi vires habet perennes, & immutabiles, pereunt enim destructa organizatione corporis, nec eas habet, cum suis similibus, nimirum cum aliis animabus, cum quibus idcirco nec impenetrabilitatem habet, nec illos nexus cohæsionum, ex quibus materiæ sensibilitas oritur. Atque ex iis tam multis discriminibus, & tam insignibus, satis luculenter patet, quam longe hæc etiam lex pertinens ad unionem animæ cum corpore a materiali mechanismo distet, & penitus remota sit.

Ubi sit sedes animæ, ex puris phænomenis sciri non posse.

534. Ubi sit animæ sedes, ex *puris phænomenis certo nosse* omnino non possumus : an nimirum ea sit præsens certo cuidam punctorum numero, & toti spatio intermedio habens virtualem illam extensionem, quam num. 84 in primis materiæ clementis rejecimus, an compenetretur cum uno aliquo puncto materiæ, cui unita secum ferat & necessarios illos, & liberos nexus, ut vel illud punctum cum aliis etiam legibus agat in alia puncta quædam, vel ut, enatis certis quibusdam in eo motibus, cætera fiant per virium legem toti materiæ communem ; an ipsa existat in unico puncto spatii, quod a nullo materiæ puncto occupetur, & inde nexum habeat cum certis punctis, respectu quorum habeat omnes illas motuum localium, & animasticorum leges, quas diximus ; id sane *ex puris Naturæ phænomenis*, & vero etiam, ut arbitror, ex reflexione, & meditatione quavis, quæ fiat *circa ipsa phænomena*, nunquam nobis innotescet.

Demonstratur id ipsum producendo, quid oporteret nosse ad resolvendam ejusmodi quæstionem ex phænomenis.

535. Nam ad id determinandum ex phænomenis utcunque consideratis, oporteret nosse, an ea phænomena possint haberi eadem quovis ex iis modis, an potins requiratur aliquis ex iis determinatus ut conjunctio, localis etiam, animæ cum magna corporis parte,

but also those by which sensations are excited. From which it is also clear that the instruments which we employ to produce the two different kinds of motions must be different. Also, although in the second kind of these laws it may happen that there is, even in it, some sort of necessary connection, yet it is not a mutual connection. Thus, the whole of our power of free action consists of the excitation of acts of the will, & by means of these of ideas of the mind also; once these have been excited by a free & intrinsic motion of the mind, owing to a law of this second kind there must immediately arise certain local motions in that part of the body which is the prime instrument of free motions; but there may be no motions of any part of the body, no motions of the mind, which determine the mind to this rather than to that free act of the will. It may happen, possibly, that by a certain law there is an inclination to one thing & that the motions produce some acts more easily than others; & yet, because there always remains in the mind & that faculty of it which we call the will a perfectly free power of choosing even that thing against which it is naturally inclined, there will even be a power of bringing it about that, due merely to its own determination, the thing, which independently of this determination would have the less force, will preponderate. However, in this same kind of law, there will be also certain connections of the necessary type between the local motions of the body & the ideas of the mind, together with some involuntary affections of the mind; & how many of these laws there may be, & how different they may be, & whether all the several kinds can be reduced to a single law of fair generality, is indeed, at least up till now, quite impossible to determine.

The points in which the third kind of law agrees with the mutual connection between points of matter: and those in which it is most different from it.

533. The third kind of law agrees with the mutual law of points in the fact that it pertains to local motion of the mind itself, to a definite position which it has with regard to the body, & to the definite arrangement of the organs. Thus, while the arrangement persists, upon which life depends, the mind must of necessity change its position, as the body changes its position, & that on account of some connection of the necessary type, & not a free connection. For, if the body rushes headlong through its own gravity, or is vigorously impelled by another, or if it is borne on a ship, or if it progresses through the will of the mind itself, in every case the mind also must necessarily move along with the body, & keep to its seat with respect to the body, & accompany the body everywhere. But if this connection of the organic instruments is dissolved, straightway it goes off & leaves the body which is now useless for its purposes. But this law of forces governing the local motion of the mind differs greatly from the law of forces between points of matter in this, that it does not extend to infinity, but only to a fairly small distance, & that it does not contain that great alternation of propensity for approach & recession, going with as many limit-points; or at least we have no indication of these things. Perhaps too, even at very small distances from any point of matter, it has no propensity for recession, since it seems rather to have a power of compenetration with matter. For, I do not think that it can with certainty be decided from phenomena, whether there is compenetration with any point of matter or not. Secondly, it has no lasting & unvarying forces of this kind; for they are destroyed as soon as the organization of the body is destroyed; nor are there forces with things like itself, that is to say other minds, & so there can be no impenetrability existing between them; nor can there be those connections of cohesion from which the sensibility of matter arises. From the number of these differences & special characteristics, it is fairly evident how far even this law pertaining to the union of the mind with the body differs from a material mechanism, & that it is something of quite a different nature.

It is not possible from phenomena alone to determine the position of the seat of the mind.

534. We are quite unable to *ascertain with any certainty from phenomena alone* the position of the seat of the mind. That is to say, we cannot ascertain whether it is present in any definite number of points, & has such a virtual extension through the whole of the intermediate space, as, in Art. 84, we rejected in the case of the primary elements of matter. It cannot be ascertained whether it has compenetration with some one point of matter, &, united with this, bears along with itself those necessary & free motions, so that either this point acts on certain other points with even other laws, or so that, certain definite motions being produced in this point, others take place on account of the law of forces that is common to the whole of matter. It cannot be ascertained whether it exists in a single point of space, which is unoccupied by any point of matter, & on that account has a connection with certain definite points, with respect to which it has all those laws of local & mental motions, of which we have spoken. We can never become acquainted with any of these points from *the phenomena of Nature alone certainly*, & indeed, as I think, neither can we by reflection or any consideration whatever, that may be made *with regard to these phenomena.*

This is proved by setting forth what would have to be known in order to obtain a solution of this problem from phenomena.

535. For, in order to determine it from any consideration of phenomena in any way, it would be necessary to know whether these phenomena could happen in any of these ways, or rather some particular one of them is required, determined as a conjunction, also

vel etiam cum toto corpore. Ad id autem cognoscendum oporteret distinctam habere notitiam earum legum, quas secum trahit conjunctio animæ cum corpore, & totius dispositionis punctorum omnium, quæ corpus constituunt, ac legis vitium mutuarum inter materiæ puncta, tum etiam ha-[253]-bere tantam Geometriæ vim, quanta opus est ad determinandos omnes motus, qui ex sola mechanica distributione eorundem punctorum oriri possint. Iis omnibus opus esset ad videndum, an ex motibus, quos anima imperio suæ voluntatis, vel necessitate suæ naturæ induceret in unicum punctum, vel in aliqua determinata puncta, consequi deinde possent per solam legem virium communem punctis materiæ omnes reliqui spirituum, & nervorum motus, qui habentur in motibus nostris spontaneis, & omnes motus tot particularum corporis, ex quibus pendent secretiones, nutritio, respiratio, ac alii nostri motus non liberi. At illa omnia nobis incognita sunt, nec ad illud adeo sublime Geometriæ genus adspirare nobis licet, qui nondum penitus determinare potuimus motus omnes trium etiam massularum, quæ certis viribus in se invicem agant.

Falsitas plurium opinionum de ejus sede: non probari, eam non extendi per totum corpus.

536. Fuerunt, qui animam concluserint intra certam aliquam exiguam corporis nostri particulam, ut Cartesius intra glandulam pinealem: at deinde compertum est, ea parte sola non contineri: nam ea parte dempta, vita superfuit: sic sine pineali glandula aliquando vitam perdurasse, compertum jam est, ut animalia aliqua etiam sine cerebro vitam produxerunt. Alii diffusionem animæ per totum corpus impugnant ex eo, quod aliquando homines, rescissa etiam manu, dixerint, se digitorum dolorem sentire, tanquam si adhuc haberent digitos; qui dolor cum sentiatur absque eo quod anima ibi digitis sit præsens: inde inferri posse arbitrantur, quotiescumque digitorum sentimus dolorem, illam sentiri sine præsentia animæ in digitis. At ea ratio nihil evincit: fieri enim posset, ut ad habendum prima vice sensum, quem in digitorum dolore experimur, requireretur præsentia animæ in ipsis digitis, sine qua ejus doloris idea primo excitari non possit, possit autem efformata semel per ejusmodi præsentiam excitari iterum sine ipsa per eos motus nervorum, qui cum motu fibrarum digiti in primo illo sensu conjuncti fuerant: præterquam quod adhuc remanet definiendum illud, an ad nutritionem requiratur præsentis animæ impulsus aliquis, an ea per solum mechanismum obtineri possit tota sine ulla animæ operatione.

Conclusio pro ignoratione: ubi & quomodo possit esse.

537. Hæc omnia abunde ostendunt, phænomenis rite consultis nihil satis certo definiri posse circa animæ sedem, nec ejus diffusionem per magnam aliquam corporis partem, vel etiam per totum corpus excludi. Quod si vel per ingentem partem, vel etiam per totum corpus protendatur, id ipsum etiam cum mea theoria optime conciliabitur. Poterit enim anima per illam virtualem extensionem, de qua egimus a num. 83, existere in toto spatio, quo continentur omnia puncta constituentia illam partem, vel totum corpus: atque eo pacto adhuc magis in mea theoria differet anima a materia; cum simplicia materiæ elementa non nisi in singulis spatii punctis existant singula singulis momentis temporis, anima autem licet itidem sim-[254]-plex, adhuc tamen simul existet in punctis spatii infinitis conjungens cum unico momento temporis seriem continuam punctorum spatii, cui toti simul erit præsens per illam extensionem virtualem, ut & Deus per infinitam Immensitatem suam præsens est punctis infinitis spatii (& ille quidem omnibus omnino), sive in iis materia sit, sive sint vacua.

Nunquam produci ab anima motum, nisi æqualem in partes oppositas: quid inde consequatur.

538. Et hæc quidem de sede animæ: illud autem postremo loco addendum hic censeo de legibus omnibus constituentibus ejus conjunctionem cum corpore, quod est observationibus conforme, quod diximus num. 74, & 387, nunquam ab anima produci motum in uno materiæ puncto, quin in alio aliquo æqualis motus in partem contrariam producatur, unde fit, ut nec liberi, nec necessarii materiæ motus ab animabus nostris orti perturbent actionis, & reactionis æqualitatem, conservationem ejusdem status centri communis gravitatis, & conservationem ejusdem quantitatis motus in Mundo in candem plagam computari.

Transitus ad Auctorem Naturæ, cujus perfectiones in hac Theoria elucent maxime.

539. Hæc quidem de anima: jam quod pertinet ad ipsum Divinum Naturæ Opificem, in hac Theoria elucet maxime & necessitas ipsum omnino admittendi, & summa ipsius, atque infinita Potentia, Sapientia, Providentia, quæ venerationem a nobis demississimam,

local, of the mind with a great part of the body, or even with the whole of the body. But to know this, it would be necessary to have a clear knowledge of their laws, which conjunction of the mind with the body necessitates; & also a knowledge of the entire disposition of all the points constituting the body, & the laws for the mutual forces between points of matter. In addition, there would be the necessity for as great geometrical powers, as would be enough to determine all the motions, which might be produced merely on account of the mechanical distribution of these points. All of these would be needed for perceiving whether, from the motions, which the mind could induce, by the power of its own will or the necessity of its nature, on a single point, or on certain given points, by means of the single law of forces common to points of matter, there could follow all the other motions of the spirits & nerves, such as take place in our voluntary motions; as well as all those different motions of particles of the body upon which depend secretions, nutrition, respiration, & other motions of ours that are not voluntary. But all these are unknown to us; nor may we aspire to such a sublime kind of geometry, for as yet we cannot altogether determine all the motions of even three little masses, which act upon one another with forces that are known.

536. There have been some who would confine the mind to some very small portion of the body; for instance, Descartes suggested the pineal gland. But, later, it was discovered that it could not be contained in that part alone; for, if that part were removed, life still went on. It has been already discovered that life endured for some time without the pineal gland, just as some animals produced life even without a brain. Others argued against the diffusion of the mind throughout the whole of the body, from the fact that sometimes men, after the hand had been cut off, said that they could still feel the pain in the fingers, as if they still had fingers; & since this pain is felt, although in this case there is not the fact that the mind is present in the fingers, they thought that it could be inferred that, as often as we feel a pain in the fingers, we feel it without the presence of the mind in the fingers. But such argument proves nothing at all; for it might happen that, in order that there should be in the first place that feeling, which we experience of pain in the fingers, there were required the presence of the mind in the fingers, without which it would be impossible that an idea of the pain could be excited in the first place; but, once this idea had been formed, it might be possible that it could once more be excited, without the presence of the mind in the fingers, by the motions of the nerves, which had been conjoined with a motion of the fibres of the finger when the pain was first felt. Besides, it still remains to be decided whether any impulse of a present mind is required for nutrition, or whether this can be obtained wholly without any operation of the mind, by means of a mere mechanism alone.

Falsity of several opinions as to the seat of the mind: it cannot be proved that it does not extend throughout the whole of the body.

537. All these things show fully that nothing certain can be stated with regard to the seat of the mind from a due consideration of phenomena; nor that its diffusion throughout any great part of the body, or even throughout the whole body, is excluded. But if it should extend throughout a great part, or even the whole, of the body, that also would fit in excellently with my Theory. For, by means of such virtual extension as we discussed in Art. 83, the mind might exist in the whole of the space containing all the points which form that part of the body, or that form the whole body. With this idea, in my Theory, the mind will differ still more from matter; for the simple elements of matter cannot exist except in single points of space at single instants of time, each to each, while the mind can also be one-fold, & yet exist at one & the same time in an infinite number of points of space, conjoining with a single instant of time a continuous series of points of space; & to the whole of this series it will at one & the same time be present owing to the virtual extension it possesses; just as God also, by means of His own infinite Immensity, is present in an infinite number of points of space (& He indeed in His entirety in every single one), whether they are occupied by matter, or whether they are empty.

Conclusion that the seat of the mind is unknown; where & in what manner it may be.

538. These things indeed relate to the seat of the mind; but I think there should be added here in the last place, concerning all the laws governing its conjunction with the body, that which is in conformity with the observations that I made in Art. 74 & Art. 387; namely, that motion can never be produced by the mind in a point of matter, without producing an equal motion in some other point in the opposite direction. Whence it comes about that neither the necessary nor the free motions of matter produced by our minds can disturb the equality of action & reaction, the conservation of the same state of the centre of gravity, & the conservation of the same quantity of motion in the Universe, reckoned in the same direction.

Motion can never be produced by the mind, unless it is equal in opposite directions; what follows from this.

539. So much for the mind; now, as regards the Divine Founder of Nature Himself, there shines forth very clearly in my Theory, not only the necessity of admitting His existence in every way, but also His excellent & infinite Power, Wisdom, & Foresight; which demand from us the most humble veneration, along with a grateful heart, & loving affection. The

Transition to the Author of Nature, the perfections of Whom shine forth very clearly in this Theory.

& simul gratum animum, atque amorem exposcant : ac vanissima illorum somnia corruunt penitus, qui Mundum vel casu quodam fortuito putant, vel fatali quadam necessitate potuisse condi, vel per se ipsum existere ab æterno suis necessariis legibus consistentem.

Error tribuentium Mundi originem casui fortuito: casum esse vocem vanam sine re.

540. Et primo quidem quod ad casum pertinet, sic ratiocinantur : finiti terminorum numeri combinationes numero finitas habent, combinationes autem per totam infinitam æternitatem debent extitisse numero infinitæ ; etiamsi nomine combinationum assumamus totam seriem pertinentem ad quotcunque millenos annos. Quamobrem in fortuita atomorum agitatione, si omnia se æqualiter habuerint, ut in longa fortuitorum serie semper accidit, debuit quævis ex ipsis redire infinitis vicibus, adeoque infinities major est probabilitas pro reditu hujus individuæ combinationis, quam habemus, quocunque finito numero vicium redeuntis mero casu, quam pro non reditu. Hi quidem inprimis in eo errant, quod putent esse aliquid, quod in se ipso revera fortuitum sit ; cum omnia determinatas habeant in Natura causas, ex quibus profluunt, & idcirco a nobis fortuita dicantur quædam, quia causas, a quibus eorum existentia determinatur, ignoramus.

Numerum combinationum in terminis etiam numero finitis esse infinitum : si rite omnia expendantur.

541. Sed eo omisso, falsissimum est, numerum combinationum esse finitum in terminis numero finitis : si omnia, quæ ad Mundi constitutionem necessaria sunt, perpendantur. Est quidem finitus numerus combinationum, si nomine combinationis assumatur tantummodo ordo quidam, quo alii termini post alios jacent : hine ultro agnosco illud : si omnes litteræ, quæ [255] Virgilii poema componunt, versentur temere in sacco aliquo, tum extrahantur, & ordinentur omnes litteræ, aliæ post alias, atque ejusmodi operatio continuetur in infinitum, redituram & ipsam combinationem Virgilianam numero vicium quenvis determinatum numerum superante. At ad Mundi constitutionem habetur inprimis dispositio punctorum materiæ in spatio patente in longum, latum, & profundum : porro rectæ in uno plano sunt infinitæ, plana in spatio sunt infinita, & pro quavis recta in quovis plano infinita sunt curvarum genera, quæ cum eadem ex dato puncto directione oriantur, in quarum singularum classibus infinities plures sunt, quæ per datum punctorum numerum non transeant. Quare ubi seligenda sit curva, quæ transeat per omnia materiæ puncta, jam habemus infinitum saltem ordinis tertii. Præterea, determinata ejusmodi curva, potest variari in infinitum distantia puncti cujusvis a sibi proximo : quamobrem numerus dispositionum possibilium pro quovis puncto materiæ adhuc ceteris manentibus est infinitus, adeoque is numerus ex omnium mutationibus possibilibus est infinitus ordinis expositi a numero punctorum aucto saltem ternario. Iterum velocitas, quam habet dato tempore punctum quodvis, potest variari in infinitum, & directio motus potest variari in infinitum ordinis secundi ob directiones infinitas in eodem plano, & plana infinita in spatio. Quare cum constitutio Mundi, & sequentium phænomenorum series pendeat ab ipsa velocitate, & directione motus ; numerus, qui exprimit gradum infiniti, ad quem assurgit numerus casuum diversorum, debet multiplicari ter per numerum punctorum materiæ.

Cujus ordinis infinitus sit : nimirum altissimi, & in immensum altioris numero momentorum temporis in tota æternitate

542. Est igitur numerus casuum diversorum non finitus, sed infinitus ordinis expositi a quarta potentia numeri punctorum aucta saltem ternario, atque id etiam determinata curva virium, quæ potest itidem infinitis modis variari. Quamobrem numerus combinationum relativarum ad Mundi constitutionem non est finitus pro dato quovis momento temporis, sed infinitus ordinis altissimi, respectu infiniti ejus generis, cujus generis est infinitum numeri punctorum spatii in recta quapiam, quæ concipiatur utrinque in infinitum producta. At huic infinito est analogum infinitum momentorum temporis in tota utraque æternitate, cum unicam dimensionem habeat tempus. Igitur numerus combinationum est infinitus ordinis in immensum altioris ordine infiniti momentorum temporis, adeoque non solum non omnes combinationes non debent redire infinities : sed ratio numeri earum, quæ non redeunt, est infinita ordinis altissimi, quam nimirum exponit quarta potentia numeri punctorum aueta saltem binario, vel, si libeat variare virium leges, saltem ternario. Quamobrem ruit futile ejusmodi, atque inane argumentum.

In ipso immenso combinationum numero in immensum plures esse combinationes inordinatas, quam ordinatas.

543. Sed inde etiam illud eruitur, in immenso isto com-[256]-binationum numero infinities esse plures pro quovis genere combinationes inordinatas, quæ exhibeant incertum chaos, & massam temere volitantium punctorum, quam quæ exhibeant Mundum ordinatum, & certis constantem perpetuis legibus. Sic ex. gr. ad efformandas particulas, quæ constanter suam formam retineant, requiritur collocatio in punctis illis, in quibus sunt limites, &

truly groundless dreams of those, who think that the Universe could have been founded either by some fortuitous chance or some necessity of fate, or that it existed of itself from all eternity dependent on necessary laws of its own, all these must altogether come to nothing.

540. Now first of all, the argument that it is due to chance is as follows. The combinations of a finite number of terms are finite in number; but the combinations throughout the whole of infinite eternity must have been infinite in number, even if we assume that what is understood by the name of combinations is the whole series pertaining to so many thousands of years. Hence, in a fortuitous agitation of the atoms, if all cases happen equally, as is always the ca^e in a long series of fortuitous things, one of them is bound to recur an infinite number of times in turn. Thus, the probability of the recurrence of this individual combination, which we have, is infinitely more probable, in any finite number of succeeding returns by mere chance, than of its non-recurrence. Here, first of all, they err in the fact that they consider that there is anything that is in itself truly fortuitous; for, all things have definite causes in Nature, from which they arise; & therefore some things are called by us fortuitous, simply because we are ignorant of the causes by which their existence is determined.

The error made by those who consider that the Universe was produced by fortuitous chance; 'chance' is an empty phrase without a thing to correspond to it.

541. But, leaving that out of account, it is quite false to say that the number of combinations from a finite number of terms is finite, if all things that are necessary to the constitution of the Universe are considered. The number of combinations is indeed finite, if by the term combination there is implied merely a certain order, in which some of the terms follow the others. I readily acknowledge this much; that, if all the letters that go to form a poem of Virgil are shaken haphazard in a bag, & then taken out of it, & all the letters are set in order, one after the other, & this operation is carried on indefinitely, that combination which formed the poem of Virgil will return after a number of times, if this number is greater than some definite number. But, for the constitution of the Universe, we have first of all the arrangement of the points of matter, in a space that extends in length, breadth & depth; further, there are an infinite number of straight lines in any one plane, an infinite number of planes in space, & for any straight line in any plane there are an infinite number of classes of curves, which will start from a given point in the same direction as the straight line; & in every one of these classes there are infinitely more which do not pass through a given number of points. Hence, when a curve has to be selected which shall pass through all points of matter, we now have an infinity of at least the third order. Besides, after any curve has been chosen, the distance of each point from the one next to it can be varied indefinitely; hence the number of possible arrangements for any one point of matter, while the rest remain fixed, is infinite. Therefore it follows that the number derived from the possible changes in all of these things is infinite, of the order determined by the number of points increased at least three times. Again, the velocity which any point has at a given time can be varied indefinitely; & the direction of motion can be varied to an infinity of the second order, on account of the infinity of directions in the same plane & the infinity of planes in space. Hence, since the constitution of the Universe, & the series of consequent phenomena, depend on the velocity & the direction of motion; the number, which expresses the degree of infinity to which the number of different cases mounts up, must be multiplied three times by the number of points of matter.

The number of combinations amongst terms that are even finite in number are infinite, if they are all rightly considered.

542. Therefore the number of cases is not finite, but infinite of the order expressed by the fourth power of the number of points increased threefold at least; & that is so, even if there is a definite curve of forces which also can be varied in an infinity of ways. Hence the number of relative combinations necessary to the formation of the Universe is not finite for any given instant of time; but it is infinite, of an exceedingly high order with respect to an infinity of the kind to which belongs the infinity of the number of points of space in any straight line, which is conceived to be produced to infinity in both directions. To this infinity the infinity of the instants in the whole of eternity past & present is analogous; for time has but one dimension. Hence, the number of combinations is infinite of an order that is immensely higher than the order of the infinity of instants of time; & thus, not only does it follow that not all the combinations are not bound to return an infinite number of times, but the ratio even of those that do not return is infinite, of a very high order, namely that which is expressed by the fourth power of the number of points increased twofold at least, or threefold at least if we choose to vary the laws of forces. Hence, the arguments of this sort that are brought forward are futile & worthless.

The order of the infinity; it is exceedingly high, immensely higher than the number of instants of time in the whole of eternity.

543. Moreover from this it also follows that, in this immense number of combinations, there will be, for any kind, infinitely more irregular combinations, such as represent indefinite chaos & a mass of points flying about haphazard, than there are of those that exhibit the regular combinations of the Universe, which follow definite & everlasting laws. For instance, in order to form particles which continually maintain their form, there is required their

In this immense number of combinations even, there are immensely more of them that are irregular than there are regular.

quorum numerus debet esse infinities minor, quam numerus punctorum sitorum extra ipsos : nam intersectiones curvæ cum axe debent fieri in certis punctis, & inter ipsa debent intercedere segmenta axis continua, habentia puncta spatii infinita. Quamobrem nisi sit aliquis, qui ex omnibus æque per se possibilibus seligat unam ex ordinatis ; infinities probabilius est, infinitate ordinis admodum elevati, obventuram inordinatam combinationum seriem, & chaos, non ordinatam, & Mundum, quem cernimus, & admiramur. Atque ad vincendam determinate eam infinitam improbabilitatem, requiritur infinita vis Conditoris Supremi seligentis unam ex iis infinitis.

Non determinari ab homine individuum: sed eo determinante intra limites, ad quos cognitio pertingit, reliquam indeterminationem vinci ab Ente in infinitum libero.

544. Nec vero illud objici potest, etiam hominem, qui statuam aliquam effingat, finita vi eligere illam individuam formam, quam illi dat, inter infinitas, quæ haberi possunt. Nam imprimis ille eam individuam non eligit, sed determinat modo admodum confuso figuram quandam, & individua illa oritur ex Naturæ legibus, & Mundi constitutione illa individua, quam naturæ Opifex Infinitus infinitam indeterminationem superans determinavit, per quam ab ejus voluntatis actu oriuntur illi certi motus in ejus brachiis, & ab hisce motus instrumentorum. Quin etiam in genere idcirco tam multi Philosophi determinationem ad individuum, & determinationem ad omnes illos gradus, ad quos cognitio creati determinantis non pertingit, rejecerunt in Deum infinita cognoscendi, & discernendi vi præditum, necessaria ad determinandum unum individuum casum ex infinitis ad idem genus pertinentibus ; cum creatæ mentis cognitio ad finitum tantummodo graduum diversorum numerum distincte percipiendum extendi possit : sine ullo autem determinante ex casibus infinitis, & quidem tanto infinitatis gradu, individuus unus præ aliis per se, aut per fortuitam eventualitatem prodire omnino non potest.

Hunc ordinem non posse dici per se necessarium : prima impugnatio a nullo nexu, qui videtur haberi inter distantiam, & vim, quæ idcirco liberum determinantem requirunt.

545. Sed nec dici potest, hunc ipsum ordinem necessarium esse, & æternum ac per se subsistere, casu quovis sequente determinato a proxime præcedente, & a lege virium intrinseca, & necessaria iis individuis punctis, & non aliis. Nam contra hoc ipsum miserum sane effugium quamplurima sunt, quæ opponi possunt. Inprimis admodum difficile est, ut homo sibi serio persuadeat, hanc unam virium legem, quam habet hoc individuum punctum respectu hujus individui puncti, fuisse possibilem, & necessariam, ut nimirum in hac individua [257] distantia se potius attrahant, quam repellant, & se attrahant tanta potius attractione, quam alia. Nulla appaıet sane connexio inter distantiam tantam, & tantam talis speciei vim, ut ibi non potuerit esse alia quævis, & ut hanc potius, quam aliam pro hisce punctis non selegerit arbitrium cntis habentis infinitam determinativam potentiam, vel pro hisce punctis id, si libeat, ex natura sua petentibus, non posuerit alia puncta illam aliam petentia ex sua itidem natura.

Secunda a numero punctorum finito, qui determinantem voluntatem poscit.

546. Præterea cum & infinitum, & infinite parvum in se determinatum, & in se tale, in creatis sit impossibile (quod de infinito in extensione demonstravi (t) pluribus in locis, nec una tantum demonstratione, ut in dissertatione *De Natura, & usu infinitorum, & infinite parvorum*, ac in dissertatione adjecta meis *Sectionum Conicarum Elementis*, Element. tom. 3) ; finitus est numerus punctorum materiæ, vel saltem in communi etiam sententia finita est materiæ existentis massa, quæ finitum spatium occupare debet, & non in infinitum

(t) *En unam ex ejusmodi demonstrationibus. Sit in fig.* 71 *spatium a* C *versus* AE *infinitum, & in eo angulus rectilineus* ACE *bifariam sectus per rectam* CD. *Sit autem* GH *parallela* CA, *quæ occurrat* CD *in* H, *ac producatur ita, ut* HF *fiat dupla* GH, *ducaturque* CF, *& omnes* CA, CB, CD, CE *in infinitum producantur. Inprimis totum spatium infinitum* ECD *debet esse æquale infinito* ACD : *nam ob angulum* ACE *bifariam sectum sibi invicem congruerent. Deinde triangulum* HCF *est duplum* HCG, *ob* FH *duplum* HG. *Eodem pacto ductis aliis* ghf *ipsi parallelis,* hCf *erit duplam* hCg, *adeoque & area* FHhf *dupla* HGgh. *Quare & summa omnium* FHhf *dupla summæ omnium* HGgh, *nimirum tota area infinita* BCD *dupla infinitæ* DCE, *adeoque dupla* ACD, *nimirum pars dupla totius, quod est absurdum. Porro absurdum oritur ab ipsa infinitate, si enim sint arcus circulares* GMI, gmi *centro* C ; *sector* ICM *erit æqualis* GCM, *& triangulum* FCH *duplum* GCH. *Donec sumus in quantitatibus finitis, res bene procedit, qui a* FCH *non est pars* ICM, *sicut* BCD *est pars* ACD, *nec* MCG, *&* HCG *sunt unum, & idem, ut* DCE *est unicum infinitum absolutum contentum cruribus* CD, CE. *Absurdum oritur tantummodo, ubi sublatis prorsus limitibus, a quibus oriuntur discrimina spatiorum inclusorum iisdem angulis ad* C, *sit suppositio infiniti absoluti, quæ contradictionem involvit.*

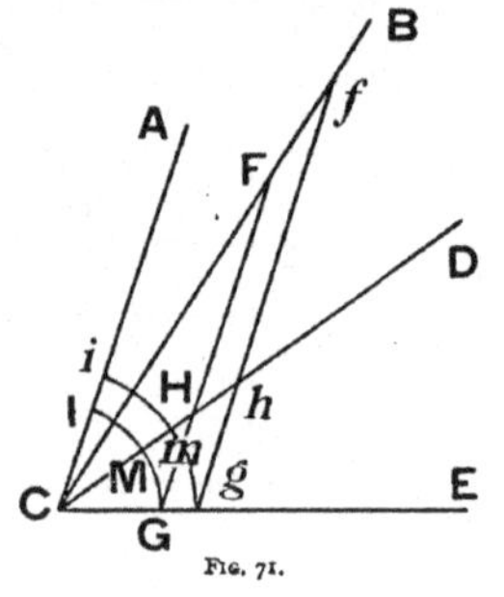

Fig. 71.

grouping together in those points in which there are limit-points; & of these the number must be infinitely less than the number of points situated without them. For the intersections of the curve with the axis must take place in certain points; & between these points there must lie continuous segments of the axis, having on them an infinite number of points of space. Hence, unless there were One to select, from among all the combinations that are equally possible in themselves, one of the regular combinations, it would be infinitely more probable, the infinity being of a very high order, that there would happen an irregular series of combinations & chaos, rather than one that was regular, & such an Universe as we see & wonder at. Then, to overcome definitely this infinite improbability, there would be required the infinite power of a Supreme Founder selecting one from among those infinite combinations.

The individual is not determined by man; but, when it has been determined within the limits to which man's knowledge attains, the rest that is undetermined is overcome by a Being who is infinitely free.

544. Nor can the argument be raised that even man, when he fashions a statue, with but a finite force selects that individual form which he gives to it, from among an infinite number which are possible. For, first of all, the man does not select that individual form; he determines in a very confused way a certain shape, & that individual form arises from the laws of Nature, & from that individual constitution of the Universe which the Infinite Founder of Nature, overcoming the infinite lack of determination, has determined; through which, by an act of his will, arise those definite motions in the arms of the man, & from these the motions of his tools. Moreover, in general, on this account, so many philosophers have thrown back individual determination, & a determination for all those stages to which the knowledge of a determining created thing cannot attain, upon a God endowed with an infinite power of knowing & distinguishing, such as is necessary for the task of determining one individual case from among an infinite number pertaining to the same class. For the knowledge of a created mind can only be extended to perceiving distinctly a finite number of different stages. But, unless there is someone to determine it, one individual cannot of itself, or through fortuitous happening, possibly come forth in preference to others, from among an infinite number of cases, let alone from an infinity of such a high degree.

This regularity cannot be said to be necessary in itself; first, because of the apparent absence of any connection between distance & force; the latter therefore requires a free determination.

545. No more can it be said that this very regularity is necessary, everlasting, & self-sustained, any one case following the one next before it & determined by it, & by a law of forces that is intrinsic & necessary to those individual points & to no others. For against this really worthless subterfuge there are very many arguments that can be brought forward. First of all, it is very difficult to see how a man can seriously persuade himself that one particular law of forces, which one particular point has with regard to another particular point, should be possible & necessary, so that, for instance, at one particular distance the points should attract one another rather than repel one another, & attract one another with an attraction that is so much greater than that with which they attract others. In truth, there is apparently no connection between so great a distance & so great a force of such a sort, that there could not be any other in the circumstances; & that the will of a Being having infinite determinative power should not select one in particular rather than another for these points; or should not substitute, for these points that from their very nature, if you like to say so, require the first, other points that also from their nature require that other connection.

Second argument derived from the finite number of points, which requires a determining will.

546. Besides, the infinite & the infinitely small, self-determined & such of themselves, is impossible in created things; as I proved concerning the infinite in extension (t) in several places, & with more than one proof, for instance, in the dissertation *De Natura, & usu infinitorum, & infinite parvorum*, & in a dissertation added to my *Sectionum Conicarum Elementa*, Elem. Vol. 3. It therefore follows that the number of points of matter is finite; or at least, even in the commonly accepted opinion, the mass of existing matter is finite;

(t) *Here is one of these proofs. In Fig.* 71, *let the space from* C *in the direction of* A, E *be infinite; & in this space, let the rectilineal angle* ACE *be bisected by the straight line* CD. *Also let* GH *be parallel to* CA, *meeting* CD *in* H; *& let it be produced so that* HF *is double* GH; *join* CF, *& let all the straight lines* CA, CB, CD, CE *be produced to infinity. Now, first of all, the whole of the infinite space* ECD *must be equal to the infinite space* ACD; *for, on account of the bisection of the angle* ACE, *they will be congruent with one another. Secondly, the triangle* HCF *is double the triangle* HCG, *since* FH *is double* HG. *In the same way, if other parallels like* ghf *are drawn*, hCf *will be double* hCg; *& thus the area* FHhf *will be double* HGgh. *Hence, the sum of all such areas as* FHhf *will be double the sum of all such as* HGgh; *that is to say, the whole of the infinite area* BCD *will be double the infinite area* DCE, *& therefore double* ACD; *the part double the whole, which is impossible. Further, the impossibility springs from the supposition of infinity; for, if* GMI, gmi *are circular arcs whose centre is* C, *the sector* ICM *will be equal to* GCM, *& the triangle* FCH *will be double* GCH. *So long as we are dealing with finite quantities, the matter goes on quite correctly, because* FCH *is not a part of* ICM, *as* BCD *is a part of* ACD, *nor are* MCG *&* HCG *one & the same, as* DCE *is the unique infinite absolute content of the arms* CD, DE. *The impossibility only arises when, all limits being taken away, from which arise the differences between the spaces included by the same angles at* C, *the supposition is made of absolute infinity, which involves the contradiction.*

protendi. Porro cur hic sit potins numerus punctorum, hæc potius massæ quantitas in Natura, quam alia ; nulla sane ratio esse potest, nisi arbitrium entis infinita determinativa potentia præditi, & nemo sanus sibi facile serio persuadebit, in quodam determinato numero punctorum haberi necessitatem existentiæ potius, quam in alio quovis.

Tertia ab æternitate, per quam durassent motus, cum linea necessario infinita : ejus impossibilitas.

547. Accedit illud, quod si Mundus cum hisce legibus fuisset ab æterno ; extitissent jam motus æterni, & lineæ a singulis punctis descriptæ debuissent fuisse jam in infinitum productæ : nam in se ipsas non redeunt sine arbitrio cntis infinitam improbabilitatem vincentis, cum demonstraverim supra pluribus in locis infinities improbabilius esse, [258] aliquod punctum redire aliquando ad locum, quem alio temporis momento occupaverit, quam nullum redire unquam. Porro infinitum in extensione impossibile prorsus esse, ego quidem demonstravi, uti monui, & illa impossibilitas pertinere debet ad omne genus linearum, quæ in infinitum productæ sint. Potest utique motus continuari in infinitum per æternitatem futuram, quia si aliquando coepit, nunquam habebitur momentum temporis, in quo jam fuerit existentia infinitæ lineæ : secus vero, si per æternitatem præcedentem jam extiterit : nec in eo futuram æternitatem cum præterita prorsus analogam esse censeo, ut illud indefinitum futuræ non sit verum quoddam infinitum præteritæ. Quod si linea infinita non fuerit, & quies est infinities adhuc improbabilior, quam regressus pro unico temporis momento ad idem spatii punctum, ac multo magis æterna quies : utique nec motum habuit æternum materia, nec existere potuit ab æterno, cum sine & quiete, & motu existere non potuerit, adeoque creatione omnino, & Creatore fuit opus, qui idcirco infinitam haberet effectivam potentiam, ut omnem creare posset materiam, ac infinitam determinativam vim, ut libero arbitrio suo utens ex omnibus infinitis possibilibus momentis totius æternitatis in utramque partem indefinitæ illud posset seligere individuum momentum, in quo materiam crearet, ac ex omnibus infinitis illis possibilibus statibus, & quidem tam sublimi infinitatis gradu, seligere illum individuum statum, complectentem unam ex illis curvis per omnia puncta dato ordine accepta transeuntibus, ac in ea determinatas illas distantias, ac determinatas motuum velocitates, & directiones.

Validissima ab impossibilitate seriei infinitæ terminorum, in qua alii ab aliis determinentur ad existendum sine extrinseco determinante ; ea hic demonstratur.

548. Verum hisce omnibus etiam omissis, est illud a determinatione itidem necessaria repetitum, in quavis Theoria validissimum, sed adhuc magis in mea, in qua omnia phænomena pendent a curva virium, & inertiæ vi. Nimirum materia licet ponatur ejusmodi, ut habeat necessariam, & sibi essentialem vim inertiæ, & virium activarum legem ; adhuc ut quovis dato tempore posteriore habeat determinatum statum, quem habet, debet determinari ad ipsum a statu præcedenti, qui si fuisset diversus, diversus esset & subsequens ; neque enim lapis, qui sequenti tempore est in Tellure, ibi esset ; si immediate antecedenti fuisset in Luna. Quare status ille, qui habetur tempore sequenti, nec a se ipso, nec a materia, nec ab ullo ente materiali tum existente, habet determinationem ad existendum, & proprietates, quas habet materia perennes, indifferentiam per se continent, nec ullam determinationem inducunt. Determinationem igitur, quam habet ille status ad existendum, accipit a statu præcedenti. Porro status præcedens non potest determinare sequentem, nisi quatenus ipse determinate existit. Ipse autem nullam itidem in se habet determinationem ad existendum, sed illam accipit a præcedente. [259] Ergo nihil habemus adhuc in ipso secundum se considerato determinationis ad existendum pro postremo illo statu. Quod de secundo diximus, dicendum de tertio præcedente, qui determinationem debet accipere a quarto, adeoque in se nullam habet determinationem pro existentia sui, nec idcirco ullam pro existentia postremi. Verum eodem pacto progrediendo in infinitum, habemus infinitam seriem statuum, in quorum singulis habemus merum nihil in ordine ad determinatam existentiam postremi status. Summa autem omnium nihilorum utcunque numero infinitorum est nihil ; jam diu enim constitit, illum Guidonis Grandi, utut summi Geometræ, paralogismum fuisse, quo ex expressione seriei parallelæ ortæ per divisionem $\frac{1}{1+1}$ intulit summam infinitorum *zero* esse revera æqualem dimidio. Non potest igitur illa series per se determinare existentiam cujuscunque certi sui termini, adeoque nec tota ipsa potest determinate existere, nisi ab ente extra ipsam posito determinetur.

In quo hoc argumentum differat a communi adhibente impossibilitatem seriei contingentium sine ente necessario.

549. Hoc quidem argumento jam ab annis multis uti soleo, quod & cum aliis pluribus communicavi, neque id ab usitato argumento, quo rejicitur series contingentium infinita sine ente extrinseco dante existentiam seriei toti, in alio differt, quam in eo, quod a

& this must occupy finite space & cannot extend indefinitely. Now, there is truly no possible reason why there should be this finite number of points, or this quantity of matter in Nature, rather than that; except the will of a Being possessed of infinite determinative power. No one in his right senses will easily persuade himself seriously that there is any necessity for existence in any one number of points, rather than in any other.

A third argument derived from eternity, during which motions have lasted, because a line is necessarily infinite; the impossibility of this.

547. In addition, if the Universe had gone on with these laws from eternity, then already there would have been eternal motions, & straight lines described by the several points would already have been produced to infinity. For they do not re-enter themselves, except by the will of a Being who overcomes the infinite improbability; since I have proved above in several places that it is infinitely more improbable that any point should return at some time to the same place as it had occupied at some other instant, than that no point should ever return. Moreover, I have proved that infinity in extension is quite impossible, as I have already observed; & this impossibility must pertain to every kind of lines that have been produced indefinitely. Anyhow, the motion can be continued indefinitely throughout future eternity; for, if it commenced at any one instant there never would be an instant of time, in which there has already been the existence of an infinite line; but otherwise, if it has already existed throughout past eternity. However, in this connection, I do not think that future eternity is quite analogous with past eternity; so that this indefinite of the future is not really the same thing as an infinite of the past. But if there has not been an infinite line (& absolute rest is still more infinitely improbable than a return for a single instant to the same point of space, & eternal rest is even more improbable still), then it certainly follows that matter cannot have had eternal motion, nor can it have existed from eternity. For, it could not have existed without both rest & motion; & thus, there was altogether a need for creation, & a Creator, & therefore He would have an infinite effective power, so that He could create all matter, & an infinite determinative force; so that, out of all the possible instants, infinite in number, in the whole of eternity indefinitely prolonged in either direction, He could choose of His Own untrammelled will that particular instant in which to create matter; & out of all the infinite number of possible states, & this to such a high degree of infinity, He could select that one particular state, which involves one of those curves passing through all the points taken in a certain order; & in it could choose those determinate distances, & the determinate velocities & directions of the motions.

A very strong argument derived from the impossibility of an infinite series of terms, in which the determination to exist of one comes from that of another without something determining existence from without; proof of the impossibility here given.

548. But, leaving all these things out of the question, there is a very strong argument in any Theory, derived also from a necessity for determination; but especially strong in my Theory, where all phenomena depend on a curve of forces, & the force of inertia. Thus, although matter may be assumed to be of such a nature as to have a necessary & essential force of inertia & a law of active forces; yet, in order that at any subsequent time it may have the determinate state, which it actually has, it must be determined to that state, from the state just preceding; & if this preceding state had been different, the subsequent state would also have been different. For a stone, which at a subsequent instant is on the Earth, would not have been there at the instant, if at the instant immediately preceding it had been on the Moon. Hence the state which occurs at the subsequent instant, neither of itself, nor from matter, nor from any material entity then existing, has any determination to exist; & the properties, which matter has unvarying, contain of themselves indifference nor do they lead to any determination. The determination, then, which that state has to exist, is derived from the state preceding it. Further, a preceding state cannot determine the one which follows it, except in so far as it itself has existed determinately. Moreover, this preceding state also has no determination in itself to exist, but derives it from one that precedes it. Consequently, we have as yet nothing in this, considered by itself, yielding determination to exist for that last state. What has been said with regard to this second state, is to be said also about the third preceding state; this must receive its determination from a fourth, & so in itself has no determination for its own existence, nor on that account has it any for the existence of the last state. Now, going on indefinitely in the same manner, we have an infinite series of states, in each of which we have absolutely nothing for the purpose of determining the existence of the last state. Moreover, the sum of all these nothings, no matter how infinite the number of them, is nothing also. For, it has been long ago made clear that Guido Grandi, although a very eminent geometer, enunciated a fallacy when, from an expression of a parallel series derived by division of 1 by (1 + 1), he deduced that the sum of an infinite number of *zeros* was really equal to $\frac{1}{2}$. Therefore, that series of states cannot determine the existence of any particular one of its terms, & so neither can the whole of it exist determinately, unless it be determined by a Being situated without itself.

In what this argument differs from the usual one, depending upon the impossibility of a series of events without a necessary being.

549. I have employed this argument for many years past, & I have communicated it to several others; it does not differ from the usual argument employed, which denies the possibility of an infinite series of contingents without an outside Being giving existence

contingentia res ad determinationem est translata, & a defectu determinationis pro sua cujusque existentia res est translata ad defectum determinationis pro existentia unius determinati status assumpti pro postremo; id autem præstiti, ne eludatur argumentum dicendo, in tota serie haberi determinationem ad ipsam totam, cum pro quovis termino habeatur determinatio intra candem seriem, nimirum in termino præcedente. Illa reductione ad vim determinativam existentiæ postremi quæsitam per omnem seriem, devenitur ad seriem nihilorum respectu ipsius quorum summa adhuc est nihilum.

Attributa, quæ Ens extrinsecum habere debet.

550. Jam vero hoc ens extrinsecum seriei ipsi, quod hanc seriem elegit præ seriebus aliis infinitis ejusdem generis, infinitam habere debet determinativam, & electivam vim, ut unam illam ex infinitis seligat. Idem autem & cognitionem habere, debuit, & sapientiam, ut hanc seriem ordinatam inter inordinatas selegerit: si enim sine cognitione, & electione egisset, infinities probabilius fuisset, ab illo determinatum iri aliquam ex inordinatis, quam unam ex ordinatis, ut hanc; cum nimirum ratio inordinatarum ad ordinatas sit infinita, & quidem ordinis altissimi, adeoque & excessus probabilitatis pro cognitione, & sapientia, ac libera electione supra probabilitatem pro cæco agendi modo, fatalismo, & necessitate, sit infinitus, qui idcirco certitudinem inducit.

Infinita improbabilitas, quæ hic occurrit, a quo uno vinci possit: nimirum a solo infinite libero.

551. Atque hic notandum & illud, pro quovis indivi-[260]-duo statu respondente cuivis momento temporis, & multo magis pro quavis individua serie respondente cuivis continuo tempori, improbabilitas determinatæ ipsius existentiæ est infinita, & nos deberemus esse certi de ejus non existentia, nisi determinaretur ab infinito determinantes & nisi ejus determinationis notitiam nos haberemus. Sic si in urna sint nomina centum, & unum, & agatur de uno determinato, an extractum inde prodierit, centuplo major est improbabilitas ipsi contraria: si mille, & unum, millecupla: si numerus sit infinitus; improbabilitas erit infinita, quæ in certitudinem transit: sed si quis viderit extractionem, & nobis nunciet; tota improbabilitas illa repente corruit. Verum & in hoc exemplo individua illa determinatio a creato agente non habebitur inter infinitas possibiles, nisi ex legibus ab infinito determinante jam determinatis in Natura, & ab ejusdem determinatione ad individuum, uti paullo ante dicebamus de individuæ figuræ electione pro statua.

Quanta sapientia opus fuerit ad seligendum numerum, & ordinem punctorum, & legem virium.

552. Porro qui aliquanto diligentius perpenderit vel illa pauca, quæ adnotavimus necessaria in distributione punctorum ad efformanda diversa particularum genera, quæ exhibeant diversa corpora, videbit sane quanta sapientia, & potentia sit opus ad ea omnia perspicienda, eligenda, præstanda. Quid vero, ubi cogitet, quanta altissimorum Problematum indeterminatio occurrat in infinito illo combinationum possibilium numero, & quanta cognitione opus fuerit ad eligendas illas potissimum, quæ necessariæ erant ad hanc usque adeo inter se connexorum phænomenorum seriem exhibendam? Cogitet, quid una lux præstare debeat, ut se propaget sine occursu, ut diversam pro diversis coloribus refrangibilitatem habeat, & diversa vicium intervalla, ut calorem & igneas fermentationes excitet: interea vero aptandus fuit corporum textus, & laminarum crassitudo ad ea potissimum remittenda radiorum genera, quæ illos determinatos colores exhiberent sine ceterarum & alterationum, & transformationum jactura, disponendæ oculorum partes, ut imago pingeretur in fundo, & propagaretur ad cerebrum, ac simul nutritioni daretur locus, & alia ejusmodi præstanda sexcenta. Quid unus aer, qui simul pro sono, pro respiratione, & vero etiam nutritione animalium, pro diurni caloris conservatione per noctem, pro ventis ad navigationem, pro vaporibus continendis ad pluvias, pro innumeris aliis usibus est conditus? Quid gravitas, qua perennes fiunt planetarum motus, & cometarum, qua omnia compacta, & coadunata in ipsorum globis, qua una suis maria continentur littoribus, & currunt fluvii, imber in terram decidit, & eam irrigat, ac fœcundat, sua mole ædificia consistunt, temporis mensuram exhibent pendulorum oscillationes? [261] si ea repente deficeret; quo noster incessus, quo situs viscerum, quo aer ipse sua elasticitate dissiliens? Homo hominem arreptum a Tellure, & utcunque exigua impulsum vi, vel uno etiam oris flatu impetitum, ab hominum omnium commercio in infinitum expelleret, nunquam per totam æternitatem rediturum.

Congeries eorum, quæ evincunt in eligente potentiam, sapientiam, providentiam, immensas,

553. Sed quid ego hæc singularia persequor? quanta Geometria opus fuit ad eas com

to the whole series, except in the detail that the matter is altered from a contingence to a determination, & from a lack of determination of the existence of any thing in itself the question is transferred to a lack of determination for the existence of one determined state assumed as the last of the series. But my argument is superior to the usual one, in that it cannot be evaded by saying that there is in the whole series a determination to the series as a whole; since for any term there is a determination within the same series, namely one derivable from the preceding term. By my reduction to a force determining existence of the last term throughout the whole series, the result is a series of zeros with regard to this last term, & the sum of these is still zero.

The necessary attributes of the external Being.

550. Now, the Being external to the series, which chooses this series in preference to all others of the infinite number in the same class, must have infinite determinative & elective force, in order that He may select this one out of an infinite number. Also He must have knowledge & wisdom, in order to select this regular series from among the irregular series; for, if He had acted without knowledge & selection, it would have been infinitely more probable that there would have been a determination by Him of one of the irregular series, than of one of the regular series, such as the one in question. For the ratio of the number of irregular series to the number of regular series is infinite, & that too of a very high order; & thus, the excess of the probability in favour of knowledge, wisdom, & arbitrary selection is infinitely greater than the probability in favour of blind choice, fatalism, & necessity; & this therefore leads to a certainty.

The sort of Being who could overcome the infinite improbability which here occurs: it could be accomplished only by One Who is infinitely free.

551. Here also it is to be observed that for any individual state corresponding to any given instant of time, & much more for any particular series corresponding to a given continuous time, the improbability of a self-determined existence is infinite; & we ought to be certain of its non-existence, unless it were determined by an infinite determinator, & we had evidence of the determination. Thus, if in an urn there are a hundred & one names, & it is a question with regard to one determined name, whether it has been drawn from the urn, the improbability is a hundredfold to the contrary; & if there were a thousand & one names, a thousandfold; if the number of names is infinite, the improbability will be infinite; & this passes into a certainty. But if anyone should have seen the drawing & give us information, then the whole of the improbability would immediately be destroyed. Again, in this example, the particular determination by a created agent will not be from among an infinite number of possibles, except on account of laws already determined in Nature by an infinite Determinator and from the determination to the individual by the same power; as I said, a little earlier, when speaking of the selection of a particular form for a statue.

How great the wisdom would have to be to select the number & arrangement of the points, & the law of forces.

552. Now, if anyone will consider a little more carefully even the few things I have mentioned as necessary in the arrangement of the points for the formation of the different kinds of particles, which different bodies exhibit, he must perceive how great the wisdom & power must needs be, to comprehend, select & establish all these things. What then, when he considers how great an indeterminateness in problems of very high degree occurs through the infinite number of possible combinations; & how great the knowledge would have to be to select those of them especially, which were necessary to yield this series of phenomena so far connected with one another? Let him consider what properties the single substance called light must exhibit, such that it is propagated without collision, that it has different refrangibilities for different colours, & different intervals between its fits, that it should excite heat, & fiery fermentations. At the same time the texture of bodies & the thickness of plates had to be made suitable for the giving forth of those kinds of rays especially, which were to exhibit determinate colours, without sacrificing other alterations and transformations; the arrangement of parts of the eyes, so that an image is depicted at the back & propagated to the brain; & at the same time place should be given to nutrition, & thousands of other things of the same sort to be settled. What the properties of the single substance called air, which at one & the same time is suitable for sound, for breathing, even for the nutrition of animals, for the preservation during the night of the heat received during the day, for holding rain-clouds, & innumerable other uses. What those of gravity, through which the motions of the planets & comets go on unchanged, through which all things became compacted & united together within their spheres, through which each sea is contained within its own bounds, & rivers flow, the rain falls upon the earth & irrigates it, & fertilizes it, houses stand up owing to their own mass, & the oscillations of pendulums yield the measure of time. Consider, if gravity were taken away suddenly, what would become of our walking, of the arrangement of our viscera, of the air itself, which would fly off in all directions through its own elasticity. A man could pick up another from the Earth, & impel him with ever so slight a force, or even but blow upon him with his breath, & drive him from intercourse with all humanity away to infinity, nevermore to return throughout all eternity.

Groups of points, which prove conclusively the immeasurability of the power, wisdom and foresight employed in selection.

553. But why do I enumerate these separate things? Consider how much geometry

binationes inveniendas, quæ tot organica nobis corpora exhiberent, tot arbores, & flores educerent, tot brutis animantibus, & hominibus tam multa vitæ instrumenta subministarent ? Pro fronde unica efformanda quanta cognitione opus fuit, & providentia, ut motus omnes per tot sæcula perdurantes, & cum omnibus aliis motibus tam arcte connexi illas individuas materiæ particulas eo adducerent, ut illam demum, illo determinato tempore frondem illius determinatæ curvaturæ producerent ? quid autem hoc ipsum respectu eorum, ad quæ nulli nostri sensus pervadunt, quæ longissime supra telescopiorum, & infra microscopiorum potestatem latent ? Quid respectu eorum, quæ nulla possumus contemplatione assequi, quorum nobis nullam omnino licet, ne levissimam quidem conjecturam adipisci, de quibus idcirco, ut phrasi utar, quam alibi ad aliquid ejusdem generis exprimendum adhibui, de quibus inquam, hoc ipsum, ignorari ea a nobis, ignoramus ? Ille profecto unus immensam Divini Creatoris potentiam, sapientiam, providentiam humanæ mentis captum omnem longissime superantes, ignorare potest, qui penitus mente cæcutit, vel sibi ipsi oculos eruit, & omnem mentis obtundit vim, qui Natura altissimis undique inclamante vocibus aures occludit sibi, ne quid audiat, vel potius (nam occludere non est satis) & cochleam, & tympanum, & quidquid ad auditum utcunque confert, proscindit, dilacerat, eruit, ac a se longissime projectum amovet.

Quid prospiciendum fuerit pro nostra existentia, & nostris commodis: quantum ipsi inde simus obstricti.

554. Sed in hac tanta eligentis, ac omnia providentis Supremi Conditoris sapientia, atque exsequentis potentia, quam admirari debemus perpetuo, & venerari, illud adhuc magis cogitandum est nobis, quantum inde in nostros etiam usus promanarit, quos utique respexit ille, qui videt omnia, & fines sibi istos omnes constituit, qui per ea omnia & nostræ ipsi existentiæ viam stravit, ac nos præ infinitis aliis hominibus, qui existere utique poterant, elegit ab ipso Mundi exordio, motus omnes, ad horum, quibus utimur, organorum formationem disposuit, præter ea tam multa quæ ad tuendam, & conservandam hanc vitam, ad tot commoda, & vero etiam voluptates conducerent. Nam illud omnino credendum firmissime, non solum ea omnia vidisse unico intuitu Auctorem Naturæ, sed omnes eos animo sibi constitutos habuisse fines, ad quos conducunt media, quæ videmus adhibita.

Mundum non esse possibilium optimum, cum in possibilibus nullus terminus sit ultimus: nec officere sapientiæ, ac bonitati infinitæ, quod non fecerit, nec potentiæ, quod non potuerit id facere.

555. Haud ego quidem Leibnitianis, & aliis quibuscunque [262] Optimismi defensoribus assentior, qui Mundum hunc, in quo vivimus, & cujus pars sumus, omnium perfectissimum esse arbitrantur, ac Deum faciunt natura sua determinatum ad id creandum quod perfectissimum sit, ac eo ordine, qui perfectissimus sit. Id sane nec fieri posse arbitror : cum nimirum in quovis possibilium genere seriem agnoscam finitorum tantummodo, quanquam in infinitum productam, ut num. 90 exposui, in qua, ut in distantiis duorum punctorum nulla est minima, nulla maxima ; ita ibidem nulla sit perfectionis maximæ, nulla minimæ, sed quavis finita perfectione utcunque magna, vel parva, sit alia perfectio major, vel minor : unde fit, ut quancunque seligat Naturæ Auctor, necessario debeat alias majores omittere : nec vero ejus potentiæ illud officit, quod creare non possit optimum, aut maximum, ut nec officit, quod non possit simul creare totum, quodcunque creare potest : nam id eo evadit, ut non possit se in eum statum redigere, in quo nihil melius, aut majus, vel absolute nihil aliud creare possit : nec officit aut sapientiæ, aut bonitati infinitæ, quod optimum non seligat, ubi optimum est nullum.

Quam multa pessima consectaria secum trahat sententia Mundi perfectissimi.

556. Ex alia parte determinatio illa ad optimum, & libertatem Divinam tollit, & contingentiam rerum omnium, cum, quæ existunt, necessaria fiant, quæ non existunt, evadant impossibilia ; ac præterea nobis quodammodo in illa hypothesi debemus, quod existimus, non illi. Qui enim potuit non existere id, quod habuit pro sua existentia rationem prævalentem, quam Naturæ Auctor cum viderit, non potuerit non sequi, nec vero potuerit non videre ? Qui existere potuit id, quod eandem habuit non existendi necessitatem ? Quid vero illi pro nostra existentia debeamus, qui nos condidit idcirco, quia in nobis invenit meritum majus, quam in iis, quos omisit, & a sua ipsius natura necessario determinatus fuit, & adactus ad obsequendum ipsi huic nostro intrinseco, & essentiali merito prævalenti ? Distinguendum est inter hæc duo : unum esse alio melius, & esse melius creare potius unum, quam aliud. Illud primum habetur ubique, hoc secundum nusquam, sed æque bonum est creare, vel non creare quodcunque, quod physicam bonitatem quancunque habeat, utcunque majorem, vel minorem alio quovis omisso : solum enim

was needed to discover those combinations which were to display to us so many organic bodies, produce so many trees & flowers, & supply so many instruments of life to living brutes & men. For the formation of a single leaf, how great was the need for knowledge & foresight, in order that all those motions, lasting for so many ages, & so closely connected with all other motions, should so bring together those particular particles of matter, that at length, at a certain determinate time, they should produce that leaf with that determinate curvature. What is this in comparison with those things to which none of our senses can penetrate, things that lie hidden far & away beyond the power of telescopes, & too small for the microscope? What of those which we can never understand no matter how hard we think about them, of which we can never attain not even the slightest idea; concerning which therefore, to use a phrase I have elsewhere employed to express something of the same sort, of which I say this:—"We do not know the very fact of our ignorance." Undoubtedly he alone can be ignorant of the immeasurable power, wisdom & foresight of the Divine Creator, far surpassing all comprehension of the human intellect, whose mind is altogether blind, or who tears out his eyes, & dulls every mental power, who shuts his ears to Nature, so that he shall not hear her as she proclaims in accents loud on every side, or rather (for to shut them is not enough) cuts away, tears up & destroys, & hurls far from him the cochlea & the tympanum & anything else that helps him to hear.

How our existence and our conveniences would have to be provided for; what a debt we are therefore under to Him.

554. But, in this great wisdom of selection & universal foresight on the part of the Supreme Founder, & the power of carrying it out, there is still another thing for us to consider; namely, how much proceeds from it to meet the needs of us, who are all under the care of Him Who sees all things, & has imposed on Himself the accomplishment of all those purposes; Who has smoothed the path of our existence with them all, & from the commencement of the Universe has chosen us in preference to an infinite number of other human beings that might have existed; Who has planned all the motions necessary for the formation of the organs we employ, besides all the many things that should conduce towards the protection & preservation of this life, to its many conveniences, nay, even to its pleasures. For, it cannot be but a matter of the firmest belief, not only that the Author of Nature saw all these things with a single intuition, but also that He had settled in his mind all those purposes, to which the means that we see employed conduce.

The Universe is not 'the best of all possibles': for amongst possibles there is no last term: nor is it an argument against infinite wisdom & goodness, because He did not make it so; nor against His power, because He was unable to make it so.

555. I do not indeed agree with the followers of Leibniz, or with any of the upholders of Optimism, who consider that this Universe, in which we live & of which we are part, is the most perfect of all; & who thus make God determined by His own nature for the creation of that which is the most perfect, & in that order which is the most perfect. In truth, I think that such a thing would be impossible; for, I recognize, in any kind of possibles, a series of finites only, although prolonged to infinity, as I explained in Art. 90; & in this series, just as in the case of the distances between two points, there is no greatest or least, here also there is no case of greatest or of least perfection; but, for any finite perfection, however great or small, there is another perfection that is greater or smaller. Hence it comes about that, whatever the Author of Nature should select, He would have to omit some that were of greater perfection. But, neither is it an argument against His power, that He cannot create the best or the greatest; nor similarly is it an argument against His power that, whatever He could create, He could not create it as a whole at one & the same time. For, it would come to this, that He would put Himself in the position where He could create nothing better, nothing greater, or absolutely nothing else. Similarly, it is no argument against His infinite wisdom & goodness, that He did not select the best, when there is no best.

The number of gross imperfections involved in the idea of a most perfect Universe.

556. On the other hand, that determination for the best takes away altogether the freedom of God, & the contingency of all things; for, those things which exist become necessary, & those that do not are impossible. Besides, on that hypothesis, we should be under some sort of obligation to ourselves, & not to Him, for the fact that we exist. For how was it possible that a thing should not exist, which had a powerful reason for its existence; for, when the Author of Nature saw this reason, He could not fail to follow it, nor indeed could He fail to see it? How could a thing exist which had a like need for non-existence? For what should we have to thank Him, if He had created us for the simple reason that in us He found a greater merit than in those whom He omitted, if He was necessarily determined by His own nature, & driven by it to submit to our mere intrinsic & essential overpowering merit? We must mark the distinction between the two dictums:—(1) this thing is better than that, (2) it would be better to create this thing than to create that. There is a possibility of the first in all cases, but never any of the second. It is an equally good thing to create or not to create anything whatever, which has any physical goodness, however much greater or less than anything else which has been omitted. The exercise of Divine freedom alone is infinitely more perfect than

Divinæ libertatis exercitium infinities perfectius est quavis perfectione creata, quæ idcirco nullum potest offerre Divinæ libertati meritum determinativum ad se creandum.

Media tamen idonea necessario eligi ab ipso Auctore Naturæ ad fines sibi propositos : quantum illi debeamus.

557. Cum ea infinita libertate Divina componitur tamen illud, quod ad sapientiam pertinet, ut ad eos fines, quos sibi pro liberrimo suo arbitrio præfixit Deus, media semper apta debeat seligere, quæ finem propositum frustrari non sinant. Porro hæc media etiam in nostrum bonum selegit plurima, dum totam Naturam conderet, quod quem a nobis exigat beneficiorum memorem, & gratum animum, quem etiam tan-[263]-tæ beneficentiae respondentem amorem cum ingenti illa admiratione, & veneratione conjunctum, nemo non videt.

Deduci nos inde ad revelationem, quæ tamen huc non pertineat, ad opus nimirum pure philosophicum.

558. Superest & illud innuendum, neminem sanæ mentis hominem dubitare posse, quin, qui tantam in ordinanda Natura providentiam ostendit, tantam erga nos in nobis seligendis, in consulendo nostris & indigentiis, & commodis beneficentiam, illud etiam præstare voluerit, ut cum adeo imbecilla sit, & hebes mens nostra, & ad ipsius cognitionem per sese vix quidquam possit, se ipse nobis per aliquam revelationem voluerit multo uberius præbere cognoscendum, colendum, amandum ; quo ubi devenerimus, quæ inter tam multas falso jactatas absurdissimas revelationes unica vera sit perspiciemus utique admodum facile. Sed ea jam Philosophiæ Naturalis fines excedunt, cujus in hoc opere Theoriam meam exposui, & ex qua uberes hosce, & solidos demum fructus percepi.

any perfection created; & the latter can therefore offer no determinative merit to the freedom of God in favour of its own creation.

557. With this infinite Divine liberty is bound up all that relates to wisdom; for, God, to those purposes which he of His own unfettered will had designed, was always bound to select suitable means, such as would not allow these purposes to be frustrated. Further, He has selected these means for the most part suitable for our welfare, whilst he founded the whole of Nature; & this demands from us a remembrance of His favours & a thankful heart, nay, even a love that shall correspond to such great beneficence together with a mighty wonder & admiration, as every one will see.

Fit means, however, had necessarily to be selected by the Author of Nature Himself to carry out the purposes He has designed for Himself; how much we owe to Him.

558. It now remains but to mention that there is no man of sound mind who could possibly doubt that One, Who has shown such great foresight in the arrangement of Nature, such great beneficence towards us in selecting us, & in looking after both our needs & our comforts, would not also wish to accomplish this also; namely that, since our mind is so weak & dull that it can scarcely of itself arrive at any sort of knowledge about Him, He would have wished to present Himself to us through some kind of revelation much more fully to be known, honoured & loved. This being done, we should indeed quite easily perceive which was the only true one, from amongst so many of those absurdities falsely brought forward as revelations. But such things as this already exceed the scope of a Natural Philosophy, of which in this work I have explained my Theory, & from which I have finally gathered such ripe & solid fruit.

We are thus led to revelation, which however does not come within the scope of such a work as this, which is purely philosophical.

[264] SUPPLEMENTA

§ I

De Spatio, ac Tempore ([a])

Argumentum: quæ spatii attributa.

1. Ego materiæ extensionem prorsus continuam non admitto, sed eam constituo punctis prorsus indivisibilibus, & inextensis a se invicem disjunctis aliquo intervallo, & connexis per vires quasdam jam attractivas, jam repulsivas pendentes a mutuis ipsorum distantiis. Videndum hic, quid mihi sit in hac sententia spatium, ac tempus, quomodo utrumque dici possit continuum, divisibile in infinitum, æternum, immensum, immobile, necessarium, licet neutrum, ut in ipsa adnotatione ostendi, suam habeat naturam realem ejusmodi proprietatibus præditam.

Necessario ab omnibus admitti debere reales modos existendi locales & temporarios.

2. Inprimis illud mihi videtur evidens, tam eos, qui spatium admittunt absolutum, natura sua reali, continuum, æternum, immensum, tam eos, qui cum Leibnitianis, & Cartesianis ponunt spatium ipsum in ordine, quem habent inter se res, quæ existunt, præter ipsas res, quæ existunt, debere admittere modum aliquem non pure imaginarium, sed realem existendi, per quem ibi sint, ubi sunt, & qui existat tum, cum ibi sunt, pereat cum ibi esse desierint, ubi erant. Nam admissso etiam in prima sententia spatio illo, si hoc, quod est esse rem aliquam in ea parte spatii, haberetur tantummodo per rem, & spatium ; quotiescunque existeret res, & spatium, haberetur hoc, quod est rem illam in ea spatii parte collocari. Rursus si in posteriore sententia ordo ille, qui locum constituit, haberetur per ipsas tantummodo res, quæ ordinem illum habent, quotiescunque res illæ existerent, eodem semper existerent ordine illo, nec proinde unquam locum mutarent. Atque id, quod de loco dixi, dicendum pariter de tempore.

Quocunque is modus nomine appelletur.

3. Necessario igitur admittendus est realis aliquis existendi modus, per quem res est ibi, ubi est, & tum, cum est. Sive is modus dicatur res, sive modus rei, sive aliquid, sive nonnihil ; is extra nostram imaginationem esse debet, & res ipsum mutare potest, habens jam alium ejusmodi existendi modum, jam alium.

Modi reales, qui sint reale spatium, & tempus.

4. Ego igitur pro singulis materiæ punctis, ut de his [265] loquar, e quibus ad res etiam immateriales eadem omnia facile transferri possunt, admitto bina realia modorum existendi genera, quorum alii ad locum pertineant, alii ad tempus, & illi locales, hi dicantur temporarii. Quodlibet punctum habet modum realem existendi, per quem est ibi, ubi est, & alium, per quem est tum, cum est. Hi reales existendi modi sunt mihi reale tempus, & spatium : horum possibilitas a nobis indefinite cognita est mihi spatium vacuum, & tempus itidem, ut ita dicam, vacuum, sive etiam spatium imaginarium, & tempus imaginarium.

Eorum natura, & relationes.

5. Modi illi reales singuli & oriuntur, ac pereunt, & indivisibiles prorsus mihi sunt, ac inextensi, & immobiles, ac in suo ordine immutabiles. Ii & sua ipsorum loca sunt realia, ac tempora, & punctorum, ad quæ pertinent. Fundamentum præbent realis relationis distantiæ, sive localis inter duo puncta, sive temporariæ inter duos eventus. Nec aliud est in se, quod illam determinatam distantiam habeant illa duo materiæ puncta, quam quod illos determinatos habeant existendi modos, quos necessario mutent, ubi eam mutent distantiam. Eos modos, qui in ordine ad locum sunt, dico puncta loci realia, qui in ordine ad tempus, momenta, quæ partibus carent singula, ac omni illa quidem extensione, hæc duratione, utraque divisibilitate destituuntur.

Contiguitas punctorum spatii impossibilis.

6. Porro punctum materiæ prorsus indivisibile, & inextensum, alteri puncto materiæ contiguum esse non potest : si nullam habent distantiam ; prorsus coeunt : si non coeunt penitus ; distantiam aliquam habent. Neque enim, cum nullum habeant partium genus,

(a) *Hic, & sequens paragraphus habentur in Supplementis tomi* I. Philosophiæ Recentioris *Benedicti Stay*, § 6, & 7.

SUPPLEMENTS

§ I

Of Space and Time (a)

1. I do not admit perfectly continuous extension of matter; I consider it to be made up of perfectly indivisible points, which are non-extended, set apart from one another by a certain interval, & connected together by certain forces that are at one time attractive & at another time repulsive, depending on their mutual distances. Here it is to be seen, with this theory, what is my idea of space, & of time, how each of them may be said to be continuous, infinitely divisible, eternal, immense, immovable, necessary, although neither of them, as I have shown in a note, have a real nature of their own that is possessed of these properties.

The theme; what are the attributes of space?

2. First of all it seems clear to me that not only those who admit absolute space, which is of its own real nature continuous, eternal & immense, but also those who, following Leibniz & Descartes, consider space itself to be the relative arrangement which exists amongst things that exist, over and above these existent things; it seems to me, I say, that all must admit some mode of existence that is real & not purely imaginary; through which they are where they are, & this mode exists when they are there, & perishes when they cease to be where they were. For, such a space being admitted in the first theory, if the fact that there is some thing in that part of space depends on the thing & space alone; then, as often as the thing existed, & space, we should have the fact that that thing was situated in that part of space. Again, if, in the second theory, the arrangement, which constitutes position, depended only on the things themselves that have that arrangement; then, as often as these things should exist, they would exist in the same arrangement, & could never change their position. What I have said with regard to space applies equally to time.

Real local and temporal modes of existence must of necessity be admitted by every one.

3. Therefore it needs must be admitted that there is some real mode of existence, due to which a thing is where it is, & exists then, when it does exist. Whether this mode is called the thing, or the mode of the thing, or something or nothing, it is bound to be beyond our imagination; & the thing may change this kind of mode, having one mode at one time & another at another time.

The name by which this mode is known is immaterial.

4. Hence, for each of the points of matter (to consider these, from which all I say can be easily transferred to immaterial things), I admit two real kinds of modes of existence, of which some pertain to space & others to time; & these will be called local & temporal modes respectively. Any point has a real mode of existence, through which it is where it is; & another, due to which it exists at the time when it does exist. These real modes of existence are to me real time & space; the possibility of these modes, hazily apprehended by us, is, to my mind, empty space & again empty time, so to speak; in other words, imaginary space & imaginary time.

Real modes; what real space & real time may be.

5. These several real modes are produced & perish, and are in my opinion quite indivisible, non-extended, immovable & unvarying in their order. They, as well as the positions & times of them, & of the points to which they belong, are real. They afford the foundation of a real relation of distance, which is either a local relation between two points, or a temporal relation between two events. Nor is the fact that those two points of matter have that determinated distance anything essentially different from the fact that they have those determinated modes of existence, which necessarily alter when they change the distance. Those modes which are descriptive of position I call real points of position; & those that are descriptive of time I call instants; & they are without parts, & the former lack any kind of extension, while the latter lack duration; both are indivisible.

Their nature & relations.

6. Further, a point of matter that is perfectly indivisible & non-extended cannot be contiguous to any other point of matter; if they have no distance from one another, they coincide completely; if they do not coincide completely, they have some distance between

Contiguity of points of space is impossible.

(a) *This & the following section are to be found in the* Philosophiæ Recentior, *by Benedict Stay, Vol. I,* § 6, 7.

possunt ex parte coire tantummodo, & ex parte altera se contingere, ex altera mutuo aversari. Præjudicium est quoddam ab infantia, & ideis ortum per sensus acquisitis, a debita reflexione destitutis, qui nimirum nobis massas semper ex partibus a se invicem distantibus compositas exhibuerunt, cum videmur nobis puncta etiam invisibilia, & inextensa posse punctis adjungere ita, ut se contingant, & oblongam quandam seriem constituant. Globulos re ipsa nobis confingimus, nec abstrahimus animum ab extensione illa, & partibus, quas voce, & ore secludimus.

Posse binis punctis addi alia in directum ad distantias æquales, interseri alia in infinitum.

7. Porro ubi bina materiæ puncta a se invicem distant, semper aliud materiæ punctum potest collocari in directum ultra utrumque ad eandem distantiam, & alterum ultra hoc, & ita porro, ut patet, sine ullo fine. Potest itidem inter utrumque collocari in medio aliud punctum, quod neutrum continget : si enim alterum contigeret, utrumque contingeret, adeoque cum utroque congrueret, & illa etiam congruerent, non distarent, contra hypothesim. Dividi igitur poterit illud intervallum in partes duas, ac eodem argumento illa itidem duo in alias quatuor, & ita porro sine ullo fine. Quamobrem, utcunque ingens fuerit binorum punctorum intervallum, semper [266] aliud haberi poterit majus, utcunque id fuerit parvum, semper aliud haberi poterit minus, sine ullo limite, & fine.

Existentia puncta spatii semper fore finita numero, & in finitis distantiis : in possibilibus nullum finem.

8. Hinc ultra, & inter bina loci puncta realia quæcunque alia loci puncta realia possibilia sunt, quæ ab iis recedant, vel ad ipsa accedant sine ullo limite determinato, & divisibilitas realis intervalli inter duo puncta in infinitum est, ut ita dicam, interseribilitas punctorum realium sine ullo fine. Quotiescunque illa puncta loci realia interposita fuerint, interpositis punctis materiæ realibus, finitus erit eorum numerus, finitus intervallorum numerus illo priore interceptorum, & ipsi simul æqualium : at numerus ejusmodi partium possibilium finem habebit nullum. Illorum singulorum magnitudo certa erir, ac finita : horum magnitudo minuetur ultra quoscunque limites, sine ullo determinato hiatu, qui adjectis novis intermediis punctis imminui adhuc non possit ; licet nec possit actuali divisione, sive interpositione exhauriri.

Quomodo inde spatium infinitum, continuum, necessarium æternum, immobile per cognitionem præcisivam

9. Hinc vero dum concipimus possibilia hæc loci puncta, spatii infinitatem, & continuitatem habemus, cum divisibilitate in infinitum. In existentibus limes est semper certus, certus punctorum numerus, certus intervallorum : in possibilibus nullus est finis. Possibilium abstracta cognitio, excludens limitem a possibili augmento intervalli, & diminutione, ac hiatu, infinitatem lineæ imaginaræ, & continuitatem constituit, quæ partes actu existentes non habet, sed tantummodo possibiles. Cumque ea possibilitas & æterna sit, & necessaria, ab æterno enim, & necessario verum fuit, posse illa puncta cum illis modis existere ; spatium hujusmodi imaginarium continuum, infinitum, simul etiam æternum fuit, & necessarium, sed non est aliquid existens, sed aliquid tantummodo potens existere, & a nobis indefinite conceptum : immobilitas autem ipsius spatii a singulorum punctorum immobilitate orietur.

In momentis eadem, quæ in punctis : post primum nullum secundum, aut ultimum : sed in tempore unica dimensio, in spatio triplex.

10. Atque hæc omnia, quæ hucusque de loci punctis sunt dicta, ad temporis momenta eodem modo admodum facile transferuntur, inter quæ ingens quædam habetur analogia. Nam & punctum a puncto, & momentum a momento quovis determinato certam distantiam habet, nisi coeunt, qua major, & minor haberi alia potest sine ullo limite. In quovis intervallo spatii imaginarii, ac temporis adest primum punctum, vel momentum, & ultimum, secundum vero, & penultimum habetur nullum : quovis enim assumpto pro secundo, vel penultimo, cum non coeat cum primo, vel ultimo, debet ab eo distare, & in eo intervallo alia itidem possibilia puncta vel momenta interjacent. Nec punctum continuæ lineæ, nec momentum continui temporis, pars est, sed limes & terminus. Linea continua, & tempus continuum generari intelligentur non repetitione puncti, vel momenti, sed ductu continuo, in quo intervalla alia aliorum sint partes, non ipsa puncta, vel momenta, quæ continuo ducuntur. Illud unicum crit [267] discrimen, quod hic ductus in spatio fieri poterit, non in unica directione tantum per lineam, sed in infinitis per planum, quod concipietur ductu continuo in latus lineæ jam conceptæ, & iterum in infinitis per solidum, quod concipietur ductu continuo plani jam concepti, in tempore autem unicus ductus durationis habebitur, quod idcirco soli lineæ erit analogum, & dum spatii imaginarii extensio

them. For, since they have no kind of parts, they cannot coincide partly only; that is, they cannot touch one another on one side, & on the other side be separated. It is but a prejudice acquired from infancy, & born of ideas obtained through the senses, which have not been considered with proper care; & these ideas picture masses to us as always being composed of parts at a distance from one another. It is owing to this prejudice that we seem to ourselves to be able to bring even indivisible and non-extended points so close to other points that they touch them & constitute a sort of lengthy series. We imagine a series of little spheres, in fact; & we do not put out of mind that extension, & the parts, which we verbally exclude.

Given two points, it is possible to add others in the same straight line at equal distances apart; & it is possible to insert others between them; to any extent in either case.

7. Again, where two points of matter are at a distance from one another, another point of matter can always be placed in the same straight line with them, on the far side of either, at an equal distance; & another beyond that, & so on without end, as is evident. Also another point can be placed halfway between the two points, so as to touch neither of them; for, if it touched either of them it would touch them both, & thus would coincide with both; hence the two points would coincide with one another & could not be separate points, which is contrary to the hypothesis. Therefore that interval can be divided into two parts; & therefore, by the same argument, those two can be divided into four others, & so on without any end. Hence it follows that, however great the interval between two points, we could always obtain another that is greater; &, however small the interval might be, we could always obtain another that is smaller; &, in either case, without any limit or end.

The number of points existing in space will always be finite, & the distances between them finite; there is no end to the possible cases.

8. Hence beyond & between two real points of position of any sort there are other real points of position possible; & these recede from them & approach them respectively, without any determinate limit. There will be a real divisibility to an infinite extent of the interval between two points, or, if I may call it so, an endless 'insertibility' of real points. However often such real points of position are interpolated, by real points of matter being interposed, their number will always be finite, the number of intervals intercepted on the first interval, & at the same time constituting that interval, will be finite; but the number of possible parts of this sort will be endless. The magnitude of each of the former will be definite & finite; the magnitude of the latter will be diminished without any limit whatever; & there will be no gap that cannot be diminished by adding fresh points in between; although it cannot be completely removed either by division or by interposition of points.

Hence, the manner in which we arrive at space that is finite, continuous, necessary, eternal & immovable, by means of an abstract concept.

9. In this way, so long as we conceive as possibles these points of position, we have infinity of space, & continuity, together with infinite divisibility. With existing things there is always a definite limit, a definite number of points, a definite number of intervals; with possibles, there is none that is finite. The abstract concept of possibles, excluding as it does a limit due to a possible increase of the interval, a decrease or a gap, gives us the infinity of an imaginary line, & continuity; such a line has not actually any existing parts, but only possible ones. Also, since this possibility is eternal, in that it was true from eternity & of necessity that such points might exist in conjunction with such modes, space of this kind, imaginary, continuous & infinite, was also at the same time eternal & necessary; but it is not anything that exists, but something that is merely capable of existing, & an indefinite concept of our minds. Moreover, immobility of this space will come from immobility of the several points of position.

The same things hold for instants of time; as for points, after the first there is no second or last; in time, however, there is but one dimension, while in space there are three.

10. Everything, that has so far been said with regard to points of position, can quite easily in the same way be applied to instants of time; & indeed there is a very great analogy of a sort between the two. For, a point from a given point, or an instant from a given instant, has a definite distance, unless they coincide; & another distance can be found either greater or less than the first, without any limit whatever. In any interval of imaginary space or time, there is a first point or instant, & a last; but there is no second, or last but one. For, if any particular one is supposed to be the second, then, since it does not coincide with the first, it must be at some distance from it; & in the interval between, other possible points or instants intervene. Again, a point is not a part of a continuous line, or an instant a part of a continuous time; but a limit & a boundary. A continuous line, or a continuous time is understood to be generated, not by repetition of points or instants, but by a continuous progressive motion, in which some intervals are parts of other intervals; the points themselves, or the instants, which are continually progressing, are not parts of the intervals. There is but one difference, namely, that this progressive motion can be accomplished in space, not only in a single direction along a line, but in infinite directions over a plane which is conceived from the continuous motion of the line already conceived in the direction of its breadth; & further, in infinite directions throughout a solid, which is conceived from the continuous motion of the plane already conceived. Whereas, in time there will be had but one progressive motion, that of duration; & therefore this will be analogous

habetur triplex in longum, latum, & profundum, temporis habetur unica in longum, vel diuturnum tantummodo. In triplici tamen spatii, & unico temporis genere, punctum, ac momentum erit principium quoddam, a quo ductu illo suo hæc ipsa generata intelligentur.

Quodvis punctum materiæ habere integrum spatium, ac tempus imaginarium suum: quid sit compenetratio.

11. Illud jam hic diligenter notandum: non solum ubi duo puncta materiæ existunt, & aliquam distantiam habent, existere duos modos, qui relationis illius distantiæ fundamentum præbeant, & sint bina diversa puncta loci realia, quorum possibilitas a nobis concepta, exhibeat bina puncta spatii imaginarii, adeoque infinitis numero possibilibus materiæ punctis respondere infinitos numero possibiles existendi modos, sed cuivis puncto materiæ respondere itidem infinitos possibiles existendi modos, qui sint omnia ipsius puncti possibilia loca. Hæc omnia satis sunt ad totum spatium imaginarium habendum, & quodvis materiæ punctum habet suum spatium imaginarium immobile, infinitum, continuum, quæ tamen omnia spatia pertinentia ad omnia puncta sibi invicem congruunt, & habentur pro unico. Nam si assumatur unum punctum reale loci ad unum materiæ punctum pertinens, & conferatur cum omnibus punctis realibus loci pertinentibus ad aliud punctum materiæ; est unum inter hæc posteriora, quod si cum illo priore coexistat, relationem inducet distantiæ nullius, quam compenetrationem appellamus. Unde patet punctorum, quæ existunt, distantiam nullam non esse nihil, sed relationem inductam a binis quibusdam existendi modis. Reliquorum quivis cum illo eodem priore induceret relationem aliam, quam dicimus cujusdam determinatæ distantiæ, & positionis. Porro illa loci puncta, quæ nullius distantiæ relationem inducunt, pro eodem accipimus, & quenvis ex infinitis hujusmodi punctis ad infinita puncta materiæ pertinentibus pro eodem accipimus, ac ejusdem loci nomine intelligimus. Ea autem haberi debere pro quovis punctorum binario, sic patet. Si tertium punctum ubicunque collocetur, habebit aliquam distantiam, & positionem respectu primi. Summoto primo, poterit secundum collocari ita, ut habeat candem illam distantiam, & positionem, respectu tertii, quam habebat primum. Igitur modus hic, quo existit, pro eodem habetur, ac modus, quo existebat illud primum, & si hi bini modi simul existerent, nullius distantiæ relationem inducerent inter primum, ac secundum: & hæc pariter, quæ hic de spatii punctis dicta sunt, æque temporis momentis conveniunt.

Plura momenta ejusdem puncti non posse coexistere

[268] 12. An autem possint simul existere, id vero pertinet ad relationem, quam habent puncta loci cum momentis temporis, sive spectetur unicum materiæ punctum, sive plura. Inprimis plura momenta ejusdem puncti materiæ coexistere non possunt, sed alia necessario post alia, sic itidem bina puncta localia ejusdem puncti materiæ conjungi non possunt, sed alia jacere debent extra alia, atque id ipsum ex eorum natura, & ut ajunt, essentia.

Combinationes quatuor spatii, & temporis pro unico puncto materiæ quatuor pro binis maxime notabiles: idea singularis spatii alterius alibi positi.

13. Deinde considerentur conjunctiones variæ punctorum loci, & momentorum. Quodvis punctum materiæ, si existit, conjungit aliquod punctum spatii cum aliquo momento temporis. Nam necessario alicubi existit, & aliquando existit; ac si solum existat, semper suum habet, & localem, & temporarium existendi modum, per quod, si aliud quodpiam existat, quod suos itidem habebit modus, distantiæ & localis, & temporis relationem ad ipsum acquiret. Id saltem omnino accidet, si omnium, quæ existunt, vel existere possunt, commune est spatium, ut puncta localia unius, punctis localibus alterius perfecte congruant, singula singulis. Quid enim, si alia sunt rerum genera, vel a nostris dissimilium, vel nostris etiam prorsus similium, quæ aliud, ut ita dicam, infinitum spatium habeant, quod a nostro itidem infinito non per intervallum quoddam finitum, vel infinitum distet, sed ita alienum sit, ita, ut ita dicam, alibi positum, ut nullum cum hoc nostro commercium habeat, nullam relationem distantiæ inducat. Atque id ipsum de tempore etiam dici posset extra omne nostrum æternum tempus collocato. At id menti, ipsum conanti concipere, vim summam infert, ac a cogitatione directa admitti vel nullo modo potest, vel saltem vix potest. Quamobrem iis rebus, vel rerum spatiis, & temporibus, quæ ad nos nihil pertinere possent, prorsus omissis, agamus de notris hisce. Si igitur primo idem punctum materiæ conjungat idem punctum spatii, cum pluribus momentis temporis aliquo a se invicem intervallo disjunctis; habebitur regressus ad eundem locum. Si secundo id conjungat cum serie continua momentorum temporis continui; habebitur quies, quæ requirit tempus aliquod continuum cum eodem loci puncto, sine qua conjunctione habetur continuus motus, succedentibus sibi aliis, atque aliis loci punctis, pro aliis, atque aliis

to a single line. Thus, while for imaginary space there is extension in three dimensions, length, breadth & depth, there is only one for time, namely length or duration only. Nevertheless, in the threefold class of space, & in the onefold class of time, the point & the instant will be respectively the element, from which, by its progression, motion, space & time will be understood to be generated.

Every point of matter is possessed of the whole of imaginary space, & time; the nature of compenetration.

11. Now here there is one thing that must be carefully noted. Not only when two points of matter exist, & have a distance from one another, do two modes exist which give the foundation of the relation of this distance; & there are two different real points of position, the possibility of which, as conceived by us, will yield two points of imaginary space; & thus, to the infinite number of possible points of matter there will correspond an infinite number of possible modes of existence. But also to any one point of matter there will correspond the infinite possible modes of existing, which are all the possible positions of that point. All of these taken together are sufficient for the possession of the whole of imaginary space; & any point of matter has its own imaginary space, immovable, infinite & continuous; nevertheless, all these spaces, belonging to all points coincide with one another, & are considered to be one & the same. For if we take one real point of position belonging to one point of matter, & associate it with all the real points of position belonging to another point of matter, there is one among the latter, which, if it coexist with the former, will induce a relation of no-distance, which we call compenetration. From this it is clear that, for points which exist, no-distance is not nothing, but a relation induced by some two modes of existence. Any of the others would induce, with that same former point of position, another relation of some determinate distance & position, as we say. Further, those points of position, which induce a relation of no-distance, we consider to be the same; & we consider any of the infinite number of such points belonging to the infinite number of points of matter to be the same; & mean them when we speak of the 'same position.' Moreover this is evidently bound to be true for any pair of points. If now a third point is situated anywhere, it will have some distance & position with respect to the first. If the first is removed, the second can be so situated that it has the same distance & position with respect to the third as the first had. Hence the mode, in which it exists, will be taken to be the same in this case as the mode in which the first point was existing; & if these two modes were existing together, they would induce a relation of no-distance between the first point & the second. All that has been said above with regard to points of space applies equally well to instants of time.

Several instants belonging to the same point cannot coexist.

12. Now, whether they can coexist is a question that pertains to the relation between points of position & instants of time, whether we consider a single point of matter or several of them. In the first place, several instants of time belonging to the same point of matter cannot coexist; but they must necessarily come one after the other; & similarly, two points of position belonging to the same point of matter cannot be conjoined, but must lie one outside the other; & this comes from the nature of points of this kind, & is essential to them, to use a common phrase.

Four combinations of space & time for a single point of matter; four worth considering for two points; extraordinary idea of another space situated elsewhere.

13. Next, we have to consider the different kinds of combinations of points of space & instants of time. Any point of matter, if it exists, connects together some point of space & some instant of time; for it is bound to exist somewhere & sometime. Even if it exists alone, it always has its own mode of existence, both local & temporal; & by this fact, if any other point of matter exists, having its own modes also, it will acquire a relation of distance, both local & temporal, with respect to the first. This at least will certainly be the case, if the space belonging to all that exist, or can possibly exist, is common; so that the points of position belonging to the one coincide perfectly with those belonging to the other, each to each. But, what if there are other kinds of things, either different from those about us, or even exactly similar to ours, which have, so to speak, another infinite space, which is distant from this our infinite space by no interval either finite or infinite, but is so foreign to it, situated, so to speak, elsewhere in such a way that it has no communication with this space of ours; & thus will induce no relation of distance. The same remark can be made with regard to a time situated outside the whole of our eternity. But such an idea requires an intellect of the greatest power to try to grasp it; & it cannot be admitted by direct consideration, in any way, or at least with difficulty. Hence, omitting altogether such things, or the spaces & times of such things which are no concern of ours, let us consider the things that have to do with us. If therefore, firstly, the same point of matter connects the same point of space with several instants of time separated from one another by any interval, there will be return to the same place. If, secondly, it connects the point of space to a continuous series of instants of continuous time, there will be rest, which requires a certain continuous time to be connected with the same point of position; without this connection there will be continuous motion, points of position succeeding one another corresponding to instants of time, one after the other. Thirdly,

momentis temporis. Si tertio idem punctum materiæ conjungat idem momentum temporis cum pluribus punctis loci a se invicem distantibus aliquo intervallo; habebitur illa, quam dicimus replicationem. Si quarto id conjungat cum serie continua punctorum loci aliquo intervallo continuo contentorum, habebitur quædam quam plures Peripatetici admiserunt, virtualem appellantes extensionem, qua indivisibilis, & partibus omnino destituta materiæ particula spatium divisibile occuparet. Sunt aliæ quatuor combinationes, ubi plura materiæ pun-[269]-cta considerentur. Nimirum quinto si conjungant idem momentum temporis cum pluribus punctis loci, in quo sita est coexistentia. Sexto si conjungant idem punctum spatii cum diversis momentis temporis, quod fieret in successivo appulsu diversorum punctorum materiæ ad eundem locum. Septimo si conjungant idem momentum temporis cum eodem puncto spatii, in quo sita esset compenetratio. Octavo si nec momentum ullum, nec punctum spatii commune habeant, quod haberetur, si nec coexisterent, nec ea loca occuparent, quæ ab aliis occupata fuissent aliquando.

Relationes earum ad se invicem: quæ, & quomodo possibiles.

14. Ex hisce octo casibus primo respondet tertius, secundo quartus, quinto sextus, septimo octavus. Tertium casum, nimirum replicationem, communitur censent naturaliter haberi non posse. Quartum censent multi habere animam rationalem, quam putant esse in spatio aliquo divisibili, ut plures Peripatetici in toto corpore, alii Philosophi in quadam cerebri parte, vel in aliquo nervorum succo ita, ut cum indivisibilis sit, tota sit in toto spatio, & tota in quavis spatii parte, quemadmodum eadem indivisibilis Divina Natura est in toto spatio, & tota in qualibet spatii parte, ubique necessario præsens, & omnibus creatarum rerum realibus locis coexistens, ac adstans. Eundem alii casum in materia admittunt, cujus particulas eodem pacto extendi putant, ut diximus; licet simplices sint, licet partibus expertes, non modo actu separatis, sed etiam distinctis, ac tantummodo separabilibus. Eam sententiam amplectendam esse non censeo idcirco, quod ubicunque materiam loca distincta occupantem sensu percipimus, separabilem etiam, ingenti saltem adhibita vi, videmus; sejunctis partibus, quæ distabant: nec vero alio ullo argumento excludimus a Natura replicationem, nisi quia nullam materiæ partem, quantum sensu percipere possumus, videmus, bina simul occupare loca. Virtualis illa extensio materiæ infinities ulterius progreditur ultra simplicem replicationem.

Quietem, & regressum ad eundem locum in Natura esse in infinitum improbabiles, & inde ingens analogia.

15. Si secundus casus quietis, & primus casus regressus ad eundem locum naturaliter haberi possent, esset is quidem defectus quidam analogiæ inter spatium, & tempus. At mihi videor probare illud posse, neutrum unquam in Natura contingere, adeoque naturaliter haberi non posse. Id autem evinco hoc argumento. Sit punctum materiæ quodam momento in quodam spatii puncto, & pro quovis alio momento ignorantes, ubi sit, quæramus, quanto probabilius sit, ipsum alibi esses, quam ibidem. Tanto crit probabilius illud, quam hoc, quanto plura sunt alia spatii puncta, quam illud unicum. Hæc in quavis linea sunt infinita, infinitus in quovis plano linearum numerus, infinitus in toto spatio planorum numerus. Quare numerus aliorum punctorum est infinitus tertii generis, adeoque illa probabilitas major infinities tertii generis infinitate, ubi de quovis alio determinato momento agitur. Agatur jam inde-[269]-finite de omnibus momentis temporis infiniti, decrescet prior probabilitas in ea ratione, qua momenta crescunt, in quorum aliquo saltem posset ibidem esse punctum. Sunt autem momenta numero infinita infinitate ejusdem generis, cujus puncta possibilia in linea infinita. Igitur adhuc agendo de omnibus momentis infiniti temporis indefinite, est infinities infinite improbabilius, quod punctum in eodem illo priore sit loco, quam quod sit alibi. Consideretur jam non unicum punctum loci determinato unico momento occupatum, sed quodvis punctum loci, quovis indefinite momento occupatum, & adhuc probabilitas regressus ad aliquod ex iis crescet, ut crescit horum loci punctorum numerus, qui infinito etiam tempore est infinitus ejusdem ordinis, cujus est numerus linearum, in quovis plano. Quare improbabilitas casus, quo determinatum quodpiam materiæ punctum redeat, quovis indefinite momento temporis, ad quodvis indefinite punctum loci, in quo alio quovis fuit momento temporis indefinite sumpto, remanet infinita primi ordinis. Eadem autem pro omnibus materiæ punctis, quæ numero finita sunt, decrescit in ratione finita ejus numeri ad unitatem (quod secus accidit in communi sententia, in qua punctorum materiæ numerus est infinitus ordinis tertii). Quare

if the same point of matter connects the same instant of time with several points of position distant from one another by some interval, then we shall have replication. Fourthly, if it connects the instant with a continuous series of points of position contained within some continuous interval, we shall have something which several of the Peripatetics admitted, calling it virtual extension; by virtue of which an indivisible particle of matter, quite without parts, could occupy divisible space. There are four other combinations, when several points are considered. That is to say, fifthly, if several points connect the same instant of time with several points of position; in this is involved coexistence. Sixthly, if they connect the same point of space with several instants of time; as would be the case when different points of matter were forced successively into the same position. Seventhly, if they connect the same point of space with the same instant of time; in this is involved compenetration. Eighthly, if they have no instant of time, & no point of space, common to them; as would be the case, if they did not coexist, nor, any of them, occupied the positions that had been occupied by any of the others at any time.

The relations of these cases to one another; which of them are Possible, & how.

14. Out of these eight cases, the third corresponds to the first, the fourth to the second, the sixth to the fifth, the eighth to the seventh. The third case, namely replication, is usually considered to be naturally impossible. Many think that the fourth case holds good for the rational soul, which they consider to have its seat in some divisible space; for instance, the Peripatetics think that it pervades the whole of the body, other philosophers think it is situated in a certain part of the brain, or in some juice of the nerves; so that, since it is indivisible, the whole of it must be in the whole of the space, & the whole of it in any part of the space. Just in the same way as the same indivisible Divine Nature is as a whole in the whole of space, & as a whole in any part of space, being necessarily present everywhere, & coexisting with & accompanying created things wherever created things are. Others admit this same case for matter, & consider that particles of matter can be extended in a similar manner, as we have said; although they are simple, & although they are devoid of parts, not only parts that are really separated, but also such as are distinct & only separable. I do not consider that this supposition can be entertained, for the reason that, whenever we perceive with our senses matter occupying positions distinct from one another, we see that it is also separable, although we may have to use a very great force; here, parts are separated which were at a distance from one another. Indeed, by no other argument can we exclude replication from Nature, than that we never see any portion of matter, as far as can be perceived by the senses, occupying two positions at the same time. The idea of Virtual extension of matter goes infinitely further beyond the idea of simple replication.

Rest & return to the same position are infinitely improbable in Nature; hence arises a very great analogy between them.

15. If the second case of rest, & the first case of return to the same position could be obtained naturally, then indeed there would be a certain defect in the analogy between space & time. But it seems to me that I can prove that neither ever happens in Nature; & so they cannot be obtained naturally; this is my argument. If a point of matter at any instant of time is at a certain point of space, & we do not know where it is at some other instant, let us inquire how much more probable it is that it should be somewhere else than at the same point as before. The former will be more probable than the latter in the proportion of the number of all the other points of space to that single point. There are an infinite number of these points in any straight line, the number of lines in any plane is infinite, & the number of planes in the whole of space is infinite. Hence, the number of other points of space is an infinity of the third order; & thus the probability is infinitely greater with an infinity of the third order, when we are concerned with any other particular instant of time. Now let us deal indefinitely with all the instants of infinite time; then the first probability will decrease in proportion as the number of instants increases, at any of which it might at least be possible that the point was in the same place as before. Moreover, there are an infinite number of instants, the infinity being of the same order as that of the number of possible points in an infinite line. Hence, still considering indefinitely all the instants of infinite time, it is infinitely more improbable that the point should be in the same position as before, than that it should be somewhere else. Now consider, not a single point of position occupied at a single particular instant, but any point of position occupied at any indefinite instant; then still the probability of return to any one of these points of position will increase as the number of them increases; & this number, in a time that is also infinite, is an infinity of the same order as the number of lines in any plane. Hence the improbability of this case, in which any particular point of matter returns at some indefinite instant of time to some indefinite point of position, in which it was assumed to be at some other indefinite instant of time, remains an infinity of the first order. Moreover, this, for all points of matter, which are finite in number, will decrease in the finite ratio of this number to infinity (which would not be the case with the usual theory, in which the number of points of matter is taken to be an infinity of the third order). Hence we are still left with

adhuc remanet infinita improbabilitas regressus puncti materiæ cujusvis indefinite, ad punctum loci quodvis, occupatum quovis momento præcedenti indefinite, regressus inquam, habendi quovis indefinite momento sequenti temporis, qui regressus idcirco sine ullo erroris metu debet excludi, cum infinitam improbabilitatem in relativam quandam impossibilitatem migrare censendum sit : quæ quidem Theoria communi sententiæ applicari non potest. Quamobrem eo pacto, patet, in mea materiæ punctorum Theoria e Natura tolli & quietem, quam etiam supra exclusimus, & vero etiam regressum ad idem loci punctum, in quo semel ipsum punctum materiæ extitit : unde fit, ut omnes illi primi quatuor casus excludantur ex Natura, & in iis accurata temporis, & spatii servetur analogia.

Nullum punctum materiæ advenire ad ullum punctum spatii, in quo aliquando fuerit aliud punctum quodvis. In sola coexistentia respondente huic adventui lædi analogiam.

16. Quin imo si quæratur, an aliquod materiæ punctum occupare debeat quopiam momento punctum loci, quod alio momento aliquo aliud materiæ punctum occupavit ; adhuc improbabilitas erit infinities infinita. Nam numerus punctorum materiæ existentium est finitus, adeoque si pro regressu puncti cujusvis ad puncta loci a se occupata adhibeatur regressus ad puncta occupata a quovis alio, numerus casuum crescit in ratione unitatis ad numerum punctorum finitum utique, nimirum in ratione finita tantummodo. Hinc improbabilitas appulsus alicujus puncti materiæ indefinite sumpti ad punctum spatii aliquando ab alio quovis puncto occupati adhuc est infinita, & ipse appulsus habendus pro impossibili, quo quidem pacto excluditur & sextus casus, qui in eo ipso situs erat regressu, & multo magis septimus, qui binorum punctorum mate-**[271]**-riæ simultaneum appulsum continet ad idem aliquod loci punctum, sive compenetrationem. Octavus autem pro materia excluditur, cum tota simul creata perpetuo duret tota, adeoque semper idem momentum habeat commune.(b) Solus quintus casus, quo plura materiæ puncta idem momentum temporis cum diversis punctis loci conjungant, non modo possibilis est, sed etiam necessarius pro omnibus materiæ punctis, coexistentibus nimirum : fieri enim non potest, ut septimus, & octavus excludantur ; nisi continuo ob id ipsum includatur quintus ille, ut consideranti patebit facile. Quamobrem in eo analogia deficit, quod possint plura materiæ puncta conjungere diversa puncta spatii cum eodem momento temporis, qui est hic casus quintus, non autem possit idem punctum spatii, cum pluribus momentis temporis, qui est casus tertius, quem defectum necessario inducit exclusio septimi, & octavi, quorum altero incluso, excludi posset hic quintus, ut si possent materiæ puncta, quæ simul creata sunt, nec pereunt, non coexistere, tum enim idem momentum cum diversis loci punctis nequaquam conjungeretur.

Qui casus sint possibiles per Divinam Omnipotentiam: usus superioris theorematis in Impenetrabilitate.

17. Ex illis 7 casibus videntur omnino 6 per Divinam Omnipotentiam possibiles, dempta nimirum virtuali illa materiæ extensione, de qua dubium esse poterit, quia deberet simul existere numerus absolute infinitus punctorum illorum loci realium, quod impossibile est ; si infinitum numero actu existens repugnat in modis ipsis. Quoniam autem possunt omnia existere alia post alia puncta loci in quavis linea constituta, in motu nimirum continuo, & possunt itidem momenta omnia temporis continui, alia itidem post alia in rei cujusvis duratione ; ambigi poterit, an possint & omnia simul ipsa loci puncta, quam quæstionem definire non ausim. Illud unum moneo, sententiam hanc meam de spatii natura, & continuitate præcipuas omnes difficultates, quibus premuntur reliquæ, peni-**[272]**-tus evitare, & ad omnia, quæ huc pertinent, explicanda commodissimam esse. Tum illud addo, excluso appulsu puncti cujusvis materiæ ad punctum loci, ad quod punctum quodvis materiæ quovis momento appellit, & inde compenetratione, veram impenetrabilitatem materiæ necessario consequi, quod in decimo nobis libro (c) plurimum proderit. Nimirum

(b) *Hic casus nusquam itidem haberetur ; si duratio non esset quid continenter permanens, sed loco ipsius admitteretur quædam existentia, ut ita dicam, saltitans, nimirum si quodvis materiæ punctum (& idem potest transferri ad quævis creata entia) existeret tantum in momentis indivisibilibus a se invicem remotis, in omnibus vero intermediis possibilibus omnino non existeret. Eo casu coexistentia esset infinite improbabilis eodem fere argumento, quo adventus unius puncti materiæ ad punctum spatii, in quo aliud quodvis punctum unquam fuerit. In eodem nullum haberetur reale continuum ne in motu quidem ; diversæ celeritates multo melius explicarentur : multo magis pateret, quomodo vita insecti brevissima possit æquivalere vitæ cuivis longissimæ, per eundem nimirum numerum existentiarum inter extrema momenta. Verum & exclusio cujusvis coexistentiæ abriperet secum omnes prorsus influxus physicos immediatos, ac determinationes, & deberet haberi continua reproductio, immo creatio nova perpetua, & alia ejusmodi, quæ admitti, non possunt, haberentur.*

(c) *Stayanæ nimirum philosophiæ, in quo Auctor elegantissimus, & doctissimus hanc meam Philosophiam exponit. Hunc ejus theorematis fructum jam cepimus hic supra, ubi in ipso opere de impenetrabilitate egimus, & de apparenti compenetratione, quæ sine viribus mutuis haberetur a num. 360.*

an infinite improbability of the return of any indefinitely chosen point of matter to any point of position, occupied at any previous instant of time indefinitely, of a return, I say, taking place at any indefinite instant of subsequent time; hence, such a return must be excluded, without any fear as to error, since it must be considered that an infinite improbability merges into a sort of relative impossibility. This Theory indeed cannot be applied to the ordinary view. Hence, in this way it is clear, in my Theory of points of matter, there must be excluded from Nature both rest, which also we excluded above, & even return to the same point of position in which that point of matter once was situated. Therefore it comes about that all those first four cases will be excluded from Nature, & in them the analogy of time & space will be preserved accurately.

No Point of matter can come into any point of space that was once occupied by any other point; it is only in coexistence, which corresponds to this that the analogy is broken.

16. Finally, if we seek to find whether any point of matter is bound to occupy at some instant a point of position which was occupied by some other point of matter at some other instant, still the improbability will be infinitely infinite. For the number of existing points of matter is finite; & thus, if instead of the return of any point to points of position occupied by itself we consider the return to points that have been occupied by another, the number of cases increases in the ratio of unity to a number of points that is in every case finite, that is to say, in a finite ratio only. Hence, the improbability of the arrival of any point of matter indefinitely taken at a point of space that has been occupied at some time by any other point is still infinite; & this arrival must therefore be taken to be impossible. In this way, indeed, the sixth case, which depended on this return, is excluded; & much more so the seventh case, which involves the simultaneous arrival of a pair of points of matter at any the same point of position, that is to say, compenetration. The eighth case also is excluded for matter; for all things created together as a whole will continually last as a whole, & so will always have a common instant of time.(*b*) Only the fifth case, in which several points of matter connect the same instant of time with different points of position remains; & this is not only possible, but also necessary for all points of matter, seeing that they coexist. For it cannot be the case that the seventh & the eighth are excluded, unless straightway, on that very account, the fifth is included, as will be easily seen on consideration. Therefore in this point the analogy fails, namely, in that several points of matter can connect different points of space with the same instant of time, which is the fifth case; whereas it is impossible for the same point of space to be connected with several instants of time, which is the third case. This defect is necessarily induced by the exclusion of the seventh & eighth cases; for if either of the latter is included, the fifth might be excluded; just as if it were possible for points of matter, which had been created together, & do not perish, not to coexist; for then the same instant of time would in no way be connected with different points of position.

Which of the cases are possible through Divine Omnipotence; use of a theorem given above on impenetrability.

17. At least six of the seven cases seem to be possible through Divine Omnipotence, that is to say, omitting the virtual extension of matter, about which there may be possibly some doubt; for in this case there must exist at the same time an absolutely infinite number of those real points of position; & this is impossible, if an existing thing that is infinite in number is contradictory in the modes. Moreover, since all points of position can exist one after another, arranged along any line, for instance, in continuous motion, & so can also all instants of continuous time, one after another in the duration of any thing, there will be reason for doubt as to whether all those points of position can also exist at the same time. This is a matter upon which I dare not make a definite statement. All I say is that this theory of mine with regard to the nature of space & continuity completely avoids all the chief difficulties that are obstacles in other theories; & that it is very suitable for the explanation of everything in connection with this matter. I will also add the remark that, if the arrival of any point of matter at a point of position, at which any point of matter has arrived at any instant, is excluded, & along with it compenetration is thus excluded, then real impenetrability of matter must necessarily follow, which will be of great service to us in our tenth book (*c*). That is, unless repulsive forces prevent such a thing, any

(b) *This case also would never happen, if the duration were not something continuously permanent; in place of it, we should have to admit a kind of, so to speak, skipping existence; that is to say, as if any point of matter (and the same thing applies to all created entities) existed only in indivisible instants remote from one another, and in all intermediate instants possible did not exist at all. Coexistence, in this case, would be infinitely improbable, the argument being nearly the same, as in the case of the arrival of one point of matter at a point of space in which some other point had once been. In this case too, there would be no real continuum even in motion; different velocities could be explained much more easily; it would be much more evident in what way the very short life of an insect can be equivalent to the longest of lives, by means of the same number of existences coming in between the first & last instants. Indeed the exclusion of any coexistence would carry away with it all immediate physical influence altogether, & determinations; indeed, a continually fresh creation, & other inadmissible things of that sort, would be obtained.*

(c) *The reference is to Stay's "Philosophy," in which that most refined & learned author expounds my Philosophy. On what I have said above, I have plucked the fruit of the theorem, in which, in Art.* 360 *of this work, I dealt with impenetrability, & the apparent compenetration that would result, if there were no mutual forces.*

nisi vires repulsivæ prohiberent; liberrime massa quævis per quamvis aliam massam permearet, sine ullo periculo occursus ullius puncti cum alio quovis, ubi haberetur apparens quædam compenetratio similis penetrationi luminis per crystalla, olei per ligna, & marmora, sine ulla reali compenetratione punctorum. In massis crassioribus, & minori celeritate præditis vires repulsivæ motum ulteriorem plerumque impediunt sine ullo impactu, & sensibilem etiam illam, ac apparentem compenetrationem excludunt: in tenuissimis, & celerrimis, ut in luminis radiis per homogeneas substantias, vel per alios radios propagatis, evitatur per celeritatem ipsam, actionum exigua inæqualitas, ex circumjacentium punctorum inæquali distantia orta, ac liberrimus habetur progressus in omnes plagas sine ullo occursus periculo, quod summam, & unicam difficultatem propagationis luminis per substantiam emissam, & progredientem, penitus amovet. Sed de his jam satis.

free mass will permeate through any other mass, without there being any danger
ɩion of one point with another. Here there would be an apparent compenetration
the penetration of light through crystals, oils through wood, & marble, without
compenetration of the points. In denser masses, & those endowed with a smaller
the repulsive forces for the most part prevent further motion without any impact;
so excludes sensible as well as apparent compenetration. In very tenuous masses
with very great velocities, as rays of light propagated through homogeneous
s, or through other rays, the very slight inequality of the actions, derived from
ual distances of the circumjacent points, will be prevented by the high velocity;
tly free progress will take place in all directions without any danger of collisions.
oves altogether the greatest & only real difficulty in the idea of the propagation
y means of a substance that is emitted & travels forward. But I have now said
ough upon this matter.

[273] § II

De Spatio, & Tempore, ut a nobis cognoscuntur

Nos nec modos existendi locales posse absolute cognoscere, nec absolute distantias, & magnitudines.

18. Diximus in superiore Supplemento de spatio, ac tempore, ut sunt in se ipsis: superest, ut illud attingam, quod pertinet ad ipsa, ut cognoscuntur. Nos nequaquam immediate cognoscimus per sensus illos existendi modos reales, nec discernere possumus alios ab aliis. Sentimus quidem a discrimine idearum, quæ per sensus excitantur in animo, relationem determinatam distantiæ, & positionis, quæ e binis quibusque localibus existendi modis exoritur, sed eadem idea oriri potest ex innumeris modorum, sive punctorum realium loci binariis, quæ inducant relationes æqualium distantiarum, & similium positionum tam inter se, quam ad nostra organa, & ad reliqua circumjacentia corpora. Nam bina materiæ puncta, quæ alicubi datam habent distantiam, & positionem inductam a binis quibusdam existendi modis, alibi possunt per alios binos existendi modos habere relationem distantiæ æqualis, & positionis similis, distantiis nimirum ipsis existentibus parallelis. Si illa puncta, & nos, & omnia circumjacentia corpora mutent loca realia, ita tamen, ut omnes distantiæ æquales maneant, & prioribus parallelæ; nos easdem prorsus habebimus ideas, quin imo easdem ideas habebimus; si manentibus distantiarum magnitudinibus, directiones omnes in æquali angulo converterentur, adeoque æque ad se invicem inclinarentur ac prius. Et si minuerentur etiam distantiæ illæ omnes, manentibus angulis, & manente illarum ratione ad se invicem, vires autem ex ea distantiarum mutatione non mutarentur, rite mutata virium scala illa, nimirum curva illa linea, per cujus ordinatas ipsæ vires exprimuntur; nullam nos in nostris ideis mutationem haberemus.

Motum communem nobis, & Mundo non posse a nobis cognosci, nec si ipse in quavis ratione augeatur, vel minuatur totus.

19. Hinc autem consequitur illud, si totus hic Mundus nobis conspicuus motu parallelo promoveatur in plagam quamvis, & simul in quovis angulo convertatur, nos illum motum, & conversionem sentire non posse. Sic si cubiculi, in quo sumus, & camporum, ac montium tractus omnis motu aliquo Telluris communi ad sensum simul convertatur; motum ejusmodi sentire non possumus: ideæ enim eædem ad sensum excitantur in animo. Fieri autem posset, ut totus itidem Mundus nobis conspicuus in dies contraheretur, vel produceretur, scala virium tantundem contracta, vel producta; quod si fieret; nulla in animo nostro idearum mutatio haberetur, adeoque nullus ejusmodi mutationis sensus.

Mutata positione nostra, & omnium, quæ videmus, non mutari nostras ideas, & idcirco nos motum nec nobis adscribere, nec reliquis.

20. Ubi vel objecta externa, vel nostra organa mutant illos suos existendi modos ita, ut prior illa æqualitas, [274] vel similitudo non maneat, tum vero mutantur ideæ, & mutationis habetur sensus, sed ideæ eædem omnino sunt, sive objecta externa mutationem subeant, sive nostra organa, sive utrumque inæqualiter. Semper ideæ nostræ differentiam novi status a priore referent, non absolutam mutationem, quæ sub sensus non cadit. Sic sive astra circa Terram moveantur, sive Terra motu contrario circa se ipsam nobiscum; eædem sunt ideæ, idem sensus. Mutationes absolutas nunquam sentire possumus, discrimen a priori forma sentimus. Cum autem nihil adest, quod nos de nostrorum organorum mutatione commoneat; tum vero nos ipsos pro immotis habemus communi præjudicio habendi pro nullis in se, quæ nulla sunt in nostra mente, cum non cognoscantur, & mutationem omnem objectis extra nos sitis tribuimus. Sic errat, qui in navi clausus se immotum censet, littora autem, & montes, ac ipsam undam moveri arbitratur.

§ II

Of Space & Time, as we know them

We cannot obtain an absolute knowledge of local modes of existence; nor yet of absolute distances or magnitudes.

18. We have spoken, in the preceding Supplement, of Space & Time, as they are in themselves; it remains for us to say a few words on matters that pertain to them, in so far as they come within our knowledge. We can in no direct way obtain a knowledge through the senses of those real modes of existence, nor can we discern one of them from another. We do indeed perceive, by a difference of ideas excited in the mind by means of the senses, a determinate relation of distance & position, such as arises from any two local modes of existence; but the same idea may be produced by innumerable pairs of modes or real points of position; these induce the relations of equal distances & like positions, both amongst themselves & with regard to our organs, & to the rest of the circumjacent bodies. For, two points of matter, which anywhere have a given distance & position induced by some two modes of existence, may somewhere else on account of two other modes of existence have a relation of equal distance & like position, for instance if the distances exist parallel to one another. If those points, we, & all the circumjacent bodies change their real positions, & yet do so in such a manner that all the distances remain equal & parallel to what they were at the start, we shall get exactly the same ideas. Nay, we shall get the same ideas, if, while the magnitudes of the distances remain the same, all their directions are turned through any the same angle, & thus make the same angles with one another as before. Even if all these distances were diminished, while the angles remained constant, & the ratio of the distances to one another also remained constant, but the forces did not change owing to that change of distance; then if the scale of forces is correctly altered, that is to say, that curved line, whose ordinates express the forces; then there would be no change in our ideas.

The motion, if any, common to us & the Universe could not come within our knowledge; nor could we know it, if it were increased in any ratio, or diminished, as a whole.

19. Hence it follows that, if the whole Universe within our sight were moved by a parallel motion in any direction, & at the same time rotated through any angle, we could never be aware of the motion or the rotation. Similarly, if the whole region containing the room in which we are, the plains & the hills, were simultaneously turned round by some approximately common motion of the Earth, we should not be aware of such a motion; for practically the same ideas would be excited in the mind. Moreover, it might be the case that the whole Universe within our sight should daily contract or expand, while the scale of forces contracted or expanded in the same ratio; if such a thing did happen, there would be no change of ideas in our mind, & so we should have no feeling that such a change was taking place.

Since, if our position & that of everything we see is changed, our ideas are not changed; therefore we can ascribe no motion to ourselves or to anything else.

20. When either objects external to us, or our organs change their modes of existence in such a way that that first equality or similitude does not remain constant, then indeed the ideas are altered, & there is a feeling of change; but the ideas are the same exactly, whether the external objects suffer the change, or our organs, or both of them unequally. In every case our ideas refer to the difference between the new state & the old, & not to the absolute change, which does not come within the scope of our senses. Thus, whether the stars move round the Earth, or the Earth & ourselves move in the opposite direction round them, the ideas are the same, & there is the same sensation. We can never perceive absolute changes; we can only perceive the difference from the former configuration that has arisen. Further, when there is nothing at hand to warn us as to the change of our organs, then indeed we shall count ourselves to have been unmoved, owing to a general prejudice for counting as nothing those things that are nothing in our mind; for we cannot know of this change, & we attribute the whole of the change to objects situated outside of ourselves. In such manner any one would be mistaken in thinking, when on board ship, that he himself was motionless, while the shore, the hills & even the sea were in motion.

Quomodo judicemus de æqualitate duorum, ex æqualitate cum tertio: nunquam haberi congruentiam in longitudine, ut nec in tempore, sed inferri a causis.

21. Illud autem notandum inprimis ex hoc principio immutabilitatis eorum, quorum mutationem per sensum non cognoscimus, oriri etiam methodum, quam adhibemus in comparandis intervallorum magnitudinibus inter se, ubi id, quod pro mensura assumimus, habemus pro immutabili. Utimur autem hoc principio, *quæ sunt æqualia eidem, sunt æqualia inter se*, ex quo deducitur hoc aliud, ad ipsum pertinens, *quæ sunt æque multipla, vel submutipla alterius, sunt itidem inter se æqualia*, & hoc alio, *quæ congruant, æqualia sunt.* Assumimus ligneam, vel ferream decempedam, quam uni intervallo semel, vel centies applicatam si inveniamus congruentem, tum alteri intervallo applicatam itidem semel, vel centies itidem congruentem, illa intervalla æqualia dicimus. Porro illam ligneam, vel ferream decempedam habemus pro eodem comparationis termino post translationem. Si ea constaret ex materia prorsus continua, & solida, haberi posset pro eodem comparationis termino ; at in mea punctorum a se invicem distantium sententia, omnia illins decempedæ puncta, dum transferuntur, perpetuo distantiam revera mutant. Distantia enim constituitur per illos reales existendi modos, qui mutantur perpetuo. Si mutentur ita, ut qui modi succedunt, fundent reales æqualium distantiarum relationes ; terminus comparationis non crit idem, adhuc tamen æqualis erit, & æqualitas mensuratorum intervallorum rite colligetur. Longitudinem decempedæ in priore situ per illos priores reales modos constitutæ, cum longitudine in posteriore situ constituta per hosce posteriores, immediate inter se conferre nihilo magis possumus, quam illa ipsa intervalla, quæ mensurando conferimus. Sed quia nullam in translatione mutationem sentimus, quæ longitudinis relationem nobis ostendat, idcirco pro eadem habemus longitudinem ipsam. At ea revera semper in ipsa translatione non nihil mutabitur. Fieri posset, ut ingentem etiam mutationem aliquam subiret [275] & ipsa, & nostri sensus, quam nos non sentiremus, & ad priorem restituta locum ad priori æqualem, vel similem statum rediret. Exigua tamen aliqua mutatio habetur omnino idcirco, quod vires, quæ illa materiæ puncta inter se nectunt, mutata positione ad omnia reliquarum Mundi partium puncta, non nihil immutantur. Idem autem & in communi sententia accidit. Nullum enim corpus spatiolis vacat interjectis, & omnis penitus compressionis, ac dilatationis est incapax, quæ quidem dilatatio, & compressio saltem exigua in omni translatione omnino habetur. Nos tamen mensuram illam pro eadem habemus, cum, ut monui, nullam mutationem sentiamus.

Conclusio : discrimen vulgi a Philosophis in judicando.

22. Ex his omnibus consequitur, nos absolutas distantias nec immediate cognoscere omnino posse, nec per terminum communem inter se comparare, sed æstimare magnitudines ab ideis, per quas eas cognoscimus, & mensuras habere pro communibus terminis, in quibus nullam mutationem factam esse vulgus censet. Philosophi autem mutationem quidem debent agnoscere, sed cum nullam violatæ notabili mutatione æqualitatis causam agnoscant, mutationem ipsam pro æqualiter facta habent.

Licet translata decempeda, mutentur modi, qui intervalli relationem constituunt ; tamen intervalla æqualia haberi pro eodem ex causis.

23. Porro licet, ubi puncta materiæ locum mutant, ut in decempeda translata, mutetur revera distantia, mutatis iis modis realibus, quæ ipsam constituunt ; tamen si mutatio ita fiat, ut posterior illa distantia æqualis prorsus priori sit, ipsam appellabimus candem, & nihil mutatam ita, ut eorundem terminorum æquales distantiæ dicantur distantia eadem, & magnitudo dicatur eadem, quæ per eas æquales distantias definitur, ut itidem ejusdem directionis nomine intelligantur binæ etiam directiones parallelæ ; nec mutari distantiam, vel directionem dicemus in sequentibus, nisi distantiæ magnitudo, vel parallelismus mutetur.

Eadem ad tempus transferenda, sed in eo etiam vulgo notum esse, intervallum temporarium non posse transferri idem pro comparatione duorum : errari ab eo circa spatium.

24. Quæ de spatii mensura diximus, haud difficulter ad tempus transferentur, in quo itidem nullam habemus certam, & constantem mensuram. Desuminus a motu illam, quam possumus, sed nullum habemus motum prorsus æquabilem. Multa, quæ huc pertinent, & quæ ad idearum ipsarum naturam, & successionem spectant, diximus in notis. Unum hic addo, in mensura temporis, ne vulgus quidem censere ab uno tempore ad aliud tempus eandem temporis mensuram transferri. Videt aliam esse, sed æqualem supponit ob motum suppositum æqualem. In mensura locali æque in mea sententia, ac in mensura temporaria impossibile est certam longitudinem, ut certam durationem e sua sede abducere in alterius sedem, ut binorum comparatio habeatur per tertium. Utrobique alia longitudo, ut alia duratio substituitur, quæ priori illi æqualis censetur, nimirum nova realia

21. Again, it is to be observed first of all that from this principle of the unchangeability of those things, of which we cannot perceive the change through our senses, there comes forth the method that we use for comparing the magnitudes of intervals with one another; here, that, which is taken as a measure, is assumed to be unchangeable. Also we make use of the axiom, *things that are equal to the same thing are equal to one another*; & from this is deduced another one pertaining to the same thing, namely, *things that are equal multiples, or submultiples, of each, are also equal to one another*; & also this, *things that coincide are equal*. We take a wooden or iron ten-foot rod; & if we find that this is congruent with one given interval when applied to it either once or a hundred times, & also congruent to another interval when applied to it either once or a hundred times, then we say that these intervals are equal. Further, we consider the wooden or iron ten-foot rod to be the same standard of comparison after translation. Now, if it consisted of perfectly continuous & solid matter, we might hold it to be exactly the same standard of comparison; but in my theory of points at a distance from one another, all the points of the ten-foot rod, while they are being transferred, really change the distance continually. For the distance is constituted by those real modes of existence, & these are continually changing. But if they are changed in such a manner that the modes which follow establish real relations of equal distances, the standard of comparison will not be identically the same; & yet it will still be an equal one, & the equality of the measured intervals will be correctly determined. We can no more transfer the length of the ten-foot rod, constituted in its first position by the first real modes, to the place of the length constituted in its second position by the second real modes, than we are able to do so for intervals themselves, which we compare by measurement. But, because we perceive none of this change during the translation, such as may demonstrate to us a relation of length, therefore we take that length to be the same. But really in this translation it will always suffer some slight change. It might happen that it underwent even some very great change, common to it & our senses, so that we should not perceive the change; & that, when restored to its former position, it would return to a state equal & similar to that which it had at first. However, there always is some slight change, owing to the fact that the forces which connect the points of matter, will be changed to some slight extent, if its position is altered with respect to all the rest of the Universe. Indeed, the same is the case in the ordinary theory. For no body is quite without little spaces interspersed within it, altogether incapable of being compressed or dilated; & this dilatation & compression undoubtedly occurs in every case of translation, at least to a slight extent. We, however, consider the measure to be the same so long as we do not perceive any alteration, as I have already remarked.

The manner in which we are to judge of the equality of two things from their equality with a third; there never can be congruence in length, any more than there can be in time; the matter is to be inferred from causes.

22. The consequence of all this is that we are quite unable to obtain a direct knowledge of absolute distances; & we cannot compare them with one another by a common standard. We have to estimate magnitudes by the ideas through which we recognize them; & to take as common standards those measures which ordinary people think suffer no change. But philosophers should recognize that there is a change; but, since they know of no case in which the equality is destroyed by a perceptible change, they consider that the change is made equally.

Conclusion reached; the difference between ordinary people & philosophers in the matter of judgment.

23. Further, although the distance is really changed when, as in the case of the translation of the ten-foot rod, the position of the points of matter is altered, those real modes which constitute the distance being altered; nevertheless if the change takes place in such a way that the second distance is exactly equal to the first, we shall call it the same, & say that it is altered in no way, so that the equal distances between the same ends will be said to be the same distance & the magnitude will be said to be the same; & this is defined by means of these equal distances, just as also two parallel directions will be also included under the name of the same direction. In what follows we shall say that the distance is not changed, or the direction, unless the magnitude of the distance, or the parallelism, is altered.

Although, when the ten-foot rod is moved in position, those modes that constitute the relations of the interval are also altered, yet equal intervals are reckoned as same for the reasons stated.

24. What has been said with regard to the measurement of space, without difficulty can be applied to time; in this also we have no definite & constant measurement. We obtain all that is possible from motion; but we cannot get a motion that is perfectly uniform. We have remarked on many things that belong to this subject, & bear upon the nature & succession of these ideas, in our notes. I will but add here, that, in the measurement of time, not even ordinary people think that the same standard measure of time can be translated from one time to another time. They see that it is another, consider that it is an equal, on account of some assumed uniform motion. Just as with the measurement of time, so in my theory with the measurement of space it is impossible to transfer a fixed length from its place to some other, just as it is impossible to transfer a fixed interval of time, so that it can be used for the purpose of comparing two of them by means of a third. In both cases, a second length, or a second duration is substituted, which is supposed to be equal to the first; that is to say, fresh real positions of the points of the same ten-foot

The same observations apply equally to Time; but in it, it is well known, even to ordinary people, that the same temporal interval cannot be translated for the purpose of comparing two intervals; it is because of this that they fall into error with regard to space.

mea Theoria eadem prorsus utrobique habetur analog
tummodo in mensura locali eundem haberi putat co
ceteri fere omnes eundem saltem haberi posse per mensu
in tempore tantummodo æqualem : ego vero utrobique
eandem.

w circuit made by the same rod, or a
o ends. In my Theory, there is in each
ime. Ordinary people think that it is
f measurement is the same ; almost all
ın at least be considered to be the same
ntinuous, but that in time there is only
case the equality, & never the identity.

[277] § III

Solutio analytica Problematis determinantis naturam Legis Virium (d)

Denominatio, ac præparatio.

25. Ut hasce conditiones impleamus, formulam inveniemus algebraicam, quæ ipsam continebit legem nostram, sed hic elementa communia vulgaris Cartesianæ algebræ supponemus ut nota, sine quibus res omnino confici nequaquam potest. Dicatur autem ordinata y, abscissa x, ac ponatur $xx = z$. Capiantur omnium AE, AG, AI &c. valores cum signo negativo, & summa quadratorum omnium ejusmodi valorum dicatur a, summa productorum e binis quibusque quadratis b, summa productorum e ternis c, & ita porro; productum, autem ex omnibus dicatur f. Numerus eorundem valorum dicatur m. His positis ponatur $z^m + az^{m-1} + bz^{m-2} + cz^{m-3}$ &c . . . $+ f = \mathrm{P}$. Si ponatur $\mathrm{P} = 0$, patet æquationis ejus omnes radices fore reales, & positivas, nimirum sola illa quadrata quantitatum AE, AG, AI &c, qui crunt valores ipsius z; adeoque cum ob $xx = z$, sit $x = \pm \sqrt{z}$, patet, valores x fore tam AE, AG, AI positivas, quam AE′, AG′, &c negativas.

Assumptio cujusdam valoris ad rem idonei.

26. Deinde sumatur quæcunque quantitas data per z, & constantes quomodocunque, dummodo non habeat ullum divisorem communem cum P, ne evanescente z, eadem evanescat, ac facta x infinitesima ordinis primi, evadat infinitesima ordinis ejusdem, vel inferioris, ut crit quæcunque formula $z^r + gz^{r-1} + hz^{r-2}$ &c $+ l$, quæ posita $= 0$ habeat radices quotcunque imaginarias, & quotcunque, & quascunque reales (dummodo earum nulla sit ex iis AE, AG, AI &c, sive positiva, sive negativa), si deinde tota multiplicetur per z. Ea dicatur Q.

Formula continens æquationem quæsitum.

27. Si jam fiat $\mathrm{P} - \mathrm{Q}y = 0$; dico, hanc æquationem satisfacere reliquis omnibus hujus curvæ conditionibus, & rite determinato valore Q, posse infinitis modis satisfieri etiam postremæ conditioni expositæ sexto loco.

Aequationem fore simplicem non resolubilem in plures.

[278] 28. Nam inprimis, quoniam valores P, & Q positi $= 0$, nullam habent radicem communem, nullum habebunt divisorem communem. Hinc hæc æquatio non potest per divisionem reduci ad binas, adeoque non est composita ex binis æquationibus, sed simplex, & proinde simplicem quandam curvam continuam exhibet, quæ ex aliis non componitur. Quod erat primum.

Exhibituram datum numerum intersectionum curvæ, in datis punctis.

29. Deinde curva hujusmodi secabit axem C′AC in iis omnibus, & solis punctis, E, G, I, &c, E′, G′, &c. Nam ea secabit axem C′AC solum in iis punctis, in quibus $y = 0$, & secabit in omnibus. Porro ubi fuerit $y = 0$, erit & $\mathrm{Q}y = 0$, adeoque ob $\mathrm{P} - \mathrm{Q}y = 0$; erit $\mathrm{P} = 0$. Id autem continget solum in iis punctis, in quibus z fuerit una e radicibus æquationis $\mathrm{P} = 0$, nimirum, ut supra vidimus, in punctis E, G, I, vel E′, G′, &c. Quare solum in his punctis evanescet y, & curva axem secabit. Secaturam autem in his omnibus patet ex eo, quod in his omnibus punctis erit $\mathrm{P} = 0$. Quare crit etiam $\mathrm{Q}\,y = 0$. Non crit autem $\mathrm{Q} = 0$; cum nulla sit radix communis æquationum $\mathrm{P} = 0$, & $\mathrm{Q} = 0$. Quare crit $y = 0$, & curva axem secabit. Quod erat secundum.

Singulas ordinatas responsuras singulis abscissis.

30. Præterea cum sit $\mathrm{P} - \mathrm{Q}x = 0$, erit $y = \frac{\mathrm{P}}{\mathrm{Q}}$; determinata autem utcunque abscissa x, habebitur determinata quædam z, adeoque & P, Q crunt unicæ, & determinatæ. Erit igitur etiam y unica, & determinata; ac proinde respondebunt singulis abscissis z singulæ tantum ordinatæ y. Quod erat tertium.

(d) *Hæc solutio excerpta est ex dissertatione* De Lege Virium in Natura existentium. *Accedit iis, quæ inde sunt eruta, scholium* 3 *primo adjectum in hac editione Veneta prima. Ipsum problema hic solvendum habetur in ipso hoc Opere parte* 1 *num.* 117, *ac ejus conditiones num.* 118.

§ III

Analytical Solution of the Problem to determine the nature of the Law of Forces (d)

25. To fulfil these conditions, we will find an algebraical formula, such as will represent our law; to do so, we shall take it that the first principles of the ordinary Cartesian algebra are known; for, without that, the thing can in no way be accomplished. Suppose that y is the ordinate, x the abscissa, & let $x^2 = z$. Take the values of AE, AG, AI, &c., all with a negative sign, & let a be the sum of the squares of all such values, b the sum of the products of all these squares two at a time, c the sum of the products three at a time, & so on; & let the product of them all together be called f; suppose that the number of these values is m. Then suppose P to stand for

Statement, & preparation.

$$z^m + az^{m-1} + bz^{m-2} + cz^{m-3} + \quad \ldots \quad + f.$$

If P is put equal to zero, it is plain that all the roots of this equation will be real & positive, namely, only the squares of the quantities AE, AG, AI, &c.; & these will be the values of z. Hence, since $x^2 = z$, & therefore $x = \pm \sqrt{z}$, it is evident that the values of x will be AE, AG, AI, positive, & AE′, AG′, &c., negative. [See Fig. 1.]

26. Next, assume some quantity that is given by z, & constants, in any manner, so long as it has not got any common measure with P, nor vanishes when z vanishes; also, if x is made an infinitesimal of the first order, let the quantity become an infinitesimal of the same order, or of a lower order. Such a formula will be any one such as

Assumption of some value suitable to the matter.

$$z^r + gz^{r-1} + hz^{r-2} + \ldots\ldots\ldots + l$$

(if this is put equal to zero, it will have a number of imaginary, & a number of real roots of ome kind; but none of them will be equal to AE, AG, AI, &c., whether positive or negative) ·f we multiply the whole by z. Call the product Q.

27. If now we put $P - Qy = 0$, I say that this equation will satisfy all the remaining conditions of the curve; & if Q is correctly determined, it can satisfy in an infinite number of ways the last condition also, given as sixthly.

Formula containing the equation required.

28. For, first of all, since the values, P & Q, when separately put equal to zero, have no common root, they cannot have a common divisor. Hence this equation cannot by division be reduced to two; & therefore it is not a composite equation formed from two equations, but is simple. Hence, it will represent some simple continuous curve, which is not made up of others. This was the first condition.

The equation will be simple, that is, it cannot be resolved into several others.

29. Next, this curve will cut the axis C′AC in all those points, & in them only, such as E, G, I, &c., E′, G′, &c. For it will cut the axis C′AC in those points only, for which $y = 0$, & it will cut it in all of them. Further, when $y = 0$, we have also $Qy = 0$; & therefore, since $P - Qy = 0$, we have $P = 0$. Now this happens only at those points for which z would be one of the roots of the equation $P = 0$; that is to say, as we saw above, at the points E, G, I, &c., E′, G′, &c. Hence it is only at these points that y will vanish, & the curve will cut the axis. It is clear that it will cut the axis at all these points, from the fact that at all these points we have $P = 0$. Hence also $Qy = 0$. But Q is not equal to zero, since there is no root common to the equations $P = 0$, $Q = 0$. Hence $y = 0$, & the curve will cut the axis. This was the second condition.

It will represent a given number of intersections of the curve at given points.

30. Further, since $P - Qy = 0$, it follows that $y = P/Q$; hence, for any determinate abscissa x, there will be a determinate z; & thus P & Q will be uniquely determinate. Therefore also y will be uniquely determinate; hence, to each abscissa x there will correspond one ordinate, y, & only one. This was the third condition.

To each abscissa there will correspond one ordinate & one only.

(d) *This solution is abstracted from my dissertation* De Lege Virium in Natura existentium. *In addition to these things that have been taken from that dissertation, there has been added a third scholium, which appears for the first time in this Venetian edition. The problem here set for solution will be found in Art.* 117 *of the first part of this work, & the conditions in Art.* 118.

Abscissis hinc inde æqualibus responsuras æquales ordinatas.

31. Rursus sive x assumatur positiva, sive negativa, dummodo ejusdem longitudinis sit, semper valor $z = xx$ erit idem ; ac proinde valores tam P, quam Q erunt semper iidem. Quare semper eadem y. Sumptis igitur abscissis z æqualibus hinc, & inde ab A, altera positiva, altera negativa, respondebunt ordinatæ æquales. Quod erat quartum.

Primum arcum fore crus asymptoticum cum area infinita.

32. Si autem x minuatur in infinitum, sive ea positiva sit, sive negativa ; semper z minuetur in infinitum, & evadet infinitesima ordinis secundi. Quare in valore P decrescent in infinitum omnes termini præter f, quia omnes præter eum multiplicantur per z, adeoque valor P erit adhuc finitus. Valor autem Q, qui habet formulam ductam in z totam, minuetur in infinitum, eritque infinitesimus ordinis secundi. Igitur $\frac{P}{Q} = y$ augebitur in infinitum ita, ut evadat infinita ordinis secundi. Quare curva habebit pro asymptoto rectam AB, & area BAED excrescet in infinitum, & si ordinatæ y assumantur ad partes AB, & exprimant vires repulsivas, arcus asymptoticus ED jacebit ad partes ipsas AB. Quod erat quintum.

Post eas conditiones remanere indeterminationem parem cuicunque accessui ad quasvis curvas in punctis datis quibusvis.

[279] 33. Patet igitur, utcunque assumpto Q cum datis conditionibus, satisfieri primis quinque conditionibus curvæ. Jam vero potest valor Q variari infinitis modis ita, ut adhuc impleat semper conditiones, cum quibus assumptus est. Ac proinde arcus curvæ intercepti intersectionibus poterunt infinitis modis variari ita, ut primæ quinque ipsius curvæ conditiones impleantur ; unde fit, ut possint etiam variari ita, ut sextam conditionem impleant.

Quid requiratur ut transeat per quævis earum puncta.

34. Si enim dentur quotcunque, & quicunque arcus, quarumcunque curvarum, modo sint ejusmodi, ut ab asymptoto AB perpetuo recedant, adeoque nulla recta ipsi asymptoto parallela eos arcus secet in pluribus, quam in unico puncto, & in iis assumantur puncta quotcunque, utcunque inter se proxima ; poterit admodum facile assumi valor P ita, ut curva per omnia ejusmodi puncta transeat, & idem poterit infinitis modis variari ita, ut adhuc semper curva transeat per eadem illa puncta.

Quomodo id præstandum.

35. Sit enim numerus punctorum assumptorum quicunque $= r$, & a singulis ejusmodi punctis demittantur rectæ parallelæ AB usque ad axem C′AC, quæ debent esse ordinatæ curvæ quæsitæ, & singulæ abscissæ ab A usque ad ejusmodi ordinatas dicantur M1. M2, M3, &c, singulæ autem ordinatæ N′1, N′2, N′3, &c. Assumatur autem quædam quantitas $Az^r + Bz^{r-1} + Cz^{r-2} + \ldots + Gz$, quæ ponatur $=$ R. Tum alia assumatur quantitas T ejusmodi, ut evanescente z evanescat quivis ejus terminus, & ut nullus sit divisor communis valoris P, & valoris R + T, quod facile fiet, cum innotescant omnes divisores quantitatis P. Ponatur autem Q = R + T, & jam æquatio ad curvam crit $P - Ry - Ty = 0$. Ponantur in hac æquatione successive M1, M2, M3, &c, pro x, & N′1, N′2, N′3, &c pro y. Habebuntur æquationes numero r, quæ singulæ continebunt valores A, B, C, . . . G, unius tantum dimensionis singulos, numero pariter r, & præterea datos valores M1, M2, M3, &c, N1, N2, N3, &c, ac valores arbitrarios, qui in T sunt coefficientes ipsius z.

Progressus ulterior.

36. Per illas æquationes numero r admodum facile determinabuntur illi valores A, B, C, . . . G, qui sunt pariter numero r, assumendo in prima æquatione, juxta methodos notissimas, & elementares valorem A, & eum substituendo in æquationibus omnibus sequentibus, quo pacto habebuntur æquationes $r - 1$. Hæ autem ejecto valore B reducentur ad $r - 2$, & ita porro, donec ad unicam ventum fuerit, in qua determinato valore G, per ipsum ordine retrogrado determinabuntur valores omnes præcedentes, singuli in singulis æquationibus.

Conclusio, & cohærentia cum omnibus præcedentibus conditionibus.

37. Determinatis hoc pacto valoribus A, B, C, . . . G [280] in æquatione $P - Ry - Ty = 0$, sive $P - Qy = 0$, patet positis successive pro x valoribus M1, M2, M3, &c, debere valores ordinatæ y esse successive N1, N2, N3, &c ; ac proinde debere curvam transire per data illa puncta in datis illis curvis : & tamen valor Q adhuc habebit omnes conditiones præcedentes. Nam imminuta z ultra quoscunque limites, minuentur singuli ejus termini ultra quoscunque limites, cum minuantur termini singuli valoris T, qui ita assumpti sunt, & minuantur pariter termini valoris R, qui omnes sunt ducti in z, & præterea nullus crit communis divisor quantitatum P, & Q, cum nullus sit quantitatum P, & R + T.

Inde contactus, oscula, accessus quivis.

38. Porro si bina proxima ex punctis assumptis in arcubus curvarum ad candem axis partem concipiantur accedere ad se invicem ultra quoscunque limites, & tandem congruere, factis nimirum binis M æqualibus, & pariter æqualibus binis N ; jam curva quæsita ibidem

31. Again, whether x is taken positive or negative, so long as its length is the same, the value of z, or x^2, will be the same. Hence the values of both P & Q will be the same. Hence y will always be the same for either. Hence, if equal abscissæ x are taken one on either side of A, the one positive & the other negative, the corresponding ordinates will be equal. This was the fourth condition.

To equal abscissæ, therefore, there will correspond equal ordinates, on either side of the origin.

32. Now, if x is diminished indefinitely, whether it is positive or negative, z will be also diminished indefinitely, & will become an infinitesimal of the second order. Hence, every term in the value of P, except f, will diminish indefinitely; for each of them except this one has a factor z. Thus the value of P will remain finite. But the value of Q, in which the whole expression was multiplied by z, will diminish indefinitely; & it will become an infinitesimal of the second order. Hence y, which is equal to P/Q, will be increased indefinitely, so that it becomes an infinity of the second order. Therefore, the curve will have the straight line AB as an asymptote, & the area BAED will become infinite; also, if AB is taken to be the positive direction for the ordinates y, these will represent repulsive forces, & the asymptotic arc ED will fall in the direction given by AB. This was the fifth condition.

The first arc will be an asymptotic branch with an infinite area.

33. Hence, it is clear that, however Q is chosen subject to the given conditions, the first five conditions for our curve will be satisfied. Now, the value of Q can be varied in an infinite number of ways, such that it will still fulfil the conditions under which it was assumed. Then the arcs of the curves intercepted between the intersections with the axis could be varied in an infinite number of ways, such that the first five conditions for the curve are satisfied. Hence it follows that they can be varied also, in such a way that the sixth condition is satisfied.

After these conditions have been fulfilled, there still remains an equal indetermination for approach to any given curves at any given points.

34. Now, if any number of arcs of any kind, belonging to any curves, are given; so long as these are such that they continually recede from the asymptote AB, & therefore such that no straight line parallel to this asymptote will cut any of them in more than one point; & if in these arcs there are taken any number of points, no matter how close they are together, a value of P can be obtained quite easily, such that the curve will pass through all these points. Moreover, this can be done in an infinite number of ways, such that the curve will still pass through all these points in every case.

The conditions that it should pass through given points of these curves.

35. For, let the number of points taken be any number r. From each of these points, let a straight line be drawn parallel to AB, to meet the axis C′AC; these must be ordinates of the curve required. Let the several abscissæ measured from A to these ordinates be M_1, M_2, M_3, &c.; & let the corresponding ordinates be N_1, N_2, N_3, &c. Then assume some quantity $Az^r + B\,z^{r-1} + Cz^{r-2} + \ldots + Gz$, & suppose that this is R. Next, take another quantity, T, of such a kind that, when z vanishes, each term of T vanishes, & there is no common divisor of P & R + T. This can easily be done, since the divisors of the quantity P are known. Now, suppose that Q = R + T; the equation to the curve, will then be $P - Ry - Ty = O$. In this equation, substitute in succession M_1, M_2, M_3 &c. for x, & N_1, N_2, N_3 &c. for y. Then we shall have r equations, each of which will contain the values A, B, C, ... , G, which are also r in number; & these will all appear linearly. The equations will also contain, in addition, the given values M_1, M_2, M_3, &c., N_1, N_2, N_3, &c., & the arbitrary values which appear as the coefficients of z in the expression T.

How this can be managed.

36. From these equations, r in number, the values of A, B, C, . . . , G, which are also r in number, can quite easily be determined. Thus, from the first equation, according to well-known elementary methods, obtain the value of A in terms of the rest, & substitute this value in each of the other equations. In this way we shall obtain $r - 1$ equations. Eliminating B from these, we shall get $r - 2$ equations; & so on, until at last we shall come to a single equation. Having determined from this the value of G, we can determine, by retracing our steps, the preceding values in succession, one value from each set of equations.

Further progress.

37. The values of A, B, C, ... , G, in the equation $P - Ry - Ty = 0$, or $P - Qy = 0$, having been thus found, it is clear that, if the values M_1, M_2, M_3, &c., are substituted for x in succession, the values of y will be N_1, N_2, N_3, &c. Hence, the curve must pass through the given points on the given arcs; & still the value of Q will satisfy all the preceding conditions. For, if z is diminished beyond all limits, each of its terms will be diminished beyond all limits; since each of the terms of the value of T, according to the supposition made, will be so diminished, & likewise each of the terms of R, which all contain a factor z. In addition, there will be no common divisor of P & Q, since there is none for the quantities P & R + T.

Conclusion; agreement with all the preceding conditions.

38. Again, if two of the chosen points, next to one another in the arcs of the curves, are supposed to approach one another on the same side of the axis beyond all limits, & finally to coincide with one another, namely, by making two values of M equal to one another, & therefore also the corresponding values of N, then also the required curve will

Hence contacts, osculations, & approach of any kind.

tanget arcum curvæ datæ : & si tria ejusmodi puncta congruant, eam osculabitur : quin immo illud præstari poterit, ut coeant quot libuerit puncta, ubi libuerit, & habeantur oscula ordinis cujus libuerit, & ut libuerit sibi invicem proxima, areu curvæ datæ accedente, ut libuerit, & in quibus libuerit distantiis ad arcus, quos libuerit curvarum, quarum libuerit, & tamen ipsa curva servante omnes illas sex conditiones requisitas ad exponendam legem illam virium repulsivarum, ac attractivarum, & datos limites.

Adhuc indeterminatio relicta pro infinitis modis.

39. Cum vero adhuc infinitis modis variari possit valor T; infinitis modis idem præstari poterit : ac proinde infinitis modis inveniri poterit curva simplex datis conditionibus satisfaciens. Q.E.F.

Posse & axem contingere, osculari, &c.

40. *Coroll.* 1. Curva poterit contingere axem C′AC in quot libuerit punctis, & contingere simul, ac secare in iisdem, ac proinde eum osculari quocunque osculi genere. Nam si binæ quævis e distantiis limitum fiant æquales; curva continget rectam C′A, evanescente arcu inter binos limites; ut si punctum I abiret in L, evanescente arcu IKL; haberetur contactus in L, repulsio per arcum HI perpetuo decresceret, & in ipso contactu IL evanesceret, tum non transiret in attractionem, sed iterum cresceret repulsio ipsa per arcum LM. Idem autem accideret attractioni, si coeuntibus punctis LN, evanesceret arcus repulsivus LMN.

Posse contingere simul, & secare.

41. Si autem tria puncta coirent, ut LNP; curva contingeret simul axem C′AC, & ab eodem simul secaretur, ac proinde haberet in eodem puncto contactus flexum contrarium. Haberetur autem ibidem transitus ab attractione ad repulsionem, vel vice versa, adeoque verus limes.

Quid congruentia intersectionum plurium.

42. Eodem pacto possunt congruere puncta quatuor, quinque, quotcunque : & si congruat numerus punctorum par; habebitur contactus : si impar; contactus simul, & sectio. Sed quo plura puncta coibunt; eo magis curva accedet ad [281] axem C′AC in ipso limite, eumque osculabitur osculo arctiore.

Posse axem secari in quibuscunque angulis, & a quavis magnitudine arcuum.

43. *Coroll.* 2. In iis limitibus, in quibus curva secat axem C′AC, potest ipsa curva secare eundem in quibuscunque angulis ita tamen, ut angulus, quem efficit ad partes A arcus curvæ in perpetuo recessu ab asymptoto appellens ad axem C′AC non sit major recto, & ibidem potest aut axem, aut rectam axi perpendicularem contingere, aut osculari, quocunque contactus, aut osculi genere, nimirum habendo in utrolibet casu radium osculi magnitudinis cujuscunque, & vel utcunque evanescentem, vel utcunque abeuntem in infinitum.

Demonstratio: limitatio necessaria.

44. Nam pro illis punctis datis in arcubus curvarum quarumcunque, quas curva inventa potest vel contingere, vel osculari quocunque osculi genere, ex quibus definitus est valor R, possunt assumi arcus curvarum quarumcunque secantium axem C′AC, in angulis quibuscunque : solum quoniam semper arcus curvæ, ut *t*N*y* debet ab asymptoto recedere, non poterit punctum ullum *t* præcedens limitem N jacere ultra rectam axi perpendicularem erectam ex N, vel punctum *y* sequens ipsum N jacere citra; ac proinde non poterit angulus AN*t*, quem efficit ad partes A arcus *t*N in perpetuo recessu ab asymptoto appellens ad axem C′AC, esse major recto.

Quid possint arcus curvarum assumptarum: omnia posse & inventam.

45. Possunt autem arcus curvarum assumptarum in iisdem punctis aut axem, aut rectam axi perpendicularem contingere, aut osculari, quocunque contactus, aut osculi genere, ut nimirum sit radius osculi magnitudinis cujuscunque, & vel utcunque evanescens, vel utcunque abiens in infinitum. Quare idem accidere poterit ut innuimus, & arcui curvæ inventæ, quæ ad eos arcus potest accedere, quantum libuerit, & eos contingere, vel osculari quocunque osculi genere in iis ipsis punctis.

Conditio necessaria, ex hujus curvæ natura.

46. Solum si curva inventa tetigerit in ipso limite rectam axi C′AC perpendicularem, debebit simul ibidem candem secare; cum debeat semper recedere ab asymptoto, adeoque debebit ibidem habere flexum contrarium.

Corol. 1 includi in corol. 2.

47. *Scholium* 1. Corollarium 1 est casus particularis hujus corollarii secundi, ut patet : sed libuit ipsum seorsum diversa methodo, & faciliore prius eruere.

Quid ubivis etiam extra limites.

48. *Coroll.* 3. Arcus curvæ etiam extra limites potest habere tangentem in quovis angulo inclinatam ad axem, vel ei parallelam, vel perpendicularem cum iisdem contactuum, & osculorum conditionibus, quæ habentur in corollario 2.

Demonstratio eadem.

49. Demonstratio est prorsus eadem : nam arcus curvarum dati, ad quos arcus curvæ inventæ potest accedere ubicunque, quantum libuerit, possunt habere ejusmodi conditiones.

touch the arc of the given curve at this point. If three such points coincide with one another, it will osculate the given curve. Indeed, it can be brought about that any number of points desired shall coincide, & thus osculations of any order desired can be obtained. These may be as close together as desired, the arc approaching the given curve to any desired degree of closeness; or they may be at any distances from any of the arcs of any of the curves, as desired. Yet the curve will observe all those six conditions, which are required for representing the law of repulsive & attractive forces, as well as the limit-points.

39. Now, since the value of T can still be varied in an infinite number of ways, this can be brought about in an infinite number of ways. Hence, in an infinite number of ways, a simple curve can be found satisfying the given conditions. Q . E . F .

There is still left indetermination in countless ways.

40. *Cor.* 1. The curve may touch the axis C′AC in any desired number of points; or at the same time touch & cut it at the same points; & hence it may osculate the axis with any kind of osculation. For, if any two of the distances for the limit-points become equal, the curve will touch the straight line C′A, the arc between these two limit-points vanishing. Thus, if the point I should go off to L, the arc IKL vanishing, we should have contact at L, & repulsion would continually decrease along the arc HI, vanish at the point of contact IL; after that it would not become an attraction, but the repulsion would continually increase along the arc LM. The same thing would also happen in the case of attraction, if, owing to the points L,N coinciding, the repulsive arc LMN should vanish.

It is possible also for the curve to touch the axis, or to osculate it, etc.

41. Again, if three points, say L,N,P, should coincide, the curve would at the same time touch the axis C′AC & intersect it; thus, at that point of contact there would be contrary flexure. Also, there would be there a passage from attraction to repulsion, or vice versa, & therefore a true limit-point.

It is possible that there may be simultaneous contact & section of the axis.

42. In the same way, four points may coincide, or five, or any number. If the number of points that coincide is even, there will be touching contact; if the number is odd, there will be contact & intersection at the same time. The greater the number of the points that coincide, the more the curve will approach to coincidence with the axis C′AC at that limit-point; & thus the higher the order of the osculation.

The result of the coincidence of several intersections.

43. *Cor.* 2. At these limit-points, where the curve cuts the axis C′AC, the curve may cut it at any angle; but in such a way that the angle, which the arc of the curve, in its continuous recession from the asymptote, makes with the direction of A as it comes up to the axis C′AC, is not greater than a right angle; & it may touch either the axis or the straight line at right angles to the axis, or osculate the axis; the contact or the osculation being of any order. That is to say, it may have in either case a radius of osculation of any magnitude whatever, either vanishing or becoming infinite, in any way whatever.

The axis may be cut by the curve at any angle, & by arcs of any size.

44. For, we may take as our chosen points in the arcs of any curves, which the curve of forces is found to touch or to osculate with an osculation of any order, from which the value of R is determined, arcs of any curves cutting the axis C′AC at any angles. Except that, since the arc of the curve, such as *t*N*y*, must always recede from the asymptote, it would not be possible for any point such as *t*, which precedes the limit-point N, to lie on the far side of the straight line perpendicular to the axis erected at N; or for the point *y*, which follows N, to lie on the near side of this perpendicular. Thus, the angle AN*t*, which it makes with the direction of A, as the arc *t*N continually recedes from the asymptote, as it comes up to the axis C′AC, cannot be greater than a right angle.

Demonstration; necessary limitation.

45. Again, the arcs of the assumed curves may, at these points either touch the axis or the straight line perpendicular to the axis, or they may osculate, the contact or the osculation being of any order; that is, the radius of osculation may be of any magnitude whatever, either vanishing or becoming infinite, in any way. Hence, as I said, this may also be the case for an arc of the curve that has been found; for it can be made to approximate as closely as desired to these curves, so as to touch them or osculate them, with any order of osculation, at these points.

What the arcs of the assumed curves may be; the same properties may all be possessed by the curve found.

46. Except that, if the curve should touch at the limit-point the straight line perpendicular to the axis C′AC, it must at the same time cut it at that point; for the curve must always recede from the asymptote, & thus is bound to have contrary flexure at the point.

Necessary condition, arising from the nature of the curve.

47. *Scholium* 1. The first corollary is a particular case of the second, as is evident. But I preferred to take it first, with an independent proof by a different & an easier method.

The first corollary is included in the second.

48. *Cor.* 3. Even beyond the limit-points, the arc of the curve can have a tangent inclined at any angle to the axis, or parallel to it, or perpendicular to it; with the same conditions as to contact or osculation as we had in the second corollary.

What happens also at any point beyond the limit-points.

49. The proof is exactly the same as before; for, the given arcs of the curves, to which the arc of the curve that is found can be made to approximate as closely as desired, may have the conditions stated.

Proof the same as before.

Mutationem abscissæ posse habere ad mutationem ordinatæ relationem quancunque.

50. *Coroll.* 5. Mutata abscissa per quodcunque intervallum datum, potest ordinata mutari per aliud quodcunque datum utcunque minus, vel majus ipsa mutatione abscissæ, & ut-[282]-cunque majus quantitate quacunque data; ac si differentia abscissæ sit infinitesima, & dicatur ordinis primi; poterit differentia ordinatæ esse ordinis cujuscunque, vel utcunque inferioris, vel intermedii, inter quantitates finitas, & quantitates ordinis primi.

Demonstratur pro ratione finita.

51. Patet primum ex eo, quod, ubi determinatur valor R, potest curva transire per quotcunque, & quæcunque puncta, adeoque per puncta, ex quibus ductæ ordinatæ sint utcunque inter se proximæ, & utcunque inæquales.

Itidem pro quovis infinitesimorum ordine.

52. Patet secundum: quia in curvis, ad quas accedit arcus curvæ inventæ vel quas osculatur quocunque osculi genere, potest differentia abscissæ ad differentiam ordinatæ esse pro diversa curvarum natura in datis earum punctis in quavis ratione, quantitatis infinitesimæ ordinis cujuscunque ad infinitesimam cujuscunque alterius.

Relationem ejusmodi pendere a positione tangentis.

53. *Scholium* 2. Illud notandum, ubicunque fuerit tangens curvæ inventæ inclinata in angulo finito ad axem, fore differentiam abscissæ ejusdem ordinis, ac est differentia ordinatæ: ubi tangens fuerit parallela axi, fore differentiam ordinatæ ordinis inferioris, quam sit differentia abscissæ, & vice versa, ubi tangens fuerit perpendicularis axi.

Quid, ubi abscissa, terminetur in limite.

54. Præterea notandum: si abscissa fuerit ipsa distantia limitis, quæ vel augeatur, vel minuatur utcunque; differentia ordinatæ erit ipsa ordinata integra; cum nimirum in limite ordinata sit nihilo æqualis.

Posse arcus utcunque recedere ab axe.

55. *Coroll.* 5. Arcus repulsionum, vel attractionum intercepti binis limitibus quibuscunque, possunt recedere ab axe, quantum libuerit, adeoque fieri potest, ut alii propiores asymptoto recedant minus, quam alii remotiores, vel ut quodam ordine eo minus recedant ab axe, quo sunt remotiores ab asymptoto, vel ut post aliquot arcus minus recedentes aliquis arcus longissime recedat.

Demonstratio.

56. Omnia manifesto consequuntur ex eo, quod curva possit transire per quævis data puncta.

Posse haberi postremum crus asymptoticum, & alia crura asymptotica.

57. *Coroll.* 6. Potest curva ipsum axem C′AC habere pro asymptoto ad partes C′, & C ita, ut arcus asymptoticus sit vel repulsivus, vel attractivus; & potest arcus quivis binis limitibus quibuscunque interceptus abire in infinitum, ac habere pro asymptoto rectam axi perpendicularem, utcunque proximam utrilibet limiti, vel ab eo remotam.

Ratio præstandi primum.

58. Nam si concipiatur, binos postremos limites coire, abeuntibus binis intersectionibus in contactum, tum concipiatur, ipsam distantiam contactus excrescere in infinitum; jam axis æquivalet rectæ curvam tangenti in puncto infinite remoto, adeoque evadit asymptotus: & si arcus evanescens inter postremos duos limites coeuntes fuerit arcus repulsionis; postremus arcus asymptoticus erit arcus attractionis. Contra vero, si arcus evanescens fuerit arcus attractionis.

Ratio præstandi & reliquum.

[285] 59. Eodem pacto si concipiatur, quamvis ordinatam respondentem puncto cuilibet, per quod debet transire curva, abire in infinitum; jam arcus curvæ abibit in infinitum, & erit ejus asymptotus in illa ipsa ordinata in infinitum excrescens.

Legem virium hic exhiberi per functionem distantiæ, alios multos censere præferendam unicam potentiam: cur id.

60. *Scholium* 3. Ope formulæ exhibentis curvam propositam habetur lex virium expressa per functionem quandam distantiæ constantem plurimis terminis, immo per æquationem commiscentem abscissam, & ordinatam, ac utriusque potentias inter se, & cum rectis datis, non per solam ipsius distantiæ potentiam. Sunt, qui censeant expressionem per solam potentiam debere præferri expressioni per functionem aliam, quia haec sit simplicior, quam illa, & quia in illa præter distantias debeant haberi aliquæ aliæ parametri, quæ non sint solæ distantiæ; dum in formula $\frac{1}{x^m}$ exprimente x distantias, distantiæ solæ rem conficiant, videatur autem vis debere pendere a solis distantiis, potissimum si sit quædam essentialis proprietas materiæ: præterea addunt, nullam fore rationem sufficientem, cur una potius, quam alia parameter expressionem virium deberet ingredi, si parametri sint admiscendæ.

Qua occasione hæc quæstio fuerit agitata in Parisiensi Academia.

61. Hæc agitata sunt potissimum ante hos aliquot annos in Academia Parisiensi, cum censeretur, motum Apogei Lunaris observatum non cohærere cum gravitate decrescente in ratione reciproca duplicata distantiarum, & ad ipsum exhibendum adhiberetur gravitas

50. *Cor.* 4. If the abscissa is changed by any given interval, the ordinate can be changed by any other given interval however much the latter may be smaller or greater than the change of the abscissa, or however much greater than any given quantity it may be. Further, if the difference in the abscissa is infinitesimal, & we call it an infinitesimal of the first order, then the difference in the ordinate may be of any order, either of any order below the first whatever, or intermediate between finite quantities & quantities of this first order.

The change of the abscissa may bear any ratio to the change of the ordinate.

51. The first part is evident from the fact that, when the value of R is determined, a curve can be made to pass through any number of points of any sort; & thus, through points, from which ordinates are drawn as close to one another as we please, & unequal to one another in any way.

Proof for finite ratios.

52. The second part is evident, because in the curves, to which the arcs of the curve found approximates, or which it osculates with any order of osculation, the difference of the abscissa can bear any ratio to the difference of the ordinate for a different nature of the curves at given points on them; this ratio may be that of an infinitesimal quantity of any order to an infinitesimal quantity of any other order.

The same for any order of infinitesimals.

53. *Scholium* 2. It is to be observed that, whenever the tangent to the curve that has been found is inclined at a finite angle to the axis, the difference of the abscissa is of the same order as the difference of the ordinate; when the tangent is parallel to the axis, the difference of the ordinate will be of an inferior order to the difference of the abscissa; & the opposite is the case when the tangent is perpendicular to the axis.

This relation depends on the position of the tangent.

54. In addition, it is to be observed that, if the abscissa corresponds to a limit-point, & this is either increased or diminished in any way, the difference of the ordinate will be the whole ordinate itself, for at the limit-point itself the ordinate is indeed equal to zero.

The case of an abscissa terminating at one of the limit-points.

55. *Cor.* 5. The arcs of repulsion or attraction, which are intercepted between any pair of limit-points, may recede from the axis to any extent; & thus, it may happen that some that are nearer to the asymptote may recede less than others that are more remote; or that, to any order, they may recede the less, the further they are from the asymptote; or that, after a number of arcs that recede less, there may be one which recedes by a very large amount.

The arcs may recede from the axis to any extent.

56. Everything clearly follows from the fact that the curve can be made to pass through any given points.

Proof of this statement.

57. *Cor.* 6. The curve may have the axis C′AC as an asymptote in the directions of C′ & C, in such a manner that the asymptotic arc is either repulsive or attractive; also any arc intercepted between a pair of limit-points may go off to infinity, & have for an asymptote a straight line perpendicular to the axis, however near or far from either limit-point.

The curve can have its last branch asymptotic, & also other asymptotic branches.

58. For, if we suppose that the last two limit-points coincide, as the two intersections coincide & become a point where the curve touches the axis; & then suppose that the distance of this point of contact becomes infinite; then the axis will become equivalent to a straight line touching the curve at a point infinitely remote, & will thus be an asymptote. If the vanishing arc that is intercepted between those two last coincident limit-points should be an arc of repulsion, the last asymptotic arc will be an arc of attraction. But the opposite would be the case if the vanishing arc should be an arc of attraction.

The proof of the first part of this statement.

59. In the same way, if it is supposed that any ordinate corresponding to any point, through which the curve has to pass, should go off to infinity; then the arc of the curve will also go off to infinity, and that ordinate, as it increases indefinitely, will become an asymptote of the curve.

The proof of the remainder of the statement.

60. *Scholium* 3. By the help of the formula corresponding to the proposed curve, the law of forces is obtained expressed as a definite function of the distance with many terms; or rather, by means of an equation involving the abscissa & the ordinate, & powers of these, along with given straight lines, & not by a single power of the distance. There are some who think that representation by means of a single power is to be preferred to representation by another function; because the latter is simpler than the former; & because in it, besides the distances, there are bound to be other parameters that are not merely distances. Whereas, in the formula $1/x^m$, where x represents the distances, the distances alone settle the matter; & it is seen that the force must depend on the distance alone, especially if it should be an essential property of matter. Besides, they add, there is no sufficient reason why any one, rather than any other, parameter should enter the expression for the forces, if parameters are to be admitted.

The law of forces is here represented by a function of the distance; many others think that a single power of the distance is preferable; their reasons.

61. This question came in for a large amount of discussion a number of years ago in the Academy of Paris. For, it was thought that the motion of the lunar apogee, as observed, did not agree with the idea of gravity decreasing in the inverse duplicate ratio of the distances. They considered that an expression for gravity should be employed, in which it was represented

The occasion on which this question was discussed in the Academy of Paris.

expressa per binomium $\frac{a}{x^3} + \frac{b}{x^2}$, cujus pars prior in magnis, pars posterior in exiguis distantiis respectu sociæ partis evanesceret ad sensum, sed illa prior in distantia Lunæ a Terra adhuc turbaret hanc posteriorem, quantum satis erat ad eam præstandam rem. Atque eam ipsam binomii expressionem adhibuerant jam plures Physici ad deducendam simul ex eadem formula gravitatem, & majores minimarum particularum attractiones, ac multo validiorem cohæsionem, ut innuimus num. 121 : atque hæ difficultates in Parisiensi Encyclopædia inculcantur ad vocem *Attractio*, Tomo I tum edito.

Occasionem substituendi tum functionem cessasse, sed rationes contra allatas nullam habere vim : curvas omnes uniformis naturæ esse in se æque simplices.

62. Paullo post, correctis calculis innotuit, motum Apogei lunaris ea composita formula non indigere : at rationes contra id propositæ, quæ multo magis contra meam virium legem pugnarent, meo quidem judicio nullam habent vim. Nam in primis quod ad simplicitatem pertinet, hic habent locum ea omnia, quæ dicta sunt in ipso opere num. 116 de simplicitate curvarum. Formula exprimens solam potentiam quandam distantiæ designatæ per abscissam exprimit ordinatam ad locum geometricum pertinentem ad familiam, quam exhibet [284] $y = x^m$, qui quidem locus est Parabola quædam ; si m sit numerus positivus, nec sit unitas : recta ; si sit unitas, vel zero : quædam Hyperbola ; si sit numerus negativus : formula autem continens functionem aliam quamvis exprimit ordinatam ad aliam curvam, quæ crit continua, & simplex, si illa formula per divisionem non possit discerpi in alias plures. Omnes autem ejusmodi curvæ sunt æque simplices in se, & aliæ aliis sunt magis affines, aliæ minus. Nobis hominibus recta est omnium simplicissima, cum ejus naturam intueamur, & evidentissime perspiciamus, ad quam idcirco reducimus alias curvas, & prout sunt ipsi magis, vel minus affines, habemus eas pro simplicioribus, vel magis compositis ; cum tamen in se æque simplices sint omnes illæ, quæ ductum uniformem habent, & naturam ubique constantem.

Esse æque simplicem relationem ordinatarum ad abscissas : terminorum multitudinem pro ea exprimenda oriri a nostro cognoscendi modo.

63. Hine ipsa ordinata ad quamvis naturæ uniformis curvam est quidam terminus simplicissimæ relationis cujusdam, quam habet ordinata ad abscissam, cui termino impositum est generale nomen functionis continens sub se omnia functionum genera, ut etiam quamcunque solam potentiam, & si haberemus nomina ad ejusmodi functiones denominandas singillatim ; haberet nomen suum quævis ex ipsis, ut habet quadratum, cubus, potestas quævis. Si omnia curvarum genera, omnes ejusmodi relationes nostra mens intueretur immediate in se ipsis ; nulla indigeremus terminorum farragine, nec multitudine signorum ad cognoscendam, & enuntiandam ejusmodi functionem, vel ejus relationem ad abscissam.

Origo ejus modi ab intuitione, quam habemus nos homines naturæ solius rectæ, ad quas omnes curvas referimus.

64. Verum nos, quibus uti monni recta linea est omnium locorum geometricorum simplicissima, omnia referimus ad reetam, & idcirco etiam ad ea, quæ oriuntur ex recta, ut est quadratum, quod fit ducendo perpendiculariter reetam super aliam reetam æqualem, & cubus, qui fit ducendo quadratum eodem pacto per aliam rectam primæ radici æqualem, quibus & sua signa dedimus ope exponentium, & universalizando exponentes efformavimus nobis ideas jam non geometricas superiorum potentiarum, nec integrarum tantummodo, & positivarum, sed etiam fractionariarum, & negativarum : & vero etiam, abstrahendo semper magis, irrationalium. Ad basce potentias, & ad producta, quæ simili ductu concipiuntur genita, reducimus cæteras functiones omnes per relationem, quam habent ad ejusmodi potentias, & producta carum cum rectis datis, ac ad eam reductionem, sive ad expressionem illarum functionum per hasce potentias, & per hæc producta, indigemus terminis jam paucioribus, jam pluribus, & quandoque etiam, ut in functionibus transcendentalibus, serie terminorum infinita, quæ ad valorem, vel naturam functionis propositæ accedat semper magis, utut in hisce casibus eam nunquam ac-[285]-curate attingat : habemus autem pro magis, vel minus compositis eas, quæ pluribus, vel paucioribus terminis indigent, sive quæ ad solas potentias relationem habent propiorem.

Aliud mentium genus ad exprimendam relationem potentiæ necessario adhibiturum æqualem, vel majorem farraginem.

65. At si aliud mentium genus aliam curvam ita intime cognosceret, ut nos rectam ; haberet pro maxime simplici solam ejus functionem, & ad exprimendum quadratum, vel aliam potentiam, contemplaretur illam candem relationem, sed inverse assumptam ita, ut incipiendo a functione ipsa per eam, & per similes ejus functiones, ac functionum citeriorum functiones ulteriores, addendo, ac subtrahendo deveniret demum ad quæsitam. Relatio potentiæ ad functionem, & nexus mutuus compositionem habet, & multitudinem terminorum inducit : uterque relationis terminus est in se æque simplex.

Sola etiam potentiæ expressione includi etiam apud nos homines parametros plures : parameter in unitate arbitraria, & affixione certæ vis ad certam distantiam.

66. Quod pertinet ad parametros, quas dicitur includere functio, non autem potentia distantiæ, non est verum id ipsum, quod potentia parametros non includat. Formula $\frac{1}{x^m}$ includit unitatem ipsam, quæ non est aliquid in se determinatum, sed potest exprimere magnitudinem quamcunque. Et quidem ea species includit omnes species Hyperbolarum,

by the formula of two terms, $a/x^3 + b/x^2$; of this, the first part at large distances, & the last part at very small distances, would practically become evanescent with respect to the other part associated with it. But the first part, for the distance of the Moon from the Earth, would still disturb the last part sufficiently to account for the observed inequality. Already, several Physicists had employed such an expression with two terms to deduce at the same time from the one formula both gravity & the greater attractions of very small particles, & much more so the still stronger forces of cohesion, as I have mentioned in Art. 121. These difficulties are included in the *Encyclopædia Parisiensis* under the heading *Attraction*, in Vol. I published at that time.

The reasons for substituting the formula for the function, which then existed, have ceased to exist; but the arguments brought forward against it have no weight; all curves that are uniform are in themselves equally simple.

62. Shortly afterwards, the calculations were corrected & it was found that the motion of the lunar apogee did not necessitate this compound formula. But the arguments brought forward against it, which were still more in opposition to this Theory of mine with regard to the law of forces, have no weight, at any rate in my eyes. For, in the first place, as regards simplicity, all those things held good in this case, which I stated in this work, Art. 116, with regard to simplicity of curves. A formula in terms of a single power of the distance represented by an abscissa expresses the ordinate of a geometrical locus belonging to the family, represented by $y = x^m$; & this locus is a Parabola, if m is any positive number except unity; a straight line, if m is unity or zero; & a hyperbola, if m is a negative number. But a formula containing some other function expresses the ordinate of some other curve; & this will be continuous & simple, if the formula cannot be separated by division into several others. Further, all such curves are equally simple in themselves; & some of them are more, some less, of the same nature as others. To us men, a straight line is the simplest of all; for we observe its nature & understand it clearest of all. To it therefore we refer all other curves; & according as they are more or less like it in nature, we consider them to be the more or less simple. However, in themselves, all curves, which are composed of a continuous line & have a constant nature everywhere, are equally simple.

The relation between the ordinate & the abscissa is equally simple; the number of terms used to express this relation arises from our way of knowing it.

63. Hence, the ordinate to any curve of a uniform nature is some term of some very simple relation that the ordinate has to the abscissa. To this term there is given the general name, function; this name includes every kind of function, for instance, even a single power. If we had names to denote such functions singly, each of them would have its own name, just as a square, a cube, or any other power. If our minds were capable of viewing all kinds of curves, & all such relations in themselves, at a glance, then there would be no need of a medley of terms, & a multitude of signs in order to know & state such a function or its relation to the abscissa.

The origin of this method comes from the intuition which we men have of the nature of a straight line alone, to which we refer all curves.

64. But we, to whom, as I mentioned, the straight line is the simplest of all geometrical loci, refer all curves to a straight line, and therefore also to all those things that arise from a straight line; such as a square, which is formed by moving a straight line perpendicular to another straight line which is equal to it; & a cube, which is formed by moving the square in the same way all along another straight line equal to its prime root. To these we have given their own signs by the help of exponents; &, generalizing exponents, we have formed for ourselves ideas, that are not now geometrical, of higher powers; & these not integral only, & positive, but also fractional, & negative; & indeed, by continual abstraction, ever more & more, ideas of irrational powers. To these powers, & to products which may be considered to arise in a similar fashion, we reduce all other functions, by means of the relation they bear to such powers & their products with given straight lines. For this reduction, or expression of the functions by means of these powers & these products we require sometimes more, sometimes less, terms; even when, as in the case of transcendental functions, we have to use an infinite series of terms, which approximates more & more closely to the value & the nature of the given function, although in such cases it never actually reaches this value. Moreover, we consider these to be more or less composite, according as they require more or less terms, or have a nearer relation to single powers.

Another type of mind, to express the relation of a power would necessarily have to use an equal or greater medley of terms.

65. But if another type of mind knew another curve as intimately as we know the straight line, it would consider a single function of that curve to be the most simple of all; &, to express a square or another power, it would consider the self-same relation, inversely taken, so that, beginning with the function, through it & like functions of it, & of higher functions of these lower functions, by addition & subtraction, the mind would finally arrive at the function required. The relation of a power to a function, & the mutual connection, has a compositeness, & leads to a multitude of terms. Each term of the relation is in itself equally simple.

Even in the single expression of a power, we men include several parameters; a parameter in the arbitrary unity, & the combination of a certain force with some certain distance.

66. As regards the introduction of parameters, which they say are included in a function but not in a power of the distance, it is not true that a power does not include a parameter. The formula $1/x^m$ includes unity itself; & this is not something that is self-determinate, but something that can express any magnitude. Indeed, that species of formula includes all species of hyperbolas, &, if the exponent m is given

ac definito exponente *m*, exprimit unicam quidem earum speciem, sed quæ continet infinitas numero individuas Hyperbolas, quarum quælibet suam parametrum diversam habet pro diversitate unitatis assumptæ. Potest quidem quævis ex iis Hyperbolis ad arbitrium assumi ad exprimendam vim decrescentem in ea ratione reciproca; sed adhuc in ipsa expressione includitur quædam parameter, quæ determinet certam vim a certa ordinata exprimendam, sive certam vim certæ distantiæ respondentem, qua semel determinata remanent determinatæ reliquæ omnes, sed ipsa infinitis modis determinari potest, stante expressione facta per ordinatas ejusdem curvæ, sive per eandem potentiæ formulam. Ejusmodi primus nexus a sola distantia utique non pendet.

Parameter in exponente potentiæ.

67. Accedit autem alia quasi parameter in exponente potentiæ: illius numeri *m* determinatio utique non pendet a distantia, nec distantiam aliquam exprimit.

Non esse, cur vis debeat pendere a sola distantia etiam, si vis sit essentialis proprietas materiæ.

68. Sed nec illud video, cur etiam si dicatur vis esse proprietas quædam materiæ essentialis, ea debeat necessario pendere a solis distantiis. Si esset quædam virtus, quæ a materiæ puncto quovis egressa progrederetur motu uniformi, & rectilineo ad omnes circum distantias: tum quidem diffusio ejus virtutis per orbes majores æque crassos fieret in ratione reciproca duplicata distantiarum, & a distantiis solis penderet; quanquam ne tum quidem ab iis penitus solis, sed ab iis, & exponente secundæ potentiæ, ac primo nexu cum arbitraria [286] unitate. At cum nulla ejusmodi virtus debeat progredi, & in progressu ipso ita attenuari; nihil est, cur determinatio ad accessum debeat pendere a solis distantiis, ac proinde solæ distantiæ ingredi formulam functionis exprimentis vim.

Etiam si vis debeat pendere a solis distantiis, ordinatas quoque in se, data curva, pendere a solis abscissis.

69. Verum admisso etiam, quod necessario vis debeat pendere a solis distantiis, nihil habetur contra expressionem factam per functionem quandam. Nam ipsa functio per se immediate pendet a distantia, & est ordinata quædam ad curvam quandam certæ naturæ, respondens abscissæ datæ cuilibet sua. Parametri inducuntur ex eo, quod illius relationem ad abscissam exprimere debeamus per potentias abscissæ, & potentiarum producta cum aliis rectis; sed in se, uti supra diximus, ejusdem est naturæ & illa functio, ac potentia quævis, & illa, ut hæc, ordinatam immediate simplicem exhibet respondentem abscissæ ad curvam quandam uniformis, & in se simplicis curvæ.

Parametros ipsas esse "distantias": eas functionem esse ingressas, quod in datis distantiis debuerit haberi vis data, vel nulla.

70. Præterea ipsæ illæ parametri, quæ formulam functionis ingrediuntur, possunt esse certæ quædam distantiæ & assumi debere ad hoc, ut illis datis distantiis illæ datæ, & non aliæ vires respondeant. Sic ubi quæsita est formula, quæ exprimeret æquationem ad curvam quæsitam, assumpsimus quasdam distantias, in quibus curva secaret axem, nimirum in quibus, evanescente vi haberentur limites, & carum distantiarum valores ingressi sunt formulam inventam, ut quædam parametri. Possunt igitur ipsæ parametri esse distantiæ quædam; ac proinde posito, quod omnino debeat vis exprimi per solas distantias, potest adhuc exprimi per functionem continentem quotcunque parametros, & non exprimetur necessario per solam aliquam potentiam.

Argumentum contrarium a defectu rationis sufficientis.

71. Reliquum est, ut dicamus aliquid de Ratione Sufficienti, quæ dicitur parametros excludere, cum non sit ratio, cur aliæ præ aliis parametri seligantur.

Si vis sit essentialis materiæ; rationem talium parametrorum esse ipsam ejus naturam: cur hoc genus materiæ existat, rationem esse arbitrium Creatoris: idem, si ea non sit essentialis.

72. Inprimis si vis est in ipsa natura materiæ; nulla ratio ulterior requiri potest præter eam ipsam naturam, quæ determinet hanc potius, quam aliam vim pro hac potius, quam pro illa distantia, adeoque hanc potius, quam aliam parametrum. Quæri ad summum poterit, cur elegerit Naturæ Auctor eam potissimum materiam, quæ eam legem virium haberet essentialem, quam aliam: ubi ego quidem, qui summam in Auctore Naturæ libertatem agnosco, censeo, ut in aliis omnibus, nihil aliud requiri pro ratione sufficienti electionis, quam ipsam liberam determinationem Divinæ voluntatis, a cujus arbitrio pendeat tum, quod hanc potius, quam aliam eligat rem, quam condat, tum quod ea re hanc in se naturam habente, ubi jam condita fuerit, utatur ad hoc potius, quam ad illud ex tam multis, ad quæ natura quævis a tanti Artificis manu adhibita potest esse idonea. Atque hæc responsio [287] æque valet, si vis non est ipsi materiæ essentialis, sed libera Auctoris lege sancita, quo casu ipse pro libero arbitrio suo hanc huic materiæ potuit legem dare præ aliis electam.

Præter arbitrium retorsio in potentia: rationem utrobique esse fines, quos sibi ipse proposuerit, qui possunt esse nobis ignoti.

73. At si ratio etiam exhiberi debeat, quæ Auctorem Naturæ potuerit impellere ad seligendam materiam hac potissimum præditam essentiali virium lege, vel ad seligendam pro hac materia hanc legem virium; quæri primo potest, cur hunc potius exponentem potentiæ elegerit, & hanc parametrum in unitate inclusam, sive in quadam determinata

it represents one of these species; & any one of these has its own different parameter for a difference in the unity assumed. It is possible for any one of these hyperbolas to be arbitrarily chosen to represent a force which decreases in that reciprocal ratio; but still there is included in the expression a certain parameter; namely, one which determines a certain force to be represented by a certain ordinate, or a certain force to correspond with a certain distance; when once this is determined, all the rest are at the same time determined. But this can be done in an infinite number of ways, without altering the generation of the expression from the ordinates of the self-same curve, or the same formula of a power. A primary connection of this kind certainly does not depend on distance alone.

There is a parameter in the exponent of the power.

67. Besides there is another thing, that is very like a parameter, in the exponent of the power; the determination of the number m at any rate does not depend on the distance, nor does it express any distance.

There is no reason why it should depend on the distance alone, if force is an essential property of matter.

68. But, really, I do not see why, if it is said that force is some property essential to matter, it should of necessity depend on distances alone. If it were some virtue, which proceeded from any point of matter & progressed with uniform motion in a straight line to all distances round; then indeed the diffusion of this virtue through greater spheres equally thick would be as the inverse squares of the distances; & thus would depend on distance alone. Although not even then would it depend altogether on distances alone; but on them & the exponent of the second power, in addition to the prime connection with an arbitrary unity. But since no such virtue is bound to progress, & even in progression to be so attenuated, there is no reason why determination for approach should depend on distances alone; & that therefore distances alone should enter the formula of the function that expresses the force.

Even if the force did depend on the distances alone, the ordinates also, in themselves, depend on the abscissæ alone, for any given curve.

69. But even if it is admitted that force must necessarily depend on the distances alone; still there is nothing against the expression being formed of some function. For the function in itself depends directly upon distance, & is an ordinate to some curve of known nature, corresponding to its own given abscissa, which may be anything you please. Parameters are induced by the fact that we have to express the relation of the ordinate to the abscissa by means of powers of the abscissa, & the products of these powers with other straight lines. But in themselves, as I said above, both the function & any power are of the same nature; & the former, like the latter, will give a perfectly simple ordinate corresponding to the abscissa to any arc of a curve that is uniform & simple in itself.

The parameters themselves are distances: they have come into the function, because at given distances there must be a given force or none at all.

70. Besides, these very parameters, which come into the formula, may be certain known distances; & they have to be assumed for the purpose of ensuring that to these given distances those given forces, & not others, correspond. So, when we seek a formula to express the equation to the curve required, we assume certain distances in which the curve shall cut the axis; that is to say, distances for which, as the force vanishes, we shall obtain limit-points; & the values of these distances have entered the formula we have found, as certain parameters. Hence the parameters themselves may be distances. Therefore, if it is stated that force is absolutely bound to depend on distances alone, it is still possible to express the force by a function containing any number of parameters; & it is not necessarily expressed by some single power.

The argument against it from a defect of sufficient reason.

71. It only remains to say a few words with regard to Sufficient Reason; this being said to exclude parameters, because there is no reason why some parameters should be chosen in preference to others.

If force is an essential property of matter, the reason for such parameters is the very nature of matter; why such matter exists is due to the will of the Creator; the same thing if force is not an essential.

72. First of all, if force is an essential property of matter, there is no need for any other reason beside that of the very nature of matter, to determine that this, rather than another, force should correspond to this, rather than to another, distance; & therefore this parameter, rather than any other. It may be asked, & we can go no further, why the Architect of Nature chose this matter in particular, such as should have this essential law of forces, & no other. In that case, I, who believe in the supreme freedom of the Architect of Nature, think, as in all other things, that there is nothing else required for the sufficient reason for His choice beyond the free determination of the Divine will. Upon the free exercise of this depends not only the fact that He chose this thing rather than another to create; & also that, the thing having this nature in itself, when it was once created, He should use it for this purpose rather than for any other of the very many purposes, to which any nature employed by the hand of so mighty an Artificer may be suitable. This reply applies just as well, even if the force is not an essential property of matter, but established by the free law of the Author; for, in that case, He, of his own free will, could give this law to this matter, having chosen it in preference to all other laws.

There is something beyond will in the limitation of his power; the reason in both cases is the aim that He set before Himself; & this we may not know.

73. Now, if we have also to give the reason which might have forced the Author of Nature to select in particular this matter possessed of this essential law of forces, or to select for this matter this law of forces especially; it may first be asked why He should have preference for this exponent of the power, this parameter that is included in the unity,

distantia quandam determinatam vim. Quod de iis dicitur, applicari poterit parametris reliquis functionis cujusvis. Ut ille exponens, illa unitas, ille nexus potuit habere aliquid, quod cæteris præstaret ad eos obtinendos fines, quos sibi Naturæ Auctor præscripsit ; sic etiam aliquid ejusmodi habere poterant reliquæ omnes quotcunque, & qualescunque parametri.

Evolutio finis ipsius : necessitas habendi hunc nexum ab Algebra humana non exprimibilem, nisi per functionem, ad solvendum creationis problema pro hac corporum constitutione, & motuum serie.

74. Deinde rem ipsam diligenter consideranti facile patebit, ad obtinendos fines, quos sibi Naturæ Auctor debuit proponere, non fuisse aptam solam potentiam quandam distantiæ pro lege virium, sed debuisse assumi functionem, quæ ubi exprimi deberet per nostram humanum Algebram, alias quoque parametros admisceret. Si ex. gr. voluisset per candem vim & motum Planetarum ad sensum ellipticum cum Kepleriano nexu inter quadrata temporum periodicorum, & cubos distantiarum mediarum, & cohæsionem per contactum, nulla sola potentia ad utrumque præstandum finem fuisset satis, quem finem obtinuisset illa, formula $\frac{a}{x^3} + \frac{b}{x^2}$. At nec ea formula potuit ipsi sufficere, si vera est Theoria mea, cum ea formula nullam babeat in minimis distantiis vim contrariam vi in maximis, sed in omnibus distantiis eandem, nimirum in minimis attractivam, ut in maximis. Cohæsio punctorum se invicem repellentium in minimis distantiis, & attrahentium in majoribus haberi non potuit sine intersectione curvæ cum axe, quæ intersectio sine parametro aliqua non obtinetur. Verum ad omnem hanc phænomenorum seriem obtinendam multo pluribus, uti ostensum est suo loco, intersectionibus curvæ, & flexibus tam variis opus erat, quæ sine plurimis parametris obtineri non poterant. Consideretur elevatissimum inversum problema affine alteri, cujus mentio est facta num. 547, quo quæratur numerus punctorum, & lex virium mutuarum communis omnibus necessaria ab habendam ope cujusdam primæ combinationis, hanc omnem tam diuturnam, tam variam phænomenorum seriem, cujus perquam exiguam particulam nos homines intuemur, & statim patebit elevatissimum debere esse, & respectu habito ad nostros exprimendi modos complicatissimum genus curvæ ad ejusmodi problematis solutionem ne-[288]-cessarium ; quod tamen problema certas quasdam parametros in singulis saltem solutionibus suis, quæ numero fortasse infinito sunt, involveret, sola unica potentia ad tanti problematis solutionem inepta.

Id non potuisse solvi per solam potentiam : legem quadrati distantiæ non esse perfectissimam.

75. Debuit igitur Naturæ Auctor, qui hanc sibi potissimum Phænomenorum seriem proposuit, parametros quasdam seligere, & quidem plures, nec potuit solam unicam pro lege irium exprimenda distantiæ potentiam adhibere : ubi & illud præterea ad rem candem confirmandam recolendum, quod a num. 124, dictum est de ratione reciproca duplicata distantiarum, quam vidimus non esse omnium perfectissimam, nec omnino eligendam, & illud, quod sequenti horum Supplementorum paragrapho exhibetur contra vires in minimis distantiis attractivas & excrescentes in infinitum, ad quas sola potentia demum deducit.

Conclusio contra necessitatem, vel convenientiam solius potentiæ.

76. Atque hoc demum pacto, videtur mihi, dissoluta penitus omnis illa difficultas, quæ proposita fuerat, nec ulla esse ratio, cur sola potentia quædam distantiæ anteferri debuerit functioni utcunque, si nostrum exprimendi modum spectemus, complicatissimæ.

or a certain determined force for a certain determined distance. Now, what is to be said about these things, can be also applied to all the other parameters of any function. Namely, that this exponent, this unity, this connection might have had something in them, which was superior to all other things for the purpose of obtaining those aims which the Author of Nature had set before Himself. Similarly, all the other parameters might have something of the same sort, no matter how many or of what kind they are.

The evolution of this aim; the necessity for this connection which is not expressible by human algebra, unless by a function, to solve the problems of creation for this constitution of bodies, & series of motions.

74. Next, it will easily be clear to anyone, who considers the matter with care, that, for the purpose of obtaining the aims which the Author of Nature was bound to have set Himself, any single power of the distance would not have been convenient for the law of forces; but a function would have had to be taken; & this, as it was destined to be expressed in our human algebra, would bring in other parameters also. If, for instance, He had wished to make subject to the same force, both the practically elliptic motion of the planets, with the Keplerian connection between the squares of the periodic times & the cubes of the mean distances, & also cohesion by contact; then no single power would have been sufficient for the establishment of both aims; this aim would have been met by the formula $a/x^3 + b/x^2$. But this formula even would not have been sufficient, if my Theory is true; for it has not the force at very small distances in the opposite direction to the force at very great distances; but the same kind of force at all distances, that is, an attractive force at very small distances, just as at very great distances. Now, the cohesion of points that repel one another at very small distances, & attract one another at very large distances, cannot be obtained without intersection of the curve & the axis; & this intersection could not be obtained without the introduction of some parameter. Indeed, to obtain the whole series of phenomena, there was need, as has been shown in the proper place for each, of far more intersections of the curve, & for flexures of such different sorts; & these could not be obtained without introducing a large number of parameters. Just consider for a moment this most intricate problem, akin to another of which mention was made in Art. 547:—Required to find the number of points, & the law of mutual forces common to all of them, which would be necessary to obtain, by the aid of a given initial combination, the whole of this series of phenomena, of such duration & variety, of which we men behold but the very smallest of small portions. Immediately it will be evident that it is bound to be of the most intricate character, &, having regard to our methods of expressing things, that the kind of curve necessary for the solution of such a problem must be very complicated. This problem, however, would involve certain known parameters in each of its solutions at least, & the number of these might perchance be infinite; & a single power by itself would be ill-suited for the solution of so great a problem.

It could not be solved by a single power; the law of the squares of the distances is not the most perfect.

75. Hence, the Author of Nature, who decided on this series of phenomena in particular, must have selected certain parameters, & indeed a considerable number of them; nor could He have used a single power of the distance by itself for expressing the law of forces. In this connection also, we must recall to mind, for the confirmation of this matter, what, from Art. 124 onwards, has been said with regard to the inverse ratio of the squares of the distances. We saw that this ratio was not the most perfect of all, nor one to be chosen in all circumstances. Also, we must look at that which is shown, in the next section of these supplements, in opposition to forces that are attractive at very small distances, increasing indefinitely, to which a single power reduces in the end.

Conclusion against the necessity or the convenience of a single power.

76. Finally, in this way, it seems to me that the whole of the difficulty that was put forward has been quite done away with; there is no reason why any single power of the distance should be preferred to a function, no matter how complicated it may be, if regard is paid to our methods of expressing it.

[289] § IV

Contra vires in minimis distantiis attractivas, & excrescentes in infinitum (e)

Prima difficultas ex eo, quod ubi conatus deberet esse maximus in appulsu, debeat esse nullus, vel irritus.

77. At præterea contra solam attractionem plures habentur difficultates, quæ per gradus crescunt. Nam inprimis si eæ imminutis utcunque distantiis agant, augent veloci tatem usque ad contactum, ad quem ubi deventum est, incrementum velocitatis ibi per saltum abrumpitur, & ubi maxima est, ibi perpetuo incassum nituntur partes ad ulteriorem effectum habendum, & necessario irritos conatus edunt.

Secunda, si ratio sit reciproca distantiæ, a vi absolute infinita, ad quam deveniri deberet.

78. Quod si in infinitum imminuta distantia, crescant in aliqua ratione distantiarum reciproca; multæ itidem difficultates habentur, quæ nostrum oppositam sententiam confirmant. Inprimis in ea hypothesi virium deveniri potest ad contactum, in quo vis, sublata omni distantia, debet augeri in infinitum magis, quam esset in aliqua distantia. Porro nos putamus accurate demonstrari, nullas quantitates existere posse, quæ in se infinitæ sint, aut infinite parvæ. Hinc autem statim habemus absurdum, quod nimirum si vires in aliqua distantia aliquid sunt, in contactu debeant esse absolute infinitæ.

Tertia ex eo, quod, si sit major quam simplex, debeat in contactu deveniri etiam ad velocitatem infinitam.

79. Augetur difficultas, si debeat ratio reciproca esse major, quam simplex (ut ad gravitatem requiritur reciproca duplicata, ad cohæsionem adhuc major) & ad bina puncta pertineat. Nam illa puncta in ipso congressu devenient ad velocitatem absolute infinitam. Velocitas autem absolute infinita est impossibilis, cum ea requirat spatium finitum percursum momento temporis, adeoque replicationem, sive extensionem simultaneam per spatium finitum divisibile, & quovis finito tempore requirat spatium infinitum, quod cum inter bina puncta interjacere non possit, requireret ex natura sua, ut punctum ejusmodi velocitatem adeptum nusquam esset.

Alia absurda: si ratio sit duplicata, regressus a centro: saltus ab acceleratione crescente ad nullam in ingressu in superficiem sphæricam.

80. Accedunt plurima absurda, ad quæ ejusmodi leges nos deducunt. Tendat punctum aliquod in fig. 72 in centrum F in ratione reciproca duplicata distantiarum, & ex A projiciatur directione AB perpendiculari ad AF, cum velocitate satis exigua: describet Ellipsim ACDE, cujus focus erit F, & semper regredietur ad A. Decrescat velocitas AB per gradus, donec demum evanescat. Semper magis arctatur Ellipsis, & vertex D accedit ad focum F, in quem demum recidit abeunte Ellipsi in reetam AF. Videtur igitur id [290] punctum sibi relictum debere descendere ad F, tum post acquisitam ibi infinitam velocitatem, eam sine ulla contraria vi convertere in oppositam, & retro regredi. At si id punctum tendat in omnia puncta superficiei sphericæ, vel globi EGCH in eadem illa ratione; demonstratum est a Newtono, debere per AG descendere motu accelerato eodem modo, quo accelerаretur, si omnia ejusmodi puncta superficiei, vel sphæræ compenetrarentur in F: abrupta vero lege accelerationis in G, debere per GH ferri motu æquabili, viribus omnibus per contrarias actiones clisis, tum per HI tantundem procurrere motu retardato, adeoque perpetuam oscillationem peragere, velocitatis mutatione bis in singulis oscillationibus per saltum interrupta.

Regressus a centro simul, & procursus ultra ad eandem distantiam, vel saltus in tanto procursu, sine præviis minoribus.

81. In eo jam absurdum quoddam videtur esse: sed id quidem multo magis crescit; si consideretur, quid debeat accidere, ubi tota sphærica superficies, vel tota sphæra abeat in unicum punctum F. Tum itidem corpus sibi relictum, deveniet ad centrum cum infinita velocitate, sed procurret ulterius usque ad I, dum prius, ubi Ellipsis evanescebat, debebat redire retro. Nos quidem pluribus in locis alibi demonstravimus, in prima

(e) *Hæc excerpta sunt ex eadem dissertatione* De Lege Virium in Natura existentium *a num.* 59.

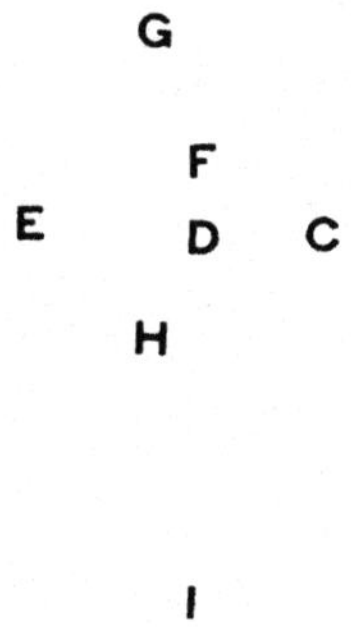

G

F

E D C

H

I

FIG. 72.

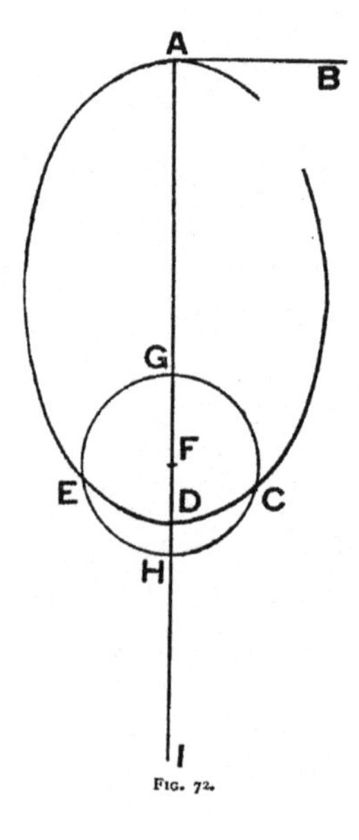

Fig. 72.

§ IV

Arguments against forces that are attractive at very small distances and increase indefinitely [e]

77. Besides, there are many difficulties in the way of attraction alone, which increase by degrees. For, first of all, if these act at diminished distances of any sort, they will increase the velocity right up to the moment of contact: & when contact is attained, the increment of the velocity will then be suddenly broken off; & when this is greatest, the parts will continually strive in vain to produce a further effect, & the efforts will necessarily turn out to be fruitless.

The first difficulty arises from the fact that, when the effort should be greatest on approach, it is bound to be either nothing or to have no effect.

78. But if, when the distances are infinitely diminished, the forces increase according to some ratio that is inversely as the distances, many difficulties will again be had, which confirm our opposite opinion. On that hypothesis of forces especially, contact may be attained, in which, as all distance is taken away, the force is bound to be increased infinitely more than it would be at a distance of some amount. Further, I think that it is rigorously proved that no quantities can possibly exist, such as are infinite in themselves or infinitely small. Hence, we immediately have an absurdity; namely, that if the forces at any distance are anything, on contact they must be absolutely infinite.

The second difficulty arises from the fact that, if the ratio is inversely as the distance, we must come to a force that is absolutely infinite.

79. The difficulty is increased, if the inverse ratio is greater than a simple ratio (as for gravity we require the inverse square, & for cohesion one that is still greater); & it has to do with a pair of points. For these points on collision will attain a velocity that is absolutely infinite. But such an absolutely infinite velocity is impossible, since it requires that a finite space should be passed over in an instant of time, that is, replication, or simultaneous extension through finite divisible space; & for any finite time it would require infinite space, which, since there cannot be such between the two points, would require of its own nature that there should not be anywhere a point that has attained such a velocity.

A third difficulty from the fact that, if the inverse ratio is greater than a simple one, we are bound also to have on contact an infinite velocity.

80. There are many more absurdities, to which such laws of forces lead us. In Fig. 72, let any point tend towards a centre F in the inverse ratio of the squares of the distances, & suppose it to be projected from the point A in a direction, AB, perpendicular to AF, with a fairly small velocity. Then it will describe the ellipse ACDE, of which F is the focus; & it will always return to A. Now let the velocity AB decrease by degrees, until finally it vanishes. Then the ellipse will continually become more & more pointed, & the vertex D will approach the focus F, & will coincide with it when the ellipse becomes the straight line AF. It seems therefore that the point, if left to itself would fall towards the focus F, then, after acquiring an infinite velocity as it reaches F, it would convert it into an equal velocity in the opposite direction without the assistance of any opposing force, & return to its original position. But if that point tended towards all the points of a spherical surface, or the sphere EGCH, in that same ratio, it was proved by Newton that it would have to descend along AG with a motion accelerated in the same manner as it would be if all such points of the surface, or the sphere, were condensed at F. Now the law of acceleration being broken at G, it will have to go on along GH with uniform velocity, all forces being counterbalanced by contrary reactions; then it will have to travel along HI for the same interval with retarded motion. Thus, there would be a continual oscillation, with the change of velocity suddenly interrupted twice in each oscillation.

Other absurdities; if the ratio is the square of the distance, there will be return from the centre; a sudden change from an acceleration that is increasing to one that is nothing on entering a spherical surface.

81. Here there is already seen to be considerable absurdity; but there is still greater to follow. For, let us consider what will necessarily happen when the whole of the spherical surface, or the whole of the sphere, becomes but a single point at F. Then indeed, the body if left to itself would arrive at the centre with infinite velocity; but it would pass through it & beyond as far as I, whereas in the former case when the ellipse vanished, it had to return to its original position. Indeed, in many places elsewhere, I have proved

Simultaneous return from the centre, & motion beyond it to an equal distance; or a sudden change in this great motion, without smaller preparatory motions.

(e) *These paragraphs are quoted from the same dissertation* De Lege Virium in Natura existentium, *starting with Art.* 59.

determinatione latere errorem, cum Ellipsi evanescente, nullæ jam adsint omnes vires, quæ agunt per arcum situm ultra F ad partes D, quæ priorem velocitatem debebant extinguere, & novam producere ipsi æqualem. Verum adhuc habetur saltus quidam, cui & Natura, & Geometria ubique repugnat. Nam donec utcunque parva est velocitas, habetur semper regressus ad A cum procursu FD eo minore, quo velocitas est minor : facta autem velocitate nulla, procursus immediate evadit FI, quin ulli intermedii minores adfuerint. Quod si quis ejus priorem determinationem tueri velit, ut punctum, quod agatur in centrum vi, quæ sit in ratione reciproca duplicata distantiarum, debeat e centro regredi retro ; tum saltus habetur similis, ubi prius in sphæricam superficiem vel sphæram tendat, quæ paullatim abeat in centrum. Donec enim aderit superficies illa, vel sphæra, habebitur semper is procursus, qui abrumpetur in illo appulsu totius superficiei ad centrum, quin habeantur prius minores procursus.

Si ratio sit triplicata pejus : annihilatio puncti in appulsu ad centrum.

82. Hæc quidem in ratione reciproca duplicata distantiarum : in reciproca triplicata habentur etiam graviora. Nam si cum debita quadam velocitate projiciatur per rectam AB fig. 73 continentem angulum acutum cum AP, mobile, quod urgeatur in P vi crescente in ratione reciproca triplicata distantiarum ; demonstratur in Mechanica, ipsum debere percurrere curvam ACDEFGH, quæ vocatur spiralis logarithmica, quæ hanc habet proprietatem, ut quævis recta, ut PF, ducta ad quodvis ejus punctum, contineat cum recta ipsam ibidem tangente angulum æqualem angulo PAB, unde illud consequitur, ut ea quidem ex una parte infinitis spiris cir-**[291]**-cumvolvatur circa punctum P, nec tamen in ipsum unquam desinat : si autem ducatur ex P recta perpendicularis ad AP, quæ tangenti AB occurrat in B, tota spiralis ACDEFGH in infinitum continuata ad mensuram longitudinis AB accedat ultra quoscunque limites, nec unquam ei æqualis fiat : velocitas autem in ejusmodi curva in continuo accessu ad centrum virium P perpetuo crescat. Quare finito tempore, & sane breviore, quam sit illud, quo velocitate initiali percurreret AB, deberet id mobile devenire ad centrum P, in quo bina gravissima absurda habentur. Primo quidem, quod haberetur tota illa spiralis, quæ in centrum desineret, contra id, quod ex ejus natura deducitur, cum nimirum in centrum cadere nequaquam possit : deinde vero, quod elapso eo finito tempore mobile illud nusquam esse deberet. Nam ea curva, ubi etiam in infinitum continuata intelligitur, nullum habet egressum e P. Et quidem formulæ analyticæ exhibent ejus locum post id tempus impossibilem, sive, ut dicimus, imaginarium ; quo quidem argumento Eulerus in sua Mechanica affirmavit illud, debere id mobile in appulsu ad centrum virium annihilari. Quanto satius fuisset inferre, eam legem virium impossibilem esse ?

Pejus in potentiis altioribus : præparatio ad demonstrandum absurdum.

83. Quanto autem majora absurda in ulterioribus potentiis, quibus vires alligatæ sint, consequentur ? Sit globus in fig. 74 ABE, & intra ipsum alius *Abe*, qui priorem contingat in A, ac in omnia utriusque puncta agant vires decrescentes in ratione reciproca quadruplicata distantiarum, vel majore, & quæratur ratio vis puncti constituti in concursu A utriusque superficiei. Concipiatur uterque resolutus in pyramides infinite arctas, quæ prodeant ex communi puncto A, ut BAD, *bAd*. In singulis autem pyramidulis divisis in partes totis proportionales sint particulæ MN, *mn* similes, & similiter positæ. Quantitas materiæ in MN, ad quantitatem in *mn* crit, ut massa totius globi majoris ad totum minorem, nimirum, ut cubus radii majoris ad cubum minoris. Cum igitur vis, qua trahitur punctum A, sit, ut quantitas materiæ directe, & ut quarta potestas distantiarum reciproce, quæ itidem distantiæ sunt, ut radii sphærarum ; erit vis in partem MN, ad vim in partem *mn* directe, ut tertia potestas radii majoris ad tertiam minoris, & reciproce, ut quarta potestas ipsius. Quare manebit ratio simplex reciproca radiorum.

Partem fore majorem toto.

84. Minor crit igitur actio singularum particularum homologarum MN, quam *mn*, in ipsa ratione radiorum, adeoque punctum A minus trahetur a tota sphæra ABE, quam a sphæra *Abe*, quod est absurdum, cum attractio in eam sphæram minorem debeat esse pars

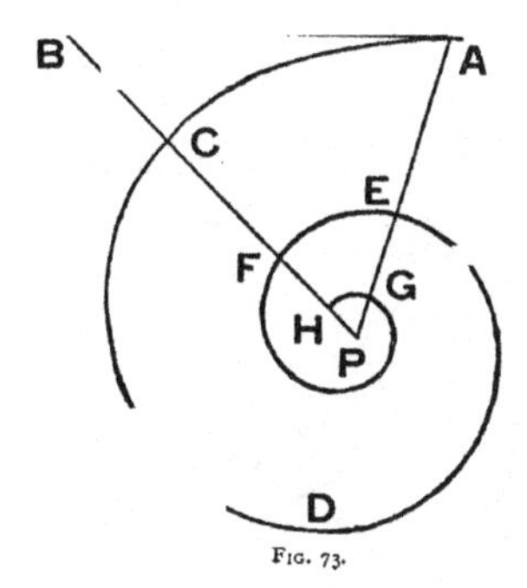

Fig. 73.

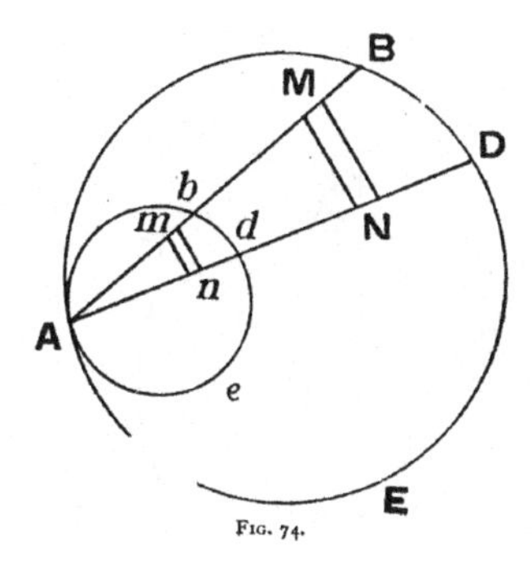

Fig. 74.

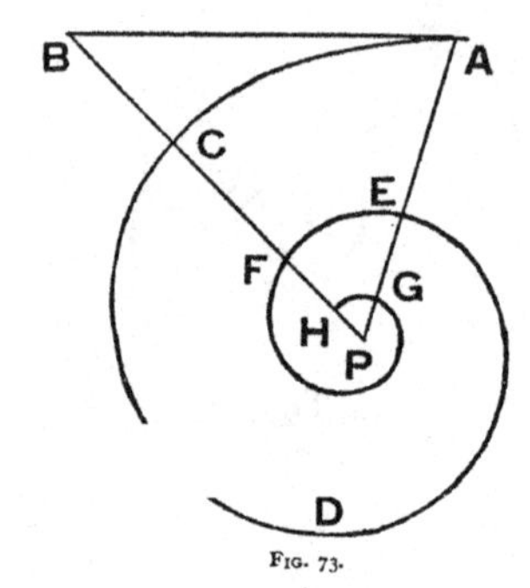

FIG. 73.

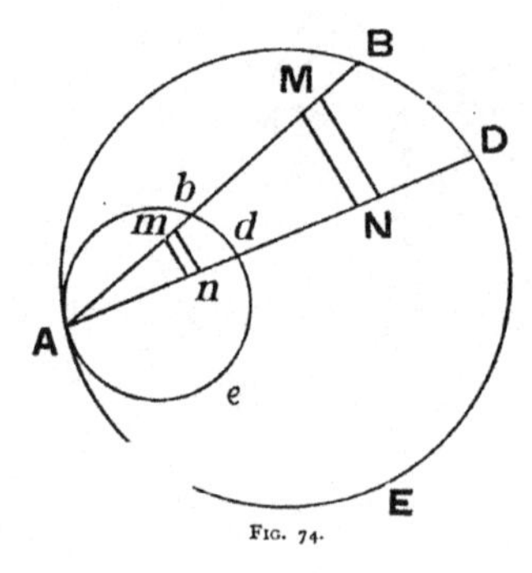

FIG. 74.

that there is an error in the first determination; for when the ellipse vanishes, there are no longer present any of all these forces, which act on the body as it goes along the arc situated beyond F in the direction of D; & these were necessary to extinguish the former velocity & to generate a new velocity equal to it. But still there is a sudden change, to which both Nature & geometry are in all cases opposed. For, so long as there is a velocity, no matter how small, we always have a return to A with a further motion beyond F, equal to FD, which is correspondingly smaller as the velocity becomes smaller; & yet, when the velocity is made nothing at all, the further motion beyond F at once becomes FI, without there being present any intermediate smaller motions. Now, if anyone would wish to adhere to the first determination of the problem, so that a point, which is attracted towards a centre by a force in the inverse ratio of the squares of the distances, is bound to return from the centre to its original position; then there too there is a sudden change of a like nature to that which took place in the first case when it tended towards a spherical surface, or a sphere, which gradually dwindled to a point at the centre. For, as long as the spherical surface, or the sphere, is there, there will always be obtained that further motion; but this is suddenly stopped on the arrival of the whole of the spherical surface, or the whole of the sphere, at the centre, without any previous smaller motions being had.

If the ratio is the triplicate, it is still worse; annihilation of the point on arrival at the centre.

82. Such indeed are the results that we obtain for the inverse ratio of the squares of the distances; for the inverse ratio of the cubes, we have even more serious difficulties. For, if a body is projected along AB, in Fig. 73, making an acute angle with AP, with a certain suitable velocity, & it is attracted towards P with a force increasing in the inverse ratio of the cubes of the distances; in that case, it is proved in Mechanics that the motion will be along a curve such as ACDEFGH, which is called the logarithmic spiral. This curve has the property that any straight line, PF, drawn from P to any point F of the curve, contains with the tangent to the curve at the point an angle equal to the angle PAB. Hence it follows that, on the one hand indeed it will rotate through an infinite number of convolutions round the point P, but will never reach that point; yet, on the other hand, if a straight line is drawn through P perpendicular to AP, to meet the tangent AB in B, then the whole length of the spiral ACDEFGH continued indefinitely will approximate to the length of AB beyond all limits, & yet never be equal to it. Further the velocity in such a curve, as it continually approaches the centre of forces P, continually increases. Hence in a finite time, & that too one that is shorter than that in which it would pass over the distance AB with the given initial velocity, the moving body would be bound to arrive at the centre P; & in this we have two very serious absurdities. The first is that the whole of the spiral, which terminates in the centre, is obtained, in opposition to the principle deduced from its nature, since truly it can never get to the centre; & secondly, that after that finite time has elapsed the moving body would have to be nowhere at all. For, the curve, even when it is understood that it is continued to infinity, has no exit through & past the point P. Indeed the analytical formulæ represent its position after the lapse of this time as impossible, or, as it is usually called imaginary. By this very argument, Euler, in his Mechanics, asserts that the moving body on approaching the centre of forces is annihilated. How much more reasonable would it be to infer that this law of forces is an impossible one?

Still worse for higher powers; preparation for demonstrating an absurdity.

83. How much greater absurdities are those that follow for higher powers, with which the forces may be connected! In Fig. 74, let ABE be a sphere, & within it let there be another one A*be*, touching the former at A; & suppose that on all points of each of them there act forces which decrease in the inverse ratio of the fourth powers of the distances, or even greater; & suppose that we require the ratio of the forces due to a point situated at the point of contact A of the two surfaces. Imagine each of the spheres to be divided into infinitely thin pyramids, proceeding from the common vertex A, such as BAD, *b*A*d*. In each of these little pyramids, which are then divided into parts proportional to the wholes, let MN & *mn* be particles that are similar & similarly situated. The quantity of matter in MN will be to the quantity of matter in *mn* as the mass of the larger sphere to the mass of the whole of the smaller; i.e., as the cube of the radius of the larger to the cube of the radius of the smaller. Hence, since the force exerted upon A varies as the quantity of matter directly, & as the fourth power of the distance inversely, & these distances also vary as the radii of the spheres. Therefore, the force on the part MN is to the force on the part *mn* directly as the third power of the radius of the larger sphere to the third power of the radius of the smaller, & inversely as the fourth powers of the same. That is, there results the simple inverse ratio of the radii.

The part greater than the whole.

84. Hence the action of each of the homologous particles MN will be less than each of the corresponding particles *mn*, in the ratio of the radii; & thus the point A will be attracted less by the whole sphere ABE than by the sphere A*be*. This is absurd; for, the attraction on the smaller sphere must be a part of the attraction on the greater sphere

attractionis in sphæram majorem, quæ continet minorem, cum magna materiæ parte sita extra ipsam usque ad superficiem sphæræ majoris, unde concluditur esse partem majorem toto, maximum nimirum absurdum. Et qui-[292]-dem in altioribus potentiis multo major est is error ; nam generaliter, si vis sit reciproce, ut R^m, posito R pro radio, & *m* pro quovis numero ternarium superante, erit attractio sphæræ eodem argumento reciproce, ut R^{m-3}, quæ eo majorem indicat vim in sphæram minorem respectu majoris ipsam continentis, quo numerus *m* est major.

Omnia absurda cessare, si in minimis distantiis habeatur repulsio, quæ appulsum impediat.

85. Hoc quidem pacto inveniuntur plurima absurda in variis generibus attractionum quæ si repulsiones, in minimis distantiis habeantur pares extinguendæ velocitati cuilibet utcunque magnæ, cessant illico omnia, cum eæ repulsiones mutuum accessum usque ad concursum penitus impediant. Inde autem manifesto iterum consequitur, repulsiones in minimis distantiis præferendas potius esse attractioni, ex quarum variis generibus tam multa absurda consequuntur.

ne, together with a great part of the matter situated beyond reater sphere; hence the conclusion is that the part is greater ogether impossible. Indeed, in still higher powers the error eral, if the force varies inversely as R^m, where R is taken as mber greater than three, then the attraction of the sphere this points to a force that is the greater on a smaller sphere ger sphere containing it, in proportion as the number m is

any absurdities in various kinds of attractions; if there are nces, sufficiently great to destroy any velocity however large, ease to be immediately, for these repulsions would prevent int of actual contact. Hence it once again manifestly follows distances are to be preferred before an attraction; for from r so many absurdities follow.

All these absurdities cease to be, if there is repulsion at very small distances, which prevents near approach.

[293] § V

De Æquilibrio binarum massarum connexarum invicem per bina alia puncta (f)

Propositio problematis de æquilibrio punctorum quatuor, quorum bina extrema habeant quascunque massa, cum viribus externis sibi proportionalibus, & altera e mediis vim a fulcro.

86. Continetur autem, quod pertinet ad momentum in vecte, & ad æquilibrium, sequentis problematis solutione. Sit in fig. 75 quivis numerus punctorum materiæ in A, qui dicatur A, in D quivis alius, qui dicatur D, & puncta ea omnia secundum directiones AZ, DX parallelas rectæ datæ CF sollicitentur simul viribus, quæ sint æquales inter omnia puncta sita in A, itidem inter omnia sita in D, licet vires in A sint utcunque diversæ a viribus in D. Sint autem in C, & B bina puncta, quæ in se invicem, & in illa puncta sita in A, & D mutuo agant, ac ejusmodi mutuis actionibus impediri debeat omnis actio virium illarum in A, & D, & omnis motus puncti B : motus autem puncti C impediri debeat actione contraria fulcri cujusdam, in quod ipsum agat secundum directionem compositam ex actionibus omnium virium, quas habet : quæritur ratio, quam habere debent summæ virium A, & D ad hoc, ut habeatur id æquilibrium, & quantitas, ac quæritur directio vis, qua fulcrum urgeri debet a puncto C.

Vis ex binis extremis in alterum e mediis.

87. Exprimant AZ, & DX vires illas parallelas singulorum punctorum positorum in A, & D. Ut ipsæ elidantur, debebunt in iis haberi vires AG, DK contrariæ, & æquales ipsis AZ, DK. Quoniam eæ debent oriri a solis actionibus punctorum C, & B agentium in A secundum rectas AC, AB, & in D secundum rectas DC, DB, ductis ex G rectis GI, GH parallelis BA, AC usque ad rectas AC, BA, & ex K rectis KM, KL parallelis BD, DC, usque ad rectas DC, BD; patet, in A vim AG debere componi ex viribus AI, AH, quarum prima quodvis punctum in A repellat a C, secunda attrahat ad B, & in D vim DK componi itidem ex viribus DM, DL, quarum prima quodvis punctum situm in D repellat a C, secunda attrahat ad B. Hinc ob actionem reactioni æqualem debebit punctum C repelli a quovis puncto sito in A secundum directionem AC vi æquali IA, & a quovis puncto sito in D secundum directio-[294]-nem DC vi æquali MD : punctum vero B debebit attrahi a quovis puncto sito in A secundum directionem BA vi æquali HA, & a quovis puncto sito in D vi æquali LD. Habebit igitur punctum C ex actione punctorum in A, & D binas vires, quarum altera aget secundum directionem AC, & erit æqualis IA ductæ in A, altera aget secundum directionem DC, & erit æqualis MD ductæ in D. Punctum vero B itidem binas, quarum altera aget secundum directionem BA, & erit æqualis HA duetæ in A, altera aget secundum directionem BD, & crit æqualis LD duetæ in D.

Vis, quam debet habere illud primum, composita e quatuor : enumeratio virium pertinentium ad omnia puncta.

88. Porro vis composita ex illis binis, quibus urgetur punctum B, elidi debet ab actione mutua inter ipsum, & C ; quare debebit habere directionem rectæ BC in casu, quem exhibet figura, in quo C jacet in angulo ABD : nam si angulus ABD hiatum obverteret ad partes oppositas, ut C jaceret extra angulum ; ea haberet directionem CB, & reliqua omnis demonstratio rediret eodem. Punctum autem C ob actionem, & reactionem æquales debebit habere vim æqualem, & contrarium illi, quam exercet B, adeoque vim æqualem, & ejusdem directionis cum vi, quam e prioribus illis binis compositam habet punctum B : nempe debebit habere binas vires æquales, & directionis ejusdem cum viribus illam componentibus, nimirum vim secundum directionem parallelam BA æqualem ipsi HA ductæ in A, & vim secundum directionem parallelam BD æqualem ipsi LD ductæ in D. Habebit

(f) *Excerpta hæc sunt ex* Synopsi Physicæ Generalis *P. Caroli Benvenuti Soc. Jesu, num.* 146, *cui hanc solutionem ibi imprimendam tradideram,*

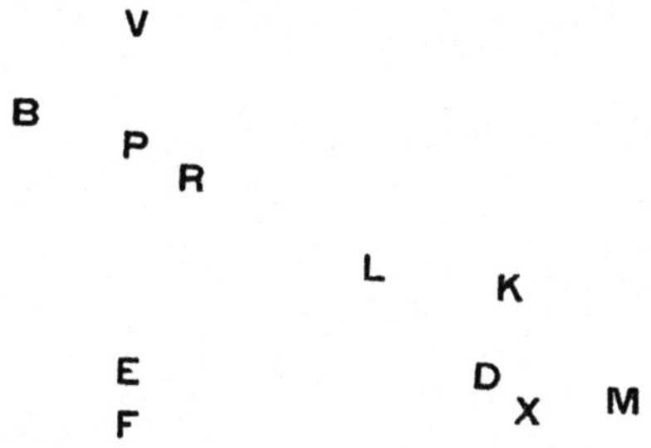

V

B

P

R

L

K

E

D

X

M

F

FIG. 75.

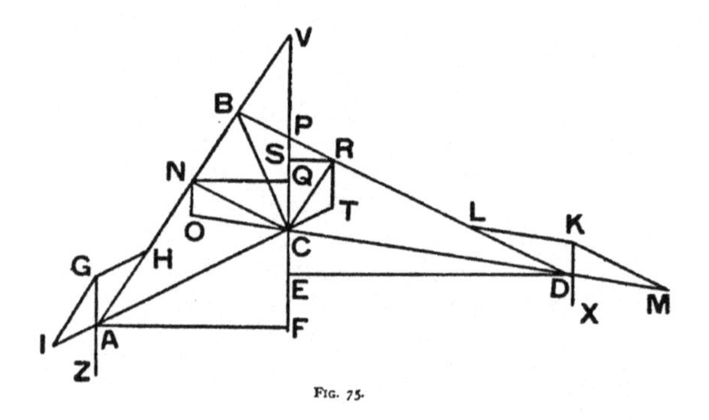

Fig. 75.

§ V

Equilibrium of two masses connected together by two other points (f)

86. All that pertains to moment in the lever, & to equilibrium is contained in the solution of the following problem. In Fig. 75, let there be any number of points of matter at the point A, & let the number be called A; similarly, any other number at D, called D; & suppose that all these are at the same time under the action of forces along the directions AZ, DX parallel to the given straight line CF, & that these forces are equal to one another for all the points situated at A, & also for all the points situated at D, although the forces at A may be altogether different from those at D. Also, at C & B, let there be two points, which act mutually upon one another & upon the points situated at A & D. Suppose that by such actions the whole of the action of the forces on A & D has to be prevented, as well as any motion of the point B. Also suppose that the motion of the point C is to be prevented by the contrary action of a fulcrum, upon which the point C acts according to the direction compounded from all the forces that act upon it. It is required to find the ratio which there must be between the forces on A & D, for the purpose of obtaining equilibrium; also to find the quantity & direction of the force to which the fulcrum must be subjected by the point C.

Enunciation of the problem of the equilibrium of four points; of which the two outside points have any masses with external forces proportional to them; & one of the inner points is subject to a force from a fulcrum.

87. Let AZ & DX represent the parallel forces of each of the points situated at A & D respectively. To cancel these, we must have acting at these points forces AG & DK, which are equal and opposite to AZ, DX. Now, these must arise purely from the actions of the points C & B, acting on A along the straight lines AC, AB, & on D along the straight lines DC & DB. Hence, if we draw through G straight lines GI, GH, parallel to BA, AC, to meet the straight lines AC, BA; & through K, straight lines KM, KL, parallel to BD, DC, to meet DC, BD; then it is plain that the force AG on A must be compounded of the forces AI, AH, of which the first will repel any one of the points at A away from C, & the second will attract it towards B; & similarly, the force DK on D must be compounded of the two forces DM, DL, of which the first repels any one of the points situated at D away from C, & the second attracts it towards B. Hence, on account of the equality of action & reaction, the point C must be repelled by every point situated at A in the direction AC by a force equal to IA, & by every point situated at D in the direction DC with a force equal to MD. Also the point B will be attracted by every point situated at A in the direction BA with a force equal to HA, & by every point at D with a force equal to LD. Therefore, the point C will have, due to the actions of the points at A & D, two forces, of which one will act in the direction AC, & be equal to IA multiplied by A, & the other will act in the direction DC & be equal to MD multiplied by D. The point B will also be under the action of two forces, one of which will act in the direction BA & be equal to HA multiplied by A, & the other will act in the direction BD & be equal to LD multiplied by D.

The force from the two extremes on either of the means.

88. Further the force composed from the two forces, which act upon the point B, must be cancelled by the mutual action between it & C; hence, this must be in the direction of the straight line BC, in the case given by the figure, where C lies within the angle ABD; for, if the angle ABD should turn its opening in the other direction, so that C should lie outside the angle, then the force would have the direction CB, & all the rest of the proof would come to the same thing. Now, the point C, on account of the equality of action & reaction, must have a force that is equal & opposite to that exerted by B; & thus, a force that is equal to, & in the same direction as, the force which B has, compounded of those first two forces. That is to say, it must have two forces that are equal to, & in the same direction as, the two forces that compose it; namely, a force in a direction parallel to BA & equal to HA multiplied by A, & a force in a direction parallel to BD & equal to

The force, which that first point, C, must have, is composed out of four; enumeration of the forces pertaining to all the points.

(f) *These are quoted from the* Synopsis Physicæ Generalis *of Fr. Carolus Benvenuto, S.J., Art.* 146, *to which author I gave this solution to print in that work.*

igitur quodvis punctum A binas vires AI, AH, quodvis punctum D binas vires DM, DL, punctum B binas vires, quarum altera dirigetur ad A, & æquabitur HA ductæ in A, altera dirigetur ad D, & æquabitur LD duetæ in D, ex quibus componi debet vis agens secundum rectam BC: & demum habebit punctum C vires quatuor, quarum prima dirigetur ad partes AC, & erit æqualis IA ductæ in A, secunda ad partes DC, & erit æqualis MD duetæ in D, tertia habebit directionem parallelam BA, & crit æqualis HA ductæ in A; quarta habebit directionem BD, & erit æqualis LD duetæ in D: ac ipsum punctum C urgebit fulcrum vi composita ex illis quatuor, quæ omnia, si habeatur ratio directionis rectarum secundum ordinem, quo enunciantur per literas, huc reducuntur:

Quodvis punctum A habebit vires binas	AI, AH
Quodvis punctum D vires binas	DM, DL
Punctum B binas	$A \times HA$, $D \times DL$
Punctum C quatuor .	$A \times IA$, $D \times MD$, $A \times HA$, $D \times LD$

Constructio præparatoria pro solutione.

89. Exprimat jam recta BC magnitudinem vis compositæ e binis CN, CR parallelis DB, AB; expriment BN, BR magnitudinem virium illarum componentium, cum exprimant [295] earum directiones, adeoque RC, NC ipsis æquales, & parallelæ expriment vires illas tertiam, & quartam puncti C. Producantur autem DC, AC donec occurrant in O, & T rectis ex N, & R parallelis ipsi CF, sive ipsis GAZ, KDX, & demittantur AF, DE, NQ, RS perpendicula in ipsam FC productam, qua opus est, quæ occurrat rectis AB, DB in V, P.

Vires sub nova expressione inde resultante.

90. Inprimis ob singula latera singulis lateribus parallela crunt similia triangula IAG, CTR, & triangula MDK, CON. Quare crit ut IG, sive AH, ad CR, sive NB, vel $A \times AH$, nimirum ut 1 ad A, ita AG ad TR, & ita AI ad TC. Erit igitur TR æqualis GA, sive AZ duetæ in A, & CT æqualis IA ductæ in A; adeoque illa exprimet summam omnium virium AZ omnium punctorum in A, hæc vim illam primam puncti C, nimirum $A \times IA$. Eodem prorsus argumento, cum sit MK, sive DL ad CN, sive RB, vel $D \times DL$, nimirum 1 ad D, ita DK ad ON, & ita DM ad OC; crit NO æqualis KD, sive DX ductæ in D, & OC æqualis MD ductæ in D, adeoque illa exprimet summam omnium virium DX omnium punctorum in D, hæc vim illam secundam puncti C, nimirum $D \times DM$. Quare jam erunt

Summa virium parallelarum in A	TR
Summa virium parallelarum in D	NO
Binæ vires in B	BN, BR
Quatuor vires in C	CT, OC, RC, NC

Vis in fulcrum cui æqualis.

91. Jam vero patet, ex tertia RC, & prima CT componi vim RT æqualem summæ virium parallelarum A: & ex quarta NC, ac secunda OC componi vim NO æqualem summæ virium parallelarum in D. Quare patet, ab unico puncto C fulcrum urgeri vi, quæ candem directionem habeat, quam habent vires parallelæ in A, & D, & æquetur carum summæ, nimirum urgeri eodem modo, quo urgeretur, si omnia illa puncta, quæ sunt in D, & A, cum his viribus essent in C, & fulcrum per se ipsa immediate urgerent.

Proportio, quæ vectem exhibet.

92. Præterea ob parallelismum itidem omnium laterum similia crunt triangula 1° CNO, DPC: 2° CNQ, PDE: 3° CPR, VCN: 4° CRS, VNQ: 5° CVA, TCR: 6° VAF, CRS. Ea exhibent sequentes sex proportiones, quarum binæ singulis versibus continentur.

ON . CP : : NC . PD : : NQ . DE
CP . CV : : CR . NV : : RS . NQ
CV . RT : : VA . RC : : AF . RS

Porro ex iis componendo primas, & postremas, ac demendo in illis CP, CV; in his QN, RS communes tam antecedentibus, quam consequentibus, fit ex æqualitate nimirum perturbata ON . RT : : AF. DE. Nempe summa omnium virium parallelarum in D, cui æquatur ON, ad summam om-[296]-nium in A, cui æquatur RT, ut e contrario distantia harum perpendicularis AF a recta CF ducta per fulcrum directioni virium earumdem parallela, ad illarum perpendicularem distantiam ab eadem. Quare habetur determinatio corum omnium quæ quærebantur (g).

(g) *Porro applicatio ad vectem est similis illi, quæ habetur hic post æquilibrium trium massarum num.* 326.

V

P R

L K

E D X M
F

Fig. 75.

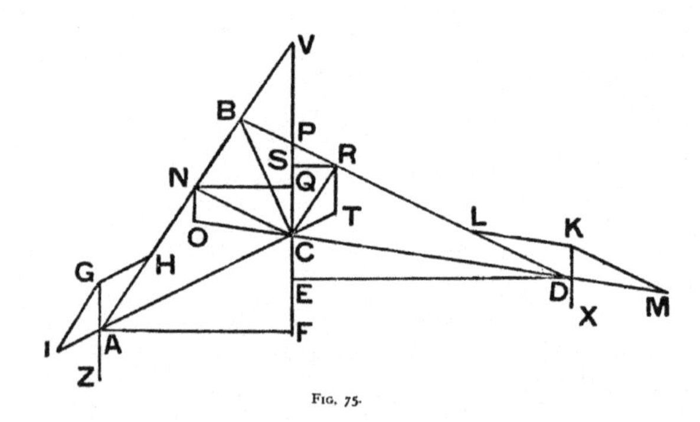

Fig. 75.

LD multiplied by D. Hence, any point at A will have two forces, AI, AH ; any point at D two forces DM, DL ; the point B two forces, of which one is directed towards A & is equal to HA multiplied by A, & the other is directed towards D & is equal to LD multiplied by D ; & lastly, the point C will have four forces, of which the first is directed along AC and is equal to IA multiplied by A, the second along DC and equal to MD multiplied by D, the third has a direction parallel to BA and is equal to HA multiplied by A, and the fourth has a direction parallel to BD and is equal to LD multiplied by D. The point C will exert on the fulcrum a force compounded from all four forces ; and all of these, if the sense of the direction of the straight lines is considered to be that given by the order of the letters by which they are named, will be as follows :—

Any point at A will have two forces . . .	AI, AH
Any point at D, two forces .	DM, DL
The point B, two	A × HA, D × LD
The point C, four . .	A × IA, D × MD, A × HA, D × LD

Construction necessary for the solution.

89. Now let BC represent the magnitude of the force compounded from the two forces CN, CR, parallel to DB, AB : then BN, BR will represent the magnitude of the component forces, since they represent their directions, and thus RC, NC, which are equal and parallel to them, will represent the third and fourth forces on the point C. Also let DC & AC be produced, until they meet in O and T respectively the straight lines drawn through N & R parallel to CF, i.e., to GAZ & KDX ; & let AF, DE, NQ, RS be drawn perpendicular to CF, produced if necessary ; & let CF meet AB, DB in V & P.

The forces that result from this new method of representation.

90. First of all, on account of their corresponding sides being parallel, the triangles IAG, CTR are similar, & so also are the triangles MDK, CON. Hence, as IG, or AH, is to CR, or NB, i.e., A × AH, in other words, as 1 is to A, so is AG to TR, or as IA to TC. Hence, TR will be equal to GA or AZ multiplied by A, & CT will be equal to IA multiplied by A. Therefore the former will represent the sum of all the forces AZ on all the points at A, & the latter that first force on the point C, i.e., A × IA. With precisely the same argument, since MK, or DL, is to CN, or RB, i.e., D × DL, in other words, as 1 to D, so is DK to ON, or DM to OC ; therefore NO will be equal to KD or DX multiplied by D, & CO equal to MD multiplied by D ; & therefore the former will represent the sum of all the forces DX for all the points at D, & the latter that second force on the point C namely, D × DM. Hence, we now have :

The sum of the parallel forces on A .	TR
The sum of the parallel forces on D .	NO
The two forces on B	BN, BR
The four forces on C	CT, OC, RC, NC

The force on the fulcrum : to what it is equal.

91. And now it is plain that, from the third force RC, & the first, CT, we have a resultant force RT which is equal to the sum of the parallel forces at A ; & from the fourth, NC, & the second, OC, we get a resultant force NO, which is equal to the sum of all the parallel forces at D. Therefore, it is evident that the fulcrum at C is subject to but a single force, which has the same direction as that of the parallel forces on the points at A & D, & that its magnitude is equal to their sum. In other words, the force acting upon it is exactly the same as if all those points which are at A & D were transferred together with the forces acting upon them to the point C, & there acted upon the fulcrum directly.

The proportion, which represents the law of the lever.

92. In addition, on account of all sides being parallel, the following pairs of triangles are similar :—(1) CNO, DPC ; (2) CNQ, PDE ; (3) CPR, VCN ; (4) CRS, VNQ ; (5) CVA, TCR ; (6) VAF, CRS. These will give the following six proportions, two of which are contained in each of the following lines :—

$$ON : CP = NC : PD = NQ : DE$$
$$CP : CV = CR : NV = RS : NQ$$
$$CV : RT = VA : RC = AF : RS.$$

Further, by compounding together the first & last of these, & removing from the antecedents the ratio CP : CV, & from the consequents the ratio QN : RS, we are left with the proportion, ON : RT = AF : DE. That is to say, the sum of all the parallel forces on D, to which ON is equal, is to the sum of all those on A, to which RT is equal, as the opposite perpendicular distance AF from the straight line CF drawn through the fulcrum in a direction parallel to that of these forces, is to the perpendicular distance of the former from the same straight line. Hence, we have obtained a solution of all that was required (g).

(g) *Moreover, the application to the lever is similar to that given in this work, after the equilibrium of three points in Art. 326.*

[297] § VI

EPISTOLA AUCTORIS AD P. CAROLUM SCHERFFER SOCIETATIS JESU

Occasio, & argumentum epistolæ.

93. In meo discessu Vienna reliqui apud Reverentiam Vestram imprimendum opus, cujus conscribendi occasionem præbuit Systema trium massarum, quarum vires mutuæ Theoremata exhibuerunt & elegantia, & fœcunda, pertinentia tam ad directionem, quam ad rationem virium compositarum e binis in massis singulis. Ex iis Theorematis evolvi nonnulla, quæ in ipso primo inventionis æstu, & scriptionis fervore quodam, atque impetu se se obtulerunt. Sunt autem & alia, potissimum nonnulla ad centrum percussionis pertinentia ibi attactum potins, quam pertractatum, quæ mihi deinde occurrerunt & in itinere, & hic in Hetruria, ubi me negotia mihi commissa detinuerunt hucusque, quæ quidem ad Reverentiam Vestram transmittenda censui, ut si forte satis mature advenerint, ad calcem operis addi possint; pertinent enim ad complementum corum, quæ ibidem exposui, & ad alias sublimiores, ac utilissimas perquisitiones viam sternunt.

Translatio Theoriæ centri oscillationis a massis jacentibus intra idem planum, ad ubicunque positas affirmata in opere, hic demonstranda.

94. Inprimis ego quidem ibi consideravi directiones virium in eodem illo plano, in quo jacent tres massæ, & idcirco ubi Theoremata applicavi ad centrum æquilibrii, & oscillationis pro pluribus etiam massis, restrinxi Theoriam ad casum, in quo omnes massæ jaceant in eodem plano perpendiculari ad axem conversionis. In nonnullis Scholiis tantummodo innui, posse rem transferri ad massas, utcunque dispersas, si eæ reducantur ad id planum per rectas perpendiculares plano eidem; sed ejus applicationis per ejusmodi reductionem nullam exhibui demonstrationem, & affirmavi, requiri systema quatuor massarum ad rem generaliter pertractandam.

Viribus trium massarum in eodem plano, in quo jacent, translatis ad aliud, rem obtineri.

95. At admodum facile demonstratur ejusmodi reductionem rite fieri, & sine nova peculiari Theoria massarum quatuor generalis habetur applicatio tenui extensione Theoriæ massarum trium. Nimirum si concipiatur planum quodvis, & vires singulæ resolvantur in duas, alteram perpendicularem plano ipsi, alteram parallelam; priorum summa elidetur, cum oriantur e viribus mutuis contrariis, & æqualibus, quæ ad quamcunque datam directionem redactæ æquales itidem remanent, & con-**[298]**-trariæ, evanescente (*h*) summa: posteriores autem componentur eodem prorsus pacto, quo componerentur; si massæ per illas perpendiculares vires reducerentur ad illud planum, & in eo essent, ibique vires haberent æquales redactas ad directionem ejusdem plani, quarum oppositio & æqualitas redderet candem figuram, & eadem Theoremata, quæ in opere demonstrata sunt pro viribus jacentibus in eodem plano, in quo sunt massæ. Porro hæc consideratio extendet Theoriam æquilibrii, & centri oscillationis ad omnes casus, in quibus systema quodvis concipitur connexum cum unico puncto axis rotationis, ut ubi globus, vel systema quotcunque massarum invicem connexarum oscillat suspensum per punctum unicum.

Si massæ sint quatuor, reducendas omnes ad planum perpendiculare rectæ jungenti duas: inde transitus ad massas quotcunque.

96. Quod si sint quatuor massæ, & concipiatur planum perpendiculare rectæ transeunti per binas ex iis, ac fiat resolutio eadem, quæ superius; res iterum eodem recidet: nam illæ binæ massæ ita in illud planum projectæ, coalescent in massam unicam, & vires ad

---

(h) *Hæc tum quidem in hac epistola. Addi potest illud, ubi nulla externa vis in ea directione agens, & in contraria applicetur diversis partibus ipsius systematis, debere vim hujusmodi in singulis etiam ipsius systematis punctis esse nullam. Nam per mutuum nexum impeditur mutatio positionis mutuæ, que utique induceretur, si in aliquibus tantummodo ejus partibus remaneret vis externis viribus non impedita. Porro ubi agitur de centro oscillationis, & percussionis, ac etiam de æquilibrio, nulla supponitur vis, externa agens secundum directionem axis rotationis, seu conversionis. Quare in iis casibus, pro quibus hæc theoria hic extenditur, satis est considerare reliquas illas vires, quæ agunt secundum directionem plani perpendicularis eidem axi, quod hic præstatur in iis, quæ consequuntur.*

§ VI

A LETTER FROM THE AUTHOR
TO
FR. CAROLUS SCHERFFER, S.J.

The occasion for, & the contents of, the letter.

93. When I departed from Vienna, I left with Your Reverence to be printed a work, which I had written as an outcome of the consideration of a system of three masses; the mutual forces between these brought out several theorems that were both elegant & fruitful, with regard to the direction & the ratio of the forces on each of the masses compounded from the other two. From these theorems I worked out certain results, which, in the first surge of discovery, & a certain fervour & impetus of writing, had forced themselves on my attention. But there are also other matters, especially some relating to the centre of percussion that are in it merely touched upon rather than dealt with thoroughly; these came to me later, some during my journey, & some here in Tuscany, where the business entrusted to me has kept me up till now. These matters I thought should be sent to Your Reverence, so that, if perchance they should reach you soon enough, they might be added at the end of the work; for they deal with the further development of those things which I have expounded therein, & open the road to more sublime & useful matters for inquiry.

The transition to the theory of the centre of oscillation from the case of masses all lying in the same plane to masses lying anywhere, merely asserted in the work itself, is here to be proved.

94. First of all, I there indeed considered the directions of the forces in the same plane as that in which the masses were situated; &, therefore, when I applied the theorems to the centre of equilibrium & oscillation even for several masses, I restricted the Theory to the case in which all the masses were lying in the same plane, perpendicular to the axis of rotation. Only in some notes did I mention that the matter could be developed for masses that were disposed in any manner, if these were reduced to that plane by perpendiculars to the plane. But I gave no demonstration of this application by means of such a reduction; & I asserted that the consideration of a system of four masses would be necessary before the matter could be dealt with thoroughly, & in general.

The forces for three masses in the same plane as that in which they lie, being transferred to another, the thing is done.

95. But it is quite easily proved that such a reduction can be correctly made; & a general application, without any special fresh theory for four masses, can take place, with a very slight extension of the theory for three masses. Thus, if any plane is taken & each force is resolved into two forces, of which one is perpendicular & the other parallel to the plane; then the sum of all the first will be eliminated, since they arise from mutual forces that are equal & opposite to one another; for, these when reduced to any given direction whatever will still remain equal & opposite to one another, & their sum will vanish (*h*). Also the latter will be compounded in exactly the same manner as they would have been compounded, if the masses, by means of those perpendicular forces, had been reduced to that plane, & were really in it, & had there equal forces reduced to the direction of that plane: the equal & opposite nature of these forces would give the same figure, & the same theorems as were proved in the work itself for forces in the same plane as that in which the masses were lying. Further, this way of looking at the matter will extend the Theory of the centre of equilibrium & oscillation to all cases, in which any system is supposed to be connected with a single point on the axis of rotation, as when a sphere, or a system of any number of masses connected together oscillates under suspension from a single point.

If there are four masses, they are all to be reduced to a plane perpendicular to the straight line joining two of them; hence the transition to any number of masses.

96. Now if there are four masses, & a plane is taken perpendicular to the straight line joining any two of them, & the same resolution is made as in the preceding paragraph; then, the matter will again come to the same thing. For, those two masses, being thus thrown into the same plane, will coalesce into a single mass; & the forces belonging to

(h) *This is what I said in the letter. To it may be added the point that, when no external force is applied acting in one direction on one part, & the opposite direction on another part, of the system, this kind of force must also be zero for each of the points of the system. For, a change of mutual position is prevented by the mutual connection; & at any rate this would be induced, if in any of the parts of it there but remained a force that was not checked by external forces. Further, when dealing with the centre of oscillation, & of percussion, & with equilibrium, no external force is supposed to act in the direction of the axis of rotation or conversion. Hence, in these cases, for which the theory is here extended, it is sufficient to consider these other forces, which act in the direction of the plane perpendicular to the axis; & this is done in what follows.*

reliquas binas massas pertinentes habebunt ad se invicem eas rationes, quæ pro systemate trium massarum deductæ sunt. Hinc ubi systema massarum utcunque dispersarum converti debet circa axem aliquem, sive de æquilibrii centro agatur, sive de centro oscillationis, sive de centro percussionis, licebit considerare massas singulas connexas cum binis punctis utcunque assumptis in axe, & cum alio puncto, vel massa quavis utcunque assumpta, vel concepta intra idem systema, & habebitur omnium massarum nexus mutuus, ac applicatio ad omnia ejusmodi centra habebitur eadem, concipiendo tantummodo massas singulas redactas ad planum perpendiculare per rectas ipsi axi parallelas.

Applicatio ad centri oscillationis generalem determinationem.

97. Sic ex. gr. ubi agitur de centro oscillationis, quæ pro massis existentibus in unico plano perpendiculari ad axem rotationis proposui, ac demonstravi respectu puncti suspensionis, centri gravitatis, traducentur ad massas quascunque, utcunque dispersas respectu axis, & respectu rectæ parallelæ axi ductæ per centrum gravitatis, quam rectam Hugenius appellat axem gravitatis. Nimirum centrum oscillationis jacebit in recta perpendiculari axi rotationis transeunte per centrum gravitatis, ac ad babendam ejus distantiam ab axe eodem, si-[299]-ve longitudinem penduli isochroni, satis erit ducere massas singulas in quadrata suarum distantiarum perpendicularium ab eodem axe, & productorum summam dividere per factum ex summa massarum, & distantia perpendiculari centri gravitatis communis ab ipso axe. Rectangulum autem sub binis distantiis centri gravitatis ab axe conversionis, & a centro oscillationis crit æquale summæ omnium productorum, quæ habentur, si massæ singulæ ducantur in quadrata suarum distantiarum perpendicularium ab axe gravitatis, divisæ per summam massarum. Si enim omnes massæ reducantur ad unicum planum perpendiculare axi conversionis, abit is totus axis in punctum suspensionis, totus axis gravitatis in centrum gravitatis, & singulæ distantiæ perpendiculares ab iis axibus evadunt distantiæ ab iis punctis : unde patet generalem Theoriam reddi omnem per solam applicationem systematis massarum trium rite adhibitam.

Aliud utile corollarium pertinens ad centrum oscillationis.

98. Quod ad centrum oscillationis pertinet, erui potest aliud Corollarium, præter illa, quæ proposui, quod summo sæpe usui esse potest : est autem ejusmodi. *Si plurium partium systematis compositarum ex massis quotcunque, utcunque dispersis inventa fuerint seorsim centra gravitatis, & centra oscillationis respondentia dato puncto suspensionis, vel dato axi conversionis ; inveniri poterit centrum oscillationis commune, ducendo singularum partium massas in distantias perpendiculares sui cujusque centri gravitatis ab axe conversionis, & centri oscillationis cujusvis ab eodem, & dividendo productorum summam per massam totius systematis ductam in distantiam centri gravitatis communis ab eodem axe.* Hoc corollarium deducitur ex formula generali eruta in ipso opere num. 334 pro centro oscillationis, quæ respondet figuræ 63 exprimenti unicam massam A ex pluribus quotcunque, quæ concipi possint ubicunque : exprimit autem ibidem P punctum suspensionis, vel axem conversionis, G centrum gravitatis, Q centrum oscillationis, M summam massarum $A + B + C$ &c, & formula est $PQ = \frac{A \times AP^2 + B \times BP^2 + \&c.}{M \times GP}$.

Ejus demonstratio.

99. Nam ex ejusmodi formula est $M \times GP \times PQ = A \times AP^2 + B \times BP^2$ &c. Quare si singularum partium massæ M ducantur in suas binas distantias GP, PQ ; habetur in singulis summa omnium $A \times AP^2 + B \times BP^2$ &c. Summa autem omnium ejusmodi summarum debet esse numerator pro formula pertinente ad totum systema, cum oporteat singulas totius systematis massas ducere in sua cujusque quadrata distantiarum ab axe. Igitur patet numeratorem ipsum rite haberi per summam productorum $M \times GP \times PQ$ pertinentium ad singulas systematis partes, uti in hoc novo Corollario enunciatur.

Usus pro longitudine penduli composito tisochroni facilius invenienda.

100. Usus hujus Corollarii facile patehit. Pendeat ex. gr. globus aliquis suspensus per filum quoddam. Pro globo jam constat centrum gravitatis esse in ipso centro globi, & constat [300] itidem, ac e superioribus etiam Theorematis facile deducitur, centrum oscillationis jacere infra centrum globi, per $\frac{2}{5}$ tertiæ proportionalis post distantiam puncti suspensionis a centro globi, & radium ; pro filo autem considerato ut recta quadam habetur centrum gravitatis in medio ipso filo, & centrum oscillationis, suspensione facta per fili extremum est in fine secundi trientis longitudinis ejusdem fili, quod itidem ex formula

the other two masses will have to one another those ratios that have already been determined for a system of three masses. Hence, when a system of masses arranged in any manner must rotate about some axis, whether it is a question of the centre of equilibrium, or of the centre of oscillation, or of the centre of percussion, we may consider each of the masses as being connected with a pair of points chosen anywhere on the axis, & with some other point, whether this is some mass taken in any manner or assumed to be within the same system ; & then, there will be a mutual connection between all the masses, & the same application can be made to all such centres, by merely considering that each of the masses is reduced to a perpendicular plane by means of straight lines parallel to the axis.

Application to the general determination of the centre of oscillation.

97. Thus, for example, when we are concerned with the centre of oscillation, the results which I enunciated for masses existing in a single plane perpendicular to the axis of rotation, and proved, with respect to the point of suspension & the centre of gravity, may be applied to any masses, however disposed with respect to the axis, & with respect to a straight line drawn parallel to the axis through the centre of gravity ; this straight line is called the axis of gravity by Huyghens. That is to say, the centre of oscillation will lie in a straight line perpendicular to the axis of rotation drawn through the centre of gravity ; & to obtain the distance of this centre of oscillation from the axis, or the length of the isochronous pendulum, it will be sufficient to multiply each of the masses by the square of its distance measured perpendicular to the same axis, & to divide the sum of the products by the product of the sum of the masses & the perpendicular distance of the common centre of gravity from the axis. Also the rectangle contained by the two distances of the centre of gravity from the axis of rotation & the centre of oscillation will be equal to the sum of all the products, which are obtained by multiplying each of the masses by the square of its perpendicular distance from the axis of gravity, divided by the sum of the masses. For, if all the masses are reduced to a single plane perpendicular to the axis of rotation, the whole axis merely becomes the point of suspension, the whole axis of gravity becomes the centre of gravity, & each of the perpendicular distances from these axes becomes a distance from these points. Thus, it will be clear that the whole of the general theory is obtained by the application of the system of three masses alone, if this is correctly done.

Another useful corollary pertaining to the centre of oscillation.

98. As regards the centre of oscillation, there can be derived another corollary, besides the one that I have enunciated ; & this has often been of great service to me ; it is as follows. *If, for two or more parts of a system composed of any number of masses, situated in any manner, the centres of gravity, & the centres of oscillation corresponding to a given point of suspension, or a given axis of rotation, have been separately determined ; then, the common centre of oscillation can be determined by multiplying the mass of each of the parts by the perpendicular distance of its centre of gravity from the axis of rotation, & the perpendicular distance of the centre of oscillation from the same axis ; & dividing the sum of these products by the mass of the whole system, & the distance of the common centre of gravity from the same axis.* This corollary is derived from the general formula derived in the work itself, Art. 334, for the centre of oscillation, which corresponds to Fig. 63, showing a single mass A out of any number whatever that might be conceived anywhere ; also in the same diagram, the point P is the point of suspension, or the axis of rotation, G the centre of gravity, Q the centre of oscillation, M the sum of the masses A + B + C, &c, and the formula is

$$PQ = \frac{A \times AP^2 + B \times BP^2 + \&c.}{M \times GP}$$

Demonstration of this corollary.

99. Thus, from the formula given, we have

$$M \times GP \times PQ = A \times AP^2 + B \times BP^2 + \&c.$$

Hence, if the mass, M, of each of the parts is multiplied each by its own two distances GP, PQ, we have for each the total sum $A \times AP^2 + B \times BP^2 + \&c.$ But the sum of all such sums as these must be the numerator belonging to the formula for the whole system, since we have to multiply each of the masses of the whole system by the square of its distance from the axis. Therefore, it is plain that the numerator can be correctly taken to be the sum of the products $M \times GP \times PQ$ belonging to the several parts of the system, as we have stated in this new corollary.

Its use in providing an easy determination of the length of a pendulum isochronous with a given composite pendulum.

100. The use of this corollary will be easily seen. For example, suppose we have a sphere suspended by a thin rod. For a sphere, it is well-known that the centre of gravity is at the centre of the sphere ; and it is also well-known, & indeed it can be easily deduced from the theorems given above, that the centre of oscillation lies below the centre of the sphere, at a distance from it equal to two-fifths of the third proportional to the distance of the point of suspension from the centre & the radius. For the rod, considered as a straight line, the centre of gravity is at the middle point of the rod ; & the centre of oscillation, when the suspension is made from one end of the rod, is two-thirds of the length of the rod from that end ; & this can also be deduced quite easily from the general formula. Hence

generali facillime deducitur. Inde centrum oscillationis commune globi, & fili nullo negotio definietur per corollarium superius.

Calculus & formula pro pendulo globi pendentis e filo.

101. Sit Longitudo fili a, massa seu pondus b, radius globi r, massa seu pondus p: erit distantia centri gravitatis fili ab axe conversionis crit $\frac{1}{2}a$, distantia centri oscillationis ejusdem $\frac{2}{3}a$. Quare productum illud pertinens ad filum erit $\frac{1}{3}\ a^2b$. Pro globo erit distantia centri gravitatis $a + r$, quæ ponatur $= m$; Distantia centri oscillationis erit $m + \frac{2}{5} \times \frac{rr}{m}$. Quare productum pertinens ad globum erit $m^2p + \frac{2}{5}\ rrp$. Horum summa est $m^2p + \frac{2}{5}\ rrp + \frac{1}{3}\ a^2b$. Porro cum centra gravitatis fili, & globi jaceant in directum cum puncto suspensionis, ad habendam distantiam centri gravitatis communis ductam in summam massarum satis erit ducere singularum partium massas in suorum centrorum distantias, ac habebitur $mp + \frac{1}{2}\ ab$. Quare formula pro centro oscillationis utriusque simul, erit

$$\frac{m^2p + \frac{2}{5}\ rrp + \frac{1}{3}a^2b}{mp + \frac{1}{2}ab}.$$

Non licere hic concipere massas singulas ut collectas in suis centris oscillationis, aut gravitatis, aut aliis intermediis documentum utile.

102. Hic autem notandum illud, ad centrum oscillationis commune habendum non licere singularem partium massas concipere, ut collectas in suis singulas aut centris oscillationis, aut centris gravitatis. In primo casu numerator colligeretur ex summa omnium productorum, quæ fierent ducendo singulas massas in quadrata distantiarum centri oscillationis sui; in secundo in quadrata distantiarum sui centri gravitatis. In illo nimirum haberetur plus justo, in hoc minus. Sed nec possunt concipi ut collectæ in aliquo puncto intermedio, cujus distantia sit media continue proportionalis inter illas distantias; nam in eo casu numerator maneret idem, at denominator non esset idem, qui ut idem perseveraret, oporteret concipere massas singulas collectas in suis centris gravitatis, non ultra ipsa. Inde autem patet, non semper licere concipere massas ingentes in suo gravitatis centro, & idcirco, ubi in Theoria centri oscillationis, vel percussionis dico massam existentem in quodam puncto, intelligi debet, ut monui in ipso opere, tota massa ibi compenetrata vel concipi massula extensionis infinitesimæ ut massæ compenetratæ in unico suo puncto æquivaleat.

Transitus ad centrum percussionis: ejus notiones haberi posse plures.

[301] 103. Quod attinet ad centrum percussionis, id attigi tantummodo determinando punctum systematis massarum jacentium in recta quadam, & libere gyrantis, cujus puncti impedito motu sistitur motus totius systematis. Porro æque facile determinatur centrum percussionis in eo sensu acceptum pro quovis systemate massarum utcunque dispositarum, & res itidem facile perficitur, si aliæ diversæ etiam centri percussionis ideæ adhibeantur. Rem hic paullo diligentius persequar.

Initium a notione adhibita in Opere: centri gravitatis status conservatus in motu libero.

104. Inprimis ut agamus de eadem centri percussionis notione, moveatur libere systema quodcunque ita inter se connexum, ut ejus partes mutare non possint distantias a se invicem. Centrum gravitatis totius systematis vel quiescet, vel movebitur uniformiter in directum, cum per theorema inventum a Newtono, & a me demonstratum in ipso Opere num. 250, actiones mutuæ non turbent statum ipsius: systema autem totum sibi relictum vel movebitur motu eodem parallelo, vel convertetur motu æquali circa axem datum transeuntem per ipsum centrum gravitatis, & vel quiescentem cum ipso centro, vel ejusdem uniformi motu parallelo delatum simul, quod itidem demonstrari potest haud difficulter.

Inde erui, in systemate translato cum rotatione, fore rectam cum eo connexam immobilem quovis tempusculo suam; quæ facile definiri possit.

105. Inde autem colligitur illud, in motu totius systematis composito ex motu uniformi in directum, & ex rotatione circulari circa axem itidem translatum haberi semper rectam quandam pertinentem ad systema, nimirum cum eo connexam, pro quovis tempusculo suam, quæ illo tempusculo maneat immota, & circa quam, ut circa quendam axem immotum convertatur eo tempusculo totum systema. Concipiatur enim planum quodvis transiens per axem rotationis circularis, & in eo plano sit recta quævis axi parallela; ea convertetur circa axem velocitate eo majore, quo magis ab ipso distat. Erit igitur aliqua distantia ejus rectæ ejusmodi, ut velocitas conversionis æquetur ibi velocitati, quam habet centrum gravitatis cum axe translato; & in altero e binis appulsibus ipsius rectæ parallelæ gyrantis

the common centre of oscillation for the sphere & the rod together can with little difficulty be determined from the corollary given above.

101. Let the length of the rod be a, its mass or weight b, the radius of the sphere r, and p its mass or weight. The distance of the centre of gravity of the rod from the axis of rotation will be $\frac{1}{2}a$, & the distance of its centre of oscillation will be $\frac{2}{3}a$. Hence, the product required in the case of the rod is $\frac{1}{3}a^2b$. For the sphere, the distance of the centre of gravity will be $a + r$; call this m. Then the distance of the centre of oscillation will be $m + \frac{2}{5} \times \frac{r^2}{m}$. Hence, the product for the sphere will be $m^2p + \frac{2}{5}r^2p$. The sum of these is $m^2p + \frac{2}{5}r^2p + \frac{1}{3}a^2b$. Further, since the centres of gravity of the rod & of the sphere lie in a straight line through the point of suspension, to obtain the distance of the common centre of gravity multiplied by the sum of the masses, it is enough to multiply the mass of each part by the distance of its own centre; in this way we obtain $mp + \frac{1}{2}ab$. Hence the formula for the centre of oscillation for both together will be

Calculation giving the formula for a pendulum formed of a sphere hanging at the end of a thin rod.

$$\frac{m^2p + \frac{2}{5}r^2p + \frac{1}{3}a^2b.}{mp + \frac{1}{2}ab}$$

102. Now, here we have to observe that, in order to find the common centre of oscillation, it will not be permissible to suppose that the mass of each part is condensed at either its centre of oscillation or its centre of gravity. In the first case, the numerator would be formed of the sum of all the products, obtained by multiplying each mass by the square of the distance of its centre of oscillation; & in the second case, by multiplying by the square of the distance of its centre of gravity. Thus, in the former, the numerator found would be greater than it ought to be; & in the latter, less. Further, the masses cannot be considered to be condensed in any point intermediate to these centres, such that its distance is some term of a continued proportion between their distances. For, in that case, the numerator would remain the same when the denominator was not the same; for, in order that the latter should remain the same, it would be necessary to suppose that each mass was condensed at its centre of gravity, & not beyond it. From this it is also evident that it is not always permissible to suppose that huge masses can be at their centre of gravity; &, on this account, when in the theory of the centre of oscillation or percussion I say that there is a mass at a certain point, it must be understood, as I mentioned in the work itself, that the whole mass is compenetrated at the point, or supposed to be a small mass of infinitesimal extension, so as to be equivalent to a mass compenetrated at a single point.

We cannot in this consider each mass as being condensed at either its centre of oscillation or its centre of gravity, or other points intermediate; a serviceable warning, to be taken from the example above.

103. Now, as regards the centre of percussion, I merely touched upon this point, when I determined its position for the case of a system of masses lying in a straight line & gyrating freely; using the idea that the point was such that, if its motion was prevented, the whole system was brought to rest. Further, the centre of percussion is determined with equal facility, when considered in this way, for any system of masses no matter how they are arranged. The matter is also easily accomplished, even if diverse other ideas of the centre of percussion are adopted. In what follows here, I will investigate the matter a little more carefully.

Passing on to the centre of percussion; several different ideas of this point are possible.

104. First of all, to use the same notion of the centre of percussion as above, let the system be in free motion of any sort so long as it is so self-connected that its parts cannot change their distances from one another. Then, the centre of gravity of the whole system will either be at rest, or will move uniformly in a straight line; for, according to a theorem, discovered by Newton, and demonstrated by myself in Art. 250 of the work, the mutual actions will not disturb the state of the centre of gravity. Also the whole system, if left to itself, will either move with the same parallel motion, or will rotate with uniform motion about a given axis passing through the centre of gravity; this axis either remains at rest along with the centre of gravity, or moves together with it with the same parallel uniform motion, as also can be proved without much difficulty.

We will start with the same idea as that used in the work itself; the state of the centre of gravity is conserved in free motion.

105. Also from this it can be deduced that, in a motion of the whole system, compounded of an uniform motion in a straight line and a circular motion about an axis that is also translated, there will always be found a certain straight line belonging to the system, that is to say, connected with it, corresponding to every small interval of time; & this straight line for that small interval of time remains motionless, and about it, as about an immovable axis, the whole system is turned in that short interval of time. For, let any plane be taken passing through the axis of circular motion, and in that plane take any straight line parallel to the axis; then this straight line will be turned about the axis with a velocity that is greater in proportion as its distance from the axis is increased. There will therefore be some distance for such a straight line, such that in that position the velocity of turning will be equal to that velocity of the centre of gravity & the axis carried along with it; & in one or other of the two positions of the parallel straight line, gyrating with the system, when it

Hence we derive the fact that, when a system is translated with rotation, there will be, corresponding to any short interval of time, a certain straight line connected with the system, which is motionless; & this straight line can easily be determined.

cum systemate ad planum perpendiculare ei plano, quod axis uniformiter progrediens describit, ejus rectæ motus circularis fiet contrarius motui axis ipsius, adeoque motui, quo ipsa axem comitatur, cui cum ibi & æqualis sit, motu altero per alterum eliso, ea recta quiescet illo tempusculo, & systema totum motu composito gyrabit circa ipsam. Nec erit difficile dato motu centri gravitatis, & binarum massarum non jacentium in eodem plano transeunte per axem rotationis, invenire positionem axis, & hujus rectæ immotæ pro quovis dato momento temporis.

Propositio problematis, & præparatio ad solutionem.

106. Quæratur jam in ejusmodi systemate punctum aliquod, cujus motus, si per aliquam vim externam impediatur, debeat mutuis actionibus sisti motus totius systematis, quod punctum, si uspiam fuerit, dicatur centrum percussionis. Concipiantur autem massæ omnes translatæ per rectas parallelas rectæ [302] illi manenti immotæ tempusculo, quo motus sistitur, quam rectam hic appellabimus axem rotationis, in planum ipsi perpendiculare transiens per centrum gravitatis, & in figura 64 exprimatur id planum ipso plano schematis : sit autem ibidem P centrum rotationis, per quod transeat axis ille, sit G centrum gravitatis, & A una ex massis. Consideretur quoddam punctum Q assumptum in ipsa recta PG, & aliud extra ipsam, ac singularum massarum motus concipiatur resolutus in duos, alterum perpendicularem rectæ PQ agentem directione Aa, alterum ipsi parallelum agentem directione PG, ac velocitas absoluta puncta Q dicatur V

Definitio velocitatis absoluutæ, & relativarum cujusvis massæ.

107. Erit PQ . PA : : V . $\frac{PA \times V}{PQ}$, quæ erit velocitas absoluta massæ A. Erit autem PA . Pa · : $\frac{PA}{QA} \times V$. $\frac{Pa}{QA} \times V$, quæ erit velocitas secundum directionem Aa, & PA . Aa · : $\frac{PA}{PQ} \times V$. $\frac{Aa}{PQ} \times V$, quæ erit velocitas secundum directionem PG. Nam in compositione, & resolutione motuum, si rectæ perpendiculares directionibus motus compositi, & binorum componentium constituant triangulum, sunt motus ipsi, ut latera ejus trianguli ipsis respondentia, velocitas autem absoluta est perpendicularis ad AP. Inde vero bini motus secundum eas duas directiones crunt

$$\frac{Pa}{PQ} \times A \times V, \ \& \ \frac{Aa}{PQ} \times A \times V.$$

Evanescentia summæ determinans problema.

108. Jam vero summa $\frac{Aa}{PQ} \times A \times V$ est *zero*, cum ob naturam centri gravitatis summa omnium Aa × A sit æqualis *zero*, & $\frac{V}{PQ}$ sit quantitas data. Quare si per vim externam applicatam cuidam puncto Q, & mutuas actiones sistatur summa omnium motuum $\frac{Pa}{PQ} \times A \times V$, sistetur totus systematis motus, reliqua summa elisa per solas vires mutuas, quarum nimirum summa est itidem *zero*.

Inventio summæ ipsius æquandæ nihilo.

109. Ut habeatur id ipsum punctum Q, concipiatur quævis massa A connexa cum eo, & cum puncto P, vel cum massis ibidem conceptis, & summa omnium motuum, qui ex nexu derivantur in Q, dum extinguitur is motus in omnibus A, debet elidi per vim externam, summa vero omnium provenientium in P, ubi nulla vis externa agit, debet elidi per sese. Hæc igitur posterior summa erit investiganda, & ponenda = o.

Calculus, & formula derivata.

[303] 110. Porro posito radio = 1, est ex Theoremate trium massarum ut P × PQ × 1 ad A × AQ × *sin* QAa, sive ut P × PQ ad A × Qa, ita actio in A perpendicularis ad PQ = $\frac{Pa}{PQ} \times V$ ad actionem in P secundum eandem directionem, quæ evadit $\frac{A \times Qa \times Pa}{P \times PQ^2} \times V$: nimirum ob Q$a$ = PQ − Pa, crit actio in P = $\frac{A \times PQ \times Pa - A \times Pa^2}{P \times PQ^2} \times V$. Cum harum summa debat æquari *zero* demptis communibus $\frac{V}{P \times PQ^2}$, æquabuntur positiva negativis, nimirum posita pro characteristica summæ, habebitur $ʃ.A \times PQ \times Pa = ʃ.A \times Pa^2$, sive $PQ = \frac{ʃ.A \times Pa^2}{ʃ.A \times Pa}$, vel ob $ʃ.\ A \times Pa = M \times PG$, posito ut prius M pro summa massarum, fiet $PQ = \frac{ʃ.A \times Pa^2}{M \times PG}$, qui valor datur ob datas omnes massas A, datas omnes rectas Pa, datam PG. Q.E.F.

arrives in a plane perpendicular to the plane which the uniformly progressing axis describes, the circular motion of the straight line will be in the opposite direction to that of the axis itself, and thus of the motion with which it accompanies the axis; & since it is also equal to it there, the one motion cancels the other, & the straight line will be at rest for the small interval of time, & the whole system will gyrate about it with a compound motion. Nor will it be difficult, given the motion of the centre of gravity, & of two masses not lying in the same plane passing through the axis of rotation, to find the position of this axis & that of the motionless straight line for any given instant of time.

Enunciation of a problem & preparation for the solution.

106. Now let it be required to find in such a system a point, such that, if its motion is prevented by some external force, the motion of the whole system is thereby checked by mutual actions; this point, if there is one, will be called the centre of percussion. Suppose all the masses to be translated along straight lines parallel to the straight line that remains motionless for the small interval of time in which the motion is checked; this straight line we will now call the axis of rotation; & suppose that by this translation they are all brought into a plane perpendicular to the axis of rotation & passing through the centre of gravity. In Fig. 64, let this plane be represented by the plane of the diagram; & there also let P stand for the centre of rotation through which the axis passes; let G be the centre of gravity, & A one of the masses. Consider any point Q, taken in the straight line PG, & another point that is not on this line; & let the motion of each mass be resolved into two, of which one is perpendicular to the straight line PG & acts in the direction A*a*, & the other is parallel to it & acts in the direction PG; let the absolute velocity of the point Q be called V.

Determination of the absolute velocity, & also the relative velocities, of any mass.

107. If v is the absolute velocity of the mass A, we have $PQ : PA = V : v$; therefore $v = V \times PA/PQ$. Similarly, since we have $PA : Pa = V \times PA/PQ : V \times Pa/PQ$; therefore $V \times Pa/PQ$ will be the velocity in the direction A*a*. Also, since we have $PA : Aa = V \times PA/PQ : V \times Aa/PQ$; hence, $V \times Aa/PQ$ will be the velocity in the direction PG. For, in composition and resolution of motion, if straight lines perpendicular to the directions of the resultant motion & its two components form a triangle, then the motions are proportional to the corresponding sides of the triangle; & the absolute velocity is perpendicular to AP. Hence, the two motions in these two directions will be equal to $\frac{Pa}{PQ} \times A \times V$, and $\frac{Aa}{PQ} \times A \times V$.

Evanescence of this sum which determines the problem.

108. Now, the sum of all such as $\frac{Aa}{PQ} \times A \times V$ is equal to *zero*, since, on account of the nature of the centre of gravity, the sum of all such as $Aa \times A$ is equal to *zero*, and V/PQ is a given quantity. Hence, if by means of an external force applied at any point Q, & the mutual actions, the sum of all the motions $\frac{Pa}{PQ} \times A \times V$ is checked, then the whole motion of the system is checked also; for the remaining sum is cancelled by the mutual forces only, of which indeed the sum is also *zero*.

The determination of the sum which is to be equated to zero.

109. In order to find the point Q, take any mass A connected with it & the point P, or with masses supposed to be situated at these points; then the sum of all the motions, which are derived from the connection for Q, when this motion is destroyed for every A, must be cancelled by the external force; but the sum of all these that arise for P, upon which no external force acts, must cancel one another. Hence it is the latter sum that will have to be investigated & put equal to zero.

The calculation, and the formula obtained.

110. Now, if the radius is made the unit, then, from the theorem for three masses, we have the ratio of $P \times PQ \times 1$ to $A \times AQ \times \sin QAa$, or $P \times PQ$ to $A \times Qa$, equal to the ratio of the action at A perpendicular to PQ (which is equal to $\frac{Pa}{PQ} \times V$) to the action at P in the same direction; & therefore the latter is equal to $\frac{A \times Qa \times Pa}{P \times PQ^2} \times V$, that is to say, since $Qa = PQ - Pa$, the action at $P = \frac{A \times PQ \times Pa - A \times Pa^2}{P \times PQ^2} \times V$. Since the sum of all of these has to be equated to *zero*, on cancelling the common factor $V/(P \times PQ^2)$, the positives will be equal to the negatives; hence, using the symbol $\int$ as the characteristic of a sum, we have $\int . A \times PQ \times Pa = \int . A \times Pa^2$; that is, $PQ = \int . A \times Pa^2/(\int . A \times Pa)$. Now, if as before we put M for the sum of all the masses, then $\int . A \times Pa = M \times PG$, & we have $PQ = \int . A \times Pa^2/(M \times PG)$. This value can be determined; for all the masses like A are given, also all the straight lines such as P*a* are given, & PG is given. Q.E.F.

Theorema erutum ex formula.

111. *Corollarium* I. Quoniam aP æquatur distantiæ perpendiculari A a plano transeunte per P perpendiculari ad reetam PG, habebitur hujusmodi Theorema. *Distantia centri percussionis ab axe rotationis in recta ipsi axi perpendiculari transeunte per centrum gravitatis habebitur, ducendo singulas massas in quadrata suarum distantiarum perpendicularium a plano perpendiculari eidem rectæ transeunte per axem ipsum rotationis, ac dividendo summam omnium ejusmodi productorum per factum ex summa massarum in distantiam perpendicularem centri gravitatis communis ab eodem plano.*(i)

Deductio casus, quo jaceant omnes massæ in eodem plano.

[304] 112. *Corollarium* II. Si massæ jaceant in eodem unico plano quovis transeunte per axem ; A, & a congruunt, adeoque distantiæ Pa sunt ipsæ distantiæ ab axe. Quamobrem in hoc casu formula hæc inventa pro centro percussionis congruit prorsus cum formula inventa pro centro oscillationis, & ea duo centra sunt idem punctum, si axis rotationis sit idem, adeoque *in eo casu transferenda sunt ad centrum percussionis, quæcunque pro centro oscillationis sunt demonstrata.*

Si qua massa sit extra : discrimen centri oscillationis, a centro percussionis.

113. *Corollarium* III. Si aliqua massa jaceat extra ejusmodi planum pertinens ad aliam quampiam ; erit ibi Pa minor, quam PA, adeoque *centrum percussionis distabit minus ab axe rotationis, quam distet centrum oscillationis.*

Formulæ deductæ pro pluribus aliis theorematis.

114. *Corollarium* IV. In formula generali $PG = \frac{f . A \times Pa^2}{M \times GP}$ habetur $Pa^2 = PG^2 + Ga^2 - 2PQ \times Ga$. Porro $f . A \times 2PQ \times Ga$ evanescit ob evanescentem $f . A \times Ga$, & $\frac{f . A \times PG^2}{M \times PG}$ est PG. Quare fit $PQ = PG + \frac{f . A \times Ga^2}{M \times PG}$, & $GQ = \frac{f . A \times Ga^2}{M \times PG}$. Inde autem deducuntur sequentia Theoremata affinia similibus pertinentibus ad centrum oscillationis deductis in ipso opere.

Theorema de positione centri gravitatis.

115. *Si impressio ad sistendum motum fiat in recta perpendiculari axi rotationis transeunte per centrum gravitatis, centrum gravitatis facet inter centrum percussionis, & axem rotationis.* Nam PQ evasit major quam PG.

Theorema de duarum distantiarum producto.

116. *Productum sub binis distantiis illius ab his est constans, ubi axis rotationis sit in eodem plano quovis transeunte per centrum gravitatis cum eadem directione in quacunque distantia ab ipso centro gravitatis.* Nam ob $GQ = \frac{f . A \times Ga^2}{M \times PG}$ erit $GQ \times PG = \frac{f . A \times Ga^2}{M}$.

Corollarium inde deductum.

117. *In eo casu punctum axis pertinens ad id planum, & centrum percussionis reciprocantur ; cum nimirum productum sub binis eorum distantiis a constanti centro gravitatis sit constans.*

Axe rotationis abeunte in infinitum, centrum percussionis abire in centrum gravitatis.

118. *Abeunte axe rotationis in infinitum, ubi nimirum totum systema movetur tantummodo motu parallelo, centrum percussionis abit in centrum gravitatis.* Nam altera e binis distantiis excrescente in infinitum, debet altera evanescere. Porro is casus accidit semper etiam, ubi omnes massæ abeunt in unum punctum, quod crit tum ipsum gravitatis centrum to-[305]-tius systematis, & progredietur sine rotatione ante percussionem.

Si axis rotationis transeat per centrum gravitatis, motum sisti non posse.

119. *Abeunte axe rotationis in centrum gravitatis, nimirum quiescente ipso gravitatis centro, centrum percussionis abit in infinitum, nec ulla percussione applicata unico puncto motus sisti potest.* Nam e contrario altera distantia evanescente, altera abit in infinitum.

Centri percussionis positio notabilis.

120. Corollarium V. *Centrum percussionis debet facere in recta perpendiculari ad axem rotationis transeunte per centrum gravitatis.* Id evincitur per quartum e superioribus Theorematis. Solutio problematis adhibita exhibet solam distantiam centri percussionis ab axe illo rotationis. Nam demonstratio manet eadem, ad quodcunque planum perpendiculare

(i) *Facile deducitur ex hoc primo corollario, ad habendum centrum percussionis massarum utcunque dispersarum satis esse singulas massas reducere ad rectam transeuntem per centrum gravitatis, & perpendicularem axi rotationis per rectas ipsi axi perpendiculares, & invenire massarum ita reductarum centrum oscillationis, habito puncto rotationis pro puncto suspensionis ; id enim erit ipsum centrum percussionis quæsitum. Nam distantiæ ab ipso plano perpendiculari illi rectæ, quarum distantiarum fit mentio in hoc corollario, manent eædem in ejusmodi translatione massarum, & evadunt distantiæ a puncto suspensionis. Theorema autem post substitutionem distantiarum a puncto suspensionis pro iis ipsis distantiis ab illo plano exhibet ipsam formulam distantiæ centri oscillationis a puncto suspensionis, quæ habetur num.* 334. *Hinc autem consequitur generalis reciprocatio puncti rotationis, & centri percussionis, ac alia plura in sequentibus deducta multo immediatius deducuntur e proprietatibus centri oscillationis jam demonstratis.*

A Theorem derived from this formula.

111. *Collorary I.* Since aP is equal to the perpendicular distance of A from a plane passing through P perpendicular to the straight line PG, we have the following theorem. *The distance of the centre of percussion from the axis of rotation in a straight line perpendicular to it passing through the centre of gravity, will be obtained by multiplying each mass by the square of its perpendicular distance from a plane passing through the axis of rotation, & perpendicular to the straight line; & then dividing the sum of all such products by the product of the sum of all the masses multiplied by the perpendicular distance of the common centre of gravity from the same plane.*(i)

Deduction of the case in which all the masses lie in the same plane.

112. *Corollary II.* If the masses lie in any the same single plane passing through the axis, A & a coincide, & therefore the distances Pa become the distances of the masses from the axis. Hence, in this case, the formula here found for the centre of percussion agrees in every way with the formula found for the centre of oscillation; thus the two centres are the same point, if the axis of rotation is the same. Hence, *in this case, everything that has been proved for the centre of oscillation, holds good for the centre of percussion.*

If any mass does not lie within the plane, there is a distinction between the centre of oscillation & the centre of percussion.

113. *Corollary III.* If any mass lies outside the plane belonging to any other, then Pa will be less than PA; hence, *the centre of percussion will be at a less distance from the axis of rotation than the centre of oscillation.*

Formulæ deduced for several other theorems.

114. *Corollary IV.* In the general formula $\mathrm{PQ} = \dfrac{\int.\mathrm{A}\times \mathrm{P}a^2}{\mathrm{M}\times \mathrm{GP}}$, we have $\mathrm{P}a^2 = \mathrm{PG}^2 + \mathrm{G}a^2 - 2\,\mathrm{PQ}\times \mathrm{G}a$. Also, the sum $\int.\mathrm{A}\times 2\,\mathrm{PQ}\times \mathrm{G}a$ vanishes, since $\int.\mathrm{A}\times \mathrm{G}a$ vanishes; & $\int.\mathrm{A}\times \mathrm{PG}^2/(\mathrm{M}\times \mathrm{PG}) = \mathrm{PG}$. Hence we have

$$\mathrm{PQ} = \mathrm{PG} + \frac{\int.\mathrm{A}\times \mathrm{G}a^2}{\mathrm{M}\times \mathrm{PG}},\ \&\ \mathrm{GQ} = \frac{\int.\mathrm{A}\times \mathrm{G}a^2}{\mathrm{M}\times \mathrm{PG}}.$$

From this can be deduced the following theorems like to similar theorems pertaining to the centre of oscillation deduced in the work itself.

Theorem concerning the position of the centre of gravity.

115. *If the impressed force applied for the purpose of checking motion is in a straight line perpendicular to the axis of rotation & passing through the centre of gravity, the centre of gravity will lie between the centre of percussion & the axis of rotation.* For PQ is greater than PG.

Theorem concerning the product of the two distances.

116. *The product of the two distances of the former from the two latter is constant, when the axis of rotation is in any the same plane passing through the centre of gravity, the direction of measurement being the same for any distance from the centre of gravity.* For, since

$$\mathrm{CQ} = \frac{\int.\mathrm{A}\times \mathrm{G}a^2}{\mathrm{M}\times \mathrm{PG}},\ \text{therefore}\ \mathrm{GQ}\times \mathrm{PG} = \frac{\int.\mathrm{A}\times \mathrm{G}a^2}{\mathrm{M}}.$$

Corollary derived from this.

117. *In that case, the point on the axis corresponding to the plane & the centre of percussion will be interchangeable; for, the product of their two distances from a constant centre of gravity is constant.*

If the axis of rotation goes off to infinity, the centre of percussion will become coincident with the centre of gravity.

118. *If the axis of rotation goes off to infinity, that is to say, when the whole system is translated with simply a parallel motion, the centre of percussion will become coincident with the centre of gravity.* For, if one of the two distances increases indefinitely, the other must become evanescent. Also, this will always happen, when all the masses coincide at a single point; this point will then be the centre of gravity of the whole system, & it will be moving without rotation before percussion.

If the axis of rotation passes through the centre of gravity, the motion cannot be checked.

119. *If the axis of rotation passes through the centre of gravity, the centre of percussion passes off to infinity, & the motion cannot be checked by any blow applied at a single point.* For, on the contrary, when the finite distance vanishes, the other distance must become infinite.

A noteworthy position of the centre of percussion.

120. *Corollary V. The centre of percussion must lie in the straight line perpendicular to the axis of rotation & passing through the centre of gravity.* This is proved by the fourth of the theorems given above. The method of solution of the problem that was employed shows the unique distance of the centre of percussion from the axis of rotation. For, the demonstration remains the same, no matter to what plane perpendicular to the axis all the

(i) *It is easily deduced from this first corollary that, in order to obtain the centre of percussion of any masses however arranged, it is sufficient to reduce each of the masses to a straight line passing through the centre of gravity & perpendicular to the axis of rotation, by means of straight lines perpendicular to the axis; & then to find the centre of oscillation of the masses thus reduced, the point of rotation being taken as the point of suspension. This will be the centre of percussion required. For, the distances from the plane perpendicular to the straight line, such as are mentioned in this corollary, remain the same in this kind of translation of the masses & become the distances from the point of suspension. Moreover, the theorem, after the substitution of the distances from the point of suspension for the distances from the plane, gives the same formula for the distance of the centre of oscillation from the point of suspension, which was obtained in Art. 334. From it also there follows the general reciprocity of the point of rotation & the centre of percussion; & many other things deduced in what follows can be more easily derived from the properties of the centre of oscillation already proved.*

axi reducantur per rectas ipsi axi parallelas & massæ omnes, & ipsum centrum gravitatis commune, adeoque inde non haberetur unicum centrum percussionis, sed series eorum continua parallela axi ipsi, quæ abeunte axe rotationis ejus directionis in infinitum, nimirum cessante conversione respectu ejus directionis, transit per centrum gravitatis juxta id Theorema. Porro si concipiatur planum quodvis perpendiculare axi rotationis, omnes massæ respectu rectarum perpendicularium axi priori in eo jacentium rotationem nullam habent, cum distantiam ab eo plano non mutent, sed ferantur secundum ejus directionem, adeoque respectu omnium directionum priori axi perpendicularium jacentium in eo plano res eodem modo se habet, ac si axis rotationis cujusdam ipsas respicientis in infinitum distet ab earum singulis, & proinde respectu ipsarum debet centrum percussionis abire ad distantiam, in qua est centrum gravitatis, nimirum jacere in eo planorum parallelorum omnes ejusmodi directiones continentium, quod transit per ipsum centrum gravitatis : adeoque ad sistendum penitus omnem motum, & ne pars altera procurrat ultra alteram, & eam vincat, debet centrum percussionis jacere in plano perpendiculari ad axem transeunte per centrum gravitatis, & debent in solutione problematis omnes massæ reduci ad id ipsum planum, ut præstitimus, non ad aliud quodpiam ipsi parallelum : ac eo pacto habebitur æquilibrium massarum, hinc & inde positarum, quarum ductarum in suas distantias ab eodem plano summæ hinc, & inde acceptæ æquabuntur inter se. Porro eo plano ad solutionem adhibito, patet ex ipsa solutione, centrum percussionis jacere in recta perpendiculari axi ducta per centrum gravitatis : jacet enim in recta, quæ a centro gravitatis ducitur ad illud punctum in quo axis id planum sceat, quæ recta ipsi axi perpendicularis toti illi plano perpendicularis esse debet.

Impactus in centrum percussionis qui sit.

121. Corollarium VI. *Impactus in centro percussionis in corpus externa vi ejus motum sistens est idem, qui esset, si singulæ massæ incurrerent in ipsum cum suis velocitatibus respecti-*[306]*-vis redactis ad directionem perpendicularem plano transeunti per axem rotationis, & centrum gravitatis, sive si massarum summa in ipsum incurreret directione, & velocitate motus, qua fertur centrum gravitatis.*

Demonstratio primæ partis.

122. Patet primum, quia debet in Q haberi vis contraria directioni illins motus perpendicularis plano transeunti per axem, & PG, par extinguendis omnibus omnium massarum velocitatibus ad eam directionem redactis, quæ vis itidem requireretur, si omnes massæ eo immediate devenirent cum ejusmodi velocitatibus.

Demonstratio secundæ.

123. Patet secundum ex eo, quod velocitas illa pro massa A sit $\frac{Pa}{PQ} \times V$, adeoque motus $\frac{A \times Pa}{PQ} \times V$, quorum motuum summa est $\frac{M \times PG}{PQ} \times V$. Est autem $\frac{PG}{PQ} \times V$, velocitas puncti G, quod punctum movetur solo motu perpendiculari ad PG, adeoque si massa totalis M incurrat in Q cum directione, & celeritate, qua fertur centrum gravitatis G, faciet impressionem candem.

Impressio ubi fieri possit extra centrum percussionis cum eodem effectu.

124. Corollarium VII. *Potest motus sisti impressione facta etiam extra rectam PG, seu extra planum transiens per axem rotationis, & centrum gravitatis, nimirum si impressio fiat in quodvis punctum rectæ eidem plano perpendicularis, & transeuntis per* Q, *directione rectæ ipsius.* Nam per nexum inter id punctum, & Q statim impressio per eam rectam transfertur ab eo puncto ad ipsum Q.

Motus communicatus quovis impactu systemati quiescenti.

125. Corollarium VIII. *Contra vero si imprimatur dato cuidam puncto systematis quiescentis vis quædam motrix ; invenietur facile motus inde communicandus ipsi systemati.* Nam ejusmodi motus crit is, qui contrario æquali impactu sisteretur. Determinatio autem regressu facto per ipsam problematis solutionem erit hujusmodi. Centrum gravitatis commune movebitur directione, qua egit vis, & velocitate, quam ea potest imprimere massæ totius systematis, quæ ad eam, quam potest imprimere massæ cuivis, est ut hæc posterior massa ad illam priorem, & si vis ipsa applicata fuerit ad centrum gravitatis, vel immediate, vel per rectam tendentem ad ipsum ; systema sine ulla rotatione movebitur eadem velocitate : sin autem applicetur ad aliud punctum quodvis directione non tendente ad ipsum centrum gravitatis, præterea habebitur conversio, cujus axis, & celeritas sic invenietur. Per centrum gravitatis G agatur planum perpendiculare rectæ, secundum quam fit impactus, & notetur punctum Q, in quo eidem plano occurrit eadem recta. Per ipsum punctum G ducatur in eo plano recta perpendicularis ad QG, quæ crit axis quæsitus. Per punctum Q concipiatur alterum planum perpendiculare rectæ GQ, ca-[307]-piantur omnes distantiæ perpendiculares omnium massarum A ab ejusmodi plano, æquales nimirum suis aQ :

masses & their common centre of gravity are reduced by straight lines parallel to the axis. Thus, from it, we should not obtain a single centre of percussion, but a continuous series of them parallel to the axis; & this, when the axis of rotation goes off to infinity for this direction, that is, when turning ceases for this direction, will pass through the centre of gravity, according to the theorem. Further, if any plane perpendicular to the axis of rotation is taken, all the masses have no rotation with regard to straight lines perpendicular to the former axis which lie in the plane; for they will not change their distances from that plane, but are carried in its direction. Hence, with regard to all directions perpendicular to the former axis which lie in that plane, the matter comes out in the same way; &, if the axis of rotation for any one of the former is infinitely distant from each of the latter, and therefore with respect to the former, the centre of percussion has to pass to that distance at which is the centre of gravity, that is to say, has to lie in that one of the parallel planes containing all such directions, which passes through the centre of gravity. Thus, to stop all motion entirely, & to prevent one part outrunning another part & overcoming it, the centre of percussion must lie in a plane perpendicular to the axis & passing through the centre of gravity; & in the solution of the problem, all the masses are bound to be reduced to that plane, as we have shown, & not to any other that is parallel to it. In this way, we shall obtain equilibrium of the masses, situated on either side of it; & the sums of these multiplied by their distances from this plane, taken together on one side & on the other, will be equal to one another. Moreover, if this plane is used for the solution, it is clear from the solution itself, that the centre of percussion lies in a straight line perpendicular to the axis, drawn through the centre of gravity. For, it will lie in the straight line that is drawn from the centre of gravity to that point in which the axis cuts the plane, & this straight line must be perpendicular to the axis, since the axis is perpendicular to the whole of the plane.

The nature of the impact at the centre of percussion.

121. *Corollary VI. The impact at the centre of percussion on a body by an external force, which checks its motion, is the same as we should have, if each mass were to collide with it with its velocity resolved in the direction perpendicular to the plane passing through the axis of rotation & the centre of gravity; or if the sum of the masses collided with it with the direction & velocity of motion, with which the centre of gravity is moving.*

Proof of the first part.

122. The first part is evident, because there must be at Q a force opposite in direction to the motion perpendicular to the plane passing through the axis & PG, capable of destroying all the velocities of all the masses resolved in that direction; & this force would also be required, if all the masses collided with it directly with such velocities.

Proof of the second part.

123. The second part is evident from the fact that the velocity for the mass A is $\frac{Pa}{PQ} \times V$; & thus, the motion is $\frac{A \times Pa}{PQ} \times V$; & the sum of these motions is $\frac{M \times PG}{PQ} \times V$. But $\frac{PG}{PQ} \times V$ is the velocity of the point G, & the sole motion of this point is perpendicular to PG; & thus, if the total mass M collided with Q with the direction & speed with which the centre of gravity G moves, it would produce the same effect.

When the blow can be applied beyond the centre of percussion with the same effect.

124. *Corollary VII. The motion may be checked even by a blow applied without the straight line PG, or without the plane passing through the axis of rotation & the centre of gravity; that is, if it is applied at any point of a straight line perpendicular to the same plane, & passing through Q, in the direction of this straight line.* For, through the connection between that point & Q, the blow is immediately transferred along the straight line from the point to Q itself.

Motion communicated by a blow to a system at rest.

125. *Corollary VIII. On the other hand, if any motive force is impressed upon any given point of a system at rest, it is easy to find the motion thereby communicated to the system.* For such motion will be that which would be checked by an equal & opposite blow. The determination of the motion, made by retracing our steps through the solution of that problem, would proceed as follows. The common centre of gravity will be moved in the direction in which the force acts, & with a velocity which it can give to the mass of the whole system; this velocity is to that which it could give to any mass as the latter mass is to the former. If the force were applied at the centre of gravity, either directly, or along a straight line tending to it, then the system, without any rotation, would move with the same velocity. But if it were applied at any other point in a direction not tending towards the centre of gravity, we should have in addition a rotation, of which the axis & the velocity will be found thus. Let a plane be drawn through the centre of gravity G perpendicular to the straight line along which the blow is impressed, & let the point in which the straight line meets this plane be denoted as the point Q. Through G draw in this plane a straight line perpendicular to QG; this will be the axis required. Draw another plane through the point Q, perpendicular to the straight line QG; take all the perpendicular distances of all the masses A

singularum quadrata ducantur in suas massas, & factorum summa dividatur per summam massarum, tum in recta GQ producta capiantur GP æqualis ; ei quoto diviso per ipsam QG, & celeritas puncti P revolventis circa axem inventum in circulo, cujus radius GP, erit æqualis celeritati inventæ centri gravitatis, directio autem motus contraria eidem. Unde habetur directio, & celeritas motus punctorum reliquorum systematis.

Demonstratio.

126. Patet constructio ex eo, quod ita motu composito movebitur systema circa axem immotum transeuntem per P, qui motus regressu facto a constructione tradita ad inventionem præmissam centri percussionis sisteretur impressione contraria, & æquali impressioni datæ.

Aditus ad perquisitiones ulteriores motu impresso systemati moto.

127. *Scholium.* Hoc postremo corollario definitur motus vi externa impressus systemati quiescenti. Quod si jam systema habuerit aliquem motum progressivum, & circularem, novus motus externa vi inductus juxta corollarium ipsum componendus erit cum priore, quod, quo pacto fieri debeat, hic non inquiram, ubi centrum percussionis persequor tantummodo. Ea perquisitio ex iisdem principiis perfici potest, & ejus ope patet, aperiri aditum ad inquirendas etiam mutationes, quæ ab inæquali actione Solis, & Lunæ in partes supra globi formam extantes inducuntur in diurnum motum, adeoque ad definiendam ex genuinis principiis præcessionem æquinoctiorum, & nutationem axis sed ea investigatio peculiarem tractionem requirit.

Transitus ad aliam notionem ejus centri.

128. Interea gradum hic faciam ad aliam notionem quandam centri percussionis, nihilo minus, imo etiam magis aptam ipsi nomini. Ad eam perquisitionem sic progrediar.

Problema continens hanc ideam.

129. Problema. *Si systema datum gyrans data velocitate circa axem datum externa vi immotum incurrat in dato suo puncto in massam datam, delatam velocitate data in directione motus puncti ejusdem, quam massam debeat abripere secum ; quæritur velocitas, quam ei massæ imprimet, & ipsum systema retinebit post impactum.*

Solutio : formulæ continentes motum massæ in quam incidit, & suum reliquum.

130 Concipiatur totum systema projectum in planum perpendiculare axi rotationis transiens per centrum gravitatis G, in quo plano punctum conversionis sit P, massa autem in recta PG in Q. Velocitas puncti cujusvis systematis, quod distet ab axe per intervallum $= 1$, ante incursum sit $= a$, velocitas ab eodem amissa sit $= x$, adeoque velocitas post impactum $= a - x$, velocitas autem massæ Q ante impactum sit $= PQ \times b$. Erit ut 1 ad AP, ita x ad velocitatem amissam a massa A, quæ crit $AP \times x$. Erit autem ut 1 ad $a - x$ ita PQ ad velocitatem residuam in puncto systematis Q, quæ fiet $PQ \times (a - x)$, & ea crit itidem velocitas massæ Q post [308] impactum, adeoque massa Q acquiret velocitatem $PQ \times (a - b - x)$, sive posito $a - b = c$, habebitur $PQ \times (c - x)$. Porro ex mutuo nexu massæ A cum P, & Q crit $Q \times PQ$ ad $A \times AP$, ut effectus ad velocitatem pertinens in $A = AP \times x$ ad effectum in $Q = \frac{A \times AP^2}{Q \times QP} \times x$. Summa horum effectuum provenientium e massis omnibus erit æqualis velocitati acquisitæ in Q. Nimirum $\frac{\int . A \times AP^2}{Q \times QP} \times x = QP \times c - QP \times x$, sive $\frac{\int . A \times AP^2 + Q \times QP^2}{Q \times QP} \times x = QP \times c$, & $x = \frac{Q \times QP^2}{\int . A \times AP^2 + Q \times QP^2} \times c$. Dato autem x datur $a - x$, & is valor ductus in distantiam puncti cujusvis systematis, vel etiam massæ Q, exhibebit velocitatem quæsitam. Q.E.F.

Casus particulares, ad quos applicari potest.

131. *Scholium.* Formula habet locum etiam pro casu, quo massa Q quiescat, vel quo feratur contra motum systematis, dummodo in primo casu fiat $b = 0$, & $c = a$, ac in secundo valor b mutetur in negativum, adeoque sit $c = a + b$. Posset etiam facile applicari ad casum, quo in conflictu ageret elasticas perfecta vel imperfecta. Determinatio tradita exhiberet partem effectus in collisione facti tempore amissæ figuræ, ex quo effectus debitus

from this plane, each equal to the corresponding aQ; multiply the square of each of these by the corresponding mass, & divide the sum of all the products by the sum of the masses. Then in the straight line QG produced take GP equal to this quotient divided by QG. The velocity of the point P rotating in a circle about the axis which has been found, of which the radius is GP, will be equal to the velocity of the centre of gravity which has also been found, but the direction of the motion will be in the opposite direction. From this, we have the direction & the velocity for all the other points of the system.

Demonstration.

126. The correctness of the construction is evident from the fact that in this way the system will move with a compound motion in a circle about a motionless axis passing through P; & this motion, by retracing our steps from the construction for finding the centre of percussion, already given, would be checked by a blow equal & opposite to the given blow.

The way is open for further investigations when motion is impressed on a moving system.

127. *Scholium.* In the last corollary the motion impressed by an external force on a system at rest is determined. But if now the system should have some motion, progressive & circular, the new motion induced by the external force in accordance with the corollary will have to be compounded with what it already has. I do not inquire here, how this will happen, for here I am only concerned with the centre of percussion. The investigation can be carried out by means of the very same principles; & by the help of this investigation, it is clear that the door would be opened also for the investigation of the variations which are induced in the daily motion by the unequal actions of the Sun, & of the Moon, on parts of the Earth that jut out beyond the figure of the sphere; & thus for determining from real principles the precession of the equinoxes & the nutation of the axis. But this investigation requires a special treatise.

Passing on to another idea of this centre.

128. Meanwhile, I will now go on to another idea of the centre of percussion, which is no less, nay it is even more, fit to have that name given to it. To this investigation I proceed in the following manner.

Problem embodying the idea.

129. Problem. *If a given system, gyrating with given velocity about a given axis, not acted upon by an external force, collides at a given point of itself with a given mass, which is moving with a given velocity in the direction of the motion of this point, the mass being of necessity borne along with the system; it is required to find the velocity impressed on the mass, & retained by the system after impact.*

Solution; formulæ containing the motion of the mass with which it collides, and the motion left in itself.

130. Suppose that the whole system is projected on a plane perpendicular to the axis of rotation passing through the centre of gravity G; in this plane let the point of rotation be P, & let the mass be in the straight line PG at Q. Let the velocity of any point of the system, whose distance from the axis is unity, before the impact be a, & let the velocity lost by it be x; & thus, the velocity after impact will be $a - x$. Also let the velocity of the mass at Q before impact be $PQ \times b$. Then, as 1 is to AP so is x to the velocity lost by the mass at A, which will therefore be $AP \times x$. Also, as 1 is to $a - x$ so is PQ to the velocity that remains in the point Q of the system; & therefore this is $PQ \times (a - x)$; this will also be the velocity of the mass Q after impact. Hence, the mass Q will acquire a velocity $PQ \times (a - b - x)$; or, if we put $a - b = c$, it will be $PQ \times (c - x)$. Further, from the mutual connection between the mass A & P & Q, we shall have the ratio of $Q \times PQ$ to $A \times AP$ equal to that of the effect pertaining to the velocity at A, which is equal to $AP \times x$, to the effect at Q, which is therefore equal to $\frac{A \times AP^2}{Q \times QP} \times x$. The sum of these effects, arising from all the masses, will be equal to the velocity acquired at Q. That is to say, we have

$$\frac{.A \times AP^2}{Q \times QP} \times x = QP \times c - QP \times x,$$

$$\text{or } \frac{\int. A \times AP^2 + Q \times QP^2}{Q \times QP} \times x = QP \times c;$$

$$\text{and } x = \frac{Q \times QP^2}{\int. A \times AP^2 + Q \times QP^2} \times c.$$

But, if we are given x, we are also given $a - x$; and this value, multiplied by the distance of any point of the system, or also that of the mass Q, will give the velocity required. Q.E.F.

Particular cases to which it can be applied.

131. *Scholium.* The formula holds good even when the mass Q is at rest, or when it moves in the opposite direction to the system; so long as, in the first case, b is made equal to zero, or $c = a$; & in the second case, the value of b is changed from positive to negative, so that $c = a + b$. It might also easily be applied to the case in which elasticity, either perfect or imperfect, would take a part in the collision. The determination given would represent that part of the effect of the collision which was produced during the interval of time corresponding to loss of shape; & from this the proper effect for the whole

tempori totus collisionis usque ad finem recuperatæ figuræ colligitur facile, duplicando priorem, vel augendo in ratione data uti fit in collisionibus.

Ejusdem ulterior extensio.

132. Itidem locum habet pro casu, quo massa nova non jaceat in Q in recta PG, sed in quovis alio puncto plani perpendicularis axi transeuntis per G, ex quo si intelligatur perpendiculum in PG ei occurrens in Q; idem prorsus erit impactus ibi, qui esset in Q, translata actione per illam systematis rectam. Qui imo si Q non jaceat in eo plano perpendiculari ad axem, quod transit per centrum gravitatis, sed ubivis extra, res eodem redit, dummodo per id punctum concipiatur planum perpendiculare axi illi immoto per vim externam ad quod planum reducatur centrum gravitatis, & quævis massa A; vel si ipsa massa Q cum reliquis reducatur ad quodvis aliud planum perpendiculare axi. Omnia eodem recidunt ob id ipsum, quod axis externa vi immotus sit. Sed jam ex generali solutione problematis deducimus plura Corollaria.

Relatio ad centrum oscillationis.

133. *Corollarium* I. Si distantia centri oscillationis totius systematis ab axe P dicatur R, distantia centri gravitatis G, massa tota M, habebitur

$$x = \frac{Q \times PQ^2}{M \times G \times R + Q \times PQ^2} \times c, \text{ \& [309] } \frac{c}{x} = \frac{M \times G \times R}{Q \times PQ^2} + 1.$$

Patet ex eo, quod ex natura centri oscillationis habetur $R = \frac{\int . A \times AP^2}{M \times G}$, adeoque $\int . A \times AP^2 = M \times G \times R$.

Expressio velocitatis in massa simplicior ope illius.

134. *Corollarium* II. Velocitas acquisita a massa Q erit $\frac{M \times G \times R \times PQ}{M \times G \times R + Q \times PQ^2} \times c$.

Est enim ea velocitas $PQ \times (c - x)$, sive $PQ\ (c - \frac{Q \times PQ^2}{M \times G \times R + Q \times PQ^2} \times c)$, quod reductum ad eundem denominatorem elisis terminis contrariis eo redit.

Ubi colligendum esset totum systema ad eandem velocitatem imprimendam massæ.

135. *Corollarium* III. Si manente velocitate circulari systematis tota ejus massa concipiatur collecta in unico puncto jacente inter centra gravitatis, & oscillationis, cujus distantia a puncto conversionis sit media geometrice proportionalis inter distantias reliquorum punctorum, vel in eadem distantia ex parte opposita; velocitas eadem imprimeretur novæ massæ in quovis puncto sitæ. Tunc enim abiret in illud punctum utrumque centrum, & valor $G \times R$ esset idem, ac prius, nimirum æqualis quadrato ejus distantiæ ab axe, quod quadratum est positivum etiam, si distantia accepta ex parte opposita fiat negativa.

In quot, & quibus distantiis ab axe massa eandem ex impactu velocitatem acquireret: ubi maximam.

136. *Corollarium* IV. Si capiatur hinc, vel inde in PG segmentum, quod ad distantiam ejus puncti ab axe sit in subduplicata ratione massæ totius systematis ad massam Q; ipsa massa Q in quatuor distantiis ab axe, binis hinc, & binis inde, quarum binarum producta æquentur singula quadrato ejus segmenti, acquiret velocitatem in omnibus candem magnitudine, licet in binis directionis contrariæ, & ea fiet maxima, ubi ipsa massa sit in fine ejus segmenti ex parte axis ultralibet. Erit enim velocitas acquisita directe ut $\frac{M \times G \times R \times PQ}{M \times G \times R + Q \times PQ^2} \times c$, vel dividendo per constantem $\frac{M \times G \times R}{Q} \times c$, & ponendo illud segmentum $= \pm T$, cujus quadratum T^2 debet esse $= \frac{M}{Q} \times G \times R$, erit directe ut $\frac{PQ}{T^2 + PQ^2}$, adeoque reciproce ut $\frac{T^2}{PQ^2} + PQ$. Is autem [310] valor manet idem, si pro PQ ponantur bini valores, quorum productum æquatur T^2, migrante tantummodo altera binomii parte in alteram Si enim alter valor sit m, erit alter $\frac{T^2}{m}$; & posito illo pro PQ: habetur $\frac{T^2}{m} + m$, posito hoc habetur $\frac{T^2 m}{T^2} + \frac{T^2}{m}$, sive $m + \frac{T^2}{m}$. Sed cum eæ distantiæ abeunt ad partes oppositas, fiunt $-m$, & $\frac{T^2}{m}$, migrante in negativum etiam

time of collision, up to the end of recovery of shape could be easily derived, by doubling in the first case, & by increasing in a given ratio in the second case; just as was done when we considered collisions.

132. The formula also holds good for the case in which the new mass does not lie at the point Q in the straight line PG, but at some other point of a plane perpendicular to the axis & passing through G; if from this point a perpendicular is supposed to be drawn to PG, meeting it in Q, then the effect will be exactly the same as if the impact had been at Q, the action being transferred by this straight line of the system. Indeed, if Q does not lie in the plane perpendicular to the axis, which passes through the centre of gravity, but somewhere without it, it all comes to the same thing, so long as through that point a plane is supposed to be drawn perpendicular to the axis that is unmoved by the external force, and the centre of gravity is reduced to this plane, together with any mass A; or if the mass Q, together with the rest, is reduced to any plane perpendicular to the axis. It all comes to the same thing, on account of the fact that there is an axis that is unmoved by the external force. But now we will deduce several corollaries from the general solution of the problem.

Further extension of this idea.

133. *Corollary I.* If the distance of the centre of oscillation of the whole system from the axis P is denoted by R, the distance of the centre of gravity by G, & the total mass by M, then we have $x = \frac{Q \times PQ^2}{M \times G \times R + Q \times QP^2} \times c$; & $\frac{c}{x} = \frac{M \times G \times R}{Q \times PQ^2} + 1$. It is evident from the fact that, from the nature of the centre of oscillation, we have

$$R = \frac{\int . A \times AP^2}{M \times G}; \text{ \& thus } \int . A \times AP^2 = M \times G \times R.$$

Relation to the centre of oscillation.

134. *Corollary II.* The velocity acquired by the mass Q will be

$$\frac{M \times G \times R \times PQ}{M \times G \times R + Q \times PQ^2} \times c;$$

for, this is the velocity PQ $(c - x)$, or PQ $\left(c - \frac{Q \times PQ^2}{M \times G \times R + Q \times PQ^2} \times c\right)$; and this, when reduced to the same denominator, comes to that which was given, after cancelling terms of opposite sign.

A simpler expression for the velocity in the mass by its help.

135. *Corollary III.* If, while the circular velocity remained unaltered, the whole mass of the system is supposed to be collected at a single point lying between the centres of gravity & oscillation, the distance of which from the point of rotation is a geometrical mean between the distances of the other points, or at the same distance on the other side of the point of rotation; then, the same velocity would be impressed on the new mass situated at any point. For, in that case, each centre would coincide with that point, & the value of G × R would be the same as before, namely, equal to the square of its distance from the axis; & this square is positive, even if the distance, when taken on the other side of the point of rotation, is negative.

The point in which the whole system would have to be collected in order to impress the same velocity on the mass.

136. If, on one side or the other, in PG a segment is taken, which is to the distance of the point from the axis in the subduplicate ratio of the whole mass of the system to the mass Q; then, the mass Q, if placed at one of four distances from the axis, two on one side & two on the other, so that the products for each pair should be equal to the square of the segment, would at each distance acquire a velocity of the same magnitude although in opposite directions for the two pairs. Also this velocity would be greatest, when the mass was placed at the end of the segment on either side of the axis. For, the velocity acquired varies directly as $\frac{M \times G \times R \times PQ}{M \times G \times R + Q \times PQ^2} \times c$; dividing this by the constant $\frac{M \times G \times R}{Q} \times c$, and denoting the segment by $\pm$ T, of which the square, T^2, must be equal to $\frac{M}{Q} \times G \times R$, the velocity will vary directly as $\frac{PQ}{T^2 + PQ^2}$, & therefore, inversely as $\frac{T^2}{PQ} + PQ$. Now, this value remains the same, if for PQ we substitute either of the pair of values whose product is T^2, the first part of the binomial expression merely interchanging with the second. For, if either value is denoted by m, the other will be T^2/m; &, if the former is substituted for PQ, we get $T^2/m + m$; or, if the latter, we have $T^2m/T^2 + T^2/m$, i.e., $m + T^2/m$. But, when these distances are taken on the opposite side, they become $-m$ & $-T^2/m$, & the value also of the formula becomes negative; this shows that the direction of the motion is opposite to what it was before; in

The number of points, and their distances from the axis, for which the mass would acquire the same velocities from the impact; where the velocity would be greatest.

valore formulæ, quod ostendit directionem motus contrariam priori, systemate nimirum hinc, & inde ab axe in partibus oppositis habente directiones motuum oppositas.

Demonstratio determinationis maximi.

137. Quoniam autem assumpto quovis valore finito pro PQ, formula $\frac{T^2}{PQ} + PQ$ est finita, & evadit infinita facto PQ tam infinito, quam $= 0$; patet in hisce postremis duobus casibus velocitatem e contrario evanescere, in reliquis esse finitam, adeoque alicubi debere esse maximam. Non potest autem esse maxima, nisi ubi ad candem magnitudinem redit, quod accidit in transitu PQ per utrumvis valorem $\pm$ T, circa quem hinc & inde valores æquales sunt. Ibi igitur id habetur maximum.

Maximi determinatio per calculum differentialem.

138. *Scholium* 2. Libuit sine calculo differentiali invenire illud maximum, quod ope calculi ipsius admodum facile definitur. Ponantur $T = t$, & $PQ = z$. Fiet formula $\frac{t^2}{z} + z$, & differentiando $-\frac{t\,dz}{zz} + dz = 0$, sive $-t^2 + z^2 = 0$, vel $z^2 = t^2$, & $z = \pm t$, sive $PQ = \pm T$, ut in corollario 4 inventum est.

Duæ aliæ acceptiones centri percussionis, & ejus determinatio ex superioribus.

139. Licebit autem jam ex postremis duobus corollariis deducere alias duas notiones centri percussionis, cum suis eorundem determinationibus. Potest primo appellari centrum percussionis illud punctum, in quo tota systematis massa collecta eandem velocitatem imprimeret massæ eidem incurrendo in eam eodem suo puncto cum eadem velocitate, quæ videtur omnium aptissima centri percussionis notio. Centrum percussionis in ea acceptione determinatur admodum eleganter ope corollarii 3: jacet nimirum inter centrum gravitatis, & centrum oscillationis ita, ut ejus distantia ab axe rotationis sit media geometrice proportionalis inter illorum distantias, vel ubivis in recta axi parallela ducta per punctum ita inventum. Potest secundo appellari centrum percussionis illud punctum, per quod si fiat percussio, imprimitur velocitas omnium maxima massæ, in [311] quam incurritur. In hac acceptione centrum percussionis itidem eleganter determinatur per corollarium quartum, mutando eam distantiam in ratione subduplicata massæ, in quam incurritur, ad massam totius systematis.

A quo ita consideratum, & pro particulari casu determinatum.

140. In hoc secundo sensu acceptum, & investigatum esse centrum percussionis a summo Geometra Celeberrimo Pisano Professore Perrellio, nuper mihi significavit Vir itidem Doctissimus, & geometra insignis Eques Mozzius, qui & suam mihi ejus centri determinationem exhibuit pro casu systematis continentis unicam massam in rectilinea virga inflexili.

Hic generalius, & aliter determinatum ad fœcunditatem Theoriæ ostendendam.

141. Libuit rem longe alia methodo hic erutam generaliter, & cum superioribus omnibus conspirantem, ac ex iis sponte propemodum profluentem proponere, ut innotescat mira sane fœcunditas Theorematis simplicissimi pertinentis ad rationem virium compositarum in systemate massarum trium. Sed de his omnibus jam satis.

DABAM FLORENTIAE, 17 *Junii*, 1758.

FINIS.

other words, the system has opposite directions for motions of opposite parts on either side of the axis.

137. Now, since, for any assumed finite value of PQ, the formula $T^2/PQ + PQ$ is finite, & comes out infinite both when PQ is made infinite & when it is made zero, it is clear that the velocity, which varies inversely as the formula, must vanish in these two extreme cases, & be finite in all other cases; hence, at some time there must be a maximum. But it cannot be a maximum, except when the two parts of the formula become equal; & this happens as PQ passes through either of the values $\pm$ T, about which, on either side, the values are equal. Hence there is a maximum there.

Demonstration that the maximum is correctly given.

138. *Scholium 2.* I have preferred to find this maximum without the help of the differential calculus; but with the help of the calculus, it can be determined very easily. Put $T = t$, & $PQ = z$; then the formula becomes $t^2/z + z$. Differentiating, we have $-t\,dz/z^2 + dz = 0$, or $-t^2 + z^2 = 0$, or $z^2 = t^2$; & $z = \pm t$, or $PQ = \pm T$, as was found in corollary IV.

Determination of the maximum by means of the differential calculus.

139. We may now, from the last two corollaries, deduce two other ideas of the centre of percussion, together with the determination of each. In the first place, we may call the centre of percussion that point which is such that if the whole mass of the system were collected therein, it would impress the same velocity on the same mass by colliding with it with this same point of itself with the same velocity; & it seems that this is the most apt idea of all for the centre of percussion. The centre of percussion, in this acceptation, is determined in a very elegant manner by the aid of corollary III. Thus, it will lie between the centre of gravity & the centre of oscillation, in such a manner that its distance from the axis of rotation is a geometrical mean between those two distances, or anywhere in a straight line parallel to the axis drawn through the point thus found. Again, the name centre of percussion may be given to that point which is such that, if the blow is delivered through it, it will give to the mass on which it falls the greatest possible velocity. In this acceptation, the centre of percussion is also elegantly determined by the fourth corollary, by changing the distance in the subduplicate ratio of the mass struck to the whole mass of the system.

Two other acceptations of the term centre of percussion; and its determination by means of what has been given above.

140. That learned man & fine geometer, Signor Mozzi, has but lately acquainted me with the fact that the centre of percussion was taken, in this second sense, & investigated by that excellent geometer, the well-known Professor at Pisa, Perrelli; & Mozzi also showed me his own determination for the case of a system consisting of a single mass in the form of a rectilinear inflexible rod.

By whom so considered, and determined in a particular case.

141. I have preferred to set forth the matter here derived in general in a far different manner, agreeing as it does with all that has gone before, & arising from it almost automatically, so as to make known the truly wonderful fertility of that very simple theorem dealing with the ratio of the composite forces in a system of three masses. But now I have said enough about all these things.

Here a more general determination from other principles has been given, in order to show the fertility of the theory.

FLORENCE,
17th June, 1758.

THE END.

INDEX

PARS I

PARS II

Applicatio Theoriæ ad Mechanicam

PARS III

Applicatio Theoriæ ad Physicam

INDEX

PART I

PART II

Application of the Theory to Mechanics

PART III

Application of the Theory to Physics

APPENDIX

Ad Metaphysicam pertinens

SUPPLEMENTA

NOI RIFORMATORI

Dello Studio di Padova.

AVENDO veduto per la Fede di Revisione, ed Approvazione del P. *F. Gio. Paolo Zapparella*, Inquisitor Generale del Santo Officio di *Venizia*, del Libro intitolato *Philosophiæ Naturalis Theoria redacta ad unicam legem virium in natura existentium, Auctore P. Rogerio Josepho Boscovich &c.* non v'esser cosa alcuna contro la Santa Fede Cattolica, e parimente per attestato del Segretario Nostro, niente contro Principi, e buoni costumi concediamo licenza a *Giambattista Remondini* Stampator *di Venezia*, che possa essere stampato, osservando gli ordini in materia di stampe, e presentando le solite Copie alle Publiche Librerie di Venizia, e di Padova.

Dat. li 7. Settembre 1758.
(Gio. Emo, Procurator, Rif.
(Z. Alvise Mocenigo, Rif.
(

Registrato in Libro a carte 47. al num. 383.

Gio. Girolamo Zuccato, Segretario.

Adi 18 Settembre 1758.

Registrato nel Magistr. Eccellentiss. degli Esec. contro la Bestemmia.

Gio. Pietro Dolfin, Segretario.

APPENDIX

Relating to Metaphysics

SUPPLEMENTS

WE, as Censors of the College of Padua, having seen, through trust in the revision & approval of Father *F. Gio. Paolo Zapparella,* Inquisitor General of the Holy Office in Venice, that there is nothing in the book, entitled *Philosophiæ Naturalis Theoria redacta ad unicam legem virium in natura existentium,* by P. Rogerius Josephus Boscovich, that is contrary to the Holy Catholic Faith; & also, on the testimony of our Secretary, that there is nothing contrary to our Rules, according to good usance, give leave to *Giambattista Remondinus,* printer in Venice, to print the book; provided that he observe the regulations governing the press, & present the usual copies to the Public Libraries of Venice & Padua.

Given this 7th of September, 1758.

Gig. Emo, Procurator, Censor.
Z. Alvise Mocenigo, Censor.
?? *

Registered in Book, p. 47, no. 383.

September 18th, 1758. *Gio. Girolamo Zuccato, Secretary.*

Registered in the High Court for the Prevention of Blasphemy.

Gio. Pietro Dolfin, Secretary.

* There is here a space for another name that was not filled in.

CATALOGUS OPERUM
P. ROGERII JOSEPHI BOSCOVICH, S.J.
impressorum usque ad initium anni 1763.

Annus primæ edition.

Opera, & opuscula justæ molis.

Sopra il Turbine, che la notte tra gli 11, e 12 Giugno del 1749 danneggiò una gran parte di Roma. Dissertazione del P. Ruggiero Giuseppe Boscovich della Comp. di Gesù. In Roma appresso Nicolò, e Marco Pagliarini, in 8. — 1749

Elementorum Matheseos tomi tres, in 4. *Prodierunt anno* 1752 *sub titulo,* Elementorum Matheseos ad usum studiosæ juventutis, tomi primi pars prima complectens Geometriam planam, Arithmeticam vulgarem, Geometriam Solidorum, & Trigonometriam cum planam, tum sphæricam. Pars altera, in qua Algebræ finitæ elementa traduntur. Romæ: excudebat Generosus Salomoni. *Iis binis tomis sine nova eorum impressione mutatus est titulus anno* 1754 *in hunc,* Elementorum Universæ Matheseos Auctore P. Rogerio Josepho Boscovich Soc. Jesu Publico Matheseos Professore Tomus I continens &c. Tomus II continens &c, *& adjectus est sequens.* — 1752

Tomus III continens Sectionum Conicarum Elementa nova quadam methodo concinnata, & Dissertationem de Transformatione locorum Geometricorum, ubi de Continuitatis lege, ac de quibusdam Infiniti mysteriis: *Typis iisdem ejusdem Generosi Salomoni omnes in* 8. *Extat eorundem impressio Veneta anni* 1758, *sed typorum mendis deformatissima.* — 1754

De Litteraria Expeditione per Pontificiam ditionem ad dimetiendos duos Meridiani gradus, & corrigendam mappam geographicam, jussu, & auspiciis Benedicti XIV. P.M. suscepta Patribus Soc. Jesu Christophoro Maire, & Rogerio Josepho Boscovich, Romæ 1755. In Typographio Palladis: excudebant Nicolaus, & Marcus Palearini, in 4. *Quidquid eo volumine continetur, est* Patris Boscovich *præter bina brevia opuscula* Patris Maire, *quæ ipse* P. Boscovich *inseruit. Prostat etiam Mappa Geographica ditionis Pontificia delineata* P. Maire *ex observationibus utrique communibus.* — 1755

De Inæqualitatibus, quas Saturnus, & Jupiter sibi mutuo videntur inducere, præsertim circa tempus conjunctionis. Opusculum ad Parisiensem Academiam transmissum, & nunc primum editum. Auctore P. Rogerio Josepho Boscovich Soc. Jesu; Romæ; ex Typographia Generosi Salomoni, in 8. — 1756

Philosophiæ Naturalis Theoria redacta ad unicam legem virium in Natura existentium Auctore P. Rogerio Jos. Boscovich S.J. publico Matheseos Professore in Collegio Romano. *Prostat Viennæ Austriæ in Officina libraria Kalivvodiana: in* 4. *In fine accedit Epistola ad* P. Carolum Scherffer Soc. Jesu. *Habetur secunda editio Viennensis paullo posterior: tertia hic exhibetur: Epistola habetur in ejus Supplementis.* — 1758

Adnotationes in aliorum Opera.

Caroli Noceti e Societate Jesu de Iride, & Aurora Boreali Carmina . . . cum notis Josephi Rogerii Boscovich ex eadem Societate. Romæ: excudebant Nicolaus, & Marcus Palearini, in 4. *Perperam nomen* Josephi *antepositum est ibi nomini* Rogerii. — 1747

Philosophiæ Recentioris a Benedicto Stay in Romano Archigymnasio Publico Eloquentiæ Professore . . . cum adnotationibus, & Supplementis P. Rogerii Josephi Boscovich S.J. in Collegio Rom. Publici Matheseos Professoris. Tomus I. Romæ: Typis, & sumptibus Nicolai, & Marci Palearini, in 8. *Duæ ejus editiones prodierunt simul.* — 1755

Tomus II Romæ: Typis, & sumptibus Nicolai, & Marci Palearini, in 8. *In singulis hisce voluminibus ea, quæ ad P. Boscovich pertinent, efficerent per se ipsa justum volumen. In solis primi Stayani tomi supplementis occurrunt* 39 *ipsius Dissertationes de variis argumentis pertinentibus potissimum ad Metaphysicam & Mechanicam.* — 1760

Dissertationes impressæ pro exercitationibus annuis, & publice propugnatæ: omnes in 4.

De Maculis Solaribus, Exercitatio Astronomica habita in Collegio Romano Soc. Jesu. Romæ: ex Typographia Komarek. — 1736

De Mercurii novissimo infra Solem transitu. Dissertatio habita in Seminario Romano. Romæ, Typis Antonii de Rubeis. — 1737

Constructio Geometrica Trigonometriæ sphæricæ. Romæ, ex Typographia Komarek. *Hujus titulus vel est hic ipse, vel parum ab hoc differt.*

De Aurora Boreali Dissertatio habita in Seminario Romano. Romæ: Typis Antonii de Rubeis. *Eadem eodem anno edita fuit etiam typis Komarek.* — 1738

De Novo Telescopii usu ad objecta cælestia determinanda. Dissertatio habenda a PP. Soc. Jesu in Collegio Romano. Romæ, ex Typographia Komarek. *Extat recusa sine ulla mutatione in Actis Lipsiensibus ad annum* 1740. — 1739

HH

Annus primæ edition.	
	De Veterum argumentis pro Telluris sphæricitate. Dissertatio habita in Seminario Romano Soc. Jesu. Romæ : Typis Antonii de Rubeis.
	Dissertatio de Telluris Figura habita in Seminario Romano Soc. Jesu. Romæ : Typis Antonii de Rubeis. *Eadem prodiit* in 8, *anno* 1744 *in opere, cui titulus* Memorie &c. In Lucca per li Salani, e Giuntini, & *in titulo additur :* nunc primum aucta, & illustrata ab ipsomet Auctore ; *sed ea editio scatet typorum erroribus, ut* & *reliqua inferius nominanda in eadem collectione inserta.*
1740	De Circulis Osculatoribus. Dissertatio habenda a PP. Societatis Jesu in Collegio Romano. Romæ : ex Typographia Komarek.
	De Motu corporum projectorum in spatio non resistente. Dissertatio habita in Seminario Romano Soc. Jesu. Romæ : Typis Antonii de Rubeis.
1741	De Natura, & usu infinitorum, & infinite parvorum. Dissertatio habita in Collegio Romano Soc. Jesu. Romæ : ex Typographia Komarek.
	De Inæqualitate gravitatis in diversis Terræ locis. Dissertatio habita in Seminario Romano Soc. Jesu. Romæ : Typis Antonii de Rubeis.
1742	De Annuis Fixarum aberrationibus. Dissertatio habita in Collegio Romano Societatis Jesu. Romæ : ex Typographia Komarek.
	De Observationibus Astronomicis, & quo pertingat earundem certitudo. Dissertatio habita in Seminario Romano Soc. Jesu. Romæ : Typis Antonii de Rubeis.
	Disquisitio in Universam Astronomiam publicæ Disputationi proposita in Collegio Romano Soc. Jesu. Romæ : ex Typographia Komarek.
1743	De Motu Corporis attracti in centrum immobile viribus decrescentibus in ratione distantiarum reciproca duplicata in spatiis non resistentibus. Dissertatio habita in Collegio Romano. Romæ : Typis Komarek. *Eadem prodiit anno* 1747 *sine ulla mutatione in Commentariis Acad. Bononiensis Tom. II. par. III.*
1744	Nova methodus adhibendi phasium observationes in Eclipsibus Lunaribus ad exercendam Geometriam, & promovendam Astronomiam. Dissertatio habita in Collegio Romano. Romæ : ex Typographia Komarek. *Eadem prodiit in* 8, *anno* 1747 *cum exigua mutatione, vel additamento in Opere superius memorato, cui titulus* Memorie &c. In Lucca per li Salani, e Giuntini.
1745	De Viribus Vivis. Dissertatio habita in Collegio Romano Soc. Jesu. Romæ : Typis Komarek. *Eadem prodiit anno* 1747 *sine ulla mutatione in Commentariis Acad. Bonon. To.* II. *par. III,* & *in Germania pluribus vicibus est recusa.*
1746	De Cometis. Dissertatio habita a PP. Soc. Jesu in Collegio Rom. Romæ : ex Typographia Komarek.
1747	De Æstu Maris. Dissertatio habita a PP. Soc. Jesu in Collegio Romano. Romæ : ex Typographia Komarek. *Ea est Dissertationis pars* 1 ; *secunda pars nunquam prodiit. Quæ pro illa fuerant destinata, habentur in Opere* De Expeditione Litteraria, & *in supplementis Philosophiæ Stayanæ tomo II.*
1748	Dissertationis de Lumine pars prima publice propugnata in Seminario Romano Soc. Jesu. Romæ : Typis Antonii de Rubeis.
	Dissertationis de Lumine pars secunda publice propugnata a PP. Soc. Jesu in Collegio Romano. Romæ : ex Typographia Komarek.
1749	De Determinanda Orbita Planetæ ope Catoptricæ, ex datis vi, celeritate, & directione motus in dato puncto. Exercitatio habita a PP. Soc. Jesu in Collegio Romano. Romæ : ex Typographia Komarek.
1751	De Centro Gravitatis. Dissertatio habita in Collegio Romano Soc. Jesu. Romæ : ex Typographia Komarek. *Eadem paullo post prodiit iterum cum sequenti titulo,* & *additamento.* De Centro Gravitatis. Dissertatio publice propugnata in Collegio Romano Soc. Jesu Auctore P. Rogerio Josepho Boscovich Societatis ejusdem. Editio altera. Accedit Disquisitio in centrum Magnitudinis, qua quædam in ea Dissertatione proposita, atque alia iis affinia demonstrantur. Romæ, Typis, & sumptibus Nicolai, & Marci Palearini.
1753	De Lunæ Atmosphæra. Dissertatio habita a PP. Soc. Jesu in Collegio Romano. Romæ : ex Typographia Generosi Salomoni. *Multa eorundem typorum exemplaria prodierunt paullo post cum nomine Auctoris in ipso titulo,* & *cum exigua unius loci mutatione.*
1754	De Continuitatis Lege, & Consectariis pertinentibus ad prima materiæ elementa, eorumque vires. Dissertatio habita a PP. Societatis Jesu in Collegio Romano. Romæ : ex Typographia Generosi Salomoni.
1755	De Lege virium in Natura existentium. Dissertatio habita a PP. Soc. Jesu in Collegio Romano. Romæ : Typis Generosi Salomoni.
	De Lentibus, & Telescopiis dioptricis. Dissertatio habita in Seminario Romano. Romæ : ex Typographia Antonii de Rubeis.

Plures ex hisce Dissertationibus prodierunt etiam iisdem typis, sed cum alio titulo, habente non locum, ubi sunt habitæ, vel propugnatæ, sed tantummodo nomen Auctoris. In hac postrema mutatæ sunt binæ paginæ, posteaquam plurima exemplaria fuerant distracta. In prioribus tribus sunt pauca quædam mutata, vel addita a P. Horatio Burgundio *adhuc Professore Matheseos in Collegio Romano, qui fuerat ejus Præceptor ; sed eo jam ad Dissertationes ejusmodi conscribendas utebatur.*

Eæ omnes, quæ pertinent ad Seminarium Romanum, habent in ipso titulo adscripta nomina Nobilium Convictorum, qui illas propugnarunt, & *sub eorum nomine referuntur plures ex iis in Actis Lipsiensibus.*

Multa pertinentia ad ipsum P. Boscovich *habentur in binis Dissertationibus, quarum tituli,* Synopsis Physicæ Generalis, & De Lumine, *quarum utraque est edita* Romæ anno 1754, Typis Antonii de Rubeis, in 4. *Id ibidem testatur earundem Auctor* (*is est* P. Carolus Benvenutus Soc. ejusdem) *affirmans, ea sibi ab eodem* P. Boscovich *fuisse communicata.*

Habetur etiam ampliatio solutionis cujusdam problematis pertinentis ad Auroram Borealem, soluti in adnotationibus ad Carmen P. Noceti, *inserta in quadam Dissertatione impressa Romæ circa annum* 1756, &

publice propugnata, cujus Auctor est P. Lunardi *Soc. Jesu, qui affirmat ibidem, se eandem acceptam ab ipso* P. Boscovich *proponere ejusdem verbis.*

Subjiciemus jam bina opuscula Italica, *quæ communi nomine PP.um Le Seur, Jacquier, ac suo conscripsit ipse* P. Boscovich. *Utrumque est sine loco impressionis, & nomine Typographi; impresserunt autem Palearini Fratres Romæ jussu Præsulis, qui tum curabat Fabricam S. Petri, a quo & publice distributa sunt per Urbem.*

Parere di tre Matematici, sopra i danni, che si sono trovati nella Cupola di S. Pietro sul fine del 1742, dato per ordine di Nostro Signore Benedetto XIV, in 4. *In fine opusculi habentur subscripta omnium tria nomina.* — 1742

Riflessioni de' PP. Tomaso Le Seur, Francesco Jacquier dell' Ordine de' Minimi, e Ruggerio Giuseppe Boscovich della Comp. di Gesù sopra alcune difficoltà spettanti i danni, e rifarcimenti della Cupola di S. Pietro proposte nella Congregazione tenutasi nel Quirinale a' 20 Gennaro 1743, e sopra alcune nuove Ispezioni fatte dopo la medesima Congregazione. — 1743

Habentur itidem Italico *sermone binæ ex iis, quas* Itali *vocant* Scritture, *pro quadam lite Ecclesiæ S. Agnetis Romanæ, pertinentes ad aquarum cursum Romæ editæ anno* 1757. — 1757

Inserta.

Nunc faciemus gradum ad inserta in Publicis Academiarum monumentis, in diariis, in collectionibus, & in privatorum Auctorum Operibus.

In Monumentis Acad. Bononiensis.

Præter reimpressionem binarum Dissertationum in To. II, de quibus supra, habetur in To. IV De Litteraria Expeditione per Pontificiam ditionem. *Est Synopsis amplioris Operis, ac habentur plura ejus exemplaria etiam seorsum impressa.* — 1757

In Romano Litteratorum diario vulgo Giornale de' Letterati appresso i Fratelli Pagliarini.

D'Un' antica villa scoperta sul dosso del Tuscolo: d'un antico Orologio a Sole, e di alcune altre rarità, che si sono tra le rovine della medesima ritrovate. Luogo di Vitruvio illustrato. *Ibi ejus schediasmatis Auctor profert, uti ipse profitetur, quæ singillatim audierat ab ipso* P. Boscovich. — 1746

Dimostrazione facile di una principale proprietà delle Sezioni Coniche, la quale non dipende da altri Teoremi conici, e disegno di un nuovo metodo di trattare questa dottrina.

Dissertazione della Tenuità della Luca Solare, Del P. Ruggiero Gius. Boscovich Matematico del Collegio Romano. — 1747

Dimostrazione di un passo spettante all' angolo massimo, e minimo dell' Iride, cavato dalla prop. IX par. 2 del libro I dell' Ottica del Newton con altre riflessioni su quel capitolo. Del P. Ruggiero Gius. Boscovich dell Comp. di Gesù.

Metodo di alzare un Infinitinomio a qualunque potenza. Del P. Ruggiero Gius. Boscovich.

Parte prima delle Riflessioni sul metodo di alzare un Infinitinomio a qualunque potenza. Del P. Ruggiero Gius. Boscovich della Comp. di Gesù. — 1748

Parte seconda &c.

Soluzione Geometrica di un Problema spettante l'ora delle alte, e basse maree, e suo confronto con una soluzione algebraica del medesimo data dal Sig. Daniele Bernoulli. Del P. Ruggiero Giuseppe Boscovich della Compagnia di Gesù.

Dialogi Pastorali V sull' Aurora Boreale del P. Ruggiero Gius. Boscovich della Comp. di Gesù.

Dimostrazione di un metodo dato dall' Eulero per dividere una frazione razionale in più frazioni più semplici con delle altre riflessioni sulla stessa materia. — 1749

Lettera del P. Ruggiero Gius. Boscovich della Comp. di Gesù al Sig. Ab. Angelo Bandini in risposta alla lettera del Sig. Ernesto Freeman sopra L'Obelisco d'Augusto. *Nomen* Freeman *est fictitium, Auctorem denotans Neapoli latentem, & aliis Operibus satis notum. Extat eadem etiam in folio.* — 1750

Altera de eodem Obelisco admodum prolixa Epistola, Italice, *& Latine scripta ad eundem* Bandinium *suo nomine ab ipso* P. Boscovich *habetur in ejusdem Bandinii Opere, cui titulus,* De Obelisco Cæsaris Augusti e Campi Martii ruderibus nuper eruto. Commentarius Auctore Angelo Maria Bandinio. Romæ apud Fratres Palearinos, in folio. *Ibidem in fine habetur alia epistola itidem admodum prolixa de eodem argumento nomine* Stuarti, *e cujus schedis relictis apud Cardinalem Valentium in ejus discessu ab Urbe eam Epistolam conscripsit, ac ejus comperta illustravit, ac auxit ipse* P. Boscovich.

Osservazioni dell' ultimo passaggio di Mercurio sotto il Sole seguito a' 6. di Maggio 1753, fatte in Roma, e raccolte dal P. Ruggiero Gius. Boscovich della Comp. di Gesù con alcune reflessioni sulle medesime. — 1753

In aliis monumentis.

In Collectione Opusculorum Lucensi cui titulus: Memorie sopra la Fisica, e Istoria naturale di diversi Valentuomini. In Lucca per li Salani, e Giuntini, in 8, *Præter binas dissertationes, de quibus supra, habetur.*

Problema Mechanicum de solido maximæ attractionis solutum a P. Rogerio Josepho Boscovich Soc. Jesu Publico Professore Matheseos in Collegio Romano: Tomo I. — 1743

De Materiæ divisibilitate, & Principiis corporum. Dissertatio conscripta jam ab anno 1748, & nunc primum edita. Auctore P. Rogerio Jos. Boscovich Soc. Jesu, To. IV. — 1757

Omnium horum quatuor Opusculorum habentur etiam exemplaria seorsum impressa.

Annus primæ edition.

1745 *In editione Elementorum Geometriæ* Patris Tacqueti *facta Romæ sumptibus* Venantii Monaldini, Typis Hieronymi Mainardi, in 8. habetur Trigonometria sphærica P. Rogerii Josephi Boscovich, *quæ deinde adhuc magis expolita prodiit Tomo I. ejus* Elementorum Matheseos. *Habetur præterea ibidem Tractatus* De Cycloide, & Logistica, *qui etiam seorsum impressus est iisdem typis.*

1752 *In Opere Comitis Zoannis Baptistæ Soardi, cui titulus* Nuovi instrumenti &c. in Brescia dalle stampe di Gio. Battista Rizzardi, in 4., *habentur binæ epistolæ Italicæ ipsius* P. Boscovich *de Curvis quibusdam, cum figuris,* & demonstrationibus.

1758 *In Optica Abbatis De la* Caille *latine reddita* a P. Carolo Scherffer Soc. Jesu, *& impressa Viennæ in Austria habetur schediasma* Patris Boscovich de Micrometro objectivo.

In postremo tomo Commentar. Academiæ Parisiensis in Historia, & in uno e tomis *Correspondentium ejusdem Academiæ, creditur esse breve aliquid pertinens ad ipsum* P. Boscovich. *Est aliquid etiam in diario Gallico* Journal des çavans, *& fortasse in Anglicanis Transactionibus, atque alibi insertum hisce itinerum annis.*

Poetica.

1753 P. Rogerii Josephi Boscovich Soc. Jesu inter Arcades Numenii Anigrei Ecloga recitata in publico Arcadum consessu primo Ludorum Olympicorum die, quo die Michæl Joseph Morejus Generalis Arcadiæ Custos illustrium Poetarum Arcadum effigies formandas jaculorum ludi substituerat. Romæ in 8. *Extat eadem iisdem Typis etiam in Collectione tum impressa omnium, quæ ea occasione sunt recitata.*

Stanislai Poloniæ Regis, Lotharingiæ, ac Barri Ducis, & inter Arcades Euthimii Aliphiraei, dum ejus effigies in publico Arcadum Cœtu erigeretur, Apotheosis. Auctore P. Rogerio Josepho Boscovich Soc. Jesu inter Arcades Numenio Anigreo. Romæ ex Typographia Generosi Salomoni, in 8. *Est poema versu heroico. Idem autem recusum fuit Nancei cum versione Gallica* Domini Cogolin.

1757 Pro Benedicto XIV. P.M. Soteria. *Est itidem poema Heroicum ejusdem* P. Boscovich *pertinens vel ad hunc, vel ad superiorem annum: est autem impressum* Romæ in 4, apud Fratres Palearinos, *occasione periculi mortis imminentis, evitati a Pontifice convalescente.*

1758 In Nuptiis Joannis Corrarii, & Andrianæ Pisauriæ e nobilissimis Venetæ Reip. Senatoriis familiis. Carmen P. Rogerii Jos. Boscovich S.J. Publici in Romano Collegio Matheseos Professoris. Romæ: ex Typographio Palladis: excudebant Nicolaus, & Marcus Palearini, in 4.

1760 De Solis, ac Lunæ defectibus libri V P. Rogerii Josephi Boscovich Societatis Jesu ad Regiam Societatem Londinensem, Londini 1760. in 4. *Non habetur nomen Typographi, qui impressit, sed Bibliopolarum quorum sumptibus est impressum: deest hic ejus editionis exemplar, ex quo ea nomina correcte describantur. Idem recusum fuit anno* 1761 *Venetiis apud Zattam in* 8°. *cum exiguo additamento in fine, & cum hoc catalogo, quem inde huc derivavimus. Habentur in adnotationibus bina Epigrammata cum versionibus Italicis, sive Sonetti.*

Est & aliud ejus poema Heroicum anno 1756 *impressum Viennæ in Austria in collectione carminum facta occasione inaugurationis novarum Academiæ Viennensis ædium.*

Sunt & epigrammata nonnulla in Collectionibus Arcadum, inter quæ unum pro recuperata valetudine Johannis V *Lusitaniæ Regis, & unum pro Rege tum utriusque Siciliæ, & nunc Hispaniæ, ac pro Regina ejus conjuge.*

Extant etiam pauca admodum exemplaria unius ex illis, quas in Italia *appellamus* Cantatine, *impressa Viterbii anno* 1750 *pro Visitatione B. Mariæ Virginis, in qua sex, quas dicimus* Ariette, *profanæ ad sacrum argumentum transferendæ erant, manente Musica, & inter se connectendæ.*

ERRATA

p. 2, l. 11, *for* ac omnem *read* ad omnem
p. 3, l. 5, *for* has been *read* should be
p. 4, l. 18, *for* Venetisis *read* Venetiis
p. 6, l. 9 from bottom, *for* exceres *read* exerces
 l. 4 from bottom, *for* eocatum *read* evocatum
p. 7, l. 18 from bottom, *after* despatched *add* to the Court of Spain
 l. 13 from bottom, *for* befits *read* befit
p. 8, l. 1, *for* publico *read* publice
 l. 13, *for* utique *read* ubique
 l. 28, *for* infiliciter *read* infeliciter
p. 10, l. 8, *for* opportunam *read* opportunum
 l. 9, *for* mediocrum *read* mediocrium
p. 12, l. 13, *for* aliquando *read* aliquanto
 l. 10 *from* bottom, *for* repulsivis *read* repulsivas
p. 14, l. 13, *for* adhibitis *read* adhibitas
 l. 24, *for* postremo *read* postrema
p. 18, l. 2, *for* alter *read* altera
p. 22, l. 15, *after* vero etiam *insert* leges
p. 28, l. 17, *for* acquiretur *read* acquireretur
 l. 28, *for* -menæ *read* -mena
p. 40, l. 22, *for* Naturam *read* Natura
 l. 23, *for* quandem *read* quandam
 l. 29, *for* recidit *read* recedit
 l. 32, *for* postquam *read* post quam
p. 47, l. 34, *for* many *read* most
p. 48, l. 18, *for* linæ *read* lineæ
 l. 29, *for* genere *read* generis
p. 50, l. 26, *for* deferendam *read* deserendam
 l. 31, *for* viderimus *read* videremus
 l. 46, *for* nominandi *read* nominando
p. 52, ll. 5, 6 of marginal note to § 7, *for* nihilmu *read* nihilum.
p. 54, l. 1, *for* exhibit *read* exhibet
 l. 3, *for* oppositæ *read* opposita
 l. 12, *for* sit *read* fit
p. 55, l. 4, *after* & *add* then
p. 56, l. 3, *for* servat *read* servant
p. 58, l. 32, *for* crederit *read* crederet
p. 60, l. 3, marg. note, Art. 46, *for* sit *read* fit
p. 64, l. 2, *for* terio *read* tertio
p. 65, l. 57, *for* fact *read* by the fact
p. 66, l. 9, *for* concipiantur *read* concipiatur
 l. 16, *for* ordinate *read* ordinatæ
p. 67, l. 48, *for* before & *read* previously
p. 68, l. 11, *for* in GM′ *read* in GM
p. 71, l. 46, *for* and this *read* and that this is found nowhere
p. 72, l. 1, *for* ejusmodi *read* hujusmodi
 l. 4 from bottom, *for* potissimuim *read* potissimum
p. 74, l. 38, *for* illo *read* illa
p. 76, l. 3 from bottom, *for* devenirent *read* devenerint
p. 81, l. 42, *for* is *read* ought to be
p. 82, l. 5, marg. note, Art. 82, *for* se *read* sed
p. 86, Art. 89, in marg. note, *for* densitatis *read* densitas
p. 88, l. 11, *for* adi *read* ad
 l. 16, *for* reliquent *read* relinquent
p. 90, l. 30, *for* diversimodo *read* diversimode
 l. 34, *for* distantia *read* distantiæ
 Art. 95, marg. note, *for* de- *read* dif-
p. 92, l. 33, *for* apparent *read* apparerent
p. 94, l. 22, *for* incurrant *read* incurrunt
p. 95, Art. 103, l. 1, *for* are *read* is ; *and in marg. note insert* in *between* and *and* what
p. 96, l. 8, *for* potissimo *read* potissimum
 l. 16, *for* præcedentum *read* præcedentem
 marg. note, Art. 105, *for* transire *read* transiri
p. 97, l. 9 from bottom, *for* quite enough *read* better
p. 99, l. 40, *insert a comma after* locus
p. 100, marg. note, Art. 112, *for* recte *read* rectæ
 l. 32, *for* ellipis *read* ellipsis
p. 106, marg. note, Art. 125, *for* perfectionum *read* perfectiorum
p. 107, l. 23, *for* off they are *read* away they go
 l. 13 from bottom, *for* have *read* has
p. 109, l. 27, *after* tantummodo *add* admitto
p. 110, l. 17, *for* expandantur *read* expendantur
 l. 24, *for* a *read* &
 bottom line, *for* distinctis *read* distinctas
p. 112, l. 27, *for* veteram *read* veterem
p. 113, l. 5 from bottom, *for* because *read* that
 l. 4 from bottom, *after* change *add* is excluded by
p. 115, l. 33, *for* and *read* et
 marg. note, Art. 139, *add at end* impugned
p. 118, l. 7 from bottom *for* ali *read* alia
p. 122, l. 26, *for* justmodi *read* ejusmodi
p. 125, l. 29, *for* ignored *read* urged in reply
p. 128, l. 31, *for* ea *read* eæ
p. 129, l. 16, *for* Principii *read* Principiis
p. 139, l. 8, *for* arm E *read* arm ED
 footnote, l. 5, *for* DP *read* OP
p. 140, l. 34, *insert* cum *before* directione
 l. 4, footnote, *for* ut in *read* ut n
p. 148, l. 10, *for* Expositas *read* Expositis, *for* curva *read* curvam
p. 156, l. 1, *for* a que *read* atque
 l. 7, *for* caculo *read* calculo
 l. 39, *for* Tam *read* Tum
p. 158, Art. 209, marg. note, *add at end* Legum multitudo & varietas
 footnote, l. 11 from bottom of page, *for* obvenerit *read* obveniret
p. 160, footnote, l. 1, *for* sit *read* fit
 l. 12, *for* ed *read* sed
p. 161, footnote, l. 20, *after* segment *add* DR
p. 162, l. 7 from bottom, *for* reflexionis *read* reflexiones
p. 167, l. 40, *for* ae *read* da
p. 168, l. 8, from bottom *for* 27C—′AC *read* 27 C′AC
p. 171, l. 4 from bottom *for* GL, or LI *read* G*I*, or IL
p. 172, l. 34, *for* compositas *read* compositis
p. 175, l. 13, for 30 *read* 27
 l. 39, *insert a comma after* approximately
p. 176, l. 7, *for* delatam *read* delatum
p. 178, marg. note, Art. 230, l. 5, *for* foco *read* focos
p. 188, l. 31, *for* summa *read* summæ
p. 195, l. 19, *insert a comma after* point P
p. 197, l. 35, *for* sum of the (*at end of line*) *read* sums of the
p. 198, Fig. 40, *insert* F *where AE cuts CD*
p. 199, l. 35, *for* ceases *read* cease
 l. 37, *after* all, & *insert I* assume
p. 202, l. 6, *for* summa *read* summam
 Art. 264, l. 3, *for* quacunque *read* quancunque *and in marg. note for* corallarium *read* corollarium
p. 205, l. 21, *for* recessions *read* recession
p. 206, l. 3, *for* globis *read* globus
 l. 2 from bottom, *insert* motu *before* quodam
p. 208, last line, *for* $\frac{m+n}{n}$ *read* $\frac{m+n}{m}$ *in each case*
p. 209, l. 11, *for* (2CQ — 2Cq) *read* (2CQ — 2cQ)
p. 210. l. 8, *for* quiescat *read* quiescit
p. 211, l. 25, *the denominator* $(Q + q)$ *should be* $(Q + q)^2$
p. 215, l. 5, *for* BP *read* BO
p. 223, l. 26, *for* 50,61,62 *read* 50,51,52
p. 227, l. 21 from bottom, *for* to *read* of
p. 228, l. 5, from bottom, *for* Angulum *read* Angulus

ERRATA

p. 233, l. 8, *for* in volute *read* evolute
  bottom line, *after* vary *insert* inversely
p. 241, marg. note, Art. 313, *add at end* This is very soon proved
p. 242, last line, *for denominator* AD *read* BD
p. 247, l. 15 from bottom, *for* A & B *read* B & C
p. 248, l. 24, *for* conversione *read* conversionem
p. 250, l. 39, *for* justa *read* juxta
p. 252, l. 5 from bottom, *for gravitas read gravitatis*
p. 256, l. 10, *for* quæsitum *read* quæsitam
p. 270, marg. note, Art. 366, l. 5, *for* magnas *read* magna
p. 278, l. 7, *for* varior *read* rarior
p. 280, l. 12 from bottom, *for* tranctanda *read* tractanda
p. 284, l. 15, *for* sit *read* fit
p. 286, l. 20, *for* multuplicetur *read* multiplicetur
p. 288, l. 21, for Solum *read* Solem
p. 292, l. 1, *for* quietam *read* quietem
  l. 3, *for* sit, *read* fit
  l. 25, *for* ullo, *read* ulla *deleting the comma*
  l. 26, *for* illa *read* ille
p. 304, l. 12 from bottom, *for* propre *read* prope
p. 310, l. 15 from bottom, *for* hæbebunt *read* hærebunt
p. 314, l. 2, *insert* & *before* mutandam
p. 319, l. 18 from bottom, *for* some repel *read* and repel
p. 324, l. 22, *for* pertinet *read* pertinent
p. 329, l. 25, *for* ethers *read* others
p. 332, l. 16, *for* æquabilis *read* æqualibus
  l. 21, *for* Benvenuti *read* Benvenutus
  l. 3 from bottom, *for* qui a *read* quia
p. 336, l. 35, *insert* utcunque *after* circunquaque
p. 346, l. 19, *for* sit *read* fit
p. 348, l. 28, *for* irregularitur *read* irregulariter
p. 350, l. 13 from bottom, *for* flexo *read* flexu
p. 355, l. 14, *for* with *read* to
p. 356, l. 9, *after* porro aliud *insert* post aliud
  l. 19, *after* accidit *insert* idem accidit
p. 366, l. 32, *for* ordores *read* odores
p. 394, l. 31, *for* imaginaræ *read* imaginariæ
p. 396, l. 19 from bottom, *after* solum *add* etiam
  l. 14 from bottom, *for* sunt *read* sint
p. 398, l. 20, *after* est *insert* tota
  l. 35, *for* esses *read* esse
p. 400, l. 33, *after* omnino *insert* saltem
p. 406, l. 6, *for* congruant *read* congruunt
p. 410, l. 8, *for* bz^{m2} *read* bz^{m-2}
  Art 27, marg. note, *for* quæsitum *read* quæsitam
p. 412, l. 11, *insert* positivæ *before* assumantur
p. 422, l. 22, *for* ab *read* ad
p. 434, Art 86, marg. note, *for* massa *read* massas
  l. 5 from bottom, for contrarium *read* contrariam
p. 444, l. 11, *insert* & *before* centri
p. 448, l. 17, *for* PQ *read* PG
  Art. 107, marg. note, *for* absoluutæ *read* absolutæ
  l. 4 from bottom, *after* posita *insert* ∫
p. 454, l. 2 from bottom, *for* elasticas *read* elasticitas
p. 456, l. 5 from bottom *for* $\frac{T^2}{PQ^2}$ *read* $\frac{T^2}{PQ}$
p. 458, l. 11, *for* $\frac{t\,dz}{zz}$ *read* $\frac{t^2dz}{zz}$
p. 459, l. 13, *for* $-tdz/z^2$ *read* $-t^2dz/z^2$
p. 464, l. 4, *for* Discrimen *read* Discrimina
  l. 10 from bottom, *for* Venizia *read* Venezia

Printed in Great Britain by Butler & Tanner, *Frome and London*

Made in the USA
Columbia, SC
16 February 2021